Handbook of
Markov Chain
Monte Carlo

Chapman & Hall/CRC
Taylor & Francis Group
6000 Broken Sound Parkway NW, Suite 300
Boca Raton, FL 33487-2742

© 2011 by Taylor and Francis Group, LLC
Chapman & Hall/CRC is an imprint of Taylor & Francis Group, an Informa business

No claim to original U.S. Government works

International Standard Book Number: 978-1-4200-7941-8 (Hardback)

Visit the Taylor & Francis Web site at
http://www.taylorandfrancis.com

and the CRC Press Web site at
http://www.crcpress.com

Contents

Part I Foundations, Methodology, and Algorithms

6. Inference from Simulations and Monitoring Convergence 163
Andrew Gelman and Kenneth Shirley

7. Implementing MCMC: Estimating with Confidence 175
James M. Flegal and Galin L. Jones

8. Perfection within Reach: Exact MCMC Sampling 199
Radu V. Craiu and Xiao-Li Meng

Part II   Applications and Case Studies

Preface

Over the past 20 years or so, Markov Chain Monte Carlo (MCMC) methods have revolutionized statistical computing. They have impacted the practice of Bayesian statistics profoundly by allowing intricate models to be posited and *used* in an astonishing array of disciplines as diverse as fisheries science and economics. Of course, Bayesians are not the only ones to benefit from using MCMC, and there continues to be increasing use of MCMC in other statistical settings. The practical importance of MCMC has also sparked expansive and deep investigation into fundamental Markov chain theory. As the use of MCMC methods mature, we see deeper theoretical questions addressed, more complex applications undertaken and their use spreading to new fields of study. It seemed to us that it was a good time to try to collect an overview of MCMC research and its applications.

This book is intended to be a reference (not a text) for a broad audience and to be of use both to developers and users of MCMC methodology. There is enough introductory material in the book to help graduate students as well as researchers new to MCMC who wish to become acquainted with the basic theory, algorithms and applications. The book should also be of particular interest to those involved in the development or application of new and advanced MCMC methods. Given the diversity of disciplines that use MCMC, it seemed prudent to have many of the chapters devoted to detailed examples and case studies of realistic scientific problems. Those wanting to see current practice in MCMC will find a wealth of material to choose from here.

Roughly speaking, we can divide the book into two parts. The first part encompasses 12 chapters concerning MCMC foundations, methodology and algorithms. The second part consists of 12 chapters which consider the use of MCMC in practical applications. Within the first part, the authors take such a wide variety of approaches that it seems pointless to try to classify the chapters into subgroups. For example, some chapters attempt to appeal to a broad audience by taking a tutorial approach while other chapters, even if introductory, are either more specialized or present more advanced material. Yet others present original research. In the second part, the focus shifts to applications. Here again, we see a variety of topics, but there are two basic approaches taken by the authors of these chapters. The first is to provide an overview of an application area with the goal of identifying best MCMC practice in the area through extended examples. The second approach is to provide detailed case studies of a given problem while clearly identifying the statistical and MCMC-related issues encountered in the application.

When we were planning this book, we quickly realized that no single source can give a truly comprehensive overview of cutting-edge MCMC research and applications—there is just too much of it and its development is moving too fast. Instead, the editorial goal was to obtain contributions of high quality that may stand the test of time. To this end, all of the contributions (including those written by members of the editorial panel) were submitted to a rigorous peer review process and many underwent several revisions. Some contributions, even after revisions, were deemed unacceptable for publication here, and we certainly welcome constructive feedback on the chapters that did survive our editorial process. We thank all the authors for their efforts and patience in this process, and we ask for understanding from those whose contributions are not included in this book. We believe the breadth and depth of the contributions to this book, including some diverse opinions expressed, imply a continuously bright and dynamic future for MCMC research. We hope this book inspires further work—theoretical, methodological, and applied—in this exciting and rich area.

Finally, no project of this magnitude could be completed with satisfactory outcome without many individuals' help. We especially want to thank Robert Calver of Chapman & Hall/CRC for his encouragements, guidelines, and particularly his patience during the entire process of editing this book. We also offer our heartfelt thanks to the numerous referees for their insightful and rigorous review, often multiple times. Of course, the ultimate appreciation for all individuals involved in this project comes from your satisfaction with the book or at least a part of it. So we thank you for reading it.

MATLAB® is a registered trademark of The MathWorks, Inc. For product information, please contact:

The MathWorks, Inc.
3 Apple Hill Drive
Natick, MA 01760-2098 USA
Tel: 508 647 7000
Fax: 508-647-7001
E-mail: info@mathworks.com
Web: www.mathworks.com

Steve Brooks
Andrew Gelman
Galin L. Jones
Xiao-Li Meng

Editors

Steve Brooks is company director of ATASS, a statistical consultancy business based in the United Kingdom. He was formerly professor of Statistics at Cambridge University and received the Royal Statistical Society Guy medal in Bronze in 2005 and the Philip Leverhulme prize in 2004. Like his co-editors, he has served on numerous professional committees both in the United Kingdom and elsewhere, as well as sitting on numerous editorial boards. He is co-author of *Bayesian Analysis for Population Ecology* (Chapman & Hall/CRC, 2009) and co-founder of the National Centre for Statistical Ecology. His research interests include the development and application of computational statistical methodology across a broad range of application areas.

Andrew Gelman is a professor of statistics and political science and director of the Applied Statistics Center at Columbia University. He has received the Outstanding Statistical Application award from the American Statistical Association, the award for best article published in the *American Political Science Review,* and the Committee of Presidents of Statistical Societies award for outstanding contributions by a person under the age of 40. His books include *Bayesian Data Analysis* (with John Carlin, Hal Stern, and Don Rubin), *Teaching Statistics: A Bag of Tricks* (with Deb Nolan), *Data Analysis Using Regression and Multilevel/Hierarchical Models* (with Jennifer Hill), and, most recently, *Red State, Blue State, Rich State, Poor State: Why Americans Vote the Way They Do* (with David Park, Boris Shor, Joe Bafumi, and Jeronimo Cortina).

Andrew has done research on a wide range of topics, including: why it is rational to vote; why campaign polls are so variable when elections are so predictable; why redistricting is good for democracy; reversals of death sentences; police stops in New York City; the statistical challenges of estimating small effects; the probability that your vote will be decisive; seats and votes in Congress; social network structure; arsenic in Bangladesh; radon in your basement; toxicology; medical imaging; and methods in surveys, experimental design, statistical inference, computation, and graphics.

Galin L. Jones is an associate professor in the School of Statistics at the University of Minnesota. He has served on many professional committees and is currently serving on the editorial board for the *Journal of Computational and Graphical Statistics.* His research interests include Markov chain Monte Carlo, Markov chains in decision theory, and applications of statistical methodology in agricultural, biological, and environmental settings.

Xiao-Li Meng is the Whipple V. N. Jones professor of statistics and chair of the Department of Statistics at Harvard University; previously he taught at the University of Chicago (1991–2001). He was the recipient of the 1997–1998 University of Chicago Faculty Award for Excellence in Graduate Teaching, the 2001 Committee of Presidents of Statistical Societies Award, the 2003 Distinguished Achievement Award and the 2008 Distinguished Service Award from the International Chinese Statistical Association, and the 2010 Medallion Lecturer from the Institute of Mathematical Statistics (IMS). He has served on numerous professional committees, including chairing the 2004 Joint Statistical Meetings and the Committee on Meetings of American Statistical Association (ASA) from 2004 until 2010. He is an elected fellow of the ASA and the IMS. He has also served on editorial boards for *The Annals of Statistics, Bayesian Analysis, Bernoulli, Biometrika, Journal of the American Statistical*

Association, as well as the coeditor of *Statistica Sinica*. Currently, he is the statistics editor for the IMS Monograph and Textbook Series. He is also a coeditor of *Applied Bayesian Modeling and Causal Inference from Incomplete-Data Perspectives* (Gelman and Meng, 2004, Wiley) and *Strength in Numbers: The Rising of Academic Statistics Departments in the U.S.* (Agresti and Meng, 2012, Springer). His research interests include inference foundations and philosophies, models of all flavors, deterministic and stochastic algorithms, signal extraction in physical, social and medical sciences, and occasionally elegant mathematical statistics.

Contributors

Susan Spear Bassett
Department of Psychiatry and
    Behavioral Sciences
Johns Hopkins Hospital
Baltimore, Maryland

John T. Behrens
Cisco Systems, Inc.
Mishawaka, Indiana

DuBois Bowman
Center for Biomedical Imaging
    Statistics (CBIS)
Emory University
Atlanta, Georgia

Steve Brooks
ATASS Ltd
Exeter, United Kingdom

Brian Caffo
Department of Biostatistics
Johns Hopkins University
Baltimore, Maryland

George Casella
Department of Statistics
University of Florida
Gainesville, Florida

Radu V. Craiu
Department of Statistics
University of Toronto
Toronto, Ontario, Canada

Francesca Dominici
Department of Biostatistics
Johns Hopkins University
Baltimore, Maryland

Lynn Eberly
Division of Biostatistics
University of Minnesota
Minneapolis, Minnesota

Yanan Fan
School of Mathematics and Statistics
University of New South Wales
Sydney, Australia

Paul Fearnhead
Department of Mathematics
    and Statistics
Lancaster University
Lancaster, United Kingdom

James M. Flegal
Department of Statistics
University of California
Riverside, California

Colin Fox
Department of Physics
University of Otago
Dunedin, New Zealand

Filiz Garip
Department of Sociology
Harvard University
Cambridge, Massachusetts

Andrew Gelman
Department of Statistics

and

Department of Political Science
Columbia University
New York, New York

Charles J. Geyer
School of Statistics
University of Minnesota
Minneapolis, Minnesota

Murali Haran
Center for Ecology and
    Environmental Statistics
Pennsylvania State University
University Park, Pennsylvania

David Higdon
Los Alamos National Laboratory
Los Alamos, New Mexico

James P. Hobert
Department of Statistics
University of Florida
Gainsville, Florida

Mark Huber
Department of Mathematical Sciences
Claremont McKenna College
Claremont, California

Galin L. Jones
School of Statistics
University of Minnesota
Minneapolis, Minnesota

Ruth King
School of Mathematics and Statistics
University of St. Andrews
St. Andrews, United Kingdom

Roy Levy
School of Social and Family Dynamics
Arizona State University
Tempe, Arizona

Thomas A. Louis
Department of Biostatistics
Johns Hopkins University
Baltimore, Maryland

Xiao-Li Meng
Department of Statistics
Harvard University
Cambridge, Massachusetts

Russell B. Millar
Department of Statistics
University of Auckland
Auckland, New Zealand

Robert J. Mislevy
Department of Measurement,
   Statistics and Evaluation
University of Maryland
Severna Park, Maryland

J. David Moulton
Los Alamos National Laboratory
Los Alamos, New Mexico

Radford M. Neal
Department of Statistics
University of Toronto
Toronto, Ontario, Canada

Jong Hee Park
Department of Political Science
University of Chicago
Chicago, Illinois

Taeyoung Park
Department of Applied Statistic
Yonsei University
Seoul, South Korea

Roger Peng
Department of Biostatistics
Johns Hopkins University
Baltimore, Maryland

C. Shane Reese
Department of Statistics
Brigham Young University
Provo, Utah

Christian Robert
CEREMADE—Université
   Paris-Dauphine
Paris, France

Jeffrey S. Rosenthal
Department of Statistics
University of Toronto
Toronto, Ontario, Canada

Kenneth Shirley
The Earth Institute
Columbia University
New York, New York

Scott A. Sisson
School of Mathematics and Statistics
University of New South Wales
Sydney, Australia

Elizabeth Thompson
Department of Statistics
University of Washington
Seattle, Washington

David A. van Dyk
Department of Statistics
University of California
Irvine, California

Jasper A. Vrugt
Center for Non-Linear Studies
Irvine, California

Bruce Western
Department of Sociology
Harvard University
Cambridge, Massachusetts

Scott Zeger
Department of Biostatistics
Johns Hopkins University
Baltimore, Maryland

Part I

Foundations, Methodology, and Algorithms

1

Introduction to Markov Chain Monte Carlo

Charles J. Geyer

1.1 History

Despite a few notable uses of simulation of random processes in the pre-computer era (Hammersley and Handscomb, 1964, Section 1.2; Stigler, 2002, Chapter 7), practical widespread use of simulation had to await the invention of computers. Almost as soon as computers were invented, they were used for simulation (Hammersley and Handscomb, 1964, Section 1.2). The name "Monte Carlo" started as cuteness—gambling was then (around 1950) illegal in most places, and the casino at Monte Carlo was the most famous in the world—but it soon became a colorless technical term for simulation of random processes.

Markov chain Monte Carlo (MCMC) was invented soon after ordinary Monte Carlo at Los Alamos, one of the few places where computers were available at the time. Metropolis et al. (1953)* simulated a liquid in equilibrium with its gas phase. The obvious way to find out about the thermodynamic equilibrium is to simulate the dynamics of the system, and let it run until it reaches equilibrium. The *tour de force* was their realization that they did not need to simulate the exact dynamics; they only needed to simulate some Markov chain having the same equilibrium distribution. Simulations following the scheme of Metropolis et al. (1953) are said to use the *Metropolis algorithm*. As computers became more widely available, the Metropolis algorithm was widely used by chemists and physicists, but it did not become widely known among statisticians until after 1990. Hastings (1970) generalized the Metropolis algorithm, and simulations following his scheme are said to use the *Metropolis–Hastings* algorithm. A special case of the Metropolis–Hastings algorithm was introduced by Geman and Geman (1984), apparently without knowledge of earlier work. Simulations following their scheme are said to use the *Gibbs sampler*. Much of Geman and Geman (1984) discusses optimization to find the posterior mode rather than simulation, and it took some time for it to be understood in the spatial statistics community that the Gibbs sampler simulated the posterior distribution, thus enabling full Bayesian inference of all kinds. A methodology that was later seen to be very similar to the Gibbs sampler was introduced by Tanner and Wong (1987), again apparently without knowledge of earlier work. To this day, some refer to the Gibbs sampler as "data augmentation" following these authors. Gelfand and Smith (1990) made the wider Bayesian community aware of the Gibbs sampler, which up to that time had been known only in the spatial statistics community. Then it took off; as of this writing, a search for Gelfand and Smith (1990) on Google Scholar yields 4003 links to other works. It was rapidly realized that most Bayesian inference could

---

* The fifth author was Edward Teller, the "father of the hydrogen bomb."

be done by MCMC, whereas very little could be done without MCMC. It took a while for researchers to properly understand the theory of MCMC (Geyer, 1992; Tierney, 1994) and that all of the aforementioned work was a special case of the notion of MCMC. Green (1995) generalized the Metropolis–Hastings algorithm, as much as it can be generalized. Although this terminology is not widely used, we say that simulations following his scheme use the *Metropolis–Hastings–Green* algorithm. MCMC is not used only for Bayesian inference. Likelihood inference in cases where the likelihood cannot be calculated explicitly due to missing data or complex dependence can also use MCMC (Geyer, 1994, 1999; Geyer and Thompson, 1992, 1995, and references cited therein).

1.2 Markov Chains

A sequence $X_1, X_2, \ldots$ of random elements of some set is a *Markov chain* if the conditional distribution of X_{n+1} given $X_1, \ldots, X_n$ depends on X_n only. The set in which the X_i take values is called the *state space* of the Markov chain.

A Markov chain has *stationary transition probabilities* if the conditional distribution of X_{n+1} given X_n does not depend on n. This is the main kind of Markov chain of interest in MCMC. Some kinds of adaptive MCMC (Chapter 4, this volume) have nonstationary transition probabilities. In this chapter we always assume stationary transition probabilities.

The joint distribution of a Markov chain is determined by

- The marginal distribution of X_1, called the *initial distribution*
- The conditional distribution of X_{n+1} given X_n, called the *transition probability distribution* (because of the assumption of stationary transition probabilities, this does not depend on n)

People introduced to Markov chains through a typical course on stochastic processes have usually only seen examples where the state space is finite or countable. If the state space is finite, written $\{x_1, \ldots, x_n\}$, then the initial distribution can be associated with a vector $\lambda = (\lambda_1, \ldots, \lambda_n)$ defined by

$$\Pr(X_1 = x_i) = \lambda_i, \quad i = 1, \ldots, n,$$

and the transition probabilities can be associated with a matrix P having elements p_{ij} defined by

$$\Pr(X_{n+1} = x_j \mid X_n = x_i) = p_{ij}, \quad i = 1, \ldots, n \quad \text{and} \quad j = 1, \ldots, n.$$

When the state space is countably infinite, we can think of an infinite vector and matrix. But most Markov chains of interest in MCMC have uncountable state space, and then we cannot think of the initial distribution as a vector or the transition probability distribution as a matrix. We must think of them as an unconditional probability distribution and a conditional probability distribution.

1.3 Computer Programs and Markov Chains

Suppose you have a computer program

```
Initialize x
repeat {
    Generate pseudorandom change to x
    Output x
}
```

If x is the entire state of the computer program exclusive of random number generator seeds (which we ignore, pretending pseudorandom is random), this is MCMC. It is important that x must be the entire state of the program. Otherwise the resulting stochastic process need not be Markov.

There is not much structure here. Most simulations can be fit into this format. Thus most simulations can be thought of as MCMC if the entire state of the computer program is considered the state of the Markov chain. Hence, MCMC is a very general simulation methodology.

1.4 Stationarity

A sequence $X_1, X_2, \ldots$ of random elements of some set is called a *stochastic process* (Markov chains are a special case). A stochastic process is *stationary* if for every positive integer k the distribution of the k-tuple

$$(X_{n+1}, \ldots, X_{n+k})$$

does not depend on n. A Markov chain is stationary if it is a stationary stochastic process. In a Markov chain, the conditional distribution of $(X_{n+2}, \ldots, X_{n+k})$ given X_{n+1} does not depend on n. It follows that a Markov chain is stationary if and only if the marginal distribution of X_n does not depend on n.

An initial distribution is said to be *stationary* or *invariant* or *equilibrium* for some transition probability distribution if the Markov chain specified by this initial distribution and transition probability distribution is stationary. We also indicate this by saying that the transition probability distribution *preserves* the initial distribution.

Stationarity implies stationary transition probabilities, but not vice versa. Consider an initial distribution concentrated at one point. The Markov chain can be stationary if and only if all iterates are concentrated at the same point, that is, $X_1 = X_2 = \ldots$, so the chain goes nowhere and does nothing. Conversely, any transition probability distribution can be combined with any initial distribution, including those concentrated at one point. Such a chain is usually not stationary (even though the transition probabilities are stationary).

Having an equilibrium distribution is an important property of a Markov chain transition probability. In Section 1.8 below, we shall see that MCMC samples the equilibrium distribution, whether the chain is stationary or not. Not all Markov chains have equilibrium distributions, but all Markov chains used in MCMC do. The Metropolis–Hastings–Green (MHG) algorithm (Sections 1.12.2, 1.17.3.2, and 1.17.4.1 below) constructs transition probability mechanisms that preserve a specified equilibrium distribution.

1.5 Reversibility

A transition probability distribution is *reversible* with respect to an initial distribution if, for the Markov chain $X_1, X_2, \ldots$ they specify, the distribution of pairs (X_i, X_{i+1}) is exchangeable.

A Markov chain is *reversible* if its transition probability is reversible with respect to its initial distribution. Reversibility implies stationarity, but not vice versa. A reversible Markov chain has the same laws running forward or backward in time, that is, for any i and k the distributions of $(X_{i+1}, \ldots, X_{i+k})$ and $(X_{i+k}, \ldots, X_{i+1})$ are the same. Hence the name.

Reversibility plays two roles in Markov chain theory. All known methods for constructing transition probability mechanisms that preserve a specified equilibrium distribution in non-toy problems are special cases of the MHG algorithm, and all of the elementary updates constructed by the MHG algorithm are reversible (which accounts for its other name, the "reversible jump" algorithm). Combining elementary updates by composition (Section 1.12.7 below) may produce a combined update mechanism that is not reversible, but this does not diminish the key role played by reversibility in constructing transition probability mechanisms for MCMC. The other role of reversibility is to simplify the Markov chain central limit theorem (CLT) and asymptotic variance estimation. In the presence of reversibility the Markov chain CLT (Kipnis and Varadhan, 1986; Roberts and Rosenthal, 1997) is much sharper and the conditions are much simpler than without reversibility. Some methods of asymptotic variance estimation (Section 1.10.2 below) only work for reversible Markov chains but are much simpler and more reliable than analogous methods for nonreversible chains.

1.6 Functionals

If $X_1, X_2, \ldots$ is a stochastic process and g is a real-valued function on its state space, then the stochastic process $g(X_1), g(X_2), \ldots$ having state space $\mathbb{R}$ is said to be a *functional* of $X_1, X_2, \ldots$.

If $X_1, X_2, \ldots$ is a Markov chain, then a functional $g(X_1), g(X_2), \ldots$ is usually not a Markov chain. The conditional distribution of X_{n+1} given $X_1, \ldots, X_n$ depends only on X_n, but this does not, in general, imply that the conditional distribution of $g(X_{n+1})$ given $g(X_1), \ldots, g(X_n)$ depends only on $g(X_n)$. Nevertheless, functionals of Markov chains have important properties not shared by other stochastic processes.

1.7 The Theory of Ordinary Monte Carlo

Ordinary Monte Carlo (OMC), also called "independent and identically distributed (i.i.d.) Monte Carlo" or "good old-fashioned Monte Carlo," is the special case of MCMC in which $X_1, X_2, \ldots$ are independent and identically distributed, in which case the Markov chain is stationary and reversible.

Suppose you wish to calculate an expectation

$$\mu = \mathrm{E}\{g(X)\},\tag{1.1}$$

where g is a real-valued function on the state space, but you cannot do it by exact methods (integration or summation using pencil and paper, a computer algebra system, or exact numerical methods). Suppose you can simulate $X_1, X_2, \ldots$ i.i.d. having the same distribution as X. Define

$$\hat{\mu}_n = \frac{1}{n} \sum_{i=1}^{n} g(X_i). \tag{1.2}$$

If we introduce the notation $Y_i = g(X_i)$, then the Y_i are i.i.d. with mean μ and variance

$$\sigma^2 = \text{var}\{g(X)\}, \tag{1.3}$$

$\hat{\mu}_n$ is the sample mean of the Y_i, and the CLT says that

$$\hat{\mu}_n \approx N\left(\mu, \frac{\sigma^2}{n}\right). \tag{1.4}$$

The variance in the CLT can be estimated by

$$\hat{\sigma}_n^2 = \frac{1}{n} \sum_{i=1}^{n} \left(g(X_i) - \hat{\mu}_n\right)^2, \tag{1.5}$$

which is the empirical variance of the Y_i. Using the terminology of Section 1.6, we can also say that $\hat{\mu}_n$ is the sample mean of the functional $g(X_1), g(X_2), \ldots$ of $X_1, X_2, \ldots$.

The theory of OMC is just elementary statistics. For example, $\hat{\mu}_n \pm 1.96 \cdot \hat{\sigma}_n/\sqrt{n}$ is an asymptotic 95% confidence interval for μ. Note that OMC obeys what an elementary statistics text (Freedman et al., 2007) calls the *square root law*: statistical accuracy is inversely proportional to the square root of the sample size. Consequently, the accuracy of Monte Carlo methods is limited. Each additional significant figure, a tenfold increase in accuracy, requires a hundredfold increase in the sample size.

The only tricky issue is that the randomness involved is the pseudorandomness of computer simulation, rather than randomness of real-world phenomena. Thus it is a good idea to use terminology that emphasizes the difference. We call Equation 1.2 the *Monte Carlo approximation* or *Monte Carlo calculation* of μ, rather than the "point estimate" or "point estimator" of μ, as we would if not doing Monte Carlo. We call n the *Monte Carlo sample size*, rather than just the "sample size." We call $\hat{\sigma}_n/\sqrt{n}$ the *Monte Carlo standard error* (MCSE), rather than just the "standard error." We also do not refer to Equation 1.1 as an unknown parameter, even though we do not know its value. It is simply the expectation we are trying to calculate, known in principle, although unknown in practice, since we do not know how to calculate it other than by Monte Carlo approximation.

It is especially important to use this terminology when applying Monte Carlo to statistics. When the expectation (Equation 1.1) arises in a statistical application, there may already be a sample size in this application, which is unrelated to the Monte Carlo sample size, and there may already be standard errors unrelated to MCSEs. It can be hopelessly confusing if these are not carefully distinguished.

1.8 The Theory of MCMC

The theory of MCMC is just like the theory of OMC, except that stochastic dependence in the Markov chain changes the standard error. We start as in OMC with an expectation (Equation 1.1) that we cannot do other than by Monte Carlo. To begin the discussion, suppose that $X_1, X_2, \ldots$ is a stationary Markov chain having initial distribution the same as the distribution of X. We assume that the Markov chain CLT (Equation 1.4) holds, where now

$$\sigma^2 = \text{var}\{g(X_i)\} + 2 \sum_{k=1}^{\infty} \text{cov}\{g(X_i), g(X_{i+k})\} \tag{1.6}$$

(this formula is correct only for stationary Markov chains; see below for nonstationary chains). Since the asymptotic variance (Equation 1.6) is more complicated than the i.i.d. case (Equation 1.3), it cannot be estimated by Equation 1.5. It can, however, be estimated in several ways discussed below (Section 1.10). Conditions for the Markov chain CLT to hold (Chan and Geyer, 1994; Jones, 2004; Roberts and Rosenthal, 1997, 2004; Tierney, 1994) are beyond the scope of this chapter.

Now we come to a somewhat confusing issue. We never use stationary Markov chains in MCMC, because if we could simulate X_1 so that it has the invariant distribution, then we could also simulate $X_2, X_3, \ldots$ in the same way and do OMC. It is a theorem, however, that, under a condition (Harris recurrence) that is easier to verify than the CLT (Chan and Geyer, 1994; Tierney, 1994), if the CLT holds for one initial distribution and transition probability, then it holds for all initial distributions and that same transition probability (Meyn and Tweedie, 1993, Proposition 17.1.6), and the asymptotic variance is the same for all initial distributions. Although the theoretical asymptotic variance formula (Equation 1.6) contains variances and covariances for the stationary Markov chain, it also gives the asymptotic variance for nonstationary Markov chains having the same transition probability distribution (but different initial distributions). In practice, this does not matter, because we can never calculate (Equation 1.6) exactly except in toy problems and must estimate it from our simulations.

1.8.1 Multivariate Theory

Suppose that we wish to approximate by Monte Carlo (Equation 1.1) where we change notation so that μ is a vector with components μ_r and $g(x)$ is a vector with components $g_r(x)$. Our Monte Carlo estimator is still given by Equation 1.2, which is now also a vector equation because each $g(X_i)$ is a vector. Then the multivariate Markov chain CLT says that

$$\hat{\mu}_n \approx N(\mu, n^{-1}\Sigma),$$

where

$$\Sigma = \text{var}\{g(X_i)\} + 2 \sum_{k=1}^{\infty} \text{cov}\{g(X_i), g(X_{i+k})\}, \tag{1.7}$$

and where, although the right-hand sides of Equations 1.6 and 1.7 are the same, they mean different things: in Equation 1.7 $\text{var}\{g(X_i)\}$ denotes the matrix with components $\text{cov}\{g_r(X_i), g_s(X_i)\}$ and $\text{cov}\{g(X_i), g(X_{i+k})\}$ denotes the matrix with components $\text{cov}\{g_r(X_i), g_s(X_{i+k})\}$.

Conditions for the multivariate CLT to hold are essentially the same as for the univariate CLT. By the Cramér–Wold theorem, the multivariate convergence in distribution $Z_n \xrightarrow{D} Z$ holds if and only if the univariate convergence in distribution $t'Z_n \xrightarrow{D} t'Z$ holds for every nonrandom vector t. Thus the multivariate CLT essentially follows from the univariate CLT, and is often not discussed. It is important, however, for users to understand that the multivariate CLT does hold and can be used when needed.

1.8.2 The Autocovariance Function

We introduce terminology for the covariances that appear in Equation 1.6:

$$\gamma_k = \mathrm{cov}\{g(X_i), g(X_{i+k})\} \tag{1.8}$$

is called the *lag-k autocovariance* of the functional $g(X_1), g(X_2), \ldots$. Recall that in Equation 1.8 as in Equation 1.6 the covariances refer to the stationary chain with the same transition probability distribution as the chain being used. The variance that appears in Equation 1.6 is then γ_0. Hence, (Equation 1.6) can be rewritten

$$\sigma^2 = \gamma_0 + 2\sum_{k=1}^{\infty} \gamma_k. \tag{1.9}$$

The function $k \mapsto \gamma_k$ is called the *autocovariance function* of the functional $g(X_1), g(X_2), \ldots$, and the function $k \mapsto \gamma_k/\gamma_0$ is called the *autocorrelation function* of this functional.

The natural estimator of the autocovariance function is

$$\hat{\gamma}_k = \frac{1}{n}\sum_{i=1}^{n-k} [g(X_i) - \hat{\mu}_n][g(X_{i+k}) - \hat{\mu}_n] \tag{1.10}$$

It might be thought that one should divide by $n - k$ instead of n, but the large k terms are already very noisy so dividing by $n - k$ only makes a bad situation worse. The function $k \mapsto \hat{\gamma}_k$ is called the *empirical autocovariance function* of the functional $g(X_1), g(X_2), \ldots$, and the function $k \mapsto \hat{\gamma}_k/\hat{\gamma}_0$ is called the *empirical autocorrelation function* of this functional.

1.9 AR(1) Example

We now look at a toy problem for which exact calculation is possible. An AR(1) process (AR stands for autoregressive) is defined recursively by

$$X_{n+1} = \rho X_n + Y_n, \tag{1.11}$$

where Y_n are i.i.d. $N(0, \tau^2)$ and X_1 may have any distribution with finite variance. From Equation 1.11 we get

$$\mathrm{cov}(X_{n+k}, X_n) = \rho \, \mathrm{cov}(X_{n+k-1}, X_n) = \ldots = \rho^{k-1} \mathrm{cov}(X_{n-1}, X_n) = \rho^k \, \mathrm{var}(X_n). \tag{1.12}$$

If the process is stationary, then

$$\mathrm{var}(X_n) = \mathrm{var}(X_{n+1}) = \rho^2 \, \mathrm{var}(X_n) + \mathrm{var}(Y_n)$$

so

$$\mathrm{var}(X_n) = \frac{\tau^2}{1 - \rho^2} \tag{1.13}$$

and since variances are nonnegative, we must have $\rho^2 < 1$. Since a linear combination of independent normal random variables is normal, we see that the normal distribution with mean zero and variance (Equation 1.13) is invariant. Define v^2 to be another notation for the right-hand side of Equation 1.13 so the invariant distribution is $N(0, v^2)$.

It can be shown that this is the unique invariant distribution and this Markov chain obeys the CLT. The variance in the CLT is

$$\sigma^2 = \mathrm{var}(X_i) + 2 \sum_{k=1}^{\infty} \mathrm{cov}(X_i, X_{i+k})$$

$$= \frac{\tau^2}{1 - \rho^2} \left(1 + 2 \sum_{k=1}^{\infty} \rho^k \right)$$

$$= \frac{\tau^2}{1 - \rho^2} \left(1 + \frac{2\rho}{1 - \rho} \right) \tag{1.14}$$

$$= \frac{\tau^2}{1 - \rho^2} \cdot \frac{1 + \rho}{1 - \rho}$$

$$= v^2 \cdot \frac{1 + \rho}{1 - \rho}.$$

1.9.1 A Digression on Toy Problems

It is hard to know what lessons to learn from a toy problem. Unless great care is taken to point out which features of the toy problem are like real applications and which unlike, readers may draw conclusions that do not apply to real-world problems.

Here we are supposed to pretend that we do not know the invariant distribution, and hence we do not know that the expectation we are trying to estimate, $\mu = E(X)$, where X has the invariant distribution, is zero.

We cannot be interested in any functional of the Markov chain other than the one induced by the identity function, because we cannot do the analog of Equation 1.14 for any function g other than the identity function, and thus would not have a closed-form expression for the variance in the Markov chain CLT, which is the whole point of this toy problem.

Observe that Equation 1.14 goes to infinity as $\rho \to 1$. Thus in order to obtain a specified accuracy for $\hat{\mu}_n$ as an approximation to μ, say $\sigma/\sqrt{n} = \epsilon$, we may need a very large Monte Carlo sample size n. How large n must be depends on how close ρ is to one. When we pretend that we do not know the asymptotic variance (Equation 1.14), which we should do because the asymptotic variance is never known in real applications, all we can conclude is that we may need the Monte Carlo sample size to be very large and have no idea how large.

We reach the same conclusion if we are only interested in approximation error relative to the standard deviation υ of the invariant distribution, because

$$\frac{\sigma^2}{\upsilon^2} = \frac{1 + \rho}{1 - \rho} \tag{1.15}$$

also goes to infinity as $\rho \to 1$.

1.9.2 Supporting Technical Report

In order to avoid including laborious details of examples while still making all examples fully reproducible, those details are relegated to a technical report (Geyer, 2010a) or the vignettes for the R package mcmc (Geyer, 2010b). All calculations in this technical report or those package vignettes are done using the R function Sweave, so all results in them are actually produced by the code shown therein and hence are fully reproducible by anyone who has R. Moreover, anyone can download the Sweave source for the technical report from the URL given in the references at the end of this chapter or find the Sweave source for the package vignettes in the doc directory of any installation of the mcmc package, separate the R from the LATEX using the Stangle function, and play with it to see how the examples work.

1.9.3 The Example

For our example, we choose $\rho = 0.99$ and Monte Carlo sample size $n = 10^4$. This makes the MCSE about 14% of the standard deviation of the invariant distribution, which is a pretty sloppy approximation. To get the relative MCSE down to 10%, we would need $n = 2 \times 10^4$. To get the relative MCSE down to 1%, we would need $n = 2 \times 10^6$.

Figure 1.1 shows a time series plot of one MCMC run for this AR(1) process. From this plot we can see that the series seems stationary—there is no obvious trend or change in spread. We can also get a rough idea of how much dependence there is in the chain by

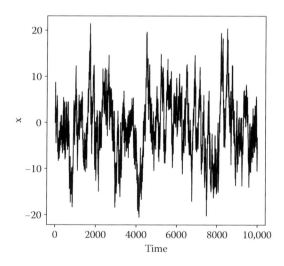

FIGURE 1.1
Time series plot for AR(1) example.

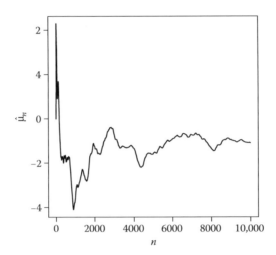

FIGURE 1.2
Running averages plot for AR(1) example.

counting large wiggles. The ratio of the variance in the CLT to the variance of the invariant distribution (Equation 1.15) is 199 for this example. Hence, this MCMC sample is about as useful as an i.i.d. sample with the same marginal distribution of sample size $10^4/199 \approx 50$.

Figure 1.2 shows a running averages plot for the same run shown in Figure 1.1. For some reason, these running averages plots seem popular among MCMC users although they provide no useful information. We know that MCMC, like OMC, obeys the square root law. A plot like Figure 1.2 does illustrate that $1/\sqrt{n}$ is a decreasing function of n, but not much else. Elementary statistics texts (Freedman et al., 2007, p. 276) often include one (and only one) figure like our Figure 1.2 to illustrate to naive students how the law of averages works. We have included Figure 1.2 only as an example of what not to do. In particular, such running averages plots should never be used to illustrate talks, since they tell the audience nothing they do not already know. Show a time series plot, like Figure 1.1, instead.

Figure 1.3 shows an autocorrelation plot for the same run shown in Figure 1.1. The black bars show the empirical autocorrelation function (ACF) defined in Section 1.8.2. We could let the domain of the ACF be zero to $n - 1$, but the R function `acf` cuts the plot at the argument `lag.max`. The `acf` function automatically adds the horizontal dashed lines, which the documentation for `plot.acf` says are 95% confidence intervals assuming white noise input. The dotted curve is the simulation truth autocorrelation function ρ^k derived from Equation 1.12. In the spirit of this toy problem, we are supposed to pretend we do not know the dotted curve, since we would not have its analog in any real application. We can see, however, how well (not very) the empirical ACF matches the theoretical ACF.

It should come as no surprise that the empirical ACF estimates the theoretical ACF less well than $\hat{\mu}_n$ estimates μ. Even in i.i.d. sampling, the mean is always much better estimated than the variance.

The ACF is well enough estimated, however, to give some idea how far significant autocorrelation extends in our Markov chain. Of course, the theoretical autocorrelation is nonzero for all lags, no matter how large, but we know (although we pretend we do not) that they decrease exponentially fast. They are not practically significantly different from zero past lag 500.

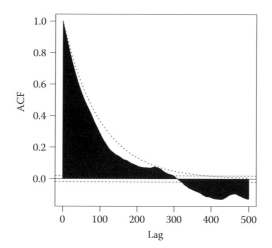

FIGURE 1.3
Autocorrelation plot for AR(1) Example. Dashed lines: 95% confidence intervals assuming white noise input. Dotted curve: simulation truth autocorrelation function.

1.10 Variance Estimation

Many methods of variance estimation have been proposed. Most come from the time series literature and are applicable to arbitrary stationary stochastic processes, not just to Markov chains. We will cover only a few very simple, but very effective, methods.

1.10.1 Nonoverlapping Batch Means

A *batch* is simply a subsequence of consecutive iterates of the Markov chain $X_{k+1}, \ldots, X_{k+b}$. The number b is called the *batch length*. If we assume the Markov chain is stationary, then all batches of the same length have the same joint distribution, and the CLT applies to each batch. The batch mean

$$\frac{1}{b} \sum_{j=1}^{b} g(X_{k+j})$$

is a Monte Carlo approximation of the expectation (Equation 1.1) we are trying to calculate, and its distribution is approximately $N(\mu, \sigma^2/b)$, where, as before, σ^2 is given by Equation 1.6. A batch of length b is just like the entire run of length n, except for length. The sample mean of a batch of length b is just like the sample mean of the entire run of length n, except that the asymptotic variance is σ^2/b instead of σ^2/n.

Suppose b divides n evenly. Divide the whole run into m nonoverlapping batches of length b. Average these batches:

$$\hat{\mu}_{b,k} = \frac{1}{b} \sum_{i=b(k-1)+1}^{bk} g(X_i). \tag{1.16}$$

Then

$$\frac{1}{m} \sum_{k=1}^{m} (\hat{\mu}_{b,k} - \hat{\mu}_n)^2 \tag{1.17}$$

estimates σ^2/b.

It is important to understand that the stochastic process $\hat{\mu}_{b,1}, \hat{\mu}_{b,2}, \ldots$ is also a functional of a Markov chain, not the original Markov chain but a different one. If S is the state space of the original Markov chain $X_1, X_2, \ldots$, then the batches

$$(X_{b(k-1)+1}, \ldots, X_{kb}), \quad k = 1, 2, \ldots$$

also form a Markov chain with state space S^b, because the conditional distribution of one batch $(X_{b(k-1)+1}, \ldots, X_{kb})$ given the past history actually depends only on $X_{b(k-1)}$, which is a component of the immediately preceding batch. The batch means are a functional of this Markov chain of batches.

Figure 1.4 shows a batch mean plot for the same run shown in Figure 1.1. The batch length is 500, the run length is 10^4, so the number of batches is 20. Like the running averages plot (Figure 1.2), we do not recommend this kind of plot for general use, because it does not show anything a sophisticated MCMC user should not already know. It is useful to show such a plot (once) in a class introducing MCMC, to illustrate the point that the stochastic process shown is a functional of a Markov chain. It is not useful for talks about MCMC.

Figure 1.5 shows the autocorrelation plot of the batch mean stochastic process for the same run shown in Figure 1.1, which shows the batches are not significantly correlated, because all of the bars except the one for lag 0 are inside the dashed lines. In this case, a confidence interval for the unknown expectation (Equation 1.1) is easily done using the R function t.test:

```
> t.test(batch)
        One Sample t-test

data:   batch
t = -1.177, df = 19, p-value = 0.2537
alternative hypothesis: true mean is not equal to 0
95 percent confidence interval:
 -2.5184770  0.7054673
sample estimates:
 mean of x
-0.9065049
```

Here, batch is the vector of batch means which is plotted in Figure 1.4.

If this plot had shown the batches to be significantly correlated, then the method of batch means should not have been used because it would have a significant downward bias. However, the time series of batches can still be used, as explained in Section 1.10.2 below.

How does one choose the batch length? The method of batch means will work well only if the batch length b is large enough so that the infinite sum in Equation 1.9 is well approximated by the partial sum of the first b terms. Hence, when the method of batch means is used blindly with no knowledge of the ACF, b should be as large as possible. The only restriction on the length of batches is that the number of batches should be enough to

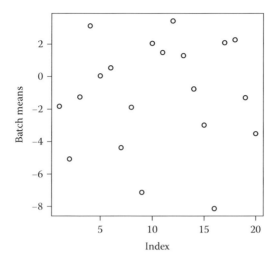

FIGURE 1.4
Batch mean plot for AR(1) example. Batch length 500.

get a reasonable estimate of variance. If one uses a *t* test, as shown above, then the *t* critical value corrects for the number of batches being small (Geyer, 1992; Schmeiser, 1982), but there is no point in the number of batches being so small that that the variance estimate is extremely unstable: 20–30 batches is a reasonable recommendation. One sometimes sees assumptions that the number of batches "goes to infinity" in theorems, but this is not necessary for simple MCSE calculation (Geyer, 1992, Section 3.2). If one is using estimated variance in a sequential stopping rule (Glynn and Whitt, 1991, 1992), then one does need the number of batches to go to infinity.

Meketon and Schmeiser (1984) pointed out that the batch means estimator of variance (Equation 1.17) is still valid if the batches are allowed to overlap, and a slight gain in efficiency is thereby achieved. For reasons explained in the following section, we do not

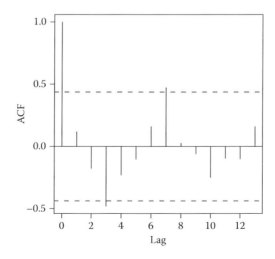

FIGURE 1.5
Autocorrelation plot of batch means for AR(1) example. Batch length 500.

recommend overlapping batch means, not because there is anything wrong with it, but because it does not fit together well with other methods we recommend.

1.10.2 Initial Sequence Methods

Another approach to variance estimation is to work directly with the representation (Equation 1.9) of the asymptotic variance. One cannot simply plug the empirical estimates (Equation 1.10) into Equation 1.9 because the variance of the high-lag terms does not decrease with lag, so as n goes to infinity an infinite amount of noise swamps the finite signal. Many solutions for this problem have been proposed in the time series literature (Geyer, 1992, Section 3.1 and references cited therein). But reversible Markov chains permit much simpler methods. Define

$$\Gamma_k = \gamma_{2k} + \gamma_{2k+1}. \tag{1.18}$$

Geyer (1992, Theorem 3.1) showed that the function $k \mapsto \Gamma_k$ is strictly positive, strictly decreasing, and strictly convex, and proposed three estimators of the asymptotic variance (Equation 1.9) that use these three properties, called the *initial positive sequence, initial monotone sequence,* and *initial convex sequence* estimators. Each is a consistent overestimate of the asymptotic variance (meaning the probability of underestimation by any fixed amount goes to zero as the Monte Carlo sample size goes to infinity) under no regularity conditions whatsoever (Geyer, 1992, Theorem 3.2). The initial convex sequence estimator is the best, because the smallest and still an asymptotic overestimate, but is a bit difficult to calculate. Fortunately, the R contributed package mcmc now has a function initseq that calculates all three estimators. We will only discuss the last. It forms

$$\widehat{\Gamma}_k = \hat{\gamma}_{2k} + \hat{\gamma}_{2k+1},$$

where $\hat{\gamma}_k$ is given by Equation 1.10, then finds the largest index m such that

$$\widehat{\Gamma}_k > 0, \quad k = 0, \ldots, m,$$

then defines $\widehat{\Gamma}_{m+1} = 0$, and then defines $k \mapsto \tilde{\Gamma}_k$ to be the greatest convex minorant of $k \mapsto \widehat{\Gamma}_k$ over the range $0, \ldots, m + 1$. Finally, it estimates

$$\hat{\sigma}^2_{\text{conv}} = -\hat{\gamma}_0 + 2 \sum_{k=0}^{m} \tilde{\Gamma}_k. \tag{1.19}$$

Figure 1.6 shows a plot of the function $k \mapsto \tilde{\Gamma}_k$ for the same run shown in Figure 1.1 compared to its theoretical value. When comparing this plot to Figure 1.3, remember that each index value in Figure 1.6 corresponds to two index values in Figure 1.3 because of the way Equation 1.18 is defined. Thus Figure 1.6 indicates significant autocorrelation out to about lag 300 (not 150).

The estimator of asymptotic variance (Equation 1.19) is calculated very simply in R:

```
> initseq(out)$var.con
[1] 7467.781
```

assuming the mcmc contributed package has already been loaded and out is the functional of the Markov chain for which the variance estimate is desired.

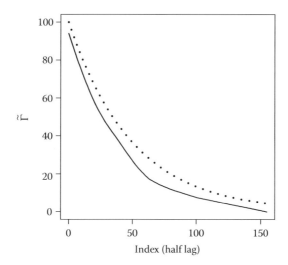

FIGURE 1.6
Plot of $\tilde{\Gamma}$ for AR(1) example. Solid line: initial convex sequence estimator of Equation 1.18. Dotted line: theoretical value.

1.10.3 Initial Sequence Methods and Batch Means

When the original Markov chain is reversible, so is the chain of batches. Hence, initial sequence methods can be applied to a sequence of nonoverlapping batch means derived from a reversible Markov chain.

This means that the method of nonoverlapping batch means can be used without testing whether the batches are large enough. Simply process them with an initial sequence method, and the result is valid regardless of the batch length.

Here is how that works. Suppose we use a batch length of 50, which is too short.

```
> blen * var(batch)
[1] 2028.515
> blen * initseq(batch)$var.con
[1] 7575.506
```

The naive batch means estimator is terrible, less than a third of the size of the initial convex sequence estimator applied to the batch means (7575.506), but this is about the same as the initial convex sequence estimator applied to the original output (7467.781). So nothing is lost when only nonoverlapping batch means are output, regardless of the batch length used.

Partly for this reason, and partly because nonoverlapping batch means are useful for reducing the size of the output, whereas overlapping batch means are not, we do not recommend overlapping batch means and will henceforth always use the term *batch means* to mean nonoverlapping batch means.

1.11 The Practice of MCMC

The practice of MCMC is simple. Set up a Markov chain having the required invariant distribution, and run it on a computer. The folklore of simulation makes this seem more complicated than it really is. None of this folklore is justified by theory and none of it actually helps users do good simulations, but, like other kinds of folklore, it persists despite its lack of validity.

1.11.1   Black Box MCMC

There is a great deal of theory about convergence of Markov chains. Unfortunately, none of it can be applied to get useful convergence information for most MCMC applications. Thus most users find themselves in the following situation we call *black box MCMC*:

1. You have a Markov chain having the required invariant distribution.
2. You know nothing other than that. The Markov chain is a "black box" that you cannot see inside. When run, it produces output. That is all you know. You know nothing about the transition probabilities of the Markov chain, nor anything else about its dynamics.
3. You know nothing about the invariant distribution except what you may learn from running the Markov chain.

Point 2 may seem extreme. You may know a lot about the particular Markov chain being used—for example, you may know that it is a Gibbs sampler—but if whatever you know is of no help in determining any convergence information about the Markov chain, then whatever knowledge you have is useless. Point 3 may seem extreme. Many examples in the MCMC literature use small problems that can be done by OMC or even by pencil and paper and for which a lot of information about the invariant distribution is available, but in complicated applications point 3 is often simply true.

1.11.2   Pseudo-Convergence

A Markov chain can appear to have converged to its equilibrium distribution when it has not. This happens when parts of the state space are poorly connected by the Markov chain dynamics: it takes many iterations to get from one part to another. When the time it takes to transition between these parts is much longer than the length of simulated Markov chain, then the Markov chain can appear to have converged but the distribution it appears to have converged to is the equilibrium distribution conditioned on the part in which the chain was started. We call this phenomenon *pseudo-convergence*.

This phenomenon has also been called "multimodality" since it may occur when the equilibrium distribution is multimodal. But multimodality does not cause pseudo-convergence when the troughs between modes are not severe. Nor does pseudo-convergence only happen when there is multimodality. Some of the most alarming cases of pseudo-convergence occur when the state space of the Markov chain is discrete and "modes" are not well defined (Geyer and Thompson, 1995). Hence pseudo-convergence is a better term.

1.11.3   One Long Run versus Many Short Runs

When you are in the black box situation, you have no idea how long runs need to be to get good mixing (convergence rather than pseudo-convergence). If you have a run that is already long enough, then an autocovariance plot like Figure 1.6 gives good information about mixing, and you know that you need to run a large multiple of the time it takes the autocovariances to decay to nearly zero. But if all the runs you have done so far are nowhere near long enough, then they provide no information about how long is long enough.

The phenomenon of pseudo-convergence has led many people to the idea of comparing multiple runs of the sampler started at different points. If the multiple runs appear to converge to the same distribution, then—according to the multistart heuristic—all is well. But this assumes that you can arrange to have at least one starting point in each part of the state space to which the sampler can pseudo-converge. If you cannot do that—and in the black box situation you never can—then the multistart heuristic is worse than useless: it can give you confidence that all is well when in fact your results are completely erroneous.

Worse, addiction to many short runs can keep one from running the sampler long enough to detect pseudo-convergence or other problems, such as bugs in the code. People who have used MCMC in complicated problems can tell stories about samplers that appeared to be converging until, after weeks of running, they discovered a new part of the state space and the distribution changed radically. If those people had thought it necessary to make hundreds of runs, none of them could have been several weeks long.

Your humble author has a dictum that the least one can do is to make an overnight run. What better way for your computer to spend its time? In many problems that are not too complicated, this is millions or billions of iterations. If you do not make runs like that, you are simply not serious about MCMC. Your humble author has another dictum (only slightly facetious) that one should start a run when the paper is submitted and keep running until the referees' reports arrive. This cannot delay the paper, and may detect pseudo-convergence.

1.11.4 Burn-In

Burn-in is a colloquial term that describes the practice of throwing away some iterations at the beginning of an MCMC run. This notion says that you start somewhere, say at x, then you run the Markov chain for n steps (the burn-in period) during which you throw away all the data (no output). After the burn-in you run normally, using each iterate in your MCMC calculations.

The name "burn-in" comes from electronics. Many electronics components fail quickly. Those that do not are a more reliable subset. So a burn-in is done at the factory to eliminate the worst ones.

Markov chains do not work the same way. Markov chain "failure" (nonconvergence or pseudo-convergence) is different from electronic component failure. Running longer may cure the first, but a dead transistor is dead forever. Thus "burn-in" is a bad term in MCMC, but there is more wrong than just the word, there is something fishy about the whole concept.

Figure 1.7 illustrates the issue that burn-in addresses. It shows an AR(1) time series with all parameters except starting position the same as Figure 1.1 so the equilibrium distribution, normal with mean zero and variance (Equation 1.13), is the same for both. In Figure 1.7 the starting position is far out in the tail of the equilibrium distribution, 10 standard deviations from the mean. In Figure 1.1 the starting position is the mean (zero). It takes several hundred iterations before the sample path in Figure 1.7 gets into the region containing the whole sample path in Figure 1.1.

The naive idea behind burn-in is that if we throw away several hundred iterations from Figure 1.7 it will be just as good as Figure 1.1. Overgeneralizing examples like Figure 1.7 leads to the idea that every MCMC run should have burn-in. Examples like Figure 1.1 show that this is not so. A Markov chain started anywhere near the center of the equilibrium distribution needs no burn-in.

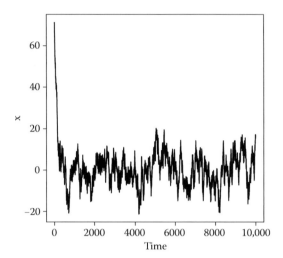

FIGURE 1.7
Time series plot for AR(1) example. Differs from Figure 1.1 only in the starting position.

> Burn-in is only one method, and not a particularly good method, of finding a good starting
> point.

There are several methods other than burn-in for finding a good starting point. One rule
that is unarguable is

> Any point you don't mind having in a sample is a good starting point.

In a typical application, one has no theoretical analysis of the Markov chain dynamics
that tells where the good starting points are (nor how much burn-in is required to get
to a good starting point). All decisions about starting points are based on the output of
some preliminary runs that appear to have "converged." Any point of the parts of these
preliminary runs one believes to be representative of the equilibrium distribution is as good
a starting point as any other.

So a good rule to follow is to start the next run where the last run ended. This is the rule
most authorities recommend for random number generator seeds and the one used by R.
It is also used by functions in the R package mcmc as discussed in Section 1.13 below.

Another method is to start at a mode of the equilibrium distribution (which can sometimes
be found by optimization before doing MCMC) if it is known to be in a region of appreciable
probability.

None of the examples in this chapter use burn-in. All use an alternative method of finding
starting points. Burn-in is mostly harmless, which is perhaps why the practice persists. But
everyone should understand that it is unnecessary, and those who do not use it are not
thereby making an error.

Burn-in has a pernicious interaction with the multistart heuristic. If one believes in multistart, then one feels the need to start at many widely dispersed, and hence bad, starting
points. Thus all of these short runs need be shortened some more by burn-in. Thus an
erroneous belief in the virtues of multistart leads to an erroneous belief in the necessity of
burn-in.

Another erroneous argument for burn-in is unbiasedness. If one could start with a
realization from the equilibrium distribution, then the Markov chain would be stationary

and the Monte Carlo approximation (Equation 1.2) would be an unbiased estimator of what it estimates (Equation 1.1). Burn-in does not produce a realization from the equilibrium distribution, hence does not produce unbiasedness. At best it produces a small bias, but the alternative methods also do that. Moreover, the bias is of order n^{-1}, where n is the Monte Carlo sample size, whereas the MCSE is of order $n^{-1/2}$, so bias is negligible in sufficiently long runs.

1.11.5 Diagnostics

Many MCMC diagnostics have been proposed in the literature. Some work with one run of a Markov chain, but tell little that cannot be seen at a glance at a time series plot like Figure 1.1 or an autocorrelation plot like Figure 1.3. Others with multiple runs of a Markov chain started at different points, what we called the multistart heuristic above. Many of these come with theorems, but the theorems never prove the property you really want a diagnostic to have. These theorems say that if the chain converges, then the diagnostic will probably say that the chain converged, but they do not say that if the chain pseudo-converges, then the diagnostic will probably say that the chain did not converge. Theorems that claim to reliably diagnose pseudo-convergence have unverifiable conditions that make them useless. For example, as we said above, it is clear that a diagnostic based on the multistart heuristic will reliably diagnose pseudo-convergence if there is at least one starting point in each part of the state space to which the sampler can pseudo-converge, but in practical applications one has no way of arranging that.

There is only one perfect MCMC diagnostic: perfect sampling (Propp and Wilson, 1996; Kendall and Møller, 2000; see also Chapter 8, this volume). This is best understood as not a method of MCMC but rather a method of Markov-chain-assisted i.i.d. sampling. Since it is guaranteed to produce an i.i.d. sample from the equilibrium distribution of the Markov chain, a sufficiently large sample is guaranteed to not miss any parts of the state space having appreciable probability. Perfect sampling is not effective as a sampling scheme. If it works, then simply running the underlying Markov chain in MCMC mode will produce more accurate results in the same amount of computer time. Thus, paradoxically, perfect sampling is most useful when it fails to produce an i.i.d. sample of the requested size in the time one is willing to wait. This shows that the underlying Markov chain is useless for sampling, MCMC or perfect.

Perfect sampling does not work on black box MCMC (Section 1.11.1 above), because it requires complicated theoretical conditions on the Markov chain dynamics. No other diagnostic ever proposed works on black box MCMC, because if you know nothing about the Markov chain dynamics or equilibrium distribution except what you learn from output of the sampler, you can always be fooled by pseudo-convergence.

> There are known knowns. These are things we know that we know. There are known unknowns. That is to say, there are things that we now know we don't know. But there are also unknown unknowns. These are things we do not know we don't know.
>
> **Donald Rumsfeld**
> *US Secretary of Defense*

Diagnostics can find the known unknowns. They cannot find the unknown unknowns. They cannot find out what a black box MCMC sampler will do eventually. Only sufficiently long runs can do that.

1.12   Elementary Theory of MCMC

We say that a bit of computer code that makes a pseudorandom change to its state is an *update mechanism*. We are interested in update mechanisms that preserve a specified distribution, that is, if the state has the specified distribution before the update, then it has the same distribution after the update. From them we can construct Markov chains to sample that distribution.

We say that an update mechanism is *elementary* if it is not made up of parts that are themselves update mechanisms preserving the specified distribution.

1.12.1   The Metropolis–Hastings Update

Suppose that the specified distribution (the desired stationary distribution of the MCMC sampler we are constructing) has *unnormalized density* h. This means that h is a positive constant times a probability density. Thus h is a nonnegative-valued function that integrates (for continuous state) or sums (for discrete state) to a value that is finite and nonzero. The *Metropolis–Hastings update* does the following:

- When the current state is x, propose a move to y, having conditional probability density given x denoted $q(x, \cdot)$.

- Calculate the *Hastings ratio*

$$r(x, y) = \frac{h(y)q(y, x)}{h(x)q(x, y)}. \tag{1.20}$$

- Accept the proposed move y with probability

$$a(x, y) = \min(1, r(x, y)), \tag{1.21}$$

  that is, the state after the update is y with probability $a(x, y)$, and the state after the update is x with probability $1 - a(x, y)$.

The last step is often called *Metropolis rejection*. The name is supposed to remind one of "rejection sampling" in OMC, but this is a misleading analogy because in OMC rejection sampling is done repeatedly until some proposal is accepted (so it always produces a new value of the state). In contrast, one Metropolis–Hastings update makes one proposal y, which is the new state with probability $a(x, y)$, but otherwise the new state the same as the old state x. Any attempt to make Metropolis rejection like OMC rejection, destroys the property that this update preserves the distribution with density h.

The Hastings ratio (Equation 1.20) is undefined if $h(x) = 0$, thus we must always arrange that $h(x) > 0$ in the initial state. There is no problem if $h(y) = 0$. All that happens is that $r(x, y) = 0$ and the proposal y is accepted with probability zero. Thus the Metropolis–Hastings update can never move to a new state x having $h(x) = 0$. Note that the proposal y must satisfy $q(x, y) > 0$ with probability one because $q(x, \cdot)$ is the conditional density of y given x. Hence, still assuming $h(x) > 0$, the denominator of the Hastings ratio is nonzero with probability one, and the Hastings ratio is well defined. Note that either term of the numerator of the Hastings ratio can be zero, so the proposal is almost surely rejected if either $h(y) = 0$ or $q(y, x) = 0$, that is, if y is an impossible value of the desired equilibrium distribution or if x is an impossible proposal when y is the current state.

We stress that nothing bad happens if the proposal y is an impossible value of the desired equilibrium distribution. The Metropolis–Hastings update automatically does the right thing, almost surely rejecting such proposals. Hence, it is not necessary to arrange that proposals are always possible values of the desired equilibrium distribution; it is only necessary to assure that one's implementation of the unnormalized density function h works when given any possible proposal as an argument and gives $h(y) = 0$ when y is impossible.

If `unifrand` is a function with no arguments that produces one $U(0, 1)$ random variate and the Hastings ratio has already been calculated and stored in a variable r, then the following computer code does the Metropolis rejection step:

```
if (unifrand() < r) {
    x = y
}
```

The variable x, which is considered the state of the Markov chain, is set to y (the proposal) when a uniform random variate is less than the Hastings ratio r and left alone otherwise.

The following computer code works with the log Hastings ratio `logr` to avoid overflow:

```
if (logr >= 0 || unifrand() < exp(logr)) {
    x = y
}
```

It uses the "short circuit" property of the `||` operator in the R or C language. Its second operand `unifrand() < exp(logr)` is only evaluated when its first operand `logr >= 0` evaluates to FALSE. Thus `exp(logr)` can never overflow.

1.12.2 The Metropolis–Hastings Theorem

We now prove that the Metropolis–Hastings update is reversible with respect to h, meaning that the transition probability that describes the update is reversible with respect to the distribution having unnormalized density h.

If X_n is the current state and Y_n is the proposal, we have $X_n = X_{n+1}$ whenever the proposal is rejected. Clearly, the distribution of (X_n, X_{n+1}) given rejection is exchangeable.

Hence, it only remains to be shown that (X_n, Y_n) is exchangeable given acceptance. We need to show that

$$E\{f(X_n, Y_n)a(X_n, Y_n)\} = E\{f(Y_n, X_n)a(X_n, Y_n)\}$$

for any function f that has expectation (assuming X_n has desired stationary distribution). That is, we must show we can interchange arguments of f in

$$\iint f(x, y)h(x)a(x, y)q(x, y)\, dx\, dy \qquad (1.22)$$

(with integrals replaced by sums if the state is discrete), and that follows if we can interchange x and y in

$$h(x)a(x, y)q(x, y) \qquad (1.23)$$

because we can exchange x and y in Equation 1.22, x and y being dummy variables. Clearly only the set of x and y such that $h(x) > 0$, $q(x, y) > 0$, and $a(x, y) > 0$ contributes to the integral or (in the discrete case) sum (Equation 1.22), and these inequalities further imply that $h(y) > 0$ and $q(y, x) > 0$. Thus we may assume these inequalities, in which case we have

$$r(y, x) = \frac{1}{r(x, y)}$$

for all such x and y.

Suppose that $r(x, y) \leq 1$, so $r(x, y) = a(x, y)$ and $a(y, x) = 1$. Then

$$h(x)a(x, y)q(x, y) = h(x)r(x, y)q(x, y)$$
$$= h(y)q(y, x)$$
$$= h(y)q(y, x)a(y, x).$$

Conversely, suppose that $r(x, y) > 1$, so $a(x, y) = 1$ and $a(y, x) = r(y, x)$. Then

$$h(x)a(x, y)q(x, y) = h(x)q(x, y)$$
$$= h(y)r(y, x)q(y, x)$$
$$= h(y)a(y, x)q(y, x).$$

In either case we can exchange x and y in Equation 1.23, and the proof is done.

1.12.3 The Metropolis Update

The special case of the Metropolis–Hastings update when $q(x, y) = q(y, x)$ for all x and y is called the *Metropolis update*. Then the Hastings ratio (Equation 1.20) simplifies to

$$r(x, y) = \frac{h(y)}{h(x)} \qquad (1.24)$$

and is called the Metropolis ratio or the odds ratio. Thus Metropolis updates save a little time in calculating $r(x, y)$ but otherwise have no advantages over Metropolis–Hastings updates.

One obvious way to arrange the symmetry property is to make proposals of the form $y = x + e$, where e is stochastically independent of x and symmetrically distributed about zero. Then $q(x, y) = f(y - x)$, where f is the density of e. Widely used proposals of this type have e normally distributed with mean zero or e uniformly distributed on a ball or a hypercube centered at zero (see Section 1.12.10 below for more on such updates).

1.12.4 The Gibbs Update

In a *Gibbs update* the proposal is from a conditional distribution of the desired equilibrium distribution. It is always accepted.

The proof of the theorem that this update is reversible with respect to the desired equilibrium distribution is trivial. Suppose that X_n has the desired stationary distribution. Suppose that the conditional distribution of X_{n+1} given $f(X_n)$ is same as the conditional distribution of X_n given $f(X_n)$. Then the pair (X_n, X_{n+1}) is conditionally exchangeable given $f(X_n)$, hence unconditionally exchangeable.

In common parlance, a Gibbs update uses the conditional distribution of one component of the state vector given the rest of the components, that is, the special case of the update described above where $f(X_n)$ is X_n with one component omitted. Conditional distributions of this form are called "full conditionals." There is no reason other than tradition why such conditional distributions should be preferred.

In fact other conditionals have been considered in the literature. If $f(X_n)$ is X_n with several components omitted, this is called "block Gibbs." Again, there is no reason other than tradition why such conditional distributions should be preferred.

If one insists that Gibbs update only apply to full conditionals, then one could call the updates described here "generalized Gibbs." But the "generalized" here is not much of a generalization. Simply do a change of variable so that $f(X_n)$ is a group of components of the new state vector and "generalized Gibbs" is "block Gibbs." Also the argument for all these updates is exactly the same.

Gibbs updates have one curious property not shared by other Metropolis–Hastings updates: they are *idempotent*, meaning the effect of multiple updates is the same as the effect of just one. This is because the update never changes $f(X_n)$, hence the result of many repetitions of the same Gibbs update results in X_{n+1} having the conditional distribution given $f(X_n)$ just like the result of a single update. In order for Gibbs elementary updates to be useful, they must be combined somehow with other updates.

1.12.5 Variable-at-a-Time Metropolis–Hastings

Gibbs updates alter only part of the state vector; when using "full conditionals" the part is a single component. Metropolis–Hastings updates can be modified to do the same.

Divide the state vector into two parts, $x = (u, v)$. Let the proposal alter u but not v. Hence, the proposal density has the form $q(x, u)$ instead of the $q(x, y)$ we had in Section 1.12.1. Again let $h(x) = h(u, v)$ be the unnormalized density of the desired equilibrium distribution. The variable-at-a-time Metropolis–Hastings update does the following:

- When the current state is $x = (u, v)$, propose a move to $y = (u^*, v)$, where u^* has conditional probability density given x denoted $q(x, \cdot) = q(u, v, \cdot)$.
- Calculate the *Hastings ratio*

$$r(x, y) = \frac{h(u^*, v)q(u^*, v, u)}{h(u, v)q(u, v, u^*)}.$$

- Accept the proposed move y with probability (Equation 1.21), that is, the state after the update is y with probability $a(x, y)$, and the state after the update is x with probability $1 - a(x, y)$.

We shall not give a proof of the validity of variable-at-a-time Metropolis–Hastings, which would look very similar to the proof in Section 1.12.2.

The term "variable-at-a-time Metropolis–Hastings" is something of a misnomer. The sampler run in Metropolis et al. (1953) was a "variable-at-a-time" sampler. For historical accuracy, the name "Metropolis algorithm" should include the updates described in Section 1.12.1 and in this section. Current usage, however, seems otherwise, naming the samplers as we have done here.

1.12.6   Gibbs Is a Special Case of Metropolis–Hastings

To see that Gibbs is a special case of Metropolis–Hastings, do a change of variable so that the new state vector can be split $x = (u, v)$ as we did in the preceding section, and v is the part of the state on which the Gibbs update conditions. Thus we are doing block Gibbs updating u from its conditional distribution given v. Factor the unnormalized density $h(u, v) = g(v)q(v, u)$, where $g(v)$ is an unnormalized marginal of v and $q(v, u)$ is the (properly normalized) conditional of u given v. Now do a Metropolis–Hastings update with q as the proposal distribution. The proposal is $y = (u^*, v)$, where u^* has the distribution $q(v, \cdot)$. The Hastings ratio is

$$r(x, y) = \frac{h(u^*, v)q(u, v)}{h(u, v)q(v, u^*)} = \frac{g(v)q(v, u^*)q(u, v)}{g(v)q(v, u)q(v, u^*)} = 1.$$

Hence the proposal is always accepted.

1.12.7   Combining Updates

1.12.7.1   Composition

Let $P_1, \ldots, P_k$ be update mechanisms (computer code) and let $P_1 P_2 \cdots P_k$ denote the composite update that consists of these updates done in that order with P_1 first and P_k last. If each P_i preserves a distribution, then obviously so does $P_1 P_2 \ldots P_k$.

If $P_1, \ldots, P_k$ are the Gibbs updates for the "full conditionals" of the desired equilibrium distribution, then the composition update is often called a *fixed scan Gibbs sampler*.

As a simple example, suppose that the desired equilibrium distribution is exchangeable and multivariate normal. Then the conditional distribution of one component of the state vector given the rest is univariate normal with mean that is a symmetric linear function of the rest of the components and constant variance. In the special case where there are just two components, the fixed scan Gibbs sampler is just consecutive pairs of an AR(1) process (Section 1.9 above).

1.12.7.2   Palindromic Composition

Note that $P_1 P_2 \ldots P_k$ is not reversible with respect to the distribution it preserves unless the transition probabilities associated with $P_1 P_2 \ldots P_k$ and $P_k P_{k-1} \ldots P_1$ are the same.

The most obvious way to arrange reversibility is to make $P_i = P_{k-i}$, for $i = 1, \ldots, k$. Then we call this composite update *palindromic*. Palindromic compositions are reversible, nonpalindromic ones need not be.

1.12.8   State-Independent Mixing

Let P_y be update mechanisms (computer code) and let $E(P_Y)$ denote the update that consists of doing a random one of these updates: generate Y from some distribution and do P_Y.

If Y is independent of the current state and each P_y preserves the same distribution, then so does $E(P_Y)$. If X_n has the desired equilibrium distribution, then it also has this distribution conditional on Y, and X_{n+1} also has this distribution conditional on Y. Since the conditional distribution of X_{n+1} does not depend on Y, these variables are independent, and X_{n+1} has the desired equilibrium distribution unconditionally.

Furthermore, the Markov chain with update $E(P_Y)$ is reversible if each P_y is reversible.

"Mixture" is used here in the sense of mixture models. The update $E(P_Y)$ is the mixture of updates P_y.

The most widely used mixtures use a finite set of y values. For example, one popular way to combine the "full conditional" Gibbs updates, one for each component of the state vector, is by state-independent mixing using the uniform distribution on the set of full conditionals as the mixing distribution. This is often called a *random scan Gibbs sampler*. The choice of the uniform distribution is arbitrary. It has no optimality properties. It does, however, make a simple default choice.

Mixing and composition can be combined. Suppose we have elementary update mechanisms $P_1, \ldots, P_k$, and let $\mathcal{Y}$ be a set of functions from $\{1, \ldots, m\}$ to $\{1, \ldots, k\}$. For $y \in \mathcal{Y}$, let Q_y denote the composition $P_{y(1)} P_{y(2)} \ldots P_{y(m)}$. Now consider the update $E(Q_Y)$, where Y is a random element of $\mathcal{Y}$ independent of the state of the Markov chain.

If $m = k$ and the P_i are the "full conditional" Gibbs updates and Y has the uniform distribution on $\mathcal{Y}$, which consists of all permutations of $1, \ldots, k$, then this mixture of compositions sampler is often called a *random sequence scan Gibbs sampler*.

We are not fond of this "scan" terminology, because it is too limiting. It focuses attention on a very few special cases of combination by composition and mixing, special cases that have no optimality properties and no reason other than tradition for their prominence.

State-independent mixing with the mixing distribution having an infinite sample space has also been used. Bélisle et al. (1993) and Chen and Schmeiser (1993) investigate the "hit and run algorithm" which uses elementary updates P_y where the state space of the Markov chain is Euclidean and y is a direction in the state space. Do a change of coordinates so that y is a coordinate direction, and do a Gibbs or other variable-at-a-time Metropolis–Hastings update of the coordinate in the y direction. The mixture update $E(P_Y)$ is called a "hit and run sampler" when Y has the uniform distribution on directions.

Again there is no particular reason to use a "hit and run" sampler. It is merely one of an infinite variety of samplers using composition and state-independent mixing.

State-dependent mixing is possible, but the argument is very different (Section 1.17.1 below).

1.12.9 Subsampling

Another topic that is not usually discussed in terms of composition and mixing, although it is another special case of them, is subsampling of Markov chains.

If P is an update mechanism, we write P^k to denote the k-fold composition of P with itself. If $X_1, X_2, \ldots$ is a Markov chain with update mechanism P, then $X_1, X_{k+1}, X_{2k+1}, \ldots$ is a Markov chain with update mechanism P^k.

The process that takes every kth element of a Markov chain $X_1, X_2, \ldots$ forming a new Markov chain $X_1, X_{k+1}, X_{2k+1}, \ldots$ is called *subsampling* the original Markov chain at *spacing* k. As we just said, the result is another Markov chain. Hence, a subsampled Markov chain is just like any other Markov chain.

According to Elizabeth Thompson, "You don't get a better answer by throwing away data." This was proved as a theorem about Markov chains by Geyer (1992) for reversible Markov chains and by MacEachern and Berliner (1994) for nonreversible Markov chains. Subsampling cannot improve the accuracy of MCMC approximation; it must make things worse.

The original motivation for subsampling appears to have been to reduce autocorrelation in the subsampled chain to a negligible level. Before 1994 the Markov chain CLT was not well understood by statisticians, so appeal was made to a non-theorem: the central limit almost-but-not-quite theorem for almost-but-not-quite i.i.d. data. Now that the Markov chain CLT is well understood, this cannot be a justification for subsampling.

Subsampling may appear to be necessary just to reduce the amount of output of a Markov chain sampler to manageable levels. Billions of iterations may be needed for convergence, but billions of iterations of output may be too much to handle, especially when using R, which chokes on very large objects. But nonoverlapping batch means (Section 1.10.1) can reduce the size of the output with no loss of accuracy of estimation. Moreover, one does not need to know the batch length necessary to make the empirical variance of the batch means a good estimate of the asymptotic variance in the Markov chain CLT in order to use batches to reduce the size of output. The method of Section 1.10.3 allows one to use batches that are too short and still obtain accurate estimates of the asymptotic variance in the Markov chain CLT. Hence, if the objective is to reduce the size of output, batching is better than subsampling.

Hence, the only reason to use subsampling is to reduce the size of output when one cannot use batching. Good MCMC code, for example the functions `metrop` and `temper` in the R contributed package `mcmc` (Geyer, 2010b), allow an arbitrary function g supplied by the user as an R function to be used in calculation of the batch means in Equation 1.16. Other MCMC code that does not allow this may not output batch means for required functionals of the Markov chain. In this case the only way to reduce the size of output and still calculate the required functionals is subsampling. Another case where one cannot use the batch means is when the required functionals are not known when the sampling is done. This occurs, for example, in Monte Carlo likelihood approximation (Geyer and Thompson, 1992).

Geyer (1992) gave another justification of subsampling based on the cost of calculating the function g in a functional (Section 1.6 above). If the cost in computing time of calculating $g(X_i)$ is much more than the cost of sampling (producing X_i given X_{i-1}), then subsampling may be justified. This is rarely the case, but it does happen.

1.12.10 Gibbs and Metropolis Revisited

Our terminology of "elementary updates" combined by "composition" or "mixing" or both is not widespread. The usual terminology for a much more limited class of samplers is the following:

- A *Gibbs sampler* is an MCMC sampler in which all of the elementary updates are Gibbs, combined either by composition (fixed scan), by mixing (random scan), or both (random sequence scan), the "scan" terminology being explained in Section 1.12.8 above.

- A *Metropolis algorithm* is an MCMC sampler in which all of the elementary updates are Metropolis, combined either by composition, mixing, or both (and the same "scan" terminology is used).

- A *Metropolis–Hastings algorithm* is an MCMC sampler in which all of the elementary updates are Metropolis–Hastings, combined either by composition, mixing, or both (and the same "scan" terminology is used).

- A *Metropolis-within-Gibbs sampler* is the same as the preceding item. This name makes no sense at all since Gibbs is a special case of Metropolis–Hastings (Section 1.12.6 above), but it is widely used.

- An *independence Metropolis–Hastings algorithm* (named by Tierney, 1994) is a special case of the Metropolis–Hastings algorithm in which the proposal distribution does not depend on the current state: $q(x, \cdot)$ does not depend on x.
- A *random-walk Metropolis–Hastings algorithm* (named by Tierney, 1994) is a special case of the Metropolis–Hastings algorithm in which the proposal has the form $x + e$, where e is stochastically independent of the current state x, so $q(x, y)$ has the form $f(y - x)$.

The Gibbs sampler became very popular after the paper of Gelfand and Smith (1990) appeared. The term MCMC had not been coined (Geyer, 1992). It was not long, however, before the limitations of the Gibbs sampler were recognized. Peter Clifford (1993), discussing Smith and Roberts (1993), Besag and Green (1993), and Gilks et al. (1993), said:

> Currently, there are many statisticians trying to reverse out of this historical *cul-de-sac*. To use the Gibbs sampler, we have to be good at manipulating conditional distributions ... this rather brings back the mystique of the statisticians.

The American translation of "reverse out of this *cul-de-sac*" is "back out of this blind alley." Despite this, many naive users still have a preference for Gibbs updates that is entirely unwarranted. If I had a nickel for every time someone had asked for help with slowly converging MCMC and the answer had been to stop using Gibbs, I would be rich. Use Gibbs updates only if the resulting sampler works well. If not, use something else.

One reason sometimes given for the use of Gibbs updates is that they are "automatic." If one chooses to use a Gibbs sampler, no other choices need be made, whereas if one uses the Metropolis–Hastings algorithm, one must choose the proposal distribution, and even if one's choice of Metropolis–Hastings algorithm is more restricted, say to normal random-walk Metropolis–Hastings, there is still the choice of the variance matrix of the normal proposal distribution. This "automaticity" of the Gibbs sampler is illusory, because even if one only knows about "scans" one still must choose between fixed and random scan. Moreover, one should consider "block Gibbs" or even the more general Gibbs updates described in Section 1.12.4 above.

Nevertheless, Gibbs does seem more automatic than Metropolis–Hastings to many users. The question is whether this lack of options is a good thing or a bad thing. It is good if it works well and bad otherwise.

1.13 A Metropolis Example

We now turn to a realistic example of MCMC, taken from the package vignette of the `mcmc` contributed R package (Geyer, 2010b). The function `metrop` in this package runs a normal random-walk Metropolis sampler in the terminology of Section 1.12.10 having equilibrium distribution for a continuous random vector specified by a user-written R function that calculates its log unnormalized density. A major design goal of this package is that there be very little opportunity for user mistakes to make the simulation incorrect. For the `metrop` function, if the user codes the log unnormalized density function correctly, then the function will run a Markov chain having the correct stationary distribution (specified by this user-written function). There is nothing other than incorrectly writing the log unnormalized density function that the user can do to make the Markov chain have the wrong stationary distribution.

It may seem that this is a very weak correctness property. There is no guarantee that the Markov chain mixes rapidly and so produces useful results in a reasonable amount of time. But nothing currently known can guarantee that for arbitrary problems. Methods of proving rapid mixing, although they are applicable in principle to arbitrary problems, are so difficult that they have actually been applied only to a few simple examples. Moreover, they are entirely pencil-and-paper proofs. There is nothing the computer can do to assure rapid mixing of Markov chains for arbitrary user-specified equilibrium distributions. Thus this weak correctness property (having the correct equilibrium distribution) is the most one can expect a computer program to assure.

Thus this "weak" correctness property is the strongest property one can reasonably assert for an MCMC program. All MCMC programs should guarantee it, but how many do? The functions in the mcmc package have been exhaustively tested using the methodology explained in Section 1.16 below and further described in the package vignette debug.pdf that comes with every installation of the package. All of the tests are in the tests directory of the source code of the package, which is available from CRAN (http://www.cran.r-project.org/).

In addition to an R function that specifies the log unnormalized density of the equilibrium distribution, the user may also provide an R function that specifies an arbitrary functional of the Markov chain to be output. If the Markov chain is $X_1, X_2, \ldots$ and this user-supplied R function codes the mathematical function g, then $g(X_1), g(X_2), \ldots$ is output. Alternatively, batch means of $g(X_1), g(X_2), \ldots$ are output.

Finally, the user must specify the variance matrix of the multivariate normal distribution used in the "random-walk" proposal. There is nothing else the user can do to affect the Markov chain simulated by the metrop function.

Let us see how it works. We use the example from the package vignette demo.pdf that comes with every installation of the package. This is a Bayesian logistic regression problem that uses the data set logit in the package. There are five variables in this data frame, the response y and four quantitative predictor variables x1, x2, x3, and x4.

A frequentist analysis of these data is done by the following R statements:

```
library(mcmc)
data(logit)
out <- glm(y ~ x1 + x2 + x3 + x4, data = logit,
    family = binomial(), x = TRUE)
summary(out)
```

We wish to do a Bayesian analysis where the prior distribution for the five regression coefficients (one for each predictor and an intercept) makes them i.i.d. normal with mean 0 and standard deviation 2.

The log unnormalized posterior (log likelihood plus log prior) density for this model is calculated by the R function lupost defined as follows:

```
x <- out$x
y <- out$y

lupost <- function(beta, x, y, ...) {
    eta <- as.numeric(x %*% beta)
```

```
    logp <- ifelse(eta < 0, eta - log1p(exp(eta)),- log1p
    (exp(- eta)))
    logq <- ifelse(eta < 0, - log1p(exp(eta)), - eta - log1p
    (exp(- eta)))
    logl <- sum(logp[y == 1]) + sum(logq[y == 0])
    return(logl - sum(beta^2) / 8)
}
```

This assumes that `out` is the result of the call to `glm` shown above, so `y` is the response vector and `x` is the model matrix for this logistic regression.

The tricky calculation of the log likelihood avoids overflow and catastrophic cancelation in calculation of $\log(p)$ and $\log(q)$, where

$$p = \frac{\exp(\eta)}{1 + \exp(\eta)} = \frac{1}{1 + \exp(-\eta)},$$

$$q = \frac{1}{1 + \exp(\eta)} = \frac{\exp(-\eta)}{1 + \exp(-\eta)},$$

so taking logs gives

$$\log(p) = \eta - \log(1 + \exp(\eta)) = -\log(1 + \exp(-\eta)),$$

$$\log(q) = -\log(1 + \exp(\eta)) = -\eta - \log(1 + \exp(-\eta)).$$

To avoid overflow, we always chose the case where the argument of exp is negative. We have also avoided catastrophic cancelation when $|\eta|$ is large. If η is large and positive, then

$$p \approx 1,$$

$$q \approx 0,$$

$$\log(p) \approx -\exp(-\eta),$$

$$\log(q) \approx -\eta - \exp(-\eta),$$

and our use of the R function `log1p`, which calculates the function $x \mapsto \log(1 + x)$ correctly for small x, avoids problems with calculating $\log(1 + \exp(-\eta))$ here. The case where η is large and negative is similar. The above definitions having been made, the following statements do an MCMC run:

```
beta.init <- as.numeric(coefficients(out))
out <- metrop(lupost, beta.init, 1e3, x = x, y = y)
```

where `beta.init` is the initial state of the Markov chain (it would be more natural to a Bayesian to use the posterior mode rather than the maximum likelihood estimate, but the starting position makes no difference so long as it is not too far out in the tails of the equilibrium distribution) and where `1e3` is the MCMC sample size. The default batch length is one, so there is no batching here. The component `out$accept` of the result gives the acceptance rate (the fraction of Metropolis updates in which the proposal is accepted) and the component `out$batch` gives the output of the Markov chain as an $n \times p$ matrix, where n is the number of iterations here where there is no batching (although in general

n is the number of batches) and where *p* is the dimension of the state space here where
no functional of the Markov chain is specified and the default is the identity functional
(although in general *p* is the dimension of the result of the user-supplied output function).

The functions in the mcmc package are designed so that if given the output of a preceding
run as their first argument, they continue the run of the Markov chain where the other run
left off. For example, if we were to say

```
out2 <- metrop(out, x = x, y = y)
```

here, then rbind(out$batch, out2$batch) would be a run of the Markov chain. The
second invocation of the metrop function starts with the seed of R's random number generator (RNG) and the state of the Markov chain set to what they were when the first invocation
finished. Thus there is no difference between rbind(out$batch, out2$batch) and the
result of one invocation starting at the same RNG seed and initial state and running for
twice as many iterations as the two shown here did.

This "restart" property obviates any need for burn-in. If the first run "converged" in the
sense that any part of the run was in a high-probability part of the state space, then the second run starts in a good place and needs no burn-in. Since the first run started at the
maximum likelihood estimate, which is in a high-probability part of the state space, the first
run needed no burn-in either.

Using this function is not quite this simple. We need to adjust the normal proposal to
achieve a reasonable acceptance rate. It is generally accepted (Gelman et al., 1996) that an
acceptance rate of about 20% is right, although this recommendation is based on the asymptotic analysis of a toy problem (simulating a multivariate normal distribution) for which
one would never use MCMC and is very unrepresentative of difficult MCMC applications.
Geyer and Thompson (1995) came to a similar conclusion, that a 20% acceptance rate is
about right, in a very different situation. But they also warned that a 20% acceptance rate
could be very wrong, and produced an example where a 20% acceptance rate was impossible and attempting to reduce the acceptance rate below 70% would keep the sampler from
ever visiting part of the state space. So the 20% magic number must be considered like
other rules of thumb we teach in introductory courses (such as *n* > 30 means the normal
approximation is valid). We know these rules of thumb can fail. There are examples in the
literature where they do fail. We keep repeating them because we want something simple
to tell beginners, and they are all right for some problems.

The scale argument to the metrop function specifies the variance matrix for the proposal. The default is the identity matrix. This results in too low an acceptance rate in this
problem (0.008). A little bit of trial and error (shown in the vignette) shows that

```
out <- metrop(out, scale = 0.4, x = x, y = y)
```

gives about 20% acceptance rate, so this scaling, which specifies proposal variance matrix
0.4 times the identity matrix, is what we use. More complicated specification of the proposal
variance is possible; see the help for the metrop function for details.

Now we do a longer run

```
out <- metrop(out, nbatch = 1e4, x = x, y = y)
```

and look at time series plots and autocorrelation plots (shown in the vignette), which show
that the Markov chain seems to mix well and that autocorrelations are negligible after

lag 25. We use batch length 100 to be safe. We are interested here in calculating both posterior means and posterior variances. Variances are not functionals of the Markov chain, but squares are, and we can use the identity $\text{var}(Z) = E(Z^2) - E(Z)^2$ to calculate variances from means and means of squares. Thus we run the following:

```
out <- metrop(out, nbatch = 1e2, blen = 100,
    outfun = function(z, ...) c(z, z^2), x = x, y = y)
```

Here the user-specified output function (argument `outfun` of the `metrop` function) maps the state `z`, a vector of length 5, to `c(z, z^2)`, a vector of length 10. So now `out$batch` is a 100×10 matrix, 100 being the number of batches (argument `nbatch`) and 10 being the length of the result of `outfun`).

Now

```
foo <- apply(out$batch, 2, mean)
foo.mcse <- apply(out$batch, 2, sd) / sqrt(out$nbatch)
```

are estimates of the posterior means of the components of the vector returned by `outfun` (the regression coefficients and their squares) and the MCSE of these estimates, respectively. The first five components are useful directly:

```
mu <- foo[1:5]
mu.mcse <- foo.mcse[1:5]
```

These are estimates of the posterior means of the regression coefficients and their MCSE (see the vignette for actual numbers).

Monte Carlo estimates of the posterior variances are found using $\text{var}(Z) = E(Z^2) - E(Z)^2$,

```
sigmasq <- foo[6:10] - foo[1:5]^2
```

but to calculate the MCSE we need the delta method. Let u_i denote the sequence of batch means for one parameter and $\bar{u}$ the grand mean of this sequence (the estimate of the posterior mean of that parameter), let v_i denote the sequence of batch means for the squares of the same parameter and $\bar{v}$ the grand mean of that sequence (the estimate of the posterior second absolute moment of that parameter), and let $\mu = E(\bar{u})$ and $v = E(\bar{v})$. Then the delta method linearizes the nonlinear function

$$g(\mu, v) = v - \mu^2$$

as

$$\Delta g(\mu, v) = \Delta v - 2\mu \Delta \mu,$$

saying that

$$g(\bar{u}, \bar{v}) - g(\mu, v)$$

has the same asymptotic normal distribution as

$$(\bar{v} - v) - 2\mu(\bar{u} - \mu)$$

which, of course, has variance 1 / out$nbatch times that of

$$(v_i - v) - 2\mu(u_i - \mu),$$

and this variance is estimated by

$$\frac{1}{n_{\text{batch}}} \sum_{i=1}^{n_{\text{batch}}} \left[(v_i - \bar{v}) - 2\bar{u}(u_i - \bar{u}) \right]^2.$$

So

```
u <- out$batch[ , 1:5]
v <- out$batch[ , 6:10]
ubar <- apply(u, 2, mean)
vbar <- apply(v, 2, mean)
deltau <- sweep(u, 2, ubar)
deltav <- sweep(v, 2, vbar)
foo <- sweep(deltau, 2, ubar, "*")
sigmasq.mcse <- sqrt(apply((deltav - 2 * foo)^2,
2, mean) / out$nbatch)
```

does the MCSE for the posterior variance (see the vignette for actual numbers).

Another application of the delta method gives MCSE for posterior standard deviations (see the vignette for details).

1.14 Checkpointing

The "restart" property of the metrop and temper functions is also useful for checkpointing. If one wants to do very long runs, they need not be done with one function invocation. Suppose that out is the result of an invocation of metrop and that the log unnormalized density function and output function (if present) do not take additional arguments, getting any additional data from the R global environment, and suppose that any such additional data has been set up. Let ncheck be the number of repetitions of out we want to make. Then

```
for (icheck in 1:ncheck) {
    out <- metrop(out)
    save(out, file = sprintf("check%03d.rda", icheck))
}
```

does them and saves them on disk, unless the computer crashes for some reason. After a crash, only the work not done and saved is left to do. Set up any required global variables and ncheck as before, and restart with

```
files <- system("ls check*.rda", intern = TRUE)
kcheck <- length(files)
```

```
load(file = files[kcheck])
if (kcheck < ncheck) {
    for (icheck in (kcheck + 1):ncheck) {
        out <- metrop(out)
        save(out, file = sprintf("check%03d.rda", icheck))
    }
}
```

(this is for UNIX, e.g., Linux or MAC OS X, and would have to be modified for Microsoft Windows). When finished collect the results with

```
files <- system("ls check*.rda", intern = TRUE)
ncheck <- length(files)
batch <- NULL
for (icheck in 1:ncheck) {
    load(file = files[icheck])
    batch <- rbind(batch, out$batch, deparse.level = 0)
}
```

and batch is the same as out$batch from one long run. This idiom allows very long runs even with unreliable computers.

1.15 Designing MCMC Code

Nothing is easier than designing MCMC algorithms. Hundreds have been introduced into the literature under various names. All that are useful in non-toy problems are special cases of the Metropolis–Hastings–Green algorithm.

When one invents a new sampler, how does one argue that it is correct? One proves a theorem: the new sampler is a special case of the MHG algorithm. The proof is usually not difficult but does require tight reasoning, like all proofs. One common error is sloppiness about what is the state of the Markov chain. Many have made the mistake of having proposals depend on some variables in the computer program that are not considered part of the state in calculating the Hastings ratio, that is, the state space is considered one thing in one part of the argument and another thing in another part—a clear error if one thinks about it.

One does not have to call this theorem a theorem, but one does need the care in proving it that any theorem requires. A few hours of careful thought about what is and what is not a special case of the MHG algorithm can save weeks or months of wasted work on a mistake. This notion that you have to prove a theorem every time you invent an MCMC algorithm came to your humble author from the experience of humbling mistakes committed by himself and others. If you think you have to prove a theorem, you will (hopefully) exercise appropriately careful argument. If you think you can use your intuition, many sad stories could be told about failure of intuition. The MHG algorithm is not difficult but is also not very intuitive.

Before one can prove a theorem, one must state the theorem, and here too care is required. The theorem must state precisely how one's MCMC algorithm works, with no vagueness.

This is very important. One cannot correctly implement an MCMC algorithm in computer code when one has to guess what the algorithm actually is. Most erroneous MCMC algorithms (just like most erroneous attempts at theorems) result from vagueness.

These general remarks having been made, we now turn to some desirable features of MCMC code that few computer packages have but the mcmc package has shown to be very useful.

The first is the "restart" property discussed in Sections 1.13 and 1.14 above and possessed by both the metrop and temper functions. This is the property that the R object output by a function doing MCMC (or the equivalent object for computer languages other than R) should contain the RNG seeds and the final state of the Markov chain, so the next run can simply continue this run. A sampler with the "restart" property needs no burn-in (Section 1.11.4 above) and is easily checkpointed (Section 1.14).

The second is the property of outputting batch means for batches of a possibly subsampled chain, also possessed by both the metrop and temper functions, specified by the arguments blen and nspac. This property allows very long runs without overly voluminous output. If nspac = 1 (the default, meaning no subsampling) is used, then no information is lost by the batching. The batches can be used for valid inference—regardless of whether the batch length is long enough for the ordinary method of batch means to work—as described in Section 1.10.3 above.

The third is the property of outputting batch means (for batches of a possibly subsampled chain) for an arbitrary functional of the Markov chain. The mcmc and temper functions do this via a user-specified function supplied as their outfun argument. This allows users to make the inferences they want without rewriting the R package. This makes statistical computer languages in which functions are not first-class objects (like they are in R) unsuitable for MCMC.

1.16   Validating and Debugging MCMC Code

Along with "black box" MCMC (Section 1.11.1) above we introduce the notion of "black box" testing of MCMC code. Black box testing is widely used terminology in software testing. It refers to tests that do not look inside the code, using only its ordinary input and output. Not looking at the code means it cannot use knowledge of the structure of the program or the values any of its internal variables. For MCMC code black box testing means you run the sampler and test that the output has the expected probability distribution.

Since goodness-of-fit testing for complicated multivariate probability distributions is very difficult, black box testing of MCMC code is highly problematic. It is even more so when the sampler is itself black box, so nothing is known about the expected equilibrium distribution except what we may learn from the sampler itself. Thus your humble author has been driven to the conclusion that black box testing of MCMC code is pointless.

Instead testing of the functions metrop and temper in the mcmc package uses a "white box" approach that exposes all important internal variables of the program when the optional argument debug = TRUE is specified. In particular, all uniform or normal random variates obtained from R's RNG system are output. This means that, assuming we can trust R's normal and uniform RNG, we can test whether metrop and temper behave properly as deterministic functions of those pseudorandom numbers obtained from R.

Testing whether a program correctly implements a deterministic function is much easier than testing whether it correctly simulates a specified probability distribution. In addition, when `debug = TRUE` these programs also output proposals, log Hastings ratios, and decisions in the Metropolis rejection step, making it easy to check whether these are correct and hence whether the Metropolis–Hastings algorithm is implemented correctly.

It must be admitted that, although this "white box" testing methodology it much superior to anything your humble author has previously used, it is not guaranteed to find conceptual problems. That is why a clearly written specification (what we called the "theorem" in the preceding section) is so important. During the writing of this chapter just such a conceptual bug was discovered in the `temper` function in versions of the mcmc package before 0.8. The terms $q(i,j)$ and $q(j,i)$ in the Hastings ratio for serial tempering (Equation 11.11 in Chapter 11, this volume) were omitted from the code, and the tests of whether the Hastings ratio was calculated correctly were implemented by looking at the code rather than the design document (the file `temper.pdf` in the `doc` directory of every installation of the mcmc package), which was correct.

Ideally, the tests should be implemented by someone other than the programmer of the code, a well-recognized principle in software testing. We know of no statistics code that conforms to this practice, perhaps because there is no tradition of refereeing computer code as opposed to papers. The most we can claim is that the "white box" testing methodology used for the mcmc would at least make such referring possible.

1.17 The Metropolis–Hastings–Green Algorithm

There are so many ideas in Green (1995) it is hard to know where to start. They include the following:

- State-dependent mixing of updates
- Measure-theoretic Metropolis–Hastings using Radon–Nikodym derivatives
- Per-update augmentation of the state space
- Metropolis–Hastings with Jacobians

any one of which would have been a major contribution by itself.

We have deferred discussion of the MHG algorithm till now because we wanted to avoid measure theory as long as we could. The MHG algorithm cannot easily be discussed without using measure-theoretic terminology and notation.

A kernel $K(x,A)$ is a generalization of regular conditional probability. For a fixed point x in the state space, $K(x, \cdot)$ is a countably-additive real signed measure on the state space. For a fixed measurable set A in the state space, $K(\cdot, A)$ is a measurable real-valued function on the state space. If

$$K(x,A) \geq 0, \quad \text{for all } x \text{ and } A,$$

then we say that K is *nonnegative*. If K is nonnegative and

$$K(x,A) \leq 1, \quad \text{for all } x \text{ and } A,$$

then we say that K is *sub-Markov*. If K is sub-Markov and

$$K(x,S) = 1, \quad \text{for all } x,$$

where S is the state space, then we say that K is *Markov*. A Markov kernel is a regular conditional probability and can be used to describe an elementary update mechanism for a Markov chain or a combined update. In widely used sloppy notation, we write

$$K(x,A) = \Pr(X_{t+1} \in A \mid X_t = x)$$

to describe the combined update (the sloppiness is the conditioning on an event of measure zero).

A kernel K is *reversible* with respect to a signed measure m if

$$\iint g(x)h(y)m(dx)K(x,dy) = \iint h(x)g(y)m(dx)K(x,dy)$$

for all measurable functions g and h such that the expectations exist. A Markov kernel P *preserves* a probability measure π if

$$\iint g(y)\pi(dx)P(x,dy) = \int g(x)\pi(dx)$$

for every bounded function g. Reversibility with respect to π implies preservation of π.

1.17.1 State-Dependent Mixing

Suppose we have a family of updates represented by Markov kernels P_i, $i \in I$. Choose one at random with probability $c_i(x)$ that depends on the current state x, and use it to update the state. The kernel that describes this combined update is

$$P(x,A) = \sum_{i \in I} c_i(x)P_i(x,A).$$

It is *not a theorem* that if each P_i preserves π, then P preserves π. The argument in Section 1.12.8 above does not work.

Define

$$K_i(x,A) = c_i(x)P_i(x,A).$$

If each K_i is reversible with respect to π, then the mixture kernel

$$P(x,A) = \sum_{i \in I} c_i(x)P_i(x,A) = \sum_{i \in I} K_i(x,A)$$

is reversible with respect to π and hence preserves π. This is how state-dependent mixing works.

It is often convenient to allow the identity kernel defined by

$$I(x,A) = \begin{cases} 1, & x \in A, \\ 0, & x \notin A, \end{cases}$$

to be among the P_i. The identity kernel is a Markov kernel that describes a do-nothing update (the state is the same before and after).

Sometimes state-dependent mixing involving the identity kernel is described differently. We insist that

$$c_i(x) \geq 0, \quad \text{for all } i \text{ and } x,$$

and

$$\sum_{i \in I} c_i(x) \leq 1, \quad \text{for all } x.$$

Then when x is the current state the mixture update chooses the ith update with probability $c_i(x)$ and performs the update described by P_i. With the remaining probability

$$1 - \sum_{i \in I} c_i(x)$$

the mixture update does nothing (which is the same as doing the update described by the identity kernel).

1.17.2   Radon–Nikodym Derivatives

Suppose that m is a finite signed measure and n a sigma-finite positive measure defined on the same space. We say that m is *dominated* by n or that m is *absolutely continuous with respect to n* if

$$n(A) = 0 \text{ implies } m(A) = 0, \quad \text{for all events } A.$$

We say that m is *concentrated* on a set C if

$$m(A) = m(A \cap C), \quad \text{for all events } A.$$

We say that measures m_1 and m_2 are *mutually singular* if they are concentrated on disjoint sets.

The Lebesgue–Radon–Nikodym theorem (Rudin, 1987, Theorem 6.10) says the following about m and n as defined above. Firstly, there exist unique finite signed measures m_a and m_s such that m_s and n are mutually singular, m_a is dominated by n, and $m = m_a + m_s$ (this is called the *Lebesgue decomposition*). Secondly, there exists a real-valued function f, which is unique up to redefinition on a set of n measure zero, such that

$$m_a(A) = \int_A f(x)n(dx), \quad \text{for all events } A. \tag{1.25}$$

We say that f is the *density* or *Radon–Nikodym derivative* of m with respect to n and write

$$f = \frac{dm}{dn}.$$

If n is Lebesgue measure and m is dominated by n, then f is an ordinary probability density function. If n is counting measure and m is dominated by n, then f is an ordinary probability mass function. Hence, the Radon–Nikodym derivative generalizes these concepts. When m is not dominated by n, we have

$$\frac{dm}{dn} = \frac{dm_a}{dn}$$

so the Radon–Nikodym derivative only determines the part of m that is absolutely continuous with respect to n and says nothing about the part of m that is singular with respect to n, but that is enough for many applications.

That the Radon–Nikodym derivative f is unique only up to redefinition on a set of n measure zero would cause a problem if we made a different choice of f every time we used it, but it causes no problem if we fix one choice of f and use it always. (The same issue arises with ordinary probability density functions.)

Radon–Nikodym derivatives are often calculated using ratios. Suppose that m and n are as above and that λ is a measure that dominates both, for example, $\lambda = m + n$. Then we have

$$\frac{dm}{dn} = \frac{dm/d\lambda}{dn/d\lambda}, \tag{1.26}$$

where the right-hand side is interpreted as ordinary division when the denominator is nonzero and an arbitrary choice when the denominator is zero.

To see this, let $f_m = dm/d\lambda$ and $f_n = dn/d\lambda$, let $C = \{ x : f_n(x) \neq 0 \}$, let h be an arbitrary measurable real-valued function, and define

$$f(x) = \begin{cases} f_m(x)/f_n(x), & x \in C, \\ h(x), & x \notin C. \end{cases}$$

By the Lebesgue–Radon–Nikodym theorem, n is concentrated on C. Define a measure m_s by

$$m_s(A) = m(A \setminus C), \quad \text{for all events } A,$$

and let $m_a = m - m_s$. It remains to be shown that m_a is dominated by n and $f = dm_a/dn$. Both are shown by verifying (Equation 1.25) as follows. For any event A,

$$m_a(A) = m(A \cap C) = \int_C f_m \, d\lambda = \int_C f \cdot f_n \, d\lambda = \int_C f \, dn = \int f \, dn$$

(the last equality being that n is concentrated on C).

1.17.3 Measure-Theoretic Metropolis–Hastings

1.17.3.1 *Metropolis–Hastings–Green Elementary Update*

We now describe the MHG elementary update with state-dependent mixing. For $i \in I$ we have proposal mechanisms described by kernels Q_i. When the current state is x, we choose the ith proposal mechanism with probability $c_i(x)$, generating a proposal y having distribution $Q_i(x, \cdot)$.

The unnormalized measure to preserve is η (the analog of the unnormalized density h in the ordinary Metropolis–Hastings algorithm). Define measures m and m_{rev} by

$$m(B) = \iint 1_B(x, y) \eta(dx) c_i(x) Q_i(x, dy), \tag{1.27a}$$

$$m_{\text{rev}}(B) = \iint 1_B(y, x) \eta(dx) c_i(x) Q_i(x, dy), \tag{1.27b}$$

where $1_B(x, y)$ is equal to one if $(x, y) \in B$ and zero otherwise, so m and m_{rev} are measures on the Cartesian product of the sample space with itself and each B is a measurable subset of that Cartesian product. Define

$$r = \frac{dm_{\text{rev}}}{dm}. \tag{1.27c}$$

Then accept the proposal with probability $\min(1, r(x, y))$.

Note the similarity of this MHG update to the Metropolis–Hastings update (Section 1.12.1 above). It differs in the incorporation of state-dependent mixing so that $c_i(x)$ appears. It also differs in that the *Green ratio* (Equation 1.27c) is actually a Radon–Nikodym derivative rather than a simple ratio like the Hastings ratio (Equation 1.20). The "Metropolis rejection" step—accept the proposal with probability $\min(1, r)$—is the same as in the Metropolis and Metropolis–Hastings algorithms.

As we saw in Equation 1.26, a Radon–Nikodym derivative is often calculated as a ratio, so the terminology "Green ratio" for Equation 1.27c is not so strange. But our main reason for introducing this terminology is the analogy between the Metropolis ratio (Equation 1.24), the Hastings ratio (Equation 1.20), and the Green ratio (Equation 1.27c). People often write things like

$$r(x, y) = \frac{c_i(y) \eta(dy) Q_i(y, dx)}{c_i(x) \eta(dx) Q_i(x, dy)} \tag{1.28}$$

as a sloppy shorthand for actual definition via Equations 1.27a through 1.27c, but Equation 1.28 has no mathematical content other than as a mnemonic for the actual definition.

Green (1995) described a specific recipe for calculating the Green ratio (Equation 1.27c) using the ratio method (Equation 1.26) in the particular case where λ is symmetric in the sense that

$$\iint 1_B(x, y) \lambda(dx, dy) = \iint 1_B(y, x) \lambda(dx, dy) \tag{1.29}$$

for any measurable set B in the Cartesian product of the state space with itself. Such λ always exist. For example, $\lambda = m + m_{\text{rev}}$ works. Then if $f = dm/d\lambda$ and

$$C = \{ (x, y) : f(x, y) \neq 0 \} \tag{1.30}$$

we have

$$r(x, y) = \begin{cases} f(y, x)/f(x, y), & x \in C, \\ 0, & x \notin C. \end{cases} \tag{1.31}$$

It does not matter whether or not we use Green's recipe for calculating (Equation 1.27c). Radon–Nikodym derivatives are unique up to redefinition on sets of measure zero, hence are the same no matter how we calculate them.

Note that the proposal distributions can be anything, described by arbitrary kernels Q_i. Thus the MHG algorithm generalizes the Metropolis–Hastings algorithm about as far as it can go. The only way your humble author can think to generalize this would be to allow state-dependent mixing over a continuum rather than a countable set of Q_i (the way state-independent mixing works; Section 1.12.8 above).

Ordinary Metropolis–Hastings samplers avoid forever the set of x such that $h(x) = 0$, where h is the unnormalized density of the equilibrium distribution (Section 1.12.1 above). Now thinking measure-theoretically, we are reminded that we may redefine h arbitrarily on sets of measure zero under the equilibrium distribution, so the set avoided depends on our choice of h. The MHG algorithm has a similar property. Suppose there is a set N that must be avoided, and $\eta(N) = 0$. Then $m_{\mathrm{rev}}(A \times N) = 0$ for any set A, and we may choose a version of the Green ratio such that $r(x, y) = 0$ for $y \in N$. Then no proposal in N can be accepted, and the chain forever avoids N.

All MCMC ideas discussed above are special cases of the MHG algorithm. Variable-at-a-time Metropolis–Hastings updates are special cases where proposals only change one coordinate. Gibbs updates are special cases where the MHG ratio is always one and the proposal is always accepted.

1.17.3.2   The MHG Theorem

Define

$$a(x, y) = \min(1, r(x, y)),$$

$$b(x) = 1 - \int a(x, y) Q_i(x, dy).$$

The kernel describing the MHG elementary update is

$$P_i(x, A) = b(x) I(x, A) + \int_A a(x, y) Q_i(x, dy),$$

and the kernel that we must verify is reversible with respect to η is

$$K_i(x, A) = c_i(x) P_i(x, A),$$

that is, we must verify that

$$\iint g(x) h(y) \eta(dx) c_i(x) P_i(x, dy)$$

is unchanged when g and h are swapped. Since

$$\iint g(x) h(y) c_i(x) \eta(dx) P_i(x, dy) = \int g(x) h(x) b(x) c_i(x) \eta(dx)$$

$$+ \iint g(x) h(y) a(x, y) c_i(x) \eta(dx) Q_i(x, dy),$$

it clearly is enough to show last term is unchanged when g and h are swapped.

Suppose we have calculated the Green ratio (Equation 1.27c) using Green's recipe (Equation 1.31) with $f = dm/d\lambda$ and λ satisfying Equation 1.29. Then

$$\iint g(x)h(y)a(x,y)c_i(x)\eta(dx)Q_i(x,dy) = \iint g(y)h(x)a(y,x)c_i(y)\eta(dy)Q_i(y,dx)$$

$$= \iint g(y)h(x)a(y,x)m_{\text{rev}}(dx,dy)$$

$$= \iint_C g(y)h(x)a(y,x)m_{\text{rev}}(dx,dy)$$

$$= \iint_C g(y)h(x)a(y,x)r(x,y)m(dx,dy)$$

$$= \iint g(y)h(x)a(y,x)r(x,y)m(dx,dy)$$

$$= \iint g(y)h(x)a(y,x)r(x,y)c_i(x)\eta(dx)Q_i(x,dy),$$

where C is defined by Equation 1.30, the first equality being the interchange of the dummy variables x and y, the second and sixth equalities being the definitions of m and m_{rev}, the third and fifth equalities being $a(y,x) = 0$ when $(x,y) \in C$, and the fourth equality being $r = dm_{\text{rev}}/dm$ and the fact that the part of m_{rev} that is dominated by m is concentrated on C, as we saw in our discussion of Equation 1.26.

Comparing the expressions at the ends of this chain of equalities, we see that it is enough to show that

$$a(y,x)r(x,y) = a(x,y), \quad \text{whenever } (x,y) \in C, \tag{1.32}$$

because the integrals are the same whether or not they are restricted to C. If $(x,y) \in C$ and $r(x,y) \leq 1$, then $a(x,y) = r(x,y)$ and $a(y,x) = 1$, in which case (1.32) holds. If $(x,y) \in C$ and $1 < r(x,y)$, then $a(x,y) = 1$ and

$$a(y,x) = r(y,x) = \frac{1}{r(x,y)}$$

by Equation 1.31, in which case (Equation 1.32) holds again.

Example: Spatial Point Processes

All of this is very abstract. That is the point! But Radon–Nikodym derivatives are nothing to be frightened of. We look at some simple examples to show how the MHG algorithm works in practice.

One only needs the MHG algorithm when proposals are singular with respect to the equilibrium distribution of the Markov chain (otherwise Metropolis–Hastings would do). This often happens when the state space is the union of sets of different dimension. One example of this is spatial point processes. Geyer and Møller (1994) proposed the sampler described here independently of Green (1995), but in hindsight it is a special case of the MHG algorithm.

A spatial point process is a random pattern of points in a region A having finite measure (length, area, volume, …), both the number of points and the positions of the points being random. A homogeneous Poisson process has a Poisson distributed number of points and the locations of the points are independent and identically and uniformly distributed conditional on the number. We consider processes having unnormalized densities h_θ with respect to the Poisson processes.

The state space of the Poisson process is

$$\mathcal{A} = \bigcup_{n=0}^{\infty} A^n,$$

where A^0 denotes a set consisting of one point, representing the spatial pattern with no points. The probability measure of the Poisson process is defined by

$$P(B) = \sum_{n=0}^{\infty} \frac{\mu^n e^{-\mu}}{n!} \cdot \frac{\lambda^n(B \cap A^n)}{\lambda(A)^n}, \quad \text{for measurable } B \subset \mathcal{A},$$

where λ is Lebesgue measure on A and μ is an adjustable parameter (the mean number of points). To say that h_θ is an unnormalized density with respect to P means that the probability measure of the non-Poisson process is defined by

$$Q_\theta(B) = \frac{1}{c(\theta)} \int_B h_\theta(x) P(dx)$$

$$= \frac{1}{c(\theta)} \sum_{n=0}^{\infty} \frac{\mu^n e^{-\mu}}{n!} \cdot \frac{1}{\lambda(A)^n} \int_{B \cap A^n} h_\theta(x) \lambda^n(dx)$$

for measurable $B \subset \mathcal{A}$, where

$$c(\theta) = \sum_{n=0}^{\infty} \frac{\mu^n e^{-\mu}}{n!} \cdot \frac{1}{\lambda(A)^n} \int h_\theta(x) \lambda^n(dx).$$

Note that the dimension of x, which is n, is different in different terms of these sums.

Let $n(x)$ denote the number of points in x. We use state-dependent mixing over a set of updates, one for each nonnegative integer i. The ith update is only valid when $n(x) = i$, in which case we propose to add one point uniformly distributed in A to the pattern, or when $n(x) = i + 1$, in which case we propose to delete a point from the pattern. (For definiteness, suppose we add or delete the last point.) The state-dependent mixing probabilities are

$$c_i(x) = \begin{cases} 1/2, & n(x) = i, \\ 1/2, & n(x) = i + 1, \\ 0, & \text{otherwise.} \end{cases}$$

For fixed x have $\sum_i c_i(x) = 1$ except when $n(x) = 0$. In that case, we do nothing (perform the identity update) with probability $1 - \sum_i c_i(x) = 1/2$ following the convention explained at the end of Section 1.17.1.

In order to apply Green's recipe for calculating Radon–Nikodym derivatives for the ith update, we need a symmetric measure on

$$(A^i \times A^{i+1}) \cup (A^{i+1} \times A^i) \tag{1.33}$$

that dominates the joint distribution m of the current state x and the proposal y or its reverse m_{rev}. This symmetric measure cannot be Lebesgue measure on Equation 1.33, because m and m_{rev} are degenerate, their first i coordinates being equal. Thus we choose the symmetric measure Λ that is the image of λ^{i+1} onto the subset of Equation 1.33 where the first i coordinates of the two parts are equal.

On the part of Equation 1.33 where $x \in A^i$ and $y \in A^{i+1}$, we have

$$f(x, y) = \frac{dm}{d\Lambda}(x, y) = \frac{\mu^i e^{-\mu} h_\theta(x)}{i! \lambda(A)^i} \cdot \frac{1}{\lambda(A)},$$

the first part on the right-hand side being the unnormalized density of the equilibrium distribution, unnormalized because we left out $c(\theta)$, which we do not know how to calculate, and the second part being the proposal density. On the part of Equation 1.33 where $x \in A^{i+1}$ and $y \in A^i$, we have

$$f(x, y) = \frac{dm}{d\Lambda}(x, y) = \frac{\mu^{i+1} e^{-\mu} h_\theta(x)}{(i+1)! \lambda(A)^{i+1}} \cdot 1,$$

the first part on the right-hand side being the unnormalized density of the equilibrium distribution, and the second part being the proposal density (which is one because deleting the last point involves no randomness). Thus the Green ratio is

$$r(x, y) = \begin{cases} \dfrac{\mu}{i+1} \cdot \dfrac{h_\theta(y)}{h_\theta(x)}, & x \in A^i \quad \text{and} \quad y \in A^{i+1}, \\[2ex] \dfrac{i+1}{\mu} \cdot \dfrac{h_\theta(y)}{h_\theta(x)}, & x \in A^{i+1} \quad \text{and} \quad y \in A^i. \end{cases}$$

We hope readers feel they could have worked this out themselves.

Since point patterns are usually considered as unordered, it is traditional to use $h_\theta(x)$ that is symmetric under exchange of points in pattern. In this case, the update that reorders the points randomly also preserves the stationary distribution. The composition of this random reordering with the update specified above (which deletes the last point) is equivalent to picking a random point to delete.

Example: Bayesian Model Selection

We consider an example done by other means in Chapter 11 of this volume. If we use MHG, there is no need for "padding" parameter vectors. We can just use the parameterization from the problem statement. If, like the ST/US sampler in Section 11.3, we only make jumps between models whose dimensions differ by one, then a very simple MHG proposal simply deletes a component of the parameter vector when moving down in dimension and adds a component distributed normally with mean zero and variance τ^2 independently of the current state when moving up in dimension. If $h(\theta)$ denotes the unnormalized posterior, then a move up in dimension from current state θ to proposed state ψ, which adds a component z to the current state, has Green ratio

$$r(\theta, \psi) = \frac{c_i(\psi) h(\psi)}{c_i(\theta) h(\theta) \phi(z/\tau)/\tau}, \tag{1.34}$$

where ϕ is the probability density function of the standard normal distribution, and a move down in dimension from current state ψ to proposed state θ, which deletes a component z from the current state, has Green ratio that is the reciprocal of the right-hand side of Equation 1.34.

1.17.4   MHG with Jacobians and Augmented State Space

Green (1995) also proposed what is in some respects a special case of MHG and in other respects an extension. We call it Metropolis–Hastings–Green with Jacobians (MHGJ). This version is so widely used that many users think that MHGJ is the general version. This form of elementary update moves between parts of the state space that are Euclidean spaces of different dimension, hence it is often called "dimension jumping"—although that name applies to other examples, such as the preceding one, that do not involve Jacobians.

Suppose that the state space is a disjoint union

$$S = \bigcup_{m \in M} S_m,$$

where S_m is a Euclidean space of dimension d_m. We assume that the equilibrium distribution of the Markov chain is specified by an unnormalized density $h(x)$ with respect to Lebesgue measure on S. MHGJ elementary updates move from one S_m to another. Say the ith elementary update moves between $S_{m(i)}$ and $S_{n(i)}$. Thus it only makes sense to have $c_i(x) > 0$ when $x \in S_{m(i)} \cup S_{n(i)}$.

Let $U_{m(i)}$ and $U_{n(i)}$ be Euclidean spaces such that $S_{m(i)} \times U_{m(i)}$ is the same dimension as $S_{n(i)} \times U_{n(i)}$. We specify a proposal density $q_i(x, \cdot)$, which describes the conditional distribution of the proposal u given the current state x such that $u \in U_{m(i)}$ when $x \in S_{m(i)}$ and $u \in U_{n(i)}$ when $x \in S_{n(i)}$. We also specify a function g_i that maps points in $S_{m(i)} \times U_{m(i)}$ to points in $S_{n(i)} \times U_{n(i)}$ and vice versa and which is its own inverse.

The MHGJ proposal is a combination of two steps. First generate a random u from the distribution $q_i(x, \cdot)$. Then propose $g_i(x, u) = (y, v)$. The MHG ratio is

$$r(x, u, y, v) = \frac{c_i(y)h(y)q_i(y, v)}{c_i(x)h(x)q_i(x, u)} \cdot \det\big(\nabla g_i(x, u)\big),$$

the last factor being the Jacobian of the mapping g_i. This is followed by the usual Metropolis rejection: accept the proposal with probability $\min(1, r)$.

For examples of the MHGJ algorithm, see Chapter 3 (this volume).

1.17.4.1   The MHGJ Theorem

The MHGJ algorithm, because of its per-update augmentation of $U_{m(i)}$ and $U_{n(i)}$, does not exactly fit in the pattern of the MHG algorithm described above. Thus we give a separate proof.

The proof starts just like the one in Section 1.17.3.2. We see that we can deal with one arbitrary elementary update, and consequently only one pair of state augmentations. Whenever one augments the state, there are two issues: what is the equilibrium distribution on the augmented state space, and how does it relate to the distribution of interest on the original state? Here the augmented state is (x, u), the equilibrium distribution on the augmented state space has unnormalized density with respect to Lebesgue measure $h(x)q_i(x, u)$. The original state is x and the distribution of interest with unnormalized density $h(x)$ is a marginal of it. The proposal $(y, v) = g(x, u)$ is deterministic.

We now determine the Radon–Nikodym derivative of the distribution of (y, v) with respect to (x, u). We use the ratio method, determining first the Radon–Nikodym derivatives of each with respect to Lebesgue measure λ on the space where (x, u) lives. We have

$$\frac{dm}{d\lambda} = c_i(x) \cdot h(x)q_i(x, u),$$

$$\frac{dm_{\text{rev}}}{d\lambda} = c_i(y) \cdot h(y)q_i(y, v) \cdot \det\big(\nabla g_i(x, u)\big),$$

where in the latter the Jacobian arises from the multivariate change-of-variable theorem, because we are differentiating with respect to (x, u) rather than (y, v).

Acknowledgments

This chapter benefited from detailed comments by Christina Knudson, Leif Johnson, Galin Jones, and Brian Shea.

References

Bélisle, C. J. P., Romeijn, H. E., and Smith, R. L. 1993. Hit-and-run algorithms for generating multivariate distributions. *Mathematics of Operations Research*, 18:255–266.

Besag, J. and Green, P. J. 1993. Spatial statistics and Bayesian computation (with discussion). *Journal of the Royal Statistical Society, Series B*, 55:25–37.

Chan, K. S. and Geyer, C. J. 1994. Discussion of the paper by Tierney (1994). *Annals of Statistics*, 22:1747–1758.

Chen, M.-H. and Schmeiser, B. 1993. Performance of the Gibbs, hit-and-run, and Metropolis samplers. *Journal of Computational and Graphical Statistics*, 2:251–272.

Clifford, P. 1993. Discussion of Smith and Roberts (1993), Besag and Green (1993), and Gilks et al. (1993). *Journal of the Royal Statistical Society, Series B*, 55:53–54.

Freedman, D., Pisani, R., and Purves, R. 2007. *Statistics*, 4th edn. W. W. Norton, New York.

Gelfand, A. E. and Smith, A. F. M. 1990. Sampling-based approaches to calculating marginal densities. *Journal of the American Statistical Association*, 85:398–409.

Gelman, A., Roberts, G. O., and Gilks, W. R. 1996. Efficient Metropolis jumping rules. In J. M. Bernardo, J. O. Berger, A. P. Dawid, and A. F. M. Smith (eds), *Bayesian Statistics 5: Proceedings of the Fifth Valencia International Meeting*, pp. 599–607. Oxford University Press, Oxford.

Geman, S. and Geman, D. 1984. Stochastic relaxation, Gibbs distribution, and the Bayesian restoration of images. *IEEE Transactions on Pattern Analysis and Machine Intelligence*, 6:721–741.

Geyer, C. J. 1992. Practical Markov chain Monte Carlo (with discussion). *Statistical Science*, 7:473–511.

Geyer, C. J. 1994. On the convergence of Monte Carlo maximum likelihood calculations. *Journal of the Royal Statistical Society, Series B*, 56:261–274.

Geyer, C. J. 1999. Likelihood inference for spatial point processes. In O. E. Barndorff-Nielsen, W. S. Kendall, and M. N. M. van Lieshout (eds), *Stochastic Geometry: Likelihood and Computation*, pp. 79–140. Chapman & Hall/CRC, Boca Raton, FL.

Geyer, C. J. 2010a. Computation for the Introduction to MCMC chapter of *Handbook of Markov Chain Monte Carlo*. Technical Report 679, School of Statistics, University of Minnesota. http://purl.umn.edu/92549.

Geyer, C. J. 2010b. *mcmc: Markov Chain Monte Carlo*. R package version 0.8, available from CRAN.

Geyer, C. J. and Møller, J. 1994. Simulation and likelihood inference for spatial point processes. *Scandinavian Journal of Statistics*, 21:359–373.

Geyer, C. J. and Thompson, E. A. 1992. Constrained Monte Carlo maximum likelihood for dependent data (with discussion). *Journal of the Royal Statistical Society, Series B*, 54:657–699.

Geyer, C. J. and Thompson, E. A. 1995. Annealing Markov chain Monte Carlo with applications to ancestral inference. *Journal of the American Statistical Association*, 90:909–920.

Gilks, W. R., Clayton, D. G., Spiegelhalter, D. J., Best, N. G., and McNeil, A. J. 1993. Modelling complexity: Applications of Gibbs sampling in medicine (with discussion). *Journal of the Royal Statistical Society, Series B*, 55:39–52.

Glynn, P. W. and Whitt, W. 1991. Estimating the asymptotic variance with batch means. *Operations Research Letters*, 10:431–435.

Glynn, P. W. and Whitt, W. 1992. The asymptotic validity of sequential stopping rules for stochastic simulations. *Annals of Applied Probability*, 2:180–198.

Green, P. J. 1995. Reversible jump Markov chain Monte Carlo computation and Bayesian model determination. *Biometrika*, 82:711–732.

Hammersley, J. M. and Handscomb, D. C. 1964. *Monte Carlo Methods*. Methuen, London.

Hastings, W. K. 1970. Monte Carlo sampling methods using Markov chains and their applications. *Biometrika*, 57:97–109.

Jones, G. L. 2004. On the Markov chain central limit theorem. *Probability Surveys*, 1:299–320.

Kendall, W. S. and Møller, J. 2000. Perfect simulation using dominating processes on ordered spaces, with application to locally stable point processes. *Advances in Applied Probability*, 32:844–865.

Kipnis, C. and Varadhan, S. R. S. 1986. Central limit theorem for additive functionals of reversible Markov processes and applications to simple exclusions. *Communications in Mathematical Physics*, 104:1–19.

MacEachern, S. N. and Berliner, L. M. 1994. Subsampling the Gibbs sampler. *American Statistician*, 48:188–190.

Meketon, M. S. and Schmeiser, B. W. 1984. Overlapping batch means: Something for nothing? In S. Sheppard, U. Pooch, and D. Pegden (eds), *Proceedings of the 1984 Winter Simulation Conference*, pp. 227–230. IEEE Press Piscataway, NJ.

Metropolis, N., Rosenbluth, A. W., Rosenbluth, M. N., Teller, A. H., and Teller, E. 1953. Equation of state calculations by fast computing machines. *Journal of Chemical Physics*, 21:1087–1092.

Meyn, S. P. and Tweedie, R. L. 1993. *Markov Chains and Stochastic Stability*. Springer, London.

Propp, J. G. and Wilson, D. B. 1996. Exact sampling with coupled Markov chains and applications to statistical mechanics. *Random Structures and Algorithms*, 9:223–252.

Roberts, G. O. and Rosenthal, J. S. 1997. Geometric ergodicity and hybrid Markov chains. *Electronic Communications in Probability*, 2:13–25.

Roberts, G. O. and Rosenthal, J. S. 2004. General state space Markov chains and MCMC algorithms. *Probability Surveys*, 1:20–71.

Rudin, W. 1987. *Real and Complex Analysis*, 3rd edn. McGraw-Hill, New York.

Schmeiser, B. 1982. Batch size effects in the analysis of simulation output. *Operations Research*, 30:556–568.

Smith, A. F. M. and Roberts, G. O. 1993. Bayesian computation via the Gibbs sampler and related Markov chain Monte Carlo methods (with discussion). *Journal of the Royal Statistical Society, Series B*, 55:3–23.

Stigler, S. M. 2002. *Statistics on the Table: The History of Statistical Concepts and Methods*. Harvard University Press, Cambridge, MA.

Tanner, M. A. and Wong, W. H. 1987. The calculation of posterior distributions by data augmentation (with discussion). *Journal of the American Statistical Association*, 82:528–550.

Tierney, L. 1994. Markov chains for exploring posterior distributions (with discussion). *Annals of Statistics*, 22:1701–1762.

2

A Short History of MCMC: Subjective Recollections from Incomplete Data

Christian Robert and George Casella

2.1 Introduction

Markov chain Monte Carlo (MCMC) methods have been around for almost as long as Monte Carlo techniques, even though their impact on statistics was not truly felt until the very early 1990s, except in the specialized fields of spatial statistics and image analysis, where those methods appeared earlier. The emergence of Markov based techniques in physics is a story that remains untold within this survey (see Landau and Binder, 2005). Also, we will not enter into a description of MCMC techniques, unless they have some historical link, as the remainder of this volume covers the technical aspects. A comprehensive treatment with further references can also be found in Robert and Casella (2004).

We will distinguish between the introduction of Metropolis–Hastings based algorithms and those related to Gibbs sampling, since they each stem from radically different origins, even though their mathematical justification via Markov chain theory is the same. Tracing the development of Monte Carlo methods, we will also briefly mention what we might call the "second-generation MCMC revolution." Starting in the mid to late 1990s, this includes the development of particle filters, reversible jump and perfect sampling, and concludes with more current work on population or sequential Monte Carlo and regeneration and the computing of "honest" standard errors.

As mentioned above, the realization that Markov chains could be used in a wide variety of situations only came (to mainstream statisticians) with Gelfand and Smith (1990), despite earlier publications in the statistical literature such as Hastings (1970), Geman and Geman (1984), and Tanner and Wong (1987). Several reasons can be advanced: lack of computing machinery (think of the computers of 1970!), or background on Markov chains, or hesitation to trust in the practicality of the method. It thus required visionary researchers like Gelfand and Smith to convince the community, supported by papers that demonstrated, through a series of applications, that the method was easy to understand, easy to implement and practical (Gelfand et al., 1990, 1992; Smith and Gelfand, 1992; Wakefield et al., 1994). The rapid emergence of the dedicated BUGS (Bayesian inference using Gibbs sampling) software as early as 1991, when a paper on BUGS was presented at the Valencia meeting, was another compelling argument for adopting, at large, MCMC algorithms.[*]

---

[*] Historically speaking, the development of BUGS can be traced back to Geman and Geman (1984) and Pearl (1987), alongside developments in the artificial intelligence community, and it pre-dates Gelfand and Smith (1990).

2.2 Before the Revolution

Monte Carlo methods were born in Los Alamos, New Mexico, during World War II, eventually resulting in the Metropolis algorithm in the early 1950s. While Monte Carlo methods were in use by that time, MCMC was brought closer to statistical practicality by the work of Hastings in the 1970s.

What can be reasonably seen as the first MCMC algorithm is what we now call the Metropolis algorithm, published by Metropolis et al. (1953). It emanates from the same group of scientists who produced the Monte Carlo method, namely the research scientists of Los Alamos, mostly physicists working on mathematical physics and the atomic bomb.

MCMC algorithms therefore date back to the same time as the development of regular (MC only) Monte Carlo methods, which are usually traced to Ulam and von Neumann in the late 1940s. Stanislaw Ulam associates the original idea with an intractable combinatorial computation he attempted in 1946 (calculating the probability of winning at the solitaire card game). This idea was enthusiastically adopted by John von Neumann for implementation with direct applications to neutron diffusion, the name "Monte Carlo" being suggested by Nicholas Metropolis. Eckhardt (1987) describes these early Monte Carlo developments, and Hitchcock (2003) gives a brief history of the Metropolis algorithm.

These occurrences very closely coincide with the appearance of the very first general-purpose digital computer, the ENIAC, which came to life in February 1946, after three years of construction. The Monte Carlo method was set up by von Neumann, who was using it on thermonuclear and fission problems as early as 1947. That same year, Ulam and von Neumann invented inversion and accept–reject techniques (also recounted in Eckhardt, 1987) to simulate from nonuniform distributions. Without computers, a rudimentary version invented by Fermi in the 1930s went unrecognized (Metropolis, 1987). Note also that, as early as 1949, a symposium on Monte Carlo was supported by Rand, the National Bureau of Standards, and the Oak Ridge laboratory and that Metropolis and Ulam (1949) published the very first paper about the Monte Carlo method.

2.2.1 The Metropolis et al. (1953) Paper

The first MCMC algorithm is associated with a second computer, called MANIAC,[*] built in Los Alamos under the direction of Metropolis in early 1952. Both a physicist and a mathematician, Nicholas Metropolis, came to Los Alamos in April 1943, and was to die there in 1999. The other members of the team also came to Los Alamos during those years, including the controversial Edward Teller. As early as 1942, this physicist became obsessed with the hydrogen bomb, which he eventually managed to design with Stanislaw Ulam, using the improved computer facilities of the early 1950s.

Published in June 1953 in the *Journal of Chemical Physics*, the primary focus of Metropolis et al. (1953) is the computation of integrals of the form

$$\mathfrak{I} = \int F(\theta) \exp\left\{\frac{-E(\theta)}{kT}\right\} d\theta \Big/ \int \exp\left\{\frac{-E(\theta)}{kT}\right\} d\theta,$$

---

[*] MANIAC stands for Mathematical Analyzer, Numerical Integrator and Computer.

on $\mathbb{R}^{2N}$, θ denoting a set of N particles on $\mathbb{R}^2$, with the energy E being defined as

$$E(\theta) = \frac{1}{2} \sum_{\substack{i=1}}^{N} \sum_{\substack{j=1 \\ j \neq i}}^{N} V(d_{ij}),$$

where V a potential function and d_{ij} the Euclidean distance between particles i and j in θ. The *Boltzmann distribution* $\exp\{-E(\theta)/kT\}$ is parameterized by the *temperature T*, k being the Boltzmann constant, with a normalization factor,

$$Z(T) = \int \exp\left\{ \frac{-E(\theta)}{kT} \right\} d\theta,$$

that is not available in closed form, except in trivial cases. Since θ is a $2N$-dimensional vector, numerical integration is impossible. Given the large dimension of the problem, even standard Monte Carlo techniques fail to correctly approximate $\mathfrak{I}$, since $\exp\{-E(\theta)/kT\}$ is very small for most realizations of the random configurations of the particle system (uniformly in the $2N$ square). In order to improve the efficiency of the Monte Carlo method, Metropolis et al. (1953) propose a random-walk modification of the N particles. That is, for each particle i ($1 \leq i \leq N$), values

$$x'_i = x_i + \sigma\xi_{1i} \quad \text{and} \quad y'_i = y_i + \sigma\xi_{2i}$$

are proposed, where both ξ_{1i} and ξ_{2i} are uniform $U(-1,1)$. The energy difference ΔE between the new configuration and the previous one is then computed and the new configuration is accepted with probability

$$\min\left\{ 1, \exp\left(\frac{-\Delta E}{kT} \right) \right\}, \tag{2.1}$$

and otherwise the previous configuration is replicated, in the sense that its counter is increased by one in the final average of the $F(\theta_t)$s over the τ moves of the random walk ($1 \leq t \leq \tau$). Note that Metropolis et al. (1953) move one particle at a time, rather than moving all of them together, which makes the initial algorithm appear a primitive kind of Gibbs sampler!

The authors of Metropolis et al. (1953) demonstrate the validity of the algorithm by first establishing irreducibility, which they call *ergodicity*, and second proving ergodicity, that is, convergence to the stationary distribution. The second part is obtained via a discretization of the space: they first note that the proposal move is reversible, then establish that $\exp\{-E/kT\}$ is invariant. The result is therefore proven in its full generality, minus the discretization. The number of iterations of the Metropolis algorithm used in the paper seems to be limited: 16 steps for burn-in and 48–64 subsequent iterations, which required 4–5 hours on the Los Alamos computer.

An interesting variation is the *simulated annealing* algorithm, developed by Kirkpatrick et al. (1983), who connected optimization with *annealing*, the cooling of a metal. Their variation is to allow the temperature T in Equation 2.1 to decrease as the algorithm runs, according to a "cooling schedule." The simulated annealing algorithm can be shown to find the global maximum with probability 1, although the analysis is quite complex due to the fact that, with varying T, the algorithm is no longer a time-homogeneous Markov chain.

2.2.2 The Hastings (1970) Paper

The Metropolis algorithm was later generalized by Hastings (1970) and his student Peskun (1973, 1981) as a statistical simulation tool that could overcome the curse of dimensionality met by regular Monte Carlo methods, a point already emphasized in Metropolis et al. (1953).[*]

In his *Biometrika* paper,[†] Hastings (1970) also defines his methodology for finite and reversible Markov chains, treating the continuous case by using a discretization analogy. The generic probability of acceptance for a move from state i to state j is

$$\alpha_{ij} = \frac{s_{ij}}{1 + \frac{\pi_i}{\pi_j} \frac{q_{ij}}{q_{ji}}},$$

where $s_{ij} = s_{ji}$ is a positive quantity ensuring that $\alpha_{ij} \leq 1$, π_i denotes the target and q_{ij} the proposal. This generic form of probability encompasses the forms of both Metropolis et al. (1953) and Barker (1965). At this stage, Hastings says that "Little is known about the relative merits of these two choices [even though] Metropolis's method may be preferable." He also warns against "high rejection rates as indicative of a poor choice of ... transition matrix," but does not mention the opposite pitfall of low rejection rates, associated with a slow exploration of the target.

The examples in the paper include a Poisson target with a ± 1 random-walk proposal and a normal target with a uniform random-walk proposal mixed with its reflection, that is, a uniform proposal centered at $-\theta_t$ rather than at the current value θ_t of the Markov chain. On a multivariate target, Hastings introduces a Gibbs sampling strategy, updating one component at a time and defining the composed transition as satisfying the stationary condition because each component does leave the target invariant. Hastings (1970) actually refers to Erhman et al. (1960) as a preliminary, if specific, instance of this sampler. More precisely, this is Metropolis-within-Gibbs except for the name. This first introduction of the Gibbs sampler has thus been completely overlooked, even though the proof of convergence is completely general, based on a composition argument as in Tierney (1994), discussed in Section 2.4.1. The remainder of the paper deals with (a) an importance sampling version of MCMC, (b) general remarks about assessment of the error, and (c) an application to random orthogonal matrices, with another example of Gibbs sampling.

Three years later, Peskun (1973) published a comparison of Metropolis' and Barker's forms of acceptance probabilities and showed in a discrete setup that the optimal choice is that of Metropolis, where optimality is to be understood in terms of the asymptotic variance of any empirical average. The proof is a direct consequence of a result by Kemeny and Snell (1960) on the asymptotic variance. Peskun also establishes that this asymptotic variance can improve upon the independently and identically distributed (i.i.d.) case if and only if the eigenvalues of $\mathbf{P} - \mathbf{A}$ are all negative, when $\mathbf{A}$ is the transition matrix corresponding to i.i.d. simulation and $\mathbf{P}$ the transition matrix corresponding to the Metropolis algorithm, but he concludes that the trace of $\mathbf{P} - \mathbf{A}$ is always positive, therefore that the uniform improvement is impossible.

---

[*] In fact, Hastings starts by mentioning a decomposition of the target distribution into a product of one-dimensional conditional distributions, but this falls short of an early Gibbs sampler.

[†] Hastings (1970) is one of the ten *Biometrika* papers reproduced in Titterington and Cox (2001).

2.3   Seeds of the Revolution

A number of early pioneers had brought forward the seeds of Gibbs sampling; in particular, Hammersley and Clifford had produced a constructive argument in 1970 to recover a joint distribution from its conditionals, a result later called the *Hammersley–Clifford* theorem by Besag (1974, 1986). Besides Hastings (1970) and Geman and Geman (1984), already mentioned, other papers that contained the seeds of Gibbs sampling are Besag and Clifford (1989), Broniatowski et al. (1984), Qian and Titterington (1990), and Tanner and Wong (1987).

2.3.1   Besag and the Fundamental (Missing) Theorem

In the early 1970s, Hammersley, Clifford, and Besag were working on the specification of joint distributions from conditional distributions and on necessary and sufficient conditions for the conditional distributions to be compatible with a joint distribution. What is now known as the *Hammersley–Clifford* theorem states that a joint distribution for a vector associated with a dependence graph (edge meaning dependence and absence of edge conditional independence) must be represented as a product of functions over the *cliques* of the graphs, that is, of functions depending only on the components indexed by the labels in the clique.*

From a historical point of view, Hammersley (1974) explains why the Hammersley–Clifford theorem was never published as such, but only through Besag (1974). The reason is that Clifford and Hammersley were dissatisfied with the positivity constraint: the joint density could be recovered from the full conditionals only when the support of the joint was made up of the product of the supports of the full conditionals. While they strived *to make the theorem independent of any positivity condition*, their graduate student published a counterexample that put a full stop to their endeavors (Moussouris, 1974).

While Besag (1974) can certainly be credited to some extent with the (re)discovery of the Gibbs sampler, Besag (1975) expressed doubt about the practicality of his method, noting that "the technique is unlikely to be particularly helpful in many other than binary situations and the Markov chain itself has no practical interpretation," clearly understating the importance of his work.

A more optimistic sentiment was expressed earlier by Hammersley and Handscomb (1964), in their textbook on Monte Carlo methods. There they cover such topics as "crude Monte Carlo," importance sampling, control variates, and "conditional Monte Carlo," which looks surprisingly like a simulation approach to missing-data models (see Section 2.3.2). Of course, they do not cover the Hammersley–Clifford theorem but they do state in the Preface: "We are convinced nevertheless that Monte Carlo methods will one day reach an impressive maturity." Well said!

2.3.2   EM and Its Simulated Versions as Precursors

Due to its connection with missing-data problems, the EM algorithm (Dempster et al., 1977) has early connections with Gibbs sampling.[†] For instance, Broniatowski et al. (1984) and Celeux and Diebolt (1985) had tried to overcome the dependence of EM methods on the

---

[*] A clique is a maximal subset of the nodes of a graphs such that every pair of nodes within the clique is connected by an edge (Cressie, 1993).

[†] This is especially relevant when considering the early introduction of a Gibbs sampler by data augmentation in Tanner and Wong (1987).

starting value by replacing the E step with a *simulation* step, the missing data z_m being generated conditionally on the observation x and on the current value of the parameter θ_m. The maximization in the M step is then carried out on the simulated complete-data likelihood, $L(\theta \mid x, z_m)$, producing a new value θ_{m+1}, and this appears as a predecessor to the Gibbs step of Gelman and King (1990) and Diebolt and Robert (1994) for mixture estimation.[*] Unfortunately, the theoretical convergence results for these methods are limited. Celeux and Diebolt (1990) have, however, solved the convergence problem of stochastic EM (SEM) by devising a hybrid version called SAEM (for *simulated annealing EM*), where the amount of randomness in the simulations decreases with the iterations, ending up with an EM algorithm.[†]

2.3.3 Gibbs and Beyond

Although somewhat removed from statistical inference in the classical sense and based on earlier techniques used in statistical physics, the landmark paper by Geman and Geman (1984) brought Gibbs sampling into the arena of statistical application. This paper is also responsible for the name *Gibbs sampling*, because it implemented this method for the Bayesian study of *Gibbs random fields* which, in turn, derive their name from the physicist Josiah Willard Gibbs (1839–1903). This original implementation of the Gibbs sampler was applied to a discrete image processing problem and did not involve completion as in Section 2.3.2. But this was one more spark that led to the explosion, as it had a clear influence on Green, Smith, Spiegelhalter, and others.

The extent to which Gibbs sampling and Metropolis algorithms were in use within the image analysis and point process communities is actually quite large, as illustrated in Ripley (1987) where Section 4.7 is entitled "Metropolis' method and random fields" and describes the implementation and validation of the Metropolis algorithm in a finite setting with an application to Markov random fields and the corresponding issue of bypassing the normalizing constant. Besag et al. (1991) is another striking example of the activity in the spatial statistics community at the end of the 1980s.

2.4 The Revolution

The gap of more than 30 years between Metropolis et al. (1953) and Gelfand and Smith (1990) can still be partially attributed to the lack of appropriate computing power, as most of the examples now processed by MCMC algorithms could not have been treated previously, even though the hundreds of dimensions processed in Metropolis et al. (1953) were quite formidable. However, by the mid 1980s, the pieces were all in place.

After Peskun, MCMC in the statistical world was dormant for about 10 years, and then several papers appeared that highlighted its usefulness in specific settings such as pattern recognition, image analysis or spatial statistics. In particular, Geman and Geman (1984) influenced Gelfand and Smith (1990) to write a paper that is the genuine starting point for an intensive use of MCMC methods by the mainstream statistical community. It sparked new

---

[*] The achievement in the former paper remained unnoticed for several years due to the low-key and off-hand use of the Gibbs sampler at a time when it was unknown to most of the community.

[†] Other and better-known connections between EM and MCMC algorithms can be found in the literature (Liu and Rubin, 1994; Meng and Rubin, 1992; Wei and Tanner, 1990), but the connection with Gibbs sampling is more tenuous in that the simulation methods there are used to approximate quantities in a Monte Carlo fashion.

interest in Bayesian methods, statistical computing, algorithms, and stochastic processes through the use of computing algorithms such as the Gibbs sampler and the Metropolis–Hastings algorithm. Casella and George (1992) wrote an elementary introduction to the Gibbs sampler* in *The American Statistician* that disseminated the technique to a wider community while explaining in simple terms why the algorithm is valid.

Interestingly, the earlier paper by Tanner and Wong (1987) had essentially the same ingredients as Gelfand and Smith (1990), namely the fact that simulating from the conditional distributions is sufficient to asymptotically simulate from the joint. This paper was considered important enough to be a discussion paper in the *Journal of the American Statistical Association*, but its impact was somehow limited, compared with Gelfand and Smith (1990). There are several reasons for this: one is that the method seemed to apply only to missing-data problems, this impression being reinforced by the name *data augmentation*; another is that the authors were more focused on approximating the posterior distribution. They suggested an MCMC approximation to the target $\pi(\theta \mid x)$ at each iteration of the sampler, based on

$$\frac{1}{m} \sum_{k=1}^{m} \pi(\theta \mid x, z^{t,k}), \quad z^{t,k} \sim \hat{\pi}_{t-1}(z \mid x), \ \ k = 1, \dots, m,$$

that is, by replicating m times the simulations from the current approximation $\hat{\pi}_{t-1}(z \mid x)$ of the marginal posterior distribution of the missing data. This focus on estimation of the posterior distribution connected the original data augmentation algorithm to EM, as pointed out by Dempster in the discussion. Although the discussion by Morris gets very close to the two-stage Gibbs sampler for hierarchical models, he is still concerned about doing m iterations, and worries about how costly that would be. Tanner and Wong mention taking $m = 1$ at the end of the paper, referring to this as an "extreme case."

In a sense, Tanner and Wong (1987) was still too close to Rubin's (1978) multiple imputation to start a new revolution. Yet another reason for this may be that the theoretical background was based on functional analysis rather than Markov chain theory, which needed, in particular, the Markov kernel to be uniformly bounded and equicontinuous. This may have discouraged potential users as requiring too much mathematics.

The authors of this review were fortunate enough to attend many focused conferences during this time, where we were able to witness the explosion of Gibbs sampling. In the summer of 1986 in Bowling Green, Ohio, Smith gave a series of ten lectures on hierarchical models. Although there was a lot of computing mentioned, the Gibbs sampler was not yet fully developed. (Interestingly, Smith commented that the limiting factor, at that time, for the full exploitation of hierarchical models in statistical problems was the inability to compute high-dimensional integrals.) In another lecture in June 1989 at a Bayesian workshop in Sherbrooke, Quebec, he revealed for the first time the generic features of Gibbs sampler, and we still remember vividly the shock induced in ourselves and in the whole audience by the sheer breadth of the method: this development of Gibbs sampling, MCMC, and the resulting seminal paper of Gelfand and Smith (1990) was an *epiphany*[†] in the world of statistics.

---

* On a humorous note, the original Technical Report of this paper was called *Gibbs for Kids*, which was changed because a referee did not appreciate the humor. However, our colleague Dan Gianola, an animal breeder at Wisconsin, liked the title. In using Gibbs sampling in his work, he gave a presentation in 1993 at the 44th Annual Meeting of the European Association for Animal Production, Aarhus, Denmark. The title: *Gibbs for Pigs*.

[†] Epiphany, *n.* A spiritual event in which the essence of a given object of manifestation appears to the subject, as in a sudden flash of recognition.

The explosion had begun, and just two years later an MCMC conference at Ohio State University, organized by Gelfand, Goel, and Smith, consisted of three full days of talks. Many of the talks were to become influential papers; including Albert and Chib (1993), Gelman and Rubin (1992), Geyer (1992), Gilks (1992), Liu et al. (1994, 1995), and Tierney (1994).

Approximately one year later, in May 1992, there was a meeting of the Royal Statistical Society on "The Gibbs sampler and other Markov chain Monte Carlo methods," where four papers were presented followed by much discussion. The papers appear in the first issue of the *Journal of the Royal Statistical Society, Series B*, in 1993, together with 49 (!) pages of discussion. The excitement is clearly evident in the writings, even though the theory and implementation were not always perfectly understood.

Looking at these meetings, we can see the paths that Gibbs sampling would lead us down. In the next two sections we will summarize some of the advances from the early to mid 1990s.

2.4.1 Advances in MCMC Theory

Perhaps the most influential MCMC theory paper of the 1990s is Tierney (1994), which carefully laid out all of the assumptions needed to analyze the Markov chains and then developed their properties, in particular, convergence of ergodic averages and central limit theorems. In one of the discussions of that paper, Chan and Geyer (1994) were able to relax a condition on Tierney's central limit theorem, and this new condition plays an important role in research today (see Section 2.5.4). A pair of very influential, and innovative, papers is the work of Liu et al. (1994, 1995), who very carefully analyzed the covariance structure of Gibbs sampling, and were able to formally establish the validity of Rao-Blackwellization in Gibbs sampling. Gelfand and Smith (1990) had used Rao-Blackwellization, but it was not justified at that time, as the original theorem was only applicable to i.i.d. sampling, which is not the case in MCMC. Another significant entry is Rosenthal (1995), who obtained one of the earliest results on exact rates of convergence.

Another paper must be singled out, namely Mengersen and Tweedie (1996), for setting the tone for the study of the speed of convergence of MCMC algorithms to the target distribution. Subsequent works in this area by Richard Tweedie, Gareth Roberts, Jeff Rosenthal and co-authors are too numerous to be mentioned here, although the paper by Roberts et al. (1997) must be cited for setting explicit targets on the acceptance rate of the random-walk Metropolis–Hastings algorithm, as well as Roberts and Rosenthal (1999) for obtaining an upper bound on the number of iterations (523) needed to approximate the target up to 1% by a slice sampler. The untimely death of Richard Tweedie in 2001 also had a major impact on the book about MCMC convergence he was contemplating with Gareth Roberts.

One pitfall arising from the widespread use of Gibbs sampling was the tendency to specify models only through their conditional distributions, almost always without referring to the positivity conditions in Section 2.3. Unfortunately, it is possible to specify a perfectly legitimate-looking set of conditionals that do not correspond to any joint distribution, and the resulting Gibbs chain cannot converge. Hobert and Casella (1996) were able to document the conditions needed for a convergent Gibbs chain, and alerted the Gibbs community to this problem, which only arises when improper priors are used, but this is a frequent occurrence.

Much other work followed, and continues to grow today. Geyer and Thompson (1995) describe how to put a "ladder" of chains together for both "hot" and "cold" exploration, followed by Neal's (1996) introduction of tempering; Athreya et al. (1996) gave more easily verifiable conditions for convergence; Meng and van Dyk (1999) and Liu and Wu (1999) developed the theory of parameter expansion in the data augmentation algorithm, leading to construction of chains with faster convergence, and to the work of Hobert and Marchev (2008), who give precise constructions and theorems to show how parameter expansion can uniformly improve over the original chain.

2.4.2 Advances in MCMC Applications

The real reason for the explosion of MCMC methods was the fact that an enormous number of problems that were deemed to be computational nightmares now cracked open like eggs. As an example, consider this very simple random effects model from Gelfand and Smith (1990). Observe

$$Y_{ij} = \theta_i + \varepsilon_{ij}, \quad i = 1, \ldots, K, \ j = 1, \ldots, J, \tag{2.2}$$

where

$$\theta_i \sim N(\mu, \sigma_\theta^2)$$

$$\varepsilon_{ij} \sim N(0, \sigma_\varepsilon^2), \quad \text{independent of } \theta_i.$$

Estimation of the variance components can be difficult for a frequentist (restricted maximum likelihood is typically preferred) but it was a nightmare for a Bayesian, as the integrals were intractable. However, with the usual priors on μ, σ_θ^2, and σ_ε^2, the full conditionals are trivial to sample from and the problem is easily solved via Gibbs sampling. Moreover, we can increase the number of variance components and the Gibbs solution remains easy to implement.

During the early 1990s, researchers found that Gibbs, or Metropolis–Hastings, algorithms would crack almost any problem that they looked at, and there was a veritable flood of papers applying MCMC to previously intractable models and getting good solutions. For example, building on Equation 2.2, it was quickly realized that Gibbs sampling was an easy route to getting estimates in the linear mixed models (Wang et al., 1993, 1994), and even generalized linear mixed models (Zeger and Karim, 1991). Building on the experience gained with the EM algorithm, similar arguments made it possible to analyze probit models using a latent variable approach in a linear mixed model (Albert and Chib, 1993) and in mixture models with Gibbs sampling (Diebolt and Robert, 1994). It progressively dawned on the community that latent variables could be artificially introduced to run the Gibbs sampler in just about every situation, as eventually published in Damien et al. (1999), the main example being the slice sampler (Neal, 2003). A very incomplete list of some other applications includes change-point analysis (Carlin et al., 1992; Stephens, 1994), genomics (Churchill, 1995; Lawrence et al., 1993; Stephens and Smith, 1993), capture–recapture (Dupuis, 1995; George and Robert, 1992), variable selection in regression (George and McCulloch, 1993), spatial statistics (Raftery and Banfield, 1991), and longitudinal studies (Lange et al., 1992).

Many of these applications were advanced though other developments such as the adaptive rejection sampling of Gilks (1992) and Gilks et al. (1995), and the simulated tempering approaches of Geyer and Thompson (1995) or Neal (1996).

2.5 After the Revolution

After the revolution comes the "second" revolution, but now we have a more mature field. The revolution has slowed, and the problems are being solved in, perhaps, deeper and more sophisticated ways, even though Gibbs sampling also offers the amateur the possibility of handling Bayesian analysis in complex models at little cost, as exhibited by the widespread use of BUGS, which mostly focuses on this approach.* But, as before, the methodology continues to expand the set of problems for which statisticians can provide meaningful solutions, and thus continues to further the impact of statistics.

2.5.1 A Brief Glimpse at Particle Systems

The realization of the possibilities of iterating importance sampling is not new: in fact, it is about as old as Monte Carlo methods themselves. It can be found in the molecular simulation literature of the 1950s, for example Hammersley and Morton (1954), Rosenbluth and Rosenbluth (1955), and Marshall (1965). Hammersley and colleagues proposed such a method to simulate a self-avoiding random walk (see Madras and Slade, 1993) on a grid, due to huge inefficiencies in regular importance sampling and rejection techniques. Although this early implementation occurred in particle physics, the use of the term "particle" only dates back to Kitagawa (1996), while Carpenter et al. (1997) coined the term "particle filter." In signal processing, early occurrences of a particle filter can be traced back to Handschin and Mayne (1969).

More in connection with our theme, the landmark paper of Gordon et al. (1993) introduced the bootstrap filter which, while formally connected with importance sampling, involves past simulations and possible MCMC steps (Gilks and Berzuini, 2001). As described in the volume edited by Doucet et al. (2001), particle filters are simulation methods adapted to sequential settings where data are collected progressively in time, as in radar detection, telecommunication correction or financial volatility estimation. Taking advantage of state–space representations of those dynamic models, particle filter methods produce Monte Carlo approximations to the posterior distributions by propagating simulated samples whose weights are actualized against the incoming observations. Since the importance weights have a tendency to degenerate, that is, all weights but one are close to zero, additional MCMC steps can be introduced at times to recover the variety and representativeness of the sample. Modern connections with MCMC in the construction of the proposal kernel are to be found, for instance, in Doucet et al. (2000) and Del Moral et al. (2006). In parallel, sequential imputation was developed in Kong et al. (1994), while Liu and Chen (1995) first formally pointed out the importance of resampling in sequential Monte Carlo, a term coined by them.

The recent literature on the topic more closely bridges the gap between sequential Monte Carlo and MCMC methods by making adaptive MCMC a possibility—see, for example, Andrieu et al. (2004) or Roberts and Rosenthal (2005).

2.5.2 Perfect Sampling

Introduced in the seminal paper of Propp and Wilson (1996), perfect sampling, namely the ability to use MCMC methods to produce an exact (or perfect) simulation from the target,

---

* BUGS now uses both Gibbs sampling and Metropolis–Hastings algorithms.

has a unique place in the history of MCMC methods. Although this exciting discovery led to an outburst of papers, in particular in the large body of work of Møller and coauthors, including the book by Møller and Waagepetersen (2004), as well as many reviews and introductory materials, such as Casella et al. (2001), Fismen (1998), and Dimakos (2001), the excitement quickly died down. The major reason for this ephemeral lifespan is that the construction of perfect samplers is most often close to impossible or impractical, despite some advances in implementation (Fill, 1998a,b).

There is, however, ongoing activity in the area of point processes and stochastic geometry, much from the work of Møller and Kendall. In particular, Kendall and Møller (2000) developed an alternative to the *coupling from the past* (CFTP) algorithm of Propp and Wilson (1996), called *horizontal CFTP*, which mainly applies to point processes and is based on continuous-time birth-and-death processes. See also Fernández et al. (1999) for another horizontal CFTP algorithm for point processes. Berthelsen and Møller (2003) exhibited a use of these algorithms for nonparametric Bayesian inference on point processes.

2.5.3 Reversible Jump and Variable Dimensions

From many viewpoints, the invention of the reversible jump algorithm in Green (1995) can be seen as the start of the second MCMC revolution: the formalization of a Markov chain that moves across models and parameter spaces allowed for the Bayesian processing of a wide variety of new models and contributed to the success of Bayesian model choice and subsequently to its adoption in other fields. There exist earlier alternative Monte Carlo solutions such as Gelfand and Dey (1994) and Carlin and Chib (1995), the latter being very close in spirit to reversible jump MCMC (as shown by the completion scheme of Brooks et al., 2003), but the definition of a proper balance condition on cross-model Markov kernels in Green (1995) gives a generic setup for exploring variable dimension spaces, even when the number of models under comparison is infinite. The impact of this new idea was clearly perceived when looking at the First European Conference on highly structured stochastic systems that took place in Rebild, Denmark, the next year, organized by Stephen Lauritzen and Jesper Møller: a large majority of the talks were aimed at direct implementations of RJMCMC to various inference problems. The application of RJMCMC to mixture order estimation in the discussion paper of Richardson and Green (1997) ensured further dissemination of the technique. More recently, Stephens (2000) proposed a continuous-time version of RJMCMC, based on earlier ideas of Geyer and Møller (1994), but with similar properties (Cappé et al., 2003), while Brooks et al. (2003) made proposals for increasing the efficiency of the moves. In retrospect, while reversible jump is somehow unavoidable in the processing of very large numbers of models under comparison, as for instance in variable selection (Marin and Robert, 2007), the implementation of a complex algorithm such as RJMCMC for the comparison of a few models is somewhat of an overkill since there exist alternative solutions based on model-specific MCMC chains (e.g. Chen et al., 2000).

2.5.4 Regeneration and the Central Limit Theorem

While the central limit theorem (CLT) is a central tool in Monte Carlo convergence assessment, its use in MCMC setups took longer to emerge, despite early signals by Geyer (1992), and it is only recently that sufficiently clear conditions emerged. We recall that the ergodic theorem (see, e.g. Robert and Casella, 2004, Theorem 6.63) states that, if $(\theta_t)_t$ is a Markov chain with stationary distribution π, and $h(\cdot)$ is a function with finite variance, then under fairly mild conditions,

$$\lim_{n \to \infty} \bar{h}_n = \int h(\theta)\pi(\theta) \, d\theta = E_\pi[h(\theta)],$$

almost everywhere, where $\bar{h}_n = (1/n) \sum_{i=1}^{n} h(\theta_i)$. For the CLT to be used to monitor this convergence,

$$\frac{\sqrt{n}(\bar{h}_n - E_\pi[h(\theta)])}{\sqrt{\text{var}[h(\theta)]}} \to N(0, 1), \tag{2.3}$$

there are two roadblocks. First, convergence to normality is strongly affected by the lack of independence. To get CLTs for Markov chains, we can use a result of Kipnis and Varadhan (1986), which requires the chain to be reversible, as is the case for Metropolis–Hastings chains, or we must delve into mixing conditions (Billingsley, 1995, Section 27), which are typically not easy to verify. However, Chan and Geyer (1994) showed how the condition of geometric ergodicity could be used to establish CLTs for Markov chains. But getting the convergence is only half of the problem. In order to use Equation 2.3, we must be able to consistently estimate the variance, which turns out to be another difficult endeavor. The "naive" estimate of the usual standard error is not consistent in the dependent case and the most promising paths for consistent variance estimates seem to be through regeneration and batch means.

The theory of regeneration uses the concept of a split chain (Athreya and Ney, 1978), and allows us to independently restart the chain while preserving the stationary distribution. These independent "tours" then allow the calculation of consistent variance estimates and honest monitoring of convergence through Equation 2.3. Early work on applying regeneration to MCMC chains was done by Mykland et al. (1995) and Robert (1995), who showed how to construct the chains and use them for variance calculations and diagnostics (see also Guihenneuc-Jouyaux and Robert, 1998), as well as deriving adaptive MCMC algorithms (Gilks et al., 1998). Rosenthal (1995) also showed how to construct and use regenerative chains, and much of this work is reviewed in Jones and Hobert (2001). The most interesting and practical developments, however, are in Hobert et al. (2002) and Jones et al. (2006), where consistent estimators are constructed for $\text{var}[h(\theta)]$, allowing valid monitoring of convergence in chains that satisfy the CLT. Interestingly, although Hobert et al. (2002) uses regeneration, Jones et al. (2006) get their consistent estimators thorough another technique, that of consistent batch means.

2.6 Conclusion

The impact of Gibbs sampling and MCMC on Bayesian statistics was to change our entire method of thinking about and attacking problems, representing a *paradigm shift* (Kuhn, 1996). Now, the collection of real problems that we could solve grew almost without bound. Markov chain Monte Carlo changed our emphasis from "closed form" solutions to algorithms, expanded our impact to solving "real" applied problems and to improving numerical algorithms using statistical ideas, and led us into a world where "exact" now means "simulated."

This has truly been a quantum leap in the evolution of the field of statistics, and the evidence is that there are no signs of a slowdown. Although the "explosion" is over, the current work is going deeper into theory and applications, and continues to expand our horizons and influence by increasing our ability to solve even bigger and more important problems. The size of the data sets, and of the models, for example in genomics or climatology, is something that could not have been conceived 60 years ago, when Ulam and von Neumann invented the Monte Carlo method. Now we continue to plod on, and hope that the advances that we make here will, in some way, help our colleagues 60 years in the future solve problems that we cannot yet conceive.

Acknowledgments

Christian Robert was supported by the Agence Nationale de la Recherche, Paris, through the 2006–2008 project ANR=05-BLAN-0299 Adap'MC. This work was partly done during a visit to the Queensland University of Technology, Brisbane, and CR is grateful to Kerrie Mengersen for her hospitality and support. George Casella was supported by National Science Foundation Grants DMS-04-05543, DMS-0631632 and SES-0631588.

We are grateful for comments and suggestions from Olivier Cappé, Alan Gelfand, Peter Green, Jun Liu, Sharon McGrayne, Peter Müller, Gareth Roberts, Adrian Smith, and David Spiegelhalter.

References

Albert, J. and Chib, S. 1993. Bayesian analysis of binary and polychotomous response data. *Journal of the American Statistical Association*, 88:669–679.

Andrieu, C., de Freitas, N., Doucet, A., and Jordan, M. 2004. An introduction to MCMC for machine learning. *Machine Learning*, 50(1–2):5–43.

Athreya, K., Doss, H., and Sethuraman, J. 1996. On the convergence of the Markov chain simulation method. *Annals of Statistics*, 24:69–100.

Athreya, K. and Ney, P. 1978. A new approach to the limit theory of recurrent Markov chains. *Transactions of the American Mathematical Society*, 245:493–501.

Barker, A. 1965. Monte Carlo calculations of the radial distribution functions for a proton–electron plasma. *Australian Journal of Physics*, 18:119–133.

Berthelsen, K. and Møller, J. 2003. Likelihood and non-parametric Bayesian MCMC inference for spatial point processes based on perfect simulation and path sampling. *Scandinavian Journal of Statistics*, 30:549–564.

Besag, J. 1974. Spatial interaction and the statistical analysis of lattice systems (with discussion). *Journal of the Royal Statistical Society*, Series B, 36:192–326.

Besag, J. 1975. Statistical analysis of non-lattice data. *The Statistician*, 24(3):179–195.

Besag, J. 1986. On the statistical analysis of dirty pictures. *Journal of the Royal Statistical Society*, Series B, 48:259–279.

Besag, J. and Clifford, P. 1989. Generalized Monte Carlo significance tests. *Biometrika*, 76:633–642.

Besag, J., York, J., and Mollie, A. 1991. Bayesian image restoration, with two applications in spatial statistics (with discussion). *Annals of the Institute of Statistical Mathematics*, 42(1):1–59.

Billingsley, P. 1995. *Probability and Measure*, 3rd edn. Wiley, New York.

Broniatowski, M., Celeux, G., and Diebolt, J. 1984. Reconnaissance de mélanges de densités par un algorithme d'apprentissage probabiliste. In E. Diday (ed.), *Data Analysis and Informatics*, Volume 3, pp. 359–373. North-Holland, Amsterdam.

Brooks, S. P., Giudici, P., and Roberts, G. O. 2003. Efficient construction of reversible jump Markov chain Monte Carlo proposal distributions (with discussion). *Journal of the Royal Statistical Society, Series B*, 65(1):3–55.

Cappé, O., Robert, C., and Rydén, T. 2003. Reversible jump, birth-and-death, and more general continuous time MCMC samplers. *Journal of the Royal Statistical Society, Series B*, 65(3): 679–700.

Carlin, B. and Chib, S. 1995. Bayesian model choice through Markov chain Monte Carlo. *Journal of the Royal Statistical Society, Series B*, 57(3):473–484.

Carlin, B., Gelfand, A., and Smith, A. 1992. Hierarchical Bayesian analysis of change point problems. *Applied Statistics*, 41:389–405.

Carpenter, J., Clifford, P., and Fernhead, P. 1997. Building robust simulation-based filters for evolving datasets. Technical report, Department of Statistics, Oxford University.

Casella, G. and George, E. 1992. An introduction to Gibbs sampling. *American Statistician*, 46:167–174.

Casella, G., Lavine, M., and Robert, C. 2001. Explaining the perfect sampler. *American Statistician*, 55(4):299–305.

Celeux, G. and Diebolt, J. 1985. The SEM algorithm: a probabilistic teacher algorithm derived from the EM algorithm for the mixture problem. *Computational Statistics Quaterly*, 2:73–82.

Celeux, G. and Diebolt, J. 1990. Une version de type recuit simulé de l'algorithme EM. *Comptes Rendus de l'Académie des Sciences (Paris)*, 310:119–124.

Chan, K. and Geyer, C. 1994. Discussion of "Markov chains for exploring posterior distribution." *Annals of Statistics*, 22:1747–1758.

Chen, M., Shao, Q., and Ibrahim, J. 2000. *Monte Carlo Methods in Bayesian Computation*. Springer, New York.

Churchill, G. 1995. Accurate restoration of DNA sequences (with discussion). In C. Gatsonis, J. S. Hodges, R. Kass, and N. Singpurwalla (eds), *Case Studies in Bayesian Statistics*, Volume 2, pp. 90–148. Springer, New York.

Cressie, N. 1993. *Spatial Statistics*. Wiley, New York.

Damien, P., Wakefield, J., and Walker, S. 1999. Gibbs sampling for Bayesian non-conjugate and hierarchical models by using auxiliary variables. *Journal of the Royal Statistical Society, Series B*, 61(2):331–344.

Del Moral, P., Doucet, A., and Jasra, A. 2006. The sequential Monte Carlo samplers. *Journal of the Royal Statistical Society, Series B*, 68(3):411–436.

Dempster, A., Laird, N., and Rubin, D. 1977. Maximum likelihood from incomplete data via the *EM* algorithm (with discussion). *Journal of the Royal Statistical Society, Series B*, 39:1–38.

Diebolt, J. and Robert, C. 1994. Estimation of finite mixture distributions by Bayesian sampling. *Journal of the Royal Statistical Society, Series B*, 56:363–375.

Dimakos, X. K. 2001. A guide to exact simulation. *International Statistical Review*, 69(1):27–48.

Doucet, A., de Freitas, N., and Gordon, N. 2001. *Sequential Monte Carlo Methods in Practice*. Springer, New York.

Doucet, A., Godsill, S., and Andrieu, C. 2000. On sequential Monte Carlo sampling methods for Bayesian filtering. *Statistics and Computing*, 10(3):197–208.

Dupuis, J. 1995. Bayesian estimation of movement probabilities in open populations using hidden Markov chains. *Biometrika*, 82(4):761–772.

Eckhardt, R. 1987. Stan Ulam, John von Neumann, and the Monte Carlo method. *Los Alamos Science*, Special Issue, pp. 131–141. Available at http://library.lanl.gov.cgi–bin/getfile?15–13.pdf.

Erhman, J., Fosdick, L., and Handscomb, D. 1960. Computation of order parameters in an Ising lattice by the Monte Carlo method. *Journal of Mathematical Physics*, 1:547–558.

Fernández, R., Ferrari, P., and Garcia, N. L. 1999. Perfect simulation for interacting point processes, loss networks and Ising models. Technical report, Laboratoire Raphael Salem, Universitè de Rouen.

Fill, J. 1998a. An interruptible algorithm for exact sampling via Markov chains. *Annals of Applied Probability*, 8:131–162.

Fill, J. 1998b. The move-to front rule: A case study for two perfect sampling algorithms. *Probability in the Engineering and Informational Sciences*, 8:131–162.

Fismen, M. 1998. Exact simulation using Markov chains. Technical Report 6/98, Institutt for Matematiske Fag, Oslo. Diploma thesis.

Gelfand, A. and Dey, D. 1994. Bayesian model choice: asymptotics and exact calculations. *Journal of the Royal Statistical Society*, Series B, 56:501–514.

Gelfand, A., Hills, S., Racine-Poon, A., and Smith, A. 1990. Illustration of Bayesian inference in normal data models using Gibbs sampling. *Journal of the American Statistical Association*, 85:972–982.

Gelfand, A. and Smith, A. 1990. Sampling based approaches to calculating marginal densities. *Journal of the American Statistical Association*, 85:398–409.

Gelfand, A., Smith, A., and Lee, T. 1992. Bayesian analysis of constrained parameters and truncated data problems using Gibbs sampling. *Journal of the American Statistical Association*, 87: 523–532.

Gelman, A. and King, G. 1990. Estimating the electoral consequences of legislative redistricting. *Journal of the American Statistical Association*, 85:274–282.

Gelman, A. and Rubin, D. 1992. Inference from iterative simulation using multiple sequences (with discussion). *Statistical Science*, 7:457–511.

Geman, S. and Geman, D. 1984. Stochastic relaxation, Gibbs distributions and the Bayesian restoration of images. *IEEE Transactions on Pattern Analysis and Machine Intelligence*, 6:721–741.

George, E. and McCulloch, R. 1993. Variable selection via Gibbs sampling. *Journal of the American Statistical Association*, 88:881–889.

George, E. and Robert, C. 1992. Calculating Bayes estimates for capture-recapture models. *Biometrika*, 79(4):677–683.

Geyer, C. 1992. Practical Monte Carlo Markov chain (with discussion). *Statistical Science*, 7:473–511.

Geyer, C. and Møller, J. 1994. Simulation procedures and likelihood inference for spatial point processes. *Scandinavian Journal of Statistics*, 21:359–373.

Geyer, C. and Thompson, E. 1995. Annealing Markov chain Monte Carlo with applications to ancestral inference. *Journal of the American Statistical Association*, 90:909–920.

Gilks, W. 1992. Derivative-free adaptive rejection sampling for Gibbs sampling. In J. Bernardo, J. Berger, A. Dawid, and A. Smith (eds), *Bayesian Statistics 4*, pp. 641–649. Oxford University Press, Oxford.

Gilks, W. and Berzuini, C. 2001. Following a moving target—Monte Carlo inference for dynamic Bayesian models. *Journal of the Royal Statistical Society*, Series B, 63(1):127–146.

Gilks, W., Best, N., and Tan, K. 1995. Adaptive rejection Metropolis sampling within Gibbs sampling. *Applied Statistics*, 44:455–472.

Gilks, W., Roberts, G. O., and Sahu, S. 1998. Adaptive Markov chain Monte Carlo. *Journal of the American Statistical Association*, 93:1045–1054.

Gordon, N., Salmond, J., and Smith, A. 1993. A novel approach to non-linear/non-Gaussian Bayesian state estimation. *IEEE Proceedings on Radar and Signal Processing*, 140:107–113.

Green, P. 1995. Reversible jump MCMC computation and Bayesian model determination. *Biometrika*, 82(4):711–732.

Guihenneuc-Jouyaux, C. and Robert, C. 1998. Finite Markov chain convergence results and MCMC convergence assessments. *Journal of the American Statistical Association*, 93:1055–1067.

Hammersley, J. 1974. Discussion of Mr Besag's paper. *Journal of the Royal Statistical Society*, Series B, 36:230–231.

Hammersley, J. and Handscomb, D. 1964. *Monte Carlo Methods*. Wiley, New York.

Hammersley, J. and Morton, K. 1954. Poor man's Monte Carlo. *Journal of the Royal Statistical Society*, Series B, 16:23–38.

Handschin, J. and Mayne, D. 1969. Monte Carlo techniques to estimate the conditional expectation in multi-stage nonlinear filtering. *International Journal of Control*, 9:547–559.

Hastings, W. 1970. Monte Carlo sampling methods using Markov chains and their application. *Biometrika*, 57:97–109.

Hitchcock, D. B. 2003. A history of the Metropolis-Hastings algorithm. *American Statistician*, 57(4): 254–257.

Hobert, J. and Casella, G. 1996. The effect of improper priors on Gibbs sampling in hierarchical linear models. *Journal of the American Statistical Association*, 91:1461–1473.

Hobert, J., Jones, G., Presnell, B., and Rosenthal, J. 2002. On the applicability of regenerative simulation in Markov chain Monte Carlo. *Biometrika*, 89(4):731–743.

Hobert, J. and Marchev, D. 2008. A theoretical comparison of the data augmentation, marginal augmentation and PX-DA algorithms. *Annals of Statistics*, 36:532–554.

Jones, G., Haran, M., Caffo, B. S., and Neath, R. 2006. Fixed-width output analysis for Markov chain Monte Carlo. *Journal of the American Statistical Association*, 101:1537–1547.

Jones, G. and Hobert, J. 2001. Honest exploration of intractable probability distributions via Markov chain Monte Carlo. *Statistical Science*, 16(4):312–334.

Kemeny, J. and Snell, J. 1960. *Finite Markov Chains*. Van Nostrand, Princeton, NJ.

Kendall, W. and Møller, J. 2000. Perfect simulation using dominating processes on ordered spaces, with application to locally stable point processes. *Advances in Applied Probability*, 32: 844–865.

Kipnis, C. and Varadhan, S. 1986. Central limit theorem for additive functionals of reversible Markov processes and applications to simple exclusions. *Communications in Mathematical Physics*, 104: 1–19.

Kirkpatrick, S., Gelatt, C., and Vecchi, M. 1983. Optimization by simulated annealing. *Science*, 220: 671–680.

Kitagawa, G. 1996. Monte Carlo filter and smoother for non-Gaussian non-linear state space models. *Journal of Computational and Graphical Statistics*, 5:1–25.

Kong, A., Liu, J., and Wong, W. 1994. Sequential imputations and Bayesian missing data problems. *Journal of the American Statistical Association*, 89:278–288.

Kuhn, T. 1996. *The Structure of scientific Revolutions*, 3rd edn. University of Chicago Press, Chicago.

Landau, D. and Binder, K. 2005. *A Guide to Monte Carlo Simulations in Statistical Physics*, 2nd edn. Cambridge University Press, Cambridge.

Lange, N., Carlin, B. P., and Gelfand, A. E. 1992. Hierarchal Bayes models for the progression of hiv infection using longitudinal cd4 t-cell numbers. *Journal of the American Statistical Association*, 87:615–626.

Lawrence, C. E., Altschul, S. F., Boguski, M. S., Liu, J. S., Neuwald, A. F., and Wootton, J. C. 1993. Detecting subtle sequence signals: a Gibbs sampling strategy for multiple alignment. *Science*, 262:208–214.

Liu, C. and Rubin, D. 1994. The ECME algorithm: a simple extension of EM and ECM with faster monotonous convergence. *Biometrika*, 81:633–648.

Liu, J. and Chen, R. 1995. Blind deconvolution via sequential imputations. *Journal of the American Statistical Association*, 90:567–576.

Liu, J., Wong, W., and Kong, A. 1994. Covariance structure of the Gibbs sampler with applications to the comparisons of estimators and sampling schemes. *Biometrika*, 81:27–40.

Liu, J., Wong, W., and Kong, A. 1995. Correlation structure and convergence rate of the Gibbs sampler with various scans. *Journal of the Royal Statistical Society, Series B*, 57:157–169.

Liu, J. and Wu, Y. N. 1999. Parameter expansion for data augmentation. *Journal of the American Statistical Association*, 94:1264–1274.

Madras, N. and Slade, G. 1993. *The Self-Avoiding Random Walk*. Birkhäuser, Boston.

Marin, J.-M. and Robert, C. 2007. *Bayesian Core*. Springer, New York.

Marshall, A. 1965. The use of multi-stage sampling schemes in Monte Carlo computations. In *Symposium on Monte Carlo Methods*. Wiley, New York.

Meng, X. and Rubin, D. 1992. Maximum likelihood estimation via the ECM algorithm: A general framework. *Biometrika*, 80:267–278.

Meng, X. and van Dyk, D. 1999. Seeking efficient data augmentation schemes via conditional and marginal augmentation. *Biometrika*, 86:301–320.

Mengersen, K. and Tweedie, R. 1996. Rates of convergence of the Hastings and Metropolis algorithms. *Annals of Statistics*, 24:101–121.

Metropolis, N. 1987. The beginning of the Monte Carlo method. *Los Alamos Science*, 15:125–130.

Metropolis, N., Rosenbluth, A., Rosenbluth, M., Teller, A., and Teller, E. 1953. Equation of state calculations by fast computing machines. *Journal of Chemical Physics*, 21(6):1087–1092.

Metropolis, N. and Ulam, S. 1949. The Monte Carlo method. *Journal of the American Statistical Association*, 44:335–341.

Møller, J. and Waagepetersen, R. 2004. *Statistical Inference and Simulation for Spatial Point Processes*. Chapman and Hall/CRC, Boca Raton, FL.

Moussouris, J. 1974. Gibbs and Markov random systems with constraints. *Journal of Statistical Physics*, 10:11–33.

Mykland, P., Tierney, L., and Yu, B. 1995. Regeneration in Markov chain samplers. *Journal of the American Statistical Association*, 90:233–241.

Neal, R. 1996. Sampling from multimodal distributions using tempered transitions. *Statistics and Computing*, 6:353–356.

Neal, R. 2003. Slice sampling (with discussion). *Annals of Statistics*, 31:705–767.

Pearl, J. 1987. Evidential reasoning using stochastic simulation in causal models. *Artificial Intelligence*, 32:247–257.

Peskun, P. 1973. Optimum Monte Carlo sampling using Markov chains. *Biometrika*, 60:607–612.

Peskun, P. 1981. Guidelines for chosing the transition matrix in Monte Carlo methods using Markov chains. *Journal of Computational Physics*, 40:327–344.

Propp, J. and Wilson, D. 1996. Exact sampling with coupled Markov chains and applications to statistical mechanics. *Random Structures and Algorithms*, 9:223–252.

Qian, W. and Titterington, D. 1990. Parameter estimation for hidden Gibbs chains. *Statistics & Probability Letters*, 10:49–58.

Raftery, A. and Banfield, J. 1991. Stopping the Gibbs sampler, the use of morphology, and other issues in spatial statistics. *Annals of the Institute of Statistical Mathematics*, 43:32–43.

Richardson, S. and Green, P. 1997. On Bayesian analysis of mixtures with an unknown number of components (with discussion). *Journal of the Royal Statistical Society, Series B*, 59:731–792.

Ripley, B. 1987. *Stochastic Simulation*. Wiley, New York.

Robert, C. 1995. Convergence control techniques for MCMC algorithms. *Statistical Science*, 10(3): 231–253.

Robert, C. and Casella, G. 2004. *Monte Carlo Statistical Methods*, 2nd edn. Springer, New York.

Roberts, G. O., Gelman, A., and Gilks, W. 1997. Weak convergence and optimal scaling of random walk Metropolis algorithms. *Annals of Applied Probability*, 7:110–120.

Roberts, G. O. and Rosenthal, J. 1999. Convergence of slice sampler Markov chains. *Journal of the Royal Statistical Society, Series B*, 61:643–660.

Roberts, G. O. and Rosenthal, J. 2005. Coupling and ergodicity of adaptive mcmc. *Journal of Applied Probability*, 44:458–475.

Rosenbluth, M. and Rosenbluth, A. 1955. Monte Carlo calculation of the average extension of molecular chains. *Journal of Chemical Physics*, 23:356–359.

Rosenthal, J. 1995. Minorization conditions and convergence rates for Markov chain Monte Carlo. *Journal of the American Statistical Association*, 90:558–566.

Rubin, D. 1978. Multiple imputation in sample surveys: a phenomenological Bayesian approach to nonresponse. In *Imputation and Editing of Faulty or Missing Survey Data*. Washington, DC: US Department of Commerce.

Smith, A. and Gelfand, A. 1992. Bayesian statistics without tears: A sampling-resampling perspective. *American Statistician*, 46:84–88.

Stephens, D. A. 1994. Bayesian retrospective multiple-changepoint identification. *Applied Statistics*, 43:159–178.

Stephens, D. A. and Smith, A. F. M. 1993. Bayesian inference in multipoint gene mapping. *Annals of Human Genetics*, 57:65–82.

Stephens, M. 2000. Bayesian analysis of mixture models with an unknown number of components—an alternative to reversible jump methods. *Annals of Statistics*, 28:40–74.

Tanner, M. and Wong, W. 1987. The calculation of posterior distributions by data augmentation. *Journal of the American Statistical Association*, 82:528–550.

Tierney, L. 1994. Markov chains for exploring posterior distributions (with discussion). *Annals of Statistics*, 22:1701–1786.

Titterington, D. and Cox, D. (eds) 2001. *Biometrika: One Hundred Years*. Oxford University Press, Oxford.

Wakefield, J., Smith, A., Racine-Poon, A., and Gelfand, A. 1994. Bayesian analysis of linear and non-linear population models using the Gibbs sampler. *Applied Statistics*, 43:201–222.

Wang, C. S., Rutledge, J. J., and Gianola, D. 1993. Marginal inferences about variance-components in a mixed linear model using Gibbs sampling. *Genetics, Selection, Evolution*, 25:41–62.

Wang, C. S., Rutledge, J. J., and Gianola, D. 1994. Bayesian analysis of mixed limear models via Gibbs sampling with an application to litter size in Iberian pigs. *Genetics, Selection, Evolution*, 26:91–115.

Wei, G. and Tanner, M. 1990. A Monte Carlo implementation of the EM algorithm and the poor man's data augmentation algorithm. *Journal of the American Statistical Association*, 85:699–704.

Zeger, S. and Karim, R. 1991. Generalized linear models with random effects; a Gibbs sampling approach. *Journal of the American Statistical Association*, 86:79–86.

3

Reversible Jump MCMC

Yanan Fan and Scott A. Sisson

3.1 Introduction

The reversible jump Markov chain Monte Carlo (MCMC) sampler (Green, 1995) provides a general framework for Markov chain Monte Carlo simulation in which the dimension of the parameter space can vary between iterates of the Markov chain. The reversible jump sampler can be viewed as an extension of the Metropolis–Hastings algorithm onto more general state spaces.

To understand this in a Bayesian modeling context, suppose that for observed data $\mathbf{x}$ we have a countable collection of candidate models $\mathcal{M} = \{\mathcal{M}_1, \mathcal{M}_2, \ldots\}$ indexed by a parameter $k \in \mathcal{K}$. The index k can be considered as an auxiliary model indicator variable, such that $\mathcal{M}_{k'}$ denotes the model where $k = k'$. Each model $\mathcal{M}_k$ has an n_k-dimensional vector of unknown parameters, $\boldsymbol{\theta}_k \in \mathbb{R}^{n_k}$, where n_k can take different values for different models $k \in \mathcal{K}$. The joint posterior distribution of $(k, \boldsymbol{\theta}_k)$ given observed data, $\mathbf{x}$, is obtained as the product of the likelihood, $L(\mathbf{x} \mid k, \boldsymbol{\theta}_k)$, and the joint prior, $p(k, \boldsymbol{\theta}_k) = p(\boldsymbol{\theta}_k \mid k)p(k)$, constructed from the prior distribution of $\boldsymbol{\theta}_k$ under model $\mathcal{M}_k$, and the prior for the model indicator k (i.e. the prior for model $\mathcal{M}_k$). Hence, the joint posterior is

$$\pi(k, \boldsymbol{\theta}_k \mid \mathbf{x}) = \frac{L(\mathbf{x} \mid k, \boldsymbol{\theta}_k)p(\boldsymbol{\theta}_k \mid k)p(k)}{\sum_{k' \in \mathcal{K}} \int_{\mathbb{R}^{n_{k'}}} L(\mathbf{x} \mid k', \boldsymbol{\theta}'_{k'})p(\boldsymbol{\theta}'_{k'} \mid k')p(k')d\boldsymbol{\theta}'_{k'}}. \tag{3.1}$$

The reversible jump algorithm uses the joint posterior distribution in Equation 3.1 as the target of an MCMC sampler over the state space $\boldsymbol{\Theta} = \bigcup_{k \in \mathcal{K}}(\{k\} \times \mathbb{R}^{n_k})$, where the states of the Markov chain are of the form $(k, \boldsymbol{\theta}_k)$, the dimension of which can vary over the state space. Accordingly, from the output of a *single* Markov chain sampler, the user is able to obtain a full probabilistic description of the posterior probabilities of each model having observed the data, $\mathbf{x}$, in addition to the posterior distributions of the individual models.

This chapter aims to provide an overview of the reversible jump sampler. We will outline the sampler's theoretical underpinnings, present the latest and most popular techniques for enhancing algorithm performance, and discuss the analysis of sampler output. Through the use of numerous worked examples it is hoped that the reader will gain a broad appreciation of the issues involved in multi-model simulation, and the confidence to implement reversible jump samplers in the course of their own studies.

3.1.1 From Metropolis–Hastings to Reversible Jump

The standard formulation of the Metropolis–Hastings algorithm (Hastings, 1970) relies on the construction of a time-reversible Markov chain via the *detailed balance* condition. This condition means that moves from state $\boldsymbol{\theta}$ to $\boldsymbol{\theta}'$ are made as often as moves from $\boldsymbol{\theta}'$ to $\boldsymbol{\theta}$ with respect to the target density. This is a simple way to ensure that the equilibrium distribution of the chain is the desired target distribution. The extension of the Metropolis–Hastings algorithm to the setting where the dimension of the parameter vector varies is more challenging theoretically, but the resulting algorithm is surprisingly simple to follow.

For the construction of a Markov chain on a general state space $\boldsymbol{\Theta}$ with invariant or stationary distribution π, the detailed balance condition can be written as

$$\int_{(\boldsymbol{\theta},\boldsymbol{\theta}')\in\mathcal{A}\times\mathcal{B}} \pi(d\boldsymbol{\theta})P(\boldsymbol{\theta},d\boldsymbol{\theta}') = \int_{(\boldsymbol{\theta},\boldsymbol{\theta}')\in\mathcal{A}\times\mathcal{B}} \pi(d\boldsymbol{\theta}')P(\boldsymbol{\theta}',d\boldsymbol{\theta}) \tag{3.2}$$

for all Borel sets $\mathcal{A}\times\mathcal{B}\subset\boldsymbol{\Theta}$, where P is a general Markov transition kernel (Green, 2001).

As with the standard Metropolis–Hastings algorithm, Markov chain transitions from a current state $\boldsymbol{\theta}=(k,\boldsymbol{\theta}_k)\in\mathcal{A}$ in model $\mathcal{M}_k$ are realized by first proposing a new state $\boldsymbol{\theta}'=(k',\boldsymbol{\theta}_{k'})\in\mathcal{B}$ in model $\mathcal{M}_{k'}$ from a proposal distribution $q(\boldsymbol{\theta},\boldsymbol{\theta}')$. The detailed balance condition (Equation 3.2) is enforced through the acceptance probability, where the move to the candidate state $\boldsymbol{\theta}'$ is accepted with probability $\alpha(\boldsymbol{\theta},\boldsymbol{\theta}')$. If rejected, the chain remains at the current state $\boldsymbol{\theta}$ in model $\mathcal{M}_k$. Under this mechanism (Green, 2001, 2003), Equation 3.2 becomes

$$\int_{(\boldsymbol{\theta},\boldsymbol{\theta}')\in\mathcal{A}\times\mathcal{B}} \pi(\boldsymbol{\theta}\mid\mathbf{x})q(\boldsymbol{\theta},\boldsymbol{\theta}')\alpha(\boldsymbol{\theta},\boldsymbol{\theta}')d\boldsymbol{\theta}\,d\boldsymbol{\theta}' = \int_{(\boldsymbol{\theta},\boldsymbol{\theta}')\in\mathcal{A}\times\mathcal{B}} \pi(\boldsymbol{\theta}'\mid\mathbf{x})q(\boldsymbol{\theta}',\boldsymbol{\theta})\alpha(\boldsymbol{\theta}',\boldsymbol{\theta})d\boldsymbol{\theta}\,d\boldsymbol{\theta}', \tag{3.3}$$

where the distributions $\pi(\boldsymbol{\theta}\mid\mathbf{x})$ and $\pi(\boldsymbol{\theta}'\mid\mathbf{x})$ are posterior distributions with respect to model $\mathcal{M}_k$ and $\mathcal{M}_{k'}$, respectively.

One way to enforce Equation 3.3 is by setting the acceptance probability as

$$\alpha(\boldsymbol{\theta},\boldsymbol{\theta}') = \min\left\{1,\frac{\pi(\boldsymbol{\theta}\mid\mathbf{x})q(\boldsymbol{\theta},\boldsymbol{\theta}')}{\pi(\boldsymbol{\theta}'\mid\mathbf{x})q(\boldsymbol{\theta}',\boldsymbol{\theta})}\right\}, \tag{3.4}$$

where $\alpha(\boldsymbol{\theta}',\boldsymbol{\theta})$ is similarly defined. This resembles the usual Metropolis–Hastings acceptance ratio (Green, 1995; Tierney, 1998). It is straightforward to observe that this formulation includes the standard Metropolis–Hastings algorithm as a special case.

Accordingly, a reversible jump sampler with N iterations is commonly constructed as follows:

Step 1. Initialize k and $\boldsymbol{\theta}_k$ at iteration $t=1$.

Step 2. For iteration $t\geq 1$ perform
- *Within-model move:* with a fixed model k, update the parameters $\boldsymbol{\theta}_k$ according to any MCMC updating scheme.
- *Between-models move:* simultaneously update model indicator k and the parameters $\boldsymbol{\theta}_k$ according to the general reversible proposal/acceptance mechanism (Equation 3.4).

Step 3. Increment iteration $t=t+1$. If $t<N$, go to Step 2.

3.1.2 Application Areas

Statistical problems in which the number of unknown model parameters is itself unknown are extensive, and as such the reversible jump sampler has been implemented in analyses

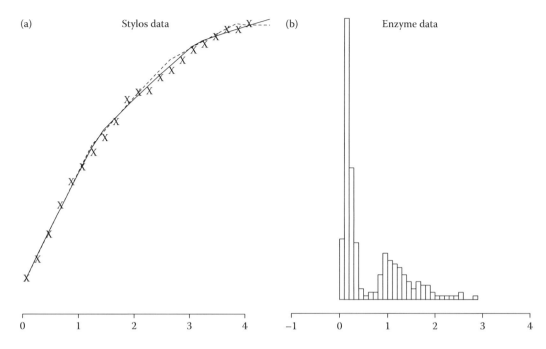

FIGURE 3.1

Examples of (a) change-point modeling and (b) mixture models. (a) With the Stylos tombs data set (crosses), a piecewise log-linear curve can be fitted between unknown change points. Illustrated are 2 (solid line) and 3 (dashed line) change points. (b) The histogram of the enzymatic activity data set suggests clear groupings of metabolizers, although the number of such groupings is not clear. (From Sisson, S. A. and Fan, Y. 2007. *Statistics and Computing*, 17:357–367. With permission.)

throughout a wide range of scientific disciplines over the last 15 years. Within the statistical literature, these predominantly concern Bayesian model determination problems (Sisson, 2005). Some of the commonly recurring models in this setting are described below.

Change-point models   One of the original applications of the reversible jump sampler was in Bayesian change-point problems, where both the number and location of change points in a system is unknown *a priori*. For example, Green (1995) analyzed mining disaster count data using a Poisson process with the rate parameter described as a step function with an unknown number and location of steps. Fan and Brooks (2000) applied the reversible jump sampler to model the shape of prehistoric tombs, where the curvature of the dome changes an unknown number of times. Figure 3.1a shows the plot of depths and radii of one of the tombs from Crete in Greece. The data appear to be piecewise log-linear, with possibly two or three change points.

Finite mixture models   Mixture models are commonly used where each data observation is generated according to some underlying categorical mechanism. This mechanism is typically unobserved, so there is uncertainty regarding which component of the resulting mixture distribution each data observation was generated from, in addition to uncertainty over the number of mixture components. A mixture model with k components for the observed data $\mathbf{x}$ takes the form

$$f(\mathbf{x} \mid \boldsymbol{\theta}_k) = \sum_{j=1}^{k} w_j f_j(\mathbf{x} \mid \boldsymbol{\phi}_j) \tag{3.5}$$

with $\theta_k = (\phi_1, \ldots, \phi_k)$, where w_j is the weight of the jth mixture component f_j, whose parameter vector is denoted by ϕ_j, and where $\sum_{j=1}^{k} w_j = 1$. The number of mixture components, k, is also unknown.

Figure 3.1b illustrates the distribution of enzymatic activity in the blood for 245 individuals. Richardson and Green (1997) analyzed these data using a mixture of normal densities to identify subgroups of slow or fast metabolizers. The multimodal nature of the data suggests the existence of such groups, but the number of distinct groupings is less clear. Tadesse et al. (2005) extend this normal mixture model for the purpose of clustering high-dimensional data.

Variable selection   The problem of variable selection arises when modeling the relationship between a response variable, Y, and p potential explanatory variables $x_1, \ldots, x_p$. The multi-model setting emerges when attempting to identify the most relevant subsets of predictors, making it a natural candidate for the reversible jump sampler. For example, under a regression model with normal errors we have

$$Y = X_\gamma \beta_\gamma + \epsilon \quad \text{with } \epsilon \sim N(0, \sigma^2 I), \tag{3.6}$$

where $\gamma = (\gamma_1, \ldots, \gamma_p)$ is a binary vector indexing the subset of $x_1, \ldots, x_p$ to be included in the linear model, X_γ is the design matrix whose columns correspond to the indexed subset given by γ, and β_γ is the corresponding subset of regression coefficients. For examples and algorithms in this setting and beyond, see, for example, George and McCulloch (1993), Smith and Kohn (1996), and Nott and Leonte (2004).

Nonparametrics   Within Bayesian nonparametrics, many authors have successfully explored the use of the reversible jump sampler as a method to automate the knot selection process when using a Pth-order spline model for curve fitting (Denison et al., 1998; DiMatteo et al., 2001). Here, a curve f is estimated by

$$f(x) = \alpha_0 + \sum_{j=1}^{P} \alpha_j x^j + \sum_{i=1}^{k} \eta_i (x - \kappa_i)_+^P, \quad x \in [a, b],$$

where $z_+ = \max(0, z)$ and κ_i, $i = 1, \ldots, k$, represent the locations of k knot points (Hastie and Tibshirani, 1990). Under this representation, fitting the curve consists of estimating the unknown number of knots k, the knot locations κ_i and the corresponding regression coefficients α_j and η_i, for $j = 0, \ldots, P$ and $i = 1, \ldots, k$.

Time series models   In the modeling of temporally dependent data, $x_1, \ldots, x_T$, multiple models naturally arise under uncertainty over the degree of dependence. For example, under a kth-order autoregressive process

$$X_t = \sum_{\tau=1}^{k} a_\tau X_{t-\tau} + \epsilon_t, \quad \text{with } t = k + 1, \ldots, T, \tag{3.7}$$

where $\epsilon_t \sim WN(0, \sigma^2)$, the order, k, of the autoregression is commonly unknown, in addition to the coefficients a_τ. Brooks et al. (2003c), Ehlers and Brooks (2003), and Vermaak et al. (2004) each detail descriptions on the use of reversible jump samplers for this class of problems.

The reversible jump algorithm has had a compelling influence in the statistical and mainstream scientific research literatures. In general, the large majority of application areas have tended to be computationally or biologically related (Sisson, 2005). Accordingly a large number of developmental and application studies can be found in the signal processing literature and the related fields of computer vision and image analysis. Epidemiological and medical studies also feature strongly.

This chapter is structured as follows: In Section 3.2 we present a detailed description of how to implement the reversible jump sampler and review methods to improve sampler performance. Section 3.3 examines post-simulation analysis, including label switching problems when identifiability is an issue, and convergence assessment. In Section 3.4 we review related sampling methods in the statistical literature, and conclude with discussion on possible future research directions for the field. Other useful reviews of reversible jump MCMC can be found in Green (2003) and Sisson (2005).

3.2 Implementation

In practice, the construction of proposal moves between different models is achieved via the concept of "dimension matching." Most simply, under a general Bayesian model determination setting, suppose that we are currently in state $(k, \boldsymbol{\theta}_k)$ in model $\mathcal{M}_k$, and we wish to propose a move to a state $(k', \boldsymbol{\theta}'_{k'})$ in model $\mathcal{M}_{k'}$, which is of a higher dimension, so that $n_{k'} > n_k$. In order to "match dimensions" between the two model states, a random vector $\mathbf{u}$ of length $d_{k \to k'} = n_{k'} - n_k$ is generated from a known density $q_{d_{k \to k'}}(\mathbf{u})$. The current state $\boldsymbol{\theta}_k$ and the random vector $\mathbf{u}$ are then mapped to the new state $\boldsymbol{\theta}'_{k'} = g_{k \to k'}(\boldsymbol{\theta}_k, \mathbf{u})$ through a one-to-one mapping function $g_{k \to k'} : \mathbb{R}^{n_k} \times \mathbb{R}^{d_k} \to \mathbb{R}^{n_{k'}}$. The acceptance probability of this proposal, combined with the joint posterior expression of Equation 3.1, becomes

$$\alpha[(k, \boldsymbol{\theta}_k), (k', \boldsymbol{\theta}'_{k'})] = \min \left\{ 1, \frac{\pi(k', \boldsymbol{\theta}'_{k'} \mid \mathbf{x}) q(k' \to k)}{\pi(k, \boldsymbol{\theta}_k \mid \mathbf{x}) q(k \to k') q_{d_{k \to k'}}(\mathbf{u})} \left| \frac{\partial g_{k \to k'}(\boldsymbol{\theta}_k, \mathbf{u})}{\partial (\boldsymbol{\theta}_k, \mathbf{u})} \right| \right\}, \tag{3.8}$$

where $q(k \to k')$ denotes the probability of proposing a move from model $\mathcal{M}_k$ to model $\mathcal{M}_{k'}$, and the final term is the determinant of the Jacobian matrix, often referred to in the reversible jump literature simply as the Jacobian. This term arises through the change of variables via the function $g_{k \to k'}$, which is required when used with respect to the integral equation (Equation 3.3). Note that the normalization constant in Equation 3.1 is not needed to evaluate the above ratio. The reverse move proposal, from model $\mathcal{M}_{k'}$ to $\mathcal{M}_k$, is made deterministically in this setting, and is accepted with probability

$$\alpha\left[(k', \boldsymbol{\theta}'_{k'}), (k, \boldsymbol{\theta}_k)\right] = \alpha\left[(k, \boldsymbol{\theta}_k), (k', \boldsymbol{\theta}'_{k'})\right]^{-1}.$$

More generally, we can relax the condition on the length of the vector $\mathbf{u}$ by allowing $d_{k \to k'} \geq n_{k'} - n_k$. In this case, nondeterministic reverse moves can be made by generating a $d_{k' \to k}$-dimensional random vector $\mathbf{u}' \sim q_{d_{k' \to k}}(\mathbf{u}')$, such that the dimension matching condition, $n_k + d_{k \to k'} = n_{k'} + d_{k' \to k}$, is satisfied. Then a reverse mapping is given by $\boldsymbol{\theta}_k = g_{k' \to k}(\boldsymbol{\theta}'_{k'}, \mathbf{u}')$, such that $\boldsymbol{\theta}_k = g_{k' \to k}(g_{k \to k'}(\boldsymbol{\theta}_k, \mathbf{u}), \mathbf{u}')$ and $\boldsymbol{\theta}'_{k'} = g_{k \to k'}(g_{k' \to k}(\boldsymbol{\theta}'_{k'}, \mathbf{u}'), \mathbf{u})$. The acceptance probability corresponding to Equation 3.8 then becomes

$$\alpha\left[(k, \boldsymbol{\theta}_k), (k', \boldsymbol{\theta}'_{k'})\right] = \min \left\{ 1, \frac{\pi(k', \boldsymbol{\theta}'_{k'} \mid \mathbf{x}) q(k' \to k) q_{d_{k' \to k}}(\mathbf{u}')}{\pi(k, \boldsymbol{\theta}_k \mid \mathbf{x}) q(k \to k') q_{d_{k \to k'}}(\mathbf{u})} \left| \frac{\partial g_{k \to k'}(\boldsymbol{\theta}_k, \mathbf{u})}{\partial (\boldsymbol{\theta}_k, \mathbf{u})} \right| \right\}. \tag{3.9}$$

Example: Dimension Matching

Consider the illustrative example given in Green (1995) and Brooks (1998). Suppose that model $\mathcal{M}_1$ has states ($k = 1, \boldsymbol{\theta}_1 \in \mathbb{R}^1$) and model $\mathcal{M}_2$ has states ($k = 2, \boldsymbol{\theta}_2 \in \mathbb{R}^2$). Let $(1, \theta^*)$ denote the current state in $\mathcal{M}_1$ and $(2, (\theta^{(1)}, \theta^{(2)}))$ denote the proposed state in $\mathcal{M}_2$. Under dimension matching, we might generate a random scalar u, and let $\theta^{(1)} = \theta^* + u$ and $\theta^{(2)} = \theta^* - u$, with the reverse move given deterministically by $\theta^* = \frac{1}{2}\left(\theta^{(1)} + \theta^{(2)}\right)$.

Example: Moment Matching in a Finite Mixture of Univariate Normals

Under the finite mixture of univariate normals model, the observed data, $\mathbf{x}$, has density given by Equation 3.5, where the jth mixture component $f_j(\mathbf{x} \mid \boldsymbol{\phi}_j) = \phi(\mathbf{x} \mid \mu_j, \sigma_j)$ is the $N(\mu_j, \sigma_j)$ density. For between-model moves, Richardson and Green (1997) implement a split (one component into two) and merge (two components into one) strategy which satisfies the dimension matching requirement. (See Dellaportas and Papageorgiou (2006) for an alternative approach.)

When two normal components j_1 and j_2 are merged into one, j^*, Richardson and Green (1997) propose a deterministic mapping which maintains the zeroth, first, and second moments:

$$w_{j^*} = w_{j_1} + w_{j_2}.$$
$$w_{j^*}\mu_{j^*} = w_{j_1}\mu_{j_1} + w_{j_2}\mu_{j_2}. \tag{3.10}$$
$$w_{j^*}\left(\mu_{j^*}^2 + \sigma_{j^*}^2\right) = w_{j_1}\left(\mu_{j_1}^2 + \sigma_{j_1}^2\right) + w_{j_2}\left(\mu_{j_2}^2 + \sigma_{j_2}^2\right).$$

The split move is proposed as

$$w_{j_1} = w_{j^*} * u_1, \quad w_{j_2} = w_{j^*} * (1 - u_1)$$

$$\mu_{j_1} = \mu_{j^*} - u_2\sigma_{j^*}\sqrt{\frac{w_{j_2}}{w_{j_1}}}$$

$$\mu_{j_2} = \mu_{j^*} + u_2\sigma_{j^*}\sqrt{\frac{w_{j_1}}{w_{j_2}}} \tag{3.11}$$

$$\sigma_{j_1}^2 = u_3\left(1 - u_2^2\right)\sigma_{j^*}^2\frac{w_{j^*}}{w_{j_1}}$$

$$\sigma_{j_2}^2 = (1 - u_3)\left(1 - u_2^2\right)\sigma_{j^*}^2\frac{w_{j^*}}{w_{j_2}},$$

where the random scalars $u_1, u_2 \sim \text{Beta}(2, 2)$ and $u_3 \sim \text{Beta}(1, 1)$. In this manner, dimension matching is satisfied, and the acceptance probability for the split move is calculated according to Equation 3.8, with the acceptance probability of the reverse merge move given by the reciprocal of this value.

3.2.1 Mapping Functions and Proposal Distributions

While the ideas behind dimension matching are conceptually simple, their implementation is complicated by the arbitrariness of the mapping function $g_{k \to k'}$ and the proposal distributions, $q_{d_{k \to k'}}(\mathbf{u})$, for the random vectors $\mathbf{u}$. Since mapping functions effectively express functional relationships between the parameters of different models, good mapping functions will clearly improve sampler performance in terms of between-model acceptance rates and chain mixing. The difficulty is that even in the simpler setting of nested models, good relationships can be hard to define, and in more general settings, parameter vectors between models may not be obviously comparable.

The only additional degree of freedom to improve between-model proposals is by choosing the form and parameters of the proposal distribution $q_{d_{k \to k'}}(\mathbf{u})$. However, there are no obvious criteria to guide this choice. Contrast this to within-model, random-walk Metropolis–Hastings moves on a continuous target density, whereby proposed moves close to the current state can have an arbitrarily large acceptance probability, and proposed moves far from the current state have low acceptance probabilities. This concept of "local" moves may be partially translated on to model space ($k \in \mathcal{K}$): proposals from $\boldsymbol{\theta}_k$ in model $\mathcal{M}_k$ to $\boldsymbol{\theta}'_{k'}$ in model $\mathcal{M}_{k'}$ will tend to have larger acceptance probabilities if their likelihood values are similar, that is, $L(\mathbf{x} \mid k, \boldsymbol{\theta}_k) \approx L(\mathbf{x} \mid k', \boldsymbol{\theta}'_{k'})$. For example, in the analysis of Bayesian mixture models, Richardson and Green (1997) propose "birth/death" and "split/merge" mappings of mixture components for the between-model move, while keeping other components unchanged. Hence, the proposed moves necessarily will have similar likelihood values to the current state. However, in general the notion of "local" move proposals does not easily extend to the parameter vectors of different models, unless considering simplified settings (e.g. nested models). In the general case, good mixing properties are achieved by the alignment of regions of high posterior probability between models.

Notwithstanding these difficulties, reversible jump MCMC is often associated with poor sampler performance. However, failure to realize acceptable sampler performance should only be considered a result of poorly constructed between-model mappings or inappropriate proposal distributions. It should even be anticipated that implementing a multi-model sampler may result in improved chain mixing, even when the inferential target distribution is a single model. In this case, sampling from a single model posterior with an "overly sophisticated" machinery is loosely analogous to the extra performance gained with augmented state space sampling methods. For example, in the case of a finite mixture of normal distributions, Richardson and Green (1997) report markedly superior sampler mixing when conditioning on there being exactly three mixture components, in comparison with the output generated by a fixed-dimension sampler. George et al. (1999) similarly obtain improved chain performance in a single model, by performing "birth-then-death" moves simultaneously so that the dimension of the model remains constant. Green (2003) presents a short study on which inferential circumstances determine whether the adoption of a multi-model sampler may be beneficial in this manner. Conversely, Han and Carlin (2001) provide an argument to suggest that multi-model sampling may have a detrimental effect on efficiency.

3.2.2 Marginalization and Augmentation

Depending on the aim or the complexity of a multi-model analysis, it may be that use of reversible jump MCMC would be somewhat heavy-handed, when reduced- or fixed-dimensional samplers may be substituted. In some Bayesian model selection settings, between-model moves can be greatly simplified or even avoided if one is prepared to make certain prior assumptions, such as conjugacy or objective prior specifications. In such cases, it may be possible to analytically integrate out some or all of the parameters $\boldsymbol{\theta}_k$ in the posterior distribution (Equation 3.1), reducing the sampler either to fixed dimensions, for example on model space $k \in \mathcal{K}$ only, or to a lower-dimensional set of model and parameter space (Berger and Pericchi, 2001; DiMatteo et al., 2001; George and McCulloch, 1993; Tadesse et al., 2005). In lower dimensions, the reversible jump sampler is often easier to implement, as the problems associated with mapping function specification are conceptually simpler to resolve.

Example: Marginalization in Variable Selection

In Bayesian variable selection for normal linear models (Equation 3.6), the vector $\gamma = (\gamma_1, \ldots, \gamma_p)$ is treated as an auxiliary (model indicator) variable, where

$$\gamma_i = \begin{cases} 1, & \text{if predictor } x_i \text{ is included in the regression,} \\ 0, & \text{otherwise.} \end{cases}$$

Under certain prior specifications for the regression coefficients β and error variance σ^2, the β coefficients can be analytically integrated out of the posterior. A Gibbs sampler directly on model space is then available for γ (George and McCulloch, 1993; Nott and Green, 2004; Smith and Kohn, 1996).

Example: Marginalization in Finite Mixture of Multivariate Normal Models

Within the context of clustering, the parameters of the normal components are usually not of interest. Tadesse et al. (2005) demonstrate that by choosing appropriate prior distributions, the parameters of the normal components can be analytically integrated out of the posterior. The reversible jump sampler may then run on a much reduced parameter space, which is simpler and more efficient.

In a general setting, Brooks et al. (2003c) proposed a class of models based on augmenting the state space of the target posterior with an auxiliary set of state-dependent variables, $\mathbf{v}_k$, so that the state space of $\pi(k, \boldsymbol{\theta}_k, \mathbf{v}_k \mid \mathbf{x}) = \pi(k, \boldsymbol{\theta}_k \mid \mathbf{x})\tau_k(\mathbf{v}_k)$ is of constant dimension for all models $\mathcal{M}_k \in \mathcal{M}$. By updating $\mathbf{v}_k$ via a (deliberately) slowly mixing Markov chain, a temporal memory is induced that persists in the $\mathbf{v}_k$ from state to state. In this manner, the motivation behind the auxiliary variables is to improve between-model proposals, in that some memory of previous model states is retained. Brooks et al. (2003c) demonstrate that this approach can significantly enhance mixing compared to an unassisted reversible jump sampler. Although the fixed dimensionality of $(k, \boldsymbol{\theta}_k, \mathbf{v}_k)$ is later relaxed, there is an obvious analogue with product space sampling frameworks (Carlin and Chib, 1995; Godsill, 2001); see Section 3.4.2.

An alternative augmented state space modification of standard MCMC is given by Liu et al. (2001). The dynamic weighting algorithm augments the original state space by a weighting factor, which permits the Markov chain to make large transitions not allowable by the standard transition rules, subject to the computation of the correct weighting factor. Inference is then made by using the weights to compute importance sampling estimates rather than simple Monte Carlo estimates. This method can be used within the reversible jump algorithm to facilitate cross-model jumps.

3.2.3   Centering and Order Methods

Brooks et al. (2003c) introduce a class of methods to achieve the automatic scaling of the proposal density, $q_{d_{k \to k'}}(\mathbf{u})$, based on "local" move proposal distributions, which are centered around the point of equal likelihood values under current and proposed models. Under this scheme, it is assumed that local mapping functions $g_{k \to k'}$ are known. For a proposed move from $(k, \boldsymbol{\theta}_k)$ in $\mathcal{M}_k$ to model $\mathcal{M}_{k'}$, the random vector "centering point"

$c_{k\to k'}(\theta_k) = g_{k\to k'}(\theta_k, \mathbf{u})$ is defined such that, for some particular choice of proposal vector $\mathbf{u}$, the current and proposed states are identical in terms of likelihood contribution, that is, $L(\mathbf{x} \mid k, \theta_k) = L(\mathbf{x} \mid k', c_{k\to k'}(\theta_k))$. For example, if $\mathcal{M}_k$ is an autoregressive model of order k (Equation 3.7) and $\mathcal{M}_{k'}$ is an autoregressive model of order $k' = k + 1$, and if $c_{k\to k'}(\theta_k) = g_{k\to k'}(\theta_k, u) = (\theta_k, u)$ (e.g. a local "birth" proposal), then we have $u = 0$ and $c_{k\to k'} = (\theta_k, 0)$, as $L(\mathbf{x} \mid k, \theta_k) = L(\mathbf{x} \mid k', (\theta_k, 0))$.

Given the centering constraint on $\mathbf{u}$, if the scaling parameter in the proposal $q_{d_{k\to k'}}(\mathbf{u})$ is a scalar, then the zeroth-order method (Brooks et al., 2003c) proposes to choose this scaling parameter such that the acceptance probability $\alpha[(k, \theta_k), (k', c_{k\to k'}(\theta_k))]$ of a move to the centering point $c_{k\to k'}(\theta_k)$ in model $\mathcal{M}_{k'}$ is exactly one. The argument is then that move proposals close to $c_{k\to k'}(\theta_k)$ will also have a large acceptance probability.

For proposal distributions, $q_{d_{k\to k'}}(\mathbf{u})$, with additional degrees of freedom, a similar method based on a series of nth-order conditions (for $n \geq 1$) requires that, for the proposed move, the nth derivative (with respect to $\mathbf{u}$) of the acceptance probability equals the zero vector at the centering point $c_{k\to k'}(\theta_k)$:

$$\nabla^n \alpha[(k, \theta_k), (k', c_{k\to k'}(\theta_k))] = \mathbf{0}. \tag{3.12}$$

That is, the m unknown parameters in the proposal distribution $q_{d_{k\to k'}}(\mathbf{u})$ are determined by solving the m simultaneous equations given by Equation 3.12 with $n = 1, \ldots, m$. The idea behind the nth-order method is that the concept of closeness to the centering point under the zeroth-order method is relaxed. By enforcing zero derivatives of $\alpha[(k, \theta_k), (k', c_{k\to k'}(\theta_k))]$, the acceptance probability will become flatter around $c_{k\to k'}(\theta_k)$. Accordingly this allows proposals further away from the centering point to still be accepted with a reasonably high probability. This will ultimately induce improved chain mixing.

With these methods, proposal distribution parameters are adapted to the current state of the chain, (k, θ_k), rather than relying on a constant proposal parameter vector for all state transitions. It can be shown that for a simple two-model case, the nth-order conditions are optimal in terms of the capacitance of the algorithm (Lawler and Sokal, 1988). See also Ehlers and Brooks (2003) for an extension to a more general setting, and Ntzoufras et al. (2003) for a centering method in the context of linear models.

One caveat with the centering schemes is that they require specification of the between-model mapping function $g_{k\to k'}$, although these methods compensate for poor choices of mapping functions by selecting the best set of parameters for the given mapping. Recently, Ehlers and Brooks (2008) suggest the posterior conditional distribution $\pi(k', \mathbf{u} \mid \theta_k)$ as the proposal for the random vector $\mathbf{u}$, side-stepping the need to construct a mapping function. In this case, the full conditionals either must be known or need to be approximated.

Example: The Zeroth-Order Method for an Autoregressive Model

Brooks et al. (2003c) consider the AR model with unknown order k (Equation 3.7), assuming Gaussian noise $\epsilon_t \sim N(0, \sigma_\epsilon^2)$ and a uniform prior on k, where $k = 1, 2, \ldots, k_{\max}$. Within each model $\mathcal{M}_k$, independent $N(0, \sigma_a^2)$ priors are adopted for the AR coefficients a_τ, $\tau = 1, \ldots, k$, with an inverse gamma prior for σ_ϵ^2. Suppose moves are made from model $\mathcal{M}_k$ to model $\mathcal{M}_{k'}$ such that $k' = k + 1$. The move from θ_k to $\theta'_{k'}$ is achieved by generating a random scalar $u \sim q(u) = N(0, 1)$, and defining the mapping function as $\theta'_{k'} = g_{k\to k'}(\theta_k, u) = (\theta_k, \sigma u)$. The centering point $c_{k\to k'}(\theta_k)$ then occurs at the point $u = 0$, or $\theta'_{k'} = (\theta_k, 0)$.

Under the mapping $g_{k \to k'}$, the Jacobian is σ, and the acceptance probability (Equation 3.8) for the move from (k, θ_k) to $(k', c_{k \to k'}(\theta_k))$ is given by $\alpha[(k, \theta_k), (k', (\theta_k, 0))] = \min(1, A)$ where

$$A = \frac{\pi(k', (\theta_k, 0) \mid \mathbf{x}) q(k' \to k) \sigma}{\pi(k, \theta_k \mid \mathbf{x}) q(k \to k') q(0)} = \frac{\left(2\pi\sigma_a^2\right)^{-1/2} q(k' \to k) \sigma}{q(k \to k')(2\pi)^{-1/2}}.$$

Note that since the likelihoods are equal at the centering point, and the priors common to both models cancel in the posterior ratio, A is only a function of the prior density for the parameter a_{k+1} evaluated at 0, the proposal distributions and the Jacobian. Hence, we solve $A = 1$ to obtain

$$\sigma^2 = \sigma_a^2 \left(\frac{q(k \to k')}{q(k' \to k)}\right)^2.$$

Thus in this case, the proposal variance is not dependent on the model parameter (θ_k) or data ($\mathbf{x}$). It depends only on the prior variance, σ_a, and the model states, k, k'.

Example: The Second-Order Method for Moment Matching

Consider the moment matching in a finite mixture of univariate normals example of Section 3.2. The mapping functions $g_{k' \to k}$ and $g_{k \to k'}$ are respectively given by Equations 3.10 and 3.11, with the random numbers u_1, u_2, u_3 drawn from independent beta distributions with unknown parameter values, so that $q_{p_i, q_i}(u_i)$: $u_i \sim \text{Beta}(p_i, q_i)$, $i = 1, 2, 3$.

Consider the split move, Equation 3.11. To apply the second-order method of Brooks et al. (2003c), we first locate a centering point, $c_{k \to k'}(\theta_k)$, achieved by setting $u_1 = 1$, $u_2 = 0$, and $u_3 \equiv u_1 = 1$ by inspection. Hence, at the centering point, the two new (split) components j_1 and j_2 will have the same location and scale as the j^* component, with new weights $w_{j_1} = w_{j*}$ and $w_{j_2} = 0$ and all observations allocated to component j_1. Accordingly this will produce identical likelihood contributions. Note that to obtain equal variances for the split proposal, substitute the expressions for w_{j_1} and w_{j_2} into those for $\sigma_{j_1}^2 = \sigma_{j_2}^2$.

Following Richardson and Green (1997), the acceptance probability of the split move evaluated at the centering point is then proportional (with respect to $\mathbf{u}$) to

$$\log A[(k, \theta_k), (k', c_{k \to k'}(\theta_k))]$$

$$\propto l_{j_1} \log(w_{j_1}) + l_{j_2} \log(w_{j_2}) - \frac{l_{j_1}}{2} \log\left(\sigma_{j_1}^2\right) - \frac{l_{j_2}}{2} \log\left(\sigma_{j_2}^2\right) - \frac{1}{2\sigma_{j_1}^2} \sum_{l=1}^{l_{j_1}} (y_l - \mu_{j_1})^2$$

$$- \frac{1}{2\sigma_{j_2}^2} \sum_{l=1}^{l_{j_2}} (y_l - \mu_{j_2})^2 + (\delta - 1 + l_{j_1}) \log(w_{j_1}) + (\delta - 1 + l_{j_2}) \log(w_{j_2})$$

$$- \left\{\frac{1}{2}\kappa[(\mu_{j_1} - \xi)^2 + (\mu_{j_2} - \xi)^2]\right\} - (\alpha + 1) \log\left(\sigma_{j_1}^2 \sigma_{j_2}^2\right) - \beta\left(\sigma_{j_1}^{-2} + \sigma_{j_2}^{-2}\right)$$

$$- \log[q_{p_1, q_1}(u_1)] - \log[q_{p_2, q_2}(u_2)] - \log[q_{p_3, q_3}(u_3)] + \log(|\mu_{j_1} - \mu_{j_2}|)$$

$$+ \log\left(\sigma_{j_1}^2\right) + \log\left(\sigma_{j_2}^2\right) - \log(u_2) - \log\left(1 - u_2^2\right) - \log(u_3) - \log(1 - u_3), \qquad (3.13)$$

where l_{j_1} and l_{j_2} respectively denote the number of observations allocated to components j_1 and j_2, and where $\delta, \alpha, \beta, \xi$ and κ are hyperparameters as defined by Richardson and Green (1997).

Thus, for example, to obtain the proposal parameter values p_1 and q_1 for u_1, we solve the first- and second-order derivatives of the acceptance probability (Equation 3.13) with respect to u_1. This yields

$$\frac{\partial \log \alpha \left[(k, \boldsymbol{\theta}_k), (k', c_{k \to k'}(\boldsymbol{\theta}_k)) \right]}{\partial u_1} = \frac{\delta + 2l_{j_1} - p_1}{u_1} + \frac{q_1 - \delta - 2l_{j_2}}{(1 - u_1)}$$

$$\frac{\partial^2 \log \alpha \left[(k, \boldsymbol{\theta}_k), (k', c_{k \to k'}(\boldsymbol{\theta}_k)) \right]}{\partial u_1^2} = -\frac{\delta + 2l_{j_1} - p_1}{u_1^2} + \frac{q_1 - \delta - 2l_{j_2}}{(1 - u_1)^2}.$$

Equating these to zero and solving for p_1 and q_1 at the centering points (with $l_{j_1} = l_{j*}$ and $l_{j_2} = 0$) gives $p_1 = \delta + 2l_{j*}$ and $q_1 = \delta$. Thus the parameter p_1 depends on the number of observations allocated to the component being split. Similar calculations to the above give solutions for p_2, q_2, p_3, and q_3.

3.2.4   Multi-Step Proposals

Green and Mira (2001) introduce a procedure for learning from rejected between-model proposals based on an extension of the splitting rejection idea of Tierney and Mira (1999). After rejecting a between-model proposal, the procedure makes a second proposal, usually under a modified proposal mechanism, and potentially dependent on the value of the rejected proposal. In this manner, a limited form of adaptive behavior may be incorporated into the proposals. The procedure is implemented via a modified Metropolis–Hastings acceptance probability, and may be extended to more than one sequential rejection (Trias et al., 2009). Delayed-rejection schemes can reduce the asymptotic variance of ergodic averages by reducing the probability of the chain remaining in the same state (Peskun, 1973; Tierney, 1998), however there is an obvious tradeoff with the extra move construction and computation required.

For clarity of exposition, in the remainder of this section we denote the current state of the Markov chain in model $\mathcal{M}_k$ by $\mathbf{x} = (k, \boldsymbol{\theta}_k)$, and the first and second stage proposed states in model $\mathcal{M}_{k'}$ by $\mathbf{y}$ and $\mathbf{z}$. Let $\mathbf{y} = g^{(1)}_{k \to k'}(\mathbf{x}, \mathbf{u}_1)$ and $\mathbf{z} = g^{(2)}_{k \to k'}(\mathbf{x}, \mathbf{u}_1, \mathbf{u}_2)$ be the mappings of the current state and random vectors $\mathbf{u}_1 \sim q^{(1)}_{d_{k \to k'}}(\mathbf{u}_1)$ and $\mathbf{u}_2 \sim q^{(2)}_{d_{k \to k'}}(\mathbf{u}_2)$ into the proposed new states. For simplicity, we again consider the framework where the dimension of model $\mathcal{M}_k$ is smaller than that of model $\mathcal{M}_{k'}$ (i.e. $n_{k'} > n_k$) and where the reverse move proposals are deterministic. The proposal from $\mathbf{x}$ to $\mathbf{y}$ is accepted with the usual acceptance probability

$$\alpha_1(\mathbf{x}, \mathbf{y}) = \min \left\{ 1, \frac{\pi(\mathbf{y}) q(k' \to k)}{\pi(\mathbf{x}) q(k \to k') q^{(1)}_{d_{k \to k'}}(\mathbf{u}_1)} \left| \frac{\partial g^{(1)}_{k \to k'}(\mathbf{x}, \mathbf{u}_1)}{\partial(\mathbf{x}, \mathbf{u}_1)} \right| \right\}.$$

If $\mathbf{y}$ is rejected, detailed balance for the move from $\mathbf{x}$ to $\mathbf{z}$ is preserved with the acceptance probability

$$\alpha_2(\mathbf{x}, \mathbf{z}) = \min \left\{ 1, \frac{\pi(\mathbf{z}) q(k' \to k)[1 - \alpha_1(\mathbf{y}^*, \mathbf{z})^{-1}]}{\pi(\mathbf{x}) q(k \to k') q^{(1)}_{d_{k \to k'}}(\mathbf{u}_1) q^{(2)}_{d_{k \to k'}}(\mathbf{u}_2)[1 - \alpha_1(\mathbf{x}, \mathbf{y})]} \left| \frac{\partial g^{(2)}_{k \to k'}(\mathbf{x}, \mathbf{u}_1, \mathbf{u}_2)}{\partial(\mathbf{x}, \mathbf{u}_1, \mathbf{u}_2)} \right| \right\},$$

where $\mathbf{y}^* = g^{(1)}_{k \to k'}(\mathbf{z}, \mathbf{u}_1)$. Note that the second stage proposal $\mathbf{z} = g^{(2)}_{k \to k'}(\mathbf{x}, \mathbf{u}_1, \mathbf{u}_2)$ is permitted to depend on the rejected first stage proposal $\mathbf{y}$ (a function of $\mathbf{x}$ and $\mathbf{u}_1$).

In a similar vein, Al-Awadhi et al. (2004) also acknowledge that an initial between-model proposal $\mathbf{x}' = g_{k \to k'}(\mathbf{x}, \mathbf{u})$ may be poor, and seek to adjust the state $\mathbf{x}'$ to a region of higher posterior probability before taking the decision to accept or reject the proposal. Specifically, Al-Awadhi et al. (2004) propose to initially evaluate the proposed move to $\mathbf{x}'$ in model $\mathcal{M}_{k'}$ through a density $\pi^*(\mathbf{x}')$ rather than the usual $\pi(\mathbf{x}')$. The authors suggest taking π^* to be some tempered distribution $\pi^* = \pi^\gamma$, $\gamma > 1$, such that the modes of π^* and π are aligned.

The algorithm then implements $\kappa \geq 1$ fixed-dimension MCMC updates, generating states $\mathbf{x}' \to \mathbf{x}^1 \to \cdots \to \mathbf{x}^\kappa = \mathbf{x}^*$, with each step satisfying detailed balance with respect to π^*. This provides an opportunity for $\mathbf{x}^*$ to move closer to the mode of π^* (and therefore π) than $\mathbf{x}'$. The move from $\mathbf{x}$ in model $\mathcal{M}_k$ to the final state $\mathbf{x}^*$ in model $\mathcal{M}_{k'}$ (with density $\pi(\mathbf{x}^*)$) is finally accepted with probability

$$\alpha(\mathbf{x}, \mathbf{x}^*) = \min \left\{ 1, \frac{\pi(\mathbf{x}^*)\pi^*(\mathbf{x}')q(k' \to k)}{\pi(\mathbf{x})\pi^*(\mathbf{x}^*)q(k \to k')q_{d_{k \to k'}}(\mathbf{u})} \left| \frac{\partial g_{k \to k'}(\mathbf{x}, \mathbf{u})}{\partial(\mathbf{x}, \mathbf{u})} \right| \right\}.$$

The implied reverse move from model $\mathcal{M}_{k'}$ to model model $\mathcal{M}_k$ is conducted by taking the κ moves with respect to π^* first, followed by the dimension-changing move.

Various extensions can easily be incorporated into this framework, such as using a sequence of π^* distributions, resulting in a slightly modified acceptance probability expression. For instance, the standard simulated annealing framework, Kirkpatrick (1984), provides an example of a sequence of distributions which encourage moves toward posterior mode. Clearly the choice of the distribution π^* can be crucial to the success of this strategy. As with all multi-step proposals, increased computational overheads are traded for potentially enhanced between-model mixing.

3.2.5 Generic Samplers

The problem of efficiently constructing between-model mapping templates, $g_{k \to k'}$, with associated random vector proposal densities, $q_{d_{k \to k'}}$, may be approached from an alternative perspective. Rather than relying on a user-specified mapping, one strategy would be to move toward a more generic proposal mechanism altogether. A clear benefit of generic between-model moves is that they may be equally be implemented for nonnested models. While the ideal of "black-box" between-model proposals is attractive, they currently remain on the research horizon. However, a number of automatic reversible jump MCMC samplers have been proposed.

Green (2003) proposed a reversible jump analogy of the random-walk Metropolis sampler of Roberts (2003). Suppose that estimates of the first- and second-order moments of $\boldsymbol{\theta}_k$ are available, for each of a small number of models, $k \in \mathcal{K}$, denoted by $\boldsymbol{\mu}_k$ and $\mathbf{B}_k \mathbf{B}_k^\top$ respectively, where $\mathbf{B}_k$ is an $n_k \times n_k$ matrix. In proposing a move from $(k, \boldsymbol{\theta}_k)$ to model $\mathcal{M}_{k'}$, a new parameter vector is proposed by

$$\boldsymbol{\theta}_{k'}' = \begin{cases} \boldsymbol{\mu}_{k'} + \mathbf{B}_{k'} \left[\mathbf{R}\mathbf{B}_k^{-1}(\boldsymbol{\theta}_k - \boldsymbol{\mu}_k) \right]_1^{n_{k'}}, & \text{if } n_{k'} < n_k, \\ \boldsymbol{\mu}_{k'} + \mathbf{B}_{k'} \mathbf{R}\mathbf{B}_k^{-1}(\boldsymbol{\theta}_k - \boldsymbol{\mu}_k), & \text{if } n_{k'} = n_k, \\ \boldsymbol{\mu}_{k'} + \mathbf{B}_{k'} \mathbf{R} \begin{pmatrix} \mathbf{B}_k^{-1}(\boldsymbol{\theta}_k - \boldsymbol{\mu}_k) \\ \mathbf{u} \end{pmatrix}, & \text{if } n_{k'} > n_k, \end{cases}$$

where $[\cdot]_1^m$ denotes the first m components of a vector, $\mathbf{R}$ is a orthogonal matrix of order $\max\{n_k, n_{k'}\}$, and $\mathbf{u} \sim q_{n_{k'}-n_k}(\mathbf{u})$ is an $(n_{k'} - n_k)$-dimensional random vector (only utilized if $n_{k'} > n_k$, or when calculating the acceptance probability of the reverse move from model $\mathcal{M}_{k'}$ to model $\mathcal{M}_k$ if $n_{k'} < n_k$). If $n_{k'} \leq n_k$, then the proposal $\boldsymbol{\theta}'_{k'}$ is deterministic and the Jacobian is trivially calculated. Hence, the acceptance probability is given by

$$\alpha\left[(k, \boldsymbol{\theta}_k), (k', \boldsymbol{\theta}'_{k'})\right] = \frac{\pi(k', \boldsymbol{\theta}'_{k'} \mid \mathbf{x})}{\pi(k, \boldsymbol{\theta}_k \mid \mathbf{x})} \frac{q(k' \to k)}{q(k \to k')} \frac{|\mathbf{B}_{k'}|}{|\mathbf{B}_k|} \times \begin{cases} q_{n_{k'} - n_k}(\mathbf{u}), & \text{for } n_{k'} < n_k, \\ 1, & \text{for } n_{k'} = n_k, \\ \dfrac{1}{q_{n_{k'} - n_k}}(\mathbf{u}), & \text{for } n_{k'} > n_k. \end{cases}$$

Accordingly, if the model-specific densities $\pi(k, \boldsymbol{\theta}_k \mid \mathbf{x})$ are unimodal with first- and second-order moments given by $\boldsymbol{\mu}_k$ and $\mathbf{B}_k\mathbf{B}_k^\top$, then high between-model acceptance probabilities may be achieved. (Unitary acceptance probabilities are available if the $\pi(k, \boldsymbol{\theta}_k \mid \mathbf{x})$ are exactly Gaussian.) Green (2003), Godsill (2003), and Hastie (2004) discuss a number of modifications to this general framework, including improving efficiency and relaxing the requirement of unimodal densities $\pi(k, \boldsymbol{\theta}_k \mid \mathbf{x})$ to realize high between-model acceptance rates. Naturally, the required knowledge of first- and second-order moments of each model density will restrict the applicability of these approaches to moderate numbers of candidate models if these require estimation (e.g. via pilot chains).

With a similar motivation to the above, Papathomas et al. (2009) put forward the multivariate normal as proposal distribution for $\boldsymbol{\theta}'_{k'}$ in the context of linear regression models, so that $\boldsymbol{\theta}'_{k'} \sim N(\boldsymbol{\mu}_{k'\mid\boldsymbol{\theta}_k}, \boldsymbol{\Sigma}_{k'\mid\boldsymbol{\theta}_k})$. The authors derive estimates for the mean $\boldsymbol{\mu}_{k'\mid\boldsymbol{\theta}_k}$ and covariance $\boldsymbol{\Sigma}_{k'\mid\boldsymbol{\theta}_k}$ such that the proposed values for $\boldsymbol{\theta}'_{k'}$ will on average produce conditional posterior values under model $\mathcal{M}_{k'}$ similar to those produced by the vector $\boldsymbol{\theta}_k$ under model $\mathcal{M}_k$. In particular, consider the normal linear model in Equation 3.6, rewriting the error covariance as V, assuming equality under the two models such that $V_k = V_{k'} = V$. The parameters of the proposal distribution for $\boldsymbol{\theta}'_{k'}$ are then given by

$$\boldsymbol{\mu}_{k'\mid\boldsymbol{\theta}_k} = (X_{\gamma'}^\top V^{-1} X_{\gamma'})^{-1} X_{\gamma'}^\top V^{-1} \left\{ Y + B^{-1} V^{-1/2} (X_\gamma \boldsymbol{\theta}_k - P_k Y) \right\},$$

$$\boldsymbol{\Sigma}_{k'\mid\boldsymbol{\theta}_k} = Q_{k',k'} - Q_{k',k} Q_{k,k}^{-1} Q_{k,k} Q_{k,k'}^{-1} Q_{k',k'} + c I_{n_{k'}},$$

where γ and γ' are indicators corresponding to models $\mathcal{M}_k$ and $\mathcal{M}_{k'}$, $B = (V + X_{\gamma'} \boldsymbol{\Sigma}_{k'\mid\boldsymbol{\theta}_k} X_{\gamma'}^\top)^{-1/2}$, $P_k = X_\gamma (X_\gamma^\top V^{-1} X_\gamma)^{-1} X_\gamma^\top V^{-1}$, $Q_{k,k'} = (X_\gamma^\top V^{-1} X_{\gamma'})^{-1}$, I_n is the $n \times n$ identity matrix and $c > 0$. Intuitively, the mean of this proposal distribution may be interpreted as the maximum likelihood estimate of $\boldsymbol{\theta}'_{k'}$ for model $\mathcal{M}_{k'}$, plus a correction term based on the distance of the current chain state $\boldsymbol{\theta}_k$ to the mode of the posterior density in model $\mathcal{M}_k$. The mapping between $\boldsymbol{\theta}'_{k'}$ and $\boldsymbol{\theta}_k$ and the random number $\mathbf{u}$ is given by

$$\boldsymbol{\theta}'_{k'} = \boldsymbol{\mu}_{k'\mid\boldsymbol{\theta}_k} + \boldsymbol{\Sigma}^{1/2}_{k'\mid\boldsymbol{\theta}_k} \mathbf{u},$$

where $\mathbf{u} \sim N(0, I_{n_{k'}})$. Accordingly the Jacobian corresponding to Equation 3.9 is given by $\left|\boldsymbol{\Sigma}^{1/2}_{k'\mid\boldsymbol{\theta}_k}\right| \left|\boldsymbol{\Sigma}^{1/2}_{k\mid\boldsymbol{\theta}_{k'}}\right|$. Under this construction, the value $c > 0$ is treated as a tuning parameter for the calibration of the acceptance probability. Quite clearly, the parameters of the between-model proposal do not require *a priori* estimation, and they adapt to the current state of the chain. The authors note that in some instances, this method produces similar results in terms of efficiency to Green (2003). One caveat is that the calculations at each proposal stage involve several inversions of matrices which can be computationally costly when the dimension is large. In addition, the method is theoretically justified for normal linear models, but can be applied to nonnormal models when transformation of data to normality is available, as demonstrated in Papathomas et al. (2009).

Fan et al. (2009) propose to construct between-model proposals based on estimating conditional marginal densities. Suppose that it is reasonable to assume some structural similarities between the parameters $\boldsymbol{\theta}_k$ and $\boldsymbol{\theta}'_{k'}$ of models $\mathcal{M}_k$ and $\mathcal{M}_{k'}$, respectively. Let c indicate the subset of the vectors $\boldsymbol{\theta}_k = (\boldsymbol{\theta}_k^c, \boldsymbol{\theta}_k^{-c})$ and $\boldsymbol{\theta}'_{k'} = (\boldsymbol{\theta}_{k'}^c, \boldsymbol{\theta}_{k'}^{-c})$ which can be kept constant between models, so that $\boldsymbol{\theta}_{k'}^c = \boldsymbol{\theta}_k^c$. The remaining r-dimensional vector $\boldsymbol{\theta}_{k'}^{-c}$ is then sampled from an estimate of the factorization of the conditional posterior of $\boldsymbol{\theta}_{k'}^{-c} = (\theta_{k'}^1, \dots, \theta_{k'}^r)$ under model $\mathcal{M}_{k'}$:

$$\pi(\boldsymbol{\theta}_{k'}^{-c} \mid \boldsymbol{\theta}_{k'}^c, \mathbf{x}) \approx \hat{\pi}_1(\theta_{k'}^1 \mid \theta_{k'}^2, \dots, \theta_{k'}^r, \boldsymbol{\theta}_{k'}^c, \mathbf{x}) \dots \hat{\pi}_{r-1}(\theta_{k'}^{r-1} \mid \theta_{k'}^r, \boldsymbol{\theta}_{k'}^c, \mathbf{x}) \hat{\pi}_r(\theta_{k'}^r \mid \boldsymbol{\theta}_{k'}^c, \mathbf{x}).$$

The proposal $\boldsymbol{\theta}_{k'}^{-c}$ is drawn by first estimating $\hat{\pi}_r(\theta_{k'}^r \mid \boldsymbol{\theta}_{k'}^c, \mathbf{x})$ and sampling $\theta_{k'}^r$, and by then estimating $\hat{\pi}_{r-1}(\theta_{k'}^{r-1} \mid \theta_{k'}^r, \boldsymbol{\theta}_{k'}^c, \mathbf{x})$ and sampling $\theta_{k'}^{r-1}$, conditioning on the previously sampled point, $\theta_{k'}^r$, and so on. Fan et al. (2009) construct the conditional marginal densities by using partial derivatives of the joint density, $\pi(k', \boldsymbol{\theta}'_{k'} \mid \mathbf{x})$, to provide gradient information within a marginal density estimator. As the conditional marginal density estimators are constructed using a combination of samples from the prior distribution and gridded values, they can be computationally expensive to construct, particularly if high-dimensional moves are attempted, for example $\boldsymbol{\theta}_{k'}^{-c} = \boldsymbol{\theta}'_{k'}$. However, this approach can be efficient, and also adapts to the current state of the sampler.

3.3 Post Simulation

3.3.1 Label Switching

The so-called "label switching" problem occurs when the posterior distribution is invariant under permutations in the labeling of the parameters. This results in the parameters having identical marginal posterior distributions. For example, in the context of a finite mixture model (Equation 3.5), the parameters of each mixture component, $\boldsymbol{\phi}_j$, are unidentifiable under a symmetric prior. This causes problems in the interpretation of the MCMC output. While this problem is general, in that it is not restricted to the multi-model case, as many applications of the reversible jump sampler encounter this type of problem, we discuss some methods of overcoming this issue below.

The conceptually simplest method of circumventing nonidentifiability is to impose artificial constraints on the parameters. For example, if μ_j denotes the mean of the jth Gaussian mixture component, then one such constraint could be $\mu_1 < \dots < \mu_k$ (Richardson and Green, 1997). However, the effectiveness of this approach is not always guaranteed (Jasra et al., 2005). One of the main problems with such constraints is that they are often artificial, being imposed for inferential convenience rather than as a result of genuine knowledge about the model. Furthermore, suitable constraints can be difficult or almost impossible to find (Frühwirth-Schnatter, 2001).

Alternative approaches to handling nonidentifiability involve the post-processing of MCMC output. Stephens (2000b) gives an inferential method based on the relabeling of components with respect to the permutation which minimizes the posterior expected loss. Celeux et al. (2000), Hurn et al. (2003), and Sisson and Hurn (2004) adopt a fully decision-theoretic approach, where for every posterior quantity of interest, an appropriate (possibly multi-model) loss function is constructed and minimized. Each of these methods can be computationally expensive.

3.3.2 Convergence Assessment

Under the assumption that an acceptably efficient method of constructing a reversible jump sampler is available, one obvious pre-requisite to inference is that the Markov chain converges to its equilibrium state. Even in fixed dimension problems, theoretical convergence bounds are in general difficult or impossible to determine. In the absence of such theoretical results, convergence diagnostics based on empirical statistics computed from the sample path of multiple chains are often the only available tool. An obvious drawback of the empirical approach is that such diagnostics invariably fail to detect a lack of convergence when parts of the target distribution are missed entirely by all replicate chains. Accordingly, these are necessary rather than sufficient indicators of chain convergence; see Mengersen et al. (1999) and Cowles and Carlin (1996) for comparative reviews under fixed dimension MCMC.

The reversible jump sampler generates additional problems in the design of suitable empirical diagnostics, since most of these depend on the identification of suitable scalar statistics of the parameters' sample paths. However, in the multi-model case, these parameters may no longer retain the same interpretation. In addition, convergence is required not only within each of a potentially large number of models, but also across models with respect to posterior model probabilities.

One obvious approach would be the implementation of independent sub-chain assessments, both within models and for the model indicator $k \in \mathcal{K}$. With focus purely on model selection, Brooks et al. (2003b) propose various diagnostics based on the sample path of the model indicator, k, including nonparametric hypothesis tests such as the χ^2 and Kolmogorov–Smirnov tests. In this manner, distributional assumptions of the models (but not the statistics) are circumvented at the price of associating marginal convergence of k with convergence of the full posterior density.

Brooks and Giudici (2000) propose the monitoring of functionals of parameters which retain their interpretations as the sampler moves between models. The deviance is suggested as a default choice in the absence of superior alternatives. A two-way ANOVA decomposition of the variance of such a functional is formed over multiple chain replications, from which the potential scale reduction factor (PSRF) (Gelman and Rubin, 1992) can be constructed and monitored. Castelloe and Zimmerman (2002) extend this approach firstly to an unbalanced (weighted) two-way ANOVA, to prevent the PRSF being dominated by a few visits to rare models, with the weights being specified in proportion to the frequency of model visits. Castelloe and Zimmerman (2002) also extend their diagnostic to the multivariate (MANOVA) setting on the observation that monitoring several functionals of marginal parameter subsets is more robust than monitoring a single statistic. This general method is clearly reliant on the identification of useful statistics to monitor, but is also sensitive to the extent of approximation induced by violations of the ANOVA assumptions of independence and normality.

Sisson and Fan (2007) propose diagnostics when the underlying model can be formulated in the marked point process framework (Diggle, 1983; Stephens, 2000a). For example, a mixture of an unknown number of univariate normal densities (Equation 3.5) can be represented as a set of k events $\xi_j = (w_j, \mu_j, \sigma_j^2)$, $j = 1, \ldots, k$, in a region $A \subset \mathbb{R}^3$. Given a reference point $v \in A$, in the same space as the events ξ_j (e.g. $v = (\omega, \mu, \sigma^2)$), then the point-to-nearest-event distance, y, is the distance from the point (v) to the nearest event (ξ_j) in A with respect to some distance measure. One can evaluate distributional aspects of the events

$\{\xi_j\}$, through y, as observed from different reference points v. A diagnostic can then be constructed based on comparisons between empirical distribution functions of the distances y, constructed from Markov chain sample paths. Intuitively, as the Markov chains converge, the distribution functions for y constructed from replicate chains should be similar.

This approach permits the direct comparison of full parameter vectors of varying dimension and, as a result, naturally incorporates a measure of across-model convergence. Due to the manner of their construction, Sisson and Fan (2007) are able to monitor an arbitrarily large number of such diagnostics. However, while this approach may have some appeal, it is limited by the need to construct the model in the marked point process setting. Common models which may be formulated in this framework include finite-mixture, change-point and regression models.

Example: Convergence Assessment for Finite Mixture Univariate Normals

We consider the reversible jump sampler of Richardson and Green (1997) implementing a finite mixture of normals model (Equation 3.5) using the enzymatic activity data set (Figure 3.1b). For the purpose of assessing performance of the sampler, we implement five independent sampler replications of length 400,000 iterations.

Figure 3.2a,b illustrates the diagnostic of Brooks et al. (2003b) which provides a test for between-chain convergence based on posterior model probabilities. The pairwise Kolmogorov–Smirnov and χ^2 (all chains simultaneously) tests assume independent realizations. Based on the estimated convergence rate (Brooks et al., 2003b), we retain every 400th iteration to obtain approximate independence. The Kolmogorov–Smirnov statistic cannot reject immediate convergence, with all pairwise chain comparisons well above the critical value of 0.05. The χ^2 statistic cannot reject convergence after the first 10,000 iterations.

Figure 3.2c illustrates the two multivariate PSRFs of Castelloe and Zimmerman (2002) using the deviance as the default statistic to monitor. The solid line shows the ratio of between- and within-chain variation; the broken line indicates the ratio of within-model variation, and the within-model, within-chain variation. The mPSRFs rapidly approach 1, suggesting convergence, beyond 166,000 iterations. This is supported by the independent analysis of Brooks and Giudici (2000) who demonstrate evidence for convergence of this sampler after around 150,000 iterations, although they caution that their chain lengths of only 200,000 iterations were too short for certainty.

Figure 3.2d, adapted from Sisson and Fan (2007), illustrates the PSRF of the distances from each of 100 randomly chosen reference points to the nearest model components, over the five replicate chains. Up to around 100,000 iterations, between-chain variation is still reducing; beyond 300,000 iterations, differences between the chains appear to have stabilized. The intervening iterations mark a gradual transition between these two states. This diagnostic appears to be the most conservative of those presented here.

This example highlights that empirical convergence assessment tools often give varying estimates of when convergence may have been achieved. As a result, it may be prudent to follow the most conservative estimates in practice. While it is undeniable that the benefits for the practitioner in implementing reversible jump sampling schemes are immense, it is arguable that the practical importance of ensuring chain convergence is often overlooked. However, it is also likely that current diagnostic methods are insufficiently advanced to permit a more rigorous default assessment of sampler convergence.

3.3.3 Estimating Bayes Factors

One of the useful by-products of the reversible jump sampler is the ease with which Bayes factors can be estimated. Explicitly expressing marginal or predictive densities of **x** under

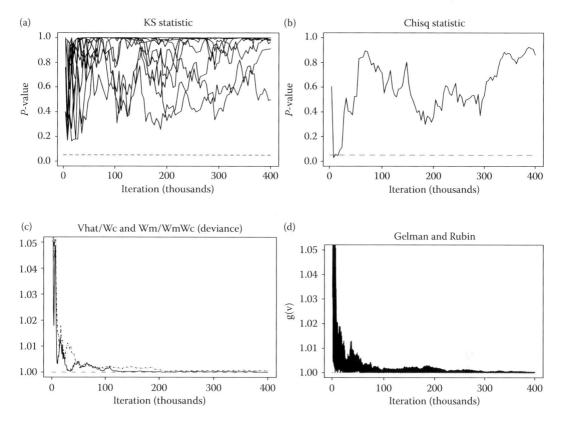

FIGURE 3.2
Convergence assessment for the enzymatic activity data set. (a) Kolmogorov–Smirnov and (b) χ^2 tests of Brooks et al. (2003b). Horizontal line denotes an $\alpha = 0.05$ significance level for test of different sampling distributions. (c) Multivariate PSRFs of Castelloe and Zimmerman (2002) and (d) PSRFvs of Sisson and Fan (2007). Horizontal lines denote the value of each statistic under equal sampling distributions. (From Sisson, S. A. and Fan, Y. 2007. *Statistics and Computing,* 17:357–367. With permission.)

model $\mathcal{M}_k$ as

$$m_k(\mathbf{x}) = \int_{\mathbb{R}^{n_k}} L(\mathbf{x} \mid k, \boldsymbol{\theta}_k) p(\boldsymbol{\theta}_k \mid k) \, d\boldsymbol{\theta}_k,$$

the normalized posterior probability of model $\mathcal{M}_k$ is given by

$$p(k \mid \mathbf{x}) = \frac{p(k)m_k(\mathbf{x})}{\sum_{k' \in \mathcal{K}} p(k')m_{k'}(\mathbf{x})} = \left(1 + \sum_{k' \neq k} \frac{p(k')}{p(k)} B_{k',k}\right)^{-1},$$

where $B_{k',k} = m_{k'}(\mathbf{x})/m_k(\mathbf{x})$ is the Bayes factor of model $\mathcal{M}_{k'}$ to $\mathcal{M}_k$, and $p(k)$ is the prior probability of model $\mathcal{M}_k$. For a discussion of Bayesian model selection techniques, see Chipman et al. (2001), Berger and Pericchi (2001), Kass and Raftery (1995), Ghosh and Samanta (2001), Berger and Pericchi (2004), and Barbieri and Berger (2004). The usual estimator of the posterior model probability, $p(k \mid \mathbf{x})$, is given by the proportion of chain iterations the reversible jump sampler spent in model $\mathcal{M}_k$.

However, when the number of candidate models $|\mathcal{M}|$ is large, the use of reversible jump MCMC algorithms to evaluate Bayes factors raises issues of efficiency. Suppose that model

$\mathcal{M}_k$ accounts for a large proportion of posterior mass. In attempting a between-model move from model $\mathcal{M}_k$, the reversible jump algorithm will tend to persist in this model and visit others models rarely. Consequently, estimates of Bayes factors based on model-visit proportions will tend to be inefficient (Bartolucci and Scaccia, 2003; Han and Carlin, 2001).

Bartolucci et al. (2006) propose enlarging the parameter space of the models under comparison with the same auxiliary variables, $\mathbf{u} \sim q_{d_{k \to k'}}(\mathbf{u})$ and $\mathbf{u}' \sim q_{d_{k' \to k}}(\mathbf{u}')$ (see Equation 3.9), defined under the between-model transitions, so that the enlarged spaces, $(\boldsymbol{\theta}_k, \mathbf{u})$ and $(\boldsymbol{\theta}_{k'}, \mathbf{u}')$, have the same dimension. In this setting, an extension to the bridge estimator for the estimation of the ratio of normalizing constants of two distributions (Meng and Wong, 1996) can be used, by integrating out the auxiliary random process (i.e. $\mathbf{u}$ and $\mathbf{u}'$) involved in the between-model moves. Accordingly, the Bayes factor of model $\mathcal{M}_{k'}$ to $\mathcal{M}_k$ can be estimated using the reversible jump acceptance probabilities as

$$\hat{B}_{k',k} = \frac{\sum_{j=1}^{J_k} \alpha^{(j)}[(k, \boldsymbol{\theta}_k), (k', \boldsymbol{\theta}'_{k'})]/J_k}{\sum_{j=1}^{J_{k'}} \alpha^{(j)}[(k', \boldsymbol{\theta}'_{k'}), (k, \boldsymbol{\theta}_k)]/J_{k'}},$$

where $\alpha^{(j)}[(k, \boldsymbol{\theta}_k), (k', \boldsymbol{\theta}'_{k'})]$ is the acceptance probability (Equation 3.9) of the jth attempt to move from model $\mathcal{M}_k$ to $\mathcal{M}_{k'}$, and where J_k and $J_{k'}$ are the number of proposed moves from model $\mathcal{M}_k$ to $\mathcal{M}_{k'}$ and vice versa during the simulation. Further manipulation is required to estimate $B_{k',k}$ if the sampler does not jump between models $\mathcal{M}_k$ and $\mathcal{M}_{k'}$ directly (Bartolucci et al., 2006). This approach can provide a more efficient way of postprocessing reversible jump MCMC with minimal computational effort.

3.4   Related Multi-Model Sampling Methods

Several alternative multi-model sampling methods are available. Some of these are closely related to the reversible jump MCMC algorithm, or include reversible jump as a special case.

3.4.1   Jump Diffusion

Before the development of the reversible jump sampler, Grenander and Miller (1994) proposed a sampling strategy based on continuous-time jump-diffusion dynamics. This process combines jumps between models at random times, and within-model updates based on a diffusion process according to a Langevin stochastic differential equation indexed by time, t, satisfying

$$d\boldsymbol{\theta}_k^t = dB_k^t + \frac{1}{2} \nabla \log \pi \left(\boldsymbol{\theta}_k^t \right) dt.$$

where dB_k^t denotes an increment of Brownian motion and ∇ the vector of partial derivatives. This method has found some application in signal processing and other Bayesian analyses (Miller et al., 1995; Phillips and Smith, 1996), but has in general been superseded by the more accessible reversible jump sampler. In practice, the continuous-time diffusion must be approximated by a discrete-time simulation. If the time discretization is corrected for via a Metropolis–Hastings acceptance probability, the jump-diffusion sampler actually results in an implementation of reversible jump MCMC (Besag, 1994).

3.4.2 Product Space Formulations

As an alternative to samplers designed for implementation on unions of model spaces, $\Theta = \bigcup_{k \in \mathcal{K}}(\{k\}, \mathbb{R}^{n_k})$, a number of "supermodel" product-space frameworks have been developed, with a state space given by $\Theta^* = \otimes_{k \in \mathcal{K}}(\{k\}, \mathbb{R}^{n_k})$. This setting encompasses all model spaces jointly, so that a sampler needs to simultaneously track $\boldsymbol{\theta}_k$ for all $k \in \mathcal{K}$. The composite parameter vector, $\boldsymbol{\theta}^* \in \Theta^*$, consisting of a concatenation of all parameters under all models, is of fixed dimension, thereby circumventing the necessity of between-model transitions. Clearly, product-space samplers are limited to situations where the dimension of $\boldsymbol{\theta}^*$ is computationally feasible. Carlin and Chib (1995) propose a posterior distribution for the composite model parameter and model indicator given by

$$\pi(k, \boldsymbol{\theta}^* \mid \mathbf{x}) \propto L(\mathbf{x} \mid k, \boldsymbol{\theta}^*_{\mathcal{I}_k}) p(\boldsymbol{\theta}^*_{\mathcal{I}_k} \mid k) p(\boldsymbol{\theta}^*_{\mathcal{I}_{-k}} \mid \boldsymbol{\theta}^*_{\mathcal{I}_k}, k) p(k),$$

where $\mathcal{I}_k$ and $\mathcal{I}_{-k}$ are index sets respectively identifying and excluding the parameters $\boldsymbol{\theta}_k$ from $\boldsymbol{\theta}^*$. Here $\mathcal{I}_k \cap \mathcal{I}_{k'} = \emptyset$ for all $k \neq k'$, so that the parameters for each model are distinct. It is easy to see that the term $p(\boldsymbol{\theta}^*_{\mathcal{I}_{-k}} \mid \boldsymbol{\theta}^*_{\mathcal{I}_k}, k)$, called a "pseudo-prior" by Carlin and Chib (1995), has no effect on the joint posterior $\pi(k, \boldsymbol{\theta}^*_{\mathcal{I}_k} \mid \mathbf{x}) = \pi(k, \boldsymbol{\theta}_k \mid \mathbf{x})$, and its form is usually chosen for convenience. However, poor choices may affect the efficiency of the sampler (Godsill, 2003; Green, 2003).

Godsill (2001) proposes a further generalization of the above by relaxing the restriction that $\mathcal{I}_k \cap \mathcal{I}_{k'} = \emptyset$ for all $k \neq k'$. That is, individual model parameter vectors are permitted to overlap arbitrarily, which is intuitive for, say, nested models. This framework can be shown to encompass the reversible jump algorithm, in addition to the setting of Carlin and Chib (1995). In theory this allows for direct comparison between the three samplers, although this has not yet been fully examined. However, one clear point is that the information contained within $\boldsymbol{\theta}^*_{\mathcal{I}_{-k}}$ would be useful in generating efficient between-model transitions when in model $\mathcal{M}_k$, under a reversible jump sampler. This idea is exploited by Brooks et al. (2003c).

3.4.3 Point Process Formulations

A different perspective on the multi-model sampler is based on spatial birth-and-death processes (Preston, 1977; Ripley, 1977). Stephens (2000a) observed that particular multi-model statistical problems can be represented as continuous-time, marked point processes (Geyer and Møller, 1994). One obvious setting is finite-mixture modeling (Equation 3.5) where the birth and death of mixture components, $\boldsymbol{\phi}_j$, indicate transitions between models. The sampler of Stephens (2000a) may be interpreted as a particular continuous-time, limiting version of a sequence of reversible jump algorithms (Cappé et al., 2003).

A number of illustrative comparisons of the reversible jump, jump diffusion, product space and point process frameworks can be found in the literature. See, for example, Andrieu et al. (2001), Dellaportas et al. (2002), Carlin and Chib (1995), Godsill (2001, 2003), Cappé et al. (2003), and Stephens (2000a).

3.4.4 Multi-Model Optimization

The reversible jump MCMC sampler may be utilized as the underlying random mechanism within a stochastic optimization framework, given its ability to traverse complex spaces efficiently (Andrieu et al., 2000; Brooks et al., 2003a). In a simulated annealing setting, the sampler would define a stationary distribution proportional to the Boltzmann distribution

$$\mathcal{B}_T(k, \boldsymbol{\theta}_k) \propto \exp\left\{\frac{-f(k, \boldsymbol{\theta}_k)}{T}\right\},$$

where $T \geq 0$ and $f(k, \boldsymbol{\theta}_k)$ is a model-ranking function to be minimized. A stochastic annealing framework will then decrease the value of T according to some schedule while using the reversible jump sampler to explore function space. Assuming adequate chain mixing, as $T \to 0$ the sampler and the Boltzmann distribution will converge to a point mass at $(k^*, \boldsymbol{\theta}_{k^*}^*) = \arg\max f(k, \boldsymbol{\theta}_k)$. Specifications for the model-ranking function may include the Akaike information criterion or Bayesian information criterion (King and Brooks, 2004; Sisson and Fan, 2009), the posterior model probability (Clyde, 1999) or a nonstandard loss function defined on variable-dimensional space (Sisson and Hurn, 2004) for the derivation of Bayes rules.

3.4.5 Population MCMC

The population Markov chain Monte Carlo method (Liang and Wong, 2001; Liu, 2001) may be extended to the reversible jump setting (Jasra et al., 2007). Motivated by simulated annealing (Geyer and Thompson, 1995), N parallel reversible jump samplers are implemented targeting a sequence of related distributions $\{\pi_i\}, i = 1, \ldots, N$, which may be tempered versions of the distribution of interest, $\pi_1 = \pi(k, \boldsymbol{\theta}_k \mid \mathbf{x})$. The chains are allowed to interact, in that the states of any two neighboring (in terms of the tempering parameter) chains may be exchanged, thereby improving the mixing across the population of samplers both within and between models. Jasra et al. (2007) demonstrate superior convergence rates over a single reversible jump sampler. For samplers that make use of tempering or parallel simulation techniques, Gramacy et al. (2010) propose efficient methods of utilizing samples from all distributions (i.e. including those not from π_1) using importance weights, for the calculation of given estimators.

3.4.6 Multi-Model Sequential Monte Carlo

The idea of running multiple samplers over a sequence of related distributions may also considered under a sequential Monte Carlo (SMC) framework (Del Moral et al., 2006). Jasra et al. (2008) propose implementing N separate SMC samplers, each targeting a different subset of model space. At some stage the samplers are allowed to interact and are combined into a single sampler. This approach permits more accurate exploration of models with lower posterior model probabilities than would be possible under a single sampler. As with population MCMC methods, the benefits gained in implementing N samplers must be weighed against the extra computational overheads.

3.5 Discussion and Future Directions

Given the degree of complexity associated with the implementation of reversible jump MCMC, a major focus for future research is in designing simple but efficient samplers, with the ultimate goal of automation. Several authors have provided new insights into the reversible jump sampler which may contribute toward achieving such goals. For example, Keith et al. (2004) present a generalized Markov sampler, which includes the reversible jump sampler as a special case. Petris and Tardella (2003) demonstrate a geometric approach for sampling from nested models, formulated by drawing from a fixed-dimension auxiliary continuous distribution on the largest model subspace, and then using transformations to recover model-specific samples. Walker (2009) has recently provided a Gibbs sampler alternative to the reversible jump MCMC, using auxiliary variables. Additionally, as noted by Sisson (2005), one does not need to work only with reversible Markov chains, and nonreversible chains may offer opportunities for sampler improvement (Diaconis et al., 2000; Mira and Geyer, 2000; Neal, 2004).

An alternative way of increasing sampler efficiency would be to explore the ideas introduced in adaptive MCMC. As with standard MCMC, any adaptations must be implemented with care—transition kernels dependent on the entire history of the Markov chain can only be used under diminishing adaptation conditions (Haario et al., 2001; Roberts and Rosenthal, 2009). Alternative schemes permit modification of the proposal distribution at regeneration times, when the next state of the Markov chain becomes completely independent of the past (Brockwell and Kadane, 2005; Gilks et al., 1998). Under the reversible jump framework, regeneration can be naturally achieved by incorporating an additional model, from which independent samples can be drawn. Under any adaptive scheme, however, how best to make use of historical chain information remains an open question. Additionally, efficiency gains through adaptations should naturally outweigh the costs of handling chain history and modification of the proposal mechanisms.

Finally, two areas remain underdeveloped in the context of reversible jump simulation. The first of these is perfect simulation, which provides an MCMC framework for producing samples exactly from the target distribution, circumventing convergence issues entirely (Propp and Wilson, 1996). Some tentative steps have been made in this area (Brooks et al., 2006; Møller and Nicholls, 1999). Secondly, while the development of "likelihood-free" MCMC has received much recent attention (Chapter 12, this volume), implementing the sampler in the multi-model setting remains a challenging problem, in terms of both computational efficiency and bias of posterior model probabilities.

Acknowledgments

This work was supported by the Australian Research Council through the Discovery Project scheme (DP0664970 and DP0877432).

References

Al-Awadhi, F., Hurn, M. A., and Jennison, C. 2004. Improving the acceptance rate of reversible jump MCMC proposals. *Statistics and Probability Letters*, 69:189–198.

Andrieu, C., de Freitas, J., and Doucet, A. 2000. Reversible jump MCMC simulated annealing for neural networks. In C. Boutilier and M. Goldszmidt (eds), *Proceedings of the 16th Conference on Uncertainty in Artificial Intelligence*, pp. 11–18. Morgan Kaufmann, San Francisco.

Andrieu, C., Djurić, P. M., and Doucet, M. 2001. Model selection by MCMC computation. *Signal Processing*, 81:19–37.

Barbieri, M. M. and Berger, J. O. 2004. Optimal predictive model selection. *Annals of Statistics*, 32:870–897.

Bartolucci, F. and Scaccia, L. 2003. A new approach for estimating the Bayes factor. Technical report, University of Perugia.

Bartolucci, F., Scaccia, L., and Mira, A. 2006. Efficient Bayes factors estimation from reversible jump output. *Biometrika*, 93(1):41–52.

Berger, J. O. and Pericchi, L. R. 2001. Objective Bayesian methods for model selection: Introduction and comparison (with discussion). In P. Lahiri (ed.), *Model Selection*, IMS Lecture Notes—Monograph Series, 38, pp. 135–207. Institute of Mathematical Statistics, Beachwood, OH.

Berger, J. O. and Pericchi, L. R. 2004. Training samples in objective Bayesian model selection. *Annals of Statistics*, 32:841–869.

Besag, J. 1994. Contribution to the discussion of a paper by Grenander and Miller. *Journal of the Royal Statistical Society, Series B*, 56:591–592.

Brockwell, A. E. and Kadane, J. B. 2005. Identification of regeneration times in MCMC simulation, with application to adaptive schemes. *Journal of Computational and Graphical Statistics*, 14(2): 436–458.

Brooks, S. P. 1998. Markov chain Monte Carlo method and its application. *The Statistician*, 47: 69–100.

Brooks, S. P. and Giudici, P. 2000. MCMC convergence assessment via two-way ANOVA. *Journal of Computational and Graphical Statistics*, 9:266–285.

Brooks, S. P., Fan, Y., and Rosenthal, J. S. 2006. Perfect forward simulation via simulated tempering. *Communications in Statistics*, 35:683–713.

Brooks, S. P., Friel, N., and King, R. 2003a. Classical model selection via simulated annealing. *Journal of the Royal Statistical Society, Series B*, 65:503–520.

Brooks, S. P., Giudici, P., and Philippe, A. 2003b. On non-parametric convergence assessment for MCMC model selection. *Journal of Computational and Graphical Statistics*, 12:1–22.

Brooks, S. P., Guidici, P., and Roberts, G. O. 2003c. Efficient construction of reversible jump Markov chain Monte Carlo proposal distributions (with discusion). *Journal of the Royal Statistical Society, Series B*, 65:3–39.

Cappé, O., Robert, C. P., and Rydén, T. 2003. Reversible jump MCMC converging to birth-and-death MCMC and more general continuous time samplers. *Journal of the Royal Statistical Society, Series B*, 65:679–700.

Carlin, B. P. and Chib, S. 1995. Bayesian model choice via Markov chain Monte Carlo. *Journal of the Royal Statistical Society, Series B*, 57:473–484.

Castelloe, J. M. and Zimmerman, D. L. 2002. Convergence assessment for reversible jump MCMC samplers. Technical Report 313, Department of Statistics and Actuarial Science, University of Iowa.

Celeux, G., Hurn, M. A., and Robert, C. P. 2000. Computational and inferential difficulties with mixture posterior distributions. *Journal of the American Statistical Association*, 95:957–970.

Chipman, H., George, E., and McCulloch, R. E. 2001. The practical implementation of Bayesian model selection (with discussion). In P. Lahiri (ed.), *Model Selection*, IMS Lecture Notes—Monograph Series, 38, pp. 67–134. Institute of Mathematical Statistics, Beachwood, OH.

Clyde, M. A. 1999. Bayesian model averaging and model search strategies. In J. M. Bernardo, J. O. Berger, A. P. Dawid, and A. F. M. Smith (eds), *Bayesian Statistics 6: Proceedings of the Sixth Valencia International Meeting*, pp. 157–185. Oxford University Press, Oxford.

Cowles, M. K. and Carlin, B. P. 1996. Markov chain Monte Carlo convergence diagnostics: A comparative review. *Journal of the American Statistical Association*, 91:883–904.

Del Moral, P., Doucet, A., and Jasra, A. 2006. Sequential Monte Carlo samplers. *Journal of the Royal Statistical Society, Series B*, 68:411–436.

Dellaportas, P. and Papageorgiou, I. 2006. Multivariate mixtures of normals with unknown number of components. *Statistics and Computing*, 16:57–68.

Dellaportas, P., Forster, J. J., and Ntzoufras, I. 2002. On Bayesian model and variable selection using MCMC. *Statistics and Computing*, 12:27–36.

Denison, D. G. T., Mallick, B. K., and Smith, A. F. M. 1998. Automatic Bayesian curve fitting. *Journal of the Royal Statistical Society, Series B*, 60:330–350.

Diaconis, P., Holmes, S., and Neal, R. M. 2000. Analysis of a non-reversible Markov chain sampler. *Annals of Applied Probability*, 10:726–752.

Diggle, P. J. 1983. *Statistical Analysis of Spatial Point Patterns*. Academic Press, London.

DiMatteo, I., Genovese, C. R., and Kass, R. E. 2001. Bayesian curve-fitting with free-knot splines. *Biometrika*, 88:1055–1071.

Ehlers, R. S. and Brooks, S. P. 2003. Constructing general efficient proposals for reversible jump MCMC. Technical report, Department of Statistics, Federal University of Paraná.

Ehlers, R. S. and Brooks, S. P. 2008. Adaptive proposal construction for reversible jump MCMC. *Scandinavian Journal of Statistics*, 35:677–690.

Fan, Y. and Brooks, S. P. 2000. Bayesian modelling of prehistoric corbelled domes. *The Statistician*, 49:339–354.

Fan, Y., Peters, G. W., and Sisson, S. A. 2009. Automating and evaluating reversible jump MCMC proposal distributions. *Statistics and Computing*, 19(4):409–421.

Frühwirth-Schnatter, S. 2001. Markov chain Monte Carlo estimation of classical and dynamic switching and mixture models. *Journal of the American Statistical Association*, 96:194–209.

Gelman, A. and Rubin, D. B. 1992. Inference from iterative simulations using multiple sequences. *Statistical Science*, 7:457–511.

George, A. W., Mengersen, K. L., and Davis, G. P. 1999. A Bayesian approach to ordering gene markers. *Biometrics*, 55:419–429.

George, E. I. and McCulloch, R. E. 1993. Variable selection via Gibbs sampling. *Journal of the American Statistical Association*, 88:881–889.

Geyer, C. J. and Møller, J. 1994. Simulation procedures and likelihood inference for spatial point processes. *Scandinavian Journal of Statistics*, 21:359–373.

Geyer, C. J. and Thompson, E. A. 1995. Annealing Markov chain Monte Carlo with applications to ancestral inference. *Journal of the American Statistical Association*, 90:909–920.

Ghosh, J. K. and Samanta, T. 2001. Model selection: An overview. *Current Science*, 80:1135–1144.

Gilks, W. R., Roberts, G. O., and Sahu, S. K. 1998. Adaptive Markov chain Monte Carlo through regeneration. *Journal of the American Statistical Association*, 93:1045–1054.

Godsill, S. 2001. On the relationship between Markov chain Monte Carlo methods for model uncertainty. *Journal of Computational and Graphical Statistics*, 10:1–19.

Godsill, S. 2003. Discussion of Trans-dimensional Markov chain Monte Carlo by P. J. Green. In P. J. Green, N. L. Hjort, and S. Richardson (eds), *Highly Structured Stochastic Systems*, pp. 199–203. Oxford University Press, Oxford.

Gramacy, R. B., Samworth, R. J., and King, R. 2010. Importance tempering. *Statistics and Computing*, 20:1–7.

Green, P. J. 1995. Reversible jump Markov chain Monte Carlo computation and Bayesian model determination. *Biometrika*, 82:711–732.

Green, P. J. 2001. A primer on Markov chain Monte Carlo. In O. E. Barndorff-Nielsen, D. R. Cox, and C. Klüppelberg (eds), *Complex Stochastic Systems*, Monographs on Statistics and Probability, 87, pp. 1–62. Chapman & Hall/CRC, Boca Raton, FL.

Green, P. J. 2003. Trans-dimensional Markov chain Monte Carlo. In P. J. Green, N. L. Hjort, and S. Richardson (eds), *Highly Structured Stochastic Systems*, pp. 179–198. Oxford University Press, Oxford.

Green, P. J. and Mira, A. 2001. Delayed rejection in reversible jump Metropolis-Hastings. *Biometrika*, 88:1035–1053.

Grenander, U. and Miller, M. I. 1994. Representations of knowledge in complex systems. *Journal of the Royal Statistical Society, Series B*, 56:549–603.

Haario, H., Saksman, E., and Tamminen, J. 2001. An adaptive Metropolis algorithm. *Bernoulli*, 7: 223–242.

Han, C. and Carlin, B. P. 2001. MCMC methods for computing Bayes factors: A comparative review. *Journal of the American Statistical Association*, 96:1122–1132.

Hastie, D. 2004. Developments in Markov chain Monte Carlo. PhD thesis, University of Bristol.

Hastie, T. J. and Tibshirani, R. J. 1990. *Generalised Additive Models*. Chapman & Hall, London.

Hastings, W. K. 1970. Monte Carlo sampling methods using Markov chains and their applications. *Biometrika*, 57:59–109.

Hurn, M., Justel, A., and Robert, C. P. 2003. Estimating mixtures of regressions. *Journal of Computational and Graphical Statistics*, 12:55–79.

Jasra, A., Doucet, A., Stephens, D. A., and Holmes, C. 2008. Interacting sequential Monte Carlo samplers for trans-dimensional simulation. *Computational Statistics and Data Analysis*, 52(4): 1765–1791.

Jasra, A., Holmes, C., and Stephens, D. A. 2005. MCMC methods and the label switching problem. *Statistical Science*, 20(1):50–67.

Jasra, A., Stephens, D. A., and Holmes, C. C. 2007. Population-based reversible jump Markov chain Monte Carlo. *Biometrika*, 94:787–807.

Kass, R. E. and Raftery, A. E. 1995. Bayes factors. *Journal of the American Statistical Association*, 90:773–796.

Keith, J. M., Kroese, D. P., and Bryant, D. 2004. A generalized Markov sampler. *Methodology and Computing in Applied Probability*, 6:29–53.

King, R. and Brooks, S. P. 2004. A classical study of catch-effort models for Hector's dolphins. *Journal of the American Statistical Association*, 99:325–333.

Kirkpatrick, S. 1984. Optimization by simulated annealing: Quantitative studies. *Journal of Statistical Physics*, 34:975–986.

Lawler, G. and Sokal, A. 1988. Bounds on the L^2 spectrum for Markov chains and Markov processes. *Transactions of the American Mathematical Society*, 309:557–580.

Liang, F. and Wong, W. H. 2001. Real parameter evolutionary Monte Carlo with applications to Bayesian mixture models. *Journal of the American Statistical Association*, 96:653–666.

Liu, J. S. 2001. *Monte Carlo Strategies in Scientific Computing*. Springer, New York.

Liu, J. S., Liang, F., and Wong, W. H. 2001. A theory for dynamic weighing in Monte Carlo computation. *Journal of the American Statistical Association*, 96(454):561–573.

Meng, X. L. and Wong, W. H. 1996. Simulating ratios of normalising constants via a simple identity: A theoretical exploration. *Statistica Sinica*, 6:831–860.

Mengersen, K. L., Robert, C. P., and Guihenneuc-Joyaux, C. 1999. MCMC convergence diagnostics: A review. In J. M. Bernardo, J. O. Berger, A. P. Dawid, and A. F. M. Smith (eds), *Bayesian Statistics 6*, pp. 415–140. Oxford University Press, Oxford.

Miller, M. I., Srivastava, A., and Grenander, U. 1995. Conditional-mean estimation via jump-diffusion processes in multiple target tracking/recognition. *IEEE Transactions on Signal Processing*, 43:2678–2690.

Mira, A. and Geyer, C. J. 2000. On non-reversible Markov chains. In N. Madras (ed.), *Monte Carlo Methods*, pp. 93–108. American Mathematical Society, Providence, RI.

Møller, J. and Nicholls, G. K. 1999. Perfect simulation for sample-based inference. Technical report, Aalborg University.

Neal, R. M. 2004. Improving asymptotic variance of MCMC estimators: Non-reversible chains are better. Technical Report 0406, Department of Statisics, University of Toronto.

Nott, D. J. and Green, P. J. 2004. Bayesian variable selection and the Swendsen-Wang algorithm. *Journal of Computational and Graphical Statistics*, 13(1):141–157.

Nott, D. J. and Leonte, D. 2004. Sampling schemes for Bayesian variable selection in generalised linear models. *Journal of Computational and Graphical Statistics*, 13(2):362–382.

Ntzoufras, I., Dellaportas, P., and Forster, J. J. 2003. Bayesian variable and link determination for generalised linear models. *Journal of Statistical Planning and Inference*, 111:165–180.

Papathomas, M., Dellaportas, P., and Vasdekis, V. G. S. 2009. A general proposal construction for reversible jump MCMC. Technical report, Athens University of Economics and Business.

Peskun, P. 1973. Optimum Monte Carlo sampling using Markov chains. *Biometrika*, 60:607–612.

Petris, G. and Tardella, L. 2003. A geometric approach to transdimensional Markov chain Monte Carlo. *Canadian Journal of Statistics*, 31.

Phillips, D. B. and Smith, A. F. M. 1996. Bayesian model comparison via jump diffusions. In W. R. Gilks, S. Richardson, and D. J, Spiegelhalter (eds), *Markov Chain Monte Carlo in Practice*, pp. 215–239. Chapman & Hall, London.

Preston, C. J. 1977. Spatial birth-and-death processes. *Bulletin of the International Statistical Institute*, 46:371–391.

Propp, J. G. and Wilson, D. B. 1996. Exact sampling with coupled Markov chains and applications to statistical mechanics. *Random Structures and Algorithms*, 9:223–252.

Richardson, S. and Green, P. J. 1997. On Bayesian analysis of mixtures with an unknown number of components (with discussion). *Journal of the Royal Statistical Society, Series B*, 59: 731–792.

Ripley, B. D. 1977. Modelling spatial patterns (with discussion). *Journal of the Royal Statistical Society, Series B*, 39:172–212.

Roberts, G. O. 2003. Linking theory and practice of MCMC. In P. J. Green, N. Hjort, and S. Richardson (eds), *Highly Structured Stochastic Systems*, pp. 145–166. Oxford University Press.

Roberts, G. O. and Rosenthal, J. S. 2009. Examples of adaptive MCMC. *Journal of Computational and Graphical Statistics*, 18:349–367.

Sisson, S. A. 2005. Trans-dimensional Markov chains: A decade of progress and future perspectives. *Journal of the American Statistical Association*, 100:1077–1089.

Sisson, S. A. and Fan, Y. 2007. A distance-based diagnostic for trans-dimensional Markov chains. *Statistics and Computing*, 17:357–367.

Sisson, S. A. and Fan, Y. 2009. Towards automating model selection for a mark-recapture-recovery analysis. *Applied Statistics*, 58(2):247–266.

Sisson, S. A. and Hurn, M. A. 2004. Bayesian point estimation of quantitative trait loci. *Biometrics*, 60:60–68.

Smith, M. and Kohn, R. 1996. Nonparametric regression using Bayesian variable selection. *Journal of Econometrics*, 75:317–344.

Stephens, M. 2000a. Bayesian analysis of mixture models with an unknown number of components— an alternative to reversible jump methods. *Annals of Statistics*, 28:40–74.

Stephens, M. 2000b. Dealing with label switching in mixture models. *Journal of the Royal Statistical Society, Series B*, 62:795–809.

Tadesse, M., Sha, N., and Vannucci, M. 2005. Bayesian variable selection in clustering high-dimensional data. *Journal of the American Statistical Association*, 100:602–617.

Tierney, L. 1998. A note on Metropolis–Hastings kernels for general state spaces. *Annals of Applied Probability*, 8:1–9.

Tierney, L. and Mira, A. 1999. Some adaptive Monte Carlo methods for Bayesian inference. *Statistics in Medicine*, 18:2507–2515.

Trias, M., Vecchio, A., and Vetich, J. 2009. Delayed rejection schemes for efficient Markov chain Monte Carlo sampling of multimodal distributions. Technical report, Universitat de les Illes Balears.

Vermaak, J., Andrieu, C., Doucet, A., and Godsill, S. J. 2004. Reversible jump Markov chain Monte Carlo strategies for Bayesian model selection in autoregressive processes. *Journal of Time Series Analysis*, 25(6):785–809.

Walker, S. G. 2009. A Gibbs sampling alternative to reversible jump MCMC. Technical report, University of Kent.

4

Optimal Proposal Distributions and Adaptive MCMC

Jeffrey S. Rosenthal

4.1 Introduction

The Metropolis–Hastings algorithm (Metropolis et al., 1953; Hastings, 1970) requires choice of proposal distributions, and it is well known that some proposals work much better than others. Determining which proposal is best for a particular target distribution is both very important and very difficult. Often this problem is attacked in an *ad hoc* manner involving much trial and error. However, it is also possible to use theory to estimate optimal proposal scalings and/or adaptive algorithms to attempt to find good proposals automatically. This chapter reviews both of these possibilities.

4.1.1 The Metropolis–Hastings Algorithm

Suppose that our target distribution has density π with respect to some reference measure (usually d-dimensional Lebesgue measure). Then, given $\mathbf{X}_n$, a "proposed value" $\mathbf{Y}_{n+1}$ is generated from some pre-specified density $q(\mathbf{X}_n, \mathbf{y})$, and is then accepted with probability

$$\alpha(\mathbf{x}, \mathbf{y}) = \begin{cases} \min\left\{ \dfrac{\pi(\mathbf{y})}{\pi(\mathbf{x})} \dfrac{q(\mathbf{y}, \mathbf{x})}{q(\mathbf{x}, \mathbf{y})}, 1 \right\}, & \pi(\mathbf{x})\, q(\mathbf{x}, \mathbf{y}) > 0, \\ 1, & \pi(\mathbf{x})\, q(\mathbf{x}, \mathbf{y}) = 0. \end{cases} \tag{4.1}$$

If the proposed value is accepted, we set $\mathbf{X}_{n+1} = \mathbf{Y}_{n+1}$; otherwise, we set $\mathbf{X}_{n+1} = \mathbf{X}_n$. The function $\alpha(\mathbf{x}, \mathbf{y})$ is chosen, of course, precisely to ensure that the Markov chain $\mathbf{X}_0, \mathbf{X}_1, \ldots$ is reversible with respect to the target density $\pi(\mathbf{y})$, so that the target density is stationary for the chain. If the proposal is *symmetric*, that is $q(\mathbf{x}, \mathbf{y}) = q(\mathbf{y}, \mathbf{x})$, then this reduces to

$$\alpha(\mathbf{x}, \mathbf{y}) = \begin{cases} \min\left\{ \dfrac{\pi(\mathbf{y})}{\pi(\mathbf{x})}, 1 \right\}, & \pi(\mathbf{x})\, q(\mathbf{x}, \mathbf{y}) > 0, \\ 1, & \pi(\mathbf{x})\, q(\mathbf{x}, \mathbf{y}) = 0. \end{cases}$$

4.1.2 Optimal Scaling

It has long been recognized that the choice of the proposal density $q(\mathbf{x}, \mathbf{y})$ is crucial to the success (e.g. rapid convergence) of the Metropolis–Hastings algorithm. Of course, the fastest-converging proposal density would be $q(\mathbf{x}, \mathbf{y}) = \pi(\mathbf{y})$ (in which case $\alpha(\mathbf{x}, \mathbf{y}) \equiv 1$, and the convergence is immediate), but in the Markov chain Monte Carlo (MCMC) context we assume that π cannot be sampled directly. Instead, the most common case (which we focus on here) involves a *symmetric random-walk Metropolis algorithm* (RMW) in which the proposal value is given by $\mathbf{Y}_{n+1} = \mathbf{X}_n + \mathbf{Z}_{n+1}$, where the increments $\{\mathbf{Z}_n\}$ are i.i.d. from some fixed symmetric distribution (e.g. $N(0, \sigma^2 I_d)$). In this case, the crucial issue becomes how to *scale* the proposal (e.g. how to choose σ): too small and the chain will move too slowly; too large and the proposals will usually be rejected. Instead, we must avoid both extremes (we sometimes refer to this as the "Goldilocks principle").

Metropolis et al. (1953) recognized this issue early on, when they considered the case $\mathbf{Z}_n \sim U[-\alpha, \alpha]$ and noted that "the maximum displacement α must be chosen with some care; if too large, most moves will be forbidden, and if too small, the configuration will not change enough. In either case it will then take longer to come to equilibrium."

In recent years, significant progress has been made in identifying optimal proposal scalings, in terms of such tangible values as asymptotic acceptance rate. Under certain conditions, these results can describe the optimal scaling precisely. These issues are discussed in Section 4.2 below.

4.1.3 Adaptive MCMC

The search for improved proposal distributions is often done manually, through trial and error, though this can be difficult, especially in high dimensions. An alternative approach is adaptive MCMC, which asks the computer to automatically "learn" better parameter values "on the fly"—that is, while an algorithm runs. Intuitively, this approach is attractive since computers are getting faster and faster, while human speed is remaining about the same.

Suppose $\{P_\gamma\}_{\gamma \in \mathcal{Y}}$ is a family of Markov chains, each having stationary distribution π. (For example, perhaps P_γ corresponds to an RWM algorithm with increment distribution $N(0, \gamma^2 I_d)$.) An adaptive MCMC algorithm would randomly update the value of γ at each iteration, in an attempt to find the best value. Adaptive MCMC has been applied in a variety of contexts (e.g. Haario et al., 2001; Giordani and Kohn, 2006; Roberts and Rosenthal, 2009), including problems in statistical genetics (Turro et al., 2007).

Counterintuitively, adaptive MCMC algorithms may not always preserve the stationarity of π. However, if the adaptations are designed to satisfy certain conditions, then stationarity is guaranteed, and significant speed-ups are possible. These issues are discussed in Section 4.3 below.

4.1.4 Comparing Markov Chains

Since much of what follows will attempt to find "better" or "best" MCMC samplers, we pause to consider what it means for one Markov chain to be better than another.

Suppose P_1 and P_2 are two Markov chains, each with the same stationary distribution π. Then P_1 *converges faster than* P_2 if $\sup_A |P_1^n(x, A) - \pi(A)| \leq \sup_A |P_2^n(x, A) - \pi(A)|$ for all n and x. This definition concerns distributional convergence (in total variation distance) as studied theoretically in, for example, Rosenthal (1995, 2002) and Roberts and Tweedie (1999).

Alternatively, P_1 *has smaller variance than* P_2 if $\mathbf{Var}\left(\frac{1}{n} \sum_{i=1}^{n} g(X_i)\right)$ is smaller when $\{X_i\}$ follows P_1 than when it follows P_2. This definition concerns the variance of a functional g, and may depend on which g is chosen, and also perhaps on n and/or the starting distribution.

Usually we assume that the Markov chain $\{X_n\}$ is in stationarity, so $\Pr(X_i \in A) = \pi(A)$, and $\Pr(X_{i+1} \in A \mid X_i = x) = P(x, A)$ where P is the Markov chain kernel being followed.

If the Markov chain $\{X_n\}$ is in stationarity, then, for large n, $\mathbf{Var}\left(\frac{1}{n} \sum_{i=1}^{n} g(X_i)\right) \approx \frac{1}{n} \mathbf{Var}_\pi(g) \, \tau_g$, where $\tau_g = \sum_{k=-\infty}^{\infty} \mathrm{corr}(g(X_0), g(X_k)) = 1 + 2 \sum_{i=1}^{\infty} \mathrm{corr}(g(X_0), g(X_i))$ is the integrated autocorrelation time. So, a related definition is that P_1 *has smaller asymptotic variance than* P_2 if τ_g is smaller under P_1 than under P_2. (Under strong conditions involving the so-called *Peskun ordering*, this improvement is sometimes uniform over choice of g; see, e.g. Mira, 2001.)

Another perspective is that a Markov chain is better if it allows for faster exploration of the state space. Thus, P_1 *mixes faster than* P_2 if $\mathrm{E}[(X_n - X_{n-1})^2]$ is larger under P_1 than under P_2, where again $\{X_n\}$ is in stationarity. (Of course, $\mathrm{E}[(X_n - X_{n-1})^2]$ would usually be estimated by $\frac{1}{n} \sum_{i=1}^{n} (X_i - X_{i-1})^2$, or perhaps by $\frac{1}{n-B} \sum_{i=B}^{n} (X_i - X_{i-1})^2$ to allow a burn-in B to approximately converge to stationarity.) Note that the evaluation of $\mathrm{E}[(X_n - X_{n-1})^2]$ is over all proposed moves, including rejected ones where $(X_n - X_{n-1})^2 = 0$. Thus, rejected moves slow down the chain, but small accepted moves do not help very much either. Best is to find reasonably large proposed moves which are reasonably likely to be accepted.

Such competing definitions of "better" Markov chain mean that the optimal choice of MCMC may depend on the specific question being asked. However, we will see in Section 4.2 that in some circumstances these different definitions are all equivalent, leading to uniformly optimal choices of algorithm parameters.

4.2 Optimal Scaling of Random-Walk Metropolis

We restrict ourselves to the RWM algorithm, where the proposals are of the form $\mathbf{Y}_{n+1} = \mathbf{X}_n + \mathbf{Z}_{n+1}$, where $\{\mathbf{Z}_i\}$ are i.i.d. with fixed symmetric density, with some scaling parameter $\sigma > 0$, for example $Z_i \sim N(0, \sigma^2 I_d)$. To avoid technicalities, we assume that the target density π is a positive, continuous function. The task is to choose σ in order to optimize the resulting MCMC algorithm.

4.2.1 Basic Principles

A first observation is that if σ is very small, then virtually all proposed moves will be accepted, but they will represent very small movements, so overall the chain will not mix well (Figure 4.1). Similarly, if σ is very large, then most moves will be rejected, so the chain will usually not move at all (Figure 4.2). What is needed is a value of σ between the two extremes, thus allowing for reasonable-sized proposal moves together with a reasonably high acceptance probability (Figure 4.3).

A simple way to avoid the extremes is to monitor the *acceptance rate* of the algorithm, that is, the fraction of proposed moves which are accepted. If this fraction is very close to 1, this suggests that σ is too small (as in Figure 4.1). If this fraction is very close to 0, this suggests that σ is too large (as in Figure 4.2). But if this fraction is far from 0 and far from 1, then we have managed to avoid both extremes (Figure 4.3).

So, this provides an easy rule of thumb for scaling RMW algorithms: choose a scaling σ so that the acceptance rate is far from 0 and far from 1. However, this still allows for a wide variety of choices. Under some conditions, much more can be said.

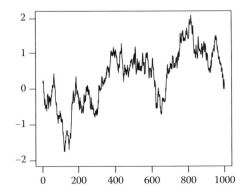

FIGURE 4.1
Trace plot with small σ, large acceptance rate, and poor mixing.

4.2.2 Optimal Acceptance Rate as $d \to \infty$

Major progress about optimal scalings was made by Roberts et al. (1997). They considered RWM on $\mathbf{R}^d$ for very special target densities, of the form

$$\pi(x_1, x_2, \ldots, x_d) = f(x_1)f(x_2)\ldots f(x_d), \tag{4.2}$$

for some one-dimensional smooth density f. That is, the target density is assumed to consist of i.i.d. components. Of course, this assumption is entirely unrealistic for MCMC, since it means that to sample from π it suffices to sample each component separately from the one-dimensional density f (which is generally easy to do numerically).

Under this restrictive assumption, and assuming proposal increment distributions of the form $N(0, \sigma^2 I_d)$, Roberts et al. (1997) proved the remarkable result that as $d \to \infty$, *the optimal acceptance rate is precisely 0.234.* This is clearly a major refinement of the general principle that the acceptance rate should be far from 0 and far from 1.

More precisely, their result is the following. Suppose that $\sigma = \ell/\sqrt{d}$ for some $\ell > 0$. Then as $d \to \infty$, if time is speeded up by a factor of d, and space is shrunk by a factor of $\sqrt{d}$, then each component of the Markov chain converges to a diffusion having stationary distribution f, and speed function given by $h(\ell) = 2\ell^2 \, \Phi\left(-\sqrt{I}\ell/2\right)$, where Φ is the cumulative distribution

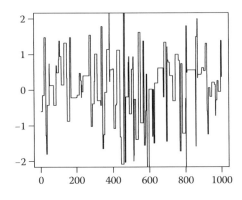

FIGURE 4.2
Trace plot with large σ, small acceptance rate, and poor mixing.

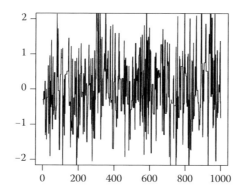

FIGURE 4.3
Trace plot with medium σ, medium acceptance rate, and good mixing.

function of a standard normal, and I is a constant depending on f, given in fact by $I =$
$$\int_{-\infty}^{\infty} \left[\left(\frac{f'(X)}{f(X)} \right)^2 \right] f(x)\, dx.$$
It follows that this diffusion is optimized (in terms of *any* of the criteria of Section 4.1.4) when ℓ is chosen to maximize $h(\ell)$. It is computed numerically that this optimal value of ℓ is given by $\ell_{\text{opt}} \doteq 2.38/\sqrt{I}$.

Furthermore, the asymptotic (stationary) acceptance rate is given by $A(\ell) = 2\,\Phi\left(-\sqrt{I}\ell/2\right)$. Hence, the optimal acceptance rate is equal to $A(\ell_{\text{opt}}) \doteq 2\,\Phi(-2.38/2) \doteq 0.234$, which is where the figure 0.234 comes from.

Figure 4.4 plots $h(\ell)$ versus ℓ, and Figure 4.5 plots $h(\ell)$ versus $A(\ell)$. (We take $I = 1$ for definiteness, but any other value of I would simply multiply all the values by a constant.) In particular, the relative speed $h(\ell)$ remains fairly close to its maximum as long as ℓ is within, say, a factor of 2 of its optimal value. Equivalently, the algorithm remains relatively efficient as long as the asymptotic acceptance rate $A(\ell)$ is between, say, 0.1 and 0.6.

Of course, the above results are all asymptotic as $d \to \infty$. Numerical studies (e.g. Gelman et al., 1996; Roberts and Rosenthal, 2001) indicate that the limiting results do seem to well approximate finite-dimensional situations for d as small as 5. On the other hand, they

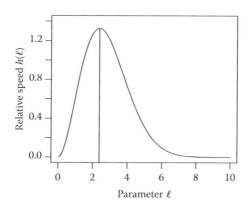

FIGURE 4.4
Algorithm relative speed $h(\ell)$ as a function of the parameter ℓ.

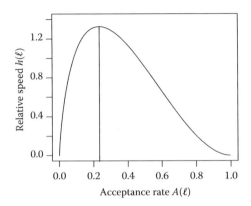

FIGURE 4.5
Algorithm relative speed $h(\ell)$ as a function of acceptance rate $A(\ell)$.

do not apply to one-dimensional increments, for example; numerical studies on normal distributions show that when $d = 1$, the optimal acceptance rate is approximately 0.44. Finally, these results are all on continuous spaces, but there have also been studies of optimal scaling for discrete Metropolis algorithms (Neal et al., 2007).

4.2.3 Inhomogeneous Target Distributions

The above result of Roberts et al. (1997) requires the strong assumption that $\pi(\mathbf{x}) = \prod_{i=1}^{d} f(x_i)$, that is, that the target distribution has i.i.d. components. In later work, this assumption was relaxed in various ways.

Roberts and Rosenthal (2001) considered inhomogeneous target densities of the form

$$\pi(\mathbf{x}) = \prod_{i=1}^{d} C_i f(C_i x_i), \tag{4.3}$$

where the $\{C_i\}$ are themselves i.i.d. from some fixed distribution. (Thus, Equation 4.2 corresponds to the special case where the C_i are constant.) They proved that in this case, the result of Roberts et al. (1997) still holds (including the optimal acceptance rate of 0.234), except that the limiting diffusion speed is divided by an "inhomogeneity factor" of $b \equiv E(C_i^2)/(E(C_i))^2 \geq 1$. In particular, the more inhomogeneous the target distribution (i.e. the greater the variability of the C_i), the slower the resulting algorithm.

As a special case, if the target distribution is $N(0, \Sigma)$ for some d-dimensional covariance matrix Σ, and the increment distribution is of the form $N(0, \Sigma_p)$, then by change of basis this is equivalent to the case of proposal increment $N(0, I_d)$ and target distribution $N(0, \Sigma \Sigma_p^{-1})$. In the corresponding eigenbasis, this target distribution is of the form (Equation 4.3) where now $C_i = \sqrt{\lambda_i}$, with $\{\lambda_i\}_{i=1}^{d}$ the eigenvalues of the matrix $\Sigma \Sigma_p^{-1}$. For large d, this approximately corresponds to the case where the $\{C_i\}$ are random with $E(C_i) = \frac{1}{d} \sum_{j=1}^{d} \sqrt{\lambda_j}$ and $E(C_i^2) = \frac{1}{d} \sum_{j=1}^{d} \lambda_j$. The inhomogeneity factor b then becomes

$$b \equiv \frac{E(C_i^2)}{(E(C_i))^2} \approx \frac{\frac{1}{d} \sum_{j=1}^{d} \lambda_j}{\left(\frac{1}{d} \sum_{j=1}^{d} \sqrt{\lambda_j}\right)^2} = d \frac{\sum_{j=1}^{d} \lambda_j}{\left(\sum_{j=1}^{d} \sqrt{\lambda_j}\right)^2}, \tag{4.4}$$

with $\{\lambda_j\}$ the eigenvalues of $\Sigma \Sigma_p^{-1}$. This expression is maximized when the $\{\lambda_j\}$ are constant, that is, when $\Sigma \Sigma_p^{-1}$ is a multiple of the identity, or in other words, when Σ_p is proportional to Σ.

We conclude that with increment distribution $N(0, \Sigma_p)$, and target distribution $N(0, \Sigma)$, it is best if Σ_p is approximately proportional to Σ, that is, $\Sigma_p \approx k\,\Sigma$ for some $k > 0$. If not, this will lead to additional slowdown by the factor b.

Once we fix $\Sigma_p = k\,\Sigma$, then we can apply the original result of Roberts et al., to conclude that the optimal constant k is then $(2.38)^2/d$. That is, it is optimal to have

$$\Sigma_p = \left[\frac{(2.38)^2}{d} \right] \Sigma. \tag{4.5}$$

In a related direction, Bédard (2007, 2008a,b; see also Bédard and Rosenthal, 2008) considered the case where the target distribution π has independent coordinates with vastly different scalings (i.e. different powers of d as $d \to \infty$). She proved that if each individual component is dominated by the sum of all components, then the optimal acceptance rate of 0.234 still holds. In cases where one component is comparable to the sum, the optimal acceptance rate is in general *less* (not more!) than 0.234. Sherlock (2006) did explicit finite-dimensional computations for the case of normal target distributions, and came to similar conclusions.

4.2.4 Metropolis-Adjusted Langevin Algorithm

Finally, Roberts and Tweedie (1996) and Roberts and Rosenthal (1998) considered the more sophisticated *Metropolis-Adjusted Langevin algorithm* (MALA). This algorithm is similar to RWM, except that the proposal increment distribution $Z_i \sim N(0, \sigma^2 I_d)$ is replaced by

$$Z_i \sim N\left(\frac{\sigma^2}{2} \nabla \log \pi(X_n), \sigma^2 I_d \right).$$

Here the extra term $\frac{\sigma^2}{2} \nabla \log \pi(X_n)$, corresponding to the discrete-time approximation to the continuous-time Langevin diffusion for π, is an attempt to move in the direction in which the (smooth) target density π is increasing.

Roberts and Rosenthal (1998) proved that in this case, under the same i.i.d. target assumption (Equation 4.2), a similar optimal scaling result holds. This time the scaling is $\sigma = \ell/d^{1/6}$ (as opposed to $\ell/\sqrt{d}$), and the optimal value ℓ_{opt} has the optimal asymptotic acceptance rate $A(\ell_{\text{opt}}) = 0.574$ (as opposed to 0.234).

This proves that the optimal proposal scaling σ and the acceptance rate are both significantly larger for MALA than for RWM, indicating that MALA an improved algorithm with faster convergence. The catch, of course, is that the gradient of π must be computed at each new state reached, which could be difficult and/or time-consuming. Thus, RWM is much more popular than MALA in practice.

4.2.5 Numerical Examples

Here we consider some simple numerical examples in dimension $d = 10$. In each case, the target density π is that of a ten-dimensional normal with some covariance matrix Σ, and we consider various forms of the RMW algorithm.

4.2.5.1 *Off-Diagonal Covariance*

Let M be the $d \times d$ matrix having diagonal elements 1, and off-diagonal elements given by the product of the row and column number divided by d^2, that is, $m_{ii} = 1$, and $m_{ij} = ij/d^2$ for $j \neq i$. Then let $\Sigma^{-1} = M^2$ (since M is symmetric, Σ is positive-definite), and let the target density π be that of $N(0, \Sigma)$. (Equivalently, π is such that $\mathbf{X} \sim \pi$ if $\mathbf{X} = MZ$, where Z is a 10-tuple of i.i.d. univariate standard normals.)

We compute numerically that the top-left entry of Σ is equal to 1.0305. So, if h is the functional equal to the square of the first coordinate, then in stationarity the mean value of h should be 1.0305.

We consider an RWM algorithm for this target $\pi(\cdot)$, with initial value $X_0 = (1, 0, 0, \ldots, 0)$, and with increment distribution given by $N(0, \sigma^2 I_d)$ for various choices of σ. For each choice of σ, we run the algorithm for 100,000 iterations, and average all the values of the square of the first coordinate to estimate its stationary mean. We repeat this 10 times for each σ, to compute a sample standard error (over the 10 independent runs) and a root mean squared error (RMSE) for each choice of σ. Our results are as follows:

σ	Mean Acc. Rate	Estimate	RMSE
0.1	0.836	0.992 ± 0.066	0.074
0.7	0.230	1.032 ± 0.019	0.018
3.0	0.002	1.000 ± 0.083	0.085

We see from this table that the value $\sigma = 0.1$ is too small, leading to an overly high acceptance rate (83.6%), a poor estimate (0.992) of the mean functional value with large standard error (0.066) and large RMSE (0.074). Similarly, the value $\sigma = 3.0$ is too high, leading to an overly low acceptance rate (0.2%), a poor estimate (1.000) of the mean functional value with large standard error (0.083) and large RMSE (0.085). On the other hand, the value $\sigma = 0.7$ is just right, leading to a nearly optimal acceptance rate (23.0%), a good estimate (1.032) of the mean functional value with smaller standard error (0.019) and smaller RMSE (0.085).

This confirms that, when scaling the increment covariance as σI_d, it is optimal to find σ to make the acceptance rate close to 0.234.

4.2.5.2 *Inhomogeneous Covariance*

To consider the effect of nondiagonal proposal increments, we again consider a case where the target density π is that of $N(0, \Sigma)$, again in dimension $d = 10$, but now we take $\Sigma = \text{diag}(1^2, 2^2, 3^2, \ldots, 10^2)$. Thus, the individual covariances are now highly variable. Since the last coordinate now has the highest variance and is thus most "interesting," we consider the functional given by the square of the last coordinate. So, the functional's true mean is now 100. We again start the algorithms with the initial value $X_0 = (1, 0, 0, \ldots, 0)$.

We first consider a usual RWM algorithm, with proposal increment distribution $N(0, \sigma^2 I_d)$, with $\sigma = 0.7$ chosen to get an acceptance rate close to the optimal value of 0.234. The result (again upon running the algorithm for 100,000 iterations, repeated 10 times to compute a sample standard error) is as follows:

σ	Mean Acc. Rate	Estimate	RMSE
0.7	0.230	114.8 ± 28.2	30.5

We thus see that, even though σ was well chosen, the resulting algorithm still converges poorly, leading to a poor estimate (114.8) with large standard error (28.2) and large RMSE (30.5).

Next we consider running the modified algorithm where now the increment proposal is equal to $N(0, \sigma^2 \Sigma)$ where Σ is the target covariance matrix as above, but otherwise the run is identical. In this case, we find the following:

σ	Mean Acc. Rate	Estimate	RMSE
0.7	0.294	100.25 ± 1.91	1.83

Comparing the two tables, we can see that the improvement from using an increment proposal covariance proportional to the target covariance (rather than the identity matrix) is very dramatic. The estimate (100.25) is much closer to the true value (100), with much smaller standard error (1.91) and much smaller RMSE (1.83). (Furthermore, the second simulation was simply run with σ = 0.7 as in the first simulation, leading to slightly too large an acceptance rate, so a slightly larger σ would make it even better.) This confirms, as shown by Roberts and Rosenthal (2001), that when running a Metropolis algorithm, it is much better to use increment proposals which mimic the covariance of the target distribution if at all possible.

Of course, in general the target covariance matrix will not be known, and it is not at all clear (especially in high dimensions) how one could arrange for proposal increment covariances to mimic the target covariance. One promising solution is adaptive MCMC, discussed in the next section. In particular, Section 4.3.2 considers the adaptive Metropolis algorithm and shows how it can successfully mimic the target covariance without any *a priori* knowledge about it, even in hundreds of dimensions.

4.2.6 Frequently Asked Questions

Isn't a larger acceptance rate always preferable?

No. For RWM, if the acceptance rate is close to 1, this means the proposal increments are so small that the algorithm is highly inefficient despite all the acceptances.

Is it essential that the acceptance rate be exactly 0.234?

No. As shown in Figure 4.5, the algorithm's efficiency remains high whenever the acceptance rate is between about 0.1 and 0.6.

Are these asymptotic results relevant to finite-dimensional problems?

Yes. While the theorems are only proven as $d \to \infty$, it appears that in many cases the asymptotics approximately apply whenever $d \geq 5$, so the infinite-dimensional results are good approximations to finite-dimensional situations.

Do these results hold for all target distributions?

No. They are only proved for very special cases involving independent target components. However, within that class they appear to be fairly robust (albeit sometimes with an even *lower* optimal acceptance rate than 0.234), and simulations seem to suggest that they approximately hold in other cases too. Furthermore, by change of basis, the results apply to all normal target distributions, too. And the general principle that the scaling should be neither too large nor too small applies much more generally, to virtually all "local" MCMC algorithms.

Do these results hold for multimodal distributions?

In principle, yes, at least for distributions with independent (though perhaps multimodal) components. However, the asymptotic acceptance rate is by definition the acceptance rate with respect to the *entire* target distribution. So, if a sampler is stuck in just one mode, it may misrepresent the asymptotic acceptance rate, leading to an incorrect estimate of the asymptotic acceptance rate, and a misapplication of the theorem.

In high dimensions, is the proposal scaling parameter σ the only quantity of interest?

No. The entire proposal distribution is of interest. In particular, it is best if the covariance of the proposal increment distribution mimics the covariance of the target distribution as much as possible. However, often significant gains can be realized simply by optimizing σ according to the theorems.

Doesn't optimality depend on which criterion is used?

Yes, in general, but these asymptotic diffusion results are valid for *any* optimality measure. That is because in the limit the processes each represent precisely the same *diffusion*, just scaled with a different speed factor. So, running a suboptimal algorithm for n steps is precisely equivalent (in the limit) to running the optimal algorithm for m steps, where $m < n$. In other words, with a suboptimal algorithm you have to run for longer to achieve precisely the same result, which is less efficient by any sensible efficiency measure at all, including all of those in Section 4.1.4.

Do these results hold for, say, Metropolis-within-Gibbs algorithms?

No, since they are proved for full-dimensional Metropolis updates only. Indeed, the Metropolis-within-Gibbs algorithm involves updating just one coordinate at a time, and thus essentially corresponds to the case $d = 1$. In that case, it appears that the optimal acceptance rate is usually closer to 0.44 than 0.234.

Isn't it too restrictive to scale σ specifically as $O(d^{-1/2})$ for RWM, or $O(d^{-1/6})$ for MALA? Wouldn't other scalings lead to other optimality results?

No, a smaller scaling would correspond to letting $\ell \to 0$, while a larger scaling would correspond to letting $\ell \to \infty$, either of which would lead to an asymptotically zero-efficiency algorithm (*cf.* Figure 4.5). The $O(d^{-1/2})$ or $O(d^{-1/6})$ scaling is the only one that leads to a nonzero limit, and thus the only scaling leading to optimality as $d \to \infty$.

4.3   Adaptive MCMC

Even if we have some idea of what criteria make an MCMC algorithm optimal, this still leaves the question of how to *find* this optimum, that is, how to run a Markov chain with (approximately) optimal characteristics. For example, even if we are convinced that an acceptance rate of 0.234 is optimal, how do we find the appropriate proposal scaling to achieve this?

One method, commonly used, is trial and error: if the acceptance rate seems too high, then we reduce the proposal scaling σ and try again (or if it seems too low, then we increase the scaling). This method is often successful, but it is generally time-consuming, requiring repeated manual intervention by the user. Furthermore, such a method cannot hope to find more complicated improvements, for example making the proposal covariance matrix Σ_p approximately proportional to the (unknown) target covariance matrix Σ as in Equation 4.5 (which requires choosing $d(d-1)/2$ separate covariance matrix entries). It is possible to use more refined versions of this, for example with increasing trial run lengths to efficiently zero in on good proposal scale and shape values (Pasarica and Gelman, 2010), but this is still not sufficient in difficult high-dimensional problems.

As an alternative, we consider algorithms which themselves try to improve the Markov chain. Specifically, let $\{P_\gamma\}_{\gamma \in \mathcal{Y}}$ be a family of Markov chain kernels, each having the same stationary distribution π. Let Γ_n be the chosen kernel choice at the nth iteration, so

$$\Pr(X_{n+1} \in A \mid X_n = x, \Gamma_n = \gamma, X_{n-1}, \ldots, X_0, \Gamma_{n-1}, \ldots, \Gamma_0) = P_\gamma(x, A),$$

for $n = 0, 1, 2, \ldots$. Here the $\{\Gamma_n\}$ are updated according to some adaptive updating algorithm. In principle, the choice of Γ_n could depend on the entire history $X_{n-1}, \ldots, X_0, \Gamma_{n-1}, \ldots, \Gamma_0$, though in practice it is often the case that the pairs process $\{(X_n, \Gamma_n)\}_{n=0}^{\infty}$ is Markovian. In general the algorithms are quite easy to implement, requiring only moderate amounts of extra computer programming—and there are even some efforts at generic adaptive software, such as Rosenthal (2007).

Whether such an adaptive scheme will improve convergence depends, obviously, on the adaptive algorithm selected. An even more fundamental question, which we now consider, is whether the adaptation might *destroy* convergence.

4.3.1 Ergodicity of Adaptive MCMC

One might think that, as long as each individual Markov chain P_γ converges to π, any adaptive mixture of the chains must also converge to π. However, this is not the case. For a simple counterexample (illustrated interactively by Rosenthal, 2004; see also Atchadé and Rosenthal, 2005; Roberts and Rosenthal, 2007), let $\mathcal{Y} = \{1, 2\}$, let $\mathcal{X} = \{1, 2, 3, 4\}$, let $\pi(1) = \pi(3) = \pi(4) = 0.333$ and $\pi(2) = 0.001$. Let each P_γ be an RWM algorithm, with proposal $Y_{n+1} \sim U\{X_n - 1, X_n + 1\}$ for P_1, or $Y_{n+1} \sim U\{X_n - 2, X_n - 1, X_n + 1, X_n + 2\}$ for P_2. (Of course, any proposed moves out of $\mathcal{X}$ are always rejected, i.e. $\pi(x) = 0$ for $x \notin \mathcal{X}$.) Define the adaptation by saying that $\Gamma_{n+1} = 2$ if the nth proposal was accepted, otherwise $\Gamma_{n+1} = 1$. Then each P_γ is reversible with respect to π. However, the adaptive algorithm can get "stuck" with $X_n = \Gamma_n = 1$ for long stretches (and only escape with probability $0.001/0.333$), so the limiting distribution of X_n is weighted too heavily toward 1 (and too lightly toward 3 and 4).

In light of such counterexamples, it is important to have sufficient conditions to guarantee convergence in distribution of $\{X_n\}$ to π. In recent years, a number of authors (Haario et al., 2001; Atchadé and Rosenthal, 2005; Andrieu and Moulines, 2006; Giordani and Kohn, 2006; Andrieu and Atchadé, 2007; Roberts and Rosenthal, 2007) have proved ergodicity of adaptive MCMC under various assumptions.

In particular, Roberts and Rosenthal (2007) proved that $\lim_{n\to\infty} \sup_{A\subseteq\mathcal{X}} \|\Pr(X_n \in A) - \pi(A)\| = 0$ (asymptotic convergence), and also $\lim_{n\to\infty} \frac{1}{n}\sum_{i=1}^{n} g(X_i) = \pi(g)$ for all bounded $g : \mathcal{X} \to \mathbf{R}$ (WLLN), assuming only the *diminishing* (a.k.a. *vanishing*) *adaptation* condition

$$\lim_{n\to\infty} \sup_{x\in\mathcal{X}} \|P_{\Gamma_{n+1}}(x,\cdot) - P_{\Gamma_n}(x,\cdot)\| = 0 \quad \text{in probability,} \tag{4.6}$$

and also the *containment* (a.k.a. *bounded convergence*) condition

$$\{M_\epsilon(X_n, \Gamma_n)\}_{n=0}^{\infty} \text{ is bounded in probability,} \quad \epsilon > 0, \tag{4.7}$$

where $M_\epsilon(x, \gamma) = \inf\{n \geq 1 : \|P_\gamma^n(x,\cdot) - \pi(\cdot)\| \leq \epsilon\}$ is the convergence time of the kernel P_γ when beginning in state $x \in \mathcal{X}$.

Now, Equation 4.7 is a technical condition which is satisfied for virtually all reasonable adaptive schemes. For example, it holds whenever $\mathcal{X} \times \mathcal{Y}$ is finite, or is compact in some topology in which either the transition kernels P_γ, or the Metropolis–Hastings proposal kernels Q_γ, have jointly continuous densities. It also holds for adaptive RWM and Metropolis-within-Gibbs algorithms under very general conditions (Bai et al., 2008). (It is, however, possible to construct pathological counterexamples, where containment does not hold; see Yang, 2008b and Bai et al., 2008.) So, in practice, the requirement (Equation 4.7) can be largely ignored.

By contrast, condition (Equation 4.6) is more fundamental. It requires that the amount of adapting at the nth iteration goes to 0 as $n \to \infty$. (Note that the *sum* of the adaptations can still be infinite, i.e. an infinite total amount of adaptation is still permissible, and it is not necessarily required that the adaptive parameters $\{\Gamma_n\}$ converge to some fixed value.) Since the user can choose the adaptive updating scheme, Equation 4.6 can be ensured directly through careful planning. For example, if the algorithm adapts at the nth iteration only with probability $p(n)$, then Equation 4.6 is automatically satisfied if $p(n) \to 0$. Alternatively, if the choice of γ depends on an empirical average over iterations 1 through n, then the influence of the nth iteration is just $O(1/n)$ and hence goes to 0.

Such results allow us to update our parameters $\{\Gamma_n\}$ in virtually any manner we wish, so long as (Equation 4.6) holds. So, what adaptations are beneficial?

4.3.2 Adaptive Metropolis

The first important modern use of adaptive MCMC was the adaptive Metropolis (AM) algorithm of Haario et al. (2001). This algorithm is motivated by the observation (Equation 4.5) that, for RWM in $\mathbf{R}^d$, at least with normal target distributions, it is optimal to have a proposal covariance matrix of the form $(2.38)^2/d$ times the target covariance matrix Σ. Since Σ is in general unknown, it is estimated by Σ_n, the empirical covariance matrix of $X_0, \ldots, X_n$.

Thus, the AM algorithm essentially uses a proposal distribution for the nth iteration given by

$$Y_{n+1} \sim N\left(X_n, \left[\frac{(2.38)^2}{d}\right]\Sigma_n\right).$$

To ensure that the proposal covariances do not simply collapse to 0 (which could violate (Equation 4.7)), Haario et al. (2001) added ϵI_d to Σ_n at each iteration, for some small

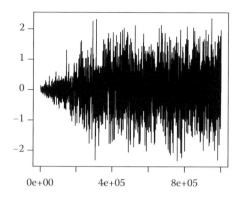

FIGURE 4.6
Trace plot of first coordinate of AM in dimension 100.

$\epsilon > 0$. Another possibility (Roberts and Rosenthal, 2009) is to instead let the proposal be a mixture distribution of the form

$$(1 - \beta)N\left(X_n, \left[\frac{(2.38)^2}{d}\right]\Sigma_n\right) + \beta N\left(X_n, \Sigma_0\right)$$

for some $0 < \beta < 1$ and some fixed nonsingular matrix Σ_0 (e.g. $\Sigma_0 = [(0.1)^2/d]\,I_d$). (With either version, it is necessary to use some alternative fixed proposal distribution for the first few iterations when the empirical covariance Σ_n is not yet well defined.)

Since empirical estimates change at the nth iteration by only $O(1/n)$, it follows that the diminishing adaptation condition (Equation 4.6) will be satisfied. Furthermore, the containment condition (Equation 4.7) will certainly be satisfied if one restricts to compact regions (Haario et al., 2001; Roberts and Rosenthal, 2009), and in fact containment still holds provided the target density π decays at least polynomially in each coordinate, a very mild assumption (Bai et al., 2008). So, AM is indeed a valid sampling algorithm.

Computer simulations (Roberts and Rosenthal, 2009) demonstrate that this AM algorithm will indeed "learn" the target covariance matrix, and approach an optimal algorithm, even in very high dimensions. While it may take many iterations before the adaptation significantly improves the algorithm, in the end it will converge considerably faster than a nonadapted RWM algorithm. For an AM run in dimension $d = 100$ (where the target was a normal distribution with an irregular and highly skewed covariance matrix), Figure 4.6 shows a trace plot of the first coordinate and Figure 4.7 a graph of the inhomogeneity factor b in Equation 4.4. These figures show that the run initially underestimates the variability of the first coordinate, which would lead to drastically incorrect estimates. However, after about 250,000 iterations, the algorithm has "found" a good proposal increment covariance matrix, so that b gets close to 1, and the trace plot correctly finds the true variability of the first coordinate. Such adaptation could never have been done manually, because of the large dimension, but the computer eventually finds a good algorithm. This shows the potential of adaptive MCMC to find good algorithms that cannot be found by hand.

4.3.3 Adaptive Metropolis-within-Gibbs

A standard alternative to the usual full-dimensional Metropolis algorithm is the "Metropolis-within-Gibbs" algorithm (arguably a misnomer, since the original work of

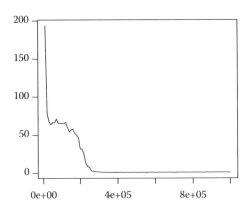

FIGURE 4.7
Trace plot of inhomogeneity factor b for AM in dimension 100.

Metropolis et al., 1953, corresponded to what we now call Metropolis-within-Gibbs). Here the variables are updated one at a time (in either systematic or random order), each using a Metropolis algorithm with a one-dimensional proposal.

To be specific, suppose that the ith coordinate is updated using a proposal increment distribution $N(0, e^{2\,ls_i})$, so ls_i is the log of the standard deviation of the increment. Obviously, we would like to find optimal values of the ls_i, which may of course be different for the different variables. We even have a rule of thumb from the end of Section 4.2.3, that each ls_i should be chosen so that the acceptance rate is approximately 0.44. However, even with this information, it is very difficult (if not impossible) in high dimensions to optimize each ls_i manually. Instead, an adaptive algorithm might be used.

One way (Roberts and Rosenthal, 2009) to adapt the ls_i values is to break up the run into "batches" of, say, 50 iterations each. After the nth batch, we update each ls_i by adding or subtracting an adaptation amount $\delta(n)$. The adapting attempts to make the acceptance rate of proposals for variable i as close as possible to 0.44. Specifically, we increase ls_i by $\delta(n)$ if the fraction of acceptances of variable i was more than 0.44 on the nth batch, or decrease ls_i by $\delta(n)$ if it was less. (A related componentwise adaptive scaling method, a one-dimensional analog of the original AM algorithm of Haario et al., 2001, is presented in Haario et al., 2005.)

To satisfy condition (Equation 4.6) we require $\delta(n) \to 0$; for example, we might take $\delta(n) = \min(0.01, n^{-1/2})$. As for Equation 4.7, it is easily seen to be satisfied if we restrict each ls_i to a finite interval $[-M, M]$. However, even this is not necessary, since it is proved by Bai et al. (2008) that Equation 4.7 is always satisfied for this algorithm, provided only that the target density π decreases at least polynomially in each direction (a very mild condition). Hence, the restriction (Equation 4.7) is once again not of practical concern.

Simulations (Roberts and Rosenthal, 2009) indicate that this adaptive Metropolis-within-Gibbs algorithm does a good job of correctly scaling the ls_i values, even in dimensions as high as 500, leading to chains which mix much faster than those with pre-chosen proposal scalings. The algorithm has recently been applied successfully to high-dimensional inference for statistical genetics (Turro et al., 2007). We believe it will be applied to many more sampling problems in the near future. Preliminary general-purpose software to implement this algorithm is now available (Rosenthal, 2007).

4.3.4 State-Dependent Proposal Scalings

Another approach involves letting the proposal scaling depend on the current state X_n, so that, for example, given $X_n = x$, we might propose $Y_{n+1} \sim N(x, \sigma_x^2)$. In this case, the acceptance probability (Equation 4.1) becomes

$$\alpha(x, y) = \min\left[1, \frac{\pi(y)}{\pi(x)}\left(\frac{\sigma_x}{\sigma_y}\right)^d \exp\left(-\frac{1}{2}(x-y)^2(\sigma_y^{-2} - \sigma_x^{-2})\right)\right]. \qquad (4.8)$$

The functional form of σ_x can be chosen and adapted in various ways to attempt to achieve efficient convergence.

For example, in many problems the target distribution becomes more spread out as we move farther from the origin. In that case, it might be appropriate to let, say, $\sigma_x = e^a(1 + |x|)^b$, where a and b are determined adaptively. For example, we could again divide the run into batches of 50 iterations as in the previous subsection. After each iteration, the algorithm updates a by adding or subtracting $\delta(n)$ in an effort to make the acceptance rate as close as possible to, for example, 0.234 or 0.44. The algorithm also adds or subtracts $\delta(n)$ to b in an effort to equalize the acceptance rates in the two regions $\{x \in \mathcal{X} : |x| > C\}$ and $\{x \in \mathcal{X} : |x| \leq C\}$ for some fixed C.

Once again, condition (Equation 4.6) is satisfied provided $\delta(n) \to 0$, and (Equation 4.7) is satisfied under very mild conditions. So, this provides a convenient way to give a useful functional form to σ_x, without knowing in advance what values of a and b might be appropriate. Simulations (Roberts and Rosenthal, 2009) indicate that this adaptive algorithm works well, at least in simple examples.

Another approach, sometimes called the regional adaptive Metropolis algorithm (RAMA), use a finite partition of the state space: $\mathcal{X} = \mathcal{X}_1 \,\dot\cup\, \ldots \,\dot\cup\, \mathcal{X}_m$. The proposal scaling is then given by $\sigma_x = e^{a_i}$ whenever $x \in \mathcal{X}_i$, with the acceptance probability (Equation 4.8) computed accordingly. Each of the values a_i is again adapted after each batch of iterations, by adding or subtracting $\delta(n)$ in an attempt to make the acceptance fraction of proposals from $\mathcal{X}_i$ close to 0.234. (As a special case, if there were no visits to $\mathcal{X}_i$ during the batch, then we always *add* $\delta(n)$ to a_i, to avoid the problem of a_i becoming so low that proposed moves to $\mathcal{X}_i$ are never accepted.) Once again, the algorithm will be valid under very mild conditions provided $\delta(n) \to 0$.

Recent work of Craiu et al. (2009) considers certain modifications of RAMA, in which multiple copies of the algorithm are run simultaneously in an effort to be sure to "learn" about *all* modes rather than getting stuck in a single mode. Their work also allows the proposal distribution to be a weighted mixture of the different $N(x, e^{2a_i})$, to allow for the possibility that the partition $\{\mathcal{X}_i\}$ was imperfectly chosen. It appears that such greater flexibility will allow for wider applicability of RAMA-type algorithms.

Of course, Langevin (MALA) algorithms may also be regarded as a type of state-dependent scaling, and it is possible to study adaptive versions of MALA as well (Atchadé, 2006).

4.3.5 Limit Theorems

Many applications of MCMC make use of such Markov chain limit theorems as the weak law of large numbers (WLLN), strong law of large numbers (SLLN), and central limit theorem (CLT), in order to guarantee good asymptotic estimates and estimate standard errors (see, e.g. Tierney, 1994; Jones and Hobert, 2001; Hobert et al., 2002; Jones, 2004; Roberts and

Rosenthal, 2004). So, it is natural to ask if such limit theorems hold for adaptive MCMC as well.

Under the assumptions of diminishing adaptation and containment, the WLLN does hold for all *bounded* functionals (Roberts and Rosenthal, 2007, Theorem 23). So, this at least means that when using adaptive MCMC for estimating means of bounded functionals, one will obtain an accurate answer with high probability if the run is sufficiently long.

For unbounded functionals, the WLLN *usually* still holds, but not always (Yang, 2008a, Theorem 2.1). Even for bounded functionals, the SLLN may not hold (Roberts and Rosenthal, 2007, Example 24), and that same example shows that a CLT might not hold as well. So, this suggests that the usual estimation of MCMC standard errors may be more challenging for adaptive MCMC if we assume only diminishing adaptation and containment.

Under stronger assumptions, more can be said. For example, Andrieu and Moulines (2006; see also Andrieu and Atchadé, 2007; Atchadé, 2007) prove various limit theorems (including CLTs) for adaptive MCMC algorithms, assuming that the adaptive parameters converge to fixed values sufficiently quickly. They also prove that such adaptive algorithms will inherit many of the asymptotic optimality properties of the corresponding fixed-parameter algorithms. Such results facilitate further applications of adaptive MCMC; however, they require various technical conditions which may be difficult to check in practice.

4.3.6 Frequently Asked Questions

Can't I adapt my MCMC algorithm any way I like, and still preserve convergence?

No. In particular, if the diminishing adaptation condition (Equation 4.6) does not hold, then there are simple counterexamples showing that adaptive MCMC can converge to the wrong answer, even though each individual Markov chain kernel would correctly converge to π.

Do I have to learn lots of technical conditions before I can apply adaptive MCMC?

Not really. As long as you satisfy diminishing adaptation (Equation 4.6), which is important but quite intuitive, then your algorithm will probably be asymptotically valid.

Have adaptive MCMC algorithms actually been used to speed up convergence on high-dimensional problems?

Yes, they have. Simulations on test problems involving hundreds of dimensions have been quite successful (Roberts and Rosenthal, 2009), and adaptive Metropolis-within-Gibbs has also been used on statistical genetics problems (Turro et al., 2007).

Does adaptation have to be designed specifically to seek out optimal parameter values?

No. The ergodicity results presented herein do *not* require that the parameters $\{\Gamma_n\}$ converge at all, only that they satisfy (Equation 4.6) which still allows for the possibility of infinite total adaptation. However, many of the specific adaptive MCMC algorithms proposed are indeed designed to attempt to converge to specific values (e.g. to proposal scalings which give an asymptotic acceptance rate of 0.234).

Why not just do the adaptation by hand, with trial runs to determine optimal parameter values, and then a long run using these values?

Well, if you can really determine optimal parameter values from a few trial runs, then that's fine. However, in high dimensions, with many parameters to choose (e.g. a large proposal covariance matrix), it is doubtful that you can find good parameter values manually.

Suppose I just have the computer adapt for some fixed, finite amount of time, and then continue the run without further adapting. Won't that guarantee asymptotic convergence to π?

Yes, it will (provided each individual kernel P_γ is ergodic), and this is a sensible method to try. However, it may be unclear how much adaptation should be done before you stop. For example, with adaptive Metropolis in 200 dimensions, it took well over a million iterations (Roberts and Rosenthal, 2009) before a truly good proposal covariance matrix was found—and it was not clear *a priori* that it would take nearly so long.

Can I use adaptation for other types of MCMC algorithms, like the Gibbs sampler?

In principle, yes. For example, an adaptive Gibbs sampler could adapt such quantities as the order of update of coordinates (for systematic scan), or the probability weights of various coordinates (for random scan), or coordinate blockings for joint updates, or such reparameterizations as rotations and centerings and so on. Only time will tell what adaptations turn out to be useful in what contexts.

Am I restricted to the specific adaptive MCMC algorithms (adaptive Metropolis, adaptive Metropolis-within-Gibbs, RAMA, ...) presented herein?

Not at all! You can make up virtually any rules for how your Markov chain parameters $\{\Gamma_n\}$ adapt over time, as long as the adaptation diminishes, and your algorithm will probably be valid. The challenge is then to find sensible/clever adaptation rules. Hopefully more and better adaptive methods will be found in the future!

Are any other methods, besides adaptive MCMC, available to help algorithms "learn" how to converge well?

Yes, there are many. For example, particle filters (e.g. Pitt and Sheppard, 1999), population Monte Carlo (e.g. Cappé et al., 2004), and sequential Monte Carlo (e.g. Del Moral et al., 2006), can all be considered as methods which attempt to "learn" faster convergence as they go. However, the details of their implementations are rather different than the adaptive MCMC algorithms presented herein.

4.4   Conclusion

We have reviewed optimal proposal scaling results, and adaptive MCMC algorithms.

While the optimal scaling theorems are all proved under very restrictive and unrealistic assumptions (e.g. target distributions with independent coordinates), they appear to provide useful guidelines much more generally. In particular, results about asymptotic acceptance rates provide useful benchmarks for Metropolis algorithms in a wide variety of settings.

Adaptive MCMC algorithms appear to provide simple, intuitive methods of finding quickly-converging Markov chains without great effort on the part of the user—aside from the initial programming, and there is even some generic software available, such as Rosenthal, (2007). While certain conditions (notably diminishing adaptation) must be satisfied to guarantee asymptotic convergence, these conditions are generally not onerous or difficult to achieve.

Overall, we feel that these results indicate the widespread applicability of both optimal scaling and adaptive MCMC algorithms to many different MCMC settings (Roberts and Rosenthal, 2009; Turro et al., 2007), including to complicated high-dimensional distributions. We hope that many MCMC users will be guided by optimal scaling results, and experiment with adaptive algorithms, in their future applications.

References

Andrieu, C. and Atchadé, Y. F. 2007. On the efficiency of adaptive MCMC algorithms. *Electronic Communications in Probability*, 12:336–349.

Andrieu, C. and Moulines, E. 2006. On the ergodicity properties of some adaptive Markov chain Monte Carlo algorithms. *Annals of Applied Probability*, 16:1462–1505.

Atchadé, Y. F. 2006. An adaptive version for the Metropolis adjusted Langevin algorithm with a truncated drift. *Methodology and Computing in Applied Probability*, 8:235–254.

Atchadé, Y. F. 2007. A cautionary tale on the efficiency of some adaptive Monte Carlo schemes. *Annals of Applied Probability*, to appear.

Atchadé, Y. F. and Rosenthal, J. S. 2005. On adaptive Markov chain Monte Carlo algorithms. *Bernoulli*, 11:815–828.

Bai, Y., Roberts, G. O., and Rosenthal, J. S. 2008. On the containment condition for adaptive Markov chain Monte Carlo algorithms. Preprint.

Bédard, M. 2007. Weak convergence of Metropolis algorithms for non-iid target distributions. *Annals of Applied Probability*, 17:1222–1244.

Bédard, M. 2008a. Efficient sampling using Metropolis algorithms: Applications of optimal scaling results. *Journal of Computational and Graphical Statistics*, 17:1–21.

Bédard, M. 2008b. Optimal acceptance rates for Metropolis algorithms: Moving beyond 0.234. *Stochastic Processes and their Applications*, 118(12):2198–2222.

Bédard, M. and Rosenthal, J. S. 2008. Optimal scaling of Metropolis algorithms: Heading towards general target distributions. *Canadian Journal of Statistics*, 36(4):483–503.

Cappé, O., Guillin, A., Marin, J. M., and Robert, C. P. 2004. Population Monte Carlo. *Journal of Computational and Graphical Statistics*, 13:907–930.

Craiu, R. V., Rosenthal, J. S., and Yang, C. 2009. Learn from thy neighbor: Parallel-chain adaptive MCMC. *Journal of the American Statistical Association*, 488:1454–1466.

Del Moral, P., Doucet, A., and Jasra, A. 2006. Sequential Monte Carlo samplers. *Journal of the Royal Statistical Society, Series B*, 68:411–436.

Gelman, A., Roberts, G. O., and Gilks, W. R. 1996. Efficient Metropolis jumping rules. In J. M. Bernardo, J. O. Berger, A. P. Dawid, and A. F. M. Smith (eds), *Bayesian Statistics 5: Proceedings of the Fifth Valencia International Meeting*, pp. 599–607. Oxford University Press, Oxford.

Giordani, P. and Kohn, R. 2006. Adaptive independent Metropolis-Hastings by fast estimation of mixtures of normals. *Journal of Computational and Graphical Statistics*, to appear.

Haario, H., Saksman, E., and Tamminen, J. 2001. An adaptive Metropolis algorithm. *Bernoulli*, 7:223–242.

Haario, H., Saksman, E., and Tamminen, J. 2005. Componentwise adaptation for high dimensional MCMC. *Computational Statistics*, 20:265–274.

Hastings, W. K. 1970. Monte Carlo sampling methods using Markov chains and their applications. *Biometrika*, 57:97–109.

Hobert, J. P., Jones, G. L., Presnell, B., and Rosenthal, J. S. 2002. On the applicability of regenerative simulation in Markov chain Monte Carlo. *Biometrika* 89:731–743.

Jones, G. L. 2004. On the Markov chain central limit theorem. *Probability Surveys*, 1:299–320.

Jones, G. L. and Hobert, J. P. 2001. Honest exploration of intractable probability distributions via Markov chain Monte Carlo. *Statistical Science*, 16(4):312–334.

Metropolis, N., Rosenbluth, A., Rosenbluth, M., Teller, A., and Teller, E. 1953. Equation of state calculations by fast computing machines. *Journal of Chemical Physics*, 21(6):1087–1092.

Mira, A. 2001. Ordering and improving the performance of Monte Carlo Markov chains. *Statistical Science*, 16:340–350.

Neal, P. J., Roberts, G. O., and Yuen, W. K. 2007. Optimal scaling of random walk Metropolis algorithms with discontinuous target densities. Preprint.

Pasarica, C. and Gelman, A. 2010. Adaptively scaling the Metropolis algorithm using expected squared jumped distance. *Statistica Sinica*, 20:343–364.

Pitt, M. K. and Shephard, N. 1999. Filtering via simulation: auxiliary particle filters. *Journal of the American Statistical Association*, 94:1403–1412.

Roberts, G. O., Gelman, A., and Gilks, W. R. 1997. Weak convergence and optimal scaling of random walk Metropolis algorithms. *Annals of Applied Probability*, 7:110–120.

Roberts, G. O. and Rosenthal, J. S. 1998. Optimal scaling of discrete approximations to Langevin diffusions. *Journal of the Royal Statistical Society, Series B*, 60:255–268.

Roberts, G. O. and Rosenthal, J. S. 2001. Optimal scaling for various Metropolis-Hastings algorithms. *Statistical Science*, 16:351–367.

Roberts, G. O. and Rosenthal, J. S. 2004. General state space Markov chains and MCMC algorithms. *Probability Surveys*, 1:20–71.

Roberts, G. O. and Rosenthal, J. S. 2007. Coupling and ergodicity of adaptive MCMC. *Journal of Applied Probability* 44:458–475.

Roberts, G. O. and Rosenthal, J. S. 2009. Examples of adaptive MCMC. *Journal of Computational and Graphical Statistics* 18(2):349–367.

Roberts, G. O. and Tweedie, R. L. 1996. Exponential convergence of Langevin diffusions and their discrete approximations. *Bernoulli*, 2:341–363.

Roberts, G. O. and Tweedie, R. L. 1999. Bounds on regeneration times and convergence rates for Markov chains. *Stochastic Processes and Their Applications*, 80: 211–229. Corrigendum (2001), 91:337–338.

Rosenthal, J. S. 1995. Minorization conditions and convergence rates for Markov chain Monte Carlo. *Journal of the American Statistical Association*, 90:558–566.

Rosenthal, J. S. 2002. Quantitative convergence rates of Markov chains: A simple account. *Electronic Communications in Probability*, 7(13):123–128.

Rosenthal, J. S. 2004. Adaptive MCMC Java applet. Available at: http://probability.ca/jeff/java/adapt.html

Rosenthal, J. S. 2007. AMCMC: An R interface for adaptive MCMC. *Computational Statistics & Data Analysis*, 51:5467–5470. (Related software available at probability.ca/amcmc.)

Sherlock, C. 2006. Methodology for inference on the Markov modulated Poisson processes and theory for optimal scaling of the random walk Metropolis. Ph.D. dissertation, Lancaster University.

Tierney, L. 1994. Markov chains for exploring posterior distributions (with discussion). *Annals of Statistics*, 22:1701–1762.

Turro, E., Bochkina, N., Hein, A. M. K., and Richardson, S. 2007. BGX: a Bioconductor package for the Bayesian integrated analysis of Affymetrix GeneChips. *BMC Bioinformatics*, 8:439–448. Available at: http://www.biomedcentral.com/1471-2105/8/439.

Yang, C. 2008a. On the weak law of large numbers for unbounded functionals for adaptive MCMC. Preprint.

Yang, C. 2008b. Recurrent and ergodic properties of adaptive MCMC. Preprint.

5

MCMC Using Hamiltonian Dynamics

Radford M. Neal

5.1 Introduction

Markov chain Monte Carlo (MCMC) originated with the classic paper of Metropolis et al. (1953), where it was used to simulate the distribution of states for a system of idealized molecules. Not long after, another approach to molecular simulation was introduced (Alder and Wainwright, 1959), in which the motion of the molecules was deterministic, following Newton's laws of motion, which have an elegant formalization as *Hamiltonian dynamics*. For finding the properties of bulk materials, these approaches are asymptotically equivalent, since even in a deterministic simulation, each local region of the material experiences effectively random influences from distant regions. Despite the large overlap in their application areas, the MCMC and molecular dynamics approaches have continued to coexist in the following decades (see Frenkel and Smit, 1996).

In 1987, a landmark paper by Duane, Kennedy, Pendleton, and Roweth united the MCMC and molecular dynamics approaches. They called their method "hybrid Monte Carlo," which abbreviates to "HMC," but the phrase "Hamiltonian Monte Carlo," retaining the abbreviation, is more specific and descriptive, and I will use it here. Duane et al. applied HMC not to molecular simulation, but to lattice field theory simulations of quantum chromodynamics. Statistical applications of HMC began with my use of it for neural network models (Neal, 1996a). I also provided a statistically-oriented tutorial on HMC in a review of MCMC methods (Neal, 1993, Chapter 5). There have been other applications of HMC to statistical problems (e.g. Ishwaran, 1999; Schmidt, 2009) and statistically-oriented reviews (e.g. Liu, 2001, Chapter 9), but HMC still seems to be underappreciated by statisticians, and perhaps also by physicists outside the lattice field theory community.

This review begins by describing Hamiltonian dynamics. Despite terminology that may be unfamiliar outside physics, the features of Hamiltonian dynamics that are needed for HMC are elementary. The differential equations of Hamiltonian dynamics must be discretized for computer implementation. The "leapfrog" scheme that is typically used is quite simple.

Following this introduction to Hamiltonian dynamics, I describe how to use it to construct an MCMC method. The first step is to define a Hamiltonian function in terms of the probability distribution we wish to sample from. In addition to the variables we are interested in (the "position" variables), we must introduce auxiliary "momentum" variables, which typically have independent Gaussian distributions. The HMC method alternates simple updates for these momentum variables with Metropolis updates in which a new state is proposed by computing a trajectory according to Hamiltonian dynamics, implemented with the leapfrog method. A state proposed in this way can be distant from the current state but nevertheless have a high probability of acceptance. This bypasses the slow exploration of the state space that occurs when Metropolis updates are done using a simple random-walk proposal distribution. (An alternative way of avoiding random walks is to use short trajectories but only partially replace the momentum variables between trajectories, so that successive trajectories tend to move in the same direction.)

After presenting the basic HMC method, I discuss practical issues of tuning the leapfrog stepsize and number of leapfrog steps, as well as theoretical results on the scaling of HMC with dimensionality. I then present a number of variations on HMC. The acceptance rate for HMC can be increased for many problems by looking at "windows" of states at the beginning and end of the trajectory. For many statistical problems, approximate computation of trajectories (e.g. using subsets of the data) may be beneficial. Tuning of HMC can be made easier using a "short-cut" in which trajectories computed with a bad choice of stepsize take little computation time. Finally, "tempering" methods may be useful when multiple isolated modes exist.

5.2 Hamiltonian Dynamics

Hamiltonian dynamics has a physical interpretation that can provide useful intuitions. In two dimensions, we can visualize the dynamics as that of a frictionless puck that slides over a surface of varying height. The state of this system consists of the *position* of the puck, given by a two-dimensional vector q, and the *momentum* of the puck (its mass times its velocity), given by a two-dimensional vector p. The *potential energy*, $U(q)$, of the puck is proportional to the height of the surface at its current position, and its *kinetic energy*, $K(p)$, is equal to $|p|^2/(2m)$, where m is the mass of the puck. On a level part of the surface, the puck moves at a constant velocity, equal to p/m. If it encounters a rising slope, the puck's momentum allows it to continue, with its kinetic energy decreasing and its potential energy increasing, until the kinetic energy (and hence p) is zero, at which point it will slide back down (with kinetic energy increasing and potential energy decreasing).

In nonphysical MCMC applications of Hamiltonian dynamics, the position will correspond to the variables of interest. The potential energy will be minus the log of the probability density for these variables. Momentum variables, one for each position variable, will be introduced artificially.

These interpretations may help motivate the exposition below, but if you find otherwise, the dynamics can also be understood as simply resulting from a certain set of differential equations.

5.2.1 Hamilton's Equations

Hamiltonian dynamics operates on a d-dimensional *position* vector, q, and a d-dimensional *momentum* vector, p, so that the full state space has $2d$ dimensions. The system is described by a function of q and p known as the *Hamiltonian*, $H(q,p)$.

5.2.1.1 Equations of Motion

The partial derivatives of the Hamiltonian determine how q and p change over time, t, according to Hamilton's equations:

$$\frac{dq_i}{dt} = \frac{\partial H}{\partial p_i}, \tag{5.1}$$

$$\frac{dp_i}{dt} = -\frac{\partial H}{\partial q_i}, \tag{5.2}$$

for $i = 1, \ldots, d$. For any time interval of duration s, these equations define a mapping, T_s, from the state at any time t to the state at time $t + s$. (Here, H, and hence T_s, are assumed to not depend on t.)

Alternatively, we can combine the vectors q and p into the vector $z = (q, p)$ with $2d$ dimensions, and write Hamilton's equations as

$$\frac{dz}{dt} = J \nabla H(z),$$

where ∇H is the gradient of H (i.e. $[\nabla H]_k = \partial H / \partial z_k$), and

$$J = \begin{bmatrix} 0_{d \times d} & I_{d \times d} \\ -I_{d \times d} & 0_{d \times d} \end{bmatrix} \tag{5.3}$$

is a $2d \times 2d$ matrix whose quadrants are defined above in terms of identity and zero matrices.

5.2.1.2 *Potential and Kinetic Energy*

For HMC we usually use Hamiltonian functions that can be written as

$$H(q, p) = U(q) + K(p). \tag{5.4}$$

Here $U(q)$ is called the *potential energy*, and will be defined to be minus the log probability density of the distribution for q that we wish to sample, plus any constant that is convenient. $K(p)$ is called the *kinetic energy*, and is usually defined as

$$K(p) = p^T M^{-1} p / 2. \tag{5.5}$$

Here M is a symmetric, positive-definite "mass matrix," which is typically diagonal, and is often a scalar multiple of the identity matrix. This form for $K(p)$ corresponds to minus the log probability density (plus a constant) of the zero-mean Gaussian distribution with covariance matrix M.

With these forms for H and K, Hamilton's equations 5.1 and 5.2 can be written as follows, for $i = 1, \ldots, d$:

$$\frac{dq_i}{dt} = [M^{-1} p]_i, \tag{5.6}$$

$$\frac{dp_i}{dt} = -\frac{\partial U}{\partial q_i}. \tag{5.7}$$

5.2.1.3   A One-Dimensional Example

Consider a simple example in one dimension (for which q and p are scalars and will be written without subscripts), in which the Hamiltonian is defined as follows:

$$H(q,p) = U(q) + K(p), \qquad U(q) = \frac{q^2}{2}, \quad K(p) = \frac{p^2}{2}. \tag{5.8}$$

As we will see later in Section 5.3.1, this corresponds to a Gaussian distribution for q with mean zero and variance one. The dynamics resulting from this Hamiltonian (following Equations 5.6 and 5.7) is

$$\frac{dq}{dt} = p, \quad \frac{dp}{dt} = -q.$$

Solutions have the following form, for some constants r and a:

$$q(t) = r \cos(a + t), \quad p(t) = -r \sin(a + t). \tag{5.9}$$

Hence, the mapping T_s is a rotation by s radians clockwise around the origin in the (q, p) plane. In higher dimensions, Hamiltonian dynamics generally does not have such a simple periodic form, but this example does illustrate some important properties that we will look at next.

5.2.2   Properties of Hamiltonian Dynamics

Several properties of Hamiltonian dynamics are crucial to its use in constructing MCMC updates.

5.2.2.1   Reversibility

First, Hamiltonian dynamics is *reversible*—the mapping T_s from the state at time t, $(q(t), p(t))$, to the state at time $t + s$, $(q(t + s), p(t + s))$, is one-to-one, and hence has an inverse, T_{-s}. This inverse mapping is obtained by simply negating the time derivatives in Equations 5.1 and 5.2. When the Hamiltonian has the form in Equation 5.4, and $K(p) = K(-p)$, as in the quadratic form for the kinetic energy of Equation 5.5, the inverse mapping can also be obtained by negating p, applying T_s, and then negating p again.

In the simple one-dimensional example of Equation 5.8, T_{-s} is just a counterclockwise rotation by s radians, undoing the clockwise rotation of T_s.

The reversibility of Hamiltonian dynamics is important for showing that MCMC updates that use the dynamics leave the desired distribution invariant, since this is most easily proved by showing reversibility of the Markov chain transitions, which requires reversibility of the dynamics used to propose a state.

5.2.2.2   Conservation of the Hamiltonian

A second property of the dynamics is that it *keeps the Hamiltonian invariant* (i.e. conserved). This is easily seen from Equations 5.1 and 5.2 as follows:

$$\frac{dH}{dt} = \sum_{i=1}^{d} \left[\frac{dq_i}{dt} \frac{\partial H}{\partial q_i} + \frac{dp_i}{dt} \frac{\partial H}{\partial p_i} \right] = \sum_{i=1}^{d} \left[\frac{\partial H}{\partial p_i} \frac{\partial H}{\partial q_i} - \frac{\partial H}{\partial q_i} \frac{\partial H}{\partial p_i} \right] = 0. \tag{5.10}$$

With the Hamiltonian of Equation 5.8, the value of the Hamiltonian is half the squared distance from the origin, and the solutions (Equation 5.9) stay at a constant distance from the origin, keeping H constant.

For Metropolis updates using a proposal found by Hamiltonian dynamics, which form part of the HMC method, the acceptance probability is one if H is kept invariant. We will see later, however, that in practice we can only make H approximately invariant, and hence we will not quite be able to achieve this.

5.2.2.3 *Volume Preservation*

A third fundamental property of Hamiltonian dynamics is that it *preserves volume* in (q, p) space (a result known as Liouville's theorem). If we apply the mapping T_s to the points in some region R of (q, p) space, with volume V, the image of R under T_s will also have volume V.

With the Hamiltonian of Equation 5.8, the solutions (Equation 5.9) are rotations, which obviously do not change the volume. Such rotations also do not change the shape of a region, but this is not so in general—Hamiltonian dynamics might stretch a region in one direction, as long as the region is squashed in some other direction so as to preserve volume.

The significance of volume preservation for MCMC is that we need not account for any change in volume in the acceptance probability for Metropolis updates. If we proposed new states using some arbitrary, non-Hamiltonian, dynamics, we would need to compute the determinant of the Jacobian matrix for the mapping the dynamics defines, which might well be infeasible.

The preservation of volume by Hamiltonian dynamics can be proved in several ways. One is to note that the divergence of the vector field defined by Equations 5.1 and 5.2 is zero, which can be seen as follows:

$$\sum_{i=1}^{d} \left[\frac{\partial}{\partial q_i} \frac{dq_i}{dt} + \frac{\partial}{\partial p_i} \frac{dp_i}{dt} \right] = \sum_{i=1}^{d} \left[\frac{\partial}{\partial q_i} \frac{\partial H}{\partial p_i} - \frac{\partial}{\partial p_i} \frac{\partial H}{\partial q_i} \right] = \sum_{i=1}^{d} \left[\frac{\partial^2 H}{\partial q_i \partial p_i} - \frac{\partial^2 H}{\partial p_i \partial q_i} \right] = 0.$$

A vector field with zero divergence can be shown to preserve volume (Arnold, 1989).

Here, I will show informally that Hamiltonian dynamics preserves volume more directly, without presuming this property of the divergence. I will, however, take as given that volume preservation is equivalent to the determinant of the Jacobian matrix of T_s having absolute value one, which is related to the well-known role of this determinant in regard to the effect of transformations on definite integrals and on probability density functions.

The $2d \times 2d$ Jacobian matrix of T_s, seen as a mapping of $z = (q, p)$, will be written as B_s. In general, B_s will depend on the values of q and p before the mapping. When B_s is diagonal, it is easy to see that the absolute values of its diagonal elements are the factors by which T_s stretches or compresses a region in each dimension, so that the product of these factors, which is equal to the absolute value of $\det(B_s)$, is the factor by which the volume of the region changes. I will not prove the general result here, but note that if we were to (say) rotate the coordinate system used, B_s would no longer be diagonal, but the determinant of B_s is invariant to such transformations, and so would still give the factor by which the volume changes.

Let us first consider volume preservation for Hamiltonian dynamics in one dimension (i.e. with $d = 1$), for which we can drop the subscripts on p and q. We can approximate T_δ

for δ near zero as follows:

$$T_\delta(q,p) = \begin{bmatrix} q \\ p \end{bmatrix} + \delta \begin{bmatrix} dq/dt \\ dp/dt \end{bmatrix} + \text{terms of order } \delta^2 \text{ or higher.}$$

Taking the time derivatives from Equations 5.1 and 5.2, the Jacobian matrix can be written as

$$B_\delta = \begin{bmatrix} 1 + \delta \dfrac{\partial^2 H}{\partial q \partial p} & \delta \dfrac{\partial^2 H}{\partial p^2} \\[2ex] -\delta \dfrac{\partial^2 H}{\partial q^2} & 1 - \delta \dfrac{\partial^2 H}{\partial p \partial q} \end{bmatrix} + \text{terms of order } \delta^2 \text{ or higher.} \tag{5.11}$$

We can then write the determinant of this matrix as

$$\det(B_\delta) = 1 + \delta \frac{\partial^2 H}{\partial q \partial p} - \delta \frac{\partial^2 H}{\partial p \partial q} + \text{terms of order } \delta^2 \text{ or higher}$$

$$= 1 + \text{terms of order } \delta^2 \text{ or higher.}$$

Since $\log(1 + x) \approx x$ for x near zero, $\log \det(B_\delta)$ is zero, except perhaps for terms of order δ^2 or higher (though we will see later that it is exactly zero). Now consider $\log \det(B_s)$ for some time interval s that is not close to zero. Setting $\delta = s/n$, for some integer n, we can write T_s as the composition of T_δ applied n times (from n points along the trajectory), so $\det(B_s)$ is the n-fold product of $\det(B_\delta)$ evaluated at these points. We then find that

$$\log \det(B_s) = \sum_{i=1}^n \log \det(B_\delta)$$

$$= \sum_{i=1}^n \left\{ \text{terms of order } 1/n^2 \text{ or smaller} \right\} \tag{5.12}$$

$$= \text{terms of order } 1/n \text{ or smaller.}$$

Note that the value of B_δ in the sum in Equation 5.12 might perhaps vary with i, since the values of q and p vary along the trajectory that produces T_s. However, assuming that trajectories are not singular, the variation in B_δ must be bounded along any particular trajectory. Taking the limit as $n \to \infty$, we conclude that $\log \det(B_s) = 0$, so $\det(B_s) = 1$, and hence T_s preserves volume.

When $d > 1$, the same argument applies. The Jacobian matrix will now have the following form (compare Equation 5.11), where each entry shown below is a $d \times d$ submatrix, with rows indexed by i and columns by j:

$$B_\delta = \begin{bmatrix} I + \delta \left[\dfrac{\partial^2 H}{\partial q_j \partial p_i} \right] & \delta \left[\dfrac{\partial^2 H}{\partial p_j \partial p_i} \right] \\[3ex] -\delta \left[\dfrac{\partial^2 H}{\partial q_j \partial q_i} \right] & I - \delta \left[\dfrac{\partial^2 H}{\partial p_j \partial q_i} \right] \end{bmatrix} + \text{terms of order } \delta^2 \text{ or higher.}$$

As for $d = 1$, the determinant of this matrix will be one plus terms of order δ^2 or higher, since all the terms of order δ cancel. The remainder of the argument above then applies without change.

5.2.2.4 Symplecticness

Volume preservation is also a consequence of Hamiltonian dynamics being *symplectic*. Letting $z = (q, p)$, and defining J as in Equation 5.3, the symplecticness condition is that the Jacobian matrix, B_s, of the mapping T_s satisfies

$$B_s^T J^{-1} B_s = J^{-1}.$$

This implies volume conservation, since $\det(B_s^T) \det(J^{-1}) \det(B_s) = \det(J^{-1})$ implies that $\det(B_s)^2$ is one. When $d > 1$, the symplecticness condition is stronger than volume preservation. Hamiltonian dynamics and the symplecticness condition can be generalized to where J is any matrix for which $J^T = -J$ and $\det(J) \neq 0$.

Crucially, reversibility, preservation of volume, and symplecticness can be maintained exactly even when, as is necessary in practice, Hamiltonian dynamics is approximated, as we will see next.

5.2.3 Discretizing Hamilton's Equations—The Leapfrog Method

For computer implementation, Hamilton's equations must be approximated by discretizing time, using some small stepsize, ε. Starting with the state at time zero, we iteratively compute (approximately) the state at times $\varepsilon, 2\varepsilon, 3\varepsilon$, etc.

In discussing how to do this, I will assume that the Hamiltonian has the form $H(q, p) = U(q) + K(p)$, as in Equation 5.4. Although the methods below can be applied with any form for the kinetic energy, I assume for simplicity that $K(p) = p^T M^{-1} p / 2$, as in Equation 5.5, and furthermore that M is diagonal, with diagonal elements $m_1, \ldots, m_d$, so that

$$K(p) = \sum_{i=1}^{d} \frac{p_i^2}{2m_i}. \tag{5.13}$$

5.2.3.1 Euler's Method

Perhaps the best-known way to approximate the solution to a system of differential equations is Euler's method. For Hamilton's equations, this method performs the following steps, for each component of position and momentum, indexed by $i = 1, \ldots, d$:

$$p_i(t + \varepsilon) = p_i(t) + \varepsilon \frac{dp_i}{dt}(t) = p_i(t) - \varepsilon \frac{\partial U}{\partial q_i}(q(t)), \tag{5.14}$$

$$q_i(t + \varepsilon) = q_i(t) + \varepsilon \frac{dq_i}{dt}(t) = q_i(t) + \varepsilon \frac{p_i(t)}{m_i}. \tag{5.15}$$

The time derivatives in Equations 5.14 and 5.15 are from the form of Hamilton's equations given by Equations 5.6 and 5.7. If we start at $t = 0$ with given values for $q_i(0)$ and $p_i(0)$, we can iterate the steps above to get a trajectory of position and momentum values at times $\varepsilon, 2\varepsilon, 3\varepsilon, \ldots$, and hence find (approximate) values for $q(\tau)$ and $p(\tau)$ after τ/ε steps (assuming τ/ε is an integer).

Figure 5.1a shows the result of using Euler's method to approximate the dynamics defined by the Hamiltonian of Equation 5.8, starting from $q(0) = 0$ and $p(0) = 1$, and using a stepsize of $\varepsilon = 0.3$ for 20 steps (i.e. to $\tau = 0.3 \times 20 = 6$). The results are not good—Euler's method produces a trajectory that diverges to infinity, but the true trajectory is a circle. Using a smaller value of ε, and correspondingly more steps, produces a more accurate result at $\tau = 6$, but although the divergence to infinity is slower, it is not eliminated.

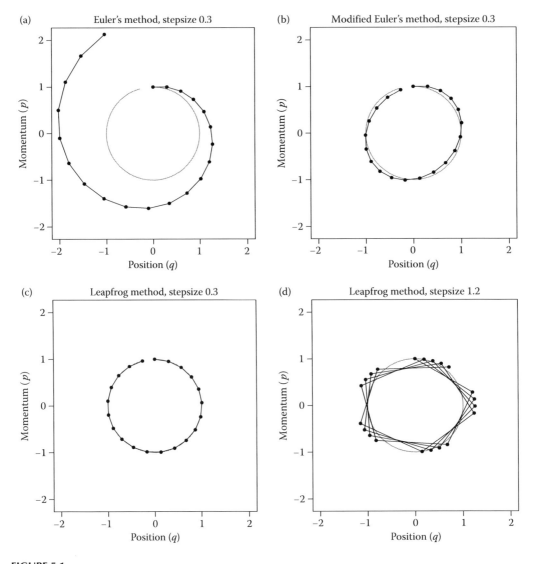

FIGURE 5.1
Results using three methods for approximating Hamiltonian dynamics, when $H(q, p) = q^2/2 + p^2/2$. The initial state was $q = 0, p = 1$. The stepsize was $\varepsilon = 0.3$ for (a), (b), and (c), and $\varepsilon = 1.2$ for (d). Twenty steps of the simulated trajectory are shown for each method, along with the true trajectory (in gray).

5.2.3.2 A Modification of Euler's Method

Much better results can be obtained by slightly modifying Euler's method, as follows:

$$p_i(t + \varepsilon) = p_i(t) - \varepsilon \frac{\partial U}{\partial q_i}(q(t)), \tag{5.16}$$

$$q_i(t + \varepsilon) = q_i(t) + \varepsilon \frac{p_i(t + \varepsilon)}{m_i}. \tag{5.17}$$

We simply use the *new* value for the momentum variables, p_i, when computing the new value for the position variables, q_i. A method with similar performance can be obtained by instead updating the q_i first and using their new values to update the p_i.

Figure 5.1b shows the results using this modification of Euler's method with $\varepsilon = 0.3$. Though not perfect, the trajectory it produces is much closer to the true trajectory than that obtained using Euler's method, with no tendency to diverge to infinity. This better performance is related to the modified method's exact preservation of volume, which helps avoid divergence to infinity or spiraling into the origin, since these would typically involve the volume expanding to infinity or contracting to zero.

To see that this modification of Euler's method preserves volume exactly despite the finite discretization of time, note that both the transformation from $(q(t), p(t))$ to $(q(t), p(t + \varepsilon))$ via Equation 5.16 and the transformation from $(q(t), p(t + \varepsilon))$ to $(q(t + \varepsilon), p(t + \varepsilon))$ via Equation 5.17 are "shear" transformations, in which only some of the variables change (either the p_i or the q_i), by amounts that depend only on the variables that do not change. Any shear transformation will preserve volume, since its Jacobian matrix will have determinant one (as the only nonzero term in the determinant will be the product of diagonal elements, which will all be one).

5.2.3.3 The Leapfrog Method

Even better results can be obtained with the *leapfrog* method, which works as follows:

$$p_i(t + \varepsilon/2) = p_i(t) - (\varepsilon/2) \frac{\partial U}{\partial q_i}(q(t)), \tag{5.18}$$

$$q_i(t + \varepsilon) = q_i(t) + \varepsilon \frac{p_i(t + \varepsilon/2)}{m_i}, \tag{5.19}$$

$$p_i(t + \varepsilon) = p_i(t + \varepsilon/2) - (\varepsilon/2) \frac{\partial U}{\partial q_i}(q(t + \varepsilon)). \tag{5.20}$$

We start with a half step for the momentum variables, then do a full step for the position variables, using the new values of the momentum variables, and finally do another half step for the momentum variables, using the new values for the position variables. An analogous scheme can be used with any kinetic energy function, with $\partial K / \partial p_i$ replacing p_i / m_i above.

When we apply Equations 5.18 through 5.20 a second time to go from time $t + \varepsilon$ to $t + 2\varepsilon$, we can combine the last half step of the first update, from $p_i(t + \varepsilon/2)$ to $p_i(t + \varepsilon)$, with the first half step of the second update, from $p_i(t + \varepsilon)$ to $p_i(t + \varepsilon + \varepsilon/2)$. The leapfrog method then looks very similar to the modification of Euler's method in Equations 5.17 and 5.16, except that leapfrog performs half steps for momentum at the very beginning and very end of the trajectory, and the time labels of the momentum values computed are shifted by $\varepsilon/2$.

The leapfrog method preserves volume exactly, since Equations 5.18 through 5.20 are shear transformations. Due to its symmetry, it is also reversible by simply negating p, applying the same number of steps again, and then negating p again.

Figure 5.1c shows the results using the leapfrog method with a stepsize of $\varepsilon = 0.3$, which are indistinguishable from the true trajectory, at the scale of this plot. In Figure 5.1d, the results of using the leapfrog method with $\varepsilon = 1.2$ are shown (still with 20 steps, so almost four cycles are seen, rather than almost one). With this larger stepsize, the approximation error is clearly visible, but the trajectory still remains stable (and will stay stable indefinitely). Only when the stepsize approaches $\varepsilon = 2$ do the trajectories become unstable.

5.2.3.4 Local and Global Error of Discretization Methods

I will briefly discuss how the error from discretizing the dynamics behaves in the limit as the stepsize, ε, goes to zero; Leimkuhler and Reich (2004) provide a much more detailed discussion. For useful methods, the error goes to zero as ε goes to zero, so that any upper limit on the error will apply (apart from a usually unknown constant factor) to any differentiable function of state—for example, if the error for (q, p) is no more than order ε^2, the error for $H(q, p)$ will also be no more than order ε^2.

The *local error* is the error after one step, that moves from time t to time $t + \varepsilon$. The *global error* is the error after simulating for some fixed time interval, s, which will require s/ε steps. If the local error is order ε^p, the global error will be order ε^{p-1}—the local errors of order ε^p accumulate over the s/ε steps to give an error of order ε^{p-1}. If we instead fix ε and consider increasing the time, s, for which the trajectory is simulated, the error can in general increase exponentially with s. Interestingly, however, this is often not what happens when simulating Hamiltonian dynamics with a symplectic method, as can be seen in Figure 5.1.

The Euler method and its modification above have order ε^2 local error and order ε global error. The leapfrog method has order ε^3 local error and order ε^2 global error. As shown by Leimkuhler and Reich (2004, Section 4.3.3), this difference is a consequence of leapfrog being reversible, since any reversible method must have global error that is of even order in ε.

5.3 MCMC from Hamiltonian Dynamics

Using Hamiltonian dynamics to sample from a distribution requires translating the density function for this distribution to a potential energy function and introducing "momentum" variables to go with the original variables of interest (now seen as "position" variables). We can then simulate a Markov chain in which each iteration resamples the momentum and then does a Metropolis update with a proposal found using Hamiltonian dynamics.

5.3.1 Probability and the Hamiltonian: Canonical Distributions

The distribution we wish to sample can be related to a potential energy function via the concept of a *canonical distribution* from statistical mechanics. Given some energy function, $E(x)$, for the state, x, of some physical system, the canonical distribution over states has probability or probability density function

$$P(x) = \frac{1}{Z} \exp\left(\frac{-E(x)}{T}\right). \tag{5.21}$$

Here, T is the temperature of the system,* and Z is the normalizing constant needed for this function to sum or integrate to one. Viewing this the opposite way, if we are interested in some distribution with density function $P(x)$, we can obtain it as a canonical distribution with $T = 1$ by setting $E(x) = -\log P(x) - \log Z$, where Z is any convenient positive constant.

The Hamiltonian is an energy function for the joint state of "position," q, and "momentum," p, and so defines a joint distribution for them as follows:

$$P(q, p) = \frac{1}{Z} \exp\left(\frac{-H(q, p)}{T}\right).$$

Note that the invariance of H under Hamiltonian dynamics means that a Hamiltonian trajectory will (if simulated exactly) move within a hypersurface of constant probability density.

If $H(q, p) = U(q) + K(p)$, the joint density is

$$P(q, p) = \frac{1}{Z} \exp\left(\frac{-U(q)}{T}\right) \exp\left(\frac{-K(p)}{T}\right), \tag{5.22}$$

and we see that q and p are independent, and each have canonical distributions, with energy functions $U(q)$ and $K(p)$. We will use q to represent the variables of interest, and introduce p just to allow Hamiltonian dynamics to operate.

In Bayesian statistics, the posterior distribution for the model parameters is the usual focus of interest, and hence these parameters will take the role of the position, q. We can express the posterior distribution as a canonical distribution (with $T = 1$) using a potential energy function defined as

$$U(q) = -\log\left[\pi(q)L(q \mid D)\right],$$

where $\pi(q)$ is the prior density, and $L(q|D)$ is the likelihood function given data D.

5.3.2 The Hamiltonian Monte Carlo Algorithm

We now have the background needed to present the Hamiltonian Monte Carlo algorithm. HMC can be used to sample only from continuous distributions on $\mathbb{R}^d$ for which the density function can be evaluated (perhaps up to an unknown normalizing constant). For the moment, I will also assume that the density is nonzero everywhere (but this is relaxed in Section 5.5.1). We must also be able to compute the partial derivatives of the log of the density function. These derivatives must therefore exist, except perhaps on a set of points with probability zero, for which some arbitrary value could be returned.

HMC samples from the canonical distribution for q and p defined by Equation 5.22, in which q has the distribution of interest, as specified using the potential energy function $U(q)$. We can choose the distribution of the momentum variables, p, which are independent of q, as we wish, specifying the distribution via the kinetic energy function, $K(p)$. Current practice with HMC is to use a quadratic kinetic energy, as in Equation 5.5, which leads p to have a zero-mean multivariate Gaussian distribution. Most often, the components of

---

* Note to physicists: I assume here that temperature is measured in units that make Boltzmann's constant unity.

p are specified to be independent, with component i having variance m_i. The kinetic energy function producing this distribution (setting $T = 1$) is

$$K(p) = \sum_{i=1}^{d} \frac{p_i^2}{2m_i}. \tag{5.23}$$

We will see in Section 5.4 how the choice for the m_i affects performance.

5.3.2.1 The Two Steps of the HMC Algorithm

Each iteration of the HMC algorithm has two steps. The first changes only the momentum; the second may change both position and momentum. Both steps leave the canonical joint distribution of (q, p) invariant, and hence their combination also leaves this distribution invariant.

In the first step, new values for the momentum variables are randomly drawn from their Gaussian distribution, independently of the current values of the position variables. For the kinetic energy of Equation 5.23, the d momentum variables are independent, with p_i having mean zero and variance m_i. Since q is not changed, and p is drawn from its correct conditional distribution given q (the same as its marginal distribution, due to independence), this step obviously leaves the canonical joint distribution invariant.

In the second step, a Metropolis update is performed, using Hamiltonian dynamics to propose a new state. Starting with the current state, (q, p), Hamiltonian dynamics is simulated for L steps using the leapfrog method (or some other reversible method that preserves volume), with a stepsize of ε. Here, L and ε are parameters of the algorithm, which need to be tuned to obtain good performance (as discussed below in Section 5.4.2). The momentum variables at the end of this L-step trajectory are then negated, giving a proposed state (q^*, p^*). This proposed state is accepted as the next state of the Markov chain with probability

$$\min\left[1, \exp(-H(q^*, p^*) + H(q, p))\right] = \min\left[1, \exp(-U(q^*) + U(q) - K(p^*) + K(p))\right].$$

If the proposed state is not accepted (i.e. it is rejected), the next state is the same as the current state (and is counted again when estimating the expectation of some function of state by its average over states of the Markov chain). The negation of the momentum variables at the end of the trajectory makes the Metropolis proposal symmetrical, as needed for the acceptance probability above to be valid. This negation need not be done in practice, since $K(p) = K(-p)$, and the momentum will be replaced before it is used again, in the first step of the next iteration. (This assumes that these HMC updates are the only ones performed.)

If we look at HMC as sampling from the joint distribution of q and p, the Metropolis step using a proposal found by Hamiltonian dynamics leaves the probability density for (q, p) unchanged or almost unchanged. Movement to (q, p) points with a different probability density is accomplished only by the first step in an HMC iteration, in which p is replaced by a new value. Fortunately, this replacement of p can change the probability density for (q, p) by a large amount, so movement to points with a different probability density is not a problem (at least not for this reason). Looked at in terms of q only, Hamiltonian dynamics for (q, p) can produce a value for q with a much different probability density (equivalently, a much different potential energy, $U(q)$). However, the resampling of the momentum variables is still crucial to obtaining the proper distribution for q. Without resampling, $H(q, p) = U(q) + K(p)$ will be (nearly) constant, and since $K(p)$ and $U(q)$ are

```
HMC = function (U, grad_U, epsilon, L, current_q)
{
  q = current_q
  p = rnorm(length(q),0,1)  # independent standard normal variates
  current_p = p

  # Make a half step for momentum at the beginning
  p = p - epsilon * grad_U(q) / 2

  # Alternate full steps for position and momentum

  for (i in 1:L)
  {
    # Make a full step for the position
    q = q + epsilon * p
    # Make a full step for the momentum, except at end of trajectory
    if (i!=L) p = p - epsilon * grad_U(q)
  }

  # Make a half step for momentum at the end.
  p = p - epsilon * grad_U(q) / 2
  # Negate momentum at end of trajectory to make the proposal symmetric
  p = -p

  # Evaluate potential and kinetic energies at start and end of trajectory

  current_U = U(current_q)
  current_K = sum(current_p^2) / 2
  proposed_U = U(q)
  proposed_K = sum(p^2) / 2

  # Accept or reject the state at end of trajectory, returning either
  # the position at the end of the trajectory or the initial position

  if (runif(1) < exp(current_U-proposed_U+current_K-proposed_K))
  {
    return (q)   # accept
  }
  else
  {
    return (current_q)  # reject
  }
}
```

FIGURE 5.2
The Hamiltonian Monte Carlo algorithm.

nonnegative, $U(q)$ could never exceed the initial value of $H(q,p)$ if no resampling for p were done.

A function that implements a single iteration of the HMC algorithm, written in the R language,* is shown in Figure 5.2. Its first two arguments are functions: U, which returns

_____
* R is available for free from www.r-project.org

the potential energy given a value for q, and grad_U, which returns the vector of partial derivatives of U given q. Other arguments are the stepsize, epsilon, for leapfrog steps; the number of leapfrog steps in the trajectory, L; and the current position, current_q, that the trajectory starts from. Momentum variables are sampled within this function, and discarded at the end, with only the next position being returned. The kinetic energy is assumed to have the simplest form, $K(p) = \sum p_i^2/2$ (i.e. all m_i are one). In this program, all components of p and of q are updated simultaneously, using vector operations. This simple implementation of HMC is available from my web page,* along with other R programs with extra features helpful for practical use, and that illustrate some of the variants of HMC in Section 5.5.

5.3.2.2   Proof That HMC Leaves the Canonical Distribution Invariant

The Metropolis update above is reversible with respect to the canonical distribution for q and p (with $T = 1$), a condition also known as "detailed balance," and which can be phrased informally as follows. Suppose that we partition the (q, p) space into regions A_k, each with the same small volume V. Let the image of A_k with respect to the operation of L leapfrog steps, plus a negation of the momentum, be B_k. Due to the reversibility of the leapfrog steps, the B_k will also partition the space, and since the leapfrog steps preserve volume (as does negation), each B_k will also have volume V. Detailed balance holds if, for all i and j,

$$P(A_i)T(B_j \mid A_i) = P(B_j)T(A_i \mid B_j), \tag{5.24}$$

where P is probability under the canonical distribution, and $T(X|Y)$ is the conditional probability of proposing and then accepting a move to region X if the current state is in region Y. Clearly, when $i \neq j$, $T(A_i \mid B_j) = T(B_j \mid A_i) = 0$ and so Equation 5.24 will be satisfied. Since the Hamiltonian is continuous almost everywhere, in the limit as the regions A_k and B_k become smaller, the Hamiltonian becomes effectively constant within each region, with value H_X in region X, and hence the canonical probability density and the transition probabilities become effectively constant within each region as well. We can now rewrite Equation 5.24 for $i = j$ (say, both equal to k) as

$$\frac{V}{Z}\exp(-H_{A_k})\min\left[1, \exp(-H_{B_k}+H_{A_k})\right] = \frac{V}{Z}\exp(-H_{B_k})\min\left[1, \exp(-H_{A_k}+H_{B_k})\right],$$

which is easily seen to be true.

Detailed balance implies that this Metropolis update leaves the canonical distribution for q and p invariant. This can be seen as follows. Let $R(X)$ be the probability that the Metropolis update for a state in the small region X leads to rejection of the proposed state. Suppose that the current state is distributed according to the canonical distribution. The probability that the next state is in a small region B_k is the sum of the probability that the current state is in B_k and the update leads to rejection, and the probability that the current state is in some region from which a move to B_k is proposed and accepted. The probability of the next state

---

* www.cs.utoronto.ca/~radford

being in B_k can therefore be written as

$$P(B_k)R(B_k) + \sum_i P(A_i)T(B_k|A_i) = P(B_k)R(B_k) + \sum_i P(B_k)T(A_i|B_k)$$

$$= P(B_k)R(B_k) + P(B_k)\sum_i T(A_i|B_k)$$

$$= P(B_k)R(B_k) + P(B_k)(1 - R(B_k))$$

$$= P(B_k).$$

The Metropolis update within HMC therefore leaves the canonical distribution invariant.

Since both the sampling of momentum variables and the Metropolis update with a proposal found by Hamiltonian dynamics leave the canonical distribution invariant, the HMC algorithm as a whole does as well.

5.3.2.3 Ergodicity of HMC

Typically, the HMC algorithm will also be "ergodic"—it will not be trapped in some subset of the state space, and hence will asymptotically converge to its (unique) invariant distribution. In an HMC iteration, any value can be sampled for the momentum variables, which can typically then affect the position variables in arbitrary ways. However, ergodicity can fail if the L leapfrog steps in a trajectory produce an exact periodicity for some function of state. For example, with the simple Hamiltonian of Equation 5.8, the exact solutions (given by Equation 5.9) are periodic with period 2π. Approximate trajectories found with L leapfrog steps with stepsize ε may return to the same position coordinate when $L\varepsilon$ is approximately 2π. HMC with such values for L and ε will not be ergodic. For nearby values of L and ε, HMC may be theoretically ergodic, but take a very long time to move about the full state space.

This potential problem of nonergodicity can be solved by randomly choosing ε or L (or both) from some fairly small interval (Mackenzie, 1989). Doing this routinely may be advisable. Although in real problems interactions between variables typically prevent any exact periodicities from occurring, near periodicities might still slow HMC considerably.

5.3.3 Illustrations of HMC and Its Benefits

I will now illustrate some practical issues with HMC, and demonstrate its potential to sample much more efficiently than simple methods such as random-walk Metropolis. I use simple Gaussian distributions for these demonstrations, so that the results can be compared with known values, but of course HMC is typically used for more complex distributions.

5.3.3.1 Trajectories for a Two-Dimensional Problem

Consider sampling from a distribution for two variables that is bivariate Gaussian, with means of zero, standard deviations of one, and correlation 0.95. We regard these as "position" variables, and introduce two corresponding "momentum" variables, defined to have a Gaussian distribution with means of zero, standard deviations of one, and zero correlation. We then define the Hamiltonian as

$$H(q,p) = q^T \Sigma^{-1} q/2 + p^T p/2, \quad \text{with } \Sigma = \begin{bmatrix} 1 & 0.95 \\ 0.95 & 1 \end{bmatrix}.$$

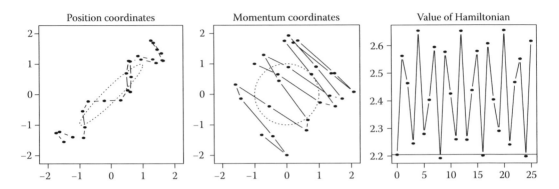

FIGURE 5.3
A trajectory for a two-dimensional Gaussian distribution, simulated using 25 leapfrog steps with a stepsize of
0.25. The ellipses plotted are one standard deviation from the means. The initial state had $q = [-1.50, -1.55]^T$ and
$p = [-1, 1]^T$.

Figure 5.3 shows a trajectory based on this Hamiltonian, such as might be used to propose
a new state in the HMC method, computed using $L = 25$ leapfrog steps, with a stepsize of
$\varepsilon = 0.25$. Since the full state space is four-dimensional, Figure 5.3 shows the two position
coordinates and the two momentum coordinates in separate plots, while the third plot
shows the value of the Hamiltonian after each leapfrog step.

Notice that this trajectory does not resemble a random walk. Instead, starting from the
lower left-hand corner, the position variables systematically move upward and to the right,
until they reach the upper right-hand corner, at which point the direction of motion is
reversed. The consistency of this motion results from the role of the momentum variables.
The projection of p in the diagonal direction will change only slowly, since the gradient
in that direction is small, and hence the direction of diagonal motion stays the same for
many leapfrog steps. While this large-scale diagonal motion is happening, smaller-scale
oscillations occur, moving back and forth across the "valley" created by the high correlation
between the variables.

The need to keep these smaller oscillations under control limits the stepsize that can
be used. As can be seen in the rightmost plot in Figure 5.3, there are also oscillations in
the value of the Hamiltonian (which would be constant if the trajectory were simulated
exactly). If a larger stepsize were used, these oscillations would be larger. At a critical
stepsize ($\varepsilon = 0.45$ in this example), the trajectory becomes unstable, and the value of the
Hamiltonian grows without bound. As long as the stepsize is less than this, however, the
error in the Hamiltonian stays bounded regardless of the number of leapfrog steps done.
This lack of growth in the error is not guaranteed for all Hamiltonians, but it does hold for
many distributions more complex than Gaussians. As can be seen, however, the error in
the Hamiltonian along the trajectory does tend to be positive more often than negative. In
this example, the error is $+0.41$ at the end of the trajectory, so if this trajectory were used
for an HMC proposal, the probability of accepting the endpoint as the next state would be
$\exp(-0.41) = 0.66$.

5.3.3.2 Sampling from a Two-Dimensional Distribution

Figures 5.4 and 5.5 show the results of using HMC and a simple random-walk Metropolis
method to sample from a bivariate Gaussian similar to the one just discussed, but with
stronger correlation of 0.98.

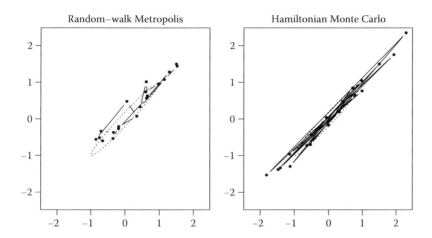

FIGURE 5.4

Twenty iterations of the random-walk Metropolis method (with 20 updates per iteration) and of the Hamiltonian Monte Carlo method (with 20 leapfrog steps per trajectory) for a two-dimensional Gaussian distribution with marginal standard deviations of one and correlation 0.98. Only the two position coordinates are plotted, with ellipses drawn one standard deviation away from the mean.

In this example, as in the previous one, HMC used a kinetic energy (defining the momentum distribution) of $K(p) = p^T p / 2$. The results of 20 HMC iterations, using trajectories of $L = 20$ leapfrog steps with stepsize $\varepsilon = 0.18$, are shown in the right plot of Figure 5.4. These values were chosen so that the trajectory length, εL, is sufficient to move to a distant point in the distribution, without being so large that the trajectory will often waste computation time by doubling back on itself. The rejection rate for these trajectories was 0.09.

Figure 5.4 also shows every 20th state from 400 iterations of random-walk Metropolis, with a bivariate Gaussian proposal distribution with the current state as mean, zero correlation, and the same standard deviation for the two coordinates. The standard deviation of the proposals for this example was 0.18, which is the same as the stepsize used for HMC proposals, so that the change in state in these random-walk proposals was comparable to that for a single leapfrog step for HMC. The rejection rate for these random-walk proposals was 0.37.

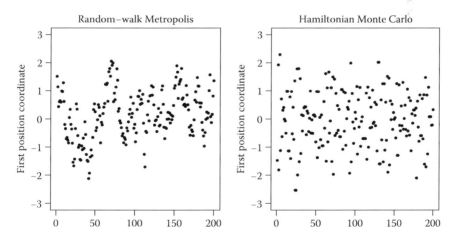

FIGURE 5.5

Two hundred iterations, starting with the 20 iterations shown above, with only the first position coordinate plotted.

One can see in Figure 5.4 how the systematic motion during an HMC trajectory (illustrated in Figure 5.3) produces larger changes in state than a corresponding number of random-walk Metropolis iterations. Figure 5.5 illustrates this difference for longer runs of 20×200 random-walk Metropolis iterations and of 200 HMC iterations.

5.3.3.3 *The Benefit of Avoiding Random Walks*

Avoidance of random-walk behavior, as illustrated above, is one major benefit of HMC. In this example, because of the high correlation between the two position variables, keeping the acceptance probability for random-walk Metropolis reasonably high requires that the changes proposed have a magnitude comparable to the standard deviation in the most constrained direction (0.14 in this example, the square root of the smallest eigenvalue of the covariance matrix). The changes produced using one Gibbs sampling scan would be of similar magnitude. The number of iterations needed to reach a state almost independent of the current state is mostly determined by how long it takes to explore the less constrained direction, which for this example has standard deviation 1.41—about ten times greater than the standard deviation in the most constrained direction. We might therefore expect that we would need around 10 iterations of random-walk Metropolis in which the proposal was accepted to move to a nearly independent state. But the number needed is actually roughly the square of this—around 100 iterations with accepted proposals—because the random-walk Metropolis proposals have no tendency to move consistently in the same direction.

To see this, note that the variance of the position after n iterations of random-walk Metropolis from some start state will grow in proportion to n (until this variance becomes comparable to the overall variance of the state), since the position is the sum of mostly independent movements for each iteration. The *standard deviation* of the amount moved (which gives the typical amount of movement) is therefore proportional to $\sqrt{n}$.

The stepsize used for the leapfrog steps is similarly limited by the most constrained direction, but the movement will be in the same direction for many steps. The distance moved after n steps will therefore tend to be proportional to n, until the distance moved becomes comparable to the overall width of the distribution. The advantage compared to movement by a random walk will be a factor roughly equal to the ratio of the standard deviations in the least confined direction and most confined direction—about 10 here.

Because avoiding a random walk is so beneficial, the optimal standard deviation for random-walk Metropolis proposals in this example is actually much larger than the value of 0.18 used here. A proposal standard deviation of 2.0 gives a very low acceptance rate (0.06), but this is more than compensated for by the large movement (to a nearly independent point) on the rare occasions when a proposal is accepted, producing a method that is about as efficient as HMC. However, this strategy of making large changes with a small acceptance rate works only when, as here, the distribution is tightly constrained in only one direction.

5.3.3.4 *Sampling from a 100-Dimensional Distribution*

More typical behavior of HMC and random-walk Metropolis is illustrated by a 100-dimensional multivariate Gaussian distribution in which the variables are independent, with means of zero, and standard deviations of $0.01, 0.02, \ldots, 0.99, 1.00$. Suppose that we have no knowledge of the details of this distribution, so we will use HMC with the same simple, rotationally symmetric kinetic energy function as above, $K(p) = p^T p / 2$, and use random-walk Metropolis proposals in which changes to each variable are independent, all with the same standard deviation. As discussed below in Section 5.4.1, the performance of both these sampling methods is invariant to rotation, so this example is illustrative of how they perform on any multivariate Gaussian distribution in which the square roots of the eigenvalues of the covariance matrix are $0.01, 0.02, \ldots, 0.99, 1.00$.

For this problem, the position coordinates, q_i, and corresponding momentum coordinates, p_i, are all independent, so the leapfrog steps used to simulate a trajectory operate independently for each (q_i, p_i) pair. However, whether the trajectory is accepted depends on the total error in the Hamiltonian due to the leapfrog discretization, which is a sum of the errors due to each (q_i, p_i) pair (for the terms in the Hamiltonian involving this pair). Keeping this error small requires limiting the leapfrog stepsize to a value roughly equal to the smallest of the standard deviations (0.01), which implies that many leapfrog steps will be needed to move a distance comparable to the largest of the standard deviations (1.00).

Consistent with this, I applied HMC to this distribution using trajectories with $L = 150$ and with ε randomly selected for each iteration, uniformly from $(0.0104, 0.0156)$, which is $0.013 \pm 20\%$. I used random-walk Metropolis with proposal standard deviation drawn uniformly from $(0.0176, 0.0264)$, which is $0.022 \pm 20\%$. These are close to optimal settings for both methods. The rejection rate was 0.13 for HMC and 0.75 for random-walk Metropolis.

Figure 5.6 shows results from runs of 1000 iterations of HMC (right) and of random-walk Metropolis (left), counting 150 random-walk Metropolis updates as one iteration, so that the computation time per iteration is comparable to that for HMC. The plot shows the last variable, with the largest standard deviation. The autocorrelation of these values is clearly much higher for random-walk Metropolis than for HMC. Figure 5.7 shows the estimates for the mean and standard deviation of each of the 100 variables obtained using the HMC and random-walk Metropolis runs (estimates were just the sample means and sample standard deviations of the values from the 1000 iterations). Except for the first few

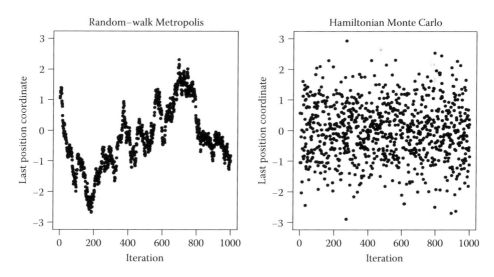

FIGURE 5.6

Values for the variable with largest standard deviation for the 100-dimensional example, from a random-walk Metropolis run and an HMC run with $L = 150$. To match computation time, 150 updates were counted as one iteration for random-walk Metropolis.

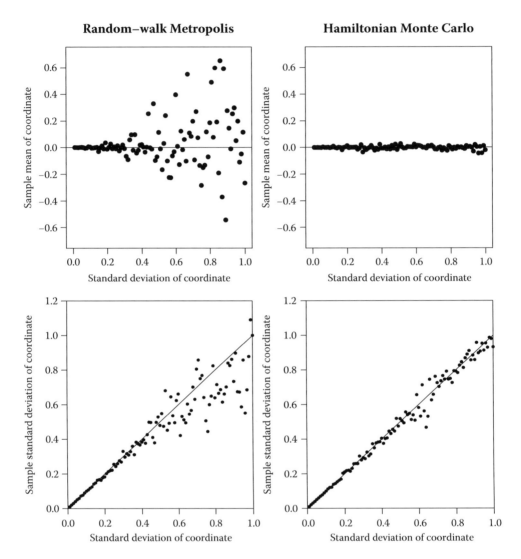

FIGURE 5.7
Estimates of means (top) and standard deviations (bottom) for the 100-dimensional example, using random-walk Metropolis (left) and HMC (right). The 100 variables are labeled on the horizontal axes by the true standard deviaton of that variable. Estimates are on the vertical axes.

variables (with smallest standard deviations), the error in the mean estimates from HMC is roughly 10 times less than the error in the mean estimates from random-walk Metropolis. The standard deviation estimates from HMC are also better.

The randomization of the leapfrog stepsize done in this example follows the advice discussed at the end of Section 5.3.2. In this example, not randomizing the stepsize (fixing $\varepsilon = 0.013$) does in fact cause problems—the variables with standard deviations near 0.31 or 0.62 change only slowly, since 150 leapfrog steps with $\epsilon = 0.013$ produces nearly a full or half cycle for these variables, so an accepted trajectory does not make much of a change in the absolute value of the variable.

5.4 HMC in Practice and Theory

Obtaining the benefits from HMC illustrated in the previous section, including random-walk avoidance, requires proper tuning of L and ε. I discuss tuning of HMC below, and also show how performance can be improved by using whatever knowledge is available regarding the scales of variables and their correlations. After briefly discussing what to do when HMC alone is not enough, I discuss an additional benefit of HMC—its better scaling with dimensionality than simple Metropolis methods.

5.4.1 Effect of Linear Transformations

Like all MCMC methods I am aware of, the performance of HMC may change if the variables being sampled are transformed by multiplication by some nonsingular matrix, A. However, performance stays the same (except perhaps in terms of computation time per iteration) if at the same time the corresponding momentum variables are multiplied by $(A^T)^{-1}$. These facts provide insight into the operation of HMC, and can help us improve performance when we have some knowledge of the scales and correlations of the variables.

Let the new variables be $q' = Aq$. The probability density for q' will be given by $P'(q') = P(A^{-1}q')/|\det(A)|$, where $P(q)$ is the density for q. If the distribution for q is the canonical distribution for a potential energy function $U(q)$ (see Section 5.3.1), we can obtain the distribution for q' as the canonical distribution for $U'(q') = U(A^{-1}q')$. (Since $|\det(A)|$ is a constant, we need not include a $\log|\det(A)|$ term in the potential energy.)

We can choose whatever distribution we wish for the corresponding momentum variables, so we could decide to use the same kinetic energy as before. Alternatively, we can choose to transform the momentum variables by $p' = (A^T)^{-1}p$, and use a new kinetic energy of $K'(p') = K(A^T p')$. If we were using a quadratic kinetic energy, $K(p) = p^T M^{-1}p/2$ (see Equation 5.5), the new kinetic energy will be

$$K'(p') = (A^T p')^T M^{-1}(A^T p')/2 = (p')^T (A M^{-1} A^T) p'/2 = (p')^T (M')^{-1} p'/2, \qquad (5.25)$$

where $M' = (A M^{-1} A^T)^{-1} = (A^{-1})^T M A^{-1}$.

If we use momentum variables transformed in this way, the dynamics for the new variables, (q', p'), essentially replicates the original dynamics for (q, p), so the performance of HMC will be the same. To see this, note that if we follow Hamiltonian dynamics for (q', p'), the result in terms of the original variables will be as follows (see Equations 5.6 and 5.7):

$$\frac{dq}{dt} = A^{-1}\frac{dq'}{dt} = A^{-1}(M')^{-1}p' = A^{-1}(A M^{-1} A^T)(A^T)^{-1}p = M^{-1}p,$$

$$\frac{dp}{dt} = A^T \frac{dp'}{dt} = -A^T \nabla U'(q') = -A^T (A^{-1})^T \nabla U(A^{-1}q') = -\nabla U(q),$$

which matches what would happen following Hamiltonian dynamics for (q, p).

If A is an orthogonal matrix (such as a rotation matrix), for which $A^{-1} = A^T$, the performance of HMC is unchanged if we transform both q and p by multiplying by A (since $(A^T)^{-1} = A$). If we chose a rotationally symmetric distribution for the momentum, with $M = mI$ (i.e. the momentum variables are independent, each having variance m), such an orthogonal transformation will not change the kinetic energy function (and hence not change the distribution of the momentum variables), since we will have $M' = (A (mI)^{-1}A^T)^{-1} = mI$.

Such an invariance to rotation holds also for a random-walk Metropolis method in which the proposal distribution is rotationally symmetric (e.g. Gaussian with covariance matrix mI). In contrast, Gibbs sampling is not rotationally invariant, nor is a scheme in which the Metropolis algorithm is used to update each variable in turn (with a proposal that changes only that variable). However, Gibbs sampling is invariant to rescaling of the variables (transformation by a diagonal matrix), which is not true for HMC or random-walk Metropolis, unless the kinetic energy or proposal distribution is transformed in a corresponding way.

Suppose that we have an estimate, Σ, of the covariance matrix for q, and suppose also that q has at least a roughly Gaussian distribution. How can we use this information to improve the performance of HMC? One way is to transform the variables so that their covariance matrix is close to the identity, by finding the Cholesky decomposition, $\Sigma = LL^T$, with L being lower-triangular, and letting $q' = L^{-1}q$. We then let our kinetic energy function be $K(p) = p^T p/2$. Since the momentum variables are independent, and the position variables are close to independent with variances close to one (if our estimate Σ and our assumption that q is close to Gaussian are good), HMC should perform well using trajectories with a small number of leapfrog steps, which will move all variables to a nearly independent point. More realistically, the estimate Σ may not be very good, but this transformation could still improve performance compared to using the same kinetic energy with the original q variables.

An equivalent way to make use of the estimated covariance Σ is to keep the original q variables, but use the kinetic energy function $K(p) = p^T \Sigma p/2$—that is, we let the momentum variables have covariance Σ^{-1}. The equivalence can be seen by transforming this kinetic energy to correspond to a transformation to $q' = L^{-1}q$ (see Equation 5.25), which gives $K(p') = (p')^T M'^{-1} p'$ with $M' = (L^{-1}(LL^T)(L^{-1})^T)^{-1} = I$.

Using such a kinetic energy function to compensate for correlations between position variables has a long history in molecular dynamics (Bennett, 1975). The usefulness of this technique is limited by the computational cost of matrix operations when the dimensionality is high.

Using a diagonal Σ can be feasible even in high-dimensional problems. Of course, this provides information only about the different scales of the variables, not their correlation. Moreover, when the actual correlations are nonzero, it is not clear what scales to use. Making an optimal choice is probably infeasible. Some approximation to the conditional standard deviation of each variable given all the others may be possible—as I have done for Bayesian neural network models (Neal, 1996a). If this also is not feasible, using approximations to the marginal standard deviations of the variables may be better than using the same scale for them all.

5.4.2 Tuning HMC

One practical impediment to the use of Hamiltonian Monte Carlo is the need to select suitable values for the leapfrog stepsize, ε, and the number of leapfrog steps, L, which together determine the length of the trajectory in fictitious time, εL. Most MCMC methods have parameters that need to be tuned, with the notable exception of Gibbs sampling when the conditional distributions are amenable to direct sampling. However, tuning HMC is more difficult in some respects than tuning a simple Metropolis method.

5.4.2.1 *Preliminary Runs and Trace Plots*

Tuning HMC will usually require preliminary runs with trial values for ε and L. In judging how well these runs work, trace plots of quantities that are thought to be indicative of overall convergence should be examined. For Bayesian inference problems, high-level hyperparameters are often among the slowest-moving quantities. The value of the potential energy function, $U(q)$, is also usually of central significance. The autocorrelation for such quantities indicates how well the Markov chain is exploring the state space. Ideally, we would like the state after one HMC iteration to be nearly independent of the previous state.

Unfortunately, preliminary runs can be misleading, if they are not long enough to have reached equilibrium. It is possible that the best choices of ε and L for reaching equilibrium are different from the best choices once equilibrium is reached, and even at equilibrium, it is possible that the best choices vary from one place to another. If necessary, at each iteration of HMC, ε and L can be chosen randomly from a selection of values that are appropriate for different parts of the state space (or these selections and can be used sequentially).

Doing several runs with different random starting states is advisable (for both preliminary and final runs), so that problems with isolated modes can be detected. Note that HMC is no less (or more) vulnerable to problems with isolated modes than other MCMC methods that make local changes to the state. If isolated modes are found to exist, something needs to be done to solve this problem—just combining runs that are each confined to a single mode is not valid. A modification of HMC with "tempering" along a trajectory (Section 5.5.7) can sometimes help with multiple modes.

5.4.2.2 What Stepsize?

Selecting a suitable leapfrog stepsize, ε, is crucial. Too large a stepsize will result in a very low acceptance rate for states proposed by simulating trajectories. Too small a stepsize will either waste computation time, by the same factor as the stepsize is too small, or (worse) will lead to slow exploration by a random walk, if the trajectory length, εL, is then too short (i.e. L is not large enough; see below).

Fortunately, as illustrated in Figure 5.3, the choice of stepsize is almost independent of how many leapfrog steps are done. The error in the value of the Hamiltonian (which will determine the rejection rate) usually does not increase with the number of leapfrog steps, *provided* that the stepsize is small enough that the dynamics is stable.

The issue of stability can be seen in a simple one-dimensional problem in which the following Hamiltonian is used:

$$H(q, p) = \frac{q^2}{2\sigma^2} + \frac{p^2}{2}.$$

The distribution for q that this defines is Gaussian with standard deviation σ. A leapfrog step for this system (as for any quadratic Hamiltonian) will be a linear mapping from $(q(t), p(t))$ to $(q(t + \varepsilon), p(t + \varepsilon))$. Referring to Equations 5.18 through 5.20, we see that this mapping can be represented by a matrix multiplication as follows:

$$\begin{bmatrix} q(t + \varepsilon) \\ p(t + \varepsilon) \end{bmatrix} = \begin{bmatrix} 1 - \varepsilon^2/2\sigma^2 & \varepsilon \\ -\varepsilon/\sigma^2 + \varepsilon^3/4\sigma^4 & 1 - \varepsilon^2/2\sigma^2 \end{bmatrix} \begin{bmatrix} q(t) \\ p(t) \end{bmatrix}.$$

Whether iterating this mapping leads to a stable trajectory, or one that diverges to infinity, depends on the magnitudes of the eigenvalues of the above matrix, which are

$$\left(1 - \frac{\varepsilon^2}{2\sigma^2}\right) \pm \left(\frac{\varepsilon}{\sigma}\right)\sqrt{\varepsilon^2/4\sigma^2 - 1}.$$

When $\varepsilon/\sigma > 2$, these eigenvalues are real, and at least one will have absolute value greater than one. Trajectories computed using the leapfrog method with this ε will therefore be unstable. When $\varepsilon/\sigma < 2$, the eigenvalues are complex, and both have squared magnitude of

$$\left(1 - \frac{\varepsilon^2}{2\sigma^2}\right)^2 + \left(\frac{\varepsilon^2}{\sigma^2}\right)\left(1 - \frac{\varepsilon^2}{4\sigma^2}\right) = 1.$$

Trajectories computed with $\varepsilon < 2\sigma$ are therefore stable.

For multidimensional problems in which the kinetic energy used is $K(p) = p^T p/2$ (as in the example above), the stability limit for ε will be determined (roughly) by the width of the distribution in the most constrained direction—for a Gaussian distribution, this would the square root of the smallest eigenvalue of the covariance matrix for q. Stability for more general quadratic Hamiltonians with $K(p) = p^T M^{-1} p/2$ can be determined by applying a linear transformation that makes $K(p') = (p')^T p'/2$, as discussed above in Section 5.4.1.

When a stepsize, ε, that produces unstable trajectories is used, the value of H grows exponentially with L, and consequently the acceptance probability will be extremely small. For low-dimensional problems, using a value for ε that is just a little below the stability limit is sufficient to produce a good acceptance rate. For high-dimensional problems, however, the stepsize may need to be reduced further than this to keep the error in H to a level that produces a good acceptance probability. This is discussed further in Section 5.4.4.

Choosing too large a value of ε can have very bad effects on the performance of HMC. In this respect, HMC is more sensitive to tuning than random-walk Metropolis. A standard deviation for proposals needs to be chosen for random-walk Metropolis, but performance degrades smoothly as this choice is made too large, without the sharp degradation seen with HMC when ε exceeds the stability limit. (However, in high-dimensional problems, the degradation in random-walk Metropolis with too large a proposal standard deviation can also be quite sharp, so this distinction becomes less clear.)

This sharp degradation in performance of HMC when the stepsize is too big would not be a serious issue if the stability limit were constant—the problem would be obvious from preliminary runs, and so could be fixed. The real danger is that the stability limit may differ for several regions of the state space that all have substantial probability. If the preliminary runs are started in a region where the stability limit is large, a choice of ε a little less than this limit might appear to be appropriate. However, if this ε is above the stability limit for some other region, the runs may never visit this region, even though it has substantial probability, producing a drastically wrong result. To see why this could happen, note that if the run ever does visit the region where the chosen ε would produce instability, it will stay there for a very long time, since the acceptance probability with that ε will be very small. Since the method nevertheless leaves the correct distribution invariant, it follows that the run only rarely moves to this region from a region where the chosen ε leads to stable trajectories. One simple context where this problem can arise is when sampling from a distribution with very light tails (lighter than a Gaussian distribution), for which the log of the density will fall faster than quadratically. In the tails, the gradient of the log density will be large, and a small stepsize will be needed for stability. See Roberts and Tweedie (1996) for a discussion of this in the context of the Langevin method (see Section 5.5.2).

This problem can be alleviated by choosing ε randomly from some distribution. Even if the mean of this distribution is too large, suitably small values for ε may be chosen occasionally. (See Section 5.3.2 for another reason to randomly vary the stepsize.) The random choice of ε should be done once at the start of a trajectory, not for every leapfrog step, since even if all the choices are below the stability limit, random changes at each step lead to a random walk in the error for H, rather than the bounded error that is illustrated in Figure 5.3.

The "short-cut" procedures described in Section 5.5.6 can be seen as ways of saving computation time when a randomly chosen stepsize is inappropriate.

5.4.2.3 What Trajectory Length?

Choosing a suitable trajectory length is crucial if HMC is to explore the state space systematically, rather than by a random walk. Many distributions are difficult to sample from because they are tightly constrained in some directions, but much less constrained in other directions. Exploring the less constrained directions is best done using trajectories that are long enough to reach a point that is far from the current point in that direction. Trajectories can be too long, however, as is illustrated in Figure 5.3. The trajectory shown on the left of that figure is a bit too long, since it reverses direction and then ends at a point that might have been reached with a trajectory about half its length. If the trajectory were a little longer, the result could be even worse, since the trajectory would not only take longer to compute, but might also end near its starting point.

For more complex problems, one cannot expect to select a suitable trajectory length by looking at plots like Figure 5.3. Finding the linear combination of variables that is least confined will be difficult, and will be impossible when, as is typical, the least confined "direction" is actually a nonlinear curve or surface.

Setting the trajectory length by trial and error therefore seems necessary. For a problem thought to be fairly difficult, a trajectory with $L = 100$ might be a suitable starting point. If preliminary runs (with a suitable ε; see above) show that HMC reaches a nearly independent point after only one iteration, a smaller value of L might be tried next. (Unless these "preliminary" runs are actually sufficient, in which case there is of course no need to do more runs.) If instead there is high autocorrelation in the run with $L = 100$, runs with $L = 1000$ might be tried next.

As discussed at the end of Sections 5.3.2 and 5.3.3, randomly varying the length of the trajectory (over a fairly small interval) may be desirable, to avoid choosing a trajectory length that happens to produce a near-periodicity for some variable or combination of variables.

5.4.2.4 Using Multiple Stepsizes

Using the results in Section 5.4.1, we can exploit information about the relative scales of variables to improve the performance of HMC. This can be done in two equivalent ways. If s_i is a suitable scale for q_i, we could transform q, by setting $q_i' = q_i/s_i$, or we could instead use a kinetic energy function of $K(p) = p^T M^{-1} p$, with M being a diagonal matrix with diagonal elements $m_i = 1/s_i^2$.

A third equivalent way to exploit this information, which is often the most convenient, is to use different stepsizes for different pairs of position and momentum variables. To see how this works, consider a leapfrog update (following Equations 5.18 through 5.20) with $m_i = 1/s_i^2$:

$$p_i(t + \varepsilon/2) = p_i(t) - (\varepsilon/2) \frac{\partial U}{\partial q_i}(q(t)),$$

$$q_i(t + \varepsilon) = q_i(t) + \varepsilon s_i^2 p_i(t + \varepsilon/2),$$

$$p_i(t + \varepsilon) = p_i(t + \varepsilon/2) - (\varepsilon/2) \frac{\partial U}{\partial q_i}(q(t + \varepsilon)).$$

Define $(q^{(0)}, p^{(0)})$ to be the state at the beginning of the leapfrog step (i.e. $(q(t), p(t))$), define $(q^{(1)}, p^{(1)})$ to be the final state (i.e. $(q(t + \varepsilon), p(t + \varepsilon))$), and define $p^{(1/2)}$ to be half-way momentum (i.e. $p(t + \varepsilon/2)$). We can now rewrite the leapfrog step above as

$$p_i^{(1/2)} = p_i^{(0)} - (\varepsilon/2) \frac{\partial U}{\partial q_i}(q^{(0)}),$$

$$q_i^{(1)} = q_i^{(0)} + \varepsilon s_i^2 p_i^{(1/2)},$$

$$p_i^{(1)} = p_i^{(1/2)} - (\varepsilon/2) \frac{\partial U}{\partial q_i}(q^{(1)}).$$

If we now define rescaled momentum variables, $\tilde{p}_i = s_i p_i$, and stepsizes $\varepsilon_i = s_i \varepsilon$, we can write the leapfrog update as

$$\tilde{p}_i^{(1/2)} = \tilde{p}_i^{(0)} - (\varepsilon_i/2) \frac{\partial U}{\partial q_i}(q^{(0)}),$$

$$q_i^{(1)} = q_i^{(0)} + \varepsilon_i \tilde{p}_i^{(1/2)},$$

$$\tilde{p}_i^{(1)} = \tilde{p}_i^{(1/2)} - (\varepsilon_i/2) \frac{\partial U}{\partial q_i}(q^{(1)}).$$

This is just like a leapfrog update with all $m_i = 1$, but with different stepsizes for different (q_i, p_i) pairs. Of course, the successive values for $(q, \tilde{p})$ can no longer be interpreted as following Hamiltonian dynamics at consistent time points, but that is of no consequence for the use of these trajectories in HMC. Note that when we sample for the momentum before each trajectory, each $\tilde{p}_i$ is drawn independently from a Gaussian distribution with mean zero and variance one, regardless of the value of s_i.

This multiple stepsize approach is often more convenient, especially when the estimated scales, s_i, are not fixed, as discussed in Section 5.4.5, and the momentum is only partially refreshed (Section 5.5.3).

5.4.3 Combining HMC with Other MCMC Updates

For some problems, MCMC using HMC alone will be impossible or undesirable. Two situations where non-HMC updates will be necessary are when some of the variables are discrete, and when the derivatives of the log probability density with respect to some of the variables are expensive or impossible to compute. HMC can then be feasibly applied only to the other variables. Another example is when special MCMC updates have been devised that may help convergence in ways that HMC does not—for example, by moving between otherwise isolated modes—but which are not a complete replacement for HMC. As discussed in Section 5.4.5 below, Bayesian hierarchical models may also be best handled with a combination of HMC and other methods such as Gibbs sampling.

In such circumstances, one or more HMC updates for all or a subset of the variables can be alternated with one or more other updates that leave the desired joint distribution of all variables invariant. The HMC updates can be viewed as either leaving this same joint distribution invariant, or as leaving invariant the conditional distribution of the variables that HMC changes, given the current values of the variables that are fixed during the HMC update. These are equivalent views, since the joint density can be factored as this conditional density times the marginal density of the variables that are fixed, which is just a constant from the point of view of a single HMC update, and hence can be left out of the potential energy function.

When both HMC and other updates are used, it may be best to use shorter trajectories for HMC than would be used if only HMC were being done. This allows the other updates to be done more often, which presumably helps sampling. Finding the optimal tradeoff is likely to be difficult, however. A variation on HMC that reduces the need for such a tradeoff is described below in Section 5.5.3.

5.4.4 Scaling with Dimensionality

In Section 5.3.3, one of the main benefits of HMC was illustrated—its ability to avoid the inefficient exploration of the state space via a random walk. This benefit is present (to at least some degree) for most practical problems. For problems in which the dimensionality is moderate to high, another benefit of HMC over simple random-walk Metropolis methods is a slower increase in the computation time needed (for a given level of accuracy) as the dimensionality increases. (Note that here I will consider only sampling performance after equilibrium is reached, not the time needed to approach equilibrium from some initial state not typical of the distribution, which is harder to analyze.)

5.4.4.1 *Creating Distributions of Increasing Dimensionality by Replication*

To talk about how performance scales with dimensionality we need to assume something about how the distribution changes with dimensionality, d.

I will assume that dimensionality increases by adding independent replicas of variables—that is, the potential energy function for $q = (q_1, \ldots, q_d)$ has the form $U(q) = \Sigma\, u_i(q_i)$, for functions u_i drawn independently from some distribution. Of course, this is not what any real practical problem is like, but it may be a reasonable model of the effect of increasing dimensionality for some problems—for instance, in statistical physics, distant regions of large systems are often nearly independent. Note that the independence assumption itself is not crucial since, as discussed in Section 5.4.1, the performance of HMC (and of simple random-walk Metropolis) does not change if independence is removed by rotating the coordinate system, provided the kinetic energy function (or random-walk proposal distribution) is rotationally symmetric.

For distributions of this form, in which the variables are independent, Gibbs sampling will perform very well (assuming it is feasible), producing an independent point after each scan of all variables. Applying Metropolis updates to each variable separately will also work well, provided the time for a single-variable update does not grow with d. However, these methods are not invariant to rotation, so this good performance may not generalize to the more interesting distributions for which we hope to obtain insight with the analysis below.

5.4.4.2 *Scaling of HMC and Random-Walk Metropolis*

Here, I discuss informally how well HMC and random-walk Metropolis scale with dimension, loosely following Creutz (1988, Section III).

To begin, Cruetz notes that the following relationship holds when any Metropolis-style algorithm is used to sample a density $P(x) = (1/Z) \exp(-E(x))$:

$$1 = \mathrm{E}\,[P(x^*)/P(x)] = \mathrm{E}\,[\exp(-(E(x^*) - E(x)))] = \mathrm{E}\,[\exp(-\Delta)], \qquad (5.26)$$

where x is the current state, assumed to be distributed according to $P(x)$, x^* is the proposed state, and $\Delta = E(x^*) - E(x)$. Jensen's inequality then implies that the expectation of the energy difference is nonnegative:

$$\mathrm{E}\,[\Delta] \geq 0.$$

The inequality will usually be strict.

When $U(q) = \Sigma\,u_i(q_i)$, and proposals are produced independently for each i, we can apply these relationships either to a single variable (or pair of variables) or to the entire state. For a single variable (or pair), I will write Δ_1 for $E(x^*) - E(x)$, with $x = q_i$ and $E(x) = u_i(q_i)$, or $x = (q_i, p_i)$ and $E(x) = u_i(q_i) + p_i^2/2$. For the entire state, I will write Δ_d for $E(x^*) - E(x)$, with $x = q$ and $E(x) = U(q)$, or $x = (q, p)$ and $E(x) = U(q) + K(p)$. For both random-walk Metropolis and HMC, increasing dimension by replicating variables will lead to increasing energy differences, since Δ_d is the sum of Δ_1 for each variable, each of which has positive mean. This will lead to a decrease in the acceptance probability—equal to $\min(1, \exp(-\Delta_d))$—unless the width of the proposal distribution or the leapfrog stepsize is decreased to compensate.

More specifically, for random-walk Metropolis with proposals that change each variable independently, the difference in potential energy between a proposed state and the current state will be the sum of independent differences for each variable. If we fix the standard deviation, ς, for each proposed change, the mean and the variance of this potential energy difference will both grow linearly with d, which will lead to a progressively lower acceptance rate. To maintain reasonable performance, ς will have to decrease as d increases. Furthermore, the number of iterations needed to reach a nearly independent point will be proportional to ς^{-2}, since exploration is via a random walk.

Similarly, when HMC is used to sample from a distribution in which the components of q are independent, using the kinetic energy $K(p) = \Sigma\,p_i^2/2$, the different (q_i, p_i) pairs do not interact during the simulation of a trajectory—each (q_i, p_i) pair follows Hamiltonian dynamics according to just the one term in the potential energy involving q_i and the one term in the kinetic energy involving p_i. There is therefore no need for the length in fictitious time of a trajectory to increase with dimensionality. However, acceptance of the endpoint of the trajectory is based on the error in H due to the leapfrog approximation, which is the sum of the errors pertaining to each (q_i, p_i) pair. For a fixed stepsize, ε, and fixed trajectory length, εL, both the mean and the variance of the error in H grow linearly with d. This will lead to a progressively lower acceptance rate as dimensionality increases, if it is not counteracted by a decrease in ε. The number of leapfrog steps needed to reach an independent point will be proportional to ε^{-1}.

To see which method scales better, we need to determine how rapidly we must reduce ς and ε as d increases, in order to maintain a reasonable acceptance rate. As d increases and ς or ε goes to zero, Δ_1 will go to zero as well. Using a second-order approximation of $\exp(-\Delta_1)$ as $1 - \Delta_1 + \Delta_1^2/2$, together with Equation 5.26, we find that

$$\mathrm{E}\,[\Delta_1] \approx \frac{\mathrm{E}\,[\Delta_1^2]}{2}. \tag{5.27}$$

It follows from this that the variance of Δ_1 is twice the mean of Δ_1 (when Δ_1 is small), which implies that the variance of Δ_d is twice the mean of Δ_d (even when Δ_d is not small). To achieve a good acceptance rate, we must therefore keep the mean of Δ_d near one, since a large mean will not be saved by a similarly large standard deviation (which would produce fairly frequent acceptances as Δ_d occasionally takes on a negative value).

For random-walk Metropolis with a symmetric proposal distribution, we can see how ς needs to scale by directly averaging Δ_1 for a proposal and its inverse. Let the proposal for one variable be $x^* = x + c$, and suppose that $c = a$ and $c = -a$ are equally likely. Approximating $U(x^*)$ to second order as $U(x) + cU'(x) + c^2U''(x)/2$, we find that the average of $\Delta_1 = U(x^*) - U(x)$ over $c = a$ and $c = -a$ is $a^2U''(x)$. Averaging this over the distribution of a, with standard deviation ς, and over the distribution of x, we see that $E[\Delta_1]$ is proportional to ς^2. It follows that $E[\Delta_d]$ is proportional to $d\varsigma^2$, so we can maintain a reasonable acceptance rate by letting ς be proportional to $d^{-1/2}$. The number of iterations needed to reach a nearly independent point will be proportional to ς^{-2}, which will be proportional to d. The amount of computation time needed will typically be proportional to d^2.

As discussed at the end of Section 5.2.3, the error in H when using the leapfrog discretization to simulate a trajectory of a fixed length is proportional to ε^2 (for sufficiently small ε). The error in H for a single (q_i, p_i) pair is the same as Δ_1, so we see that Δ_1^2 is proportional to ε^4. Equation 5.27 then implies that $E[\Delta_1]$ is also proportional to ε^4. The average total error in H for all variables, $E[\Delta_d]$, will be proportional to $d\varepsilon^4$, and hence we must make ε be proportional to $d^{-1/4}$ to maintain a reasonable acceptance rate. The number of leapfrog updates to reach a nearly independent point will therefore grow as $d^{1/4}$, and the amount of computation time will typically grow as $d^{5/4}$, which is much better than the d^2 growth for random-walk Metropolis.

5.4.4.3 Optimal Acceptance Rates

By extending the analysis above, we can determine what the acceptance rate of proposals is when the optimal choice of ς or ϵ is used. This is helpful when tuning the algorithms—provided, of course, that the distribution sampled is high-dimensional, and has properties that are adequately modeled by a distribution with replicated variables.

To find this acceptance rate, we first note that since Metropolis methods satisfy detailed balance, the probability of an accepted proposal with Δ_d negative must be equal to the probability of an accepted proposal with Δ_d positive. Since all proposals with negative Δ_d are accepted, the acceptance rate is simply twice the probability that a proposal has a negative Δ_d. For large d, the central limit theorem implies that the distribution of Δ_d is Gaussian, since it is a sum of d independent Δ_1 values. (This assumes that the variance of each Δ_1 is finite.) We saw above that the variance of Δ_d is twice its mean, $E[\Delta_d] = \mu$. The acceptance probability can therefore be written as follows (Gupta et al., 1990), for large d:

$$P(\text{accept}) = 2\,\Phi\left(\frac{(0 - \mu)}{\sqrt{2\mu}}\right) = 2\,\Phi\left(-\sqrt{\mu/2}\right) = a(\mu), \qquad (5.28)$$

where $\Phi(z)$ is the cumulative distribution function for a Gaussian variable with mean zero and variance one.

For random-walk Metropolis, the cost of obtaining an independent point will be proportional to $1/(a\varsigma^2)$, where a is the acceptance rate. We saw above that $\mu = E[\Delta_d]$ is proportional to ς^2, so the cost follows the proportionality

$$C_{\text{rw}} \propto \frac{1}{(a(\mu)\mu)}.$$

Numerical calculation shows that this is minimized when $\mu = 2.8$ and $a(\mu) = 0.23$.

For HMC, the cost of obtaining an independent point will be proportional to $1/(a\varepsilon)$, and as we saw above, μ is proportional to ε^4. From this we obtain

$$C_{\text{HMC}} \propto \frac{1}{(a(\mu)\mu^{1/4})}.$$

Numerical calculation shows that the minimum is when $\mu = 0.41$ and $a(\mu) = 0.65$.

The same optimal 23% acceptance rate for random-walk Metropolis was previously obtained using a more formal analysis by Roberts et al. (1997). The optimal 65% acceptance rate for HMC that I derive above is consistent with previous empirical results on distributions following the model here (Neal, 1994, Figure 2), and on real high-dimensional problems (Creutz, 1988, Figures 2 and 3; Sexton and Weingarten, 1992, Table 1). Kennedy and Pendleton (1991) obtained explicit and rigorous results for HMC applied to multivariate Gaussian distributions.

5.4.4.4 Exploring the Distribution of Potential Energy

The better scaling behavior of HMC seen above depends crucially on the resampling of momentum variables. We can see this by considering how well the methods explore the distribution of the potential energy, $U(q) = \Sigma u_i(q_i)$. Because $U(q)$ is a sum of d independent terms, its standard deviation will grow in proportion to $d^{1/2}$.

Following Caracciolo et al. (1994), we note that the expected change in potential energy from a single Metropolis update will be no more than order 1—intuitively, large upwards changes are unlikely to be accepted, and since Metropolis updates satisfy detailed balance, large downward changes must also be rare (in equilibrium). Because changes in U will follow a random walk (due again to detailed balance), it will take at least order $(d^{1/2}/1)^2 = d$ Metropolis updates to explore the distribution of U.

In the first step of an HMC iteration, the resampling of momentum variables will typically change the kinetic energy by an amount that is proportional to $d^{1/2}$, since the kinetic energy is also a sum of d independent terms, and hence has standard deviation that grows as $d^{1/2}$ (more precisely, its standard deviation is $d^{1/2}/2^{1/2}$). If the second step of HMC proposes a distant point, this change in kinetic energy (and hence in H) will tend, by the end of the trajectory, to have become equally split between kinetic and potential energy. If the endpoint of this trajectory is accepted, the change in potential energy from a single HMC iteration will be proportional to $d^{1/2}$, comparable to its overall range of variation. So, in contrast to random-walk Metropolis, we may hope that only a few HMC iterations will be sufficient to move to a nearly independent point, even for high-dimensional distributions.

Analyzing how well methods explore the distribution of U can also provide insight into their performance on distributions that are not well modeled by replication of variables, as we will see in the next section.

5.4.5 HMC for Hierarchical Models

Many Bayesian models are defined hierarchically. A large set of low-level parameters have prior distributions that are determined by fewer higher-level "hyperparameters," which in turn may have priors determined by yet-higher-level hyperparameters. For example, in a regression model with many predictor variables, the regression coefficients might be given Gaussian prior distributions, with a mean of zero and a variance that is a hyperparameter. This hyperparameter could be given a broad prior distribution, so that its posterior distribution is determined mostly by the data.

One could apply HMC to these models in an obvious way (after taking the logs of variance hyperparameters, so they will be unconstrained). However, it may be better to apply HMC only to the lower-level parameters, for reasons I will now discuss. (See Section 5.4.3 for general discussion of applying HMC to a subset of variables.)

I will use my work on Bayesian neural network models (Neal, 1996a) as an example. Such models typically have several groups of low-level parameters, each with an associated variance hyperparameter. The posterior distribution of these hyperparameters reflects important aspects of the data, such as which predictor variables are most relevant to the task. The efficiency with which values for these hyperparameters are sampled from the posterior distribution can often determine the overall efficiency of the MCMC method.

I use HMC only for the low-level parameters in Bayesian neural network models, with the hyperparameters being fixed during an HMC update. These HMC updates alternate with Gibbs sampling updates of the hyperparameters, which (in the simpler versions of the models) are independent given the low-level parameters, and have conditional distributions of standard form. By using HMC only for the low-level parameters, the leapfrog stepsizes used can be set using heuristics that are based on the current hyperparameter values. (I use the multiple stepsize approach described at the end of Section 5.4.2, equivalent to using different mass values, m_i, for different parameters.) For example, the size of the network "weights" on connections out of a "hidden unit" determine how sensitive the likelihood function is to changes in weights on connections into the hidden unit; the variance of the weights on these outgoing connections is therefore useful in setting the stepsize for the weights on the incoming connections. If the hyperparameters were changed by the same HMC updates as change the lower-level parameters, using them to set stepsizes would not be valid, since a reversed trajectory would use different stepsizes, and hence not retrace the original trajectory. Without a good way to set stepsizes, HMC for the low-level parameters would likely be much less efficient.

Choo (2000) bypassed this problem by using a modification of HMC in which trajectories are simulated by alternating leapfrog steps that update only the hyperparameters with leapfrog steps that update only the low-level parameters. This procedure maintains both reversibility and volume-preservation (though not necessarily symplecticness), while allowing the stepsizes for the low-level parameters to be set using the current values of the hyperparameters (and vice versa). However, performance did not improve as hoped because of a second issue with hierarchical models.

In these Bayesian neural network models, and many other hierarchical models, the joint distribution of both low-level parameters and hyperparameters is highly skewed, with the probability density varying hugely from one region of high posterior probability to another. Unless the hyperparameters controlling the variances of low-level parameters have very narrow posterior distributions, the joint posterior density for hyperparameters and low-level parameters will vary greatly from when the variance is low to when it is high.

For instance, suppose that in its region of high posterior probability, a variance hyperparameter varies by a factor of 4. If this hyperparameter controls 1000 low-level parameters, their typical prior probability density will vary by a factor of $2^{1000} = 1.07 \times 10^{301}$, corresponding to a potential energy range of $\log(2^{1000}) = 693$, with a standard deviation of $693/12^{1/2} = 200$ (since the variance of a uniform distribution is one twelfth of its range). As discussed at the end of Section 5.4.4, one HMC iteration changes the energy only through the resampling of the momentum variables, which at best leads to a change in potential energy with standard deviation of about $d^{1/2}/2^{3/2}$. For this example, with 1000 low-level parameters, this is 11.2, so about $(200/11.2)^2 = 319$ HMC iterations will be needed to reach an independent point.

One might obtain similar performance for this example using Gibbs sampling. However, for neural network models, there is no feasible way of using Gibbs sampling for the posterior distribution of the low-level parameters, but HMC can be applied to the conditional distribution of the low-level parameters given the hyperparameters. Gibbs sampling can then be used to update the hyperparameters. As we have seen, performance would not be improved by trying to update the hyperparameters with HMC as well, and updating them by Gibbs sampling is easier.

Choo (2000) tried another approach that could potentially improve on this—reparameterizing low-level parameters θ_i, all with variance $\exp(\kappa)$, by letting $\theta_i = \phi_i \exp(\kappa/2)$, and then sampling for κ and the ϕ_i using HMC. The reparameterization eliminates the extreme variation in probability density that HMC cannot efficiently sample. However, he found that it is difficult to set a suitable stepsize for κ, and that the error in H tended to grow with trajectory length, unlike the typical situation when HMC is used only for the low-level parameters. Use of "tempering" techniques (see Section 5.5.7) is another possibility.

Even though it does not eliminate all difficulties, HMC is very useful for Bayesian neural network models—indeed, without it, they might not be feasible for most applications. Using HMC for at least the low-level parameter can produce similar benefits for other hierarchical models (e.g. Ishwaran, 1999), especially when the posterior correlations of these low-level parameters are high. As in any application of HMC, however, careful tuning of the stepsize and trajectory length is generally necessary.

5.5   Extensions of and Variations on HMC

The basic HMC algorithm (Figure 5.2) can be modified in many ways, either to improve its efficiency, or to make it useful for a wider range of distributions. In this section, I will start by discussing alternatives to the leapfrog discretization of Hamilton's equations, and also show how HMC can handle distributions with constraints on the variables (e.g. variables that must be positive). I will then discuss a special case of HMC—when only one leapfrog step is done—and show how it can be extended to produce an alternative method of avoiding random walks, which may be useful when not all variables are updated by HMC. Most applications of HMC can benefit from using a variant in which "windows" of states are used to increase the acceptance probability. Another widely applicable technique is to use approximations to the Hamiltonian to compute trajectories, while still obtaining correct results by using the exact Hamiltonian when deciding whether to accept the endpoint of the trajectory. Tuning of HMC may be assisted by using a "short-cut" method that avoids computing the whole trajectory when the stepsize is inappropriate. Tempering methods have potential to help with distributions having multiple modes, or which are highly skewed.

There are many other variations that I will not be able to review here, such as the use of a "shadow Hamiltonian" that is exactly conserved by the inexact simulation of the real Hamiltonian (Izagguirre and Hampton, 2004), and the use of symplectic integration methods more sophisticated than the leapfrog method (e.g. Creutz and Gocksch, 1989), including a recent proposal by Girolami et al. (2009) to use a symplectic integrator for a nonseparable

Hamiltonian in which M in the kinetic energy of (Equation 5.5) depends on q, allowing for "adaptation" based on local information.

5.5.1 Discretization by Splitting: Handling Constraints and Other Applications

The leapfrog method is not the only discretization of Hamilton's equations that is reversible and volume-preserving, and hence can be used for HMC. Many "symplectic integration methods" have been devised, mostly for applications other than HMC (e.g. simulating the solar system for millions of years to test its stability). It is possible to devise methods that have a higher order of accuracy than the leapfrog method (see, e.g. McLachlan and Atela, 1992). Using such a method for HMC will produce asymptotically better performance than the leapfrog method, as dimensionality increases. Experience has shown, however, that the leapfrog method is hard to beat in practice.

Nevertheless, it is worth taking a more general look at how Hamiltonian dynamics can be simulated, since this also points to how constraints on the variables can be handled, as well as possible improvements such as exploiting partial analytic solutions.

5.5.1.1 Splitting the Hamiltonian

Many symplectic discretizations of Hamiltonian dynamics can be derived by "splitting" the Hamiltonian into several terms, and then, for each term in succession, simulating the dynamics defined by that term for some small time step, then repeating this procedure until the desired total simulation time is reached. If the simulation for each term can be done analytically, we obtain a symplectic approximation to the dynamics that is feasible to implement.

This general scheme is described by Leimkuhler and Reich (2004, Section 4.2) and by Sexton and Weingarten (1992). Suppose that the Hamiltonian can be written as a sum of k terms, as follows:

$$H(q, p) = H_1(q, p) + H_2(q, p) + \cdots + H_{k-1}(q, p) + H_k(q, p).$$

Suppose also that we can *exactly* implement Hamiltonian dynamics based on each H_i, for $i = 1, \ldots, k$, with $T_{i,\varepsilon}$ being the mapping defined by applying dynamics based on H_i for time ε. As shown by Leimkuhler and Reich, if the H_i are twice differentiable, the composition of these mappings, $T_{1,\varepsilon} \circ T_{2,\varepsilon} \circ \cdots \circ T_{k-1,\varepsilon} \circ T_{k,\varepsilon}$, is a valid discretization of Hamiltonian dynamics based on H, which will reproduce the exact dynamics in the limit as ε goes to zero, with global error of order ε or less.

Furthermore, this discretization will preserve volume, and will be symplectic, since these properties are satisfied by each of the $T_{i,\varepsilon}$ mappings. The discretization will also be reversible if the sequence of H_i is symmetrical—that is, $H_i(q, p) = H_{k-i+1}(q, p)$. As mentioned at the end of Section 5.2.3, any reversible method must have global error of even order in ε (Leimkuhler and Reich, 2004, Section 4.3.3), which means that the global error must be of order ε^2 or better.

We can derive the leapfrog method from a symmetrical splitting of the Hamiltonian. If $H(q, p) = U(q) + K(p)$, we can write the Hamiltonian as

$$H(q, p) = \frac{U(q)}{2} + K(p) + \frac{U(q)}{2},$$

which corresponds to a split with $H_1(q,p) = H_3(q,p) = U(q)/2$ and $H_2(q,p) = K(p)$. Hamiltonian dynamics based on H_1 is (Equations 5.1 and 5.2):

$$\frac{dq_i}{dt} = \frac{\partial H_1}{\partial p_i} = 0,$$

$$\frac{dp_i}{dt} = -\frac{\partial H_1}{\partial q_i} = -\frac{1}{2}\frac{\partial U}{\partial q_i}.$$

Applying this dynamics for time ε just adds $-(\varepsilon/2)\,\partial U/\partial q_i$ to each p_i, which is the first part of a leapfrog step (Equation 5.18). The dynamics based on H_2 is as follows:

$$\frac{dq_i}{dt} = \frac{\partial H_2}{\partial p_i} = \frac{\partial K}{\partial p_i},$$

$$\frac{dp_i}{dt} = -\frac{\partial H_2}{\partial q_i} = 0.$$

If $K(p) = \frac{1}{2}\sum p_i^2/m_i$, applying this dynamics for time ϵ results in adding $\epsilon p_i/m_i$ to each q_i, which is the second part of a leapfrog step Equation 5.19. Finally, H_3 produces the third part of a leapfrog step (Equation 5.20), which is the same as the first part, since $H_3 = H_1$.

5.5.1.2 Splitting to Exploit Partial Analytical Solutions

One situation where splitting can help is when the potential energy contains a term that can, on its own, be handled analytically. For example, the potential energy for a Bayesian posterior distribution will be the sum of minus the log prior density for the parameters and minus the log likelihood. If the prior is Gaussian, the log prior density term will be quadratic, and can be handled analytically (see the one-dimensional example at the end of Section 5.2.1).

We can modify the leapfrog method for this situation by using a modified split. Suppose that $U(q) = U_0(q) + U_1(q)$, with U_0 being analytically tractable, in conjunction with the kinetic energy, $K(p)$. We use the split

$$H(q,p) = \frac{U_1(q)}{2} + \left[U_0(q) + K(p)\right] + \frac{U_1(q)}{2}, \tag{5.29}$$

that is, $H_1(q,p) = H_3(q,p) = U_1(q)/2$ and $H_2(q,p) = U_0(q) + K(p)$. The first and last half steps for p are the same as for ordinary leapfrog, based on U_1 alone. The middle full step for q, which in ordinary leapfrog just adds εp to q, is replaced by the analytical solution for following the exact dynamics based on the Hamiltonian $U_0(q) + K(p)$ for time ε.

With this procedure, it should be possible to use a larger stepsize (and hence use fewer steps in a trajectory), since part of the potential energy has been separated out and handled exactly. The benefit of handling the prior exactly may be limited, however, since the prior is usually dominated by the likelihood.

5.5.1.3 Splitting Potential Energies with Variable Computation Costs

Splitting can also help if the potential energy function can be split into two terms, one of which requires less computation time to evaluate than the other (Sexton and Weingarten,

1992). Suppose that $U(q) = U_0(q) + U_1(q)$, with U_0 being cheaper to compute than U_1, and let the kinetic energy be $K(p)$. We can use the following split, for some $M > 1$:

$$H(q,p) = \frac{U_1(q)}{2} + \sum_{m=1}^{M} \left[\frac{U_0(q)}{2M} + \frac{K(p)}{M} + \frac{U_0(q)}{2M} \right] + \frac{U_1(q)}{2}.$$

We label the $k = 3M + 2$ terms as $H_1(q,p) = H_k(q,p) = U_1(q)/2$ and, for $i = 1, \ldots, M$, $H_{3i-1}(q,p) = H_{3i+1}(q,p) = U_0(q)/2M$ and $H_{3i}(q,p) = K(p)/M$. The resulting discretization can be seen as a nested leapfrog method. The M inner leapfrog steps involve only U_0, and use an effective stepsize of ε/M. The outer leapfrog step takes half steps for p using only U_1, and replaces the update for q in the middle with the M inner leapfrog steps.

If U_0 is much cheaper to compute than U_1, we can use a large value for M without increasing computation time by much. The stepsize, ε, that we can use will then be limited mostly by the properties of U_1, since the effective stepsize for U_0 is much smaller, ε/M. Using a bigger ε than with the standard leapfrog method will usually be possible, and hence we will need fewer steps in a trajectory, with fewer computations of U_1.

5.5.1.4 *Splitting according to Data Subsets*

When sampling from the posterior distribution for a Bayesian model of independent data points, it may be possible to save computation time by splitting the potential energy into terms for subsets of the data.

Suppose that we partition the data into subsets S_m, for $m = 1, \ldots, M$, typically of roughly equal size. We can then write the log likelihood function as $\ell(q) = \sum_{m=1}^{M} \ell_m(q)$, where ℓ_m is the log likelihood function based on the data points in S_m. If $\pi(q)$ is the prior density for the parameters, we can let $U_m(q) = -\log(\pi(q))/M - \ell_m(q)$, and split the Hamiltonian as follows:

$$H(q,p) = \sum_{m=1}^{M} \left[\frac{U_m(q)}{2} + K(p) \Big/ M + \frac{U_m(q)}{2} \right];$$

that is, we let the $k = 3M$ terms be $H_{3m-2}(q,p) = H_{3m}(q,p) = U_m(q)/2$ and $H_{3m-1}(q,p) = K(p)/m$. The resulting discretization with stepsize ε effectively performs M leapfrog steps with stepsize ε/M, with the mth step using MU_m as the potential energy function.

This scheme can be beneficial if the data set is redundant, with many data points that are similar. We then expect $MU_m(q)$ to be approximately the same as $U(q)$, and we might hope that we could set ε to be M times larger than with the standard leapfrog method, obtaining similar results with M times less computation. In practice, however, the error in H at the end of the trajectory will be larger than with standard leapfrog, so the gain will be less than this. I found (Neal, 1996a, Sections 3.5.1 and 3.5.2) that the method can be beneficial for neural network models, especially when combined with the windowed HMC procedure described below in Section 5.5.4.

Note that unlike the other examples above, this split is *not* symmetrical, and hence the resulting discretization is not reversible. However, it can still be used to produce a proposal for HMC as long as the labeling of the subsets is randomized for each iteration, so that the reverse trajectory has the same probability of being produced as the forward trajectory. (It is possible, however, that some symmetrical variation on this split might produce better results.)

5.5.1.5   Handling Constraints

An argument based on splitting shows how to handle constraints on the variables being sampled. Here, I will consider only separate constraints on some subset of the variables, with the constraint on q_i taking the form $q_i \leq u_i$, or $q_i \geq l_i$, or both. A similar scheme can handle constraints taking the form $G(q) \geq 0$, for any differentiable function G.

We can impose constraints on variables by letting the potential energy be infinite for values of q that violate any of the constraints, which will give such points probability zero. To see how to handle such infinite potential energies, we can look at a limit of potential energy functions that approach infinity, and the corresponding limit of the dynamics.

To illustrate, suppose that $U_*(q)$ is the potential energy ignoring constraints, and that q_i is constrained to be less than u_i. We can take the limit as $r \to \infty$ of the following potential energy function (which is one of many that could be used):

$$U(q) = U_*(q) + C_r(q_i, u_i), \quad \text{where } C_r(q_i, u_i) = \begin{cases} 0, & \text{if } q_i \leq u_i, \\ r^{r+1}(q_i - u_i)^r, & \text{if } q_i > u_i. \end{cases}$$

It is easy to see that $\lim_{r \to \infty} C_r(q_i, u_i)$ is zero for any $q_i \leq u_i$ and infinity for any $q_i > u_i$. For any finite $r > 1$, $U(q)$ is differentiable, so we can use it to define Hamiltonian dynamics.

To simulate the dynamics based on this $U(q)$, with a kinetic energy $K(p) = \frac{1}{2} \sum p_i^2 / m_i$, we can use the split of Equation 5.29, with $U_1(q) = U_*(q)$ and $U_0(q) = C_r(q_i, u_i)$:

$$H(q, p) = \frac{U_*(q)}{2} + \left[C_r(q_i, u_i) + K(p) \right] + \frac{U_*(q)}{2}.$$

This produces a variation on the leapfrog method in which the half steps for p (Equations 5.18 and 5.19) remain the same, but the full step for q (Equation 5.19) is modified to account for the constraint on q_i. After computing $q_i' = q_i(t) + \varepsilon p_i(t + \varepsilon/2)/m_i$, we check if $q_i' > u_i$. If not, the value of $C_r(q_i, u_i)$ must be zero all along the path from q_i to q_i', and we can set $q(t + \varepsilon)$ to q_i'. But if $q_i' > u_i$, the dynamics based on the Hamiltonian $C_r(q_i, u_i) + K(p)$ will be affected by the C_r term. This term can be seen as a steep hill, which will be climbed as q_i moves past u_i, until the point is reached where C_r is equal to the previous value of $\frac{1}{2} p_i^2 / m_i$, at which point p_i will be zero. (If r is sufficiently large, as it will be in the limit as $r \to \infty$, this point will be reached before the end of the full step.) We will then fall down the hill, with p_i taking on increasingly negative values, until we again reach $q_i = u_i$, when p_i will be just the negative of the original value of p_i. We then continue, now moving in the opposite direction, away from the upper limit.

If several variables have constraints, we must follow this procedure for each, and if a variable has both upper and lower constraints, we must repeat the procedure until neither constraint is violated. The end result is that the full step for q of Equation 5.19 is replaced by the procedure shown in Figure 5.8. Intuitively, the trajectory just bounces off the "walls" given by the constraints. If $U_*(q)$ is constant, these bounces are the only interesting aspect of the dynamics, and the procedure is sometimes referred to as "billiards" (see, e.g. Ruján, 1997).

5.5.2   Taking One Step at a Time—The Langevin Method

A special case of HMC arises when the trajectory used to propose a new state consists of only a single leapfrog step. Suppose that we use the kinetic energy $K(p) = \frac{1}{2} \sum p_i^2$. An

```
For each variable, i=1,...,d:

  1) Let p'_i= p_i(t+ε/2)
  2) Let q'_i=q_i(t)+εp'_i/m_i
  3) If q_i is constrained, repeat the following until q'_i satisfies
     all constraints:
       a)  If q_i has an upper constraint, and q'_i > u_i
             Let q'_i=u_i-(q'_i-u_i) and p'_i=-p'_i
       b)  If q_i has a lower constraint, and q'_i < l_i
             Let q'_i=l_i+(l_i-q'_i) and p'_i=-p'_i
  4) Let q_i(t+ε)=q'_i and p_i(t+ε/2)=p'_i
```

FIGURE 5.8
Modification to the leapfrog update of q (Equation 5.19) to handle constraints of the form $q_i \le u_i$ or $q_i \le l_i$.

iteration of HMC with one leapfrog step can be expressed in the following way. We sample values for the momentum variables, p, from their Gaussian distributions with mean zero and variance one, and then propose new values, q^* and p^*, as follows:

$$q_i^* = q_i - \frac{\varepsilon^2}{2}\frac{\partial U}{\partial q_i}(q) + \varepsilon p_i, \qquad (5.30)$$

$$p_i^* = p_i - \frac{\varepsilon}{2}\frac{\partial U}{\partial q_i}(q) - \frac{\varepsilon}{2}\frac{\partial U}{\partial q_i}(q^*). \qquad (5.31)$$

We accept q^* as the new state with probability

$$\min\left[1, \exp\left(-(U(q^*) - U(q)) - \frac{1}{2}\sum_i ((p_i^*)^2 - p_i^2)\right)\right], \qquad (5.32)$$

and otherwise keep q as the new state. Equation 5.30 is known in physics as one type of "Langevin equation," and this method is therefore known as *Langevin Monte Carlo* (LMC) in the lattice field theory literature (e.g. Kennedy, 1990).

One can also remove any explicit mention of momentum variables, and view this method as performing a Metropolis–Hastings update in which q^* is proposed from a Gaussian distribution where the q_i^* are independent, with means of $q_i - (\varepsilon^2/2)[\partial U/\partial q_i](q)$ and variances of ε^2. Since this proposal is not symmetrical, it must be accepted or rejected based both on the ratio of the probability densities of q^* and q and on the ratio of the probability densities for proposing q from q^* and vice versa (Hastings, 1970). To see the equivalence with HMC using one leapfrog step, we can write the Metropolis–Hastings acceptance probability as follows:

$$\min\left[1, \frac{\exp(-U(q^*))}{\exp(-U(q))} \prod_{i=1}^{d} \frac{\exp\left(-\left(q_i - q_i^* + (\varepsilon^2/2)[\partial U/\partial q_i](q^*)\right)^2/2\varepsilon^2\right)}{\exp\left(-\left(q_i^* - q_i + (\varepsilon^2/2)[\partial U/\partial q_i](q))\right)^2/2\varepsilon^2\right)}\right]. \qquad (5.33)$$

To see that this is the same as Equation 5.32, note that using Equations 5.30 and 5.31, we can write

$$p = \frac{1}{\varepsilon}\left[q_i^* - q_i + \frac{\varepsilon^2}{2}\frac{\partial U}{\partial q_i}(q)\right],$$

$$p^* = -\frac{1}{\varepsilon}\left[q_i - q_i^* + \frac{\varepsilon^2}{2}\frac{\partial U}{\partial q_i}(q^*)\right].$$

After substituting these into Equation 5.32, it is straightforward to see the equivalence to Equation 5.33.

In this Metropolis–Hastings form, the LMC method was first proposed by Rossky et al. (1978) for use in physical simulations. Approximate Langevin methods without an accept/reject step can also be used (for a discussion of this, see Neal, 1993, Section 5.3)—as, for instance, in a paper on statistical inference for complex models by Grenander and Miller (1990), where also an accept/reject step is proposed in the discussion by J. Besag (p. 591).

Although LMC can be seen as a special case of HMC, its properties are quite different. Since LMC updates are reversible, and generally make only small changes to the state (since ε typically cannot be very large), LMC will explore the distribution via an inefficient random walk, just like random-walk Metropolis updates.

However, LMC has better scaling behavior than random-walk Metropolis as dimensionality increases, as can be seen from an analysis paralleling that in Section 5.4.4 (Creutz, 1988; Kennedy, 1990). The local error of the leapfrog step is of order ε^3, so $\mathrm{E}[\Delta_1^2]$, the average squared error in H from one variable, will be of order ε^6. From Equation 5.27, $\mathrm{E}[\Delta]$ will also be of order ε^6, and with d independent variables, $\mathrm{E}[\Delta_d]$ will be of order $d\varepsilon^6$, so that ε must scale as $d^{-1/6}$ in order to maintain a reasonable acceptance rate. Since LMC explores the distribution via a random walk, the number of iterations needed to reach a nearly independent point will be proportional to ε^{-2}, which grows as $d^{1/3}$, and the computation time to reach a nearly independent point grows as $d^{4/3}$. This is better than the d^2 growth in computation time for random-walk Metropolis, but worse than the $d^{5/4}$ growth when HMC is used with trajectories that are long enough to reach a nearly independent point.

We can also find what the acceptance rate for LMC will be when the optimal ε is used, when sampling a distribution with independent variables replicated d times. As for random-walk Metropolis and HMC, the acceptance rate is given in terms of $\mu = \mathrm{E}[\Delta_d]$ by Equation 5.28. The cost of obtaining a nearly independent point using LMC is proportional to $1/(a(\mu)\varepsilon^2)$, and since μ is proportional to ε^6, we can write the cost as

$$C_{\mathrm{LMC}} \propto \frac{1}{(a(\mu)\mu^{1/3})}.$$

Numerical calculation shows that this is minimized when $a(\mu)$ is 0.57, a result obtained more formally by Roberts and Rosenthal (1998). This may be useful for tuning, if the behavior of LMC for the distribution being sampled resembles its behavior when sampling for replicated independent variables.

5.5.3   Partial Momentum Refreshment: Another Way to Avoid Random Walks

The single leapfrog step used in the LMC algorithm will usually not be sufficient to move to a nearly independent point, so LMC will explore the distribution via an inefficient random walk. This is why HMC is typically used with trajectories of many leapfrog steps. An alternative that can suppress random-walk behavior even when trajectories consist of just one leapfrog step is to only partially refresh the momentum between trajectories, as proposed by Horowitz (1991).

Suppose that the kinetic energy has the typical form $K(p) = p^T M^{-1} p/2$. The following update for p will leave invariant the distribution for the momentum (Gaussian with mean zero and covariance M):

$$p' = \alpha p + (1 - \alpha^2)^{1/2} n. \tag{5.34}$$

Here, α is any constant in the interval $[-1, +1]$, and n is a Gaussian random vector with mean zero and covariance matrix M. To see this, note that if p has the required Gaussian distribution, the distribution of p' will also be Gaussian (since it is a linear combination of independent Gaussians), with mean 0 and covariance $\alpha^2 M + (1 - \alpha^2)M = M$.

If α is only slightly less than one, p' will be similar to p, but repeated updates of this sort will eventually produce a value for the momentum variables almost independent of the initial value. When $\alpha = 0$, p' is just set to a random value drawn from its Gaussian distribution, independent of its previous value. Note that when M is diagonal, the update of each momentum variable, p_i, is independent of the updates of other momentum variables.

The partial momentum update of Equation 5.34 can be substituted for the full replacement of the momentum in the standard HMC algorithm. This gives a generalized HMC algorithm in which an iteration consists of three steps:

1. Update the momentum variables using Equation 5.34. Let the new momentum be p'.

2. Propose a new state, (q^*, p^*), by applying L leapfrog steps with stepsize ε, starting at (q, p'), and then negating the momentum. Accept (q^*, p^*) with probability

$$\min\left[1, \exp\left(-U(q^*) + U(q) - K(p^*) + K(p')\right)\right].$$

If (q^*, p^*) is accepted, let $(q'', p'') = (q^*, p^*)$; otherwise, let $(q'', p'') = (q, p')$.

3. Negate the momentum, so that the new state is $(q'', -p'')$.

The transitions in each of these steps—$(q, p) \to (q, p')$, $(q, p') \to (q'', p'')$, and $(q'', p'') \to (q'', -p'')$—leave the canonical distribution for (q, p) invariant. The entire update therefore also leaves the canonical distribution invariant. The three transitions also each satisfy detailed balance, but the sequential combination of the three does *not* satisfy detailed balance (except when $\alpha = 0$). This is crucial, since if the combination were reversible, it would still result in random-walk behavior when L is small.

Note that omitting step (3) above would result in a valid algorithm, but then, far from suppressing random walks, the method (with α close to one) would produce nearly back-and-forth motion, since the direction of motion would reverse with every trajectory accepted in step (2). With the reversal in step (3), motion continues in the same direction as long as the trajectories in step (2) are accepted, since the two negations of p will cancel. Motion reverses whenever a trajectory is rejected, so if random-walk behavior is to be suppressed, the rejection rate must be kept small.

If $\alpha = 0$, the above algorithm is the same as standard HMC, since step (1) will completely replace the momentum variables, step (2) is the same as for standard HMC, and step (3) will have no effect, since the momentum will be immediately replaced anyway, in step (1) of the next iteration.

Since this algorithm can be seen as a generalization of standard HMC, with an additional α parameter, one might think it will offer an improvement, provided that α is tuned for best performance. However, Kennedy and Pendleton (2001) show that when the method is applied to high-dimensional multivariate Gaussian distributions only a small constant factor improvement is obtained, with no better scaling with dimensionality. Best performance is obtained using long trajectories (L large), and a value for α that is not very close to one (but not zero, so the optimum choice is not standard HMC). If L is small, the need to keep the rejection rate very low (by using a small ε), as needed to suppress random walks, makes the method less advantageous than standard HMC.

It is disappointing that only a small improvement is obtained with this generalization when sampling a multivariate Gaussian, due to limitations that likely apply to other distributions as well. However, the method may be more useful than one would think from this. For reasons discussed in Sections 5.4.3 and 5.4.5, we will often combine HMC updates with other MCMC updates (perhaps for variables not changed by HMC). There may then be a tradeoff between using long trajectories to make HMC more efficient, and using shorter trajectories so that the other MCMC updates can be done more often. If shorter-than-optimal trajectories are to be used for this reason, setting α greater than zero can reduce the random-walk behavior that would otherwise result.

Furthermore, rejection rates can be reduced using the "window" method described next. An analysis of partial momentum refreshment combined with the window method might find that using trajectories of moderate length in conjunction with a value for α greater than zero produces a more substantial improvement.

5.5.4 Acceptance Using Windows of States

Figure 5.3 (right plot) shows how the error in H varies along a typical trajectory computed with the leapfrog method. Rapid oscillations occur, here with a period of between 2 and 3 leapfrog steps, due to errors in simulating the motion in the most confined direction (or directions, for higher-dimensional distributions). When a long trajectory is used to propose a state for HMC, it is essentially random whether the trajectory ends at a state where the error in H is negative or close to zero, and hence will be accepted with probability close to one, or whether it happens to end at a state with a large positive error in H, and a correspondingly lower acceptance probability. If somehow we could smooth out these oscillations, we might obtain a high probability of acceptance for all trajectories.

I introduced a method for achieving this result that uses "windows" of states at the beginning and end of the trajectory (Neal, 1994). Here, I will present the method as an application of a general technique in which we probabilistically map to a state in a different space, perform a Markov chain transition in this new space, and then probabilistically map back to our original state space (Neal, 2006).

Our original state space consists of pairs, (q, p), of position and momentum variables. We will map to a sequence of W pairs, $[(q_0, p_0), \ldots, (q_{W-1}, p_{W-1})]$, in which each (q_i, p_i) for $i > 0$ is the result of applying one leapfrog step (with some fixed stepsize, ε) to (q_{i-1}, p_{i-1}). Note that even though a point in the new space seems to consist of W times as many numbers as a point in the original space, the real dimensionality of the new space is the same as the old, since the whole sequence of W pairs is determined by (q_0, p_0).

To probabilistically map from (q, p) to a sequence of pairs, $[(q_0, p_0), \ldots, (q_{W-1}, p_{W-1})]$, we select s uniformly from $\{0, \ldots, W-1\}$, and set (q_s, p_s) in the new state to our current state

(q, p). The other (q_i, p_i) pairs in the new state are obtained using leapfrog steps from (q_s, p_s), for $i > s$, or backwards leapfrog steps (i.e. done with stepsize $-\varepsilon$) for $i < s$. It is easy to see, using the fact that leapfrog steps preserve volume, that if our original state is distributed with probability density $P(q, p)$, then the probability density of obtaining the sequence $[(q_0, p_0), \ldots, (q_{W-1}, p_{W-1})]$ by this procedure is

$$P\left([(q_0, p_0), \ldots, (q_{W-1}, p_{W-1})]\right) = \frac{1}{W} \sum_{i=0}^{W-1} P(q_i, p_i), \qquad (5.35)$$

since we can obtain this sequence from a (q, p) pair that matches any pair in the sequence, and the probability is $1/W$ that we will produce the sequence starting from each of these pairs (which happens only if the random selection of s puts the pair at the right place in the sequence).

Having mapped to a sequence of W pairs, we now perform a Metropolis update that keeps the sequence distribution defined by Equation 5.35 invariant, before mapping back to the original state space. To obtain a Metropolis proposal, we perform $L - W + 1$ leapfrog steps (for some $L \geq W-1$), starting from (q_{W-1}, p_{W-1}), producing pairs (q_W, p_W) to (q_L, p_L). We then propose the sequence $[(q_L, -p_L), \ldots, (q_{L-W+1}, -p_{L-W+1})]$. We accept or reject this proposed sequence by the usual Metropolis criterion, with the acceptance probability being

$$\min \left[1, \frac{\sum_{i=L-W+1}^{L} P(q_i, p_i)}{\sum_{i=0}^{W-1} P(q_i, p_i)} \right], \qquad (5.36)$$

with $P(q, p) \propto \exp(-H(q, p))$. (Note here that $H(q, p) = H(q, -p)$, and that starting from the proposed sequence would lead symmetrically to the original sequence being proposed.)

This Metropolis update leaves us with either the sequence $[(q_L, p_L), \ldots, (q_{L-W+1}, p_{L-W+1})]$, called the "accept window," or the sequence $[(q_0, p_0), \ldots, (q_{W-1}, p_{W-1})]$, called the "reject window." (Note that these windows will overlap if $L + 1 < 2W$.) We label the pairs in the window chosen as $[(q_0^+, p_0^+), \ldots, (q_{W-1}^+, p_{W-1}^+)]$. We now produce a final state for the windowed HMC update by probabilistically mapping from this sequence to a single pair, choosing (q_e^+, p_e^+) with probability

$$\frac{P(q_e^+, p_e^+)}{\sum_{i=0}^{W-1} P(q_i^+, p_i^+)}.$$

If the sequence in the chosen window was distributed according to Equation 5.35, the pair (q_e^+, p_e^+) chosen will be distributed according to $P(q, p) \propto \exp(-H(q, p))$, as desired. To see this, let (q_{e+n}^+, p_{e+n}^+) be the result of applying n leapfrog steps (backward ones if $n < 0$) starting at (q_e^+, p_e^+). The probability density that (q_e^+, p_e^+) will result from mapping from a sequence to a single pair can then be written as follows, considering all sequences that can contain (q_e^+, p_e^+) and their probabilities:

$$\sum_{k=e-W+1}^{e} \left[\frac{1}{W} \sum_{i=k}^{k+W-1} P(q_i^+, p_i^+) \right] \frac{P(q_e^+, p_e^+)}{\sum_{i=k}^{k+W-1} P(q_i^+, p_i^+)} = P(q_e^+, p_e^+).$$

The entire procedure therefore leaves the correct distribution invariant.

When $W > 1$, the potential problem with ergodicity discussed at the end of Section 5.3.2 does not arise, since there is a nonzero probability of moving to a state only one leapfrog step away, where q may differ arbitrarily from its value at the current state.

It might appear that the windowed HMC procedure requires saving all $2W$ states in the accept and reject windows, since any one of these states might become the new state when a state is selected from either the accept window or reject window. Actually, however, at most three states need to be saved—the start state, so that forward simulation can be resumed after the initial backward simulation, plus one state from the reject window and one state from the accept window, one of which will become the new state after one of these windows is chosen. As states in each window are produced in sequence, a decision is made whether the state just produced should replace the state presently saved for that window. Suppose that the sum of the probability densities of states seen so far is $s_i = p_1 + \cdots + p_i$. If the state just produced has probability density p_{i+1}, it replaces the previous state saved from this window with probability $p_{i+1}/(s_i + p_{i+1})$.

I showed (Neal, 1994) that, compared to standard HMC, using windows improves the performance of HMC by a factor of 2 or more, on multivariate Gaussian distributions in which the standard deviation in some directions is much larger than in other directions. This is because the acceptance probability in Equation 5.36 uses an average of probability densities over states in a window, smoothing out the oscillations in H from inexact simulation of the trajectory. Empirically, the advantage of the windowed method was found to increase with dimensionality. For high-dimensional distributions, the acceptance probability when using the optimal stepsize was approximately 0.85, larger than the theoretical value of 0.65 for HMC (see Section 5.4.4).

These results for multivariate Gaussian distributions were obtained with a window size, W, much less than the trajectory length, L. For less regular distributions, it may be advantageous to use a much larger window. When $W = L/2$, the acceptance test determines whether the new state is from the first half of the trajectory (which includes the current state) or the second half; the new state is then chosen from one half or the other with probabilities proportional to the probability densities of the states in that half. This choice of W guards against the last few states of the trajectory having low probability density (high H), as might happen if the trajectory had by then entered a region where the stepsize used was too big.

The windowed variant of HMC may make other variants of HMC more attractive. One such variant (Section 5.5.1) splits the Hamiltonian into many terms corresponding to subsets of the data, which tends to make errors in H higher (while saving computation). Errors in H have less effect when averaged over windows. As discussed in Section 5.5.3, very low rejection rates are desirable when using partial momentum refreshment. It is easier to obtain a low rejection probability using windows (i.e. a less drastic reduction in ϵ is needed), which makes partial momentum refreshment more attractive.

Qin and Liu (2001) introduced a variant on windowed HMC. In their version, L leapfrog steps are done from the start state, with the accept window consisting of the states after the last W of these steps. A state from the accept window is then selected with probabilities proportional to their probability densities. If the state selected is k states before the end, k backwards leapfrog steps are done from the start state, and the states found by these steps along with those up to $W - k - 1$ steps forward of the start state form the reject window. The state selected from the accept window then becomes the next state with probability given by the analog of Equation 5.36; otherwise the state remains the same.

Qin and Liu's procedure is quite similar to the original windowed HMC procedure. One disadvantage of Qin and Liu's procedure is that the state is unchanged when the accept window is rejected, whereas in the original procedure a state is selected from the reject window (which might be the current state, but often will not be). The only other difference is that the number of steps from the current state to an accepted state ranges from $L - W + 1$ to L (average $L - (W + 1)/2$) with Qin and Liu's procedure, versus from $L - 2W + 2$

to L (average $L - W + 1$) for the original windowed HMC procedure, while the number of leapfrog steps computed varies from L to $L + W - 1$ with Qin and Liu's procedure, and is fixed at L with the original procedure. These differences are slight if $W \ll L$. Qin and Lin claim that their procedure performs better than the original on high-dimensional multivariate Gaussian distributions, but their experiments are flawed.*

Qin and Liu (2001) also introduce the more useful idea of weighting the states in the accept and reject windows nonuniformly, which can be incorporated into the original procedure as well. When mapping from the current state to a sequence of W weighted states, the position of the current state is chosen with probabilities equal to the weights, and when computing the acceptance probability or choosing a state from the accept or reject window, the probability densities of states are multiplied by their weights. Qin and Liu use weights that favor states more distant from the current state, which could be useful by usually causing movement to a distant point, while allowing choice of a nearer point if the distant points have low probability density. Alternatively, if one sees a window as a way of smoothing the errors in H, symmetrical weights that implement a better "low pass filter" would make sense.

5.5.5   Using Approximations to Compute the Trajectory

The validity of HMC does not depend on using the correct Hamiltonian when simulating the trajectory. We can instead use some approximate Hamiltonian, as long as we simulate the dynamics based on it by a method that is reversible and volume-preserving. However, the exact Hamiltonian must be used when computing the probability of accepting the endpoint of the trajectory. There is no need to look for an approximation to the kinetic energy, when it is of a simple form such as Equation 5.13, but the potential energy is often much more complex and costly to compute—for instance, it may involve the sum of log likelihoods based on many data points, if the data cannot be summarized by a simple sufficient statistic. When using trajectories of many leapfrog steps, we can therefore save much computation time if a fast and accurate approximation to the potential energy is available, while still obtaining exact results (apart from the usual sampling variation inherent in MCMC).

Many ways of approximating the potential energy might be useful. For example, if its evaluation requires iterative numerical methods, fewer iterations might be done than are necessary to get a result accurate to machine precision. In a Bayesian statistical application, a less costly approximation to the unnormalized posterior density (whose log gives the potential energy) may be obtainable by simply looking at only a subset of the data. This may not be a good strategy in general, but I have found it useful for Gaussian process models (Neal, 1998; Rasmussen and Williams, 2006), for which computation time scales as the cube of the number of data points, so that even a small reduction in the number of points produces a useful speedup.

Rasmussen (2003) has proposed approximating the potential energy by modeling it as a Gaussian process, inferred from values of the potential energy at positions selected during an initial exploratory phase. This method assumes only a degree of smoothness of the potential energy function, and so could be widely applied. It is limited, however, by the cost of

---

* In their first comparison, their method computes an average of 55 leapfrog steps per iteration, but the original only computes 50 steps, a difference in computation time which if properly accounted for negates the slight advantage they see for their procedure. Their second comparison has a similar problem, and it is also clear from an examination of the results (in their Table I) that the sampling errors in their comparison are too large for any meaningful conclusions to be drawn.

Gaussian process inference, and so is most useful for problems of moderate dimensionality for which exact evaluation of the potential energy is very costly.

An interesting possibility, to my knowledge not yet explored, would be to express the exact potential energy as the sum of an approximate potential energy and the error in this approximation, and to then apply one of the splitting techniques described in Section 5.5.1—exploiting either the approximation's analytic tractability (e.g. for a Gaussian approximation, with quadratic potential energy), or its low computational cost, so that its dynamics can be accurately simulated at little cost using many small steps. This would reduce the number of evaluations of the gradient of the exact potential energy if the variation in the potential energy removed by the approximation term permits a large stepsize for the error term.

5.5.6 Short-Cut Trajectories: Adapting the Stepsize without Adaptation

One significant disadvantage of HMC is that, as discussed in Section 5.4.2, its performance depends critically on the settings of its tuning parameters—which consist of at least the leapfrog stepsize, ϵ, and number of leapfrog steps, L, with variations such as windowed HMC having additional tuning parameters as well. The optimal choice of trajectory length (ϵL) depends on the global extent of the distribution, so finding a good trajectory length likely requires examining a substantial number of HMC updates. In contrast, just a few leapfrog steps can reveal whether some choice of stepsize is good or bad, which leads to the possibility of trying to set the stepsize "adaptively" during an HMC run.

Recent work on adaptive MCMC methods is reviewed by Andrieu and Thoms (2008). As they explain, naively choosing a stepsize for each HMC update based on results of previous updates—for example, reducing the stepsize by 20% if the previous 10 trajectories were all rejected, and increasing it by 20% if less than two of the 10 previous trajectories were rejected—undermines proofs of correctness (in particular, the process is no longer a Markov chain), and will in general produce points from the wrong distribution. However, correct results can be obtained if the degree of adaptation declines over time. Adaptive methods of this sort could be used for HMC, in much the same way as for any other tunable MCMC method.

An alternative approach (Neal, 2005, 2007) is to perform MCMC updates with various values of the tuning parameters, set according to a schedule that is predetermined or chosen randomly without reference to the realized states, so that the usual proofs of MCMC convergence and error analysis apply, but to do this using MCMC updates that have been tweaked so that they require little computation time when the tuning parameters are not appropriate for the distribution. Most of the computation time will then be devoted to updates with appropriate values for the tuning parameters. Effectively, the tuning parameters are set adaptively from a computational point of view, but not from a mathematical point of view.

For example, trajectories that are simulated with a stepsize that is much too large can be rejected after only a few leapfrog steps, by rejecting whenever the change (either way) in the Hamiltonian due to a single step (or a short series of steps) is greater than some threshold—that is, we reject if $|H(q(t + \epsilon), p(t + \epsilon)) - H(q(t), p(t))|$ is greater than the threshold. If this happens early in the trajectory, little computation time will have been wasted on this unsuitable stepsize. Such early termination of trajectories is valid, since any MCMC update that satisfies detailed balance will still satisfy detailed balance if it is modified to eliminate transitions either way between certain pairs of states.

With this simple modification, we can randomly choose stepsizes from some distribution without wasting much time on those stepsizes that turn out to be much too large. However, if we set the threshold small enough to reject when the stepsize is only a little too large, we may terminate trajectories that would have been accepted, perhaps after a substantial amount of computation has already been done. Trying to terminate trajectories early when the stepsize is smaller than optimal carries a similar risk.

A less drastic alternative to terminating trajectories when the stepsize seems inappropriate is to instead *reverse* the trajectory. Suppose that we perform leapfrog steps in groups of k steps. Based on the changes in H over these k steps, we can test whether the stepsize is inappropriate—for example, the group may fail the test if the standard deviation of H over the $k + 1$ states is greater than some upper threshold or less than some lower threshold (any criterion that would yield the same decision for the reversed sequence is valid). When a group of k leapfrog steps fails this test, the trajectory stays at the state where this group started, rather than moving k steps forward, and the momentum variables are negated. The trajectory will now exactly retrace states previously computed (and which therefore need not be recomputed), until the initial state is reached, at which point new states will again be computed. If another group of k steps fails the test, the trajectory will again reverse, after which the whole remainder of the trajectory will traverse states already computed, allowing its endpoint to be found immediately without further computation.

This scheme behaves the same as standard HMC if no group of k leapfrog steps fails the test. If there are two failures early in the trajectory, little computation time will have been wasted on this (most likely) inappropriate stepsize. Between these extremes, it is possible that one or two reversals will occur, but not early in the trajectory; the endpoint of the trajectory will then usually not be close to the initial state, so the nonnegligible computation performed will not be wasted (as it would be if the trajectory had been terminated).

Such short-cut schemes can be effective at finding good values for a small number of tuning parameters, for which good values will be picked reasonably often when drawing them randomly. It will not be feasible for setting a large number of tuning parameters, such as the entries in the "mass matrix" of Equation 5.5 when dimensionality is high, since even if two reversals happen early on, the cost of using inappropriate values of the tuning parameters will dominate when appropriate values are chosen only very rarely.

5.5.7 Tempering during a Trajectory

Standard HMC and the variations described so far have as much difficulty moving between modes that are separated by regions of low probability as other local MCMC methods, such as random-walk Metropolis and Gibbs sampling. Several general schemes have been devised for solving problems with such isolated modes that involve sampling from a series of distributions that are more diffuse than the distribution of interest. Such schemes include parallel tempering (Geyer, 1991; Earl and Deem, 2005), simulated tempering (Marinari and Parisi, 1992), tempered transitions (Neal, 1996b), and annealed importance sampling (Neal, 2001). Most commonly, these distributions are obtained by varying a "temperature" parameter, T, as in Equation 5.21, with $T = 1$ corresponding to the distribution of interest, and larger values of T giving more diffuse distributions. Any of these "tempering" methods could be used in conjunction with HMC. However, tempering-like behavior can also be incorporated directly into the trajectory used to propose a new state in the HMC procedure.

In the simplest version of such a "tempered trajectory" scheme (Neal, 1999, Section 6), each leapfrog step in the first half of the trajectory is combined with multiplication of the momentum variables by some factor α slightly greater than one, and each leapfrog step in the second half of the trajectory is combined with division of the momentum by the same factor α. These multiplications and divisions can be done in various ways, as long as the result is reversible, and the divisions are paired exactly with multiplications. The most symmetrical scheme is to multiply the momentum by $\sqrt{\alpha}$ before the first half step for momentum (Equation 5.18) and after the second half step for momentum (Equation 5.20), for leapfrog steps in the first half of the trajectory, and correspondingly, to divide the momentum by $\sqrt{\alpha}$ before the first and after the second half steps for momentum in the second half of the trajectory. (If the trajectory has an odd number of leapfrog steps, for the middle leapfrog step of the trajectory, the momentum is multiplied by $\sqrt{\alpha}$ before the first half step for momentum, and divided by $\sqrt{\alpha}$ after the second half step for momentum.) Note that most of the multiplications and divisions by $\sqrt{\alpha}$ are preceded or followed by another such, and so can be combined into a single multiplication or division by α.

It is easy to see that the determinant of the Jacobian matrix for such a tempered trajectory is one, just as for standard HMC, so its endpoint can be used as a proposal without any need to include a Jacobian factor in the acceptance probability.

Multiplying the momentum by an α that is slightly greater than one increases the value of $H(q, p)$ slightly. If H initially had a value typical of the canonical distribution at $T = 1$, after this multiplication, H will be typical of a value of T that is slightly higher.[*] Initially, the change in $H(q, p) = K(p) + U(q)$ is due entirely to a change in $K(p)$ as p is made bigger, but subsequent dynamical steps will tend to distribute the increase in H between K and U, producing a more diffuse distribution for q than is seen when $T = 1$. After many such multiplications of p by α, values for q can be visited that are very unlikely in the distribution at $T = 1$, allowing movement between modes that are separated by low-probability regions. The divisions by α in the second half of the trajectory result in H returning to values that are typical for $T = 1$, but perhaps now in a different mode.

If α is too large, the probability of accepting the endpoint of a tempered trajectory will be small, since H at the endpoint will likely be much larger than H at the initial state. To see this, consider a trajectory consisting of only one leapfrog step. If $\epsilon = 0$, so that this step does nothing, the multiplication by $\sqrt{\alpha}$ before the first half step for momentum would be exactly canceled by the division by $\sqrt{\alpha}$ after the second half step for momentum, so H would be unchanged, and the trajectory would be accepted. Since we want something to happen, however, we will use a nonzero ϵ, which will on average result in the kinetic energy decreasing when the leapfrog step is done, as the increase in H from the multiplication by $\sqrt{\alpha}$ is redistributed from K alone to both K and U. The division of p by $\sqrt{\alpha}$ will now not cancel the multiplication by $\sqrt{\alpha}$—instead, on average, it will reduce H by less than the earlier increase. This tendency for H to be larger at the endpoint than at the initial state can be lessened by increasing the number of leapfrog steps, say by a factor of R, while reducing α to $\alpha^{1/R}$, which (roughly) maintains the effective temperature reached at the midpoint of the trajectory.

Figure 5.9 illustrates tempered trajectories used to sample from an equal mixture of two bivariate Gaussian distributions, with means of [0 0] and [10 10], and covariances of I and $2I$. Each trajectory consists of 200 leapfrog steps, done with $\epsilon = 0.3$, with tempering done as described above with $\alpha = 1.04$. The left plots show how H varies along the trajectories; the right plots show the position coordinates for the trajectories. The

---

[*] This assumes that the typical value of H is a continuous function of T, which may not be true for systems that have a "phase transition." Where there is a discontinuity (in practice, a near-discontinuity) in the expected value of H as a function of T, making small changes to H, as here, may be better than making small changes to T (which may imply big changes in the distribution).

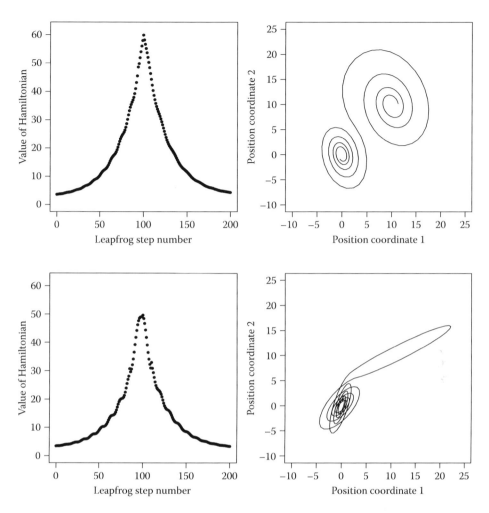

FIGURE 5.9

Illustration of tempered trajectories on a mixture of two Gaussians. The trajectory shown in the top plots moves between modes; the one shown in the bottom plots ends in the same mode.

top plots are for a trajectory starting at $q = [-0.4 \ -0.9]$ and $p = [0.7 \ -0.9]$, which has an endpoint in the other mode around $[10 \ 10]$. The bottom plots are for a trajectory starting at $q = [0.1 \ 1.0]$ and $p = [0.5 \ 0.8]$, which ends in the same mode it begins in. The change in H for the top trajectory is 0.69, so it would be accepted with probability $\exp(-0.69) = 0.50$. The change in H for the bottom trajectory is -0.15, so it would be accepted with probability one.

By using such tempered trajectories, HMC is able to sample these two well-separated modes—11% of the trajectories move to the other mode and are accepted—whereas standard HMC does very poorly, being trapped for a very long time in one of the modes. The parameters for the tempered trajectories in Figure 5.9 were chosen to produce easily interpreted pictures, and are not optimal. More efficient sampling is obtained with a much smaller number of leapfrog steps, larger stepsize, and larger α—for example, $L = 20, \epsilon = 0.6$, and $\alpha = 1.5$ give a 6% probability of moving between modes.

A fundamental limitation of the tempering method described above is that (as for standard HMC) the endpoint of the tempered trajectory is unlikely to be accepted if the value for H there is much higher that for the initial state. This corresponds to the probability density at the endpoint being much lower than at the current state. Consequently, the method will not move well between two modes with equal total probability if one mode is high and narrow and the other low and broad, especially when the dimensionality is high. (Since acceptance is based on the joint density for q and p, there is some slack for moving to a point where the density for q alone is different, but not enough to eliminate this problem.) I have proposed (Neal, 1999) a modification that addresses this, in which the point moved to can come from anywhere along the tempered trajectory, not just the endpoint. Such a point must be selected based both on its value for H and the accumulated Jacobian factor for that point, which is easily calculated, since the determinant of the Jacobian matrix for a multiplication of p by α is simply α^d, where d is the dimensionality. This modified tempering procedure can not only move between modes of differing width, but also move back and forth between the tails and the central area of a heavy-tailed distribution.

More details on these variations on HMC can be found in the R implementations available from my web page, at www.cs.utoronto.ca/~radford

Acknowledgment

This work was supported by the Natural Sciences and Engineering Research Council of Canada. The author holds a Canada Research Chair in Statistics and Machine Learning.

References

Alder, B. J. and Wainwright, T. E. 1959. Studies in molecular dynamics. I. General method. *Journal of Chemical Physics*, 31:459–466.

Andrieu, C. and Thoms, J. 2008. A tutorial on adaptive MCMC. *Statistics and Computing*, 18:343–373.

Arnold, V. I. 1989. *Mathematical Methods of Classical Mechanics*, 2nd edn. Springer, New York.

Bennett, C. H. 1975. Mass tensor molecular dynamics. *Journal of Computational Physics*, 19:267–279.

Caracciolo, S., Pelissetto, A., and Sokal, A. D. 1994. A general limitation on Monte Carlo algorithms of Metropolis type. *Physical Review Letters*, 72:179–182. Also available at http://arxiv.org/abs/hep-lat/9307021

Choo, K. 2000. Learning hyperparameters for neural network models using Hamiltonian dynamics. MSc thesis, Dept. of Computer Science, University of Toronto. Available at http://www.cs.toronto.edu/~radford/ftp/kiam-thesis.pdf

Creutz, M. 1988. Global Monte Carlo algorithms for many-fermion systems. *Physical Review D*, 38:1228–1238.

Creutz, M. and Gocksch, A. 1989. Higher order hybrid Monte Carlo algorithms. *Physical Review Letters*, 63:9–12.

Duane, S., Kennedy, A. D., Pendleton, B. J., and Roweth, D. 1987. Hybrid Monte Carlo. *Physics Letters B*, 195:216–222.

Earl, D. J. and Deem, M. W. 2005. Parallel tempering: Theory, applications, and new perspectives. *Physical Chemistry Chemical Physics*, 7:3910–3916.

Frenkel, D. and Smit, B. 1996. *Understanding Molecular Simulation: From Algorithms to Applications*. Academic Press, San Diego, CA.

Geyer, C. J. 1991. Markov chain Monte Carlo maximum likelihood for dependent data. In E. M. Keramidas (ed.), *Computing Science and Statistics: Proceedings of the 23rd Symposium on the Interface*, pp. 156–163. American Statistical Association, New York.

Girolami, M., Calderhead, B., and Chin, S. A. 2009. Riemannian manifold Hamiltonian Monte Carlo. http://arxiv.org/abs/arxiv:0907.1100, 35 pp.

Grenander, U. and Miller, M. I. 1990. Representations of knowledge in complex systems. *Physics Letters B*, 242:437–443.

Gupta, S., Irbäck, A., Karsch, F., and Petersson, B. 1990. The acceptance probability in the hybrid Monte Carlo method. *Physics Letters B*, 242:437–443.

Hastings, W. K. 1970. Monte Carlo sampling methods using Markov chains and their applications. *Biometrika*, 57:97–109.

Horowitz, A. M. 1991. A generalized guided Monte Carlo algorithm. *Physics Letters B*, 268:247–252.

Ishwaran, H. 1999. Applications of hybrid Monte Carlo to generalized linear models: quasicomplete separation and neural networks. *Journal of Computational and Graphical Statistics*, 8:779–799.

Izagguirre, J. A. and Hampton, S. S. 2004. Shadow hybrid Monte Carlo: an efficient propagator in phase space of macromolecules. *Journal of Computational Physics*, 200:581–604.

Kennedy, A. D. 1990. The theory of hybrid stochastic algorithms. In P. H. Damgaard, H. Hüffel, and A. Rosenblum (eds), *Probabilistic Methods in Quantum Field Theory and Quantum Gravity*, pp. 209–223. Plenum Press, New York.

Kennedy, A. D. and Pendleton, B. 1991. Acceptances and autocorrelations in hybrid Monte Carlo. *Nuclear Physics B (Proc. Suppl.)*, 20:118–121.

Kennedy, A. D. and Pendleton, B. 2001. Cost of the generalized hybrid Monte Carlo algorithm for free field theory. *Nuclear Physics B*, 607:456–510. Also available at http://arxiv.org/abs/hep-lat/0008020

Leimkuhler, B. and Reich, S. 2004. *Simulating Hamiltonian Dynamics*. Cambridge University Press, Cambridge.

Liu, J. S. 2001. *Monte Carlo Strategies in Scientific Computing*. Springer, New York.

Mackenzie, P. B. 1989. An improved hybrid Monte Carlo method. *Physics Letters B*, 226:369–371.

Marinari, E. and Parisi, G. 1992. Simulated tempering: A new Monte Carlo scheme. *Europhysics Letters*, 19:451–458.

McLachlan, R. I. and Atela, P. 1992. The accuracy of symplectic integrators. *Nonlinearity*, 5: 541–562.

Metropolis, N., Rosenbluth, A. W., Rosenbluth, M. N., Teller, A. H., and Teller, E. 1953. Equation of state calculations by fast computing machines. *Journal of Chemical Physics*, 21:1087–1092.

Neal, R. M. 1993. Probabilistic inference using Markov chain Monte Carlo methods. Technical Report CRG-TR-93-1, Dept. of Computer Science, University of Toronto.

Neal, R. M. 1994. An improved acceptance procedure for the hybrid Monte Carlo algorithm. *Journal of Computational Physics*, 111:194–203.

Neal, R. M. 1996a. *Bayesian Learning for Neural Networks*, Lecture Notes in Statistics, Vol. 118. Springer, New York.

Neal, R. M. 1996b. Sampling from multimodal distributions using tempered transitions. *Statistics and Computing*, 6:353–356.

Neal, R. M. 1998. Regression and classification using Gaussian process priors (with discussion). In J. M. Bernardo, J. O. Berger, A. P. Dawid, and A. F. M. Smith (eds.), *Bayesian Statistics 6*, pp. 475–501. Oxford University Press, Oxford.

Neal, R. M. 1999. Markov chain sampling using Hamiltonian dynamics. Talk at the Joint Statistical Meetings, Baltimore, MD, August. Slides are available at http://www.cs.utoronto.ca/~radford/ftp/jsm99.pdf

Neal, R. M. 2001. Annealed importance sampling. *Statistics and Computing*, 11:125–139.

Neal, R. M. 2005. The short-cut Metropolis method. Technical Report No. 0506, Department of Statistics, University of Toronto, 28 pp. Available at http://arxiv.org/abs/math.ST/0508060

Neal, R. M. 2006. Constructing efficient MCMC methods using temporary mapping and caching. Talk at Columbia University, Dept. of Statistics, December 11. Slides are available at http://www.cs.utoronto.ca/~radford/ftp/cache-map.pdf

Neal, R. M. 2007. Short-cut MCMC: An alternative to adaptation. Talk at the Third Workshop on Monte Carlo Methods, Harvard, May. Slides are available at http://www.cs.utoronto.ca/~radford/ftp/short-mcmc-talk.pdf

Qin, Z. S. and Liu, J. S. (2001). Multipoint Metropolis method with application to hybrid Monte Carlo. *Journal of Computational Physics*, 172:827–840.

Rasmussen, C. E. 2003. Gaussian processes to speed up hybrid Monte Carlo for expensive Bayesian integrals. In J. M. Bernardo, M. J. Bayarri, J. O. Berger, A. P. Dawid, D. Heckerman, A. F. M. Smith, and M. West, (eds.), *Bayesian Statistics 7. Proceedings of the Seventh Valencia International Meeting*, pp. 651–659. Oxford University Press, Oxford.

Rasmussen, C. E. and Williams, C. K. I. 2006. *Gaussian Processes for Machine Learning*. MIT Press, Cambridge, MA.

Roberts, G. O., Gelman, A., and Gilks, W. R. 1997. Weak convergence and optimal scaling of random walk Metropolis algorithms. *Annals of Applied Probability*, 7:110–120.

Roberts, G. O. and Rosenthal, J. S. 1998. Optimal scaling of discrete approximations to Langevin diffusions. *Journal of the Royal Statistical Society, Series B*, 60:255–268.

Roberts, G. O. and Tweedie, R. L. 1996. Exponential convergence of Langevin distributions and their discrete approximations. *Bernoulli*, 2:341–363.

Rossky, P. J., Doll, J. D., and Friedman, H. L. 1978. Brownian dynamics as smart Monte Carlo simulation. *Journal of Chemical Physics*, 69:4628–4633.

Ruján, P. 1997. Playing billiards in version space. *Neural Computation*, 9:99–122.

Schmidt, M. N. 2009. Function factorization using warped Gaussian processes. In *Proceedings of the Twenty-Sixth International Conference on Machine Learning*. ACM, New York.

Sexton, J. C. and Weingarten, D. H. 1992. Hamiltonian evolution for the hybrid Monte Carlo method. *Nuclear Physics B*, 380:665–677.

6

Inference from Simulations and Monitoring Convergence

Andrew Gelman and Kenneth Shirley

Constructing efficient iterative simulation algorithms can be difficult, but inference and monitoring convergence are relatively easy. We first give our recommended strategy (following Section 11.10 of Gelman et al., 2003) and then explain the reasons for our recommendations, illustrating with a relatively simple example from our recent research: a hierarchical model fit to public-opinion survey data.

6.1 Quick Summary of Recommendations

1. Simulate three or more chains in parallel. We typically obtain starting points by adding random perturbations to crude estimates based on a simpler model or approximation.

2. Check convergence by discarding the first part of the simulations—we discard the first half, although that may be overly conservative—and using within-chain analysis to monitor stationarity and between/within chains comparisons to monitor mixing.

3. Once you have reached approximate convergence, mix all the simulations from the second halves of the chains together to summarize the target distribution. For most purposes there is no longer any need to worry about autocorrelations in the chains.

4. Adaptive Markov chain Monte Carlo (MCMC)—for example, tuning the jumping distribution of a Metropolis algorithm—can often be a good idea and presents no problems for convergence if you restart after adapting. For example, if you have already run 400 iterations and have not reached approximate convergence, you can adjust your algorithm and run another 400 steps, discarding the earlier simulations. At the next step of adaptation, you can run another 400, and so forth, possibly adapting the adaptation time itself to balance the goals of rapid convergence and computational efficiency. (Newer, more sophisticated algorithms have the promise of allowing continuous adaptation and do not necessarily require discarding early iterations.)

5. If you have run your simulations for a while and they are not close to convergence, stop, look at plots of simulations from different chains, and go back and improve your algorithm, for example, by adding new kinds of jumps to get faster

mixing (see, e.g., Sections 11.8 through 11.9 of Gelman et al., 2003, for some simple approaches, or some of the chapters in this handbook for more advanced ideas for tackling harder problems). It is not generally a good idea to wait hours for convergence, and in many practical examples there is a real gain from getting an answer in ten seconds, say, rather than two minutes. Faster computation translates into the ability to fit more models and to do more real-time data analysis.

6. When all is done, compare inferences to those from simpler models or approximations. Examine discrepancies to see whether they represent programming errors, poor convergence, or actual changes in inferences as the model is expanded. (Here we are talking about using these comparisons as a way to diagnose potential problems in a simulation. Other chapters in this handbook discuss ways of combining MCMC runs from different models to perform more efficient computations, using techniques such as parallel tempering and path sampling.)

Another useful debugging technique is the fake-data check: Choose or simulate some "true values" of the parameters and simulate data given these parameters. Then run the MCMC algorithm and check that it converges to a distribution consistent with the assumed true parameter values.[*]

To illustrate the concepts in this chapter, we introduce a model fit using MCMC that comes from a political science application: modeling state-level attitudes on the death penalty over time using national survey data (Shirley and Gelman, 2010). The model is a multilevel logistic regression for the binary response representing support ($y = 1$) or opposition ($y = 0$) to the death penalty for people convicted of murder (this is how the question was phrased in repeated polls given by Gallup and the National Opinion Research Center during the time span 1953–2006). The predictors in the model include demographics such as race, sex, and age, as well as the state of residence of the respondent (nested within one of four regions of the United States), so that we can model opinion trends in different parts of the country.

6.2 Key Differences between Point Estimation and MCMC Inference

Markov chain Monte Carlo methods are wonderfully convenient and flexible but, compared to simpler methods of statistical computation, they involve two difficulties: running the Markov chains long enough for convergence, and having enough simulation draws for suitably accurate inference.

- The distribution of simulations depends on starting values of the algorithm. The user must correct for starting-value bias or else run simulations long enough that starting values have essentially been forgotten.

- Inferences are based on simulations rather than deterministic estimates; as a result the user must account for Monte Carlo error or else average over enough simulation draws that such error is negligible.

---

[*] The basic idea is that, over many simulations, 50% of the 50% posterior intervals should contain the true value, 95% of the 95% intervals should contain the true value, and so forth. Cook et al. (2006) provide a more formal procedure along these lines.

The first item above is sometimes called the problem of monitoring convergence of the sampler and is commonly assessed in two ways: by studying time trends *within* chains (thus detecting movement away from the starting points) and by examining mixing *between* chains (thus detecting influence of the starting values of the different chains).

The second item above arises because, even if all chains were started from random draws from the target distribution, we would still need to think about their speed of mixing: the iterative simulation must cycle through the distribution enough times to give us the equivalent of many independent draws. In practice, though, once chains have been run long enough that the distribution of each of them is close to the distribution of all of them mixed together, we usually have created enough simulation draws that Monte Carlo error is not a problem. So typically we simply monitor convergence and then stop. It can also be a good idea to examine movement within chains (via trace plots or time series summaries) to catch the occasional situation when a group of chains have mixed but still have not converged to a stable distribution.

Beyond this, there can be convergence problems which are essentially undetectable from output analysis alone, for example if a target distribution has multiple, well-separated modes and all the chains are started from within a single mode. Here there may be specific workarounds for particular models, but in general the only solution is the usual combination of subject-matter understanding, comparisons to previous fitted models, and mathematical analysis: the usual set of tools we use in any data analysis. In the words of Brooks et al. (2003):

> Diagnostics can only reliably be used to determine a lack of convergence and not detect convergence *per se*. For example, it is relatively easy for a sampler to become stuck in a local mode and naively applied diagnostics would not detect that the chain had not explored the majority of the model/parameter space. Therefore, it is important to use a range of techniques, preferably assessing different aspects of the chains and each based upon independent chains started at a range of different starting points. If only a single diagnostic is used and it detects no lack of convergence, then this provides only mild reassurance that the sampler has performed well. However, if a range of diagnostics can be used and each detects no lack of convergence, then we can be far more confident that we would gain reliable inference from the sampler output.

Ultimately, MCMC computation, and simulation in general, is part of a larger statistical enterprise.

In the case of our example, we aim to summarize patterns and trends in public opinion on the death penalty for political scientists by fitting a model to survey data. To see how knowledge of the problem leads to better decision-making regarding inferences via simulation, consider the situation in which we encounter multiple modes in the target distribution of some parameter, such as the time trend for the coefficient of a particular state. Given that we are modeling survey data, we might hypothesize that the multiple modes represent a mixture of distributions that correspond to different subgroups of the population in that state, and we would then want to add an interaction term in the model between state of residence and some demographic variable, such as sex, to see if the multimodality disappears. Such situations highlight that the convergence of MCMC algorithms depends strongly on whether the model actually fits the data: these are never totally separate, and convergence problems are often related to modeling issues.

6.3   Inference for Functions of the Parameters vs. Inference for Functions of the Target Distribution

It is sometimes said that simulation-based inference is all about the problem of estimating expectations $E(\theta)$ under the target distribution, $p(\theta)$.* This is not correct. There are actually *two* sorts of inferential or computational task:

Task 1. Inference about θ or, more generally, about any quantity of interest $g(\theta)$. Such inference will typically be constructed using a collection of 1000 (say) simulations of the parameter vector, perhaps summarized by a mean and standard deviation, or maybe a 95% interval using the empirical distribution of the simulations that have been saved. Even if these summaries could be computed analytically, we would in general still want simulations because these allow us directly to obtain inferences for any posterior or predictive summary.

Task 2. Computation of $E(\theta)$ or, more generally, any function of the target distribution. For example, suppose we are interested in a parameter θ and we plan to summarize our inference using a posterior mean and standard deviation. Then what we really want are $E(\theta)$ and $E(\theta^2)$, which indeed are expectations of functions of θ. Or suppose we plan to summarize our inference using a 95% central posterior interval. These can be derived from posterior expectations; for example, the lower endpoint of the interval is the value L for which $\Pr(\theta < L) = 0.025$.

The precision we need depends on our inferential goals. Consider a scalar parameter θ whose posterior distribution happens to be approximately normal with mean and standard deviation estimated at 3.47 and 1.83, respectively. Suppose you are now told that the Monte Carlo standard deviation of the mean is estimated to be 0.1. If your goal is inference for θ—Task 1 above—you can stop right there: the Monte Carlo error is trivial compared to the inherent uncertainty about θ in your posterior distribution, and further simulation will be a waste of time (at least for the purposes of estimating 50% and 95% intervals for θ).† However, if your goal is to compute $E(\theta)$—Task 2—then you might want to go further: depending on your ultimate goal, you might want to learn that $E(\theta)$ is actually 3.53 or 3.53840 or whatever.

Task 1 is by far the more common goal in Bayesian statistics, but Task 2 arises in other application areas such as statistical physics and, in statistics, the computation of normalizing constants and marginal distributions. Much of the routine use of Markov chain simulation (e.g. inferences for hierarchical models using the Bayesian software package BUGS) culminates in inferences for parameters and model predictions (i.e. Task 1). Many of the most technically demanding simulation problems have Task 2 as a goal.

---

* In Bayesian applications, the target distribution is the posterior distribution, $p(\theta \mid y)$, but more generally it can be any probability distribution. Our discussion of inference and convergence does not require that the MCMC be done for a Bayesian purpose, so we simply write the target distribution as $p(\theta)$, with the understanding that it might be conditional on data.

† In this example, the standard deviation is only estimated, not known, but our point remains. If the standard deviation is estimated at 1.83, it is highly doubtful that adding further precision to the $E(\theta)$ will tell us anything useful about θ itself. If computation is free, it is fine to run longer, but to the extent that computation time is an issue and some stopping criterion must be used, it makes sense to tie the convergence to the estimated uncertainty in θ rather than to keep going to get some arbitrary preset level of precision.

It may be that much of the confusion of the statistical literature on MCMC convergence involves methods being designed for Task 1 problems and applied to Task 2, and vice versa.

One goal of our death penalty analysis is to measure the changes in attitudes during the past half-century in each of the four regions of the United States (Northeast, South, Midwest, and West). We model changes as being linear on the logistic scale, estimating a different slope parameter for each region, β^{North}, β^{South}, $\beta^{Midwest}$, and β^{West}, and also estimating the standard deviation among regions, σ_{region}. We will take the posterior distributions of these five parameters as our target distribution of interest, and our inferential goals are of the Task 1 variety. That is, we care about basic summaries of the distributions of these parameters, and not functions of them, such as their means.

6.4 Inference from Noniterative Simulations

We first consider the simple problem of inference based on simulations taken directly from the target distribution. Let us consider specific instances of the two tasks mentioned above:

1. Inference about a parameter (or function of parameters) θ, to be represented by a set of simulations and possibly a 95% interval. We can order our simulation draws and use the 2.5% and 97.5% quantiles of these simulations.

As pointed out by Raftery and Lewis (1992), these extreme order statistics are numerically unstable. For example, Table 6.1 shows five replications of inferences from a unit normal distribution based on 100 simulations, then based on 1000 simulations. If the goal is to precisely determine the endpoints of the interval (e.g. to determine if a coefficient is statistically significant, or simply to present a replicable value for publication), then many simulations are required—even in this extremely easy problem, 1000 independent draws are not enough to pin down the interval endpoints to one decimal place. However, if the goal is to get an interval for θ with approximate 95% coverage in the target distribution, even 100 draws are reasonable.

TABLE 6.1

Simple Examples of Inference from Direct Simulation

Inferences Based on 100 Random Draws	Inferences Based on 1000 Random Draws
[−1.79, 1.69]	[−1.83, 1.97]
[−1.80, 1.85]	[−2.01, 2.04]
[−1.64, 2.15]	[−2.10, 2.13]
[−2.08, 2.38]	[−1.97, 1.95]
[−1.68, 2.10]	[−2.10, 1.97]

Simple examples of inference from direct simulation. Left column: five replications of 95% intervals for a hypothetical parameter θ that has a unit normal distribution, each based on 100 independent simulation draws. Right column: five replications of the same inference, each based on 1000 draws. For either column, the correct answer is [−1.96, 1.96]. From one perspective, these estimates are pretty bad: even with 1000 simulations, either bound can easily be off by more than 0.1, and the entire interval width can easily be off by 10%. On the other hand, for the goal of inference about θ, even the far-off estimates above aren't so bad: the interval [−2.08, 2.38] has 97% probability coverage, and [−1.79, 1.69] has 92% coverage.

Our practical advice is to use the estimated uncertainty in the target distribution to decide when simulations are sufficient; the purpose of this particular simple example is to demonstrate that an appropriate minimal number of simulations depends on inferential goals.

Many fewer draws are needed for everyday inference than for a final published result. And we see this even with direct simulations without even getting into the "Markov chain" part of MCMC.

2. Inference about the mean, $E(\theta)$. We can divide our simulations of θ into k groups, compute the sample mean for each group, and then get a standard error for the grand mean by taking the standard deviation of the k group means, divided by $\sqrt{k}$.

Dividing into groups is an arbitrary choice—presumably it would be better to use a tool such as the jackknife (Efron and Tibshirani, 1993)—but we go with a simple batch means approach here because it generalizes so naturally to MCMC with blocking and parallel chains. In any case, the appropriate number of simulation draws will depend on the inferential goal. For example, 1000 random draws from a unit normal distribution allow its mean to be estimated to within a standard error of approximately 0.03.

6.5   Burn-In

It is standard practice to discard the initial iterations of iterative simulation as they are too strongly influenced by starting values and do not provide good information about the target distribution. We follow this "burn-in" idea ourselves and generally discard the first half of simulated sequences. Thus, if we run MCMC for 100 iterations, we keep only the iterations 51–100 of each chain. If we then run another 100 iterations, we discard the 50 we have already kept, now keeping only iterations 101–200, and so forth.

Burn-in is convenient, but discarding early iterations certainly cannot be the most efficient approach; see Geyer (1998) for a general argument and Liu and Rubin (1996, 2002) for specific methods for output analysis accounting for the dependence of the simulations on the starting values. That said, we typically go with the simple burn-in approach, accepting the increased Monte Carlo error involved in discarding half the simulations.

There has been some confusion on this point, however. For example, we recently received the following question by email:

> I was wondering about MCMC burn-in and whether the oft-cited emphasis on this in the literature might not be a bit overstated. My thought was that the chain is Markovian. In a Metropolis (or Metropolis–Hastings) context, once you establish the scale of the proposal distribution(s), successful burn-in gets you only a starting location inside the posterior—nothing else is remembered, by definition! However, there is nothing really special about this particular starting point; it would have been just as valid had it been your initial guess and the burn-in would then have been superfluous. Moreover, the sampling phase will eventually reach the far outskirts of the posterior, often a lot more extreme than the sampling starting location, yet it will still (collectively) describe the posterior correctly. This implies that *any* valid starting point is just as good as any other, burn-in or no burn-in.
>
> The only circumstance that I can think of in which a burn-in would be essential is in the case in which prior support regions for the parameters are not all jointly valid (inside

the joint posterior), if that is even possible given the min/max limits set for the priors. Am I missing something?

Indeed, our correspondent was missing the point that any inference from a finite number of simulations is an approximation, and the starting point can affect the quality of the approximation. Consider an extreme example in which your target distribution is normal with mean μ and standard deviation σ; and your sampler takes independent draws directly from the target distribution; but you pick a starting value of X. The average of n simulations will then have the value, in expectation, of $(1/n)X + ((n-1)/n)\mu$, instead of the correct value of μ. If, for example, $X = 100$, $n = 100$, and $\mu = 1$, you are in trouble! But a burn-in of 1 will solve all your problems in this example. True, if you draw a few million simulations, the initial value will be forgotten, but why run a few million simulations if you do not have to? That will just take time away from your more important work.

More generally, the starting distribution will persist for a while, basically as long as it takes for your chains to converge. If your starting values persist for a time T, then these will pollute your inferences for some time of order T, by which time you can already have stopped the simulations if you had discarded some early steps. In this example, you might say that it would be fine to just start at the center of the distribution. One difficulty, though, is that you do not know where the center of the distribution is before you have done your simulations. More realistically, we start from estimates $\pm$ uncertainty as estimated from some simpler model that was easier to fit.

We illustrate with our example. Figure 6.1a contains a trace plot of β^{South}, the slope coefficient for the Southern region. We initialized three chains at values that were overdispersed relative to the estimate of this parameter from a simpler model (a linear model of the differences in the sample percentages of supporters in the South relative to the national average). The crude estimate of β^{South} from the simple model was 0.33, with a standard error of about 0.03, so we started our three chains at –0.7, 0.3, and 1.3, which are roughly centered at the crude estimate, but widely dispersed relative to the crude estimate's standard error,

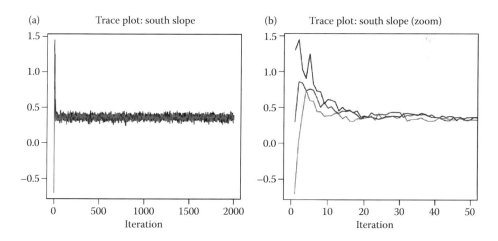

FIGURE 6.1
(a) Trace plots of β^{South}, the time trend in the logistic regression coefficients for death penalty support (per decade) for the Southern states. Three chains were initialized at –0.7, 0.3, and 1.3, respectively, and they converge to the target distribution within about 20 or 30 iterations. (b) The first 50 iterations, showing the movement away from the starting values.

so that we would be unlikely to miss a local mode in the potentially multimodal posterior distribution. The model converges very quickly, so that the initial values have been essentially forgotten after about 25 iterations (see Figure 6.1b). We ran the MCMC for 2000 iterations, because not every parameter converged as quickly as β^{South}, and also because it only took a few minutes to obtain chains of this length. Had the algorithm been much slower, we could have stopped partway through, inspected trace plots, and then made a decision about continuing the algorithm.

6.6 Monitoring Convergence Comparing between and within Chains

We never reach exact convergence; as a result it does not make sense to try to check convergence using statistical hypothesis tests of the null hypothesis of perfect mixing. Instead, we use statistical estimation—postprocessing of simulation results—to estimate how far current simulations are from perfect mixing.

We typically monitor the convergence of all the parameters and other quantities of interest separately. There have also been some methods proposed for monitoring the convergence of the entire distribution at once (see, e.g. Brooks and Gelman, 1998), but these methods may sometimes represent overkill: individual parameters can be well estimated even while approximate convergence of simulations of a multivariate distribution can take a very long time.

Our usual approach is, for each parameter or quantity of interest, to compute the variance of the simulations from each chain (after the first halves of each have been discarded, as explained in our discussion of burn-in), to average these within-chain variances, and compare this to the variances of all the chains mixed together. We take the mixture variance divided by the average within-chain variance, compute the square root of this ratio, and call it R.hat or the "potential scale reduction factor" (Gelman and Rubin, 1992, following ideas of Fosdick, 1959). R.hat is calculated in various MCMC software including BUGS (Spiegelhalter et al., 1994, 2003) and the R2WinBUGS and coda packages (Plummer et al., 2005; Sturtz et al., 2005) in R, and the underlying idea has also been applied to transdimensional simulations—mixture of models with different parameter spaces (see Brooks and Giudici, 2000; Brooks et al., 2003).

At convergence, the chains will have mixed, so that the distribution of the simulations between and within chains will be identical, and the ratio R.hat should equal 1. If R.hat is greater than 1, this implies that the chains have not fully mixed and that further simulation might increase the precision of inferences. In practice we typically go until R.hat is less than 1.1 for all parameters and quantities of interest; however, we recognize that this rule can declare convergence prematurely, which is one reason why we always recommend comparing results to estimates from simpler models. It can also be useful to check other convergence diagnostics (Cowles and Carlin, 1996). In our death penalty example, Figure 6.2a illustrates that convergence happens quickly—if we recompute R.hat every 50 iterations, discarding the first half of the iterations as burn-in in our computations, we see that it is less than 1.05 for every batch of such samples after 200 iterations and is less than 1.02 for every batch of such samples after about 500 iterations.

When problems show up, we typically look at time series plots of simulated chains to see where the poor mixing occurs and get insight into how to fix the algorithm to run more efficiently. Multivariate visual tools can make this graphical process more effective (Venna et al., 2003; Peltonen et al., 2009).

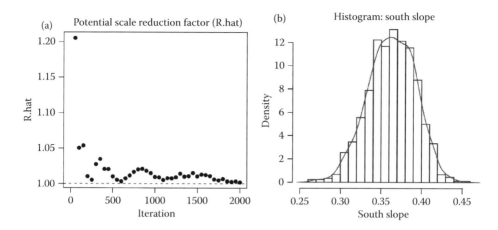

FIGURE 6.2

(a) R.hat, the potential scale reduction factor, for β^{South}, calculated repeatedly every 50 iterations using only the last half of each set of iterations. That is, it was calculated using iterations 26–50, then using iterations 51–100, then using iterations 76–150, and so on. R.hat was less than 1.05 for all batches of samples after 200 iterations and was less than 1.02 for all batches of samples after about 500 iterations. (b) The 900 posterior samples formed by taking iterations 301–600 from each of the three chains, with a density estimate overlain.

Mixing of chains can also be monitored nonparametrically, for example by computing the 80% (say) central interval from each chain and then determining its coverage with respect to the empirical distribution of all the other chains combined together (as always, after discarding the early burn-in iterations). At convergence, the average coverage of these 80% intervals should be 80%; a much lesser value indicates poor mixing. Brooks and Gelman (1998) and Brooks et al. (2003) discuss this and other methods for monitoring convergence by measuring mixing of multiple sequences, along with problems that can arise.

6.7 Inference from Simulations after Approximate Convergence

In considering inferential summaries from our simulations, we again separately consider our two tasks:

1. Inference about a parameter (or function of parameters) θ, to be summarized by a set of simulations and possibly a 95% interval.

Here we can use the collection of all our simulation draws (after discarding burn-in), or we can "thin" them by saving every nth iteration. The purpose of thinning (i.e. setting n to some integer greater than 1) is computational, not statistical. If we have a model with 2000 parameters and we are running three chains with a million iterations each, we do not want to be carrying around 6 billion numbers in our simulation. The key is to realize that, if we really needed a million iterations, they must be so highly autocorrelated that little is gained by saving them all. In practice, we find it is generally more than enough to save 1000 iterations in total, and so we thin accordingly. But ultimately this will depend on the size of the model and computational constraints.

2. Inference about an expectation, $E(\theta)$.

To estimate expectations, we can use the batch means method, dividing the chains (again, after discarding burn-in) into k batches and then computing the mean and standard error based on the average of the batch means. Each chain should be divided into batches, so that k is the number of chains multiplied by the number of batches per chain. The number of batches per chain should be set large enough that the standard error can be estimated reasonably precisely (the precision is essentially that of a chi-squared distribution with $k - 1$ degrees of freedom) while still having the batch means be approximately statistically independent (so that the $1/\sqrt{k}$ standard error formula is applicable). If necessary, the standard error can be adjusted upward to correct for any autocorrelation remaining even after batching.

In the death penalty example, our inference for β^{South}, as noted above, is of the Task 1 variety. According to Figure 6.2a, we can assume that the three chains have converged after 300 iterations, and we can use the next 300 iterations from each of the three chains to form a posterior sample of size 900 for our inference. From this we estimate the posterior mean of β^{South} to be 0.363 and the standard deviation of β^{South} to be 0.029. A 95% interval for β^{South} is obtained by taking the 2.5th and 97.5th percentiles of this sample, which are 0.31 and 0.42, respectively. Figure 6.2b contains a histogram of these 900 posterior samples, with a smooth density estimate overlain.

To estimate the uncertainty about the expectation (in this case the posterior mean, $E(\beta^{\text{South}}|y)$), we can use the batch means method. We divide the 300 samples from each chain into six batches each, and compute the standard error of these $k = 18$ batch means (where each set of six batch means per chain has autocorrelation approximately zero). This approximate standard error is about 0.002. As in the toy example from earlier, the uncertainty about the mean is tiny compared to the uncertainty in the posterior distribution of the parameter β^{South}, and we conclude that these 900 samples are sufficient for our Task 1 inference for β^{South}. We can double-check our batch means calculations by computing the effective sample size of these 900 draws (which accounts for autocorrelation) using standard methods (Kass et al., 1998), and we compute that the approximate total number of independent draws from these three sets of 300 autocorrelated samples is 169. Thus, the standard error of the mean computed this way is $0.029/\sqrt{169}$, which is about 0.002, confirming our earlier batch means calculation.

In practice, we could compute more accurate estimates of summaries of the posterior distribution of β^{South}, for example the expectation, variance, and various quantiles, by including iterations 101–2000 in our posterior sample: visual inspection of the trace plot that the chains converged by iteration 100, and in fact R.hat = 1.00 when it is computed using iterations 101–2000. But more samples will not improve our Task 1 inference in any meaningful way, and since this was our goal, we could have stopped the samplers after about 600 iterations (instead of running them for 2000 iterations as we did).

We have also worked on problems that have required tens of thousands of iterations or more to reach approximate convergence, and the same inferential principles apply.

6.8 Summary

Monitoring convergence of iterative simulation is straightforward (discard the first part of the simulations and then compare the variances of quantities of interest within and between chains) and inference given approximate convergence is even simpler (just mix the simulations together and use them as a joint distribution). Both these ideas can be and have been refined, but the basic concepts are straightforward and robust.

The hard part is knowing what to do when the simulations are slow to converge. Then it is a good idea to look at the output and put together a more efficient simulation algorithm, which sometimes can be easy enough (e.g. using redundant parameterization for the Gibbs sampler or tuning the proposal distributions for a Metropolis algorithm), sometimes can require more elaborate algorithms (such as hybrid sampling or parallel tempering), and sometimes requires development of a simulation algorithm specifically tailored to the problem at hand. Once we have an improved algorithm, we again monitor its convergence by measuring the mixing of independent chains and checking that each chain seems to have reached a stationary distribution. And then we can perform simulation-based inferences as described above.

Acknowledgments

We thank Steve Brooks, Brad Carlin, David Dunson, Hal Stern, and an anonymous reviewer for helpful comments and discussion. The National Science Foundation grants SES-1023176, ATM-0934516; National Institutes of Health; National Security Agency grant H98230-10-1-0184; Department of Energy grant DE-SC0002099; Institute for Education Sciences grants ED-GRANTS-032309-005, R305D090006-09A; and Yahoo! Research provided partial support for this research.

References

Brooks, S. and Gelman, A. 1998. General methods for monitoring convergence of iterative simulations. *Journal of Computational and Graphical Statistics*, 7:434–455.

Brooks, S. and Giudici, P. 2000. MCMC convergence assessment via two-way ANOVA. *Journal of Computational and Graphical Statistics*, 9:266–285.

Brooks, S., Giudici, P., and Phillipe, A. 2003. Nonparametric convergence assessment for MCMC model selection. *Journal of Computational and Graphical Statistics*, 12:1–22.

Cowles, M. K. and Carlin, B. P. 1996. Markov chain Monte Carlo convergence diagnostics: A comparative review. *Journal of the American Statistical Association*, 91:883–904.

Cook, S., Gelman, A., and Rubin, D. B. 2006. Validation of software for Bayesian models using posterior quantiles. *Journal of Computational and Graphical Statistics*, 15:675–692.

Efron, B. and Tibshirani, R. 1993. *An Introduction to the Bootstrap*. Chapman & Hall, New York.

Fosdick, L. D. 1959. Calculation of order parameters in a binary alloy by the Monte Carlo method. *Physical Review*, 116:565–573.

Gelman, A. and Rubin, D. B. 1992. Inference from iterative simulation using multiple sequences (with discussion). *Statistical Science*, 7:457–511.

Gelman, A., Carlin, J. B., Stern, H. S., and Rubin, D. B. 2003. *Bayesian Data Analysis*, 2nd edition. Chapman & Hall/CRC, Boca Raton, FL.

Geyer, C. 1998. Burn-in is unnecessary. www.stat.umn.edu/~charlie/mcmc/burn.html

Kass, R. E., Carlin, B. P., Gelman, A., and Neal, R. 1998. Markov chain Monte Carlo in practice: A roundtable discussion. *American Statistician*, 52:93–100.

Liu, C. and Rubin, D. B. 1996. Markov-normal analysis of iterative simulations before their convergence. *Journal of Econometrics*, 75:69–78.

Liu, C. and Rubin, D. B. 2002. Model-based analysis to improve the performance of iterative simulations. *Statistica Sinica*, 12:751–767.

Peltonen, J., Venna, J., and Kaski, S. 2009. Visualizations for assessing convergence and mixing of Markov chain Monte Carlo simulations. *Computational Statistics and Data Analysis*, 53:4453–4470.

Plummer, M., Best, N., Cowles, K., and Vines, K. 2005. Output analysis and diagnostics for MCMC: The coda package for R. http://pbil.univ-lyon1.fr/library/coda/html/00Index.html

Raftery, A. E. and Lewis, S. M. 1992. How many iterations in the Gibbs sampler? In J. M. Bernardo, J. O. Berger, A. P. Dawid, and A. F. M. Smith (eds), *Bayesian Statistics 4: Proceedings of the Fourth Valencia International Meeting*, pp. 765–776. Oxford University Press, Oxford.

Shirley, K. and Gelman, A. 2010. State-level trends in public opinion about the death penalty, 1953–2006. Technical report, Department of Statistics, Columbia University.

Spiegelhalter, D., Thomas, A., Best, N., Gilks, W., and Lunn, D. 1994, 2003. BUGS: Bayesian inference using Gibbs sampling. MRC Biostatistics Unit, Cambridge, UK. www.mrc-bsu.cam.ac.uk/bugs/

Sturtz, S., Ligges, U., and Gelman, A. 2005. R2WinBUGS: A package for running WinBUGS from R. *Journal of Statistical Software*, 12(3).

Venna, J., Kaski, S., and Peltonen, J. 2003. Visualizations for assessing convergence and mixing of MCMC. In N. Lavraè, D. Gamberger, H. Blockeel, and L. Todorovski (eds), *Machine Learning: ECML 2003*, Lecture Notes in Artificial Intelligence, Vol. 2837. Springer, Berlin.

7

Implementing MCMC: Estimating with Confidence

James M. Flegal and Galin L. Jones

7.1   Introduction

Our goal is to introduce some of the tools useful for analyzing the output of a Markov chain Monte Carlo (MCMC) simulation. In particular, we focus on methods which allow the practitioner (and others!) to have confidence in the claims put forward. The following are the main issues we will address: (1) initial graphical assessment of MCMC output; (2) using the output for estimation; (3) assessing the Monte Carlo error of estimation; and (4) terminating the simulation.

Let π be a density function with support $X \subseteq \mathbb{R}^d$ about which we wish to make an inference. This inference often is based on some feature of π. For example, if $g : X \to \mathbb{R}$ a common goal is the calculation of

$$E_{\pi}g = \int_X g(x)\pi(x)\,dx. \tag{7.1}$$

We will typically want the value of several features such as mean and variance parameters, along with quantiles and so on. As a result, the features of interest form a p-dimensional vector which we call θ_{π}. Unfortunately, in practically relevant settings we often cannot calculate any of the components of θ_{π} analytically or even numerically. Thus we are faced with a classical statistical problem: given a density π, we want to estimate several fixed, unknown features of it. For ease of exposition we focus on the case where θ_{π} is univariate, but we will come back to the general case at various points throughout.

Consider estimating an expectation as in Equation 7.1. The basic MCMC method entails constructing a Markov chain $X = \{X_0, X_1, X_2, \ldots\}$ on X having π as its invariant density. (See Chapter 1, this volume, for an introduction to MCMC algorithms.) Then we simulate X for a finite number of steps, say n, and use the observed values to estimate $E_{\pi}g$ with a sample average

$$\bar{g}_n := \frac{1}{n}\sum_{i=0}^{n-1} g(x_i).$$

The use of this estimator is justified through the Markov chain strong law of large numbers (SLLN)*: If $E_{\pi}|g| < \infty$, then $\bar{g}_n \to E_{\pi}g$ almost surely as $n \to \infty$. From a practical point

---

* This is a special case of the Birkhoff ergodic theorem (Fristedt and Gray, 1997, p. 558).

of view, this means we can obtain an accurate estimate of $E_\pi g$ with a sufficiently long simulation.

Outside of toy examples, no matter how long our simulation, there will be an unknown *Monte Carlo error*, $\bar{g}_n - E_\pi g$. While it is impossible to assess this error directly, we can obtain its approximate sampling distribution if a Markov chain central limit theorem (CLT) holds—that is, if

$$\sqrt{n}(\bar{g}_n - E_\pi g) \xrightarrow{d} N(0, \sigma_g^2) \tag{7.2}$$

as $n \to \infty$, where $\sigma_g^2 \in (0, \infty)$. It is important to note that due to the correlation present in a Markov chain, $\sigma_g^2 \neq \text{var}_\pi g$, except in trivial cases. For now, suppose we have an estimator such that $\hat{\sigma}_n^2 \to \sigma_g^2$ almost surely as $n \to \infty$ (see Section 7.4 for some suitable techniques). This allows construction of an asymptotically valid confidence interval for $E_\pi g$ with half-width

$$t_* \frac{\hat{\sigma}_n}{\sqrt{n}}, \tag{7.3}$$

where t_* is an appropriate quantile.

Most importantly, calculating and reporting the Monte Carlo standard error (MCSE), $\hat{\sigma}_n/\sqrt{n}$, allows everyone to judge the reliability of the estimates. In practice this is done in the following way. Suppose that after n simulations our estimate of $E_\pi g$ is $\bar{g}_n = 1.3$. Let h_α denote the half-width given in Equation 7.3 of a $(1 - \alpha)100\%$ interval. We can be confident in the "3" in our estimate if $1.3 \pm h_\alpha \subseteq [1.25, 1.35]$. Otherwise, reasonable values such as 1.2 or 1.4 could be obtained by rounding. If the interval is too wide for our purposes, then more simulation should be conducted. Of course, we would be satisfied with a wider interval if we only wanted to trust the "1" or the sign of our estimate. Thus the interval estimator (Equation 7.3) allows us to describe the confidence in the reported estimate, and moreover, including an MCSE with the point estimate allows others to assess its reliability. Unfortunately, this is not currently standard practice in MCMC (Flegal et al., 2008).

The rest of this chapter is organized as follows. In Section 7.2 we consider some basic techniques for graphical assessment of MCMC output, then Section 7.3 contains a discussion of various point estimators of θ_π. Next, Section 7.4 introduces techniques for constructing interval estimators of θ_π. Then Section 7.5 considers estimating marginal densities associated with π and Section 7.6 further considers stopping rules for MCMC simulations. Finally, in Section 7.7 we give conditions for ensuring the CLT (Equation 7.2). The computations presented in our examples were carried out using the R language. See Flegal and Jones (2010c) for an `Sweave` file from which the reader can reproduce all of our calculations.

7.2 Initial Examination of Output

As a first step it pays to examine the empirical finite-sample properties of the Markov chain being simulated. A few simple graphical methods are often used in the initial assessment of the simulation output. These include scatterplots, histograms, time series plots, autocorrelation plots and plots of sample means. We will content ourselves with an illustration of some of these techniques; see Chapter 1 (this volume) for further discussion. Consider the following toy example, which we will return to several times.

Example 7.1 (Normal AR(1) Markov Chains)

The normal AR(1) time series is defined by

$$X_{n+1} = \rho X_n + \epsilon_n. \tag{7.4}$$

where the ϵ_n are i.i.d. $N(0,1)$ and $\rho < 1$. This Markov chain has invariant distribution $N\left(0, 1/(1 - \rho^2)\right)$.

As a simple numerical example, consider simulating the chain (Equation 7.4) in order to estimate the mean of the invariant distribution, that is $E_\pi X = 0$. While this is a toy example, it is quite useful because ρ plays a crucial role in the behavior of this chain. Figure 7.1 contains plots based on single sample path realizations starting at $X_1 = 1$ with $\rho = 0.5$ and $\rho = 0.95$. In each figure the top plot is a time series plot of the observed sample path. The mean of the target distribution is 0 and the horizontal lines are 2 standard deviations above and below the mean. Comparing the time series plots, it is apparent that while we may be getting a representative sample from the invariant distribution, when $\rho = 0.95$ the sample is highly correlated. This is also apparent from the autocorrelation (middle) plots in both figures. When $\rho = 0.5$ the autocorrelation is negligible after about lag 4, but when $\rho = 0.95$ there is a substantial autocorrelation until about lag 30. The impact of this correlation is apparent in the bottom two plots which plot the running estimates of the mean versus iterations in the chain. The true value is displayed as the horizontal line at 0. Clearly, the more correlated sequence requires many more iterations to achieve a reasonable estimate. From these plots, we can see that the simulation with $\rho = 0.5$ may have been run long enough while the simulation with $\rho = 0.95$ likely has not.

In the example, the plots were informative because we were able to draw horizontal lines depicting the true values. In practically relevant MCMC settings—where the truth is unavailable—it is hard to know when to trust these plots. Nevertheless, they can still be useful since a Markov chain that is mixing well would tend to have time series and autocorrelation plots that look like Figure 7.1a, while time series and autocorrelation plots like the one in Figure 7.1b would indicate a potentially problematic simulation in the sense

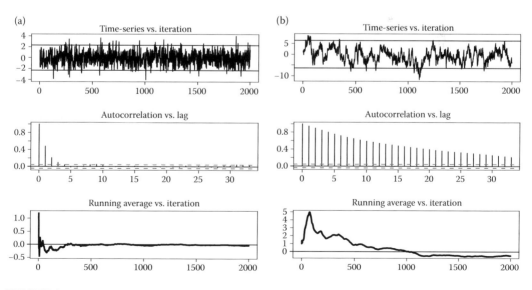

FIGURE 7.1
Initial output examination for AR(1) model: (a) $\rho = 0.5$ and (b) $\rho = 0.95$.

that a large simulation effort will be required to obtain good estimates. Also, plots of current parameter estimates (with no reference to a standard error) versus number of iterations are not as helpful since they provide little information as to the quality of estimation.

In simulating a d-dimensional Markov chain to simultaneously estimate the p-dimensional vector θ_π of features of π, p can be either greater than or less than d. When either d or p is large, the standard graphical techniques are obviously problematic. That is, even if each component's time series plot indicates good mixing, one should not necessarily infer that the chain has converged to its joint target distribution. In addition, if either d or p is large it will be impractical to look at plots of each component. These issues have received very little attention in the MCMC literature, but see Peltonen et al. (2009) and Sturtz et al. (2005) for some recent work.

7.3 Point Estimates of θ_π

In this section, we consider two specific cases of θ_π: estimating a univariate expectation $E_\pi g$; and estimating a quantile of one of the univariate marginal distributions from π.

7.3.1 Expectations

Suppose that $\theta_\pi = E_\pi g$ and assume that $E_\pi|g| < \infty$. Recall from Section 7.1 that there is an SLLN and hence it is natural to use the sample average $\bar{g}_n$ to estimate θ_π. Alternatively, we could use point estimates of θ_π obtained through the use of burn-in or averaging conditional expectations.

Consider the use of burn-in. Usually, the simulation is not started with a draw from π since otherwise we would just do ordinary Monte Carlo. It follows that marginally each $X_i \nsim \pi$ and $E_\pi g \neq E[g(X_i)]$. Thus $\bar{g}_n$ is a biased estimator of $E_\pi g$. In the current setting, we have that $X_n \overset{d}{\to} \pi$ as $n \to \infty$ so, in order to diminish the effect of this "initialization bias," an alternative estimator may be employed:

$$
\bar{g}_{n,B} = \frac{1}{n} \sum_{i=B}^{n+B-1} g(x_i),
$$

where B denotes the burn-in or amount of simulation discarded. By keeping only the draws obtained after $B - 1$ we are effectively choosing a new initial distribution that is "closer" to π. The SLLN still applies to $\bar{g}_{n,B}$ since if it holds for any initial distribution it holds for every initial distribution. Of course, one possible (perhaps even likely) consequence of using burn-in is that $\mathrm{var}(\bar{g}_{n,B}) \geq \mathrm{var}(\bar{g}_{n+B,0})$, that is, the bias decreases but the variance increases for the same total simulation effort. Obviously, this means that using burn-in could result in an estimator having larger mean-squared error than one without burn-in. Moreover, without some potentially difficult theoretical work (Jones and Hobert, 2001; Latuszynski and Niemiro, 2009; Rosenthal, 1995; Rudolf, 2009), it is not clear what value of B should be chosen. Popular approaches to determining B include simply discarding a fraction of the total run length (see Gelman and Rubin, 1992), or are based on convergence diagnostics (for a review, see Cowles and Carlin, 1996). Unfortunately, there simply is no guarantee that any of these diagnostics will detect a problem with the simulation and, in fact, using them can introduce bias (Cowles et al., 1999).

Now consider the estimator obtained by averaging conditional expectations. To motivate this discussion, suppose that the target is a function of two variables $\pi(x, y)$ and we are interested in estimating the expectation of a function of only one of the variables, say $g(x)$. Let $(X, Y) = \{(X_0, Y_0), (X_1, Y_1), (X_2, Y_2), \ldots\}$ denote the Markov chain, $m_Y(y)$ denote the marginal density, and $f_{X|Y}(x \mid y)$ denote the conditional density. Notice that

$$E_\pi g = \int \int g(x) \pi(x, y) dx \, dy = \int \left[\int g(x) f_{X|Y}(x \mid y) \, dx \right] m_Y(y) \, dy$$

so, by letting

$$h(y) = \int g(x) f_{X|Y}(x \mid y) \, dx,$$

we can appeal to the SLLN again to see that, as $n \to \infty$,

$$\bar{h}_n = \frac{1}{n} \sum_{i=0}^{n-1} h(y_i) = \frac{1}{n} \sum_{i=0}^{n-1} \int g(x) f_{X|Y}(x \mid y_i) \, dx \xrightarrow{a.s.} E_\pi g.$$

This estimator is conceptually the same as $\bar{g}_n$ in the sense that both are sample averages and the Markov chain SLLN applies to both. The estimator $\bar{h}_n$ is often called the *Rao-Blackwellized* (RB) estimator* of $E_\pi g$ (Casella and Robert, 1996). A natural question is which of $\bar{g}_n$, the sample average, or $\bar{h}_n$, the RB estimator, is better. It is obvious that $\bar{h}_n$ will sometimes be impossible to use if $f_{X|Y}(x \mid y)$ is not available in closed form or if the integral is intractable. Hence, $\bar{h}_n$ will not be as generally practical as $\bar{g}_n$. However, there are settings, such as in data augmentation (Chapter 10, this volume), where $\bar{h}_n$ is theoretically and empirically superior to $\bar{g}_n$; see Liu et al. (1994) and Geyer (1995) for theoretical investigation of these two estimators.

Example 7.2

This example is also considered in Chapter 10 (this volume). Suppose that our goal is to estimate the first two moments of a Student's t distribution with 4 degrees of freedom and having density

$$m(x) = \frac{3}{8} \left(1 + \frac{x^2}{4} \right)^{-5/2}.$$

There is nothing about this that requires MCMC since we can easily calculate that $E_m X = 0$ and $E_m X^2 = 2$. Nevertheless, we will use a data augmentation algorithm based on the joint density

$$\pi(x, y) = \frac{4}{\sqrt{2\pi}} y^{3/2} e^{-y(2 + x^2/2)},$$

so that the full conditionals are $X \mid Y \sim N(0, y^{-1})$ and $Y \mid X \sim \Gamma(5/2, 2 + x^2/2)$. Consider the Gibbs sampler that updates X then Y so that a one-step transition looks like $(x', y') \to (x, y') \to$

---

* This is an unfortunate name since it is only indirectly related to the Rao–Blackwell theorem, but the name has stuck in the literature.

(x, y). Suppose that we have obtained n observations $\{x_i, y_i; i = 0, \ldots, n-1\}$ from running the Gibbs sampler. Then the standard sample average estimates of $E_m X$ and $E_m X^2$ are

$$\frac{1}{n} \sum_{i=0}^{n-1} x_i \quad \text{and} \quad \frac{1}{n} \sum_{i=0}^{n-1} x_i^2,$$

respectively. Further, the RB estimates are easily computed. Since $X \mid Y \sim N(0, y^{-1})$ the RB estimate of $E_m X$ is 0. On the other hand, the RB estimate of $E_m X^2$ is

$$\frac{1}{n} \sum_{i=0}^{n-1} y_i^{-1}.$$

As an illustration of these estimators we simulated 2000 iterations of the Gibbs sampler and plotted the running values of the estimators in Figure 7.2. In this example, the RB averages are less variable than the standard sample averages.

It is not the case that RB estimators are always better than sample means. Whether they are better depends on the expectation being estimated as well as the properties of the MCMC sampler. In fact, Liu et al. (1994) and Geyer (1995) give an example where the RB estimator is provably worse than the sample average. RB estimators are more general than our presentation suggests. Let h be any function and set

$$f(x) = E[g(X) \mid h(X) = h(x)]$$

so that $E_\pi g = E_\pi f$. Thus, by the Markov chain SLLN with probability 1 as $n \to \infty$,

$$\frac{1}{n} \sum_{i=0}^{n-1} f(X_i) = \frac{1}{n} \sum_{i=0}^{n-1} E[g(X) \mid h(X_i) = h(x_i)] \to E_\pi g.$$

As long as the conditional distribution $X \mid h(x)$ is tractable, RB estimators are available.

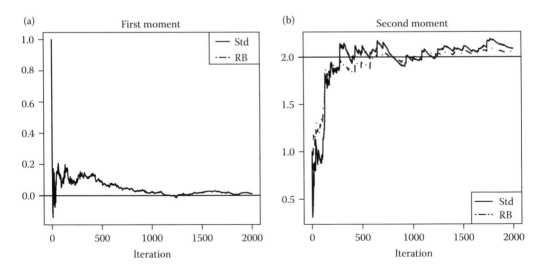

FIGURE 7.2
Estimators of the first two moments for Example 7.2. The horizontal line denotes the truth, the solid curves are the running sample averages while the dotted curves are the running RB sample averages. (a) $E_m X$ and (b) $E_m X^2$.

7.3.2 Quantiles

It is common to report estimated quantiles in addition to estimated expectations. Actually, what is nearly always reported is not a multivariate quantile but rather quantiles of the univariate marginal distributions associated with π. This is the only setting we consider. Let F be a marginal distribution function associated with π. Then the qth quantile of F is

$$\phi_q = F^{-1}(q) = \inf\{x : F(x) \geq q\}, \quad 0 < q < 1. \tag{7.5}$$

There are many potential estimates of ϕ_q, but we consider only the inverse of the empirical distribution function from the observed sequence. First define $\{X_{(1)}, \ldots, X_{(n)}\}$ as the order statistics of $\{X_0, \ldots, X_{n-1}\}$; then the estimator of ϕ_q is given by

$$\hat{\phi}_{q,n} = X_{(j+1)}, \quad \text{where } \frac{j}{n} \leq q < \frac{j+1}{n}. \tag{7.6}$$

Example 7.3 (Normal AR(1) Markov Chains)

Consider again the time series defined in Equation 7.4. Our goal in this example is to illustrate estimating the first and third quartiles, denoted Q_1 and Q_3. The true values of Q_1 and Q_3 are $\pm\Phi^{-1}(0.75)/\sqrt{1-\rho^2}$, where Φ is the cumulative distribution function of a standard normal distribution.

Using the same realization of the chain as in Example 7.1, Figure 7.3 shows plots of the running quartiles versus iteration number when $\rho = 0.5$ and $\rho = 0.95$. It is immediately apparent that estimation is more difficult when $\rho = 0.95$ and hence the simulation should continue. Also, without the horizontal lines, these plots would not be as useful. Recall that a similar conclusion was reached for estimating the mean.

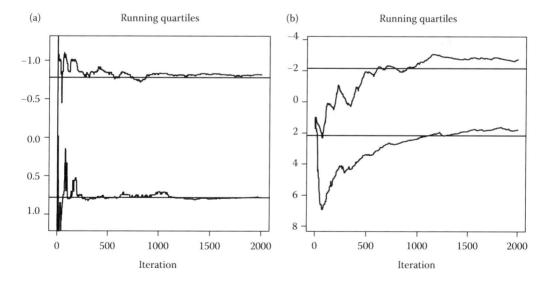

FIGURE 7.3
Plots for AR(1) model of running estimates of Q_1 and Q_3. The horizontal lines are the true quartiles. (a) $\rho = 0.5$ and (b) $\rho = 0.95$.

7.4   Interval Estimates of θ_π

In our examples we have known the truth, enabling us to draw the horizontal lines on the plots which allow us to gauge the quality of estimation. Obviously, the true parameter value is unknown in practical settings, and hence the size of the Monte Carlo error is unknown. For this reason, when reporting a point estimate of θ_π, a MCSE should be included so that the reader can assess the quality of the reported point estimates. In this section we address how to calculate MCSEs and construct asymptotically valid interval estimates of θ_π.

7.4.1   Expectations

Suppose that $\theta_\pi = E_\pi g$, which will be estimated with $\bar{g}_n = \bar{g}_{n,0}$ (i.e. with no burn-in). However, using burn-in presents no theoretical difficulties since, as with the SLLN, if the CLT holds for any initial distribution then it holds for every initial distribution. Thus the use of burn-in does not affect the existence of a CLT, but it may affect the quality of the asymptotic approximation. If $\hat{\sigma}_n^2$ is an estimate of σ_g^2, then one can form a confidence interval for $E_\pi g$ with half-width

$$t_* \frac{\hat{\sigma}_n}{\sqrt{n}}, \tag{7.7}$$

where t_* is an appropriate quantile. Thus the difficulty in finding interval estimates is in estimating σ_g^2, which requires specialized techniques to account for correlation in the Markov chain. We restrict attention to strongly consistent estimators of σ_g^2. Some interval estimation techniques do not require consistent estimation of σ_g^2 (see Schruben, 1983) but we need it for the methods presented later in Section 7.6. The methods yielding strongly consistent estimators include batch means methods, spectral variance methods and regenerative simulation. Alternatives include the initial sequence methods of Geyer (1992); however, the theoretical properties of Geyer's estimators are not well understood. We will focus on batch means as it is the most generally applicable method; for more on spectral methods, see Flegal and Jones (2010a), while Hobert et al. (2002) and Mykland et al. (1995) study regenerative simulation. There are many variants of batch means; here we emphasize overlapping batch means (OLBM).

7.4.1.1   Overlapping Batch Means

As the name suggests, in OLBM we divide the simulation into overlapping batches of length b_n, say. For example, if $b_n = 3$, then $\{X_0, X_1, X_2\}$ and $\{X_1, X_2, X_3\}$ would be the first two overlapping batches. In general, there are $n - b_n + 1$ batches of length b_n, indexed by j running from 0 to $n - b_n$. Let $\bar{Y}_j(b_n) := b_n^{-1} \sum_{i=0}^{b_n-1} g(X_{j+i})$ for $j = 0, \ldots, n - b_n$. Then the OLBM estimator of σ_g^2 is

$$\hat{\sigma}_{OLBM}^2 = \frac{nb_n}{(n - b_n)(n - b_n + 1)} \sum_{j=0}^{n-b_n} (\bar{Y}_j(b_n) - \bar{g}_n)^2. \tag{7.8}$$

Batch means estimators are not generally consistent for σ_g^2 (Glynn and Whitt, 1991). However, roughly speaking, Flegal and Jones (2010a) show that if the Markov chain mixes quickly and b_n is allowed to increase as the overall length of the simulation does, then $\hat{\sigma}^2_{OLBM}$ is a strongly consistent estimator of σ^2_g. It is often convenient to take $b_n = \lfloor n^v \rfloor$ for some $0 < v < 3/4$, and $v = 1/2$ may be a reasonable default. However, v values yielding strongly consistent estimators are dependent on the number of finite moments of g with respect to the target π and the mixing conditions of the Markov chain. These conditions are similar to those required for a Markov chain CLT, see Flegal and Jones (2010a), Jones (2004), and Jones et al. (2006). Finally, when constructing the interval (Equation 7.7), t_* is a quantile from a Student's t distribution with $n - b_n$ degrees of freedom.

Example 7.4 (Normal AR(1) Markov Chains)

Recall the AR(1) model defined in Equation 7.4. Using the same realization of the chain as in Example 7.1, that is, 2000 iterations with $\rho \in \{0.5, 0.95\}$ starting from $X_1 = 1$, we consider estimating the mean of the invariant distribution, that is, $E_\pi X = 0$. Utilizing OLBM with $b_n = \lfloor \sqrt{n} \rfloor$, we calculated an MCSE and resulting 80% confidence interval. Figure 7.4 shows the running means versus number of iterations for $\rho = 0.5$ and $\rho = 0.95$. The dashed lines correspond to upper and lower 80% confidence bounds. Notice that for the larger value of ρ it takes longer for the MCSE to stabilize and begin decreasing. After 2000 iterations for $\rho = 0.5$ we obtained an interval of -0.034 ± 0.056, while for $\rho = 0.95$ the interval is -0.507 ± 0.451.

Many of our plots are based on simulating only 2000 iterations. We chose this value strictly for illustration purposes. An obvious question is whether the simulation has been run long enough—that is, whether the interval estimates are sufficiently narrow after 2000 iterations. In the $\rho = 0.5$ case, the answer is "perhaps," while in the $\rho = 0.95$ case it is clearly "no." Consider the final interval estimate of the mean with $\rho = 0.5$, that is, $-0.034 \pm 0.056 = (-0.090, 0.022)$. If the user is satisfied with this level of precision, then 2000 iterations are sufficient. On the other hand, when $\rho = 0.95$ our interval estimate is $-0.507 \pm 0.451 = (-0.958, -0.056)$, indicating that we cannot trust any of the significant figures reported in the point estimate.

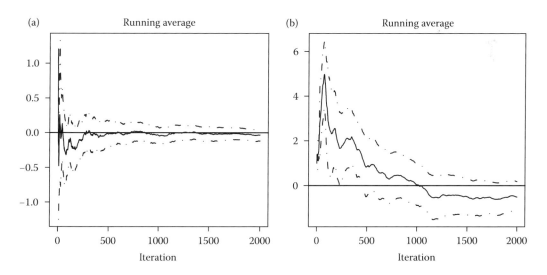

FIGURE 7.4
Plots for AR(1) model of running estimates of the mean along with confidence intervals calculated via OLBM. The horizontal line denotes the truth. (a) $\rho = 0.5$ and (b) $\rho = 0.95$.

Recall the RB estimators of Section 7.3.1. It is straightforward to use OLBM to calculate the MCSE for these estimators since the conditional expectations being averaged define the sequence of batch means $\bar{Y}_j(b_n)$.

Example 7.5

Recall Example 7.2, where the first two moments of a Student's t distribution with 4 degrees of freedom were estimated using sample average and RB estimators. Using the same Markov chain, an 80% confidence interval is calculated via OLBM with $b_n = \lfloor \sqrt{n} \rfloor$ at each iteration. Figure 7.5a shows the running estimate of $E_\pi X$ versus iteration number and includes confidence bounds for the sample average estimator. Recall that the RB estimator is exact, so there is no uncertainty in this estimate. Figure 7.5b shows the running estimate of $E_\pi X^2$ versus iteration number, with confidence bounds for both estimates. Here it is provable that the RB estimator has a smaller asymptotic variance than the sample average estimator (Geyer, 1995). This is clearly reflected by the narrower interval estimates.

7.4.1.2　Parallel Chains

To this point, the recipe for implementing MCMC seems straightforward: given a sampler, pick a starting value and run the simulation for a sufficiently long time using the SLLN and the CLT to produce a point estimate and a measure of its uncertainty. A variation of this procedure relies on simulating multiple independent, or parallel, chains. Debate between a single long run and parallel chains began in the early statistics literature on MCMC (see Gelman and Rubin, 1992; Geyer, 1992), even earlier in the operations research and physics literature (Bratley et al., 1987; Fosdick, 1959; Kelton and Law, 1984), and continues today (Alexopoulos and Goldsman, 2004; Alexopoulos et al., 2006). The main idea

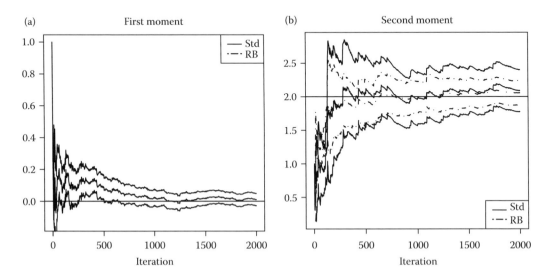

FIGURE 7.5
Estimators of the first two moments from a t distribution with 4 degrees of freedom. The horizontal line denotes the truth, the solid curves are the running sample averages with confidence bounds, while the dotted curves are the running RB sample averages with confidence bounds. (a) $E_m X$ and (b) $E_m X^2$.

of parallel chains is to run r independent chains using different starting values, where each chain is the same length and using the same burn-in. This yields r independent estimates of $E_\pi g$, namely $\bar{g}_{n,B,1}, \bar{g}_{n,B,2}, \ldots, \bar{g}_{n,B,r}$. The grand mean would then estimate $E_\pi g$—although Glynn and Heidelberger (1991) have shown that an alternative estimator may be superior—and our estimate of its performance, σ_g^2, would be the usual sample variance of the $\bar{g}_{n,B,i}$.

This approach has some intuitive appeal in that estimation avoids some of the serial correlation inherent in MCMC and it is easily implemented when more than one processor is available. Moreover, there is value in trying a variety of initial values for any MCMC experiment. It has also been argued that by choosing the r starting points in a widely dispersed manner there is a greater chance of encountering modes that one long run may have missed. Thus, for example, some argue that using independent replications results in "superior inferential validity" (Gelman and Rubin, 1992, p. 503). However, there is no agreement on this issue; indeed, Bratley et al. (1987, p. 80) "are skeptical about [the] rationale" of some proponents of independent replications. Notice that the total simulation effort using independent replications is $r(n + B)$. To obtain good estimates of σ_g^2 will require r to be large, which will require $n + B$ to be small for a given computational effort. If we use the same value of B as we would when using one long run, this means that each $\bar{g}_{n,B,i}$ will be based on a comparatively small number n of observations. Using more than one chain will also enhance the initialization bias, so that a careful choice of B can be quite important to the statistical efficiency of the estimator of $E_\pi g$ (Glynn and Heidelberger, 1991). Moreover, since each run will be comparatively short, there is a reasonable chance that a given replication will not move far from its starting value. Alexopoulos and Goldsman (2004) have shown that this can result in much poorer estimates (in terms of mean square error) of $E_\pi g$ than a single long run. On the other hand, if we can find a variety of starting values that are from a distribution very close to π, then independent replications may indeed be superior. This should not be surprising since independent draws directly from π are clearly desirable.

There is an important caveat to the above analysis. There are settings (see Chapter 20, this volume) where it is prohibitively difficult (or time-consuming) to produce a sufficiently large Monte Carlo sample without parallel computing. This has received limited attention in MCMC settings (Brockwell, 2006; Rosenthal, 2000), but perhaps deserves more.

7.4.2 Functions of Moments

Suppose that we are interested in estimating $\phi(E_\pi g)$, where ϕ is some function. If ϕ is continuous, then $\phi(\bar{g}_n) \to \phi(E_\pi g)$ with probability 1 as $n \to \infty$, making estimation of $\phi(E_\pi g)$ straightforward. Also, a valid Monte Carlo error can be obtained via the delta method (Ferguson, 1996; van der Vaart, 1998). Assuming (Equation 7.2), the delta method says that if ϕ is continuously differentiable in a neighborhood of $E_\pi g$ and $\phi'(E_\pi g) \neq 0$, then as $n \to \infty$,

$$\sqrt{n}\left(\phi(\bar{g}_n) - \phi(E_\pi g)\right) \xrightarrow{d} N\left(0, [\phi'(E_\pi g)]^2 \sigma_g^2\right).$$

If the estimator of σ_g^2, say $\hat{\sigma}_n^2$, is strongly consistent and ϕ' is continuous, then $[\phi'(\bar{g}_n)]^2 \hat{\sigma}_n^2$ is strongly consistent for $[\phi'(E_\pi g)]^2 \sigma_g^2$.

Example 7.6

Consider estimating $(E_\pi X)^2$ with $(\bar{X}_n)^2$. Let $\phi(x) = x^2$ and assume $E_\pi X \neq 0$ and a CLT as in Equation 7.2. Then as $n \to \infty$,

$$\sqrt{n}\left((\bar{X}_n)^2 - (E_\pi X)^2\right) \overset{d}{\to} N(0, 4(E_\pi X)^2 \sigma_x^2),$$

and we can use OLBM to consistently estimate σ_x^2 with $\hat{\sigma}_n^2$, which means that $4(\bar{X}_n)^2 \hat{\sigma}_n^2$ is a strongly consistent estimator of $4(E_\pi X)^2 \sigma_x^2$.

From this example we see that the univariate delta method makes it straightforward to handle powers of moments. The multivariate delta method allows us to handle more complicated functions of moments. Let T_n denote a sequence of d-dimensional random vectors and θ be a d-dimensional parameter. If, as $n \to \infty$,

$$\sqrt{n}(T_n - \theta) \overset{d}{\to} N(\mu, \Sigma)$$

and ϕ is continuously differentiable in a neighborhood of θ and $\phi'(\theta) \neq 0$, then as $n \to \infty$,

$$\sqrt{n}(\phi(T_n) - \phi(\theta)) \overset{d}{\to} N(\phi'(\theta)\mu, \phi'(\theta)\Sigma\phi'(\theta)^T).$$

Example 7.7

Consider estimating $\mathrm{var}_\pi g = E_\pi g^2 - (E_\pi g)^2$ with, setting $h = g^2$,

$$\frac{1}{n}\sum_{i=1}^n h(X_i) - \left(\frac{1}{n}\sum_{i=1}^n g(X_i)\right)^2 := \hat{v}_n.$$

Assume

$$\sqrt{n}\left(\begin{pmatrix} \bar{g}_n \\ \bar{h}_n \end{pmatrix} - \begin{pmatrix} E_\pi g \\ E_\pi g^2 \end{pmatrix}\right) \overset{d}{\to} N\left(\begin{pmatrix} 0 \\ 0 \end{pmatrix}, \begin{pmatrix} \sigma_g^2 & c \\ c & \sigma_h^2 \end{pmatrix}\right),$$

where $c = E_\pi g^3 - E_\pi g E_\pi g^2$. Let $\phi(x, y) = y - x^2$. Then as $n \to \infty$,

$$\sqrt{n}(\hat{v}_n - \mathrm{var}_\pi g) \overset{d}{\to} N(0, 4(E_\pi g)(\sigma_g^2 E_\pi g - E_\pi g^3 + E_\pi g E_\pi g^2) + \sigma_h^2).$$

Since it is easy to use OLBM to construct strongly consistent estimators of σ_g^2 and $\sigma_{h'}^2$ a strongly consistent estimator of the variance in the asymptotic normal distribution for $\hat{v}_n$ is given by

$$4(\bar{g}_n)(\hat{\sigma}_{g,n}^2 \bar{g}_n - \bar{j}_n + \bar{g}_n \bar{h}_n) + \hat{\sigma}_{h,n'}^2$$

where $j = g^3$.

7.4.3 Quantiles

Suppose that our goal is to estimate ϕ_q with $\hat{\phi}_{q,n}$ defined in Equations 7.5 and 7.6, respectively. We now turn our attention to constructing an interval estimate of ϕ_q. It is tempting to think that bootstrap methods would be appropriate for this problem. Indeed, there has been a substantial amount of research into bootstrap methods for stationary time series which would be appropriate for MCMC settings (see Bertail and Clémençon, 2006; Bühlmann, 2002; Datta and McCormick, 1993; Politis, 2003). Unfortunately, our experience has been that these methods are *extremely* computationally intensive (compared to the MCMC simulation itself) and have inferior finite-sample properties compared to the method presented below.

As above, we assume the existence of an asymptotic normal distribution for the Monte Carlo error—that is, there is a constant $\gamma_q^2 \in (0, \infty)$ such that, as $n \to \infty$,

$$\sqrt{n}(\hat{\phi}_{q,n} - \phi_q) \xrightarrow{d} N(0, \gamma_q^2). \tag{7.9}$$

Flegal and Jones (2010b) give conditions under which Equation 7.9 obtains. Just as when we were estimating an expectation, we find ourselves in the position of estimating a complicated constant γ_q^2. We focus on the use of the subsampling bootstrap method (SBM) in this context. The reader should be aware that our use of the term "subsampling" is quite different than the way it is often used in the context of MCMC, in that we are not deleting any observations of the Markov chain.

7.4.3.1 *Subsampling Bootstrap*

This section will provide a brief overview of SBM in the context of MCMC and illustrate its use for calculating the MCSE of $\hat{\phi}_{q,n}$. While this section focuses on quantiles, SBM methods apply much more generally; the interested reader is encouraged to consult Politis et al. (1999).

The main idea for SBM is similar to OLBM in that we are taking overlapping batches (or subsamples) of size b_n from the first n observations of the chain $\{X_0, X_1, \ldots, X_{n-1}\}$. There are $n - b_n + 1$ such subsamples. Let $\{X_i, \ldots, X_{i+b_n-1}\}$ be the ith subsample with corresponding ordered subsample $\{X_{(1)}^*, \ldots, X_{(b_n)}^*\}$. Then define the quantile based on the ith subsample as

$$\phi_i^* = X_{(j+1)}^* \quad \text{where} \quad \frac{j}{b_n} \leq q < \frac{j+1}{b_n} \quad \text{for } i = 0, \ldots, n - b_n.$$

The SBM estimate of γ_q^2 is then

$$\hat{\gamma}_q^2 = \frac{b_n}{n - b_n + 1} \sum_{i=0}^{n-b_n+1} (\phi_i^* - \bar{\phi}^*)^2,$$

where

$$\bar{\phi}^* = \frac{1}{n - b_n + 1} \sum_{i=0}^{n-b_n+1} \phi_i^*.$$

Politis et al. (1999) give conditions that ensure this estimator is strongly consistent, but their conditions could be difficult to check in practice. SBM implementation requires choosing b_n such that, as $n \to \infty$, we have $b_n \to \infty$ and $b_n/n \to 0$. A natural choice is $b_n = \lfloor \sqrt{n} \rfloor$.

Example 7.8 (Normal AR(1) Markov Chains)

Using the AR(1) model defined in Equation 7.4, we again consider estimating the first and third quartiles, denoted Q_1 and Q_3. Recall that the true values for the quartiles are $\pm \Phi^{-1}(0.75)/\sqrt{1 - \rho^2}$, respectively.

Figure 7.6 shows the output from the same realization of the chain used previously in Example 7.3, but this time the plot includes an interval estimate of the quartiles. Figure 7.6a shows a plot of the running quartiles versus iteration number when $\rho = 0.5$. In addition, the dashed lines show the 80% confidence interval bounds at each iteration. These intervals were produced with SBM using $b_n = \lfloor \sqrt{n} \rfloor$. At around 200 iterations, the MCSE (and hence interval estimates) seem to stabilize and begin to decrease. At 2000 iterations, the estimates for Q_1 and Q_3 are -0.817 ± 0.069 and 0.778 ± 0.065, respectively. Figure 7.6b shows the same plot when $\rho = 0.95$. At 2000 iterations, the estimates for Q_1 and Q_3 are -2.74 ± 0.481 and 1.78 ± 0.466, respectively.

Are the intervals sufficiently narrow after 2000 iterations? In both cases ($\rho = 0.5$ and $\rho = 0.95$) the answer is likely "no." Consider the narrowest interval, which is the one for Q_3 with $\rho = 0.5$, that is, $0.778 \pm 0.065 = (0.713, 0.843)$, which indicates that all we can say is that this is evidence that the true quantile is between 0.71 and 0.85. Note that in a real problem we would not have the horizontal line in the plot depicting the truth.

SBM is applicable much more generally than presented here and, in fact, essentially generalizes the method of OLBM previously discussed in the context of estimating an expectation. The subsample mean is $\bar{Y}_j(b_n)$ and the resulting estimate of σ_g^2 is

$$\hat{\sigma}^2_{\text{SBM}} = \frac{b_n}{n - b_n + 1} \sum_{j=0}^{n-b_n} (\bar{Y}_j(b_n) - \bar{Y}^*)^2, \tag{7.10}$$

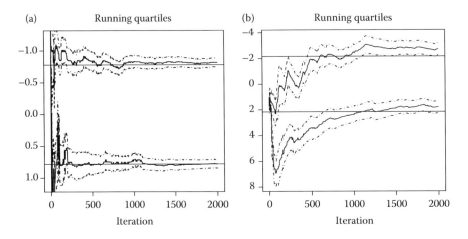

FIGURE 7.6
Plots for AR(1) model of running estimates of Q_1 and Q_3, along with 80% pointwise confidence intervals calculated via SBM. The horizontal lines denote the true values. (a) $\rho = 0.5$ and (b) $\rho = 0.95$.

where

$$\bar{Y}^* = \frac{1}{n - b_n + 1} \sum_{i=0}^{n-b_n+1} \bar{Y}_j(b_n).$$

It is straightforward to establish that the OLBM estimate defined in Equation 7.8 is asymptotically equivalent to the SBM estimate defined in Equation 7.10.

7.4.4 Multivariate Estimation

While we have largely focused on the univariate setting, recall from Section 7.1 that a typical MCMC experiment is conducted with the goal of estimating a p-dimensional vector of parameters, θ_π, associated with the d-dimensional target π. Generally, θ_π could be composed of expectations, quantiles and so on, and p could be either much larger or much smaller than d. Suppose that each component $\theta_{\pi,i}$ can be estimated with $\hat{\theta}_{n,i}$ so that $\hat{\theta}_n = (\hat{\theta}_{n,1}, \ldots, \hat{\theta}_{n,1}) \to \theta_\pi$ almost surely as $n \to \infty$. It is natural to seek to establish the existence of an asymptotic distribution of the Monte Carlo error $\hat{\theta}_n - \theta_\pi$ and then use this distribution to construct asymptotically valid confidence regions. To our knowledge this problem has not been investigated. However, it has received some attention in the case where θ_π consists only of expectations; we know of one paper in the statistics literature (Kosorok, 2000) and a few more in operations research, including Muñoz and Glynn (2001), Seila (1982), and Yang and Nelson (1992). Currently, the most common approach is to ignore the multiplicity issue and simply construct the MCSE for each component of the Monte Carlo error. If p is not too large then a Bonferroni correction could be used, but this is clearly less than optimal. This is obviously an area in MCMC output analysis that could benefit from further research.

7.5 Estimating Marginal Densities

A common inferential goal is the production of a plot of a marginal density associated with π. In this section we cover two methods for doing this. We begin with a simple graphical method, and then introduce a clever method due to Wei and Tanner (1990) that reminds us of the Rao-Blackwellization methods of Section 7.3.

A histogram approximates the true marginal by the Markov chain SLLN. Moreover, histograms are popular because they are so easy to construct with existing software. Another common approach is to report a nonparametric density estimate or smoothed histogram. It is conceptually straightforward to construct pointwise interval estimates for the smoothed histogram using SBM. However, outside of toy examples, the computational cost is typically prohibitive.

Example 7.9

Suppose that $Y_i | \mu, \theta \sim N(\mu, \theta)$ independently for $i = 1, \ldots, m$, where $m \geq 3$, and assume the standard invariant prior $\upsilon(\mu, \theta) \propto \theta^{-\frac{1}{2}}$. The resulting posterior density is

$$\pi(\mu, \theta | y) \propto \theta^{-(m+1)/2} e^{-\frac{m}{2\theta}(s^2 + (\bar{y} - \mu)^2)},$$

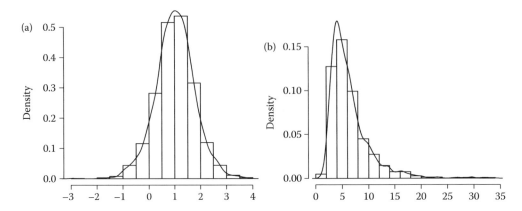

FIGURE 7.7
Histograms for estimating the marginal densities of Example 7.7: (a) marginal density of μ and (b) marginal density of θ.

where s^2 is the usual biased sample variance. It is easy to see that $\mu \mid \theta, y \sim N(\bar{y}, \theta/m)$ and that $\theta \mid \mu, y \sim \Gamma^{-1}((m-1)/2, m[s^2 + (\bar{y} - \mu)^2]/2)$, and hence a Gibbs sampler is easily implemented. We consider the Gibbs sampler that updates μ then θ so that a one-step transition is given by $(\mu', \theta') \to (\mu, \theta') \to (\mu, \theta)$ and use this sampler to estimate the marginal densities of μ and θ.

Now suppose $m = 11$, $\bar{y} = 1$ and $s^2 = 4$. We simulated 2000 realizations of the Gibbs sampler starting from $(\mu_0, \lambda_0) = (1, 1)$. The marginal density plots were created using the default settings for the `density` function in R and are shown in Figure 7.7, while an estimated bivariate density plot (created using R functions `kde2d` and `persp`) is given in Figure 7.8. It is obvious from these figures that the posterior is simple, so it is no surprise that the Gibbs sampler has been shown to converge in just a few iterations (Jones and Hobert, 2001).

A clever technique for estimating a marginal is based on the same idea as RB estimators (Wei and Tanner, 1990). To keep the notation simple, suppose that the target is a function of only two variables, $\pi(x, y)$, and let m_X and m_Y be the associated marginals. Then

$$m_X(x) = \int \pi(x, y)\, dy = \int f_{X \mid Y}(x \mid y) m_Y(y)\, dy = E_{m_Y} f_{X \mid Y}(x \mid y),$$

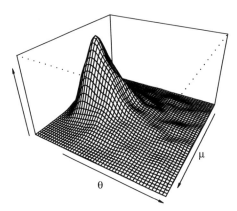

FIGURE 7.8
Estimated posterior density of Example 7.7.

suggesting that, by the Markov chain SLLN, we can get a functional approximation to m_X since for each x, as $n \to \infty$,

$$\frac{1}{n}\sum_{i=0}^{n-1} f_{X|Y}(x \mid y_i) \to m_X(x). \qquad (7.11)$$

Of course, just as with RB estimators this will only be useful when the conditionals are tractable. Note also, that it is straightforward to use OLBM to get pointwise confidence intervals for the resulting curve; that is for each x we can calculate an MCSE of the sample average in Equation 7.11.

Example 7.10

Recall the setting of Example 7.9. We will focus on estimation of the marginal posterior density of $\mu|y$, that is, $\pi(\mu|y)$. Note that

$$\pi(\mu \mid y) = \int \pi(\mu \mid \theta, y)\pi(\theta \mid y)\, d\theta,$$

so that by the Markov chain SLLN we can estimate $\pi(\mu \mid y)$ with

$$\frac{1}{n}\sum_{i=0}^{n-1} \pi(\mu \mid \theta_i, y),$$

which is straightforward to evaluate since $\mu|\theta_i, y \sim N(\bar{y}, \theta_i/m)$. Note that the resulting marginal estimate is a linear combination of normal densities. Using the same realization of the chain from Example 7.9, we estimated $\pi(\mu|y)$ using this method. Figure 7.9 shows the results with our previous estimates. One can also calculate pointwise confidence intervals using OLBM, which results in a very small Monte Carlo error (and is therefore not included in the plot). Notice that the estimate based on Equation 7.11 is a bit smoother than either the histogram or the smoothed histogram estimate, but is otherwise quite similar.

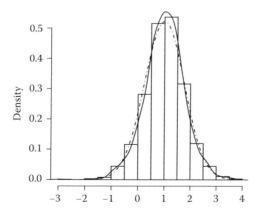

FIGURE 7.9

Estimates of the marginal density μ. The three estimates are based on a histogram, smoothed marginal densities (solid line), and the method of Wei and Tanner (1990) (dashed line).

7.6 Terminating the Simulation

A common approach to stopping an MCMC experiment is to simulate for a fixed run length. That is, the simulation is terminated using a *fixed-time* rule. Notice that this makes MCSEs crucial to understanding the reliability of the resulting estimates. There are settings where, due to the nearly prohibitive difficulty of the simulation, a fixed-time rule may be the only practical approach. However, this is not the case for many MCMC experiments.

Perhaps the most popular approach to terminating the simulation is to simulate an initial Monte Carlo sample of size n_0, say. The output is examined and if the results are found to be unsatisfactory, the simulation is continued for another n_1 steps and the output reanalyzed. If the results are still unsatisfactory, the process is repeated. Notice that this is a sequential procedure that will result in a random total simulation effort.

When implementing this sequential procedure the examination of the output can take many forms; it is often based on the use of graphical methods such as those described in Section 7.2 or on convergence diagnostics. We advocate terminating the simulation the first time the MCSE is sufficiently small. Equivalently, the simulation is terminated the first time the half-width of a confidence interval for θ_π is sufficiently small, resulting in a *fixed-width* rule. There is a substantial amount of research on fixed-width procedures in MCMC when θ_π is an expectation—see Flegal et al. (2008), Glynn and Whitt (1992), and Jones et al. (2006) and the references therein—but none that we are aware of when θ_π is not an expectation. Let $\hat{\sigma}_n^2$ be a strongly consistent estimator of σ_g^2 from Equation 7.2. Given a desired half-width ϵ the simulation terminates the first time

$$t_* \frac{\hat{\sigma}_n}{\sqrt{n}} + p(n) \leq \epsilon, \tag{7.12}$$

where t_* is the appropriate quantile and $p(n)$ is a positive function such that $p(n) = o(n^{-1/2})$ as $n \to \infty$. Letting n^* be the desired minimum simulation effort, a reasonable default is $p(n) = \epsilon I(n \leq n^*) + n^{-1}$. Glynn and Whitt (1992) established conditions ensuring that the interval at Equation 7.12 is asymptotically valid[†] in the sense that the desired coverage probability is obtained as $\epsilon \to 0$. The use of these intervals in MCMC settings has been extensively investigated and found to work well by Flegal et al. (2008), Flegal and Jones (2010a), and Jones et al. (2006).

Example 7.11 (Normal AR(1) Markov Chains)

Consider the normal AR(1) time series defined in Equation 7.4. In Example 7.4 we simulated 2000 iterations from the chain with $\rho = 0.95$ starting from $X_0 = 1$ and found that a 80% confidence interval for the mean of the invariant distribution was -0.507 ± 0.451.

Suppose that we wanted to continue our simulation until we were 80% confident that our estimate was within 0.1 of the true value after a minimum of 1000 iterations—a fixed-width procedure. If we use OLBM to estimate the variance in the asymptotic distribution, then Equation 7.12 becomes

$$t_* \frac{\hat{\sigma}_n}{\sqrt{n}} + 0.1 I(n \leq 1000) + n^{-1} \leq 0.1,$$

[†] Glynn and Whitt (1992) also provide a counterexample to show that weak consistency of $\hat{\sigma}_n^2$ for σ_g^2 is not enough to achieve asymptotic validity.

where t_* is the appropriate quantile from a Student's t distribution with $n - b_n$ degrees of freedom. It would be computationally expensive to check this criterion after each iteration, so instead we added 1000 iterations before recalculating the half-width each time. In this case, the simulation terminated after 60,000 iterations, resulting in an interval estimate of -0.0442 ± 0.100. Notice that this simple example required a relatively large simulation effort compared to what is often done in much more complicated settings, but note that ρ is large. Further, either narrowing the interval or increasing the desired confidence level will require a larger simulation effort.

7.7 Markov Chain Central Limit Theorems

Throughout we have assumed the existence of a Markov chain central limit theorem—see, for example, Equations 7.2 and 7.9. In this section we provide a brief discussion of the conditions required for these claims; the reader can find much more detail in Chan and Geyer (1994), Jones (2004), Meyn and Tweedie (1993), Roberts and Rosenthal (2004), and Tierney (1994).

Implicitly, we assumed that the Markov chain X is *Harris ergodic*, that is, Harris recurrent and aperiodic. To fully explain these conditions would require a fair amount of Markov chain theory, so we will content ourselves with providing references; the interested reader should consult Meyn and Tweedie (1993), Nummelin (1984) or Roberts and Rosenthal (2004). However, it is frequently trivial to verify Harris ergodicity (see Chapter 10, this volume; Tan and Hobert, 2009; Tierney, 1994).

Harris ergodicity alone is not sufficient for the Markov chain SLLN or a CLT. However, if X is Harris ergodic and $E_\pi|g| < \infty$, then the SLLN holds: $\bar{g}_n \to E_\pi g$ with probability 1 as $n \to \infty$. A CLT requires stronger conditions. In fact, it is important to be aware that there are simple nonpathological examples of Harris ergodic Markov chains that do not enjoy a CLT (Roberts, 1999). Let the conditional distribution of X_n given $X_0 = x$ be denoted $P^n(x, \cdot)$, that is,

$$P^n(x, A) = \Pr(X_n \in A \mid X_0 = x).$$

Then Harris ergodicity implies that, for every starting point $x \in X$,

$$\|P^n(x, \cdot) - \pi(\cdot)\| \downarrow 0 \quad \text{as } n \to \infty, \tag{7.13}$$

where $\| \cdot \|$ is the total variation norm. We will need to know the rate of the convergence in Equation 7.13 to say something about the existence of a CLT. Let $M(x)$ be a nonnegative function on X and $\gamma(n)$ be a nonnegative function on $\mathbb{Z}_+$ such that

$$\|P^n(x, \cdot) - \pi(\cdot)\| \leq M(x)\gamma(n). \tag{7.14}$$

When X is *geometrically ergodic*, $\gamma(n) = t^n$ for some $t < 1$; while *uniform ergodicity* means that X is geometrically ergodic and M is bounded. These are key sufficient conditions for the existence of an asymptotic normal distribution of the Monte Carlo error but they are not the only conditions guaranteeing a CLT. In particular, a CLT as in Equation 7.2 holds if X is geometrically ergodic and $E_\pi g^{2+\delta} < \infty$ for some $\delta > 0$, or if X is uniformly ergodic and $E_\pi g^2 < \infty$. Moreover, geometric ergodicity is a key sufficient condition for the strong consistency of the estimators of σ_g^2 from Equation 7.2. For example, Flegal and Jones (2010a) establish that when X is geometrically ergodic and $E_\pi g^{2+\delta} < \infty$ for some $\delta > 0$, the OLBM method produces a strongly consistent estimator of σ_g^2. Geometric ergodicity is also an important sufficient condition for establishing (Equation 7.9) when estimating a quantile (Flegal and Jones, 2010b).

In general, establishing (Equation 7.14) directly is apparently daunting. However, if X is finite (no matter how large), then a Harris ergodic Markov chain is uniformly ergodic. When X is a general space there are constructive methods which can be used to establish geometric or uniform ergodicity; see Chapter 10 (this volume), and Jones and Hobert (2001) for accessible introductions. These techniques have been applied to many MCMC samplers. For example, Metropolis–Hastings samplers with state-independent proposals can be uniformly ergodic (Tierney, 1994). Standard random-walk Metropolis–Hastings chains on $\mathbb{R}^d$, $d \geq 1$, cannot be uniformly ergodic but may still be geometrically ergodic; see Mengersen and Tweedie (1996). An incomplete list of other research on establishing convergence rates of Markov chains used in MCMC is given by Atchadé and Perron (2007), Christensen et al. (2001), Geyer (1999), Jarner and Hansen (2000), Meyn and Tweedie (1994), and Neath and Jones (2009) who considered Metropolis–Hastings algorithms, and Doss and Hobert (2010), Hobert and Geyer (1998), Hobert et al. (2002), Johnson and Jones (2010), Jones and Hobert (2004), Marchev and Hobert (2004), Roberts and Polson (1994), Roberts and Rosenthal (1999), Rosenthal (1995, 1996), Roy and Hobert (2007, 2010), Tan and Hobert (2009), and Tierney (1994) who examined Gibbs samplers.

7.8 Discussion

The main point of this chapter is that a MCSE should be reported along with the point estimate obtained from an MCMC experiment. At some level this probably seems obvious to most statisticians, but it is not the case in the reporting of most MCMC-based simulation experiments. In fact, Doss and Hobert (2010) recently wrote:

> Before the MCMC revolution, when classical Monte Carlo methods based on i.i.d. samples were used to estimate intractable integrals, it would have been deemed unacceptable to report a Monte Carlo estimate without an accompanying asymptotic standard error (based on the CLT). Unfortunately, this seems to have changed with the advent of MCMC.

While it is tempting to speculate on the reasons for this change, the fact remains that most currently published work in MCMC reports point estimates while failing to even acknowledge an associated MCSE; see also Flegal et al. (2008). Thus we have little ability to assess the reliability of the reported results. This is especially unfortunate since it is straightforward to compute a valid MCSE.

The only potentially difficult part of the method presented here is in establishing the existence of a Markov chain CLT. In essence, this means simulating a Markov chain known to be geometrically ergodic and checking a moment condition. Given the amount of work that has been done on establishing geometric ergodicity for standard algorithms in common statistical settings, this is not the obstacle it was in the past. However, this remains an area rich in important open research questions.

Acknowledgments

This work was supported by NSF Grant DMS-0806178. The authors would also like to thank the anonymous referees for their helpful suggestions.

References

Alexopoulos, C. and Goldsman, D. 2004. To batch or not to batch? *ACM Transactions on Modeling and Computer Simulation*, 14(1):76–114.

Alexopoulos, C., Andradóttir, S., Argon, N. T., and Goldsman, D. 2006. Replicated batch means variance estimators in the presence of an initial transient. *ACM Transactions on Modeling and Computer Simulation*, 16:317–328.

Atchadé, Y. F. and Perron, F. 2007. On the geometric ergodicity of Metropolis-Hastings algorithms. *Statistics*, 41:77–84.

Bertail, P. and Clémençon, S. 2006. Regenerative block-bootstrap for Markov chains. *Bernoulli*, 12:689–712.

Bratley, P., Fox, B. L., and Schrage, L. E. 1987. *A Guide to Simulation*. Springer, New York.

Brockwell, A. E. 2006. Parallel Markov chain Monte Carlo by pre-fetching. *Journal of Computational and Graphical Statistics*, 15:246–261.

Bühlmann, P. 2002. Bootstraps for time series. *Statistical Science*, 17:52–72.

Casella, G. and Robert, C. P. 1996. Rao-Blackwellisation of sampling schemes. *Biometrika*, 83:81–94.

Chan, K. S. and Geyer, C. J. 1994. Comment on Markov chains for exploring posterior distributions. *Annals of Statistics*, 22:1747–1758.

Christensen, O. F., Møller, J., and Waagepetersen, R. P. 2001. Geometric ergodicity of Metropolis-Hastings algorithms for conditional simulation in generalized linear mixed models. *Methodology and Computing in Applied Probability*, 3:309–327.

Cowles, M. K. and Carlin, B. P. 1996. Markov chain Monte Carlo convergence diagnostics: A comparative review. *Journal of the American Statistical Association*, 91:883–904.

Cowles, M. K., Roberts, G. O., and Rosenthal, J. S. 1999. Possible biases induced by MCMC convergence diagnostics. *Journal of Statistical Computing and Simulation*, 64:87–104.

Datta, S. and McCormick, W. P. 1993. Regeneration-based bootstrap for Markov chains. *Canadian Journal of Statistics*, 21:181–193.

Doss, H. and Hobert, J. P. 2010. Estimation of Bayes factors in a class of hierarchical random effects models using a geometrically ergodic MCMC algorithm. *Journal of Computational and Graphical Statistics*, 19(2):295—312.

Ferguson, T. S. 1996. *A Course in Large Sample Theory*. Chapman & Hall, London.

Flegal, J. M. and Jones, G. L. 2010a. Batch means and spectral variance estimators in Markov chain Monte Carlo. *Annals of Statistics*, 38:1034–1070.

Flegal, J. M. and Jones, G. L. 2010b. Quantile estimation via Markov chain Monte Carlo. Technical report, Department of Statistics, University of California, Riverside.

Flegal, J. M. and Jones, G. L. 2010c. Sweave documentation for Implementing Markov chain Monte Carlo: Estimating with confidence. http://arXiv.org/abs/1006.5690.

Flegal, J. M., Haran, M., and Jones, G. L. 2008. Markov chain Monte Carlo: Can we trust the third significant figure? *Statistical Science*, 23:250–260.

Fosdick, L. D. 1959. Calculation of order parameters in a binary alloy by the Monte Carlo method. *Physical Review*, 116:565–573.

Fristedt, B. and Gray, L. F. 1997. *A Modern Approach to Probability Theory*. Birkhäuser, Boston.

Gelman, A. and Rubin, D. B. 1992. Inference from iterative simulation using multiple sequences (with discussion). *Statistical Science*, 7:457–511.

Geyer, C. J. 1992. Practical Markov chain Monte Carlo (with discussion). *Statistical Science*, 7:473–511.

Geyer, C. J. 1995. Conditioning in Markov chain Monte Carlo. *Journal of Computational and Graphical Statistics*, 4:148–154.

Geyer, C. J. 1999. Likelihood inference for spatial point processes. In O. E. Barndorff-Nielsen, W. S. Kendall, and M. N. M. van Lieshout (eds), *Stochastic Geometry: Likelihood and Computation*, pp. 79–140. Chapman & Hall/CRC, Boca Raton, FL.

Glynn, P. W. and Heidelberger, P. 1991. Analysis of initial transient deletion for replicated steady-state simulations. *Operations Research Letters*, 10:437–443.

Glynn, P. W. and Whitt, W. 1991. Estimating the asymptotic variance with batch means. *Operations Research Letters*, 10:431–435.

Glynn, P. W. and Whitt, W. 1992. The asymptotic validity of sequential stopping rules for stochastic simulations. *Annals of Applied Probability*, 2:180–198.

Hobert, J. P. and Geyer, C. J. 1998. Geometric ergodicity of Gibbs and block Gibbs samplers for a hierarchical random effects model. *Journal of Multivariate Analysis*, 67:414–430.

Hobert, J. P., Jones, G. L., Presnell, B., and Rosenthal, J. S. 2002. On the applicability of regenerative simulation in Markov chain Monte Carlo. *Biometrika*, 89:731–743.

Jarner, S. F. and Hansen, E. 2000. Geometric ergodicity of Metropolis algorithms. *Stochastic Processes and Their Applications*, 85:341–361.

Johnson, A. A. and Jones, G. L. 2010. Gibbs sampling for a Bayesian hierarchical general linear model. *Electronic Journal of Statistics*, 4:313–333.

Jones, G. L. 2004. On the Markov chain central limit theorem. *Probability Surveys*, 1:299–320.

Jones, G. L. and Hobert, J. P. 2001. Honest exploration of intractable probability distributions via Markov chain Monte Carlo. *Statistical Science*, 16:312–334.

Jones, G. L. and Hobert, J. P. 2004. Sufficient burn-in for Gibbs samplers for a hierarchical random effects model. *Annals of Statistics*, 32:784–817.

Jones, G. L., Haran, M., Caffo, B. S., and Neath, R. 2006. Fixed-width output analysis for Markov chain Monte Carlo. *Journal of the American Statistical Association*, 101:1537–1547.

Kelton, A. M. and Law, W. D. 1984. An analytical evaluation of alternative strategies in steady-state simulation. *Operations Research*, 32:169–184.

Kosorok, M. R. 2000. Monte Carlo error estimation for multivariate Markov chains. *Statistics and Probability Letters*, 46:85–93.

Latuszynski, K. and Niemiro, W. 2009. Rigorous confidence bounds for MCMC under a geometric drift condition. http://arXiv.org/abs/0908.2098v1.

Liu, J. S., Wong, W. H., and Kong, A. 1994. Covariance structure of the Gibbs sampler with applications to the comparisons of estimators and augmentation schemes. *Biometrika*, 81:27–40.

Marchev, D. and Hobert, J. P. 2004. Geometric ergodicity of van Dyk and Meng's algorithm for the multivariate Student's *t* model. *Journal of the American Statistical Association*, 99:228–238.

Mengersen, K. and Tweedie, R. L. 1996. Rates of convergence of the Hastings and Metropolis algorithms. *Annals of Statistics*, 24:101–121.

Meyn, S. P. and Tweedie, R. L. 1993. *Markov Chains and Stochastic Stability*. Springer, London.

Meyn, S. P. and Tweedie, R. L. 1994. Computable bounds for geometric convergence rates of Markov chains. *Annals of Applied Probability*, 4:981–1011.

Muñoz, D. F. and Glynn, P. W. 2001. Multivariate standardized time series for steady-state simulation output analysis. *Operations Research*, 49:413–422.

Mykland, P., Tierney, L., and Yu, B. 1995. Regeneration in Markov chain samplers. *Journal of the American Statistical Association*, 90:233–241.

Neath, R. and Jones, G. L. 2009. Variable-at-a-time implementations of Metropolis-Hastings. Technical report, School of Statistics, University of Minnesota.

Nummelin, E. 1984. *General Irreducible Markov Chains and Non-negative Operators*. Cambridge University Press, Cambridge.

Peltonen, J., Venna, J., and Kaski, S. 2009. Visualizations for assessing convergence and mixing of Markov chain Monte Carlo simulations. *Computational Statistics and Data Analysis*, 53:4453–4470.

Politis, D. N. 2003. The impact of bootstrap methods on time series analysis. *Statistical Science*, 18:219–230.

Politis, D. N., Romano, J. P., and Wolf, M. 1999. *Subsampling*. Springer, New York.

Roberts, G. O. 1999. A note on acceptance rate criteria for CLTs for Metropolis-Hastings algorithms. *Journal of Applied Probability*, 36:1210–1217.

Roberts, G. O. and Polson, N. G. 1994. On the geometric convergence of the Gibbs sampler. *Journal of the Royal Statistical Society, Series B*, 56:377–384.

Roberts, G. O. and Rosenthal, J. S. 1999. Convergence of slice sampler Markov chains. *Journal of the Royal Statistical Society, Series B*, 61:643–660.

Roberts, G. O. and Rosenthal, J. S. 2004. General state space Markov chains and MCMC algorithms. *Probability Surveys*, 1:20–71.

Rosenthal, J. S. 1995. Minorization conditions and convergence rates for Markov chain Monte Carlo. *Journal of the American Statistical Association*, 90:558–566.

Rosenthal, J. S. 1996. Analysis of the Gibbs sampler for a model related to James-Stein estimators. *Statistics and Computing*, 6:269–275.

Rosenthal, J. S. 2000. Parallel computing and Monte Carlo algorithms. *Far East Journal of Theoretical Statistics*, 4:201–236.

Roy, V. and Hobert, J. P. 2007. Convergence rates and asymptotic standard errors for Markov chain Monte Carlo algorithms for Bayesian probit regression. *Journal of the Royal Statistical Society, Series B*, 69(4):607–623.

Roy, V. and Hobert, J. P. 2010. On Monte Carlo methods for Bayesian multivariate regression models with heavy-tailed errors. *Journal of Multivariate Analysis*, 101(5):1190–1202.

Rudolf, D. 2009. Error bounds for computing the expectation by Markov chain Monte Carlo. Preprint.

Schruben, L. 1983. Confidence interval estimation using standardized time series. *Operations Research*, 31:1090–1108.

Seila, A. F. 1982. Multivariate estimation in regenerative simulation. *Operations Research Letters*, 1:153–156.

Sturtz, S., Ligges, U., and Gelman, A. 2005. R2WinBUGS: A package for running WinBUGS from R. *Journal of Statistical Software*, 12(3).

Tan, A. and Hobert, J. P. 2009. Block Gibbs sampling for Bayesian random effects models with improper priors: Convergence and regeneration. *Journal of Computational and Graphical Statistics*, 18:861–878.

Tierney, L. 1994. Markov chains for exploring posterior distributions (with discussion). *Annals of Statistics*, 22:1701–1762.

van der Vaart, A. W. 1998. *Asymptotic Statistics*. Cambridge University Press, Cambridge.

Wei, G. C. G. and Tanner, M. A. 1990. A Monte Carlo implementation of the EM algorithm and the poor man's data augmentation algorithms. *Journal of the American Statistical Association*, 85:699–704.

Yang, W.-N. and Nelson, B. L. 1992. Multivariate batch means and control variates. *Management Science*, 38:1415–1431.

8

Perfection within Reach: Exact MCMC Sampling

Radu V. Craiu and Xiao-Li Meng

8.1 Intended Readership

The amount of research done by the Markov chain Monte Carlo (MCMC) community has been very impressive in the last two decades, as testified by this very volume. The power of MCMC has been demonstrated in countless instances in which more traditional numerical algorithms are helpless. However, one ubiquitous problem remains: the detection of convergence or lack thereof. Among the large number of procedures designed for detecting lack of convergence or for establishing convergence bounds (see, e.g. Chapters 6 and 7 in this volume), there is one class of MCMC algorithms that stands apart simply because it avoids the problem altogether. Whereas examples of such algorithms can be traced back to at least 1989 (see [56]), it is Propp and Wilson's 1996 seminal paper [48] that introduced the general scheme of *coupling from the past* (CFTP). Since then, there has been an intense search for perfect sampling or exact sampling algorithms, so named because such algorithms use Markov chains and yet obtain genuine independent and identically distributed (i.i.d.) draws—hence perfect or exact—from their limiting distributions within a finite number of iterations.

There is, of course, no free lunch. Whereas this is a class of very powerful algorithms, their construction and implementation tend to require a good deal of labor and great care. Indeed, even the most basic general themes are not entirely trivial to understand, and subtleties and traps can be overwhelming for novices. Our central goal in this chapter is therefore to provide an *intuitive* overview of some of the most basic sampling schemes developed since [48]. We do not strive for completeness, nor for mathematical generality or rigorousness. Rather, we focus on a few basic schemes and try to explain them as intuitively as we can, via figures and simple examples. The chapter is therefore not intended for the residents but rather the visitors of the MCMC kingdom who want to tour the magic land of perfect sampling. There are of course a number of other tour packages—see, for example, the list provided at http://dimacs.rutgers.edu/~dbwilson/exact.html, maintained by David Wilson. But we hope ours is one of the least expensive ones in terms of readers' mental investment, though by no means are we offering a free ride.

8.2 Coupling from the Past

8.2.1 Moving from Time-Forward to Time-Backward

The CFTP algorithm is based on an idea that is both simple and revolutionary. Suppose that we are interested in sampling from a distribution with probability law $\Pi(\cdot)$ with state space $\mathcal{S} \subset \mathbb{R}^d$. We define a Markov chain with stationary law Π using a transition kernel $K(x, \cdot)$ whose transitions can be written in a *stochastic recursive sequence* (SRS) form,

$$X_{t+1} = \phi(X_t, \xi_{t+1}), \quad t = 0, 1, 2, \ldots, \tag{8.1}$$

where ϕ is a deterministic map and ξ_{t+1} is a random variable with state space $\Lambda \subset \mathbb{R}^r$. (Sometimes it is automatically assumed that $\Lambda = (0,1)^r$, but that is not necessary here.) More precisely, the distribution of ξ is such that $P(X_{t+1} \in A) = \Pi(A) = \int K(x_t, A)\Pi(dx_t)$, that is, it guarantees that the output X_{t+1} has the same (marginal) distribution as the input X_t if $X_t \sim \Pi$.

To explain the key idea of CFTP, let us first review the usual implementation of MCMC. When the chain can be written as in Equation 8.1, we can simply compute it iteratively starting from an *arbitrary* starting point $X_0 \in \mathcal{S}$, by generating a sequence of $\xi_1, \xi_2, \ldots, \xi_t$, if we decide to run for t iterations. If the Markov chain formed by Equation 8.1 is positive recurrent and aperiodic (see Chapter 10, this volume), then we know that as $t \to \infty$, the probability law of X_t, P_t, will approach Π, regardless of the distribution of X_0. Of course, how large t needs to be before the difference between P_t and Π becomes too small to have practical consequences is the very thorny issue we try to avoid here.

The CFTP, as its name suggests, resolves this problem using an ingenious idea of running the chain *from the past* instead of *into the future*. To see this clearly, compare the following two sequences based on the same random sequence $\{\xi_1, \xi_2, \ldots, \xi_t\}$ used above:

$$\text{Forward: } X_{0 \to t}^{(x)} = \phi(\phi(\ldots \phi(\phi(x, \xi_1), \xi_2), \ldots \xi_{t-1}), \xi_t);$$

$$\text{Backward: } X_{t \to 0}^{(x)} = \phi(\phi(\ldots \phi(\phi(x, \xi_t), \xi_{t-1}), \ldots \xi_2), \xi_1). \tag{8.2}$$

The *time-forward* sequence $X_{0 \to t}^{(x)}$ is obviously identical to the X_t computed previously with $X_0 = x$. The *time-backward* sequence $X_{t \to 0}^{(x)}$ is evidently not the same as $X_{0 \to t}^{(x)}$ but clearly they have *identical distribution* whenever $\{\xi_1, \xi_2, \ldots, \xi_t\}$ are exchangeable, which certainly is the case when $\{\xi_t, t = 1, 2, \ldots\}$ are i.i.d., as in a typical implementation. (Note a slight abuse of notation: the use of t both as the length of the chain and as a generic index.) Consequently, we see that if we can somehow compute $X_{t \to 0}^{(x)}$ at its limit at $t = \infty$, then it will be a genuine draw from the desired distribution because it has the same distribution as $X_{0 \to t}^{(x)}$ at $t = \infty$.

8.2.2 Hitting the Limit

At first sight, we seem to have accomplished absolutely nothing by constructing the time-backward sequence because computing $X_{t \to 0}^{(x)}$ at $t = \infty$ surely should be as impossible as computing $X_{0 \to t}^{(x)}$ at $t = \infty$! However, a simple example reveals where the magic lies. Consider a special case where $\phi(X_t, \xi_{t+1}) = \xi_{t+1}$, that is, the original $\{X_t, t = 1, 2, \ldots\}$ already forms an i.i.d. sequence, which clearly has the distribution of ξ_1 as its stationary distribution (again, we assume $\{\xi_t, t = 1, 2, \ldots\}$ are i.i.d.). It is easy to see that in such cases, $X_{0 \to t}^{(x)} = \xi_t$, but $X_{t \to 0}^{(x)} = \xi_1$, for all t. Therefore, with $X_{0 \to \infty}^{(x)}$ we can only say that it has the *same distribution* as ξ_1, whereas for $X_{\infty \to 0}^{(x)}$ we can say *it is* ξ_1!

More generally, under regularity conditions, one can show that there exists a *stopping time* T such that $P(T < \infty) = 1$ and that the distribution of $X_{T \to 0}^{(x)}$ is exactly Π, that is, $X_{T \to 0}^{(x)}$ "hits the limit" with probability 1. Intuitively, this is possible because unlike $X_t^{(x)} \equiv X_{0 \to t}^{(x)}$, which forms a Markov chain, $\tilde{X}_t^{(x)} \equiv X_{t \to 0}^{(x)}$ depends on the entire history of $\{\tilde{X}_1, \ldots, \tilde{X}_{t-1}\}$. It is this dependence that restricts the set of possible paths $\tilde{X}_t$ can take and hence makes it possible to "hit the limit" in a finite number of steps. For a mathematical proof of the existence of such T, we refer readers to [48,53,54].

The CFTP strategy, in a nutshell, is to identify the aforementioned stopping time T via coupling. To see how it works, let us first map t to $-t$ and hence relabel $X_{T \to 0}^{(x)}$ as $X_{-T \to 0}^{(x)}$, which makes the meaning *from the past* clearer. That is, there is a chain coming from the infinite past (and hence negative time) whose value at the present time $t = 0$ is the draw from the desired stationary distribution. This is because coming from infinite past and reaching the present time is mathematically the same as starting from the present time and reaching the infinite future. However, this equivalence will remain just a mathematical statement if we really have to go into the infinite past in order to determine the current value of the chain. But the fact that the backward sequence can hit the limit in a finite number of steps suggests that, for a given infinite sequence $\{\xi_t, t = -1, -2, \ldots\}$, there exists a finite T such that, when $t \geq T$, $X_{-t \to 0}^{(x)}$ will no longer depend on x, that is, all paths determined by $\{\xi_t, t = -1, -2, \ldots\}$ will coalesce by time 0, regardless of their origins at the infinite past. It was proved in [10] that such coupling is possible if and only if the Markov chain $\{X_1, X_2, \ldots\}$ determined by ϕ is uniformly ergodic.

8.2.3 Challenges for Routine Applications

Clearly once all paths coalesce, their common value $X_0 = X_{-T \to 0}^{(x)}$ is a genuine draw from the stationary law Π. Therefore, the CFTP protocol relies on our ability to design the MCMC process given by ϕ, or more generally by the transition kernel K, such that the *coalescence of all paths* takes place for moderate values of T. This requirement poses immediate challenges in its routine applications, especially for Bayesian computation, where S typically contains many states, very often uncountably many. The brute-force way of monitoring each path is infeasible for two reasons. First, it is simply impossible to follow infinitely many paths individually. Second, when the state space is continuous, even if we manage to reduce the process to just two paths (as with the monotone coupling discussed below), the probability that these will meet is zero if they are left to run independently. Therefore, our first challenge is to design the algorithm so that the number of paths shrinks to a finite one within a few steps. A hidden obstacle in this challenge is being able to figure out exactly *which* paths will emerge from this reduction process as they are the ones that need to be monitored until coalescence. The second challenge is to find effective ways to "force" paths to meet, that is, to couple them in such a way that, at each step, the probability that they take the same value is positive.

The rest of this chapter will illustrate a variety of methods designed to address both challenges and other implementation issues. We do not know any universal method, nor do we believe it exists. But there are methods for certain classes of problems, and some of them are rather ingenious.

8.3 Coalescence Assessment

8.3.1 Illustrating Monotone Coupling

Suppose that the space S is endowed with a partial order relationship $\prec$ so that

$$x \prec y \Rightarrow \phi(x, \xi) \leq \phi(y, \xi) \tag{8.3}$$

for any $x, y \in \mathcal{S}, \xi \in \Lambda$, and where ϕ is an SRS as in Equation 8.1. If we can find the minimum and maximum states $X_{\min}, X_{\max} \in \mathcal{S}$ with respect to the order $\prec$, then we can implement this *monotone coupler*—as defined by Equation 8.3—in which it is sufficient to verify the coupling of the paths started at these two extremal points because all other states are "squeezed" between them. Therefore, the monotone coupler is an efficient way to address the first challenge discussed in Section 8.2.3. For illustration, consider the random walk with state space $\mathcal{S} = \{0.25, 0.5, 2, 4\}$, with probability p of moving up or staying if the chain is already at the ceiling state $X_t = 4$, and probability $1 - p$ of moving down or staying if already at the floor state $X_t = 0.25$. It is easy to see that this construction forms a monotone chain, expressible as $X_t = \phi(X_{t-1}, \xi_t)$, where $\xi_t \sim \text{Bernoulli}(p)$ and its value determines the direction of the walk, with one going up and zero going down.

Figure 8.1 shows a realization of the CFTP process, corresponding to

$$\{\xi_{-8}, \xi_{-7}, \ldots, \xi_{-2}, \xi_{-1}\} = \{0, 1, 0, 1, 1, 1, 1, 0\}. \tag{8.4}$$

One can see that the order between paths is preserved by ϕ. In particular, all the paths are at all times between the paths started at $X_{\min} = 0.25$ (solid line) and $X_{\max} = 4$ (dashed line), respectively. Therefore, in order to check the coalescence of all four paths, we only need to check if the top chain starting from $X = 4$ and the bottom chain starting from $X = 0.25$ have coalesced. In this toy example, the saving from checking two instead of all four is obviously insignificant, but one can easily imagine the potentially tremendous computational savings when there are many states, such as with the Ising model applications in [48].

8.3.2 Illustrating Brute-Force Coupling

This toy example also illustrates well the "brute-force" implementation of CFTP, that is, checking directly the coalescence of all paths. Figure 8.1 establishes that for any infinite binary sequences $\{\xi_t, t \le -1\}$, as long its last eight values (i.e. from $t = -8$ to $t = -1$) are the same as that given in Equation 8.4, the backward sequence given in Equation 8.2 will hit the limit $X = 2$, that is, the value of the coalesced chain at $t = 0$. Pretending that the

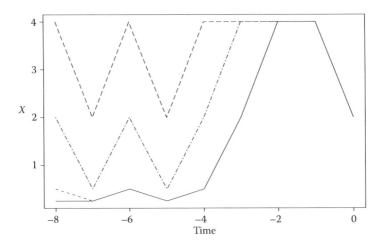

FIGURE 8.1
Illustration of a monotone SRS which preserves the natural order on the real line (i.e. paths can coalesce but never cross each other). Different lines represent sample paths started from different states.

monotone property was not noticed, we can still check the coalescence step by step for all paths. Or, more efficiently, we can use the "binary back-off" scheme proposed in [48]; that is, whenever a check fails to detect coalescence, we double the number of "backward" steps. Specifically, imagine we first made one draw of ξ, and it is zero (corresponding to $\xi_{-1} = 0$). We compute $X^{(x)}_{-1\to0}$ of Equation 8.2 for all values of $x \in \mathcal{S}$, which leads to

$$X^{(4)}_{-1\to0} = 2, \quad X^{(2)}_{-1\to0} = 0.5, \quad X^{(0.5)}_{-1\to0} = X^{(0.25)}_{-1\to0} = 0.25,$$

indicating that coalescence has not occurred. We therefore double the number of steps going back, which requires only one new draw from $\xi \sim$ Bernuolli(p), as we already have $\xi_{-1} = 0$. It is important to emphasize that we always *reuse* the draws of ξ_t that we have already made because the point here is to simply check what the coalesced value would be for a *given* infinite sequence of $\{\xi_t,\ t \le -1\}$. The device of making draws starting from $t = -1$ and going backward is the ingenious part of CFTP because it allows us to determine the property of an *infinite* sequence by revealing and examining only a *finite* number of its last elements. That is, since the remaining numbers in the (infinite) sequence cannot alter the value of the chain at $t = 0$, we do not even need to care what they are.

Now this new draw yields $\xi = 1$, and hence we have $\{\xi_{-2}, \xi_{-1}\} = \{1, 0\}$, which is then used to compute Equation 8.2 again but with $T = -2$:

$$X^{(4)}_{-2\to0} = X^{(2)}_{-2\to0} = 2, \quad X^{(0.5)}_{-2\to0} = 0.5, \quad X^{(0.25)}_{-2\to0} = 0.25,$$

hence, again, no coalescence. Once again we double the steps and go further back to $T = -4$, which means we need two more draws of ξ, and this time they both are one, yielding $\{\xi_{-4}, \xi_{-3}, \xi_{-2}, \xi_{-1}\} = \{1, 1, 1, 0\}$. Since we only need at most three consecutive upward steps to bring any state to the ceiling state $X = 4$, the $\{1, 1, 1, 0\}$ sequence immediately implies that

$$X^{(x)}_{-4\to0} = \phi(4, 0) = 2, \quad \text{for all } x \in \mathcal{S}.$$

We have therefore detected coalescence after going back to only $T = -4$. This is not in any contradiction to Figure 8.1, but points to an even stronger statement that only the last four elements in the sequence (Equation 8.4), $\{\xi_{-4}, \xi_{-3}, \xi_{-2}, \xi_{-1}\} = \{1, 1, 1, 0\}$, rather than all eight elements, are really relevant.

8.3.3 General Classes of Monotone Coupling

One may wonder when such ordering exists in more general situations and, if so, what important classes of distributions can be identified to satisfy Equation 8.3. Such questions have been investigated by [13,18] in the case of *monotone* (also called *attractive*) and *anti-monotone* (also called *repulsive*) distributions Π. Suppose that $\mathcal{S} = \mathcal{Z}^d$, for some set $\mathcal{Z} \subset \mathbb{R}$. We consider the componentwise partial order on $\mathcal{S}$ so that $x \prec y$ if and only if $x_i \le y_i$ for all $1 \le i \le d$. The probability measure P on $\mathcal{S}$ is defined to be *monotone* if, for each $1 \le i \le d$,

$$P(X_i \le s | X_{[-i]} = a) \ge P(X_i \le s | X_{[-i]} = b), \quad \text{for all } s \in \mathcal{S}, \tag{8.5}$$

whenever $a \prec b$ in $\mathcal{Z}^{d-1}$, where $X_{[-i]} = (X_1, \ldots, X_{i-1}, X_{i+1}, \ldots, X_d)$. Similarly, P is called *anti-monotone* if

$$P(X_i \leq s | X_{[-i]} = a) \leq P(X_i \leq s | X_{[-i]} = b),$$

whenever $a \prec b$ in $\mathcal{Z}^{d-1}$.

This definition of monotonicity via all full conditional distributions $P(X_i | X_{[-i]})$, $i = 1, \ldots, d$, was motivated by their use with the Gibbs sampler. In particular, Equations 8.3 and 8.5 are easily connected when the sampling from $P(X_i \leq s | X_{[-i]} = a)$ is done via the inverse CDF method. Put $F_i(s|a) = P(X_i \leq s | X_{[-i]} = a)$ and assume that the ith component is updated using $\phi_i(x, U) = (x_1, x_2, \ldots, x_{i-1}, \inf\{s : F_i(s|x_{[-i]}) = U\}, x_{i+1}, \ldots, x_d)$, with $U \sim U(0, 1)$. If we assume $x \prec y$, then from Equation 8.5 we get

$$\phi_i(x, U) \prec \phi_i(y, U) \tag{8.6}$$

because $\inf\{s : F_i(s|x_{[-i]}) = U\} \leq \inf\{s : F_i(s|y_{[-i]}) = U\}$. Applying Equation 8.6 in sequential order from $i = 1$ to $i = d$, as in a Gibbs sampler fashion, we can conclude that for $\vec{U} = \{U_1, \ldots, U_d\}$, the composite map

$$\phi(x, \vec{U}) = \phi_d(\phi_{d-1}(\ldots \phi_2(\phi_1(x, U_1), U_2), \ldots, U_{d-1}), U_d) \tag{8.7}$$

is monotone in x with respect to the same partial order $\prec$.

In the case of anti-monotone target distributions, it is not hard to see that the $\phi(x, \vec{U})$ of Equation 8.7 is also anti-monotone with respect to $\prec$ if d is odd, but monotone if d is even. Indeed, the ceiling/upper and floor/lower chains switch at each step (indexed by $i = 1$ to $i = d$), that is, the ceiling chain becomes the floor chain and vice versa. This oscillating behavior, however, still permits us to construct *bounding chains* that squeeze in between all the sample paths such that the general coalescence can be detected once the bounding chains have coalesced. See, for example, [13], which also discusses other examples of monotone target distributions; see also [6,21].

8.3.4   Bounding Chains

In a more general setup, [18] discusses the use of bounding chains without any condition of monotonicity. To better fix ideas, consider the following simple random walk with state space $\mathcal{S} = \{0.25, 0.5, 2\}$ and with transition probability matrix

$$A = \begin{pmatrix} p & 1-p & 0 \\ 0 & p & 1-p \\ p & 0 & 1-p \end{pmatrix},$$

where the $(1, 1)$ entry corresponds to the probability that the chain stays at 0.25. Unlike the previous random walk, the recursion defined by the matrix A is neither monotone nor anti-monotone with respect to the natural order on the real line. For example, with $\xi \sim \text{Bernoulli}(p)$, and if $\xi = 1$, we have $\phi(0.25, \xi) = 0.25 < \phi(0.5, \xi) = 0.5 > \phi(2, \xi) = 0.25$, where ϕ is the chain's SRS. In contrast to the previous random walk, here $\xi = 1$ can indicate both moving up or down depending on the starting position, and this is exactly what destroys monotonicity with respect to the same ordering as in the previous random-walk example. (This, of course, by no means implies that no (partial) ordering existed under

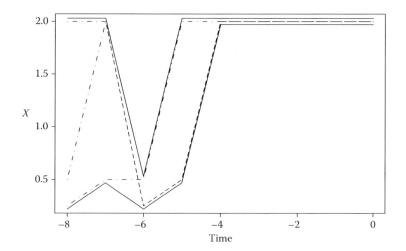

FIGURE 8.2
Nonmonotone Markov chain. The upper and lower solid lines mark the bounding processes.

which the SRS is monotone; seeking such an ordering is indeed a common implementation strategy for perfect sampling.)

In Figure 8.2, we show one run of the CFTP algorithm implemented for this simple example with $p = 0.1$, where $\{\xi_{-8}, \ldots, \xi_{-1}\} = \{0, 1, 0, 0, 0, 0, 0, 0\}$. One can see that the three paths cross multiple times and no single path remains above or below all the others at all times. A bounding chain, in the general definition introduced by [18], is a chain $\{Y_t : t \geq 0\}$ defined on $2^{\mathcal{S}}$, the set of all subsets of $\mathcal{S}$, with the property that if $X_t^{(x)} \in Y_t$ for all $x \in \mathcal{S}$ then $X_{t+1}^{(x)} \in Y_{t+1}$ for all $x \in \mathcal{S}$; evidently Y_0 needs to contain all values in S. If, at some time t, Y_t is a singleton then coalescence has occurred. Clearly, there are many ways to define the chain Y_t, but only a few are actually useful in practice and these are obtained, usually, from a careful study of the original chain X_t.

For instance, in our example we notice that after one iteration $Y_0 = \mathcal{S}$ will become either $Y_1 = \{0.25, 0.5\}$ or $Y_1 = \{0.5, 2\}$, depending on whether $\xi = 1$ or $\xi = 0$, and therefore Y_t will always be a subset of these two sets (possibly themselves). Therefore, for $t \geq 1$, the updating rule $Y_{t+1} = \Psi(Y_t, \xi)$ can be simplified to

$$
\Psi(Y_t, \xi) = \begin{cases}
Y_t, & \text{if } \xi = 1 \text{ and } Y_t = \{0.25, 0.5\}, \\
\{0.25, 0.5\} & \text{if } \xi = 1 \text{ and } Y_t = \{0.5, 2\}, \\
\{0.5, 2\}, & \text{if } \xi = 0 \text{ and } Y_t = \{0.25, 0.5\}, \\
\{2\}, & \text{if } \xi = 0 \text{ and } Y_t = \{0.5, 2\}, \\
\phi(X_t, \xi), & \text{if } Y_t = \{X_t\}.
\end{cases}
\tag{8.8}
$$

One can see then that having the ordered triplet $\{1, 0, 0\}$ in the ξ-sequence triggers coalescence, after which one simply follows the path to time zero.

Two essential requirements for an effective bounding chain are that (i) it can detect coalescence of the original chain and (ii) it requires less effort than running all original sample paths. The chain $Y_t \equiv \{\mathcal{S}\}$ for all t is a bounding chain and satisfies (ii), but clearly it is useless. As an example of bounding chains that do not satisfy (ii), consider the upper and lower solid paths in Figure 8.2. Here the upper solid path is the maximum value attained by all paths at each time t, and the lower solid path is the minimal value (both have been slighted shifted for better visualization). For each time t, the interval between the upper and lower solid paths, denoted by $\tilde{Y}_t$, clearly forms a bounding chain. But unlike Y_t in Equation 8.8, the updating function for $\tilde{Y}_t$ is not easy to define, so running $\tilde{Y}_t$ involves checking the extremes of all the paths for X_t and is thus as complicated as running all paths for X_t.

As far as general strategies go, [13] shows how to construct bounding chains when each component of the random vector X is updated via a Gibbs sampler step, whereas [18] presents a general method for constructing bounding chains and applies it to problems from statistical mechanics and graph theory.

8.4   Cost-Saving Strategies for Implementing Perfect Sampling

The plain vanilla CFTP described in Section 8.2 suffers from two main drawbacks. First, the implementation "from the past" requires the random seeds used in the backward process to be stored until coupling is observed and a random sample is obtained. Second, the impatient user cannot abandon runs that are too long without introducing sampling bias, because the coupling time T is correlated with the sample obtained at time zero. In the following two sections we provide intuitive explanations of the read-once CFTP and Fill's interruptible algorithm, designed respectively to address these two drawbacks.

8.4.1   Read-Once CFTP

Read-once CFTP (Ro-CFTP), as proposed by Wilson [56], is a clever device that turns CFTP into an equivalent "forward-moving" implementation. It collects the desired i.i.d. draws as the process moves forward and without ever needing to save any of the random numbers previously used. The method starts with a choice of a fixed block size K, such that the K-composite map

$$\phi_K(x; \vec{\xi}) = \phi(\phi(\ldots \phi(\phi(x, \xi_1), \xi_2), \ldots, \xi_{K-1}), \xi_K),$$

where $\vec{\xi} = \{\xi_1, \ldots, \xi_K\}$, has a high probability of coalescing, that is, the value of $\phi_K(x; \vec{\xi})$ will be free of x, or equivalently, all paths coalesce *within the block* defined by $\vec{\xi}$. In [56], Wilson suggests selecting K such that the probability of ϕ_K coalescing, denoted by p_K, is at least 50%. Given such a ϕ_K, we first initialize the process by generating i.i.d. $\vec{\xi}_j, j = 1, 2, \ldots,$ until we find a $\vec{\xi}_{j_0}$ such that $\phi_K(x; \vec{\xi}_{j_0})$ coalesces. Without loss of generality, in the top panel of Figure 8.3, we assumed that $j_0 = 1$; and we let $S_0 = \phi_K(x; \vec{\xi}_{j_0})$. We then repeat the same process, that is, generating i.i.d. $\vec{\xi}_j$s until $\phi_K(x; \vec{\xi})$ coalesces again. In the top panel of Figure 8.3, this occurred after three blocks. We denote the coalescent value as S_1. During this process, we follow from block to block only the *coalescence path* that goes through S_0 while all the other paths are reinitialized at the beginning of each block. The location of the coalescence path *just before* the beginning of the next coalescent composite map is a sample from the desired Π. In Figure 8.3 this implies that we retain the circled X_1 as a sample. The process then is repeated as we move forward, and this time we follow the path starting from S_1 and the next sample X_2 (not shown) is the output of this path immediately before the beginning of the next coalescent block. We continue this process to obtain i.i.d. draws X_3, X_4, and so on.

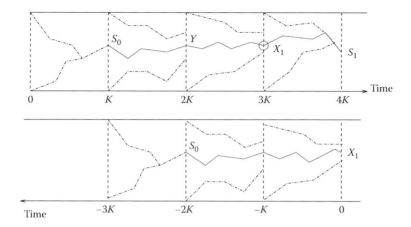

FIGURE 8.3
Top: The read-once CFTP with blocks of fixed length. Bottom: Comparison with CFTP2.

The connection between the Ro-CFTP and CFTP may not be immediately clear. Indeed, in the plain vanilla CFTP, the concept of a composite/block map is not emphasized because, although we "back off" in blocks, we do not require to have a coalescent composite map of fixed length. For instance, if we set $K = 4$, we can see that in Figure 8.1 the paths started at $-2K$ coalesce in the interval $(-K, 0)$ rather than within the block $(-2K, -K)$. However, suppose we consider a modified implementation of the plain vanilla CFTP, call it CFTP2, in which we go back from time zero block by block, each with size K, until we find a block that is coalescent, that is, all paths coalesce *within that block*. Clearly, if we trace the path from the coalescent value from that block until it reaches time zero, it will be exactly the same value as found by the original plain vanilla CFTP because once the coalescence takes place, all paths will stay together forever. The bottom panel of Figure 8.3 illustrates CFTP2, where the third block (counting backward from $t = 0$) is the coalescent block, and X_1 is our draw.

The resemblance of the bottom panel and the first three blocks in the top panel (counting forward from time $t = 0$) is intended to highlight the equivalence between Ro-CFTP and CFTP2. On its own, CFTP2 is clearly less cost-effective than CFTP because by insisting on having block coalescence, it typically requires going back further in time than does the original CFTP (since block coalescence is a more stringent detecting criterion, as discussed above). However, by sacrificing a little of the efficiency of detecting coalescence, we gain the *independence* between the block coalescent value S_0 and the entire backward search process for S_0, and hence we can reverse the order of the search without affecting the end result.

As this independence is the backbone of the Ro-CFTP, here we show how critically it depends on having *fixed-size* blocks. Intuitively, when the blocks all have the same size, they each have the same probability of being a coalescent block, and the distribution of the coalesced state given a coalescent block is the same regardless of which block it is. To confirm this intuition and see how it implies the independence, let us define the block random vector $\vec{\xi}_{-t} = (\xi_{-tK}, \xi_{-tK+1} \ldots, \xi_{-tK+K-1})$ and, for a given set $\vec{\xi}_{-t}, t = 1, 2, \ldots$, let T be the first t such that $\phi_K(x; \vec{\xi}_{-t})$ coalesces, and let $S_0 = \phi_K(x; \vec{\xi}_{-T})$ be the coalescent value. Also let $C_j = \{\phi_K(x, \vec{\xi}_{-j}) \text{ coalesces}\}$, that is, the event that the jth block map coalesces. Then $\{T = t\} = (\cap_{j=1}^{t-1} C_j^c) \cap C_t$. For notational simplicity, denote $A_j = \{\phi_K(x, \vec{\xi}_{-j}) \in A\}$ and $B_j = \{\Xi_j \in B\}$, where A and B are two arbitrary (measurable) sets on the appropriate probability spaces, and $\Xi_j = \{\vec{\xi}_{-1}, \ldots, \vec{\xi}_{-j}\}$. Then for any positive integer t,

$$P(S_0 \in A, T = t, \Xi_{T-1} \in B) = P(A_t \cap [\cap_{j=1}^{t-1} C_j^c \cap C_t] \cap B_{t-1})$$

$$= P(A_t \cap C_t)P(\cap_{j=1}^{t-1} C_j^c \cap B_{t-1})$$

$$= P(A_t|C_t)P(C_t)P(\cap_{j=1}^{t-1} C_j^c \cap B_{t-1}) \qquad (8.9)$$

$$= P(A_t|C_t)P(C_t \cap_{j=1}^{t-1} C_j^c \cap B_{t-1})$$

$$= P(A_1|C_1)P(T = t, B_{T-1}).$$

In deriving the above equalities, we have repeatedly used the fact that $\{A_t, C_t\}$ are independent of $\{A_{t-1}, B_{t-1}, C_{t-1}\}$ since they are determined respectively by $\vec{\xi}_{-t}$ and $\{\vec{\xi}_j, j = -1, \ldots, -(t-1)\}$. The last switching from $P(A_t|C_t)$ to $P(A_1|C_1)$ is due to the i.i.d. nature of the $\{A_t, C_t\}$, because all blocks have the same size K. This switching is critical in establishing the factorization in Equation 8.9, and hence the independence.

Clearly, as depicted in Figure 8.3, the output of CFTP2, namely X_1, can be expressed as $M(S_0, T, \Xi_{T-1})$, where M is a *deterministic* map. The aforementioned independence ensures that if we can find $\{\tilde{T}, \Xi_{\tilde{T}-1}\}$ such that it has the same distribution as $\{T, \Xi_{T-1}\}$ and is independent of S_0, then $\tilde{X}_1 = M(S_0, \tilde{T}, \Xi_{\tilde{T}-1})$ will have the same distribution as $X_1 = M(S_0, T, \Xi_{T-1})$, and hence it is also an exact draw from the stationary distribution Π. Because $\{\vec{\xi}_{-1}, \vec{\xi}_{-2}, \ldots, \}$ are i.i.d., obviously the distribution of $\{T, \Xi_{T-1}\}$ is invariant to the order at which we check for the block coalescence. We can therefore reverse the original backward order into a forward one and start at an arbitrary block which must be independent of S_0. This naturally leads to the Ro-CFTP, because we can start with the block immediately after a coalescence has occurred (which serves as S_0), since it is independent of S_0. Moreover, the number of blocks and all the block random numbers (i.e. ξs) needed *before* we reach the next coalescent block represents a sample from the distribution of $\{T, \Xi_{T-1}\}$. It is worth emphasizing that each coalescent composite map fulfills two roles as it marks the end of a successful run (inclusive) and the beginning of a new run (exclusive). Alternatively, Equation 8.9 implies that we can first generate T from a geometric distribution with mean $1/p_K$ (recall that p_K is the probability of coalescence within each block), and then generate $T - 1$ noncoalescent blocks, via which we then run the chain forward starting from S_0. This observation has little practical impact since p_K is usually unknown, but it is useful for understanding the connection with the splitting chain technique that will be discussed in Section 8.5. The forward implementation brought by Ro-CFTP also makes it easier to implement the efficient use of perfect sampling tours proposed by [40], which will be discussed in Section 8.6.

8.4.2 Fill's Algorithm

Fill's algorithm [8] and its extension to general chains [9] break the dependence between the backward time to coalescence and the sample obtained at time zero. In the following we use the slightly modified description from [39].

The algorithm relies on the time-reversal version of the Markov chain designed to sample from Π. If the original chain has transition kernel $K(x, \cdot)$, then the time-reversal version has kernel $\tilde{K}(z, \cdot)$, such that

$$\tilde{k}(x|z)\pi(z) = k(z \mid x)\pi(x), \quad \forall (x, z) \in \mathcal{S} \times \mathcal{S}, \qquad (8.10)$$

where, for simplicity of presentation, we have assumed that the stationary law Π has density π, and $K(x, \cdot)$ and $\tilde{K}(z, \cdot)$ have kernel densities $k(\cdot|x)$ and $\tilde{k}(\cdot \mid z)$, respectively. It also requires that, given a particular path $X_0 \rightarrow X_1 \rightarrow \cdots \rightarrow X_t$, we can sample, *conditional on the observed path*, a sample of the same length from any state in $\mathcal{S}$.

The algorithm starts by sampling a random $Z \in \mathcal{S}$ from an arbitrary distribution P_0 (with density p_0) that is absolutely continuous with respect to Π, and by selecting a positive integer T. The first stage is illustrated in the top panel of Figure 8.4: using the reversal time chain, we simulate the path $Z = X_T \rightarrow X_{T-1} \rightarrow \cdots \rightarrow X_1 \rightarrow X_0$ (note that the arrow is pointing against the direction of time). In the second stage, we sample forward from all the states in $\mathcal{S}$ conditional on the existing path $X_0 \rightarrow X_1 \rightarrow \cdots \rightarrow X_T = Z$ (note that this path is considered now in the same direction as time). If by time T all the paths have coalesced, as depicted in the middle panel of Figure 8.4 (where we used monotone coupling for simplicity of illustration, but the idea is general), we retain X_0 as a sample from π, as shown in the bottom panel of Figure 8.4, and restart with a new pair (Z, T). Otherwise, we select a new T or we restart with a new pair (Z, T).

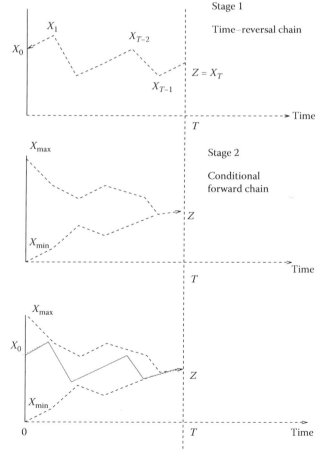

FIGURE 8.4
Illustration of Fill's algorithm.

To understand why the algorithm produces i.i.d. samples from π, we first note that Equation 8.10 holds in the more general form

$$\tilde{k}_t(x|z)\pi(z) = k_t(z \mid x)\pi(x), \quad \forall\, (x,z) \in \mathcal{S} \times \mathcal{S}, \tag{8.11}$$

where k_t is the kernel density of the t-step forward transition kernel K_t and $\tilde{k}_t$ is the corresponding time-reversal one, $\tilde{K}_t$. Fill's algorithm retains only those paths from Z to X_0 (obtained via $\tilde{K}_T$) such that the corresponding $k_T(z|x)$ is free of x—and hence it can be expressed as $h_T(z)$—due to coalescence; in this sense Fill's algorithm is a case of rejection sampling. Therefore, using Equation 8.11, the probability density for those retained X_0s is

$$p(x) = \int \tilde{k}_T(x \mid z)p_0(z)\,dz = \int \frac{\pi(x)h_T(z)}{\pi(z)}p_0(z)\,dz \propto \pi(x), \tag{8.12}$$

hence the correctness of the sampling algorithm (see also [4]). Note that here, for simplicity, we have deliberately blurred the distinction between the fixed t in Equation 8.11 and the potentially random T in Equation 8.12; in this sense Equation 8.12 is a heuristic argument for building intuition rather than a rigorous mathematical proof. In its general form, Fill's algorithm can search for the coalescence time T just as with CFTP—see [9] for a detailed treatment of the general form of Fill's algorithm, including a rigorous proof of its validity. See also [4] for an alternative proof based directly on the rejection-sampling argument, as well as for a numerical illustration.

The conditional sampling is the main difficulty encountered when implementing Fill's algorithm, but in some cases it can be straightforward. For instance, if $X_{t+1} = \phi(X_t, U_{t+1})$ then it is possible that U_{t+1} is uniquely determined once both X_t and X_{t+1} are fixed (e.g. if we generate X_{t+1} using the inverse CDF transformation). If we denote by $\{U_1, \ldots, U_T\} := \mathcal{U}^0$ the set of random deviates determined by the path $Z = X_T \to X_{T-1} \to \cdots \to X_1 \to X_0$, then at the second stage we simply run for T steps, from all the states of $\mathcal{S}$, the Markov chains using the recursive form (Equation 8.1) with the set $\mathcal{U}^0$ as random seeds.

8.5 Coupling Methods

All algorithms described so far require the coupling of a finite or infinite number of paths in finite time. This is the greatest difficulty of applying perfect sampling algorithms to continuous state spaces, especially those with unbounded spaces (which is the case for most routine applications in Bayesian computation) and this is where the greatest ingenuity is required to run perfect sampling in more realistic settings. A good coupling method must be usable in practice and it is even better if it is implementable for different models with the same degree of success. In this section, we review some of the most useful coupling techniques, which essentially belong to two different types: (i) those which induce a "common regeneration state" that all sample paths must enter with positive probability; and (ii) those which explore hidden discretization and hence effectively convert the problem into one with a finite state space.

8.5.1 Splitting Technique

A very common technique for coupling MCMC paths is initiated in [46] and discussed in detail by [54]. Consider the Markov chain X_t defined using the transition kernel K and suppose that there exist $t > 0, 0 < \epsilon < 1$, a set C (called a *small set*) and a *probability measure* ν such that

$$K_t(x, dy) \geq \epsilon \nu(dy), \quad \forall\, x \in C,$$

where K_t represents the t-step transition kernel. Thus, for any $x \in C$,

$$K_t(x, dy) = \epsilon \nu(dy) + (1 - \epsilon)\frac{K_t(x, dy) - \epsilon \nu(dy)}{1 - \epsilon} = \epsilon \nu(dy) + (1 - \epsilon)Q(x, dy), \quad (8.13)$$

where $Q(x, dy) = [K_t(x, dy) - \epsilon \nu(dy)]/(1 - \epsilon)$. The representation given by Equation 8.13 is important because with probability ϵ the updating of the chain will be done using the probability measure ν, that is, *independently* of the chain's current state. If at time t all the paths are in the set C and all the updates use the same random numbers ξ that lead to the transition into the ν component of Equation 8.13, then all paths will coalesce at time $t + 1$, even if there are uncountably many. However, for a set $C \subset S$ it will be difficult, if not impossible, to determine whether it contains all paths at a given time. This problem is alleviated in the case of CFTP where the existence of successful coupling has been shown (see [10]) to be equivalent to the uniform ergodicity of the chain X_t, in which case the small set is the whole sample space, S, so all paths are automatically within a small set at all times. An example where this idea has been brought to fruition is the *multigamma coupler* introduced by [37], following the gamma coupler of [25]. The method is further developed by [36] in the context of perfect sampling from continuous state distributions.

The multigamma coupler applies when the update kernel density $f(\cdot|x)$ of the Markov chain is known. In addition, it requires that there is a nonnegative function r such that

$$f(y \mid x) \geq r(y), \quad \forall x, y \in S. \quad (8.14)$$

If we denote $\rho = \int r(y)dy > 0$, then in line with the splitting technique discussed above we can write

$$P(X_{t+1} \leq y \mid X_t = x) = \rho R(y) + (1 - \rho)Q(y|x), \quad (8.15)$$

where $R(y) = \rho^{-1}\int_{-\infty}^{y} r(v)\, dv$ and $Q(y \mid x) = (1 - \rho)^{-1}\int_{-\infty}^{y}[f(v \mid x) - r(v)]\, dv$.

As a simple example, assume that the transition kernel has the gamma density $f(y \mid a, b_x) = y^{a-1}b_x^a \exp(-yb_x)/\Gamma(a)$, where a is fixed, and b_x depends on the previous state $X_t = x$ but is always within a fixed interval, say $b_x \in [b_0, b_1]$, where b_0 and b_1 are known constants. Then we can set $r(y) = y^{a-1}b_0^a \exp(-yb_1)/\Gamma(a)$, which yields $\rho = (b_0/b_1)^a$. At each t, we sample $\xi \sim \text{Bernoulli}(\rho)$, and if $\xi = 1$, we draw y from $\text{Gamma}(a, b_1)$, and let all paths $X_{t+1} = y$ regardless of their previous states, hence coalescence takes place. If $\xi = 0$, then we draw from the Q component in Equation 8.15 (though this step requires drawing from a nonstandard distribution).

In situations when no uniform bound can be found on S for Equation 8.14 to hold, Murdoch and Green [37] propose partitioning $S = S_1 \cup \ldots \cup S_m$ and bounding the kernel density f on each S_i with r_i and introduce a *partitioned multigamma coupler* for this setting. A more difficult coupling strategy has been described in [22] in the case of geometrically (but not necessarily uniformly) ergodic chains, though the approach has not been implemented on a wide scale.

There is a direct connection between the multigamma coupler and the Ro-CFTP in Section 8.4.1. With a block of size $K = 1$ the multigamma coupler construction implies that the probability of coalescence within the block is ρ. As described above, we can therefore sample a geometric T with success probability ρ, and start from a coalesced value, that is, an independent draw from $R(y)$ in Equation 8.15. We then run the chain forward for $T - 1$ steps conditioning on noncoalesced blocks, namely, we use the Q component of Equation 8.15 as the transition kernel. The resulting value then is an exact draw from Π [37].

There is also a close connection between the multigamma coupler and the slice sampler (see Section 8.5.4), as both can be viewed as building upon the following simple idea: For a given (not necessarily normalized) density $g(y)$, if (U, Y) is distributed uniformly on $\Omega_g = \{(u, y) : u \leq g(y)\}$, then the marginal density of Y is proportional to $g(y)$. Therefore, when $f(y|x) \geq r(y)$ for all x and y, we have

$$\Omega_r = \{(u, y) : u \leq r(y)\} \subset \Omega_{f,x} = \{(u, y) : u \leq f(y \mid x)\}, \quad \forall x \in \mathcal{S}. \tag{8.16}$$

For simplicity of illustration, let us assume that all $\Omega_{f,x}$ are contained in the unit square $[0, 1] \times [0, 1]$. Imagine now we use rejection sampling to achieve the uniform sampling on $\Omega_{f,x}$ for a particular x by drawing uniformly on the unit square. The chance that the draw (u, y) will fall into Ω_r is precisely ρ, and more importantly, if it is in Ω_r, it is an acceptable proposal for $f(y|x)$ regardless of the value of x because of Equation 8.16. This is the geometric interpretation of how the coalescence takes place for splitting coupling, which also hints at the more general idea of coupling via a common proposal, to which we now turn.

8.5.2   Coupling via a Common Proposal

The idea of using a common proposal to induce coalescence was given in [3] as a way to address the second challenge discussed in Section 8.2.3. (Note, however, that this strategy does not directly address the first challenge, namely discretizing a continuous set of paths into a finite set; that challenge is addressed by, for example, the augmentation method described in the next subsection, or by other clever methods such as the multishift coupler in [57].) Imagine that we have managed to reduce the number of paths to a finite one. In practice, it may still take a long time (possibly *too* long) before all paths coalesce into one. Intuitively, one would like to make it easier for paths that are close to each other to coalesce more quickly.

Remarkably, the description of coupling via a common proposal can be formulated in a general setting irrespective of the transition kernel used for the chain, as long as it has a density. Suppose that the chain of interest has transition kernel with the (conditional) density $f(\cdot|X_t)$. Instead of always accepting the next state as $X_{t+1} \sim f(\cdot|X_t)$, we occasionally replace it with a random draw $\tilde{Y}$ sampled from a user-defined density g. Thus, the X_{t+1} from the original chain plays the role of a proposal and is no longer guaranteed to be the next state; we therefore relabel it as $\tilde{X}_{t+1}$.

Instead, given $X_t = x$, the next state X_{t+1} is given by the updating rule

$$X_{t+1} = \begin{cases} \tilde{Y}, & \text{if } \dfrac{f(\tilde{Y}|x)g(\tilde{X}_{t+1})}{f(\tilde{X}_{t+1}|x)g(\tilde{Y})} > U, \\[2em] \tilde{X}_{t+1}, & \text{otherwise,} \end{cases} \tag{8.17}$$

where $U \sim U(0,1)$ and is independent of any other variables. In other words, the above coupler makes a choice between two independent random variables $\tilde{X}_{t+1}$ and $\tilde{Y}$ using a Metropolis–Hastings (MH) acceptance ratio. Note that the MH accept–reject move is introduced here simply to ensure that the next state of the chain has distribution density $f(\cdot|X_t)$ even if occasionally the state is "proposed" from g. The coupling via a common proposal tends to increase the propensity of coalescing paths that are close to each other. More precisely, suppose that two of the paths are close, that is, $X_t^{(1)} \approx X_t^{(2)}$. Then the ratios in Equation 8.17 will tend to be similar for the two chains, which implies that both chains will likely accept/reject $\tilde{Y}$ simultaneously.

It is also worth emphasizing that the above scheme needs a modification in order to be applicable to the MH sampler which does not admit a density with respect to the Lebesgue measure. The basic idea is to introduce the common proposal into the MH proposals themselves as in [3]. This perhaps is best seen via a toy example. Suppose that our target distribution is $N(0,1)$, and we adopt a random-walk Metropolis algorithm, that is, the proposal distribution is $q(y|x) = N(y-x)$, where $N(z)$ is the density of $N(0,1)$. Clearly, because $N(z)$ is continuous, two paths started at different points in the sample space will have zero probability of coalescing if we just let them "walk randomly." To stimulate coalescence, we follow the ideas in [3] and create an intermediate step in which the proposals used in the two processes can be coupled.

More precisely, at each time t we sample $\tilde{Z}_{t+1} \sim t_3(\cdot)$, where t_3 is the t distribution with three degrees of freedom. Suppose that the proposal for chain i at time t is $\tilde{Y}_{t+1}^{(i)}$, where $\tilde{Y}_{t+1}^{(i)} \sim N(X_t^{(i)}, 1)$. We then define

$$
\tilde{W}_{t+1}^{(i)} = \begin{cases} \tilde{Z}_{t+1}, & \text{if } \dfrac{N(\tilde{Z}_{t+1} - X_t^{(i)})t_3(\tilde{Y}_{t+1}^{(i)})}{N(\tilde{Y}_{t+1}^{(i)} - X_t^{(i)})t_3(\tilde{Z}_{t+1})} > U, \\[4mm] \tilde{Y}_{t+1}^{(i)}, & \text{otherwise}, \end{cases}
\tag{8.18}
$$

where $U \sim U(0,1)$ is independent of all the other variables. The proposal $\tilde{W}_{t+1}^{(i)}$ is accepted using the usual MH strategy because its density is still the density of the original proposal, $N(X_t^{(i)}, 1)$; the next state is then either $\tilde{W}_{t+1}^{(i)}$ (acceptance) or $X_t^{(i)}$ (rejection). What has changed is that regardless of which paths the chains have taken, their MH proposals now have a positive probability of taking on a common value $\tilde{Z}_{t+1}$ for all those chains for which the first inequality in Equation 8.18 is satisfied. This does not guarantee coupling, but it certainly makes it more likely. In Figure 8.5 we show two paths simulated using the simple model described above, where the two paths first came very close at $t = -8$ and then coalesced at $t = -7$.

8.5.3 Coupling via Discrete Data Augmentation

Data augmentation [51], also known in statistical physics as the auxiliary variable method, is a very effective method for constructing efficient MCMC algorithms; see [55] for a review. It turns out to be useful for perfect sampling as well, because we can purposely consider auxiliary variables that are discrete and therefore convenient for assessing coalescence. Specifically, suppose that our target density is $f(x)$, where x may be continuous. Suppose that we have a way to augment $f(x)$ into $f(x, l)$, where l is discrete. If we can perform Gibbs sampling via $f(x \mid l)$ and $f(l|x)$, then we will automatically have a Markov sub-chain with

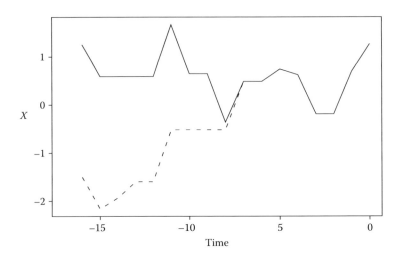

FIGURE 8.5
Illustration of coupling with proposals for two paths.

$f(l)$ as the stationary density (note the sub-chain with l only is Markovian because the Gibbs sampler here only involves two steps). Therefore, we have effectively turned the continuous problem for $f(x)$ into a discrete one because once we have an authentic draw from $f(l)$, then we can easily get a corresponding authentic draw from $f(x)$ by sampling from $f(x \mid l)$.

To illustrate, consider finite mixtures, where the obvious auxiliary variable is the indicator variable indicating the mixture component from which a sample is obtained. The coupling via augmentation has been successfully implemented by [17] in the case of two-component mixtures of distributions and by [38] in the case of Bayesian mixture priors. Below is one of the examples discussed by [17], which we recast in order to crystalize the essence of discrete data augmentation.

Consider the mixture $\alpha f_0(d) + (1 - \alpha)f_1(d)$, where only the mixture proportion α is unknown and therefore we seek its posterior density, assuming a uniform prior on $(0, 1)$. Given a sample $\{d_1, \ldots, d_n\}$ from the mixture, the posterior for α is proportional to

$$p(\alpha|\vec{d}) \propto \prod_{i=1}^{n} \{\alpha f_0(d_i) + (1 - \alpha)f_1(d_i)\}, \tag{8.19}$$

involving 2^n terms when expanded; note that here we use $\vec{d} = \{d_1, \ldots, d_n\}$ to denote the data instead of the original $\{x_1, \ldots, x_n\}$, to avoid the potential confusion of our generic notation which uses X for the sampling variable, which is α here. Let the latent variables $\vec{z} = \{z_1, \ldots, z_n\}$ be such that $z_i = 0$ if d_i has been generated from f_0 and $z_i = 1$ if d_i has been generated from f_1. Then it is easy to see that

$$P(z_i = 1|\vec{d}, \alpha) = \frac{(1 - \alpha)f_1(d_i)}{\alpha f_0(d_i) + (1 - \alpha)f_1(d_i)} := p_i \tag{8.20}$$

and

$$P(\alpha|\vec{z}) = \text{Beta}\left(n + 1 - \sum_{i=1}^{n} z_i, \sum_{i=1}^{n} z_i + 1\right). \tag{8.21}$$

This implies that we can construct the discrete augmentation as $l = \sum_i z_i$, which has a nonhomogenous binomial (NhB) distribution $\text{NhB}(n, \vec{p})$, where $\vec{p} = \{p_1, \ldots, p_n\}$. That is, l is the sum of n independent but not necessarily identically distributed Bernoulli variables. Given this data augmentation scheme $f(\alpha, l)$, the algorithm given in [17] can be reformulated as follows.

1. Because of Equation 8.21, given $l_t = l$, we generate $\alpha_{t+1} \sim \text{Beta}(n + 1 - l, l + 1)$, which can be accomplished by drawing $w_j \sim \text{Exponential}(1)$ for $j \in \{1, \ldots, n + 2\}$ and then letting

$$\alpha_{t+1} = \frac{\sum_{i=1}^{n+1-l} w_i}{\sum_{i=1}^{n+2} w_i}. \tag{8.22}$$

2. Given $\alpha_{t+1} = \alpha$, because of Equation 8.20, we need to draw l_{t+1} from $\text{NhB}(n, \vec{p}(\alpha))$, where $\vec{p}(\alpha) = \{p_1, \ldots, p_n\}$, with $p_i \equiv p_i(\alpha)$ given by the right-hand side of Equation 8.20. This draw is accomplished by generating independent $u_i \sim U(0, 1)$ and letting

$$l_{t+1} = \sum_{i=1}^{n} \mathbf{1}\{u_i \leq p_i\}, \tag{8.23}$$

where $\mathbf{1}\{A\}$ is the usual indicator function of event A.

Combining Equations 8.22 and 8.23, we see that the SRS from l_t to l_{t+1} can be written as

$$l_{t+1} \equiv \phi(l_t; \vec{u}, \vec{w}) = \sum_{i=1}^{n} \mathbf{1}\left\{ u_i \leq \left[1 + \left(\frac{\sum_{i=1}^{n+2} w_i}{\sum_{i=1}^{n+1-l_t} w_i} - 1 \right)^{-1} \frac{f_0(d_i)}{f_1(d_i)} \right]^{-1} \right\}. \tag{8.24}$$

For given $\vec{u} = \{u_1, \ldots, u_n\}$ and $\vec{w} = \{w_1, \ldots, w_n\}$, the function ϕ in Equation 8.24 is evidently increasing in l_t and thus defines, with respect to the natural integer ordering, a monotone Markov chain on the state space $S_l = \{0, \ldots, n\}$, with the ceiling and floor states given by $l = 0$ and $l = n$. Through data augmentation we have therefore converted the problem of drawing from the continuous distribution given by Equation 8.19 into one in which the sample space is the finite discrete space S_l, given by Equation 8.24, for which we only need to trace the two extreme paths starting from $l = 0$ and $l = n$.

8.5.4 Perfect Slice Sampling

Slice sampling is based on the simple observation that sampling from Π (assumed to have density π) is equivalent to sampling from the uniform distribution $g(u, x) \propto \mathbf{1}\{u \leq f(x)\}$, where f is an unnormalized version of π and is assumed known. One can easily see that the marginal distribution of x is then the desired one. In turn, the sampling from g can be performed using a Gibbs scan in which both steps involve sampling from uniform distributions:

I. Given X_t, sample $U \sim U(0, f(X_t))$.

II. Given U from Step I, sample $X_{t+1} \sim U[A(U)]$, where $A(w) = \{y : f(y) \geq w\}$.

Here, for simplicity, we assume that $A(U)$ has finite Lebesgue measure for any U; more general implementations of the slice sampler are discussed in [7,45]. The coupling for slice sampling has been designed by [30] under the assumption that there exists a minimal element $x_{\min} \in S$ with respect to the order $x \prec y \Leftrightarrow f(x) \leq f(y)$.

Specifically, the *perfect slice sampler* achieves coupling via introducing common random numbers into the implementation of Steps I and II in the following fashion. We implement Step I, regardless of the value of X_t, by drawing $\epsilon \sim U(0,1)$ and then letting $U = U(X_t) = \epsilon f(X_t)$; hence all the $U(X_t)$ share the same random number ϵ.

Given the $U = U(X_t)$ from Step I, we need to implement Step II in such a way that there is a positive (and hopefully large) probability that all X_{t+1} will take the same value regardless of the value X_t. This is achieved by forming a sequence of random variables $\mathbf{W} = \{W_j\}_{j=1,2,\dots}$, where $W_1 \sim U[A(f(x_{\min}))]$ and $W_j \sim U[A(f(W_{j-1}))]$, for any $j \geq 2$. The desired draw X_{t+1} is then the first $W_j \in A(U(X_t)) = A(\epsilon f(X_t))$, that is,

$$X_{t+1} \equiv \phi(X_t, (\epsilon, \mathbf{W})) = W_{\tau(X_t)},$$

where $\tau(x) = \inf\{j : f(W_j) \geq \epsilon f(x)\}$.

In [30] it is proven that, almost surely, only a finite number of the elements of the sequence $\mathbf{W}$ are needed in order to determine $\tau(x)$. The correctness of the algorithm is satisfied if $W_{\tau(x)} \sim U[A(\epsilon f(x))]$, and in [30] this is established by viewing it as a special case of adaptive rejection sampling. Here we provide a simple direct proof. For any given x, denote $A^{(x)} = A(\epsilon f(x))$ and $B_j^{(x)} = \{(W_1, \dots, W_j) : f(W_i) < \epsilon f(x), \ i = 1, \dots, j\}$. Then clearly, for any $k \geq 1$, $\{\tau(x) = k\} = \{W_k \in A^{(x)}\} \cap B_{k-1}^{(x)}$ (assume $B_0^{(x)} = S$ for any $x \in S$). Hence, for any (measurable) set $C \subset A^{(x)}$, we have

$$
\begin{aligned}
P(\{W_{\tau(x)} \in C\} | \tau(x) = k) &= \frac{P(\{W_k \in C \cap A^{(x)}\} \cap B_{k-1}^{(x)})}{P(\{W_k \in A^{(x)}\} \cap B_{k-1}^{(x)})} \\
&= \frac{E\left[E\left(\mathbf{1}\{W_k \in C\}\mathbf{1}\{B_{k-1}^{(x)}\} | W_1, \dots, W_{k-1}\right)\right]}{E\left[E\left(\mathbf{1}\{W_k \in A^{(x)}\}\mathbf{1}\{B_{k-1}^{(x)}\} | W_1, \dots, W_{k-1}\right)\right]} \\
&= \frac{E\left[\mathbf{1}\{B_{k-1}^{(x)}\}P(\{W_k \in C\} | W_{k-1})\right]}{E\left[\mathbf{1}\{B_{k-1}^{(x)}\}P(\{W_k \in A^{(x)}\} | W_{k-1})\right]}.
\end{aligned}
\tag{8.25}
$$

In the above derivation, we have used the fact that $\{W_1, \dots, W_k\}$ forms a Markov chain itself. Given $W_{k-1} = w$, W_k is uniform on $A(f(w))$ by construction, so $P(\{W_k \in B\} | W_{k-1} = w) = \mu(B)/\mu(A(f(w)))$, where μ is the Lebesgue measure. Consequently, the last ratio in Equation 8.25 is exactly $\mu(C)/\mu(A^{(x)})$, the uniform measure on $A^{(x)}$. It follows immediately that $W_{\tau(x)} \sim U(A^{(x)}) = U[A(\epsilon f(x))]$.

To visualize how Steps I and II achieve coupling, Figure 8.6 depicts the update for two paths in the simple case in which f is strictly decreasing with support $(0, x_{\min})$. Suppose that the two chains are currently in X_1 and X_2. Given the ϵ drawn in Step I, the monotonicity of f allows us to write $A(\epsilon f(X_1)) = (0, A_1)$ and $A(\epsilon f(X_2)) = (0, A_2)$. Step II then starts by sampling $W_1 \sim U(0, x_{\min})$ and, since it is not in either of the intervals $(0, A_1)$ or $(0, A_2)$, we follow by sampling uniformly $W_2 \sim U(0, W_1)$ which is the same as sampling $W_2 \sim U[A(f(W_1))]$ since f is decreasing. Because $W_2 \in (0, A_2)$, we have $\tau(X_2) = 2$ so X_2 is updated

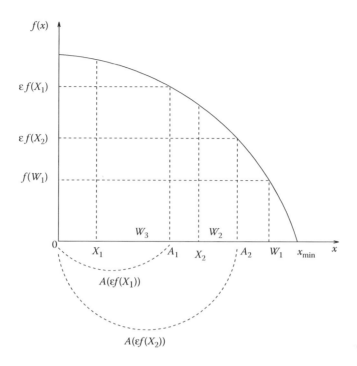

FIGURE 8.6
Illustration of perfect slice sampling.

into W_2. As $W_2 \notin (0, A_1)$ we continue by sampling $W_3 \sim U(0, W_2)$, and since $W_3 \in (0, A_1)$ we can set $\tau(X_1) = 3$. Thus, in the case illustrated by Figure 8.6, the updates are $\phi(X_1, \mathbf{W}) = W_3$ and $\phi(X_2, \mathbf{W}) = W_2$. To understand why this construction creates an opportunity for coupling, imagine that the second uniform draw, W_2, happens to be smaller than A_1. In this case, $\tau(X_1) = \tau(X_2) = 2$ so both X_1 and X_2 are updated into W_2, which means that the two paths have coalesced. In fact, for all $X \in (0, x_{\min})$ with the property that $f(X) \leq f(W_1)/\epsilon$ we have $\phi(X, \mathbf{W}) = W_1$, for all X such that $f(W_1)/\epsilon < f(X) \leq f(W_2)/\epsilon$ we have $\phi(X, \mathbf{W}) = W_2$, and so on. This shows how the continuous set of paths is discretized in only one update.

Figure 8.6 also illustrates that the density ordering $X_2 \prec X_1$ (since $f(X_2) < f(X_1)$) is consistent with the same ordering for the updates: $\phi(X_2, \mathbf{W}) = W_2 \prec \phi(X_1, \mathbf{W}) = W_3$ because $f(W_2) \leq f(W_3)$ by construction. This is true in general because if $X_2 \prec X_1$, that is, $f(X_2) \leq f(X_1)$, then $\tau(X_2) \leq \tau(X_1)$ because $A(\epsilon f(X_1)) \subset A(\epsilon f(X_2))$. Consequently, $W_{\tau(X_2)} \prec W_{\tau(X_1)}$. This property implies that we can implement the monotone CFTP as described in Section 1.3.1, if a maximal $x_{\max}$ exists. In situations in which the extremal states cannot be found, Mira et al. [30] show how to construct bounding processes for this perfect slice sampler.

8.6 Swindles

The term "swindle" has traditionally been used in the Monte Carlo literature to characterize any strategy or modification that either reduces the computational effort or increases the efficiency of the algorithm [12,50]. Usually, swindles are relatively easy-to-implement generic methods applicable to a wide class of algorithms. In the following we describe some of the swindles proposed that either are for or take advantage of perfect sampling algorithms.

8.6.1 Efficient Use of Exact Samples via Concatenation

The algorithms presented in Section 8.2 may be very slow in producing *one draw* from the distribution of interest Π, which is very uneconomical considering that the whole generation process involves a large number of random variables. Motivated by this observation, [40] proposed alternative implementations and estimators that extract more than one draw from each perfect sampling run.

One natural idea is that once a perfect draw is made, say $X_0 \sim \Pi$, then we obviously can run the chain forward for, say, k steps, all of which will be genuine draws from Π. However, this introduces serial correlation in the k samples retained for estimation. Simulations performed in [40] show that a more efficient implementation is the concatenated CFTP (CCFTP). The strategy is illustrated in Figure 8.7, in which two consecutive runs of the monotone CFTP have produced *independent* sample points $X, Y \sim \Pi$. Instead of using just the two sample points, CCFTP uses the *tour* made up of all the realizations lying on the path starting at $-T_Y$ (the time needed to detect coalescence for sampling Y) that connects X to Y, that is, the dashed line in Figure 8.7.

Since the time order of X and Y is irrelevant here (because all the random numbers are i.i.d. along the sequence), one could construct another tour using the path that starts at time $-T_X$ with state Y and ends in X; note that such a path must exist because all tours, regardless of their initial position, must coalesce after T_X iterations by design. Whereas such constructions are not hard to generalize to situations with more than two chains, it is much more straightforward to construct the tours with the Ro-CFTP algorithm. That is, in Figure 8.3, instead of using just X_1, we include in our sample the segment of the coalescence path between X_1 and X_2 (the second sample point not shown in the figure), and then between X_i and X_{i+1} for all i.

Clearly all such tours are independent, but the samples within one tour are serially correlated. If we denote the length of the ith tour by T_i and draws within the tour by X_{ij}, $j = 1, \ldots, T_i$, then obviously a consistent estimator for

$$I_g = \int g(x)\pi(dx) \tag{8.26}$$

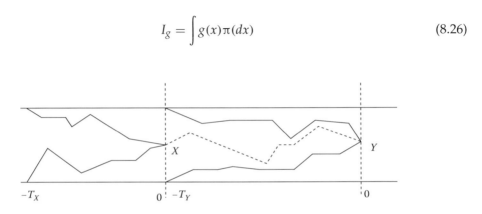

FIGURE 8.7
Illustration of a perfect tour.

that utilizes N tours is [44]

$$\hat{I}_g = \frac{\sum_{i=1}^{N} \sum_{j=1}^{T_i} g(X_{ij})}{\sum_{i=1}^{N} T_i} = \frac{\sum_{i=1}^{N} T_i \bar{g}_i}{\sum_{i=1}^{N} T_i},$$

where $\bar{g}_i$ is the sample average of $g(X_{ij})$ within the ith tour. Note, however, that this is a ratio estimator since the tour lengths are random, but the added benefit here is that since the tours are i.i.d., so are the $\{T_i, \bar{g}_i\}$. Hence, the variance of $\hat{I}_g$ can easily be estimated (when N is large) by the usual variance estimator for ratio estimators based on i.i.d. samples (see [24]), namely

$$\widehat{\mathrm{var}}(\hat{I}_g) = \frac{\sum_{i=1}^{N} T_i^2 (\bar{g}_i - \hat{I}_g)^2}{(\sum_{i=1}^{N} T_i)^2}. \tag{8.27}$$

Note that in Equation 8.27, for the sake of simplicity, we have used denominator N instead of the usual $N - 1$ in defining the cluster sample variance—we can view each tour as a cluster of draws and there are N clusters in total. The beauty of Equation 8.27 is that it entirely avoids the issue of dealing with within-tour dependence, just as in the usual cluster sampling we do not need to be concerned with intra-cluster correlations when we have genuine replications of clusters, which are the tours here.

8.6.2 Multistage Perfect Sampling

Perfect sampling offers the possibility of performing simple random sampling. It is well known in the sampling survey literature that simple random sampling is surpassed in statistical and/or cost efficiency by a number of alternative sampling protocols, for instance multistage sampling. In light of this observation [27] proposed a different approach for running a monotone CFTP algorithm. The *multistage backward coupling algorithm* is designed to perform multistage sampling within the CFTP protocol.

For the sake of simplicity, we describe first the case with two stages. Consider a partition of the sample space into m clusters $\mathcal{S} = \cup_{i=1}^{m} \mathcal{C}_i$. In the first stage we run the CFTP until *cluster coalescence* occurs, that is, all chains merge into a common cluster, say $\mathcal{C}_j$ at time 0. In the second stage, we run CFTP to sample from the conditional distribution $\Pi(\cdot|\mathcal{C}_j)$ defined via $\Pi(A|\mathcal{C}_j) = \Pi(A)/\Pi(\mathcal{C}_j)$ for any measurable $A \subset \mathcal{C}_j$. The two-stage method can easily be extended to multiple stages using a class of nested partitions—for example, each element of the partition used in the second step can in turn be partitioned, $\mathcal{C}_j = \cup_{h=1}^{K_j} \mathcal{C}_{jh}$, and sampling from $\Pi(\cdot|\mathcal{C}_j)$ can be done again in two or more stages, and so on.

The astute reader may have realized that this procedure is valid only if the detection of cluster $\mathcal{C}_j$ in the first stage guarantees that the sample we would eventually have obtained was indeed going to belong to $\mathcal{C}_j$. One way to achieve this is to restrict the proposal to $\mathcal{C}_j$ when using MH algorithm for the second stage. In general, this "foretelling" requirement can be quite stringent when implemented in brute-force ways; more effective methods need to be developed before the multistage sampling methods see general applications. Nevertheless, when the method can be implemented, empirical evidence provided in [27] demonstrates that substantial reductions (e.g. 70%) in running time are possible.

8.6.3 Antithetic Perfect Sampling

Stratified sampling is another efficient method widely used in sample surveys. An implicit way of performing stratification in Monte Carlo simulations can be implemented via *antithetic variates* [15]. Traditionally, antithetic sampling is performed in Monte Carlo using two negatively correlated copies of an unbiased estimator. Suppose that we are interested in estimating the I_g in Equation 8.26. Traditional Monte Carlo uses an i.i.d. sample $\{X_1, \ldots, X_{2n}\}$ from Π and the estimator

$$\hat{I}_{2n} = \frac{\sum_{i=1}^{2n} g(X_i)}{2n}.$$

The variance of $\hat{I}_{2n}$ can be significantly decreased if we are able to sample, for each $1 \leq j \leq n$, $X_j^{(1)}, X_j^{(2)} \sim \Pi$ such that $\text{corr}(g(X_j^{(1)}), g(X_j^{(2)})) < 0$ and use the estimator

$$\tilde{I}_{2n} = \frac{\sum_{i=1}^{n} \left[g(X_i^{(1)}) + g(X_i^{(2)}) \right]}{2n}.$$

In the case where g is monotone, there are relatively simple ways to generate the desired pairs $(X_j^{(1)}, X_j^{(2)})$. Moreover, we have shown in [6] that increasing the number of *simultaneous* negatively correlated samples can bring a significant additional variance reduction. The more complex problem of generating $k \geq 3$ random variables $\{X^{(1)}, \ldots, X^{(k)}\}$ such that any two satisfy $\text{corr}(g(X^{(i)}), g(X^{(j)})) \leq 0$ can be solved, at least for monotone g, using the concept of *negative association* (NA) introduced in [20]. The random variables $\{X_i, i = 1, \ldots, k\}$, where each X_i can be of arbitrary dimension, are said to be negatively associated (NA) if for every pair of disjoint finite subsets A_1, A_2 of $\{1, \ldots, k\}$ and for any nondecreasing functions g_1, g_2,

$$\text{cov}(g_1(X_i, i \in A_1), g_2(X_j, j \in A_2)) \leq 0,$$

whenever the above covariance function is well defined. In light of this stringent condition it is perhaps not surprising that NA is a stronger form of negative dependence which is preserved by concatenation. More precisely, if $\{X_1, \ldots, X_{k_1}\}$ and $\{Y_1, \ldots, Y_{k_2}\}$ are two independent sets of NA random variables, then their union, $\{X_1, \ldots, X_{k_1}, Y_1, \ldots, Y_{k_2}\}$, is also a set of NA random variables. A number of methods used to generate vectors of NA random deviates, especially the very promising iterative Latin hypercube sampling, are discussed in [6].

The implementation of the antithetic principle for CFTP is relatively straightforward. Given a method to generate NA $\{\xi^{(1)}, \ldots, \xi^{(k)}\}$ (where ξ is as needed in Equation 8.1), one can run k CFTP processes in parallel, the jth one using $\{\xi_t^{(j)}, t \leq 0\}$, where $\{\xi_t^{(1)}, \ldots, \xi_t^{(k)}\}$, $t \leq 0$, are i.i.d. copies of $\{\xi^{(1)}, \ldots, \xi^{(k)}\}$, as sketched in Figure 8.8. Within the jth process of CFTP all paths are positively coupled because they use the same $\{\xi_t^{(j)}, t \leq 0\}$. At each update, $\{\xi_t^{(1)}, \ldots, \xi_t^{(k)}\}$ are NA, a property that clearly does not alter the validity of each individual CFTP process.

To obtain $n = km$ draws, we repeat the above procedure *independently* m times, and collect $\{X_i^{(j)}, 1 \leq i \leq m; 1 \leq j \leq k\}$, where i indexes the replication, as our sample $\{X_1, \ldots, X_n\}$. Let

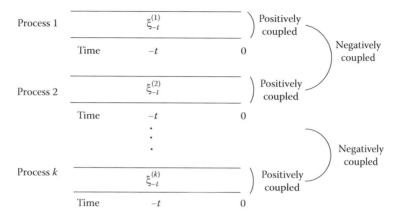

FIGURE 8.8
Parallel antithetic backward CFTP processes.

$\sigma_g^2 = \text{var}[g(X)]$ and $\rho_k^{(g)} = \text{corr}(g(X_1^{(1)}), g(X_1^{(2)}))$. Then

$$\text{var}\left(\frac{1}{n}\sum_{i=1}^{n} g(X_i)\right) = \frac{\sigma_g^2}{n}\left[1 + (k-1)\rho_k^{(g)}\right].$$

Consequently, the *variance reduction factor* (VRF), relative to the independent sampling *with the same simulation size*, is

$$S_k^{(g)} = 1 + (k-1)\rho_k^{(g)}.$$

We emphasize here the dependence of $S_k^{(g)}$ on k, and more importantly on g, and thus the actual gain in reduction can be of practical importance for some g but not for others, but $S_k^{(g)} \leq 1$ as long as $\rho_k^{(g)} \leq 0$.

8.6.4 Integrating Exact and Approximate MCMC Algorithms

It is probably clear by now to the statistician with some travel experience in the MCMC kingdom that perfect sampling may not be the vehicle that one could take on every trip. But it is possible to extend its range considerably if we couple it with more traditional MCMC methods. Here we describe such an approach devised by [33] to deal with Bayesian computation in cases where the sampling density is known only up to a constant that depends on the model parameter, and hence the likelihood function itself cannot be evaluated directly.

More precisely, consider the case in which the target of interest is the posterior density $\pi(\theta|y) \propto p(\theta)p(y\,|\,\theta)$, where $p(\theta)$ is the prior density and $p(y\,|\,\theta)$ is the sampling density of the data. There is a large spectrum of problems (e.g. Markov random fields, image analysis, Markov point processes, Gaussian graphical models, neural networks) for which $p(y\,|\,\theta)$ is known only up to a constant, that is, $p(y\,|\,\theta) = q(y\,|\,\theta)/C_\theta$, with the functional form of q known but the normalizing constant C_θ unknown, in the sense that its value at any particular θ is hard or even impossible to calculate. Obviously, for such problems, the classical MCMC approach cannot be directly implemented. For instance, a Metropolis algorithm with a symmetric proposal, moving from $\theta \rightarrow \theta'$, would require the calculation of the acceptance ratio

$$\alpha(\theta'; \theta) = \min \left\{ 1, \frac{p(\theta')q(y \mid \theta')}{p(\theta)q(y \mid \theta)} \times \frac{C_\theta}{C_{\theta'}} \right\}$$

which involves the unknown ratio of two normalizing constants, $C_\theta / C_{\theta'}$, a problem which occurs in many areas (see, e.g., [11,28,29]).

One obvious way to deal with this problem is to use Monte Carlo or other approximations to estimate each ratio needed in the implementation of Metropolis–Hastings algorithm. A more creative and "exact" solution is proposed by [33] with the help of perfect sampling. The idea is to add into the mix an auxiliary variable x such that the chain updates not only θ but (θ, x) via MH sampling with an acceptance ratio in which no unknown constant appears. Since the auxiliary variable is just a computational artifact, as long as the marginal distribution of θ is preserved there is a lot of freedom in choosing how to update x. In particular, we consider updating (θ, x) via a proposal (θ', x') in which the proposal θ' is generated as in the original chain (it does not depend on x) but $x' \mid \theta', \theta, x \sim q(\cdot \mid \theta')/C_{\theta'}$. Essentially, x' is pseudo-data simulated from the sampling distribution when the parameter is equal to the proposal, θ'. For the new chain, the acceptance ratio is then

$$\tilde{\alpha} = \min \left\{ 1, \frac{p(\theta')q(y \mid \theta')q(x \mid \theta)}{p(\theta)q(y \mid \theta)q(x' \mid \theta')} \right\}, \tag{8.28}$$

which no longer involves any unknown normalizing constant.

The perceptive reader may immediately have realized that the above scheme simply transfers one difficult problem into another, namely, simulating from the original sampling density $p(\cdot \mid \theta') = q(\cdot \mid \theta')/C_{\theta'}$. Since $C_{\theta'}$ is not available, direct methods such as inverse CDF are out of the question (even when they are applicable otherwise). We can of course apply the Metropolis–Hastings algorithm itself for this sampling, which will not require any value of C_θ (since here we sample for x, not θ). But then we would need to introduce a new proposal, and more critically we would need to worry about the convergence of this imbedded Metropolis–Hastings algorithm *within each step* of creating a proposal (θ', x') as called for by Equation 8.28. This is clearly cumbersome and, indeed, entirely defeats the purpose of introducing x' in order to have a "clean" solution to the problem without invoking any approximation (beyond the original Metropolis–Hastings algorithm for θ). This is where the perfect sampling methodologies kick in, because if we have an exact draw from $p(x' \mid \theta')$, then the acceptance ratio given in Equation 8.28 is exactly correct for implementing the Metropolis–Hastings algorithm for drawing (θ, x) and hence for θ. This is particularly fitting, since intractable likelihoods are common in inference for point processes and this is also the area where exact sampling has been most successful. For instance, in [33], the method is illustrated on the well-known Ising model which was proposed as a main application in Propp and Wilson's landmark paper [48], which is a "must" for any tourist of the magic land of perfect sampling.

In closing, we should mention that the method discussed here is only one among a number of promising attempts that have been made to couple the power of traditional MCMC to the precision of perfect sampling such as in [40,43]. See also [42] for related ideas and algorithms.

8.7 Where Are the Applications?

The most resounding successes of perfect sampling have been reported from applications involving finite state spaces, especially in statistical physics (e.g. [13,19,48,49]) and point processes (e.g. [1,14,21,23,26,31,34,35,52]). Other applications include sampling from truncated distributions (e.g. [2,47]), queuing ([41]), Bayesian inference (as in [16,32,37,38]), and mixture of distributions (see [5,17]).

Whereas applications are many, and some are exceedingly successful, much still needs to be done before perfect sampling can be applied routinely. What is gained by perfect sampling is its "perfectness," that is, once it delivers a draw, we are theoretically guaranteed that its distribution is mathematically the same as our desired distribution. The price one pays for this mathematical precision is that any perfect sampling method refuses to produce a draw unless it is absolutely perfect, much like a craftsman reputed for his fixation with perfection refuses to sell a product unless it is 100% flawless. In contrast, any "nonperfect" MCMC method can sell plenty of its "products," but it will either ask the consumers to blindly trust their qualities or leave the consumers to determine their qualities at their own risk. The perfect sampling versus nonperfect sampling is therefore a tradeoff between quality and quantity. As with anything else in life, perhaps the future lies in finding a sensible balance. Perfect quality in small quantity only excites treasure collectors, and lousy quality in abundant quantity only helps garbage collectors. The future of perfect sampling methods lies in how successfully we can strike a balance—producing many quality products at an affordable price in terms of users' implementation cost.

Acknowledgments

We would like to thank Jeff Rosenthal for helpful discussions and three exceptionally thorough anonymous referees for comments and suggestions. RVC has been partially supported by the Natural Sciences and Engineering Research Council of Canada and XLM partially by the National Science Foundation of USA.

References

1. Berthelsen, K. K. and Møller, J. 2002. A primer on perfect simulation for spatial point processes. *Bulletin of the Brazilian Mathematical Society (N.S.)*, 33(3):351–367.
2. Beskos, A. and Roberts, G. O. 2005. One-shot CFTP; application to a class of truncated Gaussian densities. *Methodology and Computing in Applied Probability*, 7(4):407–437.
3. Breyer, L. and Roberts, G. O. 2000. Some multi-step coupling constructions for Markov chains. Technical report, Lancaster University.
4. Casella, G., Lavine, M., and Robert, C. P. 2001. Explaining the perfect sampler. *American Statistician*, 55:299–305.
5. Casella, G., Mengersen, K. L., Robert, C. P., and Titterington, D. M. 2002. Perfect slice samplers for mixtures of distributions. *Journal of the Royal Statistical Society*, Series B, 64:777–790.

6. Craiu, R. V. and Meng, X. L. 2005. Multi-process parallel antithetic coupling for forward and backward MCMC. *Annals of Statistics*, 33(2):661–697.

7. Damien, P., Wakefield, J., and Walker, S. 1999. Gibbs sampling for Bayesian non-conjugate and hierarchical models by using auxiliary variables. *Journal of the Royal Statistical Society, Series B*, 61:331–344.

8. Fill, J. A. 1998. An interruptible algorithm for perfect sampling via Markov chains. *Journal of Applied Probability*, 8:131–162.

9. Fill, J. A., Machida, M., Murdoch, D. J., and Rosenthal, J. S. 1999. Extension of Fill's perfect rejection sampling algorithm to general chains. *Random Structures and Algorithms*, 17(3–4): 290–316.

10. Foss, S. G. and Tweedie, R. L. 1998. Perfect simulation and backward coupling. *Stochastic Models*, 14:187–203.

11. Gelman, A. and Meng, X.-L. 1998. Simulating normalizing constants: from importance sampling to bridge sampling to path sampling. *Statistical Science*, 13(2):163–185.

12. Gentle, J. 1998. *Random Number Generation and Monte Carlo Methods*. Springer, New York.

13. Häggström, O. and Nelander, K. 1998. Exact sampling from anti-monotone systems. *Statistica Neerlandica*, 52:360–380.

14. Häggström, O., van Lieshout, M. N. M., and Møller, J. 1999. Characterization results and Markov chain Monte Carlo algorithms including exact simulation for some spatial point processes. *Bernoulli*, 5(4):641–658.

15. Hammersley, D. C. and Morton, K. V. 1956. A new Monte Carlo technique: antithetic variates. *Mathematical Proceedings of the Cambridge Philosophical Society*, 52:449–475.

16. Haran, M. 2003. Efficient perfect and MCMC sampling methods for Bayesian spatial and components of variance models. PhD thesis, Department of Statistics, University of Minnesota.

17. Hobert, J. P., Robert, C. P., and Titterington, D. M. 1999. On perfect simulation for some mixtures of distributions. *Statistics and Computing*, 9:287–298.

18. Huber, M. 2004. Perfect sampling using bounding chains. *Annals of Applied Probability*, 14(2):734–753.

19. Huber, M. L. 2002. A bounding chain for Swendsen-Wang. *Random Structures and Algorithms*, 22:43–59.

20. Joag-Dev, K. and Proschan, F. 1983. Negative association of random variables with applications. *Annals of Statistics*, 11:286–295.

21. Kendall, W. S. 1998. Perfect simulation for the area-interaction point process. In L. Accardi and C. C. Heyde (eds), *Probability Towards 2000*, pp. 218–234. Springer, New York.

22. Kendall, W. S. 2004. Geometric ergodicity and perfect simulation. *Electronic Communications in Probability*, 9:140–151.

23. Kendall, W. S. and Møller, J. 2000. Perfect simulation using dominating processes on ordered spaces, with application to locally stable point processes. *Advances in Applied Probability*, 32(3):844–865.

24. Kish, L. 1965. *Survey Sampling*. Wiley, New York, 1965.

25. Lindvall, T. 1992. *Lectures on the Coupling Method*. Wiley, New York.

26. McKeague, I. W. and Loizeaux, M. 2002. Perfect sampling for point process cluster modelling. In A. B. Lawson and D. G. T. Denison (eds), *Spatial Cluster Modelling*, pp. 87–107. Chapman & Hall/CRC, Boca Raton, FL.

27. Meng, X.-L. 2000. Towards a more general Propp-Wilson algorithm: Multistage backward coupling. In N. Madras (ed.), *Monte Carlo Methods*, pp. 85–93. American Mathematical Society, Providence, RI.

28. Meng, X.-L. and Shilling, S. 2002. Warp bridge sampling. *Journal of Computational and Graphical Statistics*, 11(3):552–586.

29. Meng, X.-L. and Wong, W. 1996. Simulating ratios of normalizing constants via a simple identity: A theoretical exploration. *Statistica Sinica*, 6:831–860.

30. Mira, A., Møller, J., and Roberts, G. O. 2001. Perfect slice samplers. *Journal of the Royal Statistical Society, Series B*, 63(3):593–606.

31. Møller, J. 1999. Markov chain Monte Carlo and spatial point processes. In O. E. Barndorff-Nielsen, W. S. Kendall, and M. N. M. van Lieshout (eds), *Stochastic Geometry: Likelihood and Computations*, pp. 141–172. Chapman & Hall/CRC, Boca Raton, FL.

32. Møller, J. and Nicholls, G. K. 1999. Perfect simulation for sample-based inference. Preprint, Department of Mathematical Statistics, Chalmers Institute of Technology.

33. Møller, J., Pettitt, A. N., Reeves, R., and Berthelsen, K. K. 2006. An efficient Markov chain Monte Carlo method for distributions with intractable normalising constants. *Biometrika*, 93(2):451–458.

34. Møller, J. and Rasmussen, J. G. 2005. Perfect simulation of Hawkes processes. *Advances in Applied Probability*, 37(3):629–646.

35. Møller, J. and Waagepetersen, R. P. 2004. *Statistical Inference and Simulation for Spatial Point Processes*. Chapman & Hall/CRC, Boca Raton, FL.

36. Murdoch, D. 2000. Exact sampling for Bayesian inference: Unbounded state spaces. In N. Madras (ed.), *Monte Carlo Methods*, pp. 111–121. American Mathematical Society, Providence, RI.

37. Murdoch, D. J. and Green, P. J. 1998. Exact sampling from a continuous state space. *Scand. J. Stat.*, 25:483–502.

38. Murdoch, D. J. and Meng, X.-L. 2001. Towards perfect sampling for Bayesian mixture priors. In E. George (ed.), *Proceedings of ISBA 2000*, pp. 381–390. Eurostat.

39. Murdoch, D. J. and Rosenthal, J. S. 1998. An extension of Fill's exact sampling algorithm to non-monotone chains. Technical report, University of Toronto.

40. Murdoch, D. J. and Rosenthal, J. S. 2000. Efficient use of exact samples. *Statistics and Computing*, 10:237–243.

41. Murdoch, D. J. and Takahara. G. 2006. Perfect sampling for queues and network models. *ACM Transactions on Modeling and Computer Simulation*, 16:76–92.

42. Murray, I. 2005. Advances in Markov chain Monte Carlo methods. PhD thesis, University College London.

43. Murray, I., Ghahramani, Z., and MacKay, D. 2006. MCMC for doubly-intractable distributions. In *Proceedings of the 22nd Annual Conference on Uncertainty in Artificial Intelligence (UAI)*, pp. 359–366. AUAI Press, Corvallis, OR.

44. Mykland, P., Tierney, L., and Yu, B. 1995. Regeneration in Markov chain samplers. *Journal of the American Statistical Association*, 25:483–502.

45. Neal, R. M. 2003. Slice sampling (with discussion). *Annals of Statistics*, 31(3):705–767.

46. Nummelin, E. 1978. A splitting technique for Harris recurrent Markov chains. *Zeitschrift für Wahrscheinlichkeitstheorie und Verwandte Gebiete*, 43(4):309–318.

47. Philippe, A. and Robert, C. 2003. Perfect simulation of positive Gaussian distributions. *Statistics and Computing*, 13:179–186.

48. Propp, J. G. and Wilson, D. B. 1996. Exact sampling with coupled Markov chains and applications to statistical mechanics. *Random Structures and Algorithms*, 9(1–2):223–252.

49. Servidea, J. D. and Meng, X.-L. 2006. Statistical physics and statistical computing: A critical link. In J. Fan and H. Koul (eds), *Frontiers in Statistics: Dedicated to Peter John Bickel in Honor of His 65th Birthday*, pp. 327–344. Imperial College Press, London.

50. Simon, G. 1976. Computer simulation swindles, with applications to estimates of location and dispersion. *Applied Statistics*, 25:266–274.

51. Tanner, M. A. 1996. *Tools for Statistical Inference : Methods for the Exploration of Posterior Distributions and Likelihood Functions*, 3rd edn. Springer, New York.

52. Thönnes, E. 1999. Perfect simulation of some point processes for the impatient user. *Advances in Applied Probability*, 31(1):69–87.

53. Thönnes, E. 2000. A primer on perfect simulation. In K. Mecke and D. Stoyan (eds), *Statistical Physics and Spatial Statistics. The Art of Analyzing and Modeling Spatial Structures and Pattern Formation*, Lecture Notes in Physics 554, pp. 349–378. Springer, Berlin.

54. Thorisson, H. 2000. *Coupling, Stationarity, and Regeneration*. Springer, New York.

55. van Dyk, D. and Meng, X.-L. 2001. The art of data augmentation (with discussion). *Journal of Computational and Graphical Statistics*, 10:1–111.

56. Wilson, D. B. 2000a. How to couple from the past using a read-once source of randomness. *Random Structures Algorithms*, 16(1):85–113.
57. Wilson, D. B. 2000b. Layered multishift coupling for use in perfect sampling algorithms (with a primer on CFTP). In N. Madras (ed.), *Monte Carlo Methods*, pp. 141–176. American Mathematical Society, Providence, RI.

9

Spatial Point Processes

Mark Huber

9.1 Introduction

Spatial point processes arise naturally in many contexts, including population studies, forestry, epidemiology, agriculture, and material science; for more examples, see Ripley (1977), Stoyan and Stoyan (1995), and Møller and Waagepetersen (2007). Typically, statistical models for these data sets are given by densities with respect to a Poisson point process. In Section 9.2 these Poisson point processes and densities are described in detail, together with several examples. As in many applications, these densities are often unnormalized, and calculating the normalizing constant exactly is computationally unfeasible. Therefore Monte Carlo methods are used instead.

Many of these methods involve construction of a Markov chain whose stationary distribution matches the target density. There are two primary types of chains used for these point processes. In Section 9.3, the Metropolis–Hastings and reversible jump (Green, 1995) methods are described. Section 9.4 shows how to build continuous-time spatial birth and death chains for these problems.

Next, in Section 9.5 perfect sampling techniques are introduced. These methods draw samples exactly drawn from the target distribution. Acceptance/rejection methods can be used for small problems, while larger problems require methods such as Kendall and Møller's (2000) dominated coupling from the past.

Sections 9.2 through 9.5 develop techniques for sampling from the statistical model. When the interest is in sampling from the posterior, Section 9.6 goes further and shows how these methods can be modified in order to accomplish this task and carry out Bayesian inference.

Finally, Section 9.7 examines what is known about the running time of these methods, and strategies for improving the convergence of these chains.

9.2 Setup

The models considered here are described by using densities with respect to a Poisson point process. Consider a space S that is separable (so it has a countable dense subset) equipped with a set of measurable sets $\mathcal{B}$, and intensity measure λ satisfying $\lambda(S) < \infty$. Usually this intensity is proportional to Lebesgue measure; throughout this chapter we use $m(A)$ to denote the Lebesgue measure of a set A. Examples of point processes include the following:

- Spatial data. $S \subseteq \mathbb{R}^2$, $\lambda(A)$ is proportional to $m(A)$ and $\mathcal{B}$ is the usual Borel σ-algebra.
- Marked spatial data. For instance $S \subseteq \mathbb{R}^2 \times [0, \infty)$, where the $\mathbb{R}^2$ coordinate is the location of the point and the $[0, \infty)$ coordinate is the radius of a disk centered at the point. The intensity $\lambda(\cdot)$ is the cross product of $m(\cdot)$ on $\mathbb{R}^2$ and an arbitrary measure on $[0, \infty)$. For instance, Fiksel (1984) studied data of Klier on Norwegian spruces together with their trunk diameters.
- Typed spatial processes. When the mark is a finite set, it can be used to represent the type of point. Here $S \subseteq \mathbb{R}^2 \times \{1, 2, \ldots, k\}$, with intensity the cross product of $m(\cdot)$ with a measure on $\{1, 2, \ldots, k\}$. The value of the second coordinate determines the type of point. For instance, Harkness and Isham (1983) analyzed locations of ants' nests where there were two different possible types of nests.

Configurations in a Poisson point process over a space S are vectors of random length (including length 0) whose components are points, that is, elements of S. For example, if $S = [-10, 10]^2$, then $((2.78, 3.42), (1.23, -3.21))$ and $\emptyset$ are valid configurations of points. The process is governed by λ, the intensity measure on S. If X is a Poisson point process and A a measurable region, then the expected number of components of X (the points) that lie in A must equal $\lambda(A)$.

Formally, a Poisson point process can be viewed as a measure on these configurations that can be defined using an *exponential space*; see Carter and Prenter (1972) for details. A Poisson point process can also be viewed as the distribution of the output of the following procedure.

ALGORITHM 9.1   POISSON POINT PROCESS GENERATOR

Input: space S, intensity measure $\lambda(\cdot)$ with $\lambda(S) < \infty$
Output: X
1: **draw** $N \leftarrow \text{Poisson}(\lambda(S))$
2: **draw** $X_1, \ldots, X_N$ i.i.d. from $\lambda(\cdot)$
3: $X \leftarrow (X_1, \ldots, X_N)$

In line 1, N (the number of points in the configuration) is a draw from a discrete Poisson random variable with parameter $\lambda(S)$, so $P(N = i) = [\lambda(S)^i/i!] \exp(-\lambda(S))$. In line 2, each point is distributed according to the normalized intensity, so $P(X_1 \in A) = \lambda(A)/\lambda(S)$.

Note that if S is discrete or λ assigns positive measure to any point of S, then it is possible that $X_i = X_j$ for some $i < j \leq N$. That is, it is possible to have repeated points in the configuration. On the other hand, when S is a continuous space and λ is atomless, the probability of repeated points becomes 0.

Since the order of the points is immaterial, this means that the configurations can be treated as a set rather than a vector. If $x = \{x_1, \ldots, x_{\#x}\}$ is a set containing $\#x$ distinct points, then there are exactly $(\#x)!$ different vectors that give rise to the set. Therefore, if μ is the distribution of the output of this algorithm on sets of distinct points,

$$\mu(dx) = (\#x)! \frac{\lambda(S)^{\#x}}{(\#x)!} \exp(-\lambda(S)) \prod_{i=1}^{\#x} \frac{\lambda(dx_i)}{\lambda(S)} = \exp(-\lambda(S)) \prod_{i=1}^{\#x} \lambda(dx_i). \tag{9.1}$$

While it is easy to sample from the basic Poisson point process, it is far more difficult to sample from models where the target distribution is given by an unnormalized density with respect to a Poisson point process. A random variable X has density f with respect to measure μ if $P(X \in A) = \int_A f(x)\, d\mu(x)$. Typically densities either reward or penalize point processes by attaching multiplicative factors greater or less than 1.

In fact, for most models these densities are of the form

$$f(x) = \frac{g(x)}{Z}, \quad \text{where } g(x) = \alpha(\#x)h(x), \quad Z = \int g(x)\, d\mu(x),$$

and $\#x$ is the number of points in configuration x. Here the function $\alpha(n)$ controls the number of points, while $h(x)$ is a measurable function that describes the interaction structure between points of the process. Z is the normalizing constant, and usually no direct method is known for even approximating Z, let alone calculating it exactly.

Example 9.1

The *Strauss process* (Strauss, 1975) is a repulsive model that surrounds each point with a disk of radius R, and then penalizes configurations for each pair of points whose disks overlap. This process has three parameters: λ is the nonnegative activity of points, γ is the interaction parameter in $[0, 1]$, and R is the radius of the disks. Typically, $S = \mathbb{R}^2$ and the density is with respect to μ where the intensity is $m(\cdot)$:

$$g(x) = \lambda^{\#x}\gamma^{s(x)}, \quad s(x) = \#\{\{i, j\} : \text{dist}(x_i, x_j) < 2R\}, \tag{9.2}$$

where x consists of the points $x_1, \ldots, x_{\#x}$. In order to normalize g to be a probability density, set $Z(\lambda, \gamma, R) = \int g(s)\, d\mu(s)$ and $f(x) = g(x)/Z(\lambda, \gamma, R)$.

As $\gamma \to 0$, it becomes more and more unlikely to find overlapping disks. The limit of this process is called a *hard core model*, since each disk becomes a hard that cannot overlap. When $\gamma > 0$, the process can be called a soft core model, as it is possible for the cores surrounding the disks to interpenetrate. The Strauss process was generalized to pair-repulsion processes in Kelly and Ripley (1976), where the penalty factor is allowed to be a general function, rather than a constant γ.

A weakness of the Strauss process is that it can only be used to model repulsion, since if $\gamma > 1$ then the density cannot be normalized. The next example, the area interaction process, solves this problem and allows for both repulsion and attraction as one of its parameters varies from 0 to infinity.

Example 9.2

The *area interaction process* (Baddeley and van Lieshout, 1995; Widom and Rowlinson, 1970) has two parameters, λ and γ, each in $[0, \infty)$. Each point x_i is surrounded by a region A_i called the grain of x_i. The unnormalized density is

$$g(x) = \lambda^{\#x}\gamma^{-m(\cup A_i)}. \tag{9.3}$$

When $\gamma > 1$, points tend to clump closer together, making the model attractive; and when $\gamma < 1$, the points are forced farther apart, making the model repulsive. The model is written above using Lebesgue measure, but in fact is very general and can be used with any measure on the space.

9.3 Metropolis–Hastings Reversible Jump Chains

Typical discrete-time Markov chains for approximately sampling from g operate by using three types of moves: (1) shifting a point in the configuration, (2) adding a point to the configuration, (3) removing a point from the configuration. These algorithms were first described by Geyer and Møller (1994), and can be considered a special case of the reversible jump procedure of Green (1995). Reversible jump is an extension of Metropolis–Hastings chains (Hastings, 1970; Metropolis et al., 1953) for problems where the dimension of the space is not fixed.

Metropolis–Hastings chains are constructed to satisfy the reversibility condition. A Markov chain with transition kernel $\mathbf{K}(x, \cdot)$ is reversible with respect to π if

$$\pi(dx)\mathbf{K}(x, dy) = \pi(dy)\mathbf{K}(y, dx).$$

The purpose of reversibility in the design of Markov chains is that if $\mathbf{K}$ is reversible with respect to π, then π is a stationary distribution of the Markov chain.

To reach the goal of generating variates from a point process with distribution π, first a Markov chain must be constructed with π as its stationary distribution. Suppose that π has density $g(x)/Z$ with respect to a Poisson point process with intensity $\lambda(\cdot)$ (so that the underlying Poisson point process has measure $\mu(\cdot)$ as in Equation 9.1). In other words,

$$\pi(dx) = \frac{g(x)}{Z} \cdot \mu(dx). \tag{9.4}$$

For a configuration x and points $v \in S$, let $x + v$ denote the configuration that contains all the points in x and v. For $v \in x$, let $x - v$ denote the configuration x after removing the point v.

Metropolis–Hastings chains begin with a proposal kernel $\mathbf{K}'(x, \cdot)$, where $\pi(dy)\mathbf{K}'(y, dx)$ is absolutely continuous with respect to $\pi(dx)\mathbf{K}'(x, dy)$. Given the current state x, $\mathbf{K}'$ is used to generate a state y which is the proposed next state of the chain. Given x and y, the Metropolis–Hastings ratio is

$$r(x,y) = \frac{\pi(dy)\mathbf{K}'(y, dx)}{\pi(dx)\mathbf{K}'(x, dy)} = \frac{g(y)\mu(dy)\mathbf{K}'(y, dx)}{g(x)\mu(dx)\mathbf{K}'(x, dy)}. \tag{9.5}$$

Formally this is a Radon–Nikodym derivative, which is the reason for the absolute continuity requirement above. With probability $\min\{1, r(x, y)\}$, the move from x to y is accepted, and y becomes the next state in the chain. Otherwise the next state is the same as the current state. Note that the normalizing constant Z in Equation 9.4 cancels out in the ratio (Equation 9.5).

Shifting When $y = x + v - w$ for some $v, w \in S$, suppose that the proposal move shifts point w to point v. To find the ratio in Equation 9.5, recall that $\mu(dx) = \exp(-\lambda(S)) \prod_{i=1}^{\#x} \lambda(dx_i)$. The $\exp(-\lambda(S))$ factor cancels out. Since the point w is shifted to v in moving from x to y, $\mu(dy)$ contains a factor of $\lambda(dv)$ that $\mu(dx)$ does not. In the other direction $\mu(dx)$ contains a factor of $\lambda(dw)$ that $\mu(dy)$ does not. All the other factors cancel, and so Equation 9.5 becomes

$$r(x, y) = \frac{g(x + v - w)}{g(x)} \cdot \frac{\mathbf{K}'(x + v - w, dx)}{\mathbf{K}'(x, d(x + v - w))} \cdot \frac{\lambda(dv)}{\lambda(dw)}.$$

Birth   When $y = x + v$, suppose that point v is born. In this case $\#y = \#x + 1$, and so $\mu(dy)$ contains a factor of $\lambda(dv)$ that $\mu(dx)$ lacks. So the Metropolis–Hastings ratio becomes

$$r(x, y) = \frac{g(x + v)}{g(x)} \cdot \frac{\mathbf{K}'(x + v, dx)}{\mathbf{K}'(x, d(x + v))} \lambda(dv).$$

Death   When $y = x - w$, suppose that point w died. In this case $\#y = \#x - 1$, and $\mu(dx)$ contains a factor of $\lambda(dw)$ that $\mu(dy)$ lacks. The Metropolis–Hastings ratio is

$$r(x, y) = \frac{g(x - w)}{g(x)} \cdot \frac{\mathbf{K}'(x - w, dx)}{\mathbf{K}'(x, d(x - w))} \cdot \frac{1}{\lambda(dw)}.$$

One choice that simplifies these ratios is to use $P(\text{shift}) = P(\text{birth}) = P(\text{death}) = \frac{1}{3}$. For shifts or births, v is drawn from the normalized intensity measure $\lambda(\cdot)/\lambda(S)$, and for shifts or deaths, w is drawn uniformly from the set of points in x. This makes the shifting ratio

$$\frac{g(x + v - w)}{g(x)} \cdot \frac{\frac{1}{3}(\lambda(dw)/\lambda(S))(1/\#x)}{\frac{1}{3}(\lambda(dv)/\lambda(S))(1/\#x)} \cdot \frac{\lambda(dv)}{\lambda(dw)} = \frac{g(x + v - w)}{g(x)}.$$

Similarly, the birth ratio becomes

$$\frac{g(x + v)}{g(x)} \cdot \frac{\frac{1}{3}(1/(\#x + 1))}{\frac{1}{3}\lambda(dv)/\lambda(S)} \cdot \lambda(dv) = \frac{g(x + v)}{g(x)} \cdot \frac{\lambda(S)}{\#x + 1}.$$

The death ratio becomes

$$\frac{g(x - w)}{g(x)} \cdot \frac{\frac{1}{3}\lambda(dw)/\lambda(S)}{\frac{1}{3}(1/\#x)} \cdot \frac{1}{\lambda(dw)} = \frac{g(x - w)}{g(x)} \cdot \frac{\#x}{\lambda(S)}.$$

Therefore the following pseudocode takes one step in this chain:

ALGORITHM 9.2   METROPOLIS–HASTINGS STEP IN SHIFT–BIRTH–DEATH CHAIN

Input: current state x,
Output: next state x'
```
 1: draw type ← Unif({shift,birth,death}),  U ← Unif([0, 1]),  v ← λ(·),
       w ← Unif(x)
 2: if type = shift and U < g(x + v − w)/g(x) then
 3:     x' ← x + v − w
 4: else if type = birth and U < (g(x + v)/g(x))(λ(S)/[#x + 1]) then
 5:     x' ← x + v
 6: else if type = death and U < (g(x − w)/g(x))(#x/λ(S)) then
 7:     x' ← x − w
 8: else
 9:     x' ← x
10: end if
```

Calculation of $g(y)/g(x)$ can be the most difficult part in coding these algorithms. Fortunately, it is often the case that many factors in the ratio cancel.

9.3.1 Examples

Consider Example 9.1. In the Strauss process density (Equation 9.2), adding (or removing) a point changes the number of factors of λ by 1, and adds (or removes) a number of factors of γ equal to the number of other points within distance $2R$ of the changing point. For $u \in S$, let $n(x, u) = \#\{j : \mathrm{dist}(x_j, u) < 2R\}$, so

$$\frac{g(x+v)}{g(x)} = \lambda \gamma^{n(x,v)}, \quad \frac{g(x-w)}{g(x)} = \lambda^{-1} \gamma^{-n(x,w)}, \quad \frac{g(x+v-w)}{g(x)} = \gamma^{n(x,v)-n(x,w)}.$$

In Example 9.2 (the area interaction process), it is easy to write a formula for $g(x + v)/g(x)$, but difficult to compute in practice. Let $a(x, v)$ denote the area of the region that is inside the grain for v, but which is not inside the grain for any other point. Then

$$\frac{g(x+v)}{g(x)} = \lambda \gamma^{-a(x,v)}, \quad \frac{g(x-w)}{g(x)} = \lambda^{-1} \gamma^{a(x,v)}, \quad \frac{g(x+v-w)}{g(x)} = \gamma^{-a(x,v)+a(x+v,w)}.$$

So the ratio for a birth is $(g(x + v)/g(x))(\lambda(S)/(\#x + 1)) = (\lambda(S)/(\#x + 1))\lambda \gamma^{-a(x,v)}$. Even when the grain is a simple disk, calculation of $a(x, v)$ can be time-consuming, which in turn makes finding the Metropolis–Hastings ratio difficult. This problem is addressed in Section 9.4, where continuous-time Markov chains are considered.

9.3.2 Convergence

Let $\mathbf{K}^t(x, \cdot)$ be the distribution of X_t given that the starting state of the chain is x. The Metropolis–Hastings methodology gives a means for building a chain with an invariant π so that

$$\pi(A) = \int \mathbf{K}(x, A)\, \pi(dx).$$

For Monte Carlo purposes the goal is to have the limiting distribution match the stationary distribution:

$$\lim_{t \to \infty} \|\mathbf{K}^t(x, \cdot) - \pi(\cdot)\|_{TV} = 0,$$

where $\|v\|_{TV}$ is the total variation norm of a signed measure given by

$$\|v\|_{TV} = \sup_A |v(A)|.$$

The methods above build a Markov chain whose stationary distribution matches the target distribution given by g, but that is not a guarantee that the limiting distribution matches the stationary distribution.

The following definitions come from Section 5.6 of Durrett (2005). Suppose that a Markov chain is a *Harris chain* if there exist measurable sets A, B, real $\epsilon > 0$, and a probability measure ρ with $\rho(B) = 1$ where two properties hold. First, if $\tau_A := \inf\{n \geq 0 : X_n \in A\}$, then $P(\tau_A < \infty | X_0 = x) > 0$ for all x. Second, if $x \in A$ and C is a measurable subset of B then $\mathbf{K}(x, C) \geq \epsilon\rho(C)$. That is, from any starting state, there is positive probability of getting to A, and from any state in A, there is a positive chance ϵ that a simulator can ignore the value of the current state in deciding the location of the next state.

Furthermore, if $(\forall x \in A)(P(\tau_A < \infty | X_0 = x) = 1)$ (so that the probability of returning to A from any starting point in A is 1) then the chain will be *recurrent*. For a recurrent chain, for all $x \in A$, the greatest common divisor of $\{t : \mathbf{K}^t(x, A) > 0\}$ will be the same. Call this common value the *period* of A, and, if the period is 1, call the chain *aperiodic*. Then the following theorem gives sufficient conditions for the stationary distribution to be the limiting distribution (see Durrett, 2005):

Theorem 9.1

Let X_n be an aperiodic recurrent Harris chain with stationary distribution π. If $P(\tau_A < \infty | X_0 = x) = 1$, then as $t \to \infty$,

$$||\mathbf{K}^t(x, \cdot) - \pi(\cdot)|| \to 0.$$

For chains on the space of point processes, using $A = B = \emptyset$ is usually a valid choice for showing that a Harris chain is recurrent. A sufficient condition for the validity of this choice is local stability:

Definition 9.1

A density $g(x)$ is defined to be *locally stable* if there exists a constant K such that $g(x + v)/g(x) \le K$ for all x and v.

Note that both the Strauss process and the area interaction process discussed earlier are locally stable. In addition, most Markov point processes such as the saturated and triplets process (Geyer, 1999) or nearest-neighbor processes (Baddeley and Møller, 1989) are locally stable as well. The Metropolis–Hastings step given above will be Harris recurrent for any locally stable Markov chain; see Rosenthal (1995) for examples of this type of analysis.

Suppose that a chain is Harris recurrent, and that when the chain is in the empty set, there is a positive chance that a death is proposed. Since the chain stays in the same configuration when this happens, the chain also becomes aperiodic, thereby satisfying the conditions of Theorem 1.

9.4 Continuous-Time Spatial Birth–Death Chains

A Metropolis–Hastings chain stays at the current state for a number of steps before jumping to a new state. The time until the first jump is a geometric random variable. The continuous-time analog is the exponential distribution, and so continuous-time Markov chains operate by staying at the current state an exponential amount of time, and then jumping to a new state. These chains are also known as *jump processes* (see Feller, 1966, Chapter X.3).

As with discrete-time chains, reversibility is the key to designing continuous-time chains with the target distribution as their limiting distribution. Preston (1977) solved this problem by introducing jump process where, for a configuration x, the rate of births is controlled by a rate function $b(x, v)$, and the rate of deaths is controlled by a death function $d(x, w)$, where $d(x, w) > 0$ if $w \in x$.

Reversibility holds with respect to g if the rate of births balances the rate of deaths for all configurations x and points v:

$$g(x)b(x, v) = g(x + v)d(x + v, v). \tag{9.6}$$

The chain is updated as follows. For the total rate of births and of deaths we have

$$r_b(x) = \int b(x, v)\, \lambda(dv), \quad r_d(x) = \sum_{w \in x} d(x, w).$$

The time until the next birth is an exponential random variable with rate $r_b(x)$. Similarly, the time until the next death is exponential with rate $r_d(x)$. If a birth occurs, a point v is chosen according to $\lambda(\cdot)$. If a death occurs, a point $w \in x$ is chosen to be removed with probability $d(x, w)/r_d$.

ALGORITHM 9.3   CONTINUOUS-TIME BIRTH–DEATH CHAIN

Input: current time t, current state x
Output: new time t', new state x'
1: **draw** $t_b \leftarrow \mathrm{Exp}(\int b(x, v)\lambda(dv))$, $t_d \leftarrow \mathrm{Exp}(\sum_{w \in x} d(x, w))$
2: **if** $t_b < t_d$ (so a new point is born) **then**
3:     **draw** $v \leftarrow \lambda(\cdot)$
4:     $x' \leftarrow x + v$, $t' \leftarrow t + t_b$
5: **else**
6:     **draw** $w \leftarrow d(x, \cdot)$
7:     $x' \leftarrow x - w$, $t' \leftarrow t + t_d$
8: **end if**

To create birth and death rates that satisfy reversibility (Equation 9.6), a technique similar to Metropolis–Hastings is used. When a birth occurs, only accept the birth with a ratio $r(x, v)$. This thins the birth rate from $b(x, v)$ to $b'(x, v) = b(x, v)r(x, v)$. This procedure is easiest when the density is locally stable with constant K, so that $g(x + v)/g(x) \le K$ for all x and v.

To create a spatial birth–death jump process with a locally stable g as its target density, set $d(x, w) = 1$ and $b(x, v) = K$ for all configurations x and all points v and w. Hence $r_b = K\lambda(S)$ and $r_d = \#x$. Once a birth occurs, it is accepted with probability $r(x, v) = (1/K)(g(x + v)/g(x))$. Hence,

$$g(x)b'(x, v) = g(x)b(x, v)r(x, v) = g(x)(K/K)(g(x + v)/g(x)) = d(x + v, v)g(x + v),$$

and reversibility holds. The pseudocode is as follows:

ALGORITHM 9.4   PRESTON SPATIAL BIRTH–DEATH CHAIN

Input: current time t, current state x
Output: new time t', new state x'
1: **draw** $t_b \leftarrow \mathrm{Exp}(K\lambda(S))$, $t_d \leftarrow \mathrm{Exp}(\#x)$
2: **if** $t_b < t_d$ (a new point might be born) **then**
3:     **draw** $v \leftarrow \lambda(\cdot)$, $U \leftarrow \mathrm{Unif}([0, 1])$
4:     $x' \leftarrow x$, $t' \leftarrow t + t_b$
5:     **if** $U < (g(x + v)/g(x))(1/K)$ **then**
6:         $x' \leftarrow x' + v$
7:     **end if**
8: **else**
9:     **draw** $w \leftarrow \mathrm{Unif}(x)$
10:     $x' \leftarrow x - w$, $t' \leftarrow t + t_d$
11: **end if**

An exponential with rate α can be drawn by taking $-(1/\alpha)\ln U$, where U is a uniform random variable on $[0,1]$. Therefore the work needed in this algorithm is similar to that needed to take a discrete time step.

9.4.1 Examples

Consider Example 9.1, the Strauss point process. Here K is λ. For a point v that is born into a configuration x, $g(x+v)/g(x)$ is just $\lambda\gamma^{n(x,v)}$, where $n(x,v) = \#\{j : \text{dist}(x_j,v) < 2R\}$ as before. So lines 5 and 6 above become:

```
5:      if  U < γⁿ⁽ˣ, ᵛ⁾  then
6:          x′ ← x + v
```

For Example 9.2, the area interaction process, it was difficult to create a Metropolis chain because of the difficulty of calculating the acceptance ratio. As before, let $a(x,v)$ denote the area of the grain of v not already in a grain of a point of x and $m(\cdot)$ be the measure used to determine areas in the process. Then the Metropolis ratio for accepting a birth is $r_b = (m(S)/(\#x+1))\lambda\gamma^{-a(x,v)}$, which can be difficult to compute.

In contrast, the ratio for accepting a birth in the continuous-time chain is simpler. For a configuration $\{x_1, \ldots, x_{\#x}\}$ with grains $\{A_1, \ldots, A_{\#x}\}$, let A be the union of these grains (so $A = \cup_{w \in x} A_w$). Suppose that $\gamma > 1$, so $K = \lambda$ applies as the constant of local stability. Then $(g(x+v)/g(v))(1/K) = \gamma^{-a(x,v)}$. While this value is no easier to compute than it was before, a Bernoulli random variable with this parameter can be generated as follows. First, generate a Poisson point process with intensity measure $(\ln\gamma)m(\cdot)$ over A_v. Remove any points from this process that lie in A. This is called *thinning* the process.

The result of thinning is a Poisson point process over the region $A_v \cap A^C$ with intensity $(\ln\gamma)m(\cdot)$ restricted to $A_v \cap A^C$, since the expected number of points inside any subset of $A_v \cap A^C$ will be correct because it came from the Poisson point process over the larger region. Since it is a Poisson point process, the number of points in the region will have the Poisson distribution with parameter $(\ln\gamma)m(A_v \cap A^C)$. Hence the probability the process contains zero points will be exactly $\exp(-(\ln\gamma)m(A_v \cap A^C)) = \gamma^{-a(x,v)}$. So if no points remain after thinning, accept the birth of the grain, otherwise reject. In pseudocode:

```
6a:     draw Z ← as a Poisson point process on Aᵥ with intensity
            (ln γ)m(·)
6b:     if  Z ∩ Aᶜ = ∅  then
6c:         x′ ← x + v
```

Let a be the measure of a grain surrounding v. When $\gamma < 1$, adding the point v changes the density by a factor of at most $\lambda\gamma^{-a}$. Therefore the probability of accepting a birth becomes $[g(x+v)/g(x)][1/(\lambda\gamma^{-a})] = \lambda\gamma^{-a(x,v)}/(\lambda\gamma^{-a}) = (\gamma^{-1})^{-(a-a(x,v))}$. A Bernoulli with this probability can be found by first generating a Poisson point process with intensity $\ln(\gamma^{-1}) = -\ln\gamma$ inside the grain of v, and then accepting if none of the points lie in the area of the grain that is already covered by grains of points in x. Again thinning can be used to verify that the probabilities are correct:

```
6a:     draw Z ← as a Poisson point process on Aᵥ with intensity
            −(ln γ)m(·)
6b:     if  Z ∩ A = ∅  then
6c:         x′ ← x + v
```

9.4.2 Shifting Moves with Spatial Birth and Death Chains

Preston (1977) only included provisions for births and deaths in his examples, but his method can be extended to allow for shifting moves as well. Suppose that point v is added and point w is removed from the chain at rate $s(x, v, w)$. Reversibility for the jump process requires

$$g(x)s(x, v, w) = g(x + v - w)s(x + v - w, w, v).$$

Note that the shifting move rate can be multiplied by any constant without disturbing reversibility. This allows the user to make shifting moves more or less prevalent compared to births and deaths as needed to make the chain run more quickly.

9.4.3 Convergence

Kaspi and Mandelbaum (1994) studied Harris recurrence for continuous-time Markov chains. The results are essentially the same as in the discrete-time case, although, because of the exponential waiting time between jumps, there is no need for a notion of aperiodicity in this context. As long as the chain returns infinitely often to a set A that is hit from the starting state x with probability 1, then $||\mathbf{K}^t(x, \cdot) - \pi(\cdot)|| \to 0$ as $t \to \infty$. When the target density is locally stable, then the empty configuration gives such a set when used with the Preston spatial birth–death chain.

9.5 Perfect Sampling

While it may be possible to show convergence of the distribution of the state to the stationary distribution, it is far more difficult to assess how quickly this convergence occurs. This is the primary drawback to Markov chain methods for approximately sampling from distributions. Heuristics such as autocorrelation plots can show that a Markov chain is not mixing, but they cannot prove that the chain is mixing.

Perfect sampling algorithms generate samples exactly from π (up to the natural limits all Monte Carlo algorithms face: real numbers that are rounded to machine accuracy and the use of pseudorandom numbers rather than true uniforms). The drawback to perfect sampling algorithms is that they are Las Vegas type algorithms, and so their running time is itself a random variable. While the running time has finite expectation, the support is unbounded. That is:

Definition 9.2

A *perfect simulation algorithm* for π is an algorithm whose running time T is finite with probability 1, whose output is a draw from π, and where, for all t, $P(T > t) > 0$.

9.5.1 Acceptance/Rejection Method

As an illustration, consider the acceptance/rejection method for generating from g, the basic idea of which goes back to von Neumann (1951). Suppose that $g(x) \le u(x)$, where $u(x)$ is an unnormalized density from which it is possible to generate samples. Then a random variate with density proportional to $u(x)$ is drawn, and is accepted as a draw from density $g(x)$ with probability $g(x)/u(x)$. Otherwise the process begins again.

ALGORITHM 9.5   ACCEPTANCE/REJECTION

```
Input: target g(x), upper bound u(x) satisfying g(x) ≤ u(x)
Output: X ~ g(·)
  repeat
     draw X ← u(·)
  until U ≤ g(X)/u(X)
```

Theorem 9.2

The output of the above procedure is distributed according to $g(x)$.

Proof.   Let $X_1, X_2, \ldots$ be the i.i.d. draws from $u(x)$ used by the algorithm, $U_1, U_2, \ldots$ the uniform i.i.d. draws, N the number of times through the repeat loop, and A a measurable set. Then

$$P(X_N \in A) = \sum_{i=1}^{\infty} P(X_i \in A, N = i) = \sum_{i=1}^{\infty} P\left(X_i \in A, U_i \leq \frac{g(X_i)}{u(X_i)}\right) \prod_{j=1}^{i-1} P\left(U_j > \frac{g(X_j)}{u(X_j)}\right).$$

Each factor in the last product is the same (since U_i and X_i are i.i.d.), so call it $1 - p$. Then

$$P(X_N \in A) = \sum_{i=1}^{\infty} P(X_i \in A, N = i) = \sum_{i=1}^{\infty} P\left(X_i \in A, U_i \leq \frac{g(X_i)}{u(X_i)}\right)(1 - p)^{i-1}$$

$$= \sum_{i=1}^{\infty} (1 - p)^{i-1} \int_A P(U_i \leq g(X_i)/u(X_i))(u(x)/Z_u)\, d\mu(x)$$

$$= \int_A (g(x)/u(x))(u(x)/Z_u)\, d\mu(x) \sum_{i=1}^{\infty} (1 - p)^{i-1}$$

$$= \left[\int_A (g(x)/Z_u)\, d\mu(x)\right](1/p)$$

where Z_u is the normalizing constant for $u(x)$. Next calculate p and find $P(X_N \in A)$:

$$p = P(U_j \leq g(X_j)/u(X_j)) = \int (g(x)/u(x))(u(x)/Z_u)\, d\mu(x) = Z_g/Z_u,$$

$$P(X_N \in A) = \int_A (g(x)/Z_g)\, d\mu(x),$$

exactly as desired. ∎

The running time of the algorithm T (as measured by the number of times through the repeat loop) is a geometric random variable, and so $P(T > t) > 0$ for any fixed value of t. The output does come from π, but can take an arbitrarily long time to do so.

Moreover, the expected running time of $1/p = Z_u/Z_g$ is directly related to how close the upper bound density $u(x)$ is to $g(x)$. For example, in the Strauss process density (Equation 9.2), $u(x) = \lambda^{\#x}$ is a valid upper bound density, and is easy to sample from: simply generate a Poisson point process with intensity $\lambda m(\cdot)$ rather than $m(\cdot)$. Given a draw $X \sim u(\cdot)$, the probability of accepting the draw is $g(X)/u(X) = \gamma^{s(X)}$. If λ is large, there could be many pairs of points within distance $2R$ of each other, and so for (say) $\gamma = \frac{1}{2}$, the probability of accepting could be very small. For this reason, acceptance/rejection is usually only useful on small problems, and for larger instances more sophisticated techniques must be used.

9.5.2   Dominated Coupling from the Past

In this section the dominated coupling from the past (DCFTP) procedure of Kendall and Møller (2000) is described. This method extends the coupling from the past procedure of Propp and Wilson (1996) to work with chains with an unbounded number of dimensions for locally stable processes.

Consider again the Preston spatial birth–death chain of Section 9.4. It is worth recalling two important facts about exponential random variables here.

- If $A_1, A_2, \ldots, A_n$ are exponential random variables with rates $\lambda_1, \lambda_2, \ldots, \lambda_n$, then the minimum of the $A_1, \ldots, A_n$ variables will be an exponential random variable with rate $\lambda_1 + \cdots + \lambda_n$. Hence, the death clock of rate r_d can be thought of as putting individual death clocks of rate 1 on each of the points in the set, and activating them as needed.

- Exponential random variables are memoryless. Conditioned on a clock being larger than t, the remaining time on the clock will still be an exponential random variable. That is, if $T \sim \text{Exp}(\lambda_1)$, then $[T - t \mid T > t] \sim \text{Exp}(\lambda_1)$.

Using these properties, an equivalent method for simulating the Preston birth–death chain is as follows. Always keep track of the time of the next birth of a point, and the times of the deaths of the current points. When a point dies, it is removed from the process as before; when it is born, it is assigned a death time, and then checked to see whether or not it should be thinned. Call the resulting process X_t, and this process has stationarity density g.

Also keep track of the process where no points are thinned, and call this process D_t. Note that D_t is using the same birth and death events as process X_t, it is just that some of the births are thinned (and so those points do not appear) in X_t. So D_t always contains more points than X_t, and D_t is a *dominating process* for X_t. Since D_t has no thinning it has birth rate $K\lambda(S)$ and a death rate of 1 on each point. Hence D_0 is a Poisson point process with intensity measure $K\lambda(\cdot)$, and so is D_t for all times t. In order to know whether or not to thin the point $v \in D_t$ in the process X_t, each point in the dominating process will be marked with a value drawn uniformly from $[0, 1]$. That mark will be used to decide if the point should be thinned. (This mark is in addition to any other mark that might be part of the spatial point process model.) Figure 9.1 illustrates this process by showing a possible run of the marked dominating process D_t and the thinned underlying process X_t. In this figure, the line segments represent the lifespans of the points, while the squares are birth and death events. Shaded squares indicate a point in both the dominated and underlying process, while empty squares are in D_t, but were thinned at birth and so do not appear in X_t.

With a dominating process in hand, the CFTP procedure of Propp and Wilson (1996) can now be used. Here, instead of running the X_t process to larger times t, think of the

Space

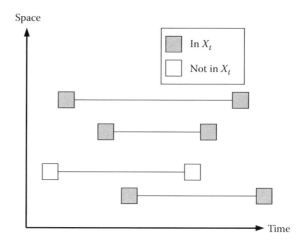

Time

FIGURE 9.1
Dominated and underlying process.

process as running over all times in $(-\infty, \infty)$ and just selecting X_0. Intuitively, since over time $(-\infty, 0]$ the process has already been running for an arbitrarily long amount of time, X_0 will be stationary. This involves running the dominating process backwards in time and then running the coupled (D_t, X_t) process forwards in time up to time 0.

First consider how to run the dominating process backwards in time. This process is time reversible, and so running backwards in time is straightforward. At time 0, D_0 is just a draw from a Poisson point process with intensity $K\lambda(S)$, which can be created using the Poisson point sampler (Algorithm 9.1).

As the process moves backwards in time, the roles of births and deaths are reversed. A "death" causes a new point to be added, and a "birth" removes an existing point from the system. Hence deaths occur backwards in time at constant rate $K\lambda(S)$, while the total birth rate for the dominating process is $|D_t|$.

Consider the following example. Let $D(-n)$ denote the state of the marked dominating process after n events (either births or deaths) backwards in time have been generated. Suppose $D_0 = D(0) = \{v_1, v_2\}$. Then the time until the first birth of v_1 or v_2 is exponential with rate 2, and the time until the first death is exponential with rate $K\lambda(S)$. Suppose that a death occurs. Then a random point v_3 is chosen by $\lambda(\cdot)$ and added to the system so that $D(-1) = \{v_1, v_2, v_3\}$. Now suppose that the next event that occurs is the birth of v_2. Then a uniform on $[0, 1]$ for v_2 is rolled for its mark (let us say it was $0.9863\ldots$), and $D(-2) = \{v_1, v_3\}$ since v_2 is removed from the system. The list of events going backwards in time for the dominating process now has two events in it, a death of v_3 followed by a birth of v_2 with mark $.09836\ldots$, so

$$\text{birth–death list} = \begin{bmatrix} \text{death} & v_3 & \\ \text{birth} & v_2 & 0.9836\ldots \end{bmatrix}.$$

In this fashion, given the current state of the marked dominating process $D(-n)$ and the list of the first n events, the list can be extended to give state $D(-n')$ and the first n' events using the following pseudocode:

ALGORITHM 9.6   DOMINATED EVENT GENERATOR

Input: state $D(-n)$, birth–death list *BDL* with n events, and the
  new size of list n'
Output: $D(-n')$, new birth–death list BDL' of size n'
  1: $D \leftarrow D(-n)$, $BDL' \leftarrow BDL$
  2: **for** i from $n+1$ to n' **do**
  3:    **draw** $t_d \leftarrow \mathrm{Exp}(K\lambda(S))$, $t_b \leftarrow \mathrm{Exp}(\#D)$
  4:    **if** $t_d < t_b$ (then next event is a death) **then**
  5:      **draw** $w \leftarrow \lambda(\cdot)$
  6:      $D \leftarrow D+w$, **add** [death w] to list BDL'
  7:    **else**
  8:      **draw** $v \leftarrow \mathrm{Unif}(D)$, $U \leftarrow \mathrm{Unif}([0, 1])$
  9:      $D \leftarrow D-v$, **add** [birth v U] to list BDL'
10:    **end if**
11: **end for**
12: $D(-n') \leftarrow D$

In the dominated event generator, the actual values of t_d and t_b are not used, they are only being used to determine if $t_d < t_b$. For two independent random variables $A_1 \sim \mathrm{Exp}(\lambda_1)$ and $A_2 \sim \mathrm{Exp}(\lambda_2)$, $P(A_1 < A_2) = \lambda_1/(\lambda_1 + \lambda_2)$, so lines 3 and 4 of the dominated event generator can be replaced with:

  draw $U \leftarrow \mathrm{Unif}([0, 1])$
  if $U < K\lambda(S)/[K\lambda(S) + \#D]$ **then**

Go back in time for n events, and suppose that the state X right before these events is known, and that the list of the next n events is given. Then all the random choices needed to calculate the state X_0 have already been made in the *BDL* list. So from the list, the state of the process up to time 0 can be calculated.

ALGORITHM 9.7   ADVANCE UNDERLYING STATE TO TIME 0

Input: state X after n events back in time, birth–death list *BDL*
  consisting of n events
Output: X, the state at time 0
  1: **for** all the events e in BDL from end to beginning **do**
  2:    **if** $e = [\text{death } w]$ **then**
  3:      $X \leftarrow X - w$
  4:    **else if** $e = [\text{birth } v \; U]$ **then**
  5:      **if** $U \le [(g(X+v)/g(X)][1/K]$ **then**
  6:        $X \leftarrow X+v$
  7:      **end if**
  8:    **end if**
  9: **end for**

Note that this forms a coupling of the marked dominated process D_t and the underlying process X_t. A coupling is a joint process (in this case (X_t, D_t)) such that the marginal processes have their original distribution. In this case X_t is evolving according to the Preston spatial birth–death chain, and D_t is evolving according to a simple spatial birth–death chain, so marginally each has the correct distribution.

The problem with this procedure is that if $D(-n) \neq \emptyset$ then $X(-n)$ is unknown, making it impossible to run the procedure! Since $P(D(-n) = \emptyset) = \exp(-K\lambda(S))$, this is unlikely except for small spaces or very small values of K. So instead a bounding chain (Huber, 2004) is used to try to find $X(0)$.

Definition 9.3

A process (L_t, M_t) is a *bounding chain* for X_t if there is a coupling (L_t, X_t, M_t) such that

$$L_t \subseteq X_t \subseteq M_t \Rightarrow L_{t'} \subseteq X_{t'} \subseteq M_{t'} \quad \text{for all } t' > t.$$

For our purposes, it is sufficient to use processes $L(-n)$ and $M(-n)$, and verify that if $L(-n) \subseteq X(-n) \subseteq M(-n)$, then $L(-n+1) \subseteq X(-n+1) \subseteq M(-n+1)$. Note that if a death occurs, then in line 4 the point is always removed from the underlying state X. However, if the point is a birth, then the probability that the point v is added to X is only $r(X) = (g(X+v)/g(X))(1/K)$. Thus a lower bound on the probability that the point is added to X is

$$r_{\min} = \min_{X : L(-n) \subseteq X \subseteq M(-n)} r(X).$$

Similarly, $r_{\max} = \max_{X : L(-n) \subseteq X \subseteq M(-n)} r(X)$ is an upper bound on the probability of adding v to X. Let X satisfy $L(-n) \subseteq X \subseteq M(-n)$. If $U \leq r_{\min}$ then v will be added to X, and so $L(-n+1) = L(-n) + v$ and $M(-n+1) = M(-n) + v$ is valid as a bounding chain step. If $U > r_{\max}$, then v is definitely not added to X, and so $L(-n+1) = L(-n)$ and $M(-n+1) = M(-n)$ is a valid bounding chain step. Finally, if $r_{\min} < U \leq r_{\max}$, setting $L(-n+1) = L(-n)$ and $M(-n+1) = M(-n) + v$ is a valid step. This is summarized in the following pseudocode:

ALGORITHM 9.8   BOUNDING CHAIN STEP

```
Input: L(-n), M(-n), event  e
Output: L(-n+1), M(-n+1)
 1: if e = [death w] then
 2:     L(-n+1) ← L(-n) - w
 3:     M(-n+1) ← M(-n) - w
 4: else if e = [birth v U] then
 5:     A ← {x : L(-n) ⊆ x ⊆ M(-n)}
 6:     r_min ← min_{x∈A}[g(x+v)/g(x)][1/K],  r_max ← max_{x∈A}[g(x+v)/g(x)][1/K]
 7:     if u < r_min then
 8:         L(-n+1) ← L(-n) + v,  M(-n+1) ← M(-n) + v
 9:     else if U < r_max then
10:         M(-n+1) ← M(-n) + v
11:     end if
12: end if
```

Now the main DCFTP loop can be created. The outline is as follows. First generate $D(0)$, and n events in the backward birth–death list. Set $L(-n) = \emptyset$ and $M(-n) = D(-n)$ so that no matter what $X(-n)$ is, $L(-n) \subseteq X(-n) \subseteq M(-n)$. Then advance forward to find $L(0)$ and $M(0)$. If $L(0) = M(0)$, then $X(0)$ is sandwiched in between and can be output as a stationary state. If they are different, then generate events even farther back in time, and begin again. This is summarized in the following pseudocode.

ALGORITHM 9.9   DOMINATED COUPLING FROM THE PAST

Input: starting value for n
Output: X a draw from the target density g
 1: **draw** $D(0) \leftarrow$ **Poisson point process sampler**$(K\lambda(\cdot))$
 2: $n_{\text{old}} \leftarrow 0$, $BDL \leftarrow \emptyset$
 3: **repeat**
 4:     $(D(-n), BDL) \leftarrow$ **Dominated event generator**$(D(-n_{\text{old}}), BDL, n_{\text{old}}, n)$
 5:     $U \leftarrow D(-n)$
 6:     **for** e the events in BDL from end of list to beginning of list **do**
 7:         $(L, U) \leftarrow$ **Bounding chain step**(L, U, e)
 8:     **end for**
 9:     $n_{\text{old}} \leftarrow n$, $n \leftarrow 2n$
10: **until** $U = L$
11: $X \leftarrow U$

It is important to note that the events generated in line 4 are being appended to the already created events in BDL. That is, suppose that the backward events 1–10 have already been created. If $n = 20$, then only events 11–20 will be added to the list, since events 1–10 are currently in the list, and must be reused on each run through the repeat loop.

In line 9, the number of events checked is doubled at each step. Let N be the smallest value of n such that $L(-n)$ and $M(-n)$, run forward in time, equal one another. Then by doubling n each time, the longest run in the repeat loop will be for $n \leq 2N$. This makes the total time at most $2N + N + N/2 + \cdots \leq 4N$, and so this method ensures that the total number of steps taken is within a factor of 4 of the optimal value N.

9.5.3   Examples

The trickiest part of a DCFTP algorithm is the creation of an efficient bounding chain. In lines 5 and 6 of the bounding chain step, the values of $r_{\min}$ and $r_{\max}$ are written as a minimum and maximum over a large set A. In many instances, it is possible to find these values without resorting to use of an optimization method.

Consider the Strauss process (Example 9.1), which is locally stable with $K = \lambda$. As before, for a configuration x and point v, let $n(x, v) = \#\{j : \text{dist}(x_j, v) < 2R\}$ so that $(g(x + v)/g(x))(1/K) = \gamma^{n(x,v)}$. Then since $\gamma \leq 1$, this ratio is smallest when $n(x, v)$ is largest, which happens when $x = M(-n)$, its upper bound. In the other direction, the ratio is largest when $n(x, v)$ is smallest, so $r_{\max}$ occurs when $x = L(-n)$. So there is no need to compute A, making the relevant line for the Strauss process in the bounding chain step:

 6:     **let** $r_{\min} \leftarrow \gamma^{n(M, v)}$, $r_{\max} \leftarrow \gamma^{n(L, v)}$.

Now consider Example 9.2, the area interaction process. In Section 9.4.1, it was shown how to take steps in the Preston birth–death chain by generating a Poisson point process Z

with intensity $|\ln \gamma| m(\cdot)$ in A_v, the grain of the proposed birth v. Let A be the region covered by the grains of points in x. If $Z \cap A^C = \emptyset$ and $\gamma > 1$, or if $Z \cap A = \emptyset$ and $\gamma < 1$, then the point is accepted. Let $M(A)$ denote the region covered by the grains of points in the upper process M, with $L(A)$ defined similarly for the lower process L. Then $L \subseteq x \subseteq M$ implies $L(A) \subseteq A \subseteq M(A)$, and $M(A)^C \subseteq A^C \subseteq L(A)^C$. For $\gamma > 1$, the equivalent to the uniform being less than $r_{\min}$ is if $Z \cap L(A)^C = \emptyset$, and the uniform being less than $r_{\max}$ is $Z \cap M(A)^c = \emptyset$. So for the bounding chain step:

```
 6:     draw Z ← Poisson point process sampler((ln γ)m(· ∩ Aᵥ))
 7:     if Z ∩ L(A)ᶜ = ∅ then
 8:         M(−n + 1) ← M(−n) + v,  L(−n + 1) ← L(−n) + v
 9:     else if Z ∩ M(A)ᶜ = ∅ then
10:         M(−n + 1) ← M(−n) + v
```

Another way to view this update is as follows. When $Z \cap L(A)^C = \emptyset$, add the birth to the lower process L. When $Z \cap M(A)^C = \emptyset$, add the birth to the upper process U. In effect, L and U are being updated exactly as though they were states in the Markov chain. When this choice of update works for the bounding chain, the update scheme is called *monotonic*.

When $\gamma < 1$, the update scheme is no longer monotonic. Now the point is least likely to be born when $X = U$, and most likely to be born when $X = L$. The pseudocode becomes:

```
 6:     draw Z ← Poisson point process sampler(−(ln γ)m(· ∩ Aᵥ))
 7:     if Z ∩ M(A) = ∅ then
 8:         M(−n + 1) ← M(−n) + v,  L(−n + 1) ← L(−n) + v
 9:     else if Z ∩ L(A) = ∅ then
10:         M(−n + 1) ← M(−n) + v
```

9.6 Monte Carlo Posterior Draws

Markov chain Monte Carlo for Bayesian analysis of spatial data creates a new set of challenges. The basic framework is as follows. First, a probabilistic model (such as the Strauss process or area interaction process) is placed on the data. These models have parameters, which are themselves treated as random variables. The distribution of these random variables is called the prior. Assuming that the data is drawn from the probabilistic model, then the distribution of the parameters conditioned on the value of the data is different from the prior, and is known as the posterior. This posterior can often be written as an unnormalized density using Bayes' rule. However, in Bayesian spatial analysis, the unnormalized density can contain normalizing constants that are themselves difficult to compute—making the problem exceptionally difficult!

Let $p(\cdot)$ denote the initial probability density of the random parameter θ (this is known as the prior for θ). Let X denote the random value of the data, and suppose that the density of X given θ is $g(\cdot|\theta)/Z_\theta$. Let π be the distribution of θ given X. From Bayes' rule,

$$\pi(da|X = x) = \frac{[g(x|\theta = a)/Z_a]p(a)\,da}{\int (g(x|\theta = b)/Z_b)p(b)\,db}.$$

So while the denominator is an unknown normalizing constant, the numerator contains such a constant as well.

Metropolis–Hastings can be used to design a chain with limiting distribution $\pi(\cdot|X=x)$ without knowing the denominator, but there is still the problem of Z_a. Suppose that $\theta = a$, and $q(\cdot|\theta = a, X = x)$ is the proposal density used to generate a proposed move to $\theta = b$. Then the probability of accepting this move is

$$r = \frac{g(x|\theta = b)/Z_b}{g(x|\theta = a)/Z_a} \cdot \frac{p(b)}{p(a)} \cdot \frac{q(a|\theta = b, X = x)}{q(b|\theta = a, X = x)}. \tag{9.7}$$

The last two factors in the ratio are typically easy to calculate, but the first represents a problem. While $g(x|\theta = b)/g(x|\theta = a)$ is straightforward, Z_a/Z_b is not feasible to compute directly.

Bognar and Cowles (2004) used importance sampling to approximate this ratio to take approximate steps in the Metropolis chain. Møller et al. (2006) suggested an auxiliary variable approach for dealing with this problem without needing to find the ratio. By adding an extra random variable to the Markov chain, and choosing the appropriate proposal density, the factor Z_a/Z_b is eliminated in the Metropolis–Hastings acceptance ratio (Equation 9.7).

Alongside the random variable θ, create a new random variable Y, conditioned on θ and $X = x$, that is itself a point process with normalized density $f(\cdot|\theta, X = x)$ with respect to a Poisson point process with intensity $m(\cdot)$. This is an example of the data augmentation method: see Chapter 10 (this volume) for more information.

Note that θ has exactly the same distribution as earlier. Therefore, in a draw from the limiting distribution of the Metropolis chain for $(\theta, Y|X = x)$, the marginal density of θ will remain as before. The good news is that unlike the Metropolis chain for $\theta|X = x$, a new Metropolis chain for $(\theta, Y|X = x)$ can be constructed where the ratios can be calculated explicitly.

Suppose that $X = x$, $\theta = a$, and $Y = y$. Using density $q(\cdot|\theta = a, X = x)$, propose a move to $\theta = b$. Now using density $g(\cdot|\theta = b)/Z_b$, propose a new state y' for Y to move to. This makes the combined density for (b, y') equal to $g(y'|\theta = b)q(b|\theta = a, X = x)/Z_b$. This makes the Metropolis–Hastings ratio for accepting the move from (a, y) to (b, y'):

$$r = \frac{g(x|\theta = b)/Z_b}{g(x|\theta = a)/Z_a} \cdot \frac{p(b)}{p(a)} \cdot \frac{f(y'|\theta = b, X = x)}{f(y|\theta = a, X = x)} \cdot \frac{g(y|\theta = a)q(a|\theta = b, X = x)/Z_a}{g(y'|\theta = b)q(b|\theta = a, X = x)/Z_b}.$$

Note that Z_a and Z_b cancel out in this ratio, and so this method can be implemented to take a Metropolis–Hastings step.

ALGORITHM 9.10   AUXILIARY VARIABLE METROPOLIS–HASTINGS STEP

Input: current state $\theta = a$, $Y = y$, data x
Output: next state (θ, Y)
    draw $b \leftarrow q(\cdot|\theta = a, X = x)$, $y' \leftarrow g(\cdot|\theta = b, X = x)$, $U \leftarrow \text{Unif}([0, 1])$
    $r \leftarrow \dfrac{g(x|\theta = b)p(b)f(y'|\theta = b, X = x)g(y|\theta = a)q(a|\theta = b, X = x)}{g(x|\theta = a)p(a)f(y|\theta = a, X = x)g(y'|\theta = b)q(b|\theta = a, X = x)}$
    if $U < r$ **then**
        $\theta \leftarrow b$, $Y \leftarrow y'$
    end if

The drawback of this method (as noted in Møller et al., 2006) is that the acceptance probability can become very low, since the ratio involves $f(\cdot|\theta = b, X = x)$. When f is very far away from g, this can lead to proposals that are only rarely accepted. This problem has been addressed by a variant of the method that uses the auxiliary variable in a different fashion: see Murray et al. (2006) and Murray (2007) for details.

9.7   Running Time Analysis

Although it is not possible to analyze the mixing time of the Markov chains or the running time of the perfect simulation methods precisely for all models, when the model is sufficiently noisy (in a sense to be described later), these methods will have a polynomial running time.

The technique used here is coupling. As mentioned in Section 9.5, a coupling (X_t, Y_t) is a paired process where both X_t and Y_t marginally have the correct distribution. Coupling relates to the mixing time of a Markov chain as follows.

Theorem 9.3

(Doeblin, 1933) *Let (X_t, Y_t) be a coupling of two processes whose marginal distributions are a Markov chain with stationary distribution π, and $Y_0 \sim \pi$. Then*

$$||P(X_t \in \cdot) - \pi(\cdot)||_{\mathrm{TV}} \leq P(X_t \neq Y_t).$$

Proof.   Let A be a measurable set. Then $P(Y_t \in A) = \pi(A)$ and

$$P(X_t \in A) = P(X_t \in A, Y_t = X_t) + P(X_t \in A, Y_t \neq X_t) \leq P(Y_t \in A) + P(Y_t \neq X_t).$$

Subtracting $\pi(A) = P(Y_t \in A)$ yields $P(X_t \in A) - \pi(A) \leq P(Y_t \neq X_t)$. Reversing X_t and Y_t in the above argument gives $\pi(A) - P(X_t \in A) \leq P(Y_t \neq X_t)$, so $|P(X_t \in A) - \pi(A)| \leq P(Y_t \neq X_t)$. This is true for all A, so $||P(X_t \in \cdot) - \pi(\cdot)||_{\mathrm{TV}} \leq P(X_t \neq Y_t)$ as desired. ∎

The processes X_t and Y_t have *coupled* if $X_t = Y_t$. Aldous (1982) used coupling to bound mixing times of Markov chains, and the following proofs are in the same style, with some differences necessary to deal with the nature of birth and death chains. Consider a process X_t started at $X_0 = \emptyset$, $Y_0 \sim \pi$, coupled as in Section 9.5 by using a dominating process D_t and sharing birth and death events. Using the Preston spatial birth–death chain, if a birth occurs in D_t, it can be thinned so that the birth does not occur in X_t. But if it survives, when the point dies in D_t, it dies in X_t as well. The same occurs for Y_t, so that a point that is born in D_t could be born in X_t or Y_t or both.

Since births come from the dominating process D_t, both X_t and Y_t are subsets of D_t. Moreover, when a death occurs in one of the processes it will also occur in the other if that point exists in the other. This tends to aid in coupling the two processes.

Births, on the other hand, might drive X_t and Y_t apart, as a point might be accepted as born into X_t but not Y_t, or vice versa. Let $W_t = X_t \oplus Y_t$ be those points in X_t or in Y_t, but not in both (this is the symmetric difference of the two configurations.) Let $w(t) = \mathrm{E}[\#W_t]$.

A birth can only change the number of points in W_t by 1, and so a differential equation for $w(t)$ can be derived by considering

$$\lim_{h \to 0} \frac{w(t+h) - w(t)}{h} = \lim_{h \to 0} E[E[\#W_{t+h}|X_t, Y_t] - \#W_t]/h.$$

Theorem 9.4

Consider a Strauss process X_t (Example 9.1) started at $X_0 = \emptyset$, and let β be the Lebesgue measure of a ball of radius 2R (this value depends on the distance metric used.) If $\beta\lambda(1 - \gamma) < 1$, then

$$\|P(X_t \in \cdot | X_0 = \emptyset) - \pi(\cdot)\|_{TV} \leq \lambda m(S) \exp(-(1 - \beta\lambda(1 - \gamma))t).$$

Proof. The idea of the proof is to couple the X process started at $X_0 = \emptyset$ with a Y process started with $Y_0 \sim \pi$. At the beginning, Y most likely contains points that X does not. But when a point in Y dies, the Y process and X process move closer together. Therefore, this is referred to as a *good event*.

The point process $W_t = X_t \oplus Y_t$ is the set of points in X but not in Y, or in Y but not in X. A good event is when one of these points dies, and so the rate at which good events occur is just $\#W_t$.

A *bad event* is when a point is born to X but not to Y, or born to Y but not to X. These bad events increase the size of W_t by 1. The goal of the proof is to find an upper bound on the rate of bad events. If the rate of bad events is smaller than the rate of good events, then the good events will dominate, and eventually, the point process W_t will lose more and more points until it reaches the empty set.

For $\#W_t$ to increase by 1 (a bad event), a birth must occur at v and must be accepted by X_t but not Y_t or vice versa. The only way that can happen is if v lies within distance 2R of at least one point in W_t. The chance of adding to X_t but not Y_t (or vice versa) is at most $1 - \gamma^{n(v)}$, where $n(v)$ denotes the number of points of W_t within distance 2R of v. Let A_i denote the measure of the region within distance 2R of exactly i points in W_t. Then $\sum_i iA_i \leq \beta\#W_t$.

The birth rate of points within 2R of i points in $\#W_t$ in the Preston birth–death chain is λA_i. The total rate at which bad events occur is then at most

$$\sum_{i=1}^{\#W_t} A_i \lambda \beta(1 - \gamma^i) \leq \sum_{i=1}^{\#W_t} iA_i \lambda \beta(1 - \gamma) \leq \beta\#W_t(1 - \gamma),$$

where the first inequality follows from

$$1 - \gamma^i = (1 - \gamma)(1 + \gamma + \cdots + \gamma^{i-1}) \leq (1 - \gamma)i.$$

Note that the rate of bad events is small when λ is small, β is small, or γ is close to 1. When the bad event rate is smaller than the good event rate, the size of $\#W_t$ will tend to 0.

To make this intuition precise, consider the probability that exactly one event (good or bad) occurs in the time interval from t to $t + h$. This will yield a differential inequality on $w(t) := E[\#W_t]$. Because this is a continuous-time Markov chain, this probability is proportional to h, and the probability that exactly n events occur in $[t, t + h]$ is $O(h^n)$. Putting this together with the known rate of good events (that decrease $\#W(t)$ by 1) and the upper bound on the rate of bad events (that increase $\#W(t)$ by 1) yields an upper bound on $w'(t)$:

$$w'(t) = \lim_{h \to 0} \frac{E[E[\#W_{t+h}|X_t, Y_t] - \#W_t]}{h}$$

$$\leq \lim_{h \to 0} \frac{E[(h)(\#W_t \beta \lambda (1 - \gamma) - \#W_t) + (\sum_{i=2}^{\infty} iO(h^i))]}{h}$$

$$= -E[\#W_t(\beta \lambda (1 - \gamma) - 1)]$$

$$= -w(t)(1 - \beta \lambda (1 - \gamma)).$$

Now $w(0) \leq E[\#W(0)] = \lambda m(S)$ so, together with the differential inequality above,

$$w(t) \leq \lambda m(S) \exp(-t(1 - \beta \lambda (1 - \gamma))).$$

The last trick is to note that $\#W_t$ is a nonnegative integer, and when $\#W_t$ does not equal 0, it is at least 1. This allows the use of Markov's inequality to say that

$$P(X_t \neq Y_t) = P(\#W_t \neq 0) = P(\#W_t \geq 1) \leq E(\#W(t)) = w(t),$$

which completes the proof. ∎

Therefore, to get the total variation distance below an arbitrary $\epsilon > 0$, when $\lambda \beta (1 - \gamma) < 1$ it suffices to take $t = (1 - \beta \lambda (1 - \gamma))^{-1}[\ln(\lambda m(S)) + \ln \epsilon^{-1}]$, and run this chain for this length of time. The number of uniforms generated in running the chain for t steps is proportional to the number of events in the dominating chain. Fortunately, this is closely related to t. The following tail bounds on Poisson random variables will be helpful.

Lemma 9.1

Let $A \sim \text{Poisson}(\alpha)$. Then $E[A] = \alpha$, and for $c \geq 1$,

$$P(A > \alpha c) \leq \exp(-\alpha(c \ln c - c + 1)), \tag{9.8}$$

$$P(A < \alpha/c) \leq \exp(-\alpha(-(\ln c)/c - (1/c) + 1)). \tag{9.9}$$

Proof. These are examples of Chernoff bounds (Chernoff, 1952). The idea is to use Markov's inequality for the moment generating function, then minimize over the argument. That is, for all $a \geq 0$,

$$P(A > c\alpha) = P(e^{aA} > e^{ac\alpha}) \leq \frac{E[\exp(aA)]}{\exp(ac\alpha)} = \exp(\alpha(e^a - 1 - ca)),$$

where the last inequality is using the fact that $E[\exp(tA)] = \exp(\alpha(e^t - 1))$. Now minimizing $e^a - 1 - ca$ gives $a = \ln c$, making the bound $\exp(\alpha(c - 1 - c \ln c))$ as in Equation 9.8.

The second inequality follows similarly, first multiplying by $-a$ where $a \geq 0$ to obtain $P(-aA > -a\alpha/c)$, and then using $a = \ln c$ as before. ∎

Theorem 9.5

Let N_t be the number of events in the Preston spatial birth–death dominating chain to move from time 0 to time t. Then $E[N_t] = 2K\lambda(S)$ and, for all $c \geq 1$,

$$P[N_t > 2cK\lambda(S)] \leq 2\exp(-2tK\lambda(S)(c\ln c - c + 1)).$$

Proof. Break the number of events N_t into the number of births N_b plus the number of deaths N_d. The birth and death processes in the chain are themselves each a one-dimensional Poisson process with rate $K\lambda(S)$, so N_b and N_d are each Poisson random variables with parameter $tK\lambda(S)$. The theorem essentially states that it is very likely that these two Poisson random variables are close to their means, where $E[N_b] = E[N_d] = tK\lambda(S)$.

If $N_t > 2ctK\lambda(S)$ then, since $N_t = N_b + N_d$, at least one of N_b and N_d is at least $ctK\lambda(S)$. So $P(N_t > 2ctK\lambda(S)) \leq P(N_b > ctK\lambda(S)) + P(N_d > ctK\lambda(S))$. The two tail probabilities on the right are bounded using Lemma 1 to complete the proof. ∎

In other words, to run for time t in the Preston birth–death chain requires evaluation of on average $2tK\lambda(S)$ events, and is unlikely to take very much more time than that.

Theorem 9.6

Consider an area interaction process X_t (Example 9.2) started at $X_0 = \emptyset$, and let a be the maximum area of a grain for any point in the space. Let $B(v)$ be the area of the region where for a point w in the region, the grains of w and v intersect, and let $\beta = \sup_{v \in S} B(v)$. If $\lambda\beta a \max\{\gamma, \gamma^{-1}\}^a < 1$, then

$$\|P(X_t \in \cdot | X_0 = \emptyset) - \pi(\cdot)\|_{TV} \leq \lambda m(S)\exp(-(1 - \lambda\beta a \max\{\gamma, \gamma^{-1}\}^a)t).$$

Proof. As in the proof of Theorem 4, what is needed is to show that the rate of bad events is smaller than the rate of good events. In the earlier proof, the rate of good events is exactly $\#W_t$, and the same rate holds here.

The rate of bad events can be bounded above by noting that a point v must be born so that its grain overlaps with a point in W_t. This birth rate is at most $\#W_t\lambda\beta$. Then the probability of accepting the birth is $\gamma^{-a(x,v)}$ for $\gamma > 1$, and $(\gamma^{-1})^{-(a-a(x,v))}$ for $\gamma < 1$. Either way, this is bounded below by $\max\{\gamma, \gamma^{-1}\}^a$, and the overall rate of bad events is $\#W_t\lambda\beta a \max\{\gamma, \gamma^{-1}\}^a$.

The rest of the proof proceeds as in Theorem 4. ∎

9.7.1 Running Time of Perfect Simulation Methods

The advantage of using a perfect simulation method is that there is no need to know the mixing time of a Markov chain—when the method runs quickly, the samples are guaranteed to come from the correct distribution.

On the other hand, it is still useful to have an *a priori* bound on the running time for a nontrivial set of parameters of the model, and these bounds can be created in a fashion similar to that used in finding the mixing time. The following theorem bounds the expectation and the tail of the number of events generated in the birth–death list in the course of creating a single random draw. The running time of dominated coupling from the past is proportional to this number of events.

Theorem 9.7

Consider perfect simulation of the Strauss process X_t using dominated coupling from the past. Let n be the smallest number of events needed in the birth–death list before convergence occurs. Let $\beta = \sup_{v \in S} m(\{w : \text{dist}(v, w) < 2R\})$ and $b = 1 - \beta\lambda(1 - \gamma)$. If $b \geq 0$, then

$$P(n \geq a) \leq \exp(-0.15a) + \lambda m(S)\exp(-2ba/[\lambda m(S)])), \quad E[n] \leq 6.2 + 1.6\lambda m(S)\lceil \lambda m(S)/b \rceil.$$

Proof. This theorem is stated in terms of the smallest number of entries n on the birth–death list needed for convergence to occur, rather than in time, since the running time of DCFTP is directly proportional to this number.

Consider a fixed $t < 0$. Then suppose that entry number a in the birth–death list comes before time t, and that, running forward from time t, the DCFTP procedure brings the lower process and upper process together. Then $n \leq a$. Taking complements and using the union bound, the probability that $n > a$ is upper-bounded by the probability that fewer than a events occur in $[t, 0]$ plus the probability the lower and upper processes fail to converge in time t.

Begin by bounding the probability that the lower process L and upper process M converge by time t. When a point dies, it is removed from both M and L, but when a point is born, there is a chance it is added to M but not L, further separating the processes.

So consider a lower process L started at $\emptyset$, and M started with a draw from a Poisson point process on S. Two types of events alter the size of $M \setminus L$. Type I events are deaths of points in $M \setminus L$. Type II events are the birth of points within distance R of $M \setminus L$ that are accepted into M but not L.

The rate of type I events is exactly $\#(M \setminus L)$. The rate of type II points can be bounded as follows. Recall that if a point is born within distance R of n_M points in M_n, $r_{\min} = \gamma^{n_M}$, while if it is born within distance R of n_L points in L_n, $r_{\max} = \gamma^{n_L}$. Therefore, the chance of adding a birth to M but not L is

$$\gamma^{n_L} - \gamma^{n_M} = \gamma^{n_L}(1 - \gamma^{n_M - n_L}) \leq 1 - \gamma^{n_M - n_L}.$$

Let A_i denote the area of points within distance R of exactly i points in $M \setminus L$. Then the total rate of births in $M \setminus L$ is bounded above by

$$\sum_{i=1}^{\#(M \setminus L)} \lambda A_i(1 - \gamma^i), \quad \text{where } A_1 + 2A_2 + \cdots + iA_i \leq \beta\#(M \setminus L).$$

Now use $1 - \gamma^i = (1 - \gamma)(1 + \gamma + \cdots + \gamma^{i-1}) \leq (1 - \gamma)i$ to say that

$$\sum_{i=1}^{\#(M \setminus L)} \lambda A_i(1 - \gamma^i) \leq \sum_{i=1}^{\#(M \setminus L)} \lambda A_i(i)(1 - \gamma) \leq \beta\lambda(1 - \gamma)\#(M \setminus L).$$

Therefore, a sufficient condition for the rate of good events to outpace the rate of bad events is that $\#(M \setminus L) \geq \beta\lambda(1 - \gamma)\#(M \setminus L)$, or equivalently, that $b = 1 - \beta\lambda(1 - \gamma) \geq 0$.

Let $w(t) := E[\#(M_t \setminus L_t)]$, where M_t and L_t are the upper and lower processes after t time has evolved. Then, as in the proof of Theorem 4, the rate computation above shows that $w(t) \leq -w'(t)(1 - \beta\lambda(1 - \gamma))$ and, taken with $w(0) = E[\#(M_0 \setminus L_0)] = \lambda m(S)$, leads to the result that $P(\#(M_t \setminus L_t) > 0) \leq \lambda m(S) \exp(-tb)$. That is, the probability of not coupling declines exponentially in the amount of time elapsed.

Now the second half of the proof starts, where it is shown that, for $t = 2a/\lambda m(S)$, the probability that fewer than a events occur in $[t, 0]$ is small. Consider birth events: they form a Poisson process with rate $\lambda m(S)$, and so the probability that fewer than a births occur in interval $[t, 0]$ is bounded by the probability that a Poisson random variable with parameter $\lambda m(S)t = 2a$ is at most a.

From Lemma 1, this probability is bounded above by $\exp(-0.15a)$. Therefore, the sum of these probabilities is

$$\exp(-0.15a) + \lambda m(S) \exp(-2ab/[\lambda m(S)]).$$

The bound on the expected value of n then follows from the tail sum formula. ∎

Some comments are in order. First, as b approaches zero, the bound approaches infinity. In fact, in this situation it is possible to obtain a bound on $E[n]$ that grows as $O([\lambda m(S)]^2)$.

Second, no effort was made to obtain tight constants in the running time because in practice they are unnecessary: this is a perfect simulation algorithm and so it can just be run and the expected running time estimated as tightly as needed. The purpose of these bounds is that under the same conditions where the Markov chain is known to be rapidly mixing, the perfect simulation algorithm is polynomial as well. But in practice, the perfect simulation algorithm could be fast over a much wider range of parameters, whereas the mixing time of the Markov chain is completely unknown.

Third, suppose that the original call to dominated coupling from the past began with $n = 10$, and after doubling twice to $n = 40$, the algorithm terminates. Then there were 40 events generated, and $40 + 20 + 10$ evaluations of how the upper and lower processes changed given those events. Because the number of calls is being doubled at each step, if n is the minimum number of steps needed, the longest call to DCFTP will run for at most $2n$ steps. Each step is evaluated twice, and so the number of steps taken in the bounding chain will be $2n + n + (n/2) + \cdots \leq 4n$. That is, the total expected number of steps will be at most $4n$, justifying the emphasis on bounding n in the preceding theorem.

A similar analysis can be done for the area interaction process.

Theorem 9.8

Consider perfect simulation of the area interaction process X_t using dominated coupling from the past. Let n be the smallest number of events needed in the birth–death list before convergence occurs. As earlier, let a be the maximum area of a grain over points in the space, and let $B(v)$ be the area of the region where, for a point w in the region, the grains of w and v intersect. Then let $\beta = \sup_{v \in S} B(v)$ and $b = 1 - [\lambda \beta a \max\{\gamma, \gamma^{-1}\}^{-a}]$. If $b \geq 0$, then

$$P(n \geq a) \leq 2\exp(-0.15a) + \lambda m(S) \exp(-2ba/(\lambda m(S))), \quad E[n] \leq 6.2 + 1.6\lambda m(S) \lceil \lambda m(S)/b \rceil.$$

The proof is the same as the previous theorem. The only difference is the definition of b, which (roughly speaking) is 1 minus the rate at which a point in $M \setminus L$ is causing new points to be born to $M \setminus L$. (The 1 measures the rate at which these points are dying.) Once b is known, the bounds on the tails of the running time, and the bound on the expected running time, follow as before.

Acknowledgment

This work was supported by NSF Grant DMS-05-48153.

References

Aldous, D. 1982. Some inequalities for reversible Markov chains. *Journal of the London Mathematical Society*, 25(2):561–576.

Baddeley, A. and Møller, J. 1989. Nearest neighbour Markov point processes and random sets. *International Statistical Review*, 57:89–121.

Baddeley, A. J. and van Lieshout, M. N. M. 1995. Area-interaction point processes. *Annals of the Institute of Statistical Mathematics*, 47:601–619.

Bognar, M. A. and Cowles, M. K. 2004. Bayesian inference for pairwise interacting point processes. *Statistics and Computing*, 14:109–117.

Carter, D. and Prenter, P. 1972. Exponential spaces and counting processes. *Zeitschrift für Wahrscheinlichkeitstheorie und Verwandte Gebiete*, 21:1–19.

Chernoff, H. 1952. A measure of asymptotic efficiency for tests of a hypothesis based on the sum of observations. *Annals of Mathematical Statistics*, 23:493–509.

Doeblin, W. 1933. Exposé de la théorie des chaînes simples constantes de Markov à un nombre fini d'états. *Revue Mathématique de l'Union Interbalkanique*, 2:77–105.

Durrett, R. 2005. *Probability: Theory and Examples*, 3rd edn. Thomson Brooks/Cole, Belmont, CA.

Feller, W. 1966. *An Introduction to Probability Theory and Its Applications, Volume II*. Wiley, New York.

Fiksel, T. 1984. Estimation of parameterized pair potentials of marked and non-marked Gibbsian point processes. *Elektronische Informationsverarbeitung and Kybernetik*, 20:270–278.

Geyer, C. 1999. Likelihood inference for spatial point processes. In O. E. Barndorff-Nielsen, W. S. Kendall, and M. N. M. van Lieshout (eds), *Stochastic Geometry: Likelihood and Computation*, pp. 79–140. Chapman & Hall/CRC, Boca Raton, FL.

Geyer, C. and Møller, J. 1994. Simulation and likelihood inference for spatial point processes. *Scandanavian Journal of Statistics*, 21:359–373.

Green, P. J. 1995. Reversible jump Markov chain Monte Carlo computation and Bayesian model determination. *Biometrika*, 82(4):711–732.

Harkness, R. D. and Isham, V. 1983. A bivariate spatial point pattern of ants' nests. *Applied Statistics*, 32:293–303.

Hastings, W. K. 1970. Monte Carlo sampling methods using Markov chains and their applications. *Biometrika*, 57:97–109.

Huber, M. 2004. Perfect sampling using bounding chains. *Annals of Applied Probability*, 14(2):734–753.

Kaspi, H. and Mandelbaum, A. 1994. On Harris recurrence in continuous time. *Mathematics of Operations Research*, 19:211–222.

Kelly, F. and Ripley, B. D. 1976. A note on Strauss's model for clustering. *Biometrika*, 63(2):357–360.

Kendall, W. and Møller, J. 2000. Perfect simulation using dominating processes on ordered spaces, with application to locally stable point processes. *Advances in Applied Probability*, 32:844–865.

Metropolis, N., Rosenbluth, A., Rosenbluth, M., Teller, A., and Teller, E. 1953. Equation of state calculations by fast computing machines. *Journal of Chemical Physics*, 21:1087–1092.

Møller, J. and Waagepetersen, R. P. 2007. Modern statistics for spatial point processes. *Scandanavian Journal of Statistics*, 34:643–684.

Møller, J., Pettitt, A., Reeves, R., and Berthelsen, K. K. 2006. An efficient Markov chain Monte Carlo method for distributions with intractable normalising constants. *Biometrika*, 93(2):451–458.

Murray, I. 2007. Advances in Markov chain Monte Carlo methods. PhD thesis, University College London.

Murray, I., Ghahramani, Z., and MacKay, D. J. C. 2006. MCMC for doubly-intractable distributions. In *Proceedings of the 22nd Annual Conference on Uncertainty in Artificial Intelligence (UAI)*, pp. 359–366. AUAI Press, Corvallis, OR.

Preston, C. 1977. Spatial birth-and-death processes. *Bulletin of the International Statistical Institute*, 46:371–391.

Propp, J. G. and Wilson, D. B. 1996. Exact sampling with coupled Markov chains and applications to statistical mechanics. *Random Structures Algorithms*, 9(1–2):223–252.

Ripley, B. D. 1977. Modelling spatial patterns (with discussion). *Journal of the Royal Statistical Society, Series B*, 39:172–212.

Rosenthal, J. 1995. Minorization conditions and convergence rates for Markov chain Monte Carlo. *Journal of the American Statistical Association*, 90(430):558–566.

Stoyan, D. and Stoyan, H. 1995. *Fractals, Random Shapes and Point Fields*. Wiley, Chichester.

Strauss, D. J. 1975. A model for clustering. *Biometrika*, 63:467–475.

von Neumann, J. 1951. Various techniques used in connection with random digits. In A. S. Householder, G. E. Forsythe, and H. H. Germond (eds), *Monte Carlo Method*, National Bureau of Standards Applied Mathematics Series, 12. US Government Printing Office, Washington, DC.

Widom, B. and Rowlinson, J. 1970. A new model for the study of liquid-vapor phase transition. *Journal of Chemical Physics*, 52:1670–1684.

10

The Data Augmentation Algorithm: Theory and Methodology

James P. Hobert

10.1 Basic Ideas and Examples

Assume that the function $f_X : \mathbb{R}^p \to [0, \infty)$ is a probability density function (pdf). Suppose that $g : \mathbb{R}^p \to \mathbb{R}$ is a function of interest and that we want to know the value of $\mathrm{E}_{f_X} g = \int_{\mathbb{R}^p} g(x) f_X(x) \, dx$, but that this integral cannot be computed analytically. There are many ways of approximating such intractable integrals, including numerical integration, analytical approximations, and Monte Carlo methods. In this chapter, we will describe a Markov chain Monte Carlo (MCMC) method called the *data augmentation (DA) algorithm*.

Here is the basic idea. In situations where classical Monte Carlo methods are not applicable because it is impossible to simulate from f_X directly, it is often possible to find a joint pdf $f : \mathbb{R}^p \times \mathbb{R}^q \to [0, \infty)$ that satisfies two properties: (i) the x-marginal is f_X, that is,

$$\int_{\mathbb{R}^q} f(x, y) \, dy = f_X(x);$$

and (ii) simulating from the associated conditional pdfs, $f_{X|Y}(x \mid y)$ and $f_{Y|X}(y \mid x)$, is straightforward. The DA algorithm is based on this joint pdf. The first property allows for the construction of a Markov chain that has f_X as an invariant pdf, and the second property provides a means of simulating this Markov chain. As long as the resulting chain is reasonably well behaved, simulations of it can be used to consistently estimate $\mathrm{E}_{f_X} g$. We now begin to fill in the details, starting with the construction of the Markov chain.

As usual, let $f_Y(y) = \int_{\mathbb{R}^p} f(x, y) \, dx$. Also, define $\mathsf{X} = \{x \in \mathbb{R}^p : f_X(x) > 0\}$ and $\mathsf{Y} = \{y \in \mathbb{R}^q : f_Y(y) > 0\}$ and assume that $f(x, y) = 0$ whenever $(x, y) \notin \mathsf{X} \times \mathsf{Y}$. Now define a function $k : \mathsf{X} \times \mathsf{X} \to [0, \infty)$ as follows:

$$k(x' \mid x) = \int_{\mathsf{Y}} f_{X|Y}(x' \mid y) f_{Y|X}(y \mid x) \, dy. \tag{10.1}$$

(We will not need to perform the integration in Equation 10.1—remember that we are still in the construction phase.) Since the integrand in Equation 10.1 is a product of conditional densities, k is never negative. Furthermore,

$$
\int_X k(x' \mid x)\, dx' = \int_X \left[\int_Y f_{X\mid Y}(x' \mid y) f_{Y\mid X}(y \mid x)\, dy \right] dx'
$$

$$
= \int_Y f_{Y\mid X}(y \mid x) \left[\int_X f_{X\mid Y}(x' \mid y)\, dx' \right] dy
$$

$$
= \int_Y f_{Y\mid X}(y \mid x)\, dy
$$

$$
= 1.
$$

Hence, for each fixed $x \in X$, $k(x' \mid x)$ is nonnegative and integrates to 1. The function k is therefore a Markov transition density (Mtd) that defines a Markov chain, $X = \{X_n\}_{n=0}^{\infty}$, with state space X. The chain evolves as follows. If the current state of the chain is $X_n = x$, then the density of the next state, X_{n+1}, is $k(\cdot \mid x)$. This Markov chain is the basis of the DA algorithm and we now describe some of its properties.

The product $k(x' \mid x) f_X(x)$ is symmetric in (x, x'). Indeed,

$$
k(x' \mid x) f_X(x) = f_X(x) \int_Y f_{X\mid Y}(x' \mid y) f_{Y\mid X}(y \mid x)\, dy = \int_Y \frac{f(x', y) f(x, y)}{f_Y(y)}\, dy.
$$

Thus, for all $x, x' \in X$,

$$
k(x' \mid x) f_X(x) = k(x \mid x') f_X(x'), \tag{10.2}
$$

which implies that the Markov chain X is *reversible* with respect to f_X (see, e.g. Ross, 1996, Section 4.7). Equation 10.2 is sometimes called the *detailed balance condition*. Integrating both sides of Equation 10.2 with respect to x yields

$$
\int_X k(x' \mid x) f_X(x)\, dx = f_X(x'), \tag{10.3}
$$

which shows that f_X is an *invariant density* for the Markov chain X. What does it mean for f_X to be invariant for X? To answer this question, note that the integrand in Equation 10.3 is the joint density of (X_0, X_1) when the starting value, X_0, is drawn from f_X. Thus, Equation 10.3 implies that, when $X_0 \sim f_X$, the marginal density of X_1 is also f_X. Actually, since X is a time homogeneous Markov chain, Equation 10.3 also implies that, if $X_n \sim f_X$, then $X_{n+1} \sim f_X$. Hence, a simple induction argument leads to the conclusion that, if $X_0 \sim f_X$, then the marginal density of X_n is f_X for all n. In other words, when $X_0 \sim f_X$, the Markov chain X is a sequence of *dependent* random vectors with density f_X. Of course, in practice, it will not be possible to start the chain by drawing X_0 from f_X. (If simulating directly from f_X is possible, then one should use classical Monte Carlo methods instead of the DA algorithm for the reasons laid out in Subsections 10.2.4 and 10.3.1.) Fortunately, as long as the Markov chain X is well behaved (see Section 10.2.1), the marginal density of X_n will *converge* to the invariant density f_X no matter how the chain is started. And, more importantly, the estimator $n^{-1} \sum_{i=0}^{n-1} g(X_i)$ will be strongly consistent for $E_{f_X} g$; that is, this estimator will converge almost surely to $E_{f_X} g$ as $n \to \infty$.

In order to keep things simple, we are considering only situations where f_X and $f(x, y)$ are densities with respect to Lebesgue measure. However, all of the results and methodology that we discuss in this chapter can easily be extended to a much more general setting. See, for example, Section 2 of Hobert and Marchev (2008).

Now consider the practical issue of simulating the Markov chain X. Given that the current state of the chain is $X_n = x$, how do we draw X_{n+1} from the Mtd $k(\cdot \mid x)$? The answer is based on a sequential simulation technique that we now describe. Suppose that we would like to simulate a random vector from some pdf $f_U(u)$, but that we cannot do this directly. Suppose further that f_U is the u-marginal of the joint pdf $f_{U,V}(u,v)$ and that we have the ability to make draws from $f_V(v)$ and from $f_{U \mid V}(u \mid v)$ for fixed v. If we draw $V \sim f_V(\cdot)$, and then, conditional on $V = v$, we draw $U \sim f_{U \mid V}(\cdot \mid v)$, then the observed pair, (u, v), is a draw from $f_{U,V}$, which means that u is a draw from f_U. This general technique will be employed many times throughout this chapter. We now explain how it is used to simulate from $k(\cdot \mid x)$.

Define

$$h(x', y \mid x) = f_{X \mid Y}(x' \mid y) f_{Y \mid X}(y \mid x),$$

and note that, for fixed $x \in \mathsf{X}$, $h(x', y \mid x)$ is a joint pdf in (x', y) with $\int_{\mathsf{Y}} h(x', y \mid x)\, dy = k(x' \mid x)$. We simply apply the technique described above with $k(\cdot \mid x)$ and $h(\cdot, \cdot \mid x)$ playing the roles of $f_U(\cdot)$ and $f_{U,V}(\cdot, \cdot)$, respectively. All we need is the y-marginal of $h(x', y \mid x)$, which is $f_{Y \mid X}(y \mid x)$, and the conditional density of X' given $Y = y$, which is

$$\frac{h(x', y \mid x)}{f_{Y \mid X}(y \mid x)} = f_{X \mid Y}(x' \mid y).$$

We now have a procedure for simulating one step of the DA algorithm. Indeed, if the current state is $X_n = x$, we simulate X_{n+1} as follows.

ONE ITERATION OF THE DA ALGORITHM

1. Draw $Y \sim f_{Y \mid X}(\cdot \mid x)$, and call the observed value y.
2. Draw $X_{n+1} \sim f_{X \mid Y}(\cdot \mid y)$.

So, as long as we can simulate from the conditional densities, $f_{X \mid Y}$ and $f_{Y \mid X}$, we can simulate the Markov chain X. (Note that, as mentioned above, we do not need $k(x' \mid x)$ in closed form.)

The name *data augmentation algorithm* comes from Tanner and Wong (1987) who used it to describe an iterative algorithm for approximating complex posterior distributions. On the last page of their paper, Tanner and Wong note that an "extreme" special case of their algorithm (in which their m is set equal to 1) yields a Markov chain whose transition density has the form (10.1). However, it does not appear to be the case that Tanner and Wong (1987) "invented" the DA algorithm (as we have defined it here), since other researchers, such as Swendsen and Wang (1987), were using it at about the same time. Here is our first example.

Example 10.1

Suppose that f_X is the standard normal density, $f_X(x) = e^{-x^2/2}/\sqrt{2\pi}$. Obviously, there is nothing intractable about this density. On the other hand, it is instructive to begin with a few simple examples in which the basic ideas of the algorithm are not overshadowed by the complexity of the target density. Take $f(x, y) = (\sqrt{2}\pi)^{-1} \exp\left\{ -(x^2 - \sqrt{2}xy + y^2) \right\}$, which is a bivariate normal density with means equal to zero, variances equal to one, and correlation equal to $1/\sqrt{2}$. The x-marginal is clearly standard normal and the two conditionals are also normal. Indeed,

$$Y \mid X = x \sim N\left(\frac{x}{\sqrt{2}}, \frac{1}{2}\right) \quad \text{and} \quad X \mid Y = y \sim N\left(\frac{y}{\sqrt{2}}, \frac{1}{2}\right).$$

Simulating from these conditionals is easy. For example, most statistically oriented programming languages, such as R (R Development Core Team, 2006), produce variates from the normal distribution and many other standard distributions. Hence, we have a viable DA algorithm that can be run by choosing an arbitrary starting value, $X_0 = x_0$, and then iterating the two-step procedure described above.

We now provide two more toy examples that will be put to good use. Two realistic examples are given in this section.

Example 10.2

Suppose that $f_X(x) = 3x^2 I_{(0,1)}(x)$. If we take $f(x, y) = 3x\, I(0 < y < x < 1)$, then the x-marginal is $f_X(x) = 3x^2 I_{(0,1)}(x)$ and the two conditional densities are given by

$$f_{Y|X}(y \mid x) = \frac{1}{x} I(0 < y < x) \quad \text{and} \quad f_{X|Y}(x \mid y) = \frac{2x}{1 - y^2} I(y < x < 1).$$

Simulating from these conditionals is straightforward. Indeed, if $U \sim U(0, 1)$, then $xU \sim f_{Y|X}(\cdot \mid x)$ and, using the probability integral transformation, $\sqrt{U(1 - y^2) + y^2} \sim f_{X|Y}(\cdot \mid y)$.

Example 10.3

Suppose that $f_X(x)$ is a Student's t density with 4 degrees of freedom,

$$f_X(x) = \frac{3}{8}\left(1 + \frac{x^2}{4}\right)^{\frac{-5}{2}}.$$

If we take

$$f(x, y) = \frac{4}{\sqrt{2\pi}} y^{\frac{3}{2}} \exp\left\{-y\left(\frac{x^2}{2} + 2\right)\right\} I_{(0,\infty)}(y),$$

then $\int_{\mathbb{R}} f(x, y)\, dy = f_X(x)$. Moreover, it is easy to show that $X|Y = y \sim N(0, y^{-1})$ and that $Y|X = x \sim \Gamma\left(\frac{5}{2}, \frac{x^2}{2} + 2\right)$. (We say that $W \sim \Gamma(\alpha, \beta)$ if its density is proportional to $w^{\alpha-1} e^{-w\beta} I(w > 0)$.)

The popularity of the DA algorithm is due in part to the fact that, given an intractable f_X, there are general techniques available for constructing a potentially useful joint density $f(x, y)$. Here is one such technique. Suppose that f_X can be factorized as $f_X(x) = q(x)l(x)$. Now define

$$f(x, y) = q(x) I_{(0, l(x))}(y),$$

and note that

$$\int_{\mathbb{R}} f(x,y) \, dy = q(x) \int_{\mathbb{R}} I_{(0,l(x))}(y) \, dy = q(x) \int_{0}^{l(x)} dy = q(x)l(x) = f_X(x).$$

A simple calculation shows that $Y \mid X = x \sim U(0, l(x))$, which is easy to sample. Thus, if it is also possible to draw from $f_{X\mid Y}(x \mid y) \propto q(x)I_{(y,\infty)}(l(x))$, then the DA algorithm can be applied. In this particular form, the DA algorithm is known as the *simple slice sampler* (Neal, 2003). The reader may verify that the DA algorithm developed in Example 10.2 is actually a simple slice sampler based on the factorization $f_X(x) = 3x^2 I_{(0,1)}(x) = \left[3xI_{(0,1)}(x)\right][x] = [q(x)][l(x)]$.

Another general technique for identifying an appropriate $f(x,y)$ involves the concept of *missing data* that underlies the EM algorithm (Dempster et al., 1977). This technique is applicable when the target, f_X, is a posterior density. Let z denote some observed data, which is assumed to be a sample from a member of a family of pdfs $\{p(z \mid \theta) : \theta \in \Theta\}$, where $\Theta \subset \mathbb{R}^p$. If $\pi(\theta)$ denotes the prior density, then the posterior density is given by $\pi(\theta \mid z) = p(z \mid \theta)\pi(\theta)/c(z)$, where $c(z) = \int_{\Theta} p(z \mid \theta)\pi(\theta) \, d\theta$ is the marginal density of the data. Assume that expectations with respect to $\pi(\theta \mid z)$ are intractable; that is, $\pi(\theta \mid z)$ is now playing the role of the problematic target $f_X(x)$.

Suppose that we can identify missing data $y \in Y \subset \mathbb{R}^q$ such that the joint density of z and y, call it $p(z, y \mid \theta)$, satisfies

$$\int_Y p(z, y \mid \theta) \, dy = p(z \mid \theta). \tag{10.4}$$

Finding such missing data is often straightforward. Indeed, the joint density $p(z, y \mid \theta)$ is precisely what is required to construct an EM algorithm for finding the maximum likelihood estimate of θ; that is, the maximizer of $p(z \mid \theta)$ over $\theta \in \Theta$ for fixed z. If such an EM algorithm already exists, we can simply use the corresponding missing data. Now define the *complete data posterior density* as

$$\pi(\theta, y \mid z) = \frac{p(z, y \mid \theta)\pi(\theta)}{\int_{\Theta} \int_Y p(z, y \mid \theta)\pi(\theta) \, dy \, d\theta} = \frac{p(z, y \mid \theta)\pi(\theta)}{\int_{\Theta} p(z \mid \theta)\pi(\theta) \, d\theta} = \frac{p(z, y \mid \theta)\pi(\theta)}{c(z)}.$$

The key feature of the complete data posterior density is that its θ-marginal is the target density, $\pi(\theta \mid z)$. Indeed,

$$\int_Y \pi(\theta, y \mid z) \, dy = \frac{\pi(\theta)}{c(z)} \int_Y p(z, y \mid \theta) \, dy = \frac{p(z \mid \theta)\pi(\theta)}{c(z)} = \pi(\theta \mid z).$$

When an EM algorithm is constructed, the missing data is chosen to make likelihood calculations under $p(z, y \mid \theta)$ much simpler than they are under the original density, $p(z \mid \theta)$. Such a choice will usually also result in conditional densities, $\pi(\theta \mid y, z)$ and $\pi(y \mid \theta, z)$, that are easy to sample. Regardless of whether or not our missing data came from a preexisting EM algorithm, as long as $\pi(\theta \mid y, z)$ and $\pi(y \mid \theta, z)$ can be straightforwardly sampled, we will have a viable DA algorithm with the complete data posterior density playing the role of $f(x,y)$. In particular, θ plays the role of x, and everything is done conditionally on the observed data z.

Example 10.4

Let $Z_1, \ldots, Z_m$ be a random sample from the location–scale Student's t density with known degrees of freedom, $v > 0$. The common density of the Z_i is given by

$$\frac{\Gamma\left(\frac{v+1}{2}\right)}{\sigma\sqrt{\pi v}\,\Gamma\left(\frac{v}{2}\right)}\left(1 + \frac{(z-\mu)^2}{v\sigma^2}\right)^{\frac{-(v+1)}{2}}.$$

Here (μ, σ^2) is playing the role of θ. The standard diffuse prior density for this location–scale problem is $\pi(\mu, \sigma^2) \propto 1/\sigma^2$. Of course, whenever an improper prior is used, it is important to check that the posterior is proper. In this case, the posterior is proper if and only if $m \geq 2$ (Fernández and Steel, 1999), and we assume this throughout. The posterior density is an intractable bivariate density that is characterized by

$$\pi(\mu, \sigma^2 \mid z) \propto \left(\sigma^2\right)^{\frac{-(m+2)}{2}} \prod_{i=1}^{m}\left(1 + \frac{(z_i-\mu)^2}{v\sigma^2}\right)^{\frac{-(v+1)}{2}},$$

where $z = (z_1, \ldots, z_m)$. Meng and van Dyk (1999) described a DA algorithm for this problem in which the missing data are based on the standard representation of a Student's t variate in terms of normal and χ^2 variates. Conditional on (μ, σ^2), let $(Z_1, Y_1), \ldots, (Z_m, Y_m)$ be independent and identically distributed (i.i.d.) pairs such that, for $i = 1, \ldots, m$,

$$Z_i \mid Y_i, \mu, \sigma^2 \sim N(\mu, \sigma^2/y_i)$$

$$Y_i \mid \mu, \sigma^2 \sim \Gamma(v/2, v/2).$$

In this case, $Y = \mathbb{R}_+^m$ where $\mathbb{R}_+ := (0, \infty)$. Letting $y = (y_1, \ldots, y_m)$, we have

$$p(z, y \mid \mu, \sigma^2) = \prod_{i=1}^{m} p(z_i \mid y_i, \mu, \sigma^2) p(y_i \mid \mu, \sigma^2)$$

$$= \prod_{i=1}^{m} \frac{\sqrt{y_i}}{\sqrt{2\pi\sigma^2}} \exp\left\{-\frac{y_i}{2\sigma^2}(z_i-\mu)^2\right\} \frac{\left(\frac{v}{2}\right)^{\frac{v}{2}}}{\Gamma\left(\frac{v}{2}\right)} y_i^{\frac{(v-2)}{2}} \exp\left\{-\frac{v y_i}{2}\right\}.$$

Now,

$$\int_Y p(z, y \mid \mu, \sigma^2)\, dy = \prod_{i=1}^{m} \int_{\mathbb{R}_+} p(z_i \mid y_i, \mu, \sigma^2) p(y_i \mid \mu, \sigma^2)\, dy_i$$

$$= \prod_{i=1}^{m} \frac{\Gamma\left(\frac{v+1}{2}\right)}{\sigma\sqrt{\pi v}\,\Gamma\left(\frac{v}{2}\right)}\left(1 + \frac{(z_i-\mu)^2}{v\sigma^2}\right)^{\frac{-(v+1)}{2}},$$

so Equation 10.4 is satisfied. The complete data posterior density is characterized by

$$\pi\left((\mu, \sigma^2), y \mid z\right) \propto \frac{1}{\sigma^2} \prod_{i=1}^{m} \frac{\sqrt{y_i}}{\sqrt{2\pi\sigma^2}} \exp\left\{-\frac{y_i}{2\sigma^2}(z_i-\mu)^2\right\} \frac{\left(\frac{v}{2}\right)^{v/2}}{\Gamma\left(\frac{v}{2}\right)} y_i^{\frac{(v-2)}{2}} \exp\left\{-\frac{v y_i}{2}\right\}. \quad (10.5)$$

In order to implement the DA algorithm, we must be able to draw from $\pi(y \mid \mu, \sigma^2, z)$ and from $\pi(\mu, \sigma^2 \mid y, z)$. Since $\pi(y \mid \mu, \sigma^2, z) \propto \pi(\mu, \sigma^2, y \mid z)$, it is clear that the y_i are conditionally independent given (μ, σ^2, z) and, in fact,

$$Y_i \mid \mu, \sigma^2, z \sim \Gamma\left(\frac{\nu+1}{2}, \frac{1}{2}\left(\frac{(z_i - \mu)^2}{\sigma^2} + \nu\right)\right). \tag{10.6}$$

We can simulate from $\pi(\mu, \sigma^2 | y, z)$ sequentially by first drawing from $\pi(\sigma^2 | y, z)$ and then from $\pi(\mu | \sigma^2, y, z)$. (Remember our sequential method of drawing from $f_{U,V}$?) Let $y. = \sum_{i=1}^{m} y_i$ and define

$$\hat{\mu} = \frac{1}{y.}\sum_{j=1}^{m} z_j y_j \quad \text{and} \quad \hat{\sigma}^2 = \frac{1}{y.}\sum_{j=1}^{m} y_j (z_j - \hat{\mu})^2.$$

Using the fact that $\pi(\mu | \sigma^2, y, z) \propto \pi(\mu, \sigma^2, y | z)$, it is straightforward to show that

$$\mu | \sigma^2, y, z \sim N\left(\hat{\mu}, \frac{\sigma^2}{y.}\right). \tag{10.7}$$

Finally, $\pi(\sigma^2 | y, z)$ is proportional to what remains when μ is integrated out of Equation 10.5. This integral can be computed in closed form and it follows that

$$\sigma^2 | y, z \sim \Gamma^{-1}\left(\frac{m+1}{2}, \frac{y.\hat{\sigma}^2}{2}\right), \tag{10.8}$$

where $\Gamma^{-1}(\alpha, \beta)$ is the distribution of $1/W$ when $W \sim \Gamma(\alpha, \beta)$. We now know how to run the DA algorithm for this problem. Given the current state, $X_n = (\mu, \sigma^2)$, we simulate the next state, $X_{n+1} = (\mu_{n+1}, \sigma^2_{n+1})$, by performing the following two steps:

1. Draw $Y_1, \ldots, Y_m$ independently according to Equation 10.6, and call the result $y = (y_1, \ldots, y_m)$.
2. Draw σ^2_{n+1} according to Equation 10.8, and then draw μ_{n+1} according to Equation 10.7 with σ^2_{n+1} in place of σ^2.

The algorithm described above is actually a special case of a more general DA algorithm developed by Meng and van Dyk (1999) that can handle observations from the *multivariate* location–scale Student's *t* density.

We end this section by describing Albert and Chib's (1993) DA algorithm for Bayesian probit regression, which is one of the most widely used DA algorithms.

Example 10.5

Let $Z_1, \ldots, Z_m$ be independent Bernoulli random variables such that $\Pr(Z_i = 1) = \Phi(v_i^T \beta)$, where v_i is a $p \times 1$ vector of known covariates associated with Z_i, β is a $p \times 1$ vector of unknown regression coefficients and $\Phi(\cdot)$ denotes the standard normal distribution function. We have

$$\Pr(Z_1 = z_1, \ldots, Z_m = z_m \mid \beta) = \prod_{i=1}^{m} \left[\Phi(v_i^T \beta)\right]^{z_i}\left[1 - \Phi(v_i^T \beta)\right]^{1-z_i},$$

where each z_i is binary—either 0 or 1. Consider a Bayesian analysis that employs a flat prior on β. Letting $z = (z_1, \ldots, z_m)$ denote the observed data, the marginal density is given by

$$c(z) = \int_{\mathbb{R}^p} \prod_{i=1}^{m} \left[\Phi(v_i^T \beta)\right]^{z_i} \left[1 - \Phi(v_i^T \beta)\right]^{1-z_i} d\beta.$$

Chen and Shao (2000) provide necessary and sufficient conditions on z and $\{v_i\}_{i=1}^m$ for propriety of the posterior; that is, for $c(z) < \infty$. We assume throughout that these conditions are satisfied. The intractable posterior density of β is given by

$$\pi(\beta \mid z) = \frac{1}{c(z)} \prod_{i=1}^{m} \left[\Phi(v_i^T \beta)\right]^{z_i} \left[1 - \Phi(v_i^T \beta)\right]^{1-z_i}.$$

Albert and Chib (1993) developed a DA algorithm for this problem. Let $\phi(u; \mu, \kappa^2)$ denote the $N(\mu, \kappa^2)$ density function evaluated at the point $u \in \mathbb{R}$. Also, let $\mathbb{R}_- = (-\infty, 0)$, let $y = (y_1, \ldots, y_m)^T \in \mathbb{R}^m$, and consider the function

$$\pi(\beta, y \mid z) = \frac{1}{c(z)} \prod_{i=1}^{m} \left\{ I_{\mathbb{R}_+}(y_i) I_{\{1\}}(z_i) + I_{\mathbb{R}_-}(y_i) I_{\{0\}}(z_i) \right\} \phi(y_i; v_i^T \beta, 1). \tag{10.9}$$

Integrating y out of $\pi(\beta, y \mid z)$, we have

$$\frac{1}{c(z)} \int_{\mathbb{R}} \int_{\mathbb{R}} \cdots \int_{\mathbb{R}} \prod_{i=1}^{m} \left\{ I_{\mathbb{R}_+}(y_i) I_{\{1\}}(z_i) + I_{\mathbb{R}_-}(y_i) I_{\{0\}}(z_i) \right\} \phi(y_i; v_i^T \beta, 1) \, dy_m \ldots dy_2 \, dy_1$$

$$= \frac{1}{c(z)} \prod_{i=1}^{m} \int_{\mathbb{R}} \left\{ I_{\mathbb{R}_+}(y_i) I_{\{1\}}(z_i) + I_{\mathbb{R}_-}(y_i) I_{\{0\}}(z_i) \right\} \phi(y_i; v_i^T \beta, 1) \, dy_i$$

$$= \frac{1}{c(z)} \prod_{i=1}^{m} \left\{ I_{\{1\}}(z_i) \int_0^{\infty} \phi(y_i; v_i^T \beta, 1) \, dy_i + I_{\{0\}}(z_i) \int_{-\infty}^0 \phi(y_i; v_i^T \beta, 1) \, dy_i \right\}$$

$$= \frac{1}{c(z)} \prod_{i=1}^{m} \left\{ I_{\{1\}}(z_i) \Phi(v_i^T \beta) + I_{\{0\}}(z_i) \left[1 - \Phi(v_i^T \beta)\right] \right\}$$

$$= \frac{1}{c(z)} \prod_{i=1}^{m} \left[\Phi(v_i^T \beta)\right]^{z_i} \left[1 - \Phi(v_i^T \beta)\right]^{1-z_i}$$

$$= \pi(\beta \mid z).$$

Hence, $\pi(\beta, y \mid z)$ is a joint density in (β, y) whose β-marginal is $\pi(\beta \mid z)$. Albert and Chib's (1993) DA algorithm is based on this joint density. We now derive the conditional densities, $\pi(\beta \mid y, z)$ and $\pi(y \mid \beta, z)$. Let V denote the $m \times p$ matrix whose ith row is v_i^T. (A necessary condition for propriety is that V have rank p.) Standard linear model-type calculations show that

$$\prod_{i=1}^{m} \phi(y_i; v_i^T \beta, 1) = (2\pi)^{-m/2} e^{-\frac{1}{2} y^T (I-H) y} \exp\left\{ -\frac{1}{2} (\beta - \hat{\beta}(y))^T V^T V (\beta - \hat{\beta}(y)) \right\}, \tag{10.10}$$

where $\hat{\beta}(y) = (V^T V)^{-1} V^T y$ and $H = V(V^T V)^{-1} V^T$. This implies that $\pi(\beta \mid y, z)$ is a p-variate normal density with mean $\hat{\beta}(y)$ and covariance matrix $(V^T V)^{-1}$.

Finally, let $TN(\mu, \kappa^2, u)$ denote a normal distribution with mean μ and variance κ^2 that is *truncated* to be positive if $u = 1$ and negative if $u = 0$. It is clear from Equation 10.9 that, given β and z, the Y_i are independent with $Y_i \sim TN(v_i^T \beta, 1, z_i)$. We now know exactly how to implement the DA algorithm. Given the current state, $X_n = \beta$, we simulate the next state, X_{n+1}, by performing the following two steps:

1. Draw $Y_1, \ldots, Y_m$ independently such that $Y_i \sim TN(v_i^T \beta, 1, z_i)$, and call the result $y = (y_1, \ldots, y_m)^T$.
2. Draw $X_{n+1} \sim N(\hat{\beta}(y), (V^T V)^{-1})$.

See Robert (1995) for an efficient method of simulating truncated normal random variables.

In the next section, we describe the theoretical properties of the Markov chain underlying the DA algorithm.

10.2   Properties of the DA Markov Chain

10.2.1   Basic Regularity Conditions

In Section 10.1 we described how to construct and simulate a Markov chain, X, that has the intractable target, f_X, as an invariant density. Unfortunately, without additional assumptions, there is no guarantee that this chain will be useful for approximating expectations with respect to f_X. Here is a simple example from Roberts and Smith (1994) that illustrates one of the potential problems.

Example 10.6

Suppose that $f_X(x) = \frac{1}{2} I_{(0,2)}(x)$. If we take

$$f(x, y) = \frac{1}{2} \Big[I_{(0,1)}(x) I_{(0,1)}(y) + I_{[1,2)}(x) I_{[1,2)}(y) \Big],$$

then $\int_{\mathbb{R}} f(x, y) \, dy = \frac{1}{2} I_{(0,2)}(x)$ and

$$f_{X|Y}(x \mid y) = f_{Y|X}(y \mid x) = I_{(0,1)}(x) I_{(0,1)}(y) + I_{[1,2)}(x) I_{[1,2)}(y).$$

Since the x-marginal of $f(x, y)$ is f_X and simulation from the conditionals is easy, there is a DA algorithm based on $f(x, y)$. However, this algorithm is useless from a practical standpoint because the underlying Markov chain is not *irreducible*. For example, suppose we start the chain at $x_0 = \frac{1}{2}$, and consider applying the two-step method to simulate X_1. First, we draw $Y \sim U(0, 1)$ and then, no matter what the result, we will draw $X_1 \sim U(0, 1)$. Continuing along these lines shows that the chain will be stuck forever in the set $(0, 1)$. Hence, there is no sense in which the chain converges to f_X.

If the Markov chain X is ψ-irreducible, aperiodic and Harris recurrent, then the DA algorithm can be employed to effectively explore the intractable target density, f_X. When X satisfies these three properties, we call it *Harris ergodic*. Unfortunately, a good bit of technical Markov chain theory must be developed before these conditions can even be formally stated (Meyn and Tweedie, 1993; Roberts and Rosenthal, 2004). To avoid a lengthy technical discussion, we simply provide one sufficient condition for Harris ergodicity of X that is easy to check and holds for all of our examples and for many other DA algorithms that are used in practice.

Define a condition (which we will refer to as condition $\mathcal{K}$) on the Mtd k as follows:

$$k(x' \mid x) > 0, \quad \text{for all } x', x \in X.$$

Condition $\mathcal{K}$ implies that the Markov chain X is Harris ergodic (see, e.g. Tan, 2008). In fact, condition $\mathcal{K}$ implies that it is possible for the chain to move from any point $x \in X$ to any "big" set in a single step. To make this precise, let λ denote Lebesgue measure on X and let $P(\cdot, \cdot)$ denote the Markov transition function of the chain; that is, for $x \in X$ and a measurable set A,

$$P(x, A) = \Pr\left(X_{n+1} \in A \mid X_n = x\right) = \int_A k(x' \mid x)\, dx'.$$

Under condition $\mathcal{K}$, if A is big in the sense that $\lambda(A) > 0$, then

$$P(x, A) = \int_A k(x' \mid x)\, dx' > 0,$$

which means that there is positive probability of moving from x to A in a single step. Recall that

$$k(x' \mid x) = \int_Y f_{X|Y}(x' \mid y) f_{Y|X}(y \mid x)\, dy.$$

Clearly, if $f(x, y)$ is strictly positive on $X \times Y$, then condition $\mathcal{K}$ holds and the Markov chain X is Harris ergodic. We now check that the Markov chains developed in the examples of Section 10.1 are indeed Harris ergodic.

Examples 10.1 and 10.3 (cont.)

In Example 10.1, we have $X = Y = \mathbb{R}$, while in Example 10.3, $X = \mathbb{R}$, $Y = \mathbb{R}_+$. In both cases, $f(x, y)$ is strictly positive on $X \times Y$. Hence, the Markov chains underlying the DA algorithms in Examples 10.1 and 10.3 are Harris ergodic.

Example 10.4 (cont.)

The role of X is played by $\Theta = \mathbb{R} \times \mathbb{R}_+$ and $Y = \mathbb{R}_+^m$. Note that the complete data posterior density (10.5) is strictly positive for all $\left((\mu, \sigma^2), y\right) \in \Theta \times Y$. Hence, the chain X is Harris ergodic.

Example 10.5 (cont.)

In this case, $X = \mathbb{R}^p$ and Y is a Cartesian product of m half-lines ($\mathbb{R}_+$ and $\mathbb{R}_-$), where the ith component is $\mathbb{R}_+$ if $z_i = 1$, and $\mathbb{R}_-$ if $z_i = 0$. It is clear that the joint density (Equation 10.9) is strictly positive on $X \times Y$, and this implies that the Markov chain underlying Albert and Chib's (1993) algorithm is Harris ergodic.

Even when $f(x, y)$ is not strictly positive on $X \times Y$, it is still often the case that condition $\mathcal{K}$ holds.

Example 10.2 (cont.)

The joint density is given by $f(x, y) = 3xI(0 < y < x < 1)$, which is not strictly positive on $X \times Y = (0, 1) \times (0, 1)$. However, we can show directly that condition $\mathcal{K}$ holds. Indeed, for fixed $x \in (0, 1)$, we have

$$k(x' \mid x) = \int_{\mathbb{R}} \frac{2x'}{x(1 - y^2)} I(0 < y < x) I(y < x' < 1) \, dy$$

$$= \frac{2x'}{x} I(0 < x' < 1) \int_0^{\min\{x, x'\}} \frac{1}{1 - y^2} \, dy$$

$$= \frac{x'}{x} \log \left(\frac{1 + \min\{x, x'\}}{1 - \min\{x, x'\}} \right) I(0 < x' < 1).$$

Hence, $k(x' \mid x)$ is strictly positive for all $x', x \in (0, 1)$ and Harris ergodicity follows. Actually, it is intuitively clear that the Markov chain has a positive probability of moving from any $x \in (0, 1)$ to any set $A \subset (0, 1)$ with $\lambda(A) > 0$ in one step. Indeed, to get from x to the new state, we first draw $Y \sim U(0, x)$, and then, given $Y = y$, the new state is drawn from a density with support $(y, 1)$. Therefore, as long as the observed y is small enough, there will be a positive (conditional) probability of the new state being in any open set in $(0, 1)$.

It is not difficult to create examples of well-behaved DA algorithms for which condition $\mathcal{K}$ fails to hold. Fortunately, there are many general results available for establishing that X is Harris ergodic in such situations; see, for example, Roberts and Smith (1994), Tierney (1994), Roberts and Rosenthal (2006) and Hobert et al. (2007). In the next subsection, we describe exactly what Harris ergodicity buys us.

10.2.2 Basic Convergence Properties

If X is Harris ergodic, then, no matter how the chain is started, the marginal distribution of X_n will converge to (the distribution associated with) f_X, and an analog of the strong law of large numbers (SLLN) holds. To make this precise, some additional notation is required. Define the n-step Markov transition function as

$$P^n(x, A) = \Pr \left(X_n \in A \mid X_0 = x \right),$$

so $P^1 \equiv P$. Also, let $\phi(\cdot)$ denote the probability measure corresponding to f_X; that is, for measurable A, $\phi(A) = \int_A f_X(x) \, dx$. If X is Harris ergodic, then the total variation distance between the probability measures $P^n(x, \cdot)$ and $\phi(\cdot)$ decreases to 0 as n gets large. In symbols,

$$\|P^n(x, \cdot) - \phi(\cdot)\| \downarrow 0 \quad \text{as } n \to \infty, \tag{10.11}$$

where

$$\|P^n(x, \cdot) - \phi(\cdot)\| := \sup_A \left| P^n(x, A) - \phi(A) \right|.$$

Harris ergodicity is also sufficient for the ergodic theorem, which is the Markov chain version of the SLLN. Let $L^1(f_X)$ denote the set of functions $h : X \to \mathbb{R}$ such that

$$\int_X |h(x)| f_X(x)\, dx < \infty,$$

and, for $h \in L^1(f_X)$, define $E_{f_X} h = \int_X h(x) f_X(x)\, dx$. The ergodic theorem implies that, if $g \in L^1(f_X)$, then, no matter what the distribution of X_0, we have

$$\bar{g}_n := \frac{1}{n} \sum_{i=0}^{n-1} g(X_i) \to E_{f_X} g$$

almost surely as $n \to \infty$; that is, $\bar{g}_n$ is a strongly consistent estimator of $E_{f_X} g$. The ergodic theorem justifies estimating $E_{f_X} g$ with $\bar{g}_n$ where X_0 is any point (or has any distribution) from which it is convenient to start the simulation. An important practical question that this basic theory does not answer is "What is an appropriate value of n?" Tools for answering this question will be presented in Section 10.3. For now, we simply point out that all rigorous methods of choosing an appropriate (Markov chain) Monte Carlo sample size are based on a central limit theorem (CLT) for $\bar{g}_n$. Assuming that $\int_X g^2(x) f_X(x)\, dx < \infty$, a simple sufficient condition for the existence of such a CLT is that the Markov chain, X, converge to its stationary distribution at a geometric rate.

10.2.3 Geometric Ergodicity

Assume that X is Harris ergodic. Note that Equation 10.11 gives no information about the *rate* at which the total variation distance converges to 0. There are important practical benefits to using a DA algorithm for which this rate is (at least) geometrically fast. Formally, the chain X is called *geometrically ergodic* if there exist a function $M : X \to [0, \infty)$ and a constant $\rho \in [0, 1)$ such that, for all $x \in X$ and all $n = 1, 2, \ldots,$

$$\|P^n(x, \cdot) - \phi(\cdot)\| \le M(x)\, \rho^n. \tag{10.12}$$

Unfortunately, Harris ergodicity does not imply geometric ergodicity. The most straightforward method of proving that the Harris ergodic chain X is geometrically Harris ergodic is by establishing a certain type of *drift condition*, which we now introduce.

A function $V : X \to [0, \infty)$ is said to be *unbounded off compact sets* if, for each $\beta \in \mathbb{R}$, the sub-level set $\{x \in X : V(x) \le \beta\}$ is compact. We say that a *geometric drift condition* holds if there exist a $V : X \to [0, \infty)$ that is unbounded off compact sets, and constants $\lambda \in [0, 1)$ and $L \in \mathbb{R}$ such that

$$E\big[V(X_{n+1}) \mid X_n = x\big] \le \lambda V(x) + L. \tag{10.13}$$

The function V is called the *drift function*. If $f(x, y) > 0$ for all $(x, y) \in X \times Y$, then the existence of a geometric drift condition implies that X is geometrically ergodic (Tan, 2008). (See Meyn and Tweedie (1993, Chapter 15) for similar results that hold when $f(x, y)$ is not strictly positive.) In practice, establishing a geometric drift condition is simply a matter of trial and error (and a lot of analysis). We now provide some pointers on calculating the expectation in Equation 10.13.

Note that the left-hand side of Equation 10.13 can be rewritten as

$$E\big[V(X_{n+1}) \mid X_n = x\big] = \int_X V(x')\,k(x' \mid x)\,dx'$$

$$= \int_X V(x')\left[\int_Y f_{X|Y}(x' \mid y) f_{Y|X}(y \mid x)\,dy\right]dx' \qquad (10.14)$$

$$= \int_Y \left[\int_X V(x') f_{X|Y}(x' \mid y)\,dx'\right] f_{Y|X}(y \mid x)\,dy.$$

Thus, the expectation can be computed (or bounded) in two steps. The first step is to compute (or bound) the expectation of $V(X')$ with respect to $f_{X|Y}(\cdot \mid y)$; call the result $e(y)$. The second step entails calculating (or bounding) the expectation of $e(Y)$ with respect to $f_{Y|X}(\cdot \mid x)$. The fact that we are able to simulate straightforwardly from $f_{X|Y}(x \mid y)$ and $f_{Y|X}(y \mid x)$ often means that these conditional densities are easy to handle from an analytical standpoint. Hence, it is usually possible to calculate (or, at least get sharp upper bounds on) expectations with respect to $f_{X|Y}(x \mid y)$ and $f_{Y|X}(y \mid x)$. We now give two simple examples illustrating how to prove that a DA algorithm is geometrically ergodic by establishing a geometric drift condition.

Example 10.3 (cont.)

Recall that $f(x, y)$ is strictly positive on $X \times Y$, so the drift technique can be used to establish geometric convergence in this example. Consider the drift function $V(x) = x^2$. For $\beta < 0$ the sublevel set $\{x \in X : V(x) \le \beta\}$ is the empty set, for $\beta = 0$ it is the set $\{0\}$, and for $\beta > 0$ it is a closed interval. Thus, V is unbounded off compact sets. Recall that $X \mid Y = y \sim N(0, y^{-1})$. Hence, the "inner expectation" in Equation 10.14 can be evaluated as follows:

$$E\big[V(X') \mid y\big] = E\big[(X')^2 \mid y\big] = \frac{1}{y}.$$

Now, using the fact that $Y \mid X = x \sim \Gamma\big(\frac{5}{2}, \frac{x^2}{2} + 2\big)$ yields

$$E\big[V(X_{n+1}) \mid X_n = x\big] = E\big[Y^{-1} \mid x\big] = \frac{1}{3}x^2 + \frac{4}{3} = \frac{1}{3}V(x) + \frac{4}{3}.$$

We have established that Equation 10.13 holds with $\lambda = \frac{1}{3}$ and $L = \frac{4}{3}$, and this shows that the Markov chain underlying this DA algorithm is geometrically ergodic.

In the toy example just considered, we were able to compute $E\big[V(X_{n+1}) \mid X_n = x\big]$ exactly and, luckily, the final expression involved the function $V(x)$ in exactly the right way. Establishing geometric drift conditions in real examples is typically much more difficult, and often involves what Fill et al. (2000) describe as "difficult theoretical analysis." Geometric drift conditions have been established for the Markov chains underlying the DA algorithms in Examples 10.4 and 10.5 (Marchev and Hobert, 2004; Roy and Hobert, 2007), but these calculations are too involved to present in this chapter. The next example is still a toy example, in the sense that the intractable target density is univariate, but it does provide a nice illustration of the type of bounding that is required in real examples.

Example 10.7

Consider a simplification of the Student's t setup in Example 10.4 where the variance is known and equal to 1. In this case, the posterior density is an intractable univariate density given by

$$\pi(\mu \mid z) \propto \prod_{i=1}^{m} \left(1 + \frac{(z_i - \mu)^2}{v} \right)^{\frac{-(v+1)}{2}},$$

where $z = (z_1, \ldots, z_m)$. Using the same missing data as before, the complete data posterior, $\pi(\mu, y \mid z)$, is proportional to the right-hand side of Equation 10.5 with σ^2 set equal to 1. Note that $\pi(\mu, y \mid z)$ is strictly positive on $X \times Y = \Theta \times Y = \mathbb{R} \times \mathbb{R}_+^m$. Of course, to run the DA algorithm, we need to be able to draw from $\pi(y \mid \mu, z)$ and from $\pi(\mu \mid y, z)$. Recall that $\hat{\mu} = \hat{\mu}(y) = \frac{1}{y.} \sum_{j=1}^{m} z_j y_j$. (Since the data, z, is fixed, we suppress this dependence in the notation.) It is easy to show that $\mu \mid y, z \sim N\left(\hat{\mu}(y), \frac{1}{y.}\right)$ and that the y_is are conditionally independent given (μ, z) with

$$Y_i \mid \mu, z \sim \Gamma\left(\frac{v+1}{2}, \frac{(z_i - \mu)^2 + v}{2} \right).$$

For notational convenience, we will denote the DA Markov chain as $\{\mu_n\}_{n=0}^{\infty}$ (instead of the usual $\{X_n\}_{n=0}^{\infty}$). The Mtd of the DA algorithm is then given by

$$k(\mu' \mid \mu) = \int_Y \pi(\mu' \mid y, z) \, \pi(y \mid \mu, z) \, dy.$$

We now show that this Markov chain is geometrically ergodic as long as $v > 1$ and $m > 1/(v-1)$. The drift function we use is $V(\mu) = \sum_{i=1}^{m} (z_i - \mu)^2$. It is easy to see that V is unbounded off compact sets. Indeed, fix $\beta \in \mathbb{R}$ and consider the sub-level set $\{\mu \in \mathbb{R} : V(\mu) \leq \beta\}$. Let $\bar{z} = m^{-1} \sum_{i=1}^{m} z_i$. If $\beta < \sum_{i=1}^{m} (z_i - \bar{z})^2$, then the sub-level set is the empty set, and if $\beta \geq \sum_{i=1}^{m} (z_i - \bar{z})^2$, the sub-level set is a closed interval.

Let z_* and z^* denote the minimum and the maximum of the z_i, respectively. Since $\hat{\mu}(y)$ is a convex combination of $z_1, \ldots, z_m$, it follows that $\hat{\mu}(y) \in [z_*, z^*]$ for all $y \in Y$. The inner expectation in Equation 10.14 can now be bounded as follows:

$$E\left[V(\mu') \mid y, z \right] = E\left[\sum_{i=1}^{m} (z_i - \mu')^2 \mid y, z \right]$$

$$= \sum_{i=1}^{m} E\left[(z_i - \mu')^2 \mid y, z \right]$$

$$= \sum_{i=1}^{m} \text{var}[(z_i - \mu') \mid y, z] + \sum_{i=1}^{m} \left\{ E[(z_i - \mu') \mid y, z] \right\}^2$$

$$= \sum_{i=1}^{m} \text{var}[\mu' \mid y, z] + \sum_{i=1}^{m} \left\{ z_i - E[\mu' \mid y, z] \right\}^2$$

$$= \frac{m}{y.} + \sum_{i=1}^{m} (z_i - \hat{\mu}(y))^2$$

$$\leq \frac{m}{y.} + m(z^* - z_*)^2.$$

Now, since the harmonic mean is less than or equal to the arithmetic mean, we have

$$\frac{m}{y_\cdot} = \frac{1}{\frac{1}{m}\sum_{i=1}^{m}\frac{1}{y_i^{-1}}} \leq \frac{1}{m}\sum_{i=1}^{m}y_i^{-1}.$$

We conclude that

$$E[V(\mu')\,|\,y, z] \leq \frac{1}{m}\sum_{i=1}^{m}y_i^{-1} + m(z^* - z_*)^2.$$

Therefore, as long as $\nu > 1$, we have

$$E[V(\mu_{n+1})\,|\,\mu_n = \mu] \leq E\left[\left(\frac{1}{m}\sum_{i=1}^{m}Y_i^{-1} + m(z^* - z_*)^2\right)\,\Big|\,\mu, z\right]$$

$$= \frac{1}{m}\sum_{i=1}^{m}E[Y_i^{-1}\,|\,\mu, z] + m(z^* - z_*)^2$$

$$= \frac{1}{m(\nu-1)}\sum_{i=1}^{m}\left[(z_i - \mu)^2 + \nu\right] + m(z^* - z_*)^2$$

$$= \frac{1}{m(\nu-1)}\sum_{i=1}^{m}(z_i - \mu)^2 + \frac{\nu}{(\nu-1)} + m(z^* - z_*)^2$$

$$= \frac{1}{m(\nu-1)}V(\mu) + \frac{\nu}{(\nu-1)} + m(z^* - z_*)^2.$$

We have established that, when $\nu > 1$, Equation 10.13 holds with $\lambda = \frac{1}{m(\nu-1)}$. Thus, the Markov chain is geometrically ergodic whenever $\nu > 1$ and $m(\nu - 1) > 1$.

Of course, the fact that our analysis did not lead to a geometric drift condition for the (extreme) situations where $\nu \leq 1$ and/or $m(\nu - 1) \leq 1$ *does not imply* that the DA chain converges at a sub-geometric rate in those cases. Indeed, it may be the case that a more delicate analysis of $E[V(\mu_{n+1})\,|\,\mu_n = \mu]$ would show that these chains are geometric as well. Or we might have to resort to changing the drift function. Unfortunately, there are currently no simple methods of proving that a DA chain is *not* geometrically ergodic.

The drift method that we have described and illustrated in this subsection provides only *qualitative* information about the rate of convergence in the sense that, once (10.13) has been established, all we can say is that *there exist* M and ρ satisfying Equation 10.12, but we cannot say what they are. There are other (more complicated) versions of this method that, in addition to establishing the existence of M and ρ, provide an upper bound on $M(x)\rho^n$ that decreases to zero geometrically in n. These methods were developed and refined in a series of papers beginning with Meyn and Tweedie (1994) and Rosenthal (1995); for an overview, see Jones and Hobert (2001). The final subsection of this chapter concerns CLTs for the estimator $\bar{g}_n$.

10.2.4 Central Limit Theorems

Harris ergodicity alone does not imply the existence of CLTs. However, as we now explain, if the DA Markov chain, X, is geometrically Harris ergodic, then there will be CLTs for square integrable functions. Let $L^2(f_X)$ denote the set of functions $h : X \to \mathbb{R}$ such that

$$\int_X h^2(x) f_X(x) \, dx < \infty.$$

Assume that $g \in L^2(f_X)$ and define $c_k = \text{cov}[g(X_0), g(X_k)]$ for $k \in \{1, 2, 3, \ldots\}$, where the covariances are calculated under the assumption that $X_0 \sim f_X$. For example,

$$c_1 = \int_X \int_X \big(g(x') - E_{f_X}g\big)\big(g(x) - E_{f_X}g\big)k(x' \mid x) f_X(x) \, dx \, dx',$$

where we have used the fact that $X_0 \sim f_X$ implies that $X_1 \sim f_X$, so the expected value of $g(X_1)$ is $E_{f_X}g$. Liu et al. (1994, Lemma 3.2) noted that this expression can be rearranged as follows:

$$
\begin{aligned}
c_1 &= \int_X \int_X \big(g(x') - E_{f_X}g\big)\big(g(x) - E_{f_X}g\big)k(x' \mid x) f_X(x) \, dx \, dx' \\
&= \int_X \int_X \big(g(x') - E_{f_X}g\big)\big(g(x) - E_{f_X}g\big)\left[\int_Y f_{X \mid Y}(x' \mid y) f_{Y \mid X}(y \mid x) \, dy\right] f_X(x) \, dx \, dx' \\
&= \int_Y \int_X \int_X \big(g(x') - E_{f_X}g\big)\big(g(x) - E_{f_X}g\big) f_{X \mid Y}(x' \mid y) f_{X \mid Y}(x \mid y) f_Y(y) \, dx \, dx' \, dy \\
&= \int_Y \left[\int_X \big(g(x) - E_{f_X}g\big) f_{X \mid Y}(x \mid y) \, dx\right]^2 f_Y(y) \, dy \\
&= \text{var}\Big\{E\big[\big(g(X') - E_{f_X}g\big) \mid Y'\big]\Big\},
\end{aligned}
$$

where $(X', Y') \sim f(x, y)$. This shows that $c_1 > 0$. In fact, this result can be used in conjunction with the reversibility of X to show that $c_k > 0$ for all $k \in \{1, 2, 3, \ldots\}$.

Assume that X is geometrically Harris ergodic and that $g \in L^2(f_X)$. As before, put $\bar{g}_n = \frac{1}{n}\sum_{i=0}^{n-1} g(X_i)$. Define $\sigma^2 = E_{f_X}g^2 - (E_{f_X}g)^2$ and $\kappa^2 = \sigma^2 + 2\sum_{k=1}^{\infty} c_k$. Results in Roberts and Rosenthal (1997) and Chan and Geyer (1994) imply that $\kappa^2 < \infty$ and that, as $n \to \infty$,

$$\sqrt{n}\big(\bar{g}_n - E_{f_X}g\big) \xrightarrow{d} N(0, \kappa^2). \tag{10.15}$$

This CLT *does not require* that $X_0 \sim f_X$—it holds for all starting distributions, including degenerate ones. We note that the reversibility of X plays a major role in the existence of the CLT (10.15). In the next section, we explain how to consistently estimate the asymptotic variance, κ^2. But first, we briefly compare the estimators of $E_{f_X}g$ based on DA and classical Monte Carlo.

Let $X_1^*, X_2^*, \ldots$ be an i.i.d. sequence from f_X. The classical Monte Carlo estimator of $E_{f_X}g$ is $\bar{g}_n^* := \frac{1}{n}\sum_{i=1}^{n} g(X_i^*)$. If $g \in L^1(f_X)$, then, by the SLLN, $\bar{g}_n^*$ is a strongly consistent estimator of $E_{f_X}g$. If, in addition, $g \in L^2(f_X)$, then standard results from i.i.d. theory tell us that, as $n \to \infty$,

$$\sqrt{n}\big(\bar{g}_n^* - E_{f_X}g\big) \xrightarrow{d} N(0, \sigma^2). \tag{10.16}$$

If $c_1 \neq 0$ (as will typically be the case), then $\kappa^2/\sigma^2 > 1$, so the asymptotic relative efficiency (ARE) of $\bar{g}_n^*$ with respect to $\bar{g}_n$ is larger than one. Therefore, if it is possible to make an i.i.d.

draw from f_X, and the computational effort of doing so is similar to the effort of simulating a single iteration of the DA algorithm, then the classical Monte Carlo estimator is to be preferred over the estimator based on the DA algorithm. In the next section, we explain how these CLTs can be used in practice to choose an appropriate Monte Carlo sample size.

10.3 Choosing the Monte Carlo Sample Size

10.3.1 Classical Monte Carlo

We begin by describing how the Monte Carlo sample size is chosen in the classical Monte Carlo context. Assume that $g \in L^2(f_X)$ and recall that the classical Monte Carlo estimator of $E_{f_X} g$ is $\bar{g}_n^* = \frac{1}{n} \sum_{i=1}^{n} g(X_i^*)$, where $X_1^*, X_2^*, \ldots$ are i.i.d. from f_X. The main motivation for using $\bar{g}_n^*$ as an estimator of $E_{f_X} g$ is that $\bar{g}_n^*$ converges to $E_{f_X} g$ almost surely as $n \to \infty$. Obviously, in practice we cannot use an infinite sample size, so we must find a finite value of n such that the error in $\bar{g}_n^*$ is (likely to be) acceptably small. To make this more precise, suppose we are willing to live with an error of size Δ. In other words, we would like to be able to assert that the interval given by $\bar{g}_n^* \pm \Delta$ is highly likely to contain the true, unknown value of $E_{f_X} g$. As we now explain, this can be accomplished through routine use of the CLT given in Equation 10.16.

Let $\hat{\sigma}_n^2$ denote the usual sample variance of the $g(X_i^*)$,

$$\hat{\sigma}_n^2 = \frac{1}{n-1} \sum_{i=1}^{n} \left(g(X_i^*) - \bar{g}_n^* \right)^2.$$

Basic asymptotic theory tell us that, since $\hat{\sigma}_n^2$ is a consistent estimator of σ^2,

$$\frac{\sqrt{n}(\bar{g}_n^* - E_{f_X} g)}{\sqrt{\hat{\sigma}_n^2}} \xrightarrow{d} N(0, 1).$$

Thus, for large n, the interval $\bar{g}_n^* \pm 2\hat{\sigma}_n/\sqrt{n}$ will contain the unknown value of $E_{f_X} g$ with probability (approximately) equal to 0.95. With this in mind, we can proceed as follows. Choose an initial sample size, say n', and make n' i.i.d. draws from f_X. (Hopefully, n' is large enough that $\hat{\sigma}_{n'}^2$ is a reasonable estimate of σ^2.) If the observed value of $2\hat{\sigma}_{n'}/\sqrt{n'}$ is less than Δ, then the current estimate of $E_{f_X} g$ is good enough and we stop. Otherwise, if $2\hat{\sigma}_{n'}/\sqrt{n'} > \Delta$, then additional simulation is required. Moreover, the current estimate of σ^2 can be used to calculate approximately how much more simulation will be necessary to achieve the stated precision. Indeed, we require an n such that $2\hat{\sigma}_n/\sqrt{n} < \Delta$, so assuming that our estimate of σ^2 has stabilized, $n > 4\hat{\sigma}_{n'}^2/\Delta^2$ should suffice.

There are two major obstacles blocking the use of a similar program for choosing n in the DA context. First, as we have already seen, even when the Markov chain X is Harris ergodic, the second moment condition, $g \in L^2(f_X)$, is not enough to guarantee that the estimator $\bar{g}_n$ satisfies a CLT. To be sure that CLTs hold for $L^2(f_X)$ functions, the practitioner must either (i) employ a DA algorithm that is known to be geometrically ergodic, or (ii) establish geometric ergodicity of the DA algorithm in question. The second problem is that, even when the CLT in Equation 10.15 is known to hold, consistent estimation of the asymptotic variance, κ^2, is a challenging problem because this variance has a fairly complex form and because the dependence among the variables in the Markov chain complicates asymptotic analysis. Consistent estimators of κ^2 have been developed using techniques from time series analysis and using the method of batch means, but these estimators are much more complicated than $\hat{\sigma}_n^2$, both practically and theoretically. Good entry points into the statistical literature on methods of estimating κ^2 are Geyer (1992), Jones et al. (2006), and Flegal et al. (2008).

There is no getting around the fact that establishing the existence of CLTs is harder for Markov chains than it is for i.i.d. sequences. However, it is possible to circumvent the difficulties associated with consistent estimation of κ^2. Indeed, there is an alternative form of the CLT in Equation 10.15 that is developed by introducing *regenerations* into the Markov chain. The advantage of this new CLT is that consistent estimation of its asymptotic variance is very simple. The price we have to pay for this added simplicity is that the user must develop a *minorization condition* for the Mtd $k(\cdot \mid \cdot)$. Fortunately, the form of k lends itself to constructing a minorization condition. Before we can fully explain regeneration and minorization, we have to introduce three new Markov chains that are all closely related to X.

10.3.2 Three Markov Chains Closely Related to *X*

Recall from Section 10.1 that, for fixed $x \in X$, the function $h(x', y' \mid x) = f_{X\mid Y}(x' \mid y')f_{Y\mid X}(y' \mid x)$ is a joint pdf in (x', y'). Now, define $\tilde{k} : (X \times Y) \times (X \times Y) \rightarrow [0, \infty)$ as

$$\tilde{k}(x', y' \mid x, y) = h(x', y' \mid x) = f_{X\mid Y}(x' \mid y')f_{Y\mid X}(y' \mid x).$$

For each fixed $(x, y) \in X \times Y$, $\tilde{k}(x', y' \mid x, y)$ is nonnegative and integrates to 1. Hence, the function $\tilde{k}$ is an Mtd that defines a Markov chain, $(X, Y) = \{(X_n, Y_n)\}_{n=0}^{\infty}$, with state space $X \times Y$. If the current state of the chain is $(X_n, Y_n) = (x, y)$, then the density of the next state, (X_{n+1}, Y_{n+1}), is $\tilde{k}(\cdot, \cdot \mid x, y)$. Furthermore, the chain (X, Y) has invariant density $f(x, y)$; indeed,

$$\int_X \int_Y \tilde{k}(x', y' \mid x, y)f(x, y)\, dy\, dx = f_{X\mid Y}(x' \mid y') \int_X f_{Y\mid X}(y' \mid x) \left[\int_Y f(x, y)\, dy \right] dx$$

$$= f_{X\mid Y}(x' \mid y') \int_X f(x, y')\, dx$$

$$= f_{X\mid Y}(x' \mid y')f_Y(y')$$

$$= f(x', y').$$

We refer to (X, Y) as the "Gibbs chain" because it is, in fact, just the Markov chain that is induced by the two-variable Gibbs sampler based on the joint density $f(x, y)$. The analogue of condition $\mathcal{K}$ for the Gibbs chain is condition $\widetilde{\mathcal{K}}$:

$$\tilde{k}(x', y' \mid x, y) > 0 \quad \text{for all } (x, y), (x', y') \in X \times Y.$$

Condition $\widetilde{\mathcal{K}}$ implies that the Gibbs chain is Harris ergodic. A sufficient condition for condition $\widetilde{\mathcal{K}}$ is that $f(x, y) > 0$ for all $(x, y) \in X \times Y$.

The reader has probably already noticed that $\tilde{k}(x', y' \mid x, y)$ does not actually depend on y. In terms of the Markov chain, this means that the future state, (X_{n+1}, Y_{n+1}), depends on the current state, (X_n, Y_n), only through X_n. This fact can be used to show that the conditional distribution of X_{n+1} given $(X_0, X_1, \ldots, X_n)$ does not depend on $(X_0, X_1, \ldots, X_{n-1})$. In other words, the sequence $X = \{X_n\}_{n=0}^{\infty}$ is itself a Markov chain on X. Moreover, its Mtd is

$$\int_Y \tilde{k}(x', y' \mid x, y) \, dy' = \int_Y f_{X|Y}(x' \mid y') f_{Y|X}(y' \mid x) \, dy' = k(x' \mid x);$$

that is, the marginal sequence $X = \{X_n\}_{n=0}^{\infty}$ from the Gibbs chain is the original DA Markov chain. (This is why it made sense to use the symbol X_n to denote the x-coordinate of Gibbs chain.) It follows that we can view our estimator, $\bar{g}_n = n^{-1} \sum_{i=0}^{n-1} g(X_i)$, as being an estimator based on the Gibbs chain. Formally, $\bar{g}_n = n^{-1} \sum_{i=0}^{n-1} \tilde{g}(X_i, Y_i)$, where $\tilde{g}(x, y) = g(x)$. This correspondence allows us work with the Gibbs chain instead of X, which turns out to be easier because, unlike k, $\tilde{k}$ is a known closed-form function.

Concerning simulation of the Gibbs chain, recall that our two-step procedure for simulating one iteration of the DA algorithm involves drawing from the joint pdf $h(x', y' \mid x)$ and throwing away the y-coordinate. In other words, the two-step procedure given in Section 10.1 actually simulates the Gibbs chain and just ignores the y-coordinates.

Not surprisingly, the marginal sequence $Y = \{Y_n\}_{n=0}^{\infty}$ from the Gibbs chain is also a Markov chain. This chain lives on Y and its Mtd is

$$k_Y(y'|y) = \int_X f_{Y|X}(y' \mid x) f_{X|Y}(x \mid y) \, dx.$$

It follows that Y can be viewed as the Markov chain underlying a DA algorithm for the target density $f_Y(y)$, and, as such, Y is reversible with respect to f_Y. There is actually an alternative estimator of $E_{f_X} g$ based on Y that we now describe. Define

$$\hat{g}(y) = \int_X g(x) f_{X|Y}(x \mid y) \, dx,$$

and note that $\int_Y \hat{g}(y) f_Y(y) \, dy = \int_X g(x) f_X(x) \, dx = E_{f_X} g$. Thus, if we can write $\hat{g}$ in closed form, which is often the case in practice, then we can compute the alternative estimator of $E_{f_X} g$ given by

$$\frac{1}{n} \sum_{i=0}^{n-1} \hat{g}(Y_i). \tag{10.17}$$

If Y is Harris ergodic, then, like $\bar{g}_n$, the estimator (10.17) is strongly consistent for $E_{f_X} g$. In fact, Liu et al. (1994) proved that, if $X_0 \sim f_X$ and $Y_0 \sim f_Y$, then the alternative estimator has a smaller (small sample) variance than $\bar{g}_n$. (Comparing variances is appropriate here since, if $X_0 \sim f_X$ and $Y_0 \sim f_Y$, then both estimators are unbiased.) We note that the methods described below for computing a valid asymptotic standard error for $\bar{g}_n$ can just as easily be applied to the estimator (10.17).

Finally, consider the Mtd given by

$$\tilde{\tilde{k}}(y', x' \mid y, x) = f_{Y|X}(y' \mid x') f_{X|Y}(x' \mid y),$$

and denote the corresponding Markov chain by $(Y', X') = \{(Y'_n, X'_n)\}_{n=0}^{\infty}$. Of course, (Y', X') is just the Markov chain induced by the two-variable Gibbs sampler for $f(x, y)$ with the variables in the opposite order. The chain (Y', X') behaves just like (X, Y). Indeed, $f(x, y)$ remains invariant and, by symmetry, the marginal sequences $\{X'_n\}_{n=0}^\infty$ and $\{Y'_n\}_{n=0}^\infty$ are equivalent (distributionally) to $X = \{X_n\}_{n=0}^\infty$ and $Y = \{Y_n\}_{n=0}^\infty$. Consequently, we can also view our estimator, $\bar{g}_n = n^{-1} \sum_{i=0}^{n-1} g(X_i)$, as being an estimator based on the chain (Y', X'); that is, $\bar{g}_n = n^{-1} \sum_{i=0}^{n-1} \tilde{g}(Y'_i, X'_i)$, where $\tilde{g}(y, x) = g(x)$. In some cases, it is more convenient to work with $\tilde{k}$ than with $\tilde{k}$. An important fact that will be used later is that all four of the Markov chains discussed in this section $(X, Y, (X, Y)$ and $(Y', X'))$ converge at exactly the same rate (Diaconis et al., 2008; Roberts and Rosenthal, 2001). Therefore, either all four chains are geometrically ergodic, or none of them is. We now describe the minorization condition and how it is used to induce regenerations, which can in turn be used to derive the alternative CLT.

10.3.3   Minorization, Regeneration and an Alternative CLT

We assume throughout this subsection that the Gibbs chain is Harris ergodic. Suppose that we can find a function $s : X \to [0, 1)$ with $E_{f_X} s > 0$ and a joint pdf $d : X \times Y \to [0, \infty)$ such that

$$\tilde{k}(x', y' \mid x, y) \geq s(x) \, d(x', y') \quad \text{for all } (x, y), (x', y') \in X \times Y. \tag{10.18}$$

Equation 10.18 is called a *minorization condition* (Jones and Hobert, 2001; Meyn and Tweedie, 1993; Roberts and Rosenthal, 2004). Here is a simple example.

Example 10.2 (cont.)

Here we have $X = Y = (0, 1)$, and we can develop a minorization condition as follows:

$$\tilde{k}(x', y' \mid x, y) = f_{X \mid Y}(x' \mid y') f_{Y \mid X}(y' \mid x)$$

$$= \frac{2x'}{1 - y'^2} I(y' < x' < 1) \frac{1}{x} I(0 < y' < x < 1)$$

$$\geq \frac{2x'}{1 - y'^2} I(y' < x' < 1) \frac{1}{x} I(0 < y' < 0.5) I(0.5 < x < 1)$$

$$= \frac{1}{x} I(0.5 < x < 1) \frac{2x'}{1 - y'^2} I(y' < x' < 1) I(0 < y' < 0.5)$$

$$= \left[\frac{1}{2x} I(0.5 < x < 1) \right] \left[\frac{4x'}{1 - y'^2} I(y' < x' < 1) I(0 < y' < 0.5) \right]$$

$$= s(x) \, d(x', y'),$$

where we have used the fact that

$$\int_0^1 \int_0^1 \frac{2x}{1 - y^2} I(y < x < 1) I(0 < y < 0.5) \, dx \, dy = \frac{1}{2}.$$

Note that the density d is not strictly positive on $X \times Y$.

The minorization condition (10.18) can be used to represent the Mtd $\tilde{k}$ as a mixture of two other Mtds. First, define

$$r(x',y' \mid x,y) = \frac{\tilde{k}(x',y' \mid x,y) - s(x)d(x',y')}{1 - s(x)},$$

and note that $r(x',y' \mid x,y)$ is an Mtd. Indeed, Equation 10.18 implies that r is nonnegative, and it is also clear that $\int_X \int_Y r(x',y' \mid x,y) \, dy' \, dx' = 1$. We can now express $\tilde{k}$ as

$$\tilde{k}(x',y' \mid x,y) = s(x)d(x',y') + \big(1 - s(x)\big)r(x',y' \mid x,y). \tag{10.19}$$

If we think of $s(x)$ and $1 - s(x)$ as two fixed numbers in $[0,1]$ whose sum is 1, then the right-hand side of Equation 10.19 can be viewed as a mixture of two Mtds, $d(x',y')$ and $r(x',y' \mid x,y)$. Since $d(x',y')$ does not depend on (x,y), the Markov chain defined by d is actually an i.i.d. sequence, and this is the key to introducing regenerations. Technically speaking, the regenerations do not occur in the Gibbs chain itself, but in an augmented Markov chain that we now describe.

For $(x,y) \in X \times Y$, let $f_1(\delta \mid (x,y))$ denote a Bernoulli$(s(x))$ probability mass function; that is, $f_1(1 \mid (x,y)) = s(x)$ and $f_1(0 \mid (x,y)) = 1 - s(x)$. Also, for $(x',y'), (x,y) \in X \times Y$ and $\delta \in \{0,1\}$, define

$$f_2\big((x',y') \mid \delta, (x,y)\big) = d(x',y') \, I(\delta = 1) + r(x',y' \mid x,y) \, I(\delta = 0). \tag{10.20}$$

Note that f_2 is a pdf in (x',y'). Finally, define

$$k_s\big((x',y'), \delta' \mid (x,y), \delta\big) = f_1\big(\delta' \mid (x,y)\big) f_2\big((x',y') \mid \delta', (x,y)\big). \tag{10.21}$$

Now, k_s is nonnegative and

$$\sum_{\delta' \in \{0,1\}} \int_Y \int_X k_s\big((x',y'), \delta' \mid (x,y), \delta\big) \, dx' \, dy' = \sum_{\delta' \in \{0,1\}} f_1\big(\delta' \mid (x,y)\big) = 1.$$

Therefore, k_s is an Mtd and the corresponding Markov chain, which we denote by $((X,Y), \delta) = \{(X_n, Y_n), \delta_n\}_{n=0}^{\infty}$, lives on $(X \times Y) \times \{0,1\}$. This is called the *split chain* (Nummelin, 1984, Section 4.4).

Before we elucidate the regeneration properties of the split chain, we describe the relationship between the split chain and the Gibbs chain. Note that k_s does not actually depend on δ. Thus, arguments similar to those used in Section 10.3.2 show that the marginal sequence $\{(X_n, Y_n)\}_{n=0}^{\infty}$ from the split chain is itself a Markov chain with Mtd given by

$$k_s\big((x',y'), 1 \mid (x,y), \delta\big) + k_s\big((x',y'), 0 \mid (x,y), \delta\big)$$
$$= f_1\big(1 \mid (x,y)\big) f_2\big((x',y') \mid 1, (x,y)\big) + f_1\big(0 \mid (x,y)\big) f_2\big((x',y') \mid 0, (x,y)\big)$$
$$= s(x)d(x',y') + \big(1 - s(x)\big)r(x',y' \mid x,y)$$
$$= \tilde{k}(x',y' \mid x,y).$$

We conclude that the marginal sequence $\{(X_n, Y_n)\}_{n=0}^{\infty}$ from the split chain is (distributionally) equivalent to the Gibbs chain. Moreover, the split chain inherits Harris ergodicity from the Gibbs chain (Nummelin, 1984, Section 4.4). As before, we can view the estimator $\bar{g}_n$ as being based on the split chain.

The split chain experiences a regeneration every time the binary component visits the set $\{1\}$. To see this, suppose that we start the split chain with $\delta_0 = 1$ and $(X_0, Y_0) \sim d(\cdot, \cdot)$. It is clear from Equations 10.21 and 10.20 that, no matter what the value of the current state, $((X_n, Y_n), \delta_n)$, if $\delta_{n+1} = 1$ then $(X_{n+1}, Y_{n+1}) \sim d(\cdot, \cdot)$ and the process stochastically restarts itself; that is, the Markov chain regenerates. We now use the regenerative structure of the split chain to recast our estimator of $E_{f_X} g$ in such a way that i.i.d. theory can be used to analyze it. This leads to an alternative CLT whose asymptotic variance is very easy to estimate.

Let $\tau_0, \tau_1, \tau_2, \ldots$ denote the *regeneration times*; that is, the random times at which the split chain regenerates. Then $\tau_0 = 0$ and, for $t = 1, 2, 3, \ldots$, we have

$$\tau_t = \min\{i > \tau_{t-1} : \delta_i = 1\}.$$

This notation allows us to identify the "tours" that the split chain takes in between regenerations:

$$\left\{ \left((X_{\tau_{t-1}}, Y_{\tau_{t-1}}), \delta_{\tau_{t-1}}\right), \ldots, \left((X_{\tau_t-1}, Y_{\tau_t-1}), \delta_{\tau_t-1}\right) : t = 1, 2, 3, \ldots \right\}.$$

These tours are independent stochastic copies of each other, and hence standard techniques from i.i.d. theory (such as the SLLN and the CLT) can be used in the asymptotic analysis of the resulting ergodic averages. In other words, the regenerative structure that we have introduced allows us to circumvent (to some extent) the complications caused by the dependence among the random vectors in the Markov chain.

Consider running the split chain for R tours; that is, the chain is started with $\delta_0 = 1$ and $(X_0, Y_0) \sim d(\cdot, \cdot)$ and is run until the Rth time that a $\delta_n = 1$. (Some practical advice concerning simulation of the split chain will be given later.) For $t = 1, 2, \ldots, R$, define $N_t = \tau_t - \tau_{t-1}$, which is the length of the tth tour, and $S_t = \sum_{i=\tau_{t-1}}^{\tau_t-1} g(X_i)$. Because the tours are independent stochastic copies of each other, the pairs $(N_1, S_1), \ldots, (N_R, S_R)$ are i.i.d. The total length of the simulation is $\sum_{t=1}^{R} N_t = \tau_R$, which is, of course, random. Our estimator of $E_{f_X} g$ will be

$$\bar{g}_R = \frac{1}{\tau_R} \sum_{i=0}^{\tau_R-1} g(X_i) = \frac{\sum_{t=1}^{R} S_t}{\sum_{t=1}^{R} N_t}.$$

Clearly, the only difference between $\bar{g}_R$ and the usual ergodic average, $\bar{g}_n$, is that here, the sample size is random. However, $\tau_R \to \infty$ almost surely as $R \to \infty$ and it follows that $\bar{g}_R$ is also strongly consistent for $E_{f_X} g$ as $R \to \infty$. The advantage of $\bar{g}_R$ over the usual ergodic average is that it can be expressed in terms of the i.i.d. pairs $\{(N_t, S_t)\}_{t=1}^{R}$. Results in Hobert et al. (2002) show that, if the Gibbs chain (or, equivalently, the DA Markov chain) is geometrically ergodic and $E_{f_X} |g|^{2+\alpha} < \infty$ for some $\alpha > 0$, then as $R \to \infty$,

$$\sqrt{R}(\bar{g}_R - E_{f_X} g) \xrightarrow{d} N(0, \gamma^2), \tag{10.22}$$

where

$$\gamma^2 = \frac{E\left[(S_1 - N_1 E_{f_X} g)^2\right]}{[E N_1]^2}.$$

Note that this asymptotic variance is written in terms of a single tour, (N_1, S_1). Results in Hobert et al. (2002) show that the geometric ergodicity of the Gibbs chain and the "$2 + \alpha$" moment condition on g together imply that EN_1^2 and ES_1^2 are both finite. Once these moments are known to be finite, routine asymptotics can be used to show that

$$\hat{\gamma}_R^2 = \frac{R \sum_{t=1}^{R} \left(S_t - \bar{g}_R N_t \right)^2}{\tau_R^2}$$

is a strongly consistent estimator of γ^2 as $R \to \infty$. Note the simple form of this estimator.

A couple of comments are in order concerning the two different CLTs (10.22 and 10.15). First, both CLTs are based on the assumption that the DA Markov chain, X, is geometrically ergodic. However, while Equation 10.15 requires only the usual second moment condition, $E_{f_X} g^2 < \infty$, Equation 10.22 requires the slightly stronger condition that $E_{f_X} |g|^{2+\alpha} < \infty$ for some $\alpha > 0$. Second, the two asymptotic variances are related via the formula $\kappa^2 = \gamma^2 / E_{f_X} s$ (Hobert et al., 2002). This makes sense intuitively because $E_{f_X} s$ is the average probability of regeneration (under stationarity) and hence $1/E_{f_X} s$ seems like a reasonable guess at the average tour length.

We conclude that, if X is geometrically ergodic and the "$2 + \alpha$" moment condition on g is satisfied, then we can employ the DA algorithm in the same way that classical Monte Carlo is used. Indeed, we can simulate R' tours of the split chain, where R' is some initial sample size. (Hopefully, R' is large enough that $\hat{\gamma}_{R'}^2$ is a reasonable estimate of γ^2.) If $2\hat{\gamma}_{R'}^2 / \sqrt{R'} \le \Delta$, then the current estimate of $E_{f_X} g$ is good enough and we stop. Otherwise, if $2\hat{\gamma}_{R'}^2 / \sqrt{R'} > \Delta$, then additional tours must be simulated.

10.3.4 Simulating the Split Chain

Exploiting the techniques described in the previous subsection in practice requires the ability to simulate the split chain. The form of k_s actually lends itself to the sequential simulation technique described in Section 10.1. If the current state is $\left((X_n, Y_n), \delta_n \right) = ((x, y), \delta)$, then the future state, $\left((X_{n+1}, Y_{n+1}), \delta_{n+1} \right)$, can be simulated as follows. First, draw $\delta_{n+1} \sim$ Bernoulli($s(x)$) and then, conditional on $\delta_{n+1} = \delta'$, draw (X_{n+1}, Y_{n+1}) from

$$f_2\left((\cdot, \cdot) \mid \delta', (x, y) \right)$$

that is, if $\delta' = 1$, draw $(X_{n+1}, Y_{n+1}) \sim d(\cdot, \cdot)$, and if $\delta' = 0$, draw $(X_{n+1}, Y_{n+1}) \sim r(\cdot, \cdot \mid x, y)$. Here is an example where this method is viable.

Example 10.2 (cont.)

Recall that we developed a minorization condition of the form (10.18) for this example earlier in this section. We now verify that it is straightforward to simulate from $d(\cdot, \cdot)$ and from $r(\cdot, \cdot \mid x, y)$. First, it is easy to show that if $(U, V) \sim d(\cdot, \cdot)$, then marginally, V is $U(0, 0.5)$, and the conditional density of U given $V = v$ is $f_{X|Y}(u \mid v) = \frac{2u}{1 - v^2} I(v < u < 1)$. Hence, simulating from d is easy. Now consider r. Since, $s(x) = 0$ when $x \in (0, 0.5)$, we must have $r(x', y' \mid x, y) = \tilde{k}(x', y' \mid x, y)$ when $x \in (0, 0.5)$. On the other hand, when $x \in (0.5, 1)$, then routine calculations show that

$$r(x', y' \mid x, y) = \frac{2x'}{(1 - y'^2)(x - 0.5)} I(y' < x' < 1) I(0.5 < y' < x),$$

and, in this case, it is easy to show that if $(U, V) \sim r(\cdot, \cdot \mid x, y)$, then marginally, V is $U(0.5, x)$, and the conditional density of U given $V = v$ is $f_{X|Y}(u \mid v)$, so it is also easy to draw from r. Note that the supports of $d(\cdot, \cdot)$ and $r(\cdot, \cdot \mid x, y)$ are mutually exclusive. We conclude that the sequential method outlined above can be used to simulate the split chain in this example.

In the toy example just considered, it is straightforward to simulate from $r(\cdot, \cdot \mid x, y)$. However, this will typically not be the case in real examples where $\tilde{k}(x', y' \mid x, y)$ is a high-dimensional, complex Mtd. Fortunately, Mykland et al. (1995) noticed a clever way of circumventing the need to draw from r. Their idea amounts to using the sequential simulation technique, but in the opposite order. Indeed, one way to draw from $(X_{n+1}, Y_{n+1}), \delta_{n+1} \mid (X_n, Y_n)$ is to draw first from $(X_{n+1}, Y_{n+1}) \mid (X_n, Y_n)$ and then from $\delta_{n+1} \mid (X_{n+1}, Y_{n+1}), (X_n, Y_n)$. A little thought reveals that these two steps are simple and do not involve drawing from r. First, we established above that $(X_{n+1}, Y_{n+1}) \mid (X_n, Y_n) = (x, y) \sim \tilde{k}(\cdot, \cdot \mid x, y)$, so this step can be accomplished by simulating a single iteration of the Gibbs chain (by drawing from $f_{Y|X}$ and then from $f_{X|Y}$). Furthermore, given (X_n, Y_n) and (X_{n+1}, Y_{n+1}), δ_{n+1} has a Bernoulli distribution with success probability given by

$$\Pr\left(\delta_{n+1} = 1 \mid X_n = x, Y_n = y, X_{n+1} = x', Y_{n+1} = y'\right) = \frac{s(x)\, d(x', y')}{\tilde{k}(x', y' \mid x, y)}. \tag{10.23}$$

Here is a summary of how Mykland et al.'s (1995) method is used to simulate the split chain. If the current state is $(X_n, Y_n) = (x, y)$, then we simply draw (X_{n+1}, Y_{n+1}) in the usual way, and then we go back and "fill in" the value of δ_{n+1} by simulating a Bernoulli with success probability (10.23). Even though we only draw from d once (at the start) and we never actually draw from r at all, there is a regeneration in the chain each time $\delta_n = 1$. In fact, we can even avoid the single draw from d (although, even in real problems, it is usually pretty easy to draw from d). Starting the chain from an arbitrary point, but then throwing away everything from the beginning up to and including the first Bernoulli that equals 1, is equivalent to drawing $(X_0, Y_0) \sim d(\cdot, \cdot)$. Finally, note the rather striking fact that the only difference between simulating the split chain and the Gibbs chain is a single Bernoulli draw per iteration! In fact, if computer code is available that runs the DA algorithm, then a few minor modifications will yield a program that runs the split chain instead. Here is an example illustrating the use of Equation 10.23.

Example 10.2 (cont.)

If the nth and $(n+1)$th states of the Gibbs chain are $(X_n, Y_n) = (x, y)$ and $(X_{n+1}, Y_{n+1}) = (x', y')$, then it must be the case that $x, x' \in (0, 1)$ and $y' \in (0, \min\{x, x'\})$. Now, applying Equation 10.23, the probability that a regeneration occurred is

$$\Pr\left(\delta_{n+1} = 1 \mid X_n = x, Y_n = y, X_{n+1} = x', Y_{n+1} = y'\right) = I(0.5 < x < 1)\, I(0 < y' < 0.5).$$

In hindsight, this formula is actually "obvious." First, if $x \notin (0.5, 1)$, then $s(x) = 0$, and regeneration could not have occurred. Likewise, if $y' \notin (0, 0.5)$, then d could not have been used to draw (X_{n+1}, Y_{n+1}) so, again, regeneration could not have occurred. On the other hand, if $x \in (0.5, 1)$ and $y' \in (0, 0.5)$, then there must have been a regeneration because $r(\cdot, \cdot \mid x, y)$ could not have been used to draw (X_{n+1}, Y_{n+1}).

In the next section, we give a general method of developing the minorization condition (Equation 10.18).

10.3.5 A General Method for Constructing the Minorization Condition

The minorization condition for Example 10.1 was derived in a somewhat *ad hoc* manner. We now describe a general recipe, due to Mykland et al. (1995), for constructing a minorization condition. This technique is most effective when $f(x, y)$ is strictly positive on $X \times Y$. Fix a "distinguished point" $x_* \in X$ and a set $D \subset Y$. Then we can write

$$
\begin{aligned}
\tilde{k}(x', y' \mid x, y) &= f_{X\mid Y}(x' \mid y') f_{Y\mid X}(y' \mid x) \\
&= \frac{f_{Y\mid X}(y' \mid x)}{f_{Y\mid X}(y' \mid x_*)} f_{X\mid Y}(x' \mid y') f_{Y\mid X}(y' \mid x_*) \\
&\geq \left[\inf_{y \in D} \frac{f_{Y\mid X}(y \mid x)}{f_{Y\mid X}(y \mid x_*)} \right] f_{X\mid Y}(x' \mid y') f_{Y\mid X}(y' \mid x_*) I_D(y') \\
&= c \left[\inf_{y \in D} \frac{f_{Y\mid X}(y \mid x)}{f_{Y\mid X}(y \mid x_*)} \right] \frac{1}{c} f_{X\mid Y}(x' \mid y') f_{Y\mid X}(y' \mid x_*) I_D(y'),
\end{aligned}
$$

where

$$
c = \int_Y \int_X f_{X\mid Y}(x \mid y) f_{Y\mid X}(y \mid x_*) I_D(y) \, dx \, dy = \int_D f_{Y\mid X}(y \mid x_*) \, dy.
$$

Thus, we have a minorization condition $\tilde{k}(x', y' \mid x, y) \geq s(x) d(x', y')$ with

$$
s(x) = c \inf_{y \in D} \frac{f_{Y\mid X}(y \mid x)}{f_{Y\mid X}(y \mid x_*)} \quad \text{and} \quad d(x', y') = \frac{1}{c} f_{X\mid Y}(x' \mid y') f_{Y\mid X}(y' \mid x_*) I_D(y').
$$

Fortunately, the value of c is not required in practice. The success probability in Equation 10.23 involves $s(x)$ and $d(x', y')$ only through their product, so c cancels out. Furthermore, it is possible to make draws from $d(x', y')$ without knowing the value of c. We first draw Y' from its marginal density, $c^{-1} f_{Y\mid X}(y' \mid x_*) I_D(y')$, by repeatedly drawing from $f_{Y\mid X}(\cdot \mid x_*)$ until the result is in the set D. Then, given $Y' = y'$, we draw X' from $f_{X\mid Y}(\cdot \mid y')$.

Since the asymptotics described in Section 10.3.3 are for large R, the more frequently the split chain regenerates, the better. Thus, in practice, one should choose the point x_* and the set D so that regenerations occur frequently. This can be done by trial and error. In applications, we have found it useful fix x_* (at a preliminary estimate of the mean of f_X) and then vary the set D. Note that, according to Equation 10.23, a regeneration could only have occurred if $y' \in D$, so it is tempting to make D large. However, as D gets larger, $s(x)$ becomes smaller, which means that the probability of regeneration becomes smaller. Hence, a balance must be struck. For examples, see Mykland et al. (1995), Jones and Hobert (2001), Roy and Hobert (2007), and Tan and Hobert (2009). We now provide two examples illustrating Mykland et al.'s (1995) method.

Example 10.3 (cont.)

Recall that $X \mid Y = y \sim N(0, y^{-1})$ and $Y \mid X = x \sim \Gamma\left(\frac{5}{2}, \frac{x^2}{2} + 2\right)$. Thus,

$$
\frac{f_{Y\mid X}(y \mid x)}{f_{Y\mid X}(y \mid x_*)} = \frac{\left[\Gamma\left(\frac{5}{2}\right)\right]^{-1} \left(\frac{x^2}{2} + 2\right)^{5/2} y^{3/2} \exp\left\{-y\left(\frac{x^2}{2} + 2\right)\right\}}{\left[\Gamma\left(\frac{5}{2}\right)\right]^{-1} \left(\frac{x_*^2}{2} + 2\right)^{5/2} y^{3/2} \exp\left\{-y\left(\frac{x_*^2}{2} + 2\right)\right\}}
$$

$$= \left(\frac{x^2 + 4}{x_*^2 + 4}\right)^{5/2} \exp\left\{-\frac{y}{2}(x^2 - x_*^2)\right\}.$$

So if we take $D = [d_1, d_2]$, where $0 < d_1 < d_2 < \infty$, we have

$$\inf_{y \in D} \frac{f_{Y|X}(y \mid x)}{f_{Y|X}(y \mid x_*)} = \left(\frac{x^2 + 4}{x_*^2 + 4}\right)^{5/2} \exp\left\{-\frac{d_2}{2}(x^2 - x_*^2)I(x^2 > x_*^2) - \frac{d_1}{2}(x^2 - x_*^2)I(x^2 \le x_*^2)\right\}.$$

Thus,

$$\Pr\left(\delta_{n+1} = 1 \mid X_n = x, Y_n = y, X_{n+1} = x', Y_{n+1} = y'\right)$$

$$= \frac{s(x)\, d(x', y')}{\tilde{k}(x', y' \mid x, y)}$$

$$= \left[\inf_{y \in [d_1, d_2]} \frac{f_{Y|X}(y \mid x)}{f_{Y|X}(y \mid x_*)}\right] \frac{f_{Y|X}(y' \mid x_*)}{f_{Y|X}(y' \mid x)} I_{[d_1, d_2]}(y')$$

$$= \exp\left\{(x^2 - x_*^2)\left[\frac{y'}{2} - \frac{d_2}{2}I(x^2 > x_*^2) - \frac{d_1}{2}I(x^2 \le x_*^2)\right]\right\} I_{[d_1, d_2]}(y').$$

A draw from $d(x', y')$ can be made by drawing a truncated gamma and then a normal.

Here is a more realistic example.

Example 10.4 (cont.)

The variable of interest is (μ, σ^2), which lives in $X = \mathbb{R} \times \mathbb{R}_+$, and the augmented data, y, live in $Y = \mathbb{R}_+^m$. In order to keep the notation under control, we use the symbol η in place of σ^2. In this example, it turns out to be more convenient to use $\tilde{k}$, which is given by

$$\tilde{k}\left(y', (\mu', \eta') \mid y, (\mu, \eta)\right) = \pi(y' \mid \mu', \eta', z)\pi(\mu', \eta' \mid y, z),$$

where the conditional densities on the right-hand side are defined in Equation 10.6 through 10.8. Fix a distinguished point $y_* \in Y$ and let $D = [d_1, d_2] \times [d_3, d_4]$ where $-\infty < d_1 < d_2 < \infty$ and $0 < d_3 < d_4 < \infty$. Now, letting y_s denote the sum of the components of y_*, we have

$$\frac{\pi(\mu, \eta \mid y, z)}{\pi(\mu, \eta \mid y_*, z)}$$

$$= \frac{\frac{\sqrt{y.}}{\sqrt{\eta 2\pi}} \exp\left\{-\frac{y.}{2\eta}(\mu - \hat{\mu}(y))^2\right\} \left(\frac{y.\hat{\sigma}^2(y)}{2}\right)^{\frac{m+1}{2}} \Gamma^{-1}\left(\frac{m+1}{2}\right) \eta^{-\frac{m+1}{2}-1} \exp\left\{-\frac{y.\hat{\sigma}^2(y)}{2\eta}\right\}}{\frac{\sqrt{y_s}}{\sqrt{\eta 2\pi}} \exp\left\{-\frac{y_s}{2\eta}(\mu - \hat{\mu}(y_*))^2\right\} \left(\frac{y_s\hat{\sigma}^2(y_*)}{2}\right)^{\frac{m+1}{2}} \Gamma^{-1}\left(\frac{m+1}{2}\right) \eta^{-\frac{m+1}{2}-1} \exp\left\{-\frac{y_s\hat{\sigma}^2(y_*)}{2\eta}\right\}}$$

$$= \frac{\sqrt{y.}}{\sqrt{y_s}}\left(\frac{y.\hat{\sigma}^2(y)}{y_s\hat{\sigma}^2(y_*)}\right)^{\frac{m+1}{2}} \exp\left\{-\frac{1}{2\eta}\left[y.(\mu - \hat{\mu}(y))^2 + y.\hat{\sigma}^2(y) - y_s(\mu - \hat{\mu}(y_*))^2 - y_s\hat{\sigma}^2(y_*)\right]\right\}$$

$$= \frac{\sqrt{y.}}{\sqrt{y_s}}\left(\frac{y.\hat{\sigma}^2(y)}{y_s\hat{\sigma}^2(y_*)}\right)^{\frac{m+1}{2}} \exp\left\{-\frac{1}{2\eta}Q(\mu; y, y_*)\right\},$$

where $Q(\mu; y, y_*)$ is a quadratic function of μ whose coefficients are determined by y and y_*. Now consider minimizing the exponential over $(\mu, \eta) \in [d_1, d_2] \times [d_3, d_4]$. Let $\tilde{\mu}$ denote the maximizer of $Q(\mu; y, y_*)$ over $\mu \in [d_1, d_2]$, which is easy to compute once y and y_* are specified. Clearly, if $Q(\tilde{\mu}; y, y_*) \geq 0$, then the exponential is minimized at $(\mu, \eta) = (\tilde{\mu}, d_3)$. On the other hand, if $Q(\tilde{\mu}; y, y_*) < 0$, then the minimizer is $(\mu, \eta) = (\tilde{\mu}, d_4)$. Let $\underline{\eta} = d_3$ if $Q(\tilde{\mu}; y, y_*) \geq 0$ and d_4 if $Q(\tilde{\mu}; y, y_*) < 0$. Then we can write

$$s(y) = c \inf_{(\mu,\eta) \in [d_1,d_2] \times [d_3,d_4]} \frac{\pi(\mu, \eta \mid y, z)}{\pi(\mu, \eta \mid y_*, z)} = c \frac{\sqrt{y_\cdot}}{\sqrt{y_s}} \left(\frac{y_\cdot \hat{\sigma}^2(y)}{y_s \hat{\sigma}^2(y_*)} \right)^{\frac{m+1}{2}} \exp\left\{ -\frac{1}{2\underline{\eta}} Q(\tilde{\mu}; y, y_*) \right\},$$

and

$$d(y', (\mu', \eta')) = \frac{1}{c} \pi(y' \mid \mu', \eta', z) \pi(\mu', \eta' \mid y_*, z) I_D(\mu', \eta').$$

Putting all of this together, if the nth and $(n + 1)$th states of the Gibbs chain are $(X_n, Y_n) = ((\mu, \eta), y)$ and $(X_{n+1}, Y_{n+1}) = ((\mu', \eta'), y')$, then the probability that a regeneration occurred (i.e. that $\delta_{n+1} = 1$) is given by

$$\frac{s(y)\, d(y', (\mu', \eta'))}{\tilde{k}(y', (\mu', \eta') \mid y, (\mu, \eta))} = \left[\inf_{(\mu,\eta) \in [d_1,d_2] \times [d_3,d_4]} \frac{\pi(\mu, \eta \mid y, z)}{\pi(\mu, \eta \mid y_*, z)} \right] \frac{\pi(\mu', \eta' \mid y_*, z)}{\pi(\mu', \eta' \mid y, z)} I_D(\mu', \eta')$$

$$= \exp\left\{ -\frac{1}{2\underline{\eta}} Q(\tilde{\mu}; y, y_*) + \frac{1}{2\eta'} Q(\mu'; y_*, y) \right\} I_D(\mu', \eta').$$

In the final section of this chapter, we describe a simple method of improving the DA algorithm.

10.4 Improving the DA Algorithm

Suppose that the current state of the DA algorithm is $X_n = x$. As we know, the move to X_{n+1} involves two steps: draw $Y \sim f_{Y|X}(\cdot \mid x)$ and then, conditional on $Y = y$, draw $X_{n+1} \sim f_{X|Y}(\cdot \mid y)$. Consider adding an extra step in between these two. Suppose that, *after* having drawn $Y = y$ but *before* drawing X_{n+1}, a random move is made from y to a new point in Y; call it Y'. Then, conditional on $Y' = y'$, draw $X_{n+1} \sim f_{X|Y}(\cdot \mid y')$. Graphically, we are changing the algorithm from $X \to Y \to X'$ to $X \to Y \to Y' \to X'$. Of course, this must all be done subject to the restriction that f_X remains invariant. Intuitively, this extra random move within Y should reduce the correlation between X_n and X_{n+1}, thereby improving the mixing properties of the DA Markov chain. On the other hand, the new algorithm requires more computational effort per iteration, which must be weighed against any improvement in mixing. In this section, we describe techniques for constructing relatively inexpensive extra moves that often result in dramatic improvements in mixing. Here is a brief description of one of these techniques.

Suppose that $G \subset \mathbb{R}^d$ and that we have a class of functions $t_g : Y \to Y$ indexed by $g \in G$. In Section 10.4.4 we show that, if this class possesses a certain group structure, then there exists a parametric family of densities on G, indexed by $y \in Y$—call it $\xi(g; y)$—that can be used to make the extra move $Y \to Y'$. It proceeds as follows. Given $Y = y$, draw $G \sim \xi(\cdot ; y)$, call the result g, and set $Y' = t_g(y)$. In other words, the extra move takes y to the random point $Y' = t_G(y)$ where G is drawn from a distribution that is constructed to ensure that f_X remains invariant. Typically, d is small, say 1 or 2, so drawing from $\xi(\cdot\,;y)$ is inexpensive. A potential downside of small d is that, for fixed y, the set $\{t_g(y) : g \in G\}$ is a low-dimensional subset of Y (that includes the point y). Thus, the potential "shakeup" resulting from the move to $Y' = t_G(y)$ may not be significant. However, it turns out that, even when $d = 1$, this shakeup often results in huge improvements. We now begin a careful development of these ideas.

10.4.1 The PX-DA and Marginal Augmentation Algorithms

Recall that our DA algorithm is based on the pdf $f(x,y)$ whose x-marginal is f_X. As above, let $G \subset \mathbb{R}^d$ and suppose that we have a class of functions $t_g : Y \to Y$ indexed by $g \in G$. Assume that, for each fixed g, $t_g(y)$ is one-to-one and differentiable in y. Let $J_g(z)$ denote the Jacobian of the transformation $z = t_g^{-1}(y)$, so, for example, in the univariate case, $J_g(z) = \frac{\partial}{\partial z} t_g(z)$. Note that

$$\int_Y f\big(x, t_g(y)\big)\, |J_g(y)|\, dy = \int_Y f(x,z)\, dz = f_X(x). \tag{10.24}$$

Now suppose that $w : G \to [0,\infty)$ is a pdf and define $f^{(w)} : X \times Y \times G \to [0,\infty)$ as follows:

$$f^{(w)}(x,y,g) = f\big(x, t_g(y)\big)\, |J_g(y)|\, w(g). \tag{10.25}$$

It is clear from Equation 10.24 that $f^{(w)}(x,y,g)$ is a pdf whose x-marginal is $f_X(x)$, and hence the pdf defined by

$$f^{(w)}(x,y) = \int_G f^{(w)}(x,y,g)\, dg$$

also has f_X as its x-marginal. Thus, if straightforward sampling from $f^{(w)}_{X|Y}(x \mid y)$ and $f^{(w)}_{Y|X}(y \mid x)$ is possible, then we have a new DA algorithm that can be compared with the one based on $f(x,y)$. (For the rest of this chapter, we assume that all Markov chains on X are Harris ergodic.) As we will see, it is often possible to choose t_g and w in such a way that there is little difference between these two DA algorithms in terms of computational effort per iteration. However, under mild regularity conditions that are described below, the new algorithm beats the original in terms of both convergence rate and ARE. The idea of introducing the extra parameter, g, to form a new DA algorithm was developed independently by Meng and van Dyk (1999), who called it *marginal augmentation*, and Liu and Wu (1999), who called it *parameter expanded-data augmentation* (or PX-DA). We find Liu and Wu's (1999) terminology a little more convenient, so we call the new DA algorithm based on $f^{(w)}(x,y)$ a *PX-DA algorithm*. Here is a simple example.

Example 10.3 (cont.)

Set $G = \mathbb{R}_+$ and let $t_g(y) = gy$. If we take $w(g)$ to be a $\Gamma(\alpha, \beta)$ pdf, then we have

$$f^{(w)}(x,y,g) = f\big(x, t_g(y)\big)\, |J_g(y)|\, w(g)$$

$$= \left[\frac{4}{\sqrt{2\pi}}(gy)^{3/2} \exp\left\{-gy\left(\frac{x^2}{2}+2\right)\right\} I_{\mathbb{R}_+}(y)\right](g)\left[\frac{\beta^\alpha}{\Gamma(\alpha)} g^{\alpha-1} \exp\{-g\beta\} I_{\mathbb{R}_+}(g)\right].$$

It follows that

$$f^{(w)}(x, y) = \int_{\mathbb{R}_+} f^{(w)}(x, y, g) \, dg = \frac{4\beta^\alpha \, \Gamma(\frac{5}{2} + \alpha)}{\Gamma(\alpha)\sqrt{2\pi}} \, y^{3/2} \left[y\left(\frac{x^2}{4} + 1\right) + \beta \right]^{-(5/2+\alpha)}.$$

According to the theory above, every $(\alpha, \beta) \in \mathbb{R}_+ \times \mathbb{R}_+$ yields a different version of $f^{(w)}(x, y)$ and every one of them has the Student's t density with 4 degrees of freedom as its x-marginal.

Now consider the conditional densities under $f^{(w)}(x, y)$. It is easy to show that $f^{(w)}_{X|Y}(\cdot \mid y)$ is a scaled Student's t density and that $f^{(w)}_{Y|X}(\cdot \mid x)$ is a scaled F density. In fact, if the current state of the PX-DA Markov chain is $X_n = x$, then the next state, X_{n+1}, can be simulated by performing the following two steps:

1. Draw U from the F distribution with 5 numerator degrees of freedom and 2α denominator degrees of freedom, and call the realized value u. Then set $y = \frac{10\beta}{\alpha(x^2+4)} u$.

2. Draw V from the Student's t distribution with $2(\alpha + 2)$ degrees of freedom, and set $X_{n+1} = \sqrt{\frac{2(y+\beta)}{y(\alpha+2)}} \, V$.

(Note that Step 2 is as difficult as drawing directly from the target pdf, f_X, which is a Student's t density, but keep in mind that this is just a toy example that we are using for illustration.) We now have infinitely many viable PX-DA algorithms—one for each (α, β) pair. This raises an obvious question. Are any of these PX-DA algorithms better than the original DA algorithm, and if so, is there a best one? These questions are answered below.

In the toy example just considered, the conditional densities $f^{(w)}_{X|Y}$ and $f^{(w)}_{Y|X}$ have standard forms. Unfortunately, in real examples, it will typically be impossible to sample directly from (or even compute) these conditionals. However, by exploiting the relationship between $f^{(w)}(x, y)$ and $f^{(w)}(x, y, g)$, it is possible to develop *indirect* methods of drawing from $f^{(w)}_{X|Y}$ and $f^{(w)}_{Y|X}$ that use only draws from $f_{X|Y}, f_{Y|X}, w(g)$ and one other density. (Recall that we have been operating since Section 10.1 under the assumption that it is easy to sample from $f_{X|Y}$ and $f_{Y|X}$.) We begin with $f^{(w)}_{Y|X}(y \mid x)$. Note that

$$\begin{aligned}
f^{(w)}_{Y|X}(y \mid x) &= \frac{\int_G f^{(w)}(x, y, g) \, dg}{f_X(x)} \\
&= \int_G \frac{f(x, t_g(y))}{f_X(x)} \, |J_g(y)| \, w(g) \, dg \qquad (10.26) \\
&= \int_G f_{Y|X}(t_g(y) \mid x) \, |J_g(y)| \, w(g) \, dg.
\end{aligned}$$

Now suppose that $Y' \sim f_{Y|X}(\cdot \mid x)$, $G \sim w(\cdot)$, and Y' and G are independent. Then the integrand in Equation 10.26 is the joint density of (G, Y) where $Y = t_G^{-1}(Y')$. Consequently, $Y = t_G^{-1}(Y')$ has density $f^{(w)}_{Y|X}(\cdot \mid x)$. This provides a simple method of drawing from $f^{(w)}_{Y|X}(\cdot \mid x)$. Indeed, we draw Y' and G independently from $f_{Y|X}(\cdot \mid x)$ and $w(\cdot)$ respectively, and then take $Y = t_G^{-1}(Y')$.

Sampling from $f^{(w)}_{X|Y}$ is a little trickier. Clearly,

$$f^{(w)}_{X|Y}(x \mid y) = \int_G f^{(w)}_{X,G|Y}(x, g \mid y) \, dg = \int_G f^{(w)}_{X|Y,G}(x \mid y, g) f^{(w)}_{G|Y}(g \mid y) \, dg.$$

Thus, we can use the sequential simulation technique from Section 10.1 to draw from $f_{X|Y}^{(w)}(x \mid y)$ as follows. First, draw $G \sim f_{G|Y}^{(w)}(\cdot \mid y)$ and then, conditional on $G = g$, draw $X \sim f_{X|Y,G}^{(w)}(\cdot \mid y, g)$. But now the question is whether we can draw from $f_{G|Y}^{(w)}$ and $f_{X|Y,G}^{(w)}$. It is actually simple to draw from $f_{X|Y,G}^{(w)}$ because

$$f_{X|Y,G}^{(w)}(x \mid y, g) = \frac{f^{(w)}(x, y, g)}{\int_X f^{(w)}(x, y, g)\, dx} = \frac{f(x, t_g(y)) \, |J_g(y)| \, w(g)}{f_Y(t_g(y)) \, |J_g(y)| \, w(g)} = f_{X|Y}(x \mid t_g(y)).$$

In other words, drawing from $f_{X|Y,G}^{(w)}(\cdot \mid y, g)$ is equivalent to drawing from $f_{X|Y}(\cdot \mid t_g(y))$. Now,

$$f_{G|Y}^{(w)}(g \mid y) = \frac{\int_X f^{(w)}(x, y, g)\, dx}{\int_G \int_X f^{(w)}(x, y, g)\, dx\, dg} \propto \int_X f^{(w)}(x, y, g)\, dx = f_Y(t_g(y)) \, |J_g(y)| \, w(g).$$

There is no simple trick for drawing from $f_{G|Y}^{(w)}$. Moreover, at first glance, sampling from the normalized version of $f_Y(t_g(y)) \, |J_g(y)| \, w(g)$ appears challenging because this function involves f_Y, from which it is impossible to sample. (Indeed, if we could draw directly from f_Y, then we could use the sequential simulation technique to get exact draws from the target, f_X, and we would not need MCMC!) Fortunately, g typically has much lower dimension than y and in such cases it is often possible to draw from $f_{G|Y}^{(w)}(g \mid y)$ despite the intractability of f_Y. Hence, our method of drawing from $f_{X|Y}^{(w)}(\cdot \mid y)$ is as follows. Draw $G \sim f_{G|Y}^{(w)}(\cdot \mid y)$ and then, conditional on $G = g$, draw $X \sim f_{X|Y}(\cdot \mid t_g(y))$.

As we know from previous sections, performing one iteration of the PX-DA algorithm entails drawing from $f_{Y|X}^{(w)}(\cdot \mid x)$ and then from $f_{X|Y}^{(w)}(\cdot \mid y)$. Liu and Wu (1999) noticed that making these two draws using the indirect techniques described above can be represented as a *three-step* procedure in which the first and third steps *are the same* as the original DA algorithm. Indeed, if the current state of the PX-DA Markov chain is $X_n = x$, then we can simulate X_{n+1} as follows.

ONE ITERATION OF THE PX-DA ALGORITHM

1. Draw $Y \sim f_{Y|X}(\cdot \mid x)$, and call the observed value y.
2. Draw $G \sim w(\cdot)$, call the result g, then draw $G' \sim f_{G|Y}^{(w)}\big(\cdot \mid t_g^{-1}(y)\big)$, call the result g', and finally set $y' = t_{g'}\big(t_g^{-1}(y)\big)$.
3. Draw $X_{n+1} \sim f_{X|Y}(\cdot \mid y')$.

Here is a recapitulation of what has been done so far in this subsection. We started with a DA algorithm for f_X based on a joint density $f(x, y)$. The density $f(x, y)$ was used to create an entire family of joint densities, $f^{(w)}(x, y)$, one for each density $w(\cdot)$. Each member of this family has f_X as its x-marginal and can therefore be used to create a new DA algorithm. We call these PX-DA algorithms. Running a PX-DA algorithm requires drawing from $f_{X|Y}^{(w)}$ and $f_{Y|X}^{(w)}$, and simple, indirect methods of making these draws were developed. Finally, we provided a representation of the PX-DA algorithm as a *three-step* algorithm in which the first and third steps are the same as the two steps of the original DA algorithm.

From a computational standpoint, the only difference between the original DA algorithm and the PX-DA algorithm is that one extra step (Step 2) must be performed at each iteration of the PX-DA algorithm. However, when g has relatively low dimension, as is usually the case in practice, the computational cost of the extra step is inconsequential compared to the cost of Steps 1 and 3. In such cases, the DA and PX-DA algorithms are (essentially) equivalent in terms of cost per iteration. What is amazing is the extent to which the mixing properties of the DA algorithm can be improved without really altering the computational complexity of the algorithm (see, e.g. Liu and Wu, 1999; Meng and van Dyk, 1999; van Dyk and Meng, 2001). Moreover, there is empirical evidence that the relative improvement of PX-DA over DA actually increases as the dimension of X increases (Meng and van Dyk, 1999). Section 10.4.3 contains a rigorous theoretical comparison of the DA and PX-DA algorithms. We end this subsection with a real example that was developed and studied in Liu and Wu (1999), van Dyk and Meng (2001), and Roy and Hobert (2007).

Example 10.5 (cont.)

In this example, $\pi(\beta, y \mid z)$ plays the role of $f(x, y)$. Take $G = \mathbb{R}_+$ and $t_g(y) = gy = (gy_1, \ldots, gy_m)$, and take w as follows:

$$w(g; \alpha, \delta) = \frac{2\delta^\alpha}{\Gamma(\alpha)} g^{2\alpha-1} e^{-g^2\delta} I_{\mathbb{R}_+}(g), \tag{10.27}$$

where $\alpha, \delta \in \mathbb{R}_+$. This is just the density of the square root of a gamma variate; that is, if $U \sim \Gamma(\alpha, \delta)$, then $G = \sqrt{U}$ has density (10.27). Substituting Equation 10.10 into Equation 10.9 and integrating with respect to β shows that

$$\pi(y \mid z) = \frac{\exp\left\{-\frac{1}{2}y^T(I-H)y\right\}}{|V^TV|^{1/2}c(z)(2\pi)^{(m-p)/2}} \prod_{i=1}^{m} \left\{ I_{\mathbb{R}_+}(y_i)I_{\{1\}}(z_i) + I_{\mathbb{R}_-}(y_i)I_{\{0\}}(z_i) \right\}.$$

Thus,

$$f_{G\mid Y}^{(w)}(g \mid y) \propto \pi(t_g(y) \mid z) \, |J_g(y)| \, w(g)$$

$$\propto \left[\exp\left\{ -\frac{1}{2}(gy)^T(I-H)(gy) \right\} \right] (g^m) \left[g^{2\alpha-1} \exp\{-g^2\delta\} I_{\mathbb{R}_+}(g) \right]$$

$$= \exp\left\{ -g^2\left[\frac{y^T(I-H)y}{2} + \delta \right] \right\} g^{m+2\alpha-1} I_{\mathbb{R}_+}(g).$$

Note that $f_{G\mid Y}^{(w)}(g \mid y)$ has the same form as $w(g; \alpha, \delta)$, which means that a draw from $f_{G\mid Y}^{(w)}(g \mid y)$ can be made be simulating a gamma and taking its square root. Putting all of this together, if the current state of the PX-DA algorithm is $X_n = \beta$, then we simulate the next state, X_{n+1}, by performing the following three steps:

1. Draw $Y_1, \ldots, Y_m$ independently such that $Y_i \sim \text{TN}(v_i^T\beta, 1, z_i)$, and call the result $y = (y_1, \ldots, y_m)^T$.

2. Draw $U \sim \Gamma(\alpha, \delta)$, call the result u, and set $\tilde{y} = y/\sqrt{u}$. Draw

$$V \sim \Gamma\left(\frac{m}{2} + \alpha, \frac{\tilde{y}^T(I - H)\tilde{y}}{2} + \delta\right),$$

call the result v, and set $y' = \sqrt{v}\tilde{y}$.

3. Draw $X_{n+1} \sim N(\hat{\beta}(y'), (V^T V)^{-1})$.

Sampling from the truncated normal distribution is typically done using an accept–reject algorithm, and Step 1 of the above procedure involves the simulation of m truncated normals. Obviously, the computational burden of Step 2, which requires only two univariate draws from the gamma distribution, is relatively minor. On the other hand, as the examples in Liu and Wu (1999) and van Dyk and Meng (2001) demonstrate, the PX-DA algorithm mixes much faster than the DA algorithm.

As a prelude to our theoretical comparison of DA and PX-DA, we introduce a bit of operator theory.

10.4.2 The Operator Associated with a Reversible Markov Chain

It is well known that techniques from spectral theory (see, e.g. Rudin, 1991, Part III) can be used to analyze reversible Markov chains. The reason for this is that every reversible Markov chain defines a self-adjoint operator on the space of functions that are square integrable with respect to the invariant density. Examples of the use of spectral theory in the analysis of reversible Markov chains can be found in Diaconis and Stroock (1991), Chan and Geyer (1994), Liu et al. (1994, 1995), Roberts and Rosenthal (1997), and Mira and Geyer (1999). Our theoretical comparison of PX-DA and DA involves ideas from this theory.

Define

$$L_0^2(f_X) = \left\{h \in L^2(f_X) : \int_X h(x) f_X(x) \, dx = 0\right\},$$

and, for $g, h \in L_0^2(f_X)$, define the inner product as $\langle g, h \rangle = \int_X g(x) h(x) f_X(x) \, dx$. The corresponding norm is given by $\|g\| = \sqrt{\langle g, g \rangle}$. Let $a : X \times X \to [0, \infty)$ denote a generic Mtd that is reversible with respect to f_X; that is, $a(x' \mid x) f_X(x) = a(x \mid x') f_X(x')$ for all $x, x' \in X$. Let $\Psi = \{\Psi_n\}_{n=0}^\infty$ denote the corresponding Markov chain and assume that Ψ is Harris ergodic. The Mtd a defines an operator, A, that maps $g \in L_0^2(f_X)$ to a new function in $L_0^2(f_X)$ given by

$$(Ag)(x) = \int_X g(x') a(x' \mid x) \, dx'.$$

Note that $(Ag)(x) = E[g(\Psi_{n+1}) \mid \Psi_n = x]$. To verify that Ag is square integrable with respect to f_X, use Jensen's inequality, Fubini's theorem, the invariance of f_X, and the fact that $g \in L_0^2(f_X)$ as follows:

$$\int_X \left[(Ag)(x)\right]^2 f_X(x) \, dx = \int_X \left[\int_X g(x') a(x' \mid x) \, dx'\right]^2 f_X(x) \, dx$$

$$\leq \int_X \left[\int_X g^2(x') a(x' \mid x) \, dx'\right] f_X(x) \, dx$$

$$= \int_X g^2(x') \left[\int_X a(x' \mid x) f_X(x) \, dx \right] dx'$$

$$= \int_X g^2(x') f_X(x') \, dx' < \infty.$$

That Ag has mean zero follows from Fubini, the invariance of f_X, and the fact that g has mean zero:

$$\int_X (Ag)(x) f_X(x) \, dx = \int_X \left[\int_X g(x') \, a(x' \mid x) \, dx' \right] f_X(x) \, dx$$

$$= \int_X g(x') \left[\int_X a(x' \mid x) f_X(x) \, dx \right] dx'$$

$$= \int_X g(x') f_X(x') \, dx' = 0.$$

We now demonstrate that the operator A is indeed self-adjoint (Rudin, 1991, Section 12). Using Fubini and the fact that $a(x' \mid x) f_X(x)$ is symmetric in (x, x'), we have, for $g, h \in L_0^2(f_X)$,

$$\langle Ag, h \rangle = \int_X (Ag)(x) \, h(x) f_X(x) \, dx$$

$$= \int_X \left[\int_X g(x') \, a(x' \mid x) \, dx' \right] h(x) f_X(x) \, dx$$

$$= \int_X \int_X g(x') \, h(x) \, a(x' \mid x) f_X(x) \, dx \, dx'$$

$$= \int_X g(x') \left[\int_X h(x) \, a(x \mid x') \, dx \right] f_X(x') \, dx'$$

$$= \int_X g(x') (Ah)(x') f_X(x') \, dx'$$

$$= \langle g, Ah \rangle.$$

The norm of the operator A is defined as

$$\|A\| = \sup_{g \in L_0^2(f_X), \, \|g\| = 1} \|Ag\|.$$

Obviously, $\|A\| \geq 0$. In fact, $\|A\| \in [0, 1]$. Indeed, $\|Ag\|^2 = \int_X \left[(Ag)(x) \right]^2 f_X(x) \, dx$ and the calculations above imply that $\|Ag\|^2 \leq \|g\|^2$. The quantity $\|A\|$ is closely related to the convergence properties of the Markov chain Ψ. For example, Ψ is geometrically ergodic if and only if $\|A\| < 1$ (Roberts and Rosenthal, 1997; Roberts and Tweedie, 2001). The closer $\|A\|$ is to 0, the faster Ψ converges to its stationary distribution (see, e.g. Rosenthal, 2003). Because of this, Monte Carlo Markov chains are sometimes ordered according to their operator norms. In particular, if there are two different chains available that are both reversible with respect to f_X, we prefer the one with the smaller operator norm (see, e.g. Liu and Wu, 1999; Liu et al., 1994; Meng and van Dyk, 1999). In the next subsection, we compare DA and PX-DA in terms of operator norms as well as performance in the CLT.

10.4.3 A Theoretical Comparison of the DA and PX-DA Algorithms

The Mtd of the PX-DA algorithm is given by

$$k_w(x' \mid x) = \int_Y f^{(w)}_{X|Y}(x' \mid y) f^{(w)}_{Y|X}(y \mid x)\, dy.$$

However, there is an alternative representation of k_w that is based on the general three-step procedure for simulating the PX-DA algorithm that was given in Section 10.4.1. This representation turns out to be much more useful for comparing DA and PX-DA. Recall that Step 2 of the three-step procedure entails making the transition $y \to y'$ by drawing Y' from a distribution that depends on y. Hence, this step can be viewed as performing a single iteration of a Markov chain whose state space is Y. If we denote the corresponding Mtd as $l_w(y' \mid y)$, then we can reexpress the Mtd of the PX-DA algorithm as

$$k_w(x' \mid x) = \int_Y \int_Y f_{X|Y}(x' \mid y')\, l_w(y' \mid y) f_{Y|X}(y \mid x)\, dy\, dy'. \tag{10.28}$$

Liu and Wu's (1999) Theorem 1 implies that f_Y is an invariant density for l_w; that is,

$$\int_Y l_w(y' \mid y) f_Y(y)\, dy = f_Y(y').$$

This invariance implies that f_X is an invariant density for $k_w(x' \mid x)$:

$$\int_X k_w(x' \mid x) f_X(x)\, dx = \int_X \left[\int_Y \int_Y f_{X|Y}(x' \mid y')\, l_w(y' \mid y) f_{Y|X}(y \mid x)\, dy\, dy' \right] f_X(x)\, dx$$

$$= \int_Y f_{X|Y}(x' \mid y') \left[\int_Y l_w(y' \mid y) f_Y(y)\, dy \right] dy'$$

$$= \int_Y f_{X|Y}(x' \mid y') f_Y(y')\, dy'$$

$$= f_X(x').$$

Of course, we did not need Equation 10.28 to conclude that f_X is invariant for $k_w(x' \mid x)$. Indeed, the fact that $k_w(x' \mid x)$ is the Mtd of a DA algorithm implies that k_w is reversible with respect to f_X, and hence that f_X is invariant for k_w. Note, however, that the previous calculation still goes through if l_w is replaced by *any* Mtd having f_Y as an invariant density. This suggests a generalization of Equation 10.28.

Let $l : Y \times Y \to [0, \infty)$ be any Mtd that has $f_Y(y)$ as an invariant density. Define the function $k_l : X \times X \to [0, \infty)$ as follows:

$$k_l(x' \mid x) = \int_Y \int_Y f_{X|Y}(x' \mid y')\, l(y' \mid y) f_{Y|X}(y \mid x)\, dy\, dy'. \tag{10.29}$$

The reader can easily verify that, for each fixed $x \in X$, $\int_X k_l(x' \mid x)\, dx' = 1$. Hence, k_l is an Mtd that defines a Markov chain on X, and the arguments above show that f_X is an invariant density for k_l. This is a generalization of Equation 10.28 in the sense that the set of Mtds having f_Y as an invariant density is much larger than the set of Mtds of the form l_w. Hobert and Marchev (2008) studied k_l and established that (under weak regularity conditions) the MCMC algorithm based on k_l is better (in terms of convergence rate and ARE) than the DA algorithm. This leads to the conclusion that every PX-DA algorithm is better than the DA algorithm upon which it is based. In order to state the results precisely, we need a couple of definitions.

If there exists a joint pdf $f^*(x, y)$ with $\int_Y f^*(x, y)\, dy = f_X(x)$ such that

$$k_l(x' \mid x) = \int_Y f^*_{X \mid Y}(x' \mid y) f^*_{Y \mid X}(y \mid x)\, dy,$$

then we say that k_l is *representable*. Clearly, if k_l is representable, then it is also reversible with respect to $f_X(x)$. (Note that, by definition, k_w is representable with $f^{(w)}(x, y)$ playing the role of $f^*(x, y)$.)

The second definition involves the CLT discussed in Section 10.2.4. Let $X = \{X_n\}_{n=0}^{\infty}$ denote the Markov chain underlying the original DA algorithm based on $f(x, y)$. Suppose that $g \in L^2(f_X)$ and, as before, let $\bar{g}_n = \frac{1}{n} \sum_{n=0}^{n-1} g(X_i)$. If $\bar{g}_n$ satisfies a CLT, then let κ_g^2 denote the corresponding asymptotic variance. If there is no CLT for $\bar{g}_n$, then set κ_g^2 equal to ∞. (Since we have not assumed that X is geometrically ergodic, a CLT for $\bar{g}_n$ may or may not exist.) Now let $X^* = \{X_n^*\}_{n=0}^{\infty}$ denote the Markov chain associated with $k_l(x' \mid x)$, and define κ_g^{*2} analogously using $\bar{g}_n^* = \frac{1}{n} \sum_{n=0}^{n-1} g(X_i^*)$ in place of $\bar{g}_n$. If $\kappa_g^{*2} \le \kappa_g^2$ for every $g \in L^2(f_X)$, then we say that k_l is *more efficient than* k.

Hobert and Marchev (2008) established two general results that facilitate comparison of the DA algorithm and the MCMC algorithm based on k_l: (i) if k_l is reversible with respect to f_X, then k_l is more efficient than k; and (ii) if k_l is representable, then $\|K_l\| \le \|K\|$, where K_l and K are the operators on $L_0^2(f_X)$ associated with k_l and k, respectively. (Hobert and Rosenthal (2007) show that, in (ii), representability can be replaced by a weaker condition at no expense.) Now, consider the implications of these results with regard to the PX-DA algorithm. Since k_w is representable, both of Hobert and Marchev's (2008) results are applicable and we may conclude that every PX-DA algorithm is better than the corresponding DA algorithm in terms of both convergence rate and ARE. (The norm comparison result was actually established in Liu and Wu (1999) and Meng and van Dyk (1999) using different techniques.)

In addition to providing information about the relative convergence rates of X and X^*, the inequality $\|K_l\| \le \|K\|$ also has a nice practical application. We know from Section 10.4.2 that a reversible Markov chain is geometrically ergodic if and only if the norm of the corresponding operator is strictly less than 1. Therefore, if we can prove that the DA Markov chain, X, is geometrically ergodic (by, say, establishing a geometric drift condition), then it follows that $\|K_l\| \le \|K\| < 1$, which implies that X^* is also geometrically ergodic. This allows one to prove that X^* is geometric without having to work directly with k_l, which, from an analytical standpoint, is much more cumbersome than k.

It is important to keep in mind that the comparison results described above are really only useful in situations where at least one of the two chains being compared is known to be geometrically ergodic. For example, if all we know is that $\|K_l\| \le \|K\|$, then it may be the case that X and X^* are both bad chains with norm 1 and neither should be used in practice. Similarly, if there are no CLTs, then the fact that k_l is more efficient than k is not very useful.

Finally, there is one very simple sufficient condition for k_l to be reversible with respect to f_X, and that is the reversibility of $l(y' \mid y)$ with respect to $f_Y(y)$. Indeed, suppose that

$l(y' \mid y) f_Y(y)$ is symmetric in (y, y'). Then

$$
\begin{aligned}
k_l(x' \mid x) f_X(x) &= f_X(x) \int_Y \int_Y f_{X \mid Y}(x' \mid y') \, l(y' \mid y) f_{Y \mid X}(y \mid x) \, dy \, dy' \\
&= \int_Y \int_Y f_{X \mid Y}(x' \mid y') \, l(y' \mid y) f(x, y) \, dy \, dy' \\
&= \int_Y \int_Y f_{X \mid Y}(x' \mid y') \, l(y' \mid y) f_Y(y) f_{X \mid Y}(x \mid y) \, dy \, dy' \\
&= \int_Y \int_Y f_{X \mid Y}(x' \mid y') \, l(y \mid y') f_Y(y') f_{X \mid Y}(x \mid y) \, dy \, dy' \\
&= \int_Y \int_Y f(x', y') \, l(y \mid y') f_{X \mid Y}(x \mid y) \, dy \, dy' \\
&= f_X(x') \int_Y \int_Y f_{X \mid Y}(x \mid y) \, l(y \mid y') f_{Y \mid X}(y' \mid x') \, dy \, dy' \\
&= k_l(x \mid x') f_X(x').
\end{aligned}
$$

There is also a simple sufficient condition on $l(y' \mid y)$ for representability of k_l (see Hobert and Marchev, 2008).

We know that each pdf $w(g)$ yields its own PX-DA algorithm. In the next subsection, we show that, under certain conditions, there is a limiting version of the PX-DA algorithm that beats all the others.

10.4.4 Is There a Best PX-DA Algorithm?

The results in the previous subsection show that every PX-DA algorithm is better than the original DA algorithm based on $f(x, y)$. This raises the question of whether there exists a particular PX-DA algorithm that beats all the others. There are actually theoretical arguments as well as empirical evidence suggesting that the PX-DA algorithm will perform better as the pdf $w(\cdot)$ becomes more "diffuse" (Liu and Wu, 1999; Meng and van Dyk, 1999; van Dyk and Meng, 2001). On the other hand, it is clear that our development of the PX-DA algorithm breaks down if w is improper. In particular, if w is improper, then Equation 10.25 is no longer a pdf. Moreover, Step 2 of the PX-DA algorithm requires a draw from w, which is obviously not possible when w is improper. However, Liu and Wu (1999) showed that, if there is a certain group structure present in the problem, then it is possible to construct a valid PX-DA-*like* algorithm using an improper *Haar density* in place of w. Moreover, the results from the previous subsection can be used to show that this *Haar PX-DA algorithm* is better than any PX-DA algorithm based on a proper w.

Suppose that the set G is a topological group; that is, a group such that the functions $(g_1, g_2) \mapsto g_1 g_2$ and $g \mapsto g^{-1}$ are continuous. (An example of such a group is the *multiplicative group*, $\mathbb{R}_+$, where the binary operation defining the group is multiplication, the identity element is 1, and $g^{-1} = 1/g$.) Let e denote the group's identity element and assume that $t_e(y) = y$ for all $y \in Y$ and that $t_{g_1 g_2}(y) = t_{g_1}(t_{g_2}(y))$ for all $g_1, g_2 \in G$ and all $y \in Y$. In other words, we are assuming that $t_g(y)$ represents G acting topologically on the left of Y (Eaton, 1989, Chapter 2).

A function $\chi : G \to \mathbb{R}_+$ is called a *multiplier* if χ is continuous and $\chi(g_1 g_2) = \chi(g_1) \chi(g_2)$ for all $g_1, g_2 \in G$. Assume that Lebesgue measure on Y is *relatively (left) invariant* with multiplier χ; that is, assume that, for any $g \in G$ and any integrable function $h : Y \to \mathbb{R}$, we have

$$\chi(g) \int_Y h\big(t_g(y)\big)\, dy = \int_Y h(y)\, dy.$$

Here is a simple example.

Example 10.5 (cont.)

Again, take $G = \mathbb{R}_+$ and $t_g(y) = gy = (gy_1, \ldots, gy_m)$. Now think of $G = \mathbb{R}_+$ as the multiplicative group and note that, for any $y \in \mathbb{R}_+^m$ and any $g_1, g_2 \in G$, we have $t_e(y) = y$ and

$$t_{g_1 g_2}(y) = g_1 g_2 y = g_1(g_2 y) = t_{g_1}\big(t_{g_2}(y)\big).$$

Hence, the compatibility conditions are satisfied. Now, for any $g \in G$, we have

$$\int_Y h\big(t_g(y)\big)\, dy = \int_{\mathbb{R}_+^m} h(gy)\, dy = g^{-m} \int_{\mathbb{R}_+^m} h(y)\, dy,$$

which shows that Lebesgue measure on $Y = \mathbb{R}_+^m$ is relatively invariant with multiplier $\chi(g) = g^m$.

Suppose that the group G has a *left-Haar measure* of the form $\nu_l(g)\, dg$, where dg denotes Lebesgue measure on G. Left-Haar measure satisfies

$$\int_G h(\tilde{g}g)\, \nu_l(g)\, dg = \int_G h(g)\, \nu_l(g)\, dg, \tag{10.30}$$

for all $\tilde{g} \in G$ and all integrable functions $h : G \to \mathbb{R}$. In most applications, this measure will be improper; that is, $\int_G \nu_l(g)\, dg = \infty$. (When the left-Haar measure is the same as the right-Haar measure, which satisfies the obvious analog of Equation 10.30, the group is called *unimodular*.) Finally, assume that

$$q(y) := \int_G f_Y\big(t_g(y)\big)\, \chi(g)\, \nu_l(g)\, dg$$

is strictly positive for all $y \in Y$ and finite for (almost) all $y \in Y$.

We now state (a generalized version of) Liu and Wu's (1999) Haar PX-DA algorithm. If the current state is $X_n^* = x$, we simulate X_{n+1}^* as follows.

ONE ITERATION OF THE HAAR PX-DA ALGORITHM

1. Draw $Y \sim f_{Y|X}(\cdot \mid x)$, and call the observed value y.
2. Draw G from the density proportional to $f_Y\big(t_g(y)\big)\, \chi(g)\, \nu_l(g)$, call the result g, and set $y' = t_g(y)$.
3. Draw $X_{n+1}^* \sim f_{X|Y}(\cdot \mid y')$.

This algorithm *is not a PX-DA algorithm*, but its Mtd does take the form (10.29). Indeed, if we let $l_H(y' \mid y)$ denote the Mtd of the Markov chain on Y that is simulated at Step 2, then we can write the Mtd of the Haar PX-DA algorithm as

$$k_H(x' \mid x) = \int_Y \int_Y f_{X|Y}(x' \mid y')\, l_H(y' \mid y)\, f_{Y|X}(y \mid x)\, dy\, dy'.$$

Hobert and Marchev (2008) show that $l_H(y' \mid y)$ is reversible with respect to f_Y, which, of course, implies that f_Y is an invariant density for $l_H(y' \mid y)$. Moreover, these authors also prove that k_H is representable. Hence, the comparison results from the previous subsection are applicable and imply that the Haar PX-DA algorithm is better than the DA algorithm in terms of both convergence rate and ARE. However, what we really want to compare is Haar PX-DA and PX-DA, and this is the subject of the rest of this section.

Hobert and Marchev (2008) show that, for any fixed proper pdf $w(\cdot)$, k_H can be reexpressed as

$$k_H(x' \mid x) = \int_Y \int_Y f^{(w)}_{X|Y}(x' \mid y')\, l^{(w)}(y' \mid y) f^{(w)}_{Y|X}(y \mid x)\, dy\, dy', \tag{10.31}$$

where $f^{(w)}_{X|Y}$ and $f^{(w)}_{Y|X}$ are as defined in Section 10.4.1, and $l^{(w)}(y' \mid y)$ is an Mtd on Y that is reversible with respect to $f^{(w)}_Y(y) := \int_Y f^{(w)}(x, y)\, dx$. Now consider the significance of Equation 10.31 in the context of the results of Section 10.4.3. In particular, we know that the PX-DA algorithm driven by $f^{(w)}(x, y)$ is itself a DA algorithm, and Equation 10.31 shows that k_H is related to k_w in exactly the same way that k_l is related to k. Therefore, since k_H is representable, we may appeal to the comparison results once more to conclude that the Haar PX-DA is better than every PX-DA algorithm in terms of both convergence rate and ARE.

Finally, note that Step 2 of the Haar PX-DA algorithm involves only one draw from a density on G, whereas the regular PX-DA algorithm calls for two such draws in its Step 2. Thus, from a computational standpoint, the Haar PX-DA algorithm is actually simpler than the PX-DA algorithm. We conclude with an application to the probit example.

Example 10.5 (cont.)

Recall that G is the multiplicative group, $\mathbb{R}_+$, and $t_g(y) = gy = (gy_1, \ldots, gy_m)$. Note that, for any $\tilde{g} \in G$, we have

$$\int_0^\infty h(\tilde{g}g)\, \frac{1}{g}\, dg = \int_0^\infty h(g)\, \frac{1}{g}\, dg,$$

which shows that $\frac{dg}{g}$ is a left-Haar measure for the multiplicative group. (This group is actually abelian and hence unimodular.) Thus,

$$\pi(t_g(y) \mid z)\, \chi(g)\, \nu_l(g) \propto g^{m-1} \exp\left\{ -g^2 \left[\frac{y^T(I - H)y}{2} \right] \right\} I_{\mathbb{R}_+}(g),$$

and it follows that

$$q(y) \propto \int_0^\infty g^{m-1} \exp\left\{ -g^2 \left[\frac{y^T(I - H)y}{2} \right] \right\} dg = \frac{2^{m/2}\Gamma\left(\frac{m}{2}\right)}{[y^T(I - H)y]^{m/2}},$$

which is clearly positive for all $y \in Y$ and finite for (almost) all $y \in Y$. We can now write down the Haar PX-DA algorithm. Given the current state, $X_n^* = \beta$, we simulate the next state, X_{n+1}^*, by performing the following three steps:

1. Draw $Y_1, \ldots, Y_m$ independently such that $Y_i \sim \mathrm{TN}(v_i^T\beta, 1, z_i)$, and call the result $y = (y_1, \ldots, y_m)^T$.

2. Draw

$$V \sim \Gamma\left(\frac{m}{2}, \frac{y^T(I - H)y}{2}\right),$$

call the result v, and set $y' = \sqrt{v}y$.

3. Draw $X_{n+1}^* \sim N\big(\hat{\beta}(y'), (V^T V)^{-1}\big)$.

In Section 10.4.1, we developed a family of PX-DA algorithms for this problem, one for each $(\alpha, \delta) \in \mathbb{R}_+ \times \mathbb{R}_+$. The results in Section 10.4.3 imply that every member of that family is better than the original DA algorithm based on $f(x, y)$. Moreover, the results described in this subsection show that the Haar PX-DA algorithm above is better than every member of that family of PX-DA algorithms.

Roy and Hobert (2007) proved that this Haar PX-DA algorithm is geometrically ergodic by establishing that the much simpler DA algorithm of Albert and Chib (1993) is geometrically ergodic, and then appealing to the fact that $\|K_H\| \leq \|K\|$. These authors also provided substantial empirical evidence suggesting that the ARE of the Haar PX-DA estimator with respect to the DA estimator is often much larger than 1.

Acknowledgments

The author is grateful to Trung Ha, Galin Jones, Aixin Tan, and an anonymous referee for helpful comments and suggestions. This work was supported by NSF Grants DMS-05-03648 and DMS-08-05860.

References

Albert, J. H. and Chib, S. 1993. Bayesian analysis of binary and polychotomous response data. *Journal of the American Statistical Association*, 88:669–679.

Chan, K. S. and Geyer, C. J. 1994. Comment on "Markov chains for exploring posterior distributions" by L. Tierney. *Annals of Statistics*, 22:1747–1758.

Chen, M.-H. and Shao, Q.-M. 2000. Propriety of posterior distribution for dichotomous quantal response models. *Proceedings of the American Mathematical Society*, 129:293–302.

Dempster, A. P., Laird, N. M., and Rubin, D. B. 1977. Maximum likelihood from incomplete data via the EM algorithm (with discussion). *Journal of the Royal Statistical Society*, Series B, 39:1–38.

Diaconis, P. and Stroock, D. 1991. Geometric bounds for eigenvalues of Markov chains. *Annals of Applied Probability*, 1:36–61.

Diaconis, P., Khare, K., and Saloff-Coste, L. 2008. Gibbs sampling, exponential families and orthogonal polynomials (with discussion). *Statistical Science*, 23:151–200.

Eaton, M. L. 1989. *Group Invariance Applications in Statistics*. Institute of Mathematical Statistics, Hayward, CA, and American Statistical Association, Alexandria, VA.

Fernández, C. and Steel, M. F. J. 1999. Multivariate Student-*t* regression models: Pitfalls and inference. *Biometrika*, 86:153–167.

Fill, J. A., Machida, M., Murdoch, D. J., and Rosenthal, J. S. 2000. Extension of Fill's perfect rejection sampling algorithm to general chains. *Random Structures and Algorithms*, 17:290–316.

Flegal, J. M., Haran, M., and Jones, G. L. 2008. Markov chain Monte Carlo: Can we trust the third significant figure? *Statistical Science*, 23:250–260.

Geyer, C. J. 1992. Practical Markov chain Monte Carlo (with discussion). *Statistical Science*, 7:473–511.

Hobert, J. P. and Marchev, D. 2008. A theoretical comparison of the data augmentation, marginal augmentation and PX-DA algorithms. *Annals of Statistics*, 36:532–554.

Hobert, J. P. and Rosenthal, J. S. 2007. Norm comparisons for data augmentation. *Advances and Applications in Statistics*, 7:291–302.

Hobert, J. P., Jones, G. L., Presnell, B., and Rosenthal, J. S. 2002. On the applicability of regenerative simulation in Markov chain Monte Carlo. *Biometrika*, 89:731–743.

Hobert, J. P., Tan, A., and Liu, R. 2007. When is Eaton's Markov chain irreducible? *Bernoulli*, 13:641–652.

Jones, G. L. and Hobert, J. P. 2001. Honest exploration of intractable probability distributions via Markov chain Monte Carlo. *Statistical Science*, 16:312–34.

Jones, G. L., Haran, M., Caffo, B. S., and Neath, R. 2006. Fixed-width output analysis for Markov chain Monte Carlo. *Journal of the American Statistical Association*, 101:1537–1547.

Liu, J. S. and Wu, Y. N. 1999. Parameter expansion for data augmentation. *Journal of the American Statistical Association*, 94:1264–1274.

Liu, J. S., Wong, W. H., and Kong, A. 1994. Covariance structure of the Gibbs sampler with applications to comparisons of estimators and augmentation schemes. *Biometrika*, 81:27–40.

Liu, J. S., Wong, W. H., and Kong, A. 1995. Covariance structure and convergence rate of the Gibbs sampler with various scans. *Journal of the Royal Statistical Society*, Series B, 57:157–169.

Marchev, D. and Hobert, J. P. 2004. Geometric ergodicity of van Dyk and Meng's algorithm for the multivariate Student's t model. *Journal of the American Statistical Association*, 99:228–238.

Meng, X.-L. and van Dyk, D. A. 1999. Seeking efficient data augmentation schemes via conditional and marginal augmentation. *Biometrika*, 86:301–320.

Meyn, S. P. and Tweedie, R. L. 1993. *Markov Chains and Stochastic Stability*. Springer, London.

Meyn, S. P. and Tweedie, R. L. 1994. Computable bounds for geometric convergence rates of Markov chains. *Annals of Applied Probability*, 4:981–1011.

Mira, A. and Geyer, C. J. 1999. Ordering Monte Carlo Markov chains. Technical Report No. 632, School of Statistics, University of Minnesota.

Mykland, P., Tierney, L., and Yu, B. 1995. Regeneration in Markov chain samplers. *Journal of the American Statistical Association*, 90:233–241.

Neal, R. M. 2003. Slice sampling (with discussion). *Annals of Statistics*, 31:705–767.

Nummelin, E. 1984. *General Irreducible Markov Chains and Non-negative Operators*. Cambridge University Press, Cambridge.

R Development Core Team 2006. *R: A Language and Environment for Statistical Computing*. R Foundation for Statistical Computing, Vienna, Austria.

Robert, C. P. 1995. Simulation of truncated normal variables. *Statistics and Computing*, 5:121–125.

Roberts, G. and Smith, A. F. M. 1994. Simple conditions for the convergence of the Gibbs sampler and Metropolis-Hastings algorithms. *Stochastic Processes and Their Applications*, 49:207–216.

Roberts, G. O. and Rosenthal, J. S. 1997. Geometric ergodicity and hybrid Markov chains. *Electronic Communications in Probability*, 2:13–25.

Roberts, G. O. and Rosenthal, J. S. 2001. Markov chains and de-initializing processes. *Scandinavian Journal of Statistics*, 28:489–504.

Roberts, G. O. and Rosenthal, J. S. 2004. General state space Markov chains and MCMC algorithms. *Probability Surveys*, 1:20–71.

Roberts, G. O. and Rosenthal, J. S. 2006. Harris recurrence of Metropolis-within-Gibbs and trans-dimensional Markov chains. *Annals of Applied Probability*, 16:2123–2139.

Roberts, G. O. and Tweedie, R. L. 2001. Geometric L^2 and L^1 convergence are equivalent for reversible Markov chains. *Journal of Applied Probability*, 38A:37–41.

Rosenthal, J. S. 1995. Minorization conditions and convergence rates for Markov chain Monte Carlo. *Journal of the American Statistical Association*, 90:558–566.

Rosenthal, J. S. 2003. Asymptotic variance and convergence rates of nearly-periodic MCMC algorithms. *Journal of the American Statistical Association*, 98:169–177.

Ross, S. M. 1996. *Stochastic Processes*, 2nd edn. Wiley, New York.

Roy, V. and Hobert, J. P. 2007. Convergence rates and asymptotic standard errors for Markov chain Monte Carlo algorithms for Bayesian probit regression. *Journal of the Royal Statistical Society, Series B*, 69:607–623.

Rudin, W. 1991. *Functional Analysis*, 2nd edn. McGraw-Hill, New York.

Swendsen, R. H. and Wang, J.-S. 1987. Nonuniversal critical dynamics in Monte Carlo simulations. *Physical Review Letters*, 58:86–88.

Tan, A. 2008. Analysis of Markov chain Monte Carlo algorithms for random effects models. PhD thesis, Department of Statistics, University of Florida.

Tan, A. and Hobert, J. P. 2009. Block Gibbs sampling for Bayesian random effects models with improper priors: Convergence and regeneration. *Journal of Computational and Graphical Statistics*, 18:861–878.

Tanner, M. A. and Wong, W. H. 1987. The calculation of posterior distributions by data augmentation (with discussion). *Journal of the American Statistical Association*, 82:528–550.

Tierney, L. 1994. Markov chains for exploring posterior distributions (with discussion). *Annals of Statistics*, 22:1701–1762.

van Dyk, D. A. and Meng, X.-L. 2001. The art of data augmentation (with discussion). *Journal of Computational and Graphical Statistics*, 10:1–50.

11

Importance Sampling, Simulated Tempering, and Umbrella Sampling

Charles J. Geyer

11.1 Importance Sampling

The importance of so-called importance sampling in Markov chain Monte Carlo (MCMC) is not what gives it that name. It is the idea that "any sample can come from any distribution" (Trotter and Tukey, 1956). Suppose that we have a Markov chain $X_1, X_2, \ldots$ having properly normalized density f for its equilibrium distribution. Let f_θ denote a parametric family of densities each absolutely continuous with respect to f. Then

$$\hat{\mu}_n(\theta) = \frac{1}{n} \sum_{i=1}^{n} g(X_i) \frac{f_\theta(X_i)}{f(X_i)} \tag{11.1}$$

is a sensible estimator of

$$\mu(\theta) = E_\theta\{g(X)\} \tag{11.2}$$

for all θ, because by the Markov chain law of large numbers (Meyn and Tweedie, 1993, Theorem 17.1.7),

$$\hat{\mu}_n(\theta) \xrightarrow{\text{a.s.}} E_f\left\{g(X)\frac{f_\theta(X)}{f(X)}\right\} = \int g(x)\frac{f_\theta(x)}{f(x)}f(x)\,dx = \int g(x)f_\theta(x)\,dx$$

(the requirement that f_θ is absolutely continuous with respect to f is required so that we divide by zero in the middle expressions with probability zero, so the value of the integral is not affected). With one sample from one distribution $f(x)$ we learn about $\mu(\theta)$ for all θ.

Monte Carlo standard errors (MCSEs) for importance sampling are straightforward: we just calculate the MCSE for the functional of the Markov chain (Equation 11.1) that gives our importance sampling estimator. This means we replace g in Equation 1.6 in Chapter 1 (this volume) by $g f_\theta/f$.

We are using here both the principle of "importance sampling" (in using the distribution with density f to learn about the distribution with density f_θ) and the principle of "common random numbers" (in using the same sample to learn about f_θ for all θ). The principle of common random numbers is very important. It means, for example, that

$$\nabla \tilde{\mu}_n(\theta) = \frac{1}{n} \sum_{i=1}^{n} g(X_i) \frac{\nabla f_\theta(X_i)}{f(X_i)}$$

is a sensible estimator of

$$\nabla \mu(\theta) = E_\theta\{\nabla g(X)\},$$

which relies on the same sample being used for all θ. Clearly, using different samples for different θ would not work at all.

The argument above relies on f and f_θ being properly normalized densities. If we replace them with unnormalized densities h and h_θ, we need a slightly different estimator (Geweke, 1989). Now we suppose that we have a Markov chain $X_1, X_2, \ldots$ having unnormalized density h for its equilibrium distribution, and we let h_θ denote a parametric family of unnormalized densities, each absolutely continuous with respect to h. Define the so-called "normalized importance weights"

$$w_\theta(x) = \frac{\dfrac{h_\theta(x)}{h(x)}}{\displaystyle\sum_{i=1}^{n} \dfrac{h_\theta(x_i)}{h(x_i)}} \tag{11.3}$$

so

$$\tilde\mu_n(\theta) = \sum_{i=1}^{n} g(X_i) w_\theta(X_i) \tag{11.4}$$

is sensible estimator of Equation 11.2 for all θ, because of the following. Define

$$d(\theta) = \int h_\theta(x)\, dx,$$

$$d = \int h\, dx,$$

so $h_\theta/d(\theta)$ and h/d are properly normalized probability densities. Then by the law of large numbers,

$$\tilde\mu_n(\theta) \xrightarrow{\text{a.s.}} \frac{E_h\left\{g(x)\dfrac{h_\theta(x)}{h(x)}\right\}}{E_h\left\{\dfrac{h_\theta(x)}{h(x)}\right\}} = \frac{\displaystyle\int g(x)\dfrac{h_\theta(x)}{h(x)}\cdot\dfrac{h(x)}{d}\,dx}{\displaystyle\int \dfrac{h_\theta(x)}{h(x)}\cdot\dfrac{h(x)}{d}\,dx} = \frac{\dfrac{d(\theta)}{d}\displaystyle\int g(x)\dfrac{h_\theta(x)}{d(\theta)}\,dx}{\dfrac{d(\theta)}{d}\displaystyle\int \dfrac{h_\theta(x)}{d(\theta)}\,dx} = E_\theta\{g(X)\}$$

(the requirement that h_θ is absolutely continuous with respect to h is required so that we divide by zero in the middle expressions with probability zero, so the value of the integral is not affected).

MCSEs for importance sampling are now a little more complicated. The estimator (Equation 11.4) is a ratio of two functionals of the Markov chain

$$\tilde\mu_n(\theta) = \frac{\dfrac{1}{n}\displaystyle\sum_{i=1}^{n} g(X_i)\dfrac{h_\theta(X_i)}{h(X_i)}}{\dfrac{1}{n}\displaystyle\sum_{i=1}^{n} \dfrac{h_\theta(X_i)}{h(X_i)}}.$$

We calculate the joint asymptotic distribution for the vector functional of the Markov chain having two components gh_θ/h and h_θ/h and then use the delta method to derive the asymptotic variance of their ratio; Geyer and Thompson (1995) give details.

The normalized importance weights trick, which uses Equation 11.3 and Equation 11.4 instead of Equation 11.1, is essential when using unnormalized densities to specify distributions. It is nice to use even when using properly normalized densities, because it makes the complement rule hold exactly rather than approximately for our Monte Carlo estimates of probabilities. If we use Equation 11.1 to estimate $\Pr(A)$ and $\Pr(A^c)$, the estimates will sum to approximately one for large n. If we use Equation 11.3 and Equation 11.4, the estimates will sum to exactly one for all n.

Even when there is only one target distribution with unnormalized density h_θ, one may do better using a different unnormalized density h as the importance sampling distribution. When there are many h_θ of interest, importance sampling is usually better than running a different sampler for every θ of interest.

> When importance sampling is allowed, it is never obvious what the equilibrium distribution of your MCMC sampler should be.

Why would one ever be interested in more than one distribution? Isn't MCMC just for Bayesian inference, and aren't Bayesians only interested in the posterior? The answers are, of course, no and no. As mentioned at the end of Section 1.1 in this volume, MCMC is also used for likelihood inference for models with complicated dependence (Geyer, 1994, 1999; Geyer and Thompson, 1992, 1995), and there one is interested in calculating the likelihood at each parameter value θ. Bayesians are often interested in inference under multiple priors (Insua and Ruggeri, 2000).

One warning is required about importance sampling. If the target distribution is not close to the importance sampling distribution, then importance sampling does not work well. Of course, it works for sufficiently large Monte Carlo sample size, but the sample sizes required may be impractical. A method for getting an importance sampling distribution close enough to target distributions of interest is umbrella sampling (Section 11.2.5 below).

11.2 Simulated Tempering

If a random-walk Metropolis sampler, as illustrated in Section 1.13, does not converge in a reasonable amount of time, the best way to detect this (other than perfect sampling, which usually one does not know how to do for one's problem) is to run a better sampler. The mcmc package provides two such sampling schemes, called parallel and serial tempering, both done by the `temper` function. Either can produce rapidly mixing samplers for problems in which no other known method works. Parallel tempering is easier to use. Serial tempering works better. Geyer and Thompson (1995) give an example in which serial tempering seems to work for a very hard problem but parallel tempering failed.

Serial tempering (Geyer and Thompson, 1995; Marinari and Parisi, 1992) runs a Markov chain whose state is (i, x), where i is a positive integer between 1 and m and x is an element of $\mathbb{R}^p$. The unnormalized density of the equilibrium distribution is $h(i, x)$. The integer i is called the *index of the component of the mixture*, and the integer m is called the *number of components of the mixture.* The reason for this terminology is that

$$h(x) = \sum_{i=1}^{m} h(i, x), \tag{11.5}$$

which is the unnormalized marginal density of x derived from the unnormalized joint density $h(i, x)$ of the equilibrium distribution of the Markov chain, is a mixture of m component distributions having unnormalized density $h(i, \cdot)$ for different i.

Parallel tempering (Geyer, 1991) runs a Markov chain whose state is $(x_1, \ldots, x_m)$ where each x_i is an element of $\mathbb{R}^p$. Thus the state is a vector whose elements are vectors, which may be thought of as an $m \times p$ matrix. The unnormalized density of the equilibrium distribution is

$$h(x_1, \ldots, x_m) = \prod_{i=1}^{m} h(i, x_i). \tag{11.6}$$

Since Equation 11.6 is the product of the unnormalized marginals $h(i, \cdot)$ for different i, this makes the x_i asymptotically independent in parallel tempering.

Parallel tempering was not so named by Geyer (1991). That name was later coined by others (Earl and Deem, 2005) to make an analogy with simulated tempering, the name Marinari and Parisi (1992) coined for their algorithm because they thought it had an analogy with simulated annealing (Kirkpatrick et al., 1983), even though the latter is a method of adaptive random search optimization rather than an MCMC method. The temper function in the mcmc package coins the name "serial tempering" for what has formerly been called "simulated tempering" on the linguistic grounds that both methods are forms of "tempering" (so say their names) and both methods "simulate" Markov chains, thus both must be forms of "simulated tempering." Fortunately, "simulated tempering" and "serial tempering" can both use the same abbreviation (ST), so there is no confusion there. The parallel–serial distinction is taken from the terminology for electrical circuits: parallel tempering (PT) simulates all the component distributions $h(i, \cdot)$ simultaneously, whereas ST simulates the component distributions $h(i, \cdot)$ one at a time.

The analogy with simulated annealing is the following. Suppose one has a function q that one wants to minimize, but the problem is hard with many local maxima, so no algorithm for finding local minima is worth trying. Define

$$h_\tau(x) = e^{-q(x)/k\tau}, \tag{11.7}$$

where k is an arbitrary positive constant and τ is an adjustable parameter. Consider the probability distribution with unnormalized density h_τ. Nothing in our setup guarantees that h_τ is actually integrable if the state space is infinite, so we assume this. As τ goes to zero, the distribution with unnormalized density h_τ converges to the distribution concentrated on the set of global minima of the function q. Conversely, as τ goes to infinity, the distribution with unnormalized density h_τ becomes more and more dispersed. Simulated annealing runs a Metropolis sampler for the distribution with unnormalized density h_τ and slowly decreases τ over the course of the run in hopes that the simulations will converge to the global minimum of the function q.

This chapter not being about optimization, we would not have bothered with the preceding point except for its relevance to simulated tempering. The physics and chemistry literature (Earl and Deem, 2005; Marinari and Parisi, 1992) seems quite taken with the annealing analogy using unnormalized densities for serial or parallel tempering of the form

$$h_i(x) = e^{-q(x)/k\tau_i}, \quad i = 1, \ldots, m, \tag{11.8}$$

by analogy with Equation 11.7. In PT we simply let $h(i, x) = h_i(x)$ so the joint distribution Equation 11.6 is the product of the distributions (Equation 11.8). In ST the definition $h(i, x) = h_i(x)$ usually does not work, and we need

$$h(i, x) = h_i(x)c_i, \tag{11.9}$$

where the c_i are constants adjusted by trial and error to make the ST chain mix well (Geyer and Thompson, 1995, call the c_i the *pseudo-prior*). This adjustment by trial and error is what makes ST harder to do than PT. As we shall see (Section 11.3 below) this adjustment by trial and error also makes ST much more useful than PT in some applications.

Geyer and Thompson (1995) point out that there is no reason to choose distributions of the form Equation 11.8 and many reasons not to. They allow the h_i to be arbitrary unnormalized densities. For example, in Bayesian problems, where sampling from the prior distribution is often easy (perhaps doable by OMCordinary Monte Carlo) and sampling from the posterior is much harder, it seems more natural to replace Equation 11.8 by

$$h_i(x) = e^{\lambda_i l(x) + p(x)}, \quad i = 1, \ldots, m, \tag{11.10}$$

where l is the log likelihood, p is the log prior, and

$$0 = \lambda_1 < \lambda_2 < \cdots < \lambda_m = 1,$$

so the sequence of tempering distributions interpolates prior and posterior. Many other schemes are possible. It is not at all clear that anyone with good insight into a particular simulation problem cannot easily invent a tempering sequence that will work better on that particular problem than any general suggestion such as Equation 11.8 or Equation 11.10. We shall see another form of tempering sequence in Section 11.3. Geyer and Thompson (1995) illustrate still other forms and also discuss the choice of m.

11.2.1 Parallel Tempering Update

Parallel tempering is a combined update. One kind of elementary update simulates a new x_i preserving the distribution with unnormalized density h_i. Since x_j, for $j \neq i$, are left fixed by such an update, this also preserves the joint distribution (Equation 11.6).

In addition, there are updates that swap states x_i and x_j of two components of the state while preserving the joint distribution (Equation 11.6). This is a Metropolis update, since a swap is its own inverse. The odds ratio is

$$r(i, j) = \frac{h_i(x_j)h_j(x_i)}{h_i(x_i)h_j(x_j)}.$$

The swap is accepted with probability $\min(1, r(i, j))$, as in any Metropolis update, and the new state is the old state with x_i and x_j swapped. Otherwise, the state is unchanged.

The combined update used by the function `temper` in the `mcmc` package is as follows. Each iteration of the Markov chain does either a within-component update or a swap update, each chosen with probability $\frac{1}{2}$. Having chosen to do a within-component update, it then chooses an i uniformly at random from $\{1, \ldots, m\}$, and then updates x_i preserving h_i using a normal-random-walk Metropolis update. Having chosen to do a swap update, it then chooses an i uniformly at random from $\{1, \ldots, m\}$, chooses a j uniformly at random from the subset of $\{1, \ldots, m\}$ that are neighbors of i, where neighborness is a user-specified symmetric irreflexive relation on $\{1, \ldots, m\}$, and then updates x_i and x_j by doing the swap update described above. This combined update is reversible.

There is no reason for this division into within-component updates and swap updates except convenience. Usually, before trying tempering, one has tried a more conventional MCMC sampler and hence already has code available for the within-component updates.

11.2.2 Serial Tempering Update

Serial tempering is a combined update. One kind of elementary update simulates a new x preserving the distribution with unnormalized density h_i. Since i is left fixed by such an update, this also preserves the joint distribution h.

In addition, there are updates that jump from one component of the mixture to another while preserving h. This is a Metropolis–Hastings update. The Hastings ratio is

$$r(i,j) = \frac{h_j(x)c_jq(j,i)}{h_i(x)c_iq(i,j)}, \tag{11.11}$$

where $q(i,j)$ is the probability of proposing a jump to j when the current state is i. The jump is accepted with probability $\min(1, r(i,j))$, as in any Metropolis–Hastings update, and the new state is (j, x). Otherwise, the state is unchanged.

The combined update used by the function `temper` in the `mcmc` package is as follows. Each iteration of the Markov chain does either a within-component update or a jump update, each chosen with probability $\frac{1}{2}$. Having chosen to do a within-component update, it updates x preserving h_i using a normal-random-walk Metropolis update. Having chosen to do a jump update, it then chooses a j uniformly at random from the subset of $\{1, \ldots, m\}$ that are neighbors of i, where neighborness is a user-specified relation as in the parallel tempering case, so

$$q(i,j) = \begin{cases} 1/n(i), & j \text{ and } i \text{ are neighbors,} \\ 0, & \text{otherwise,} \end{cases}$$

where $n(i)$ is the number of neighbors of i. This combined update is reversible.

As with PT, there is no reason for this division into within-component updates and jump updates except convenience.

11.2.3 Effectiveness of Tempering

Whether tempering works or not depends on the choice of the sequence of component distributions $h_1, \ldots, h_m$. The Gibbs distributions (Equation 11.8) are a "default" choice, but, as argued above, careful thought about one's particular problem may suggest a better choice.

Suppose h_m is the distribution we actually want to sample. The other distributions h_i in the sequence are "helpers." They should become progressively easier to sample as i decreases and neighboring distributions (however neighborness is defined) should be close enough together so that the swap (PT) or jump (ST) steps are accepted with reasonable probability (Geyer and Thompson, 1995, give recommendations for adjusting the "closeness" of these distributions). If h_1 is easy to sample, h_2 is easy to sample with help from h_1, h_3 is easy to sample with help from h_1 and h_2, and so forth, then the whole scheme will work and h_m will be (relatively) easy to sample, much easier than if one had tried to sample it directly, just repeating the mth within-component elementary update.

Your humble author has failed to invent a simple example of tempering. The stumbling block is that any toy problem is easily doable by other means, that any truly difficult problem takes a long time to explain, and—what is worse—that it is not obvious that the tempering sampler produces correct results. In a problem so difficult that nothing but tempering could possibly work, there is no way to check whether tempering actually works. In the genetics example in Geyer and Thompson (1995), we think the ST sampler worked, but cannot be sure. Thus we can do no better than the examples in the literature (Earl and Deem, 2005; Geyer and Thompson, 1995; Marinari and Parisi, 1992).

This issue applies more generally. The fact that MCMC works for toy problems which can be done by methods simpler than MCMC provides no evidence that MCMC works for problems so hard that no method other than MCMC could possibly work.

11.2.4 Tuning Serial Tempering

ST, unlike PT, has the additional issue that the user-specified constants $c_1, \ldots, c_m$ in Equation 11.9 must be correctly specified in order for the sampler to work. Define

$$d_i = \int h_i(x)\, dx$$

to be the normalizing constant for the ith component distribution (the integral replaced by a sum if the state space is discrete). Then the marginal equilibrium distribution for i in an ST sampler is given by

$$\int h(i, x)\, dx = c_i d_i, \quad i = 1, \ldots, m.$$

In order that this marginal distribution be approximately uniform, we must somehow adjust the c_i so that $c_i \approx 1/d_i$, but the d_i are unknown.

Let (I_t, X_t), $t = 1, 2, \ldots$, be the output of an ST sampler. Define

$$\hat{d}_k = \frac{1}{n c_k} \sum_{t=1}^{n} 1(I_t = k), \quad k = 1, \ldots, m, \tag{11.12}$$

where $1(I_t = k)$ is equal to one if $I_t = k$ and zero otherwise. Also define

$$d = \sum_{i=1}^{m} c_i d_i,$$

which is the normalizing constant for h. Then, by the law of large numbers,

$$\hat{d}_k \xrightarrow{\text{a.s.}} \frac{d_k}{d}$$

so the $\hat{d}_k$ estimate the unknown normalizing constants d_k up to an overall unknown constant of proportionality (which does not matter, since it does not affect the equilibrium distribution of the ST sampler).

Hence, assuming that the ST sampler already works, we can improve our choice of the c_k by setting them to $1/\hat{d}_k$. But this clearly will not work when $\hat{d}_k = 0$, which happens whenever the c_i are so badly adjusted that we have $c_k d_k/d \ll 1$. Geyer and Thompson (1995) recommend using stochastic approximation to deal with this situation, but the R function temper does not implement that. Instead we use a simple update of the tuning constants

$$c_k \leftarrow \min\left(a, \frac{1}{\hat{d}_k}\right),$$

where a is an arbitrarily chosen constant that keeps the new values of c_k finite (e^{10} was used in the vignette temper.pdf in the mcmc package) and $\leftarrow$ denotes redefinition.

A few iterations of this scheme usually suffice to adjust the c_i well enough so that the ST sampler will work well. Note that this does not require that we have $c_i d_i$ exactly the same for all i. We only need $c_i d_i$ approximately the same for all i, which is shown by having $c_i \hat{d}_i \approx 1/m$ for all i. Section 11.3 below gives an example of this trial-and-error adjustment.

11.2.5 Umbrella Sampling

The idea of sampling a mixture of distributions was not new when tempering was devised. Torrie and Valleau (1977) proposed a procedure they called umbrella sampling (US), which is exactly the same as ST but done with a different purpose. In ST one is only interested in one component of the mixture, and the rest are just helpers. In US one is interested in all the components, and wants to sample them simultaneously and efficiently.

US is very useful for importance sampling. Suppose that one is interested in target distributions with unnormalized densities h_θ for all θ in some region Θ. We need an importance sampling distribution that is "close" to all of the targets. Choose a finite set $\{\theta_1, \ldots, \theta_m\}$ and let $h_i = h_{\theta_i}$ be the components of the US mixture. If the θ_i are spread out so that each $\theta \in \Theta$ is close to some θ_i and if the parameterization $\theta \mapsto h_\theta$ is continuous enough so that closeness in θ implies closeness in h_θ, then US will be an effective importance sampling scheme. No other general method of constructing good importance sampling distributions for multiple targets is known.

Two somewhat different importance sampling schemes have been proposed. Let (I_t, X_t), $t = 1, 2, \ldots$, be the output of an ST sampler with unnormalized equilibrium density $h(i, x)$, and let $h_\theta(x)$ be a parametric family of unnormalized densities. Geyer and Thompson (1995) proposed the following scheme, which thinks of $h_\theta(x)$ as a function of i and x, giving

$$\tilde{\mu}_n(\theta) = \frac{\displaystyle\sum_{t=1}^{n} \frac{g(X_t) h_\theta(X_t)}{h(I_t, X_t)}}{\displaystyle\sum_{t=1}^{n} \frac{h_\theta(X_t)}{h(I_t, X_t)}} \tag{11.13}$$

as a sensible estimator of Equation 11.2, a special case of Equation 11.4.

Alternatively, one can think of $h_\theta(x)$ as a function of x only, in which case the importance sampling distribution must also be a function of x only, in which case

$$h_{\text{mix}}(x) = \sum_{i=1}^{m} h(i, x) = \sum_{i=1}^{m} c_i h_i(x),$$

the marginal distribution of X derived from the ST/US equilibrium distribution (Equation 11.9). Then one obtains

$$\tilde{\mu}_n(\theta) = \frac{\sum_{t=1}^{n} \dfrac{g(X_t)h_\theta(X_t)}{h_{\text{mix}}(X_t)}}{\sum_{t=1}^{n} \dfrac{h_\theta(X_t)}{h_{\text{mix}}(X_t)}} = \frac{\sum_{t=1}^{n} \dfrac{g(X_t)h_\theta(X_t)}{\sum_{i=1}^{m} c_i h_i(X_t)}}{\sum_{t=1}^{n} \dfrac{h_\theta(X_t)}{\sum_{i=1}^{m} c_i h_i(X_t)}} \tag{11.14}$$

as a sensible estimator of Equation 11.2, a different special case of Equation 11.4. This latter scheme was suggested at a meeting by someone your humble author has now forgotten as an application of the idea of "Rao-Blackwellization" (Gelfand and Smith, 1990) to Equation 11.13. It is not known whether Equation 11.14 is enough better than Equation 11.13 to justify the extra computing required, evaluating $h_i(X_t)$ for each i and t rather than just evaluating $h_{I_t}(X_t)$.

11.3 Bayes Factors and Normalizing Constants

Umbrella sampling is very useful in calculating Bayes factors and other unknown normalizing constants. Here we just illustrate Bayes factor calculation. Other unknown normalizing constant calculations are similar. We follow the example worked out in detail in the vignette `bfst.pdf` that comes with each installation of the `mcmc` package (Geyer, 2010).

11.3.1 Theory

Suppose we have m Bayesian models with data distributions $f(y \mid \theta, i)$, where i indexes the model and θ any within-model parameters, within-model priors $g(\theta \mid i)$, and prior on models $\text{pri}(i)$. For each model i the within-model prior $g(\cdot \mid i)$ must be a properly normalized density. It does not matter if the prior on models $\text{pri}(\cdot)$ is unnormalized. Each model may have a different within-model parameter. Let Θ_i denote the within-model parameter space for the ith model.

Likelihood times unnormalized prior is

$$h(i, \theta) = f(y \mid \theta, i)g(\theta \mid i)\,\text{pri}(i).$$

The unknown normalizing constant for the ith model is

$$\int_{\Theta_i} h(i, \theta)\,d\theta = \text{pri}(i)\int_{\Theta_i} f(y \mid \theta, i)g(\theta \mid i)\,d\theta,$$

and the part of this that does not involve the prior on models is called the *unnormalized Bayes factor*,

$$b(i \mid y) = \int_{\Theta_i} f(y \mid \theta, i)g(\theta \mid i)\,d\theta. \tag{11.15}$$

The properly normalized posterior on models is

$$\text{post}(i) = \frac{\text{pri}(i)b(i \mid y)}{\sum_{j=1}^{m} \text{pri}(j)b(j \mid y)}.$$

But frequently the prior on models is ignored (or left for users to decide for themselves) and one reports only the ratio of prior odds to posterior odds, the (normalized) *Bayes factor*

$$\frac{\text{post}(i)}{\text{post}(j)} \cdot \frac{\text{pri}(j)}{\text{pri}(i)} = \frac{b(i \mid y)}{b(j \mid y)}, \tag{11.16}$$

Clearly, it is enough to report the unnormalized Bayes factors (Equation 11.15) from which (Equation 11.16) are trivially calculated. Bayes factors and other unknown normalizing constants are notoriously difficult to calculate. Despite many proposals, no really effective scheme has been described in the literature. Here we show how ST/US makes the calculation easy.

In using ST/US for Bayes factors, we identify the index i on components of the umbrella distribution (Equation 11.5) with the index on Bayesian models. ST/US requires that all components have the same state space Θ but, as the problem is presented, the Bayesian models have different (within-model) state spaces Θ_i. Thus we have to do something a little more complicated. We "pad" the state vector θ so that it always has the same dimension, doing so in a way that does not interfere with the Bayes factor calculation. Write $\theta = (\theta_{\text{actual}}, \theta_{\text{pad}})$, the dimension of both parts depending on the model i. Then we insist on the conditions

$$f(y \mid \theta, i) = f(y \mid \theta_{\text{actual}}, i),$$

so the data distribution does not depend on the "padding" and

$$g(\theta \mid i) = g_{\text{actual}}(\theta_{\text{actual}} \mid i) \cdot g_{\text{pad}}(\theta_{\text{pad}} \mid i),$$

so the two parts are *a priori* independent and both parts of the prior are normalized proper priors. This ensures that

$$
\begin{aligned}
b(i \mid y) &= \int_{\Theta_i} f(y \mid \theta, i) g(\theta \mid i) \, d\theta \\
&= \iint f(y \mid \theta_{\text{actual}}, i) g_{\text{actual}}(\theta_{\text{actual}} \mid i) g_{\text{pad}}(\theta_{\text{pad}} \mid i) \, d\theta_{\text{actual}} \, d\theta_{\text{pad}} \\
&= \int_{\Theta_i} f(y \mid \theta_{\text{actual}}, i) g_{\text{actual}}(\theta_{\text{actual}} \mid i) \, d\theta_{\text{actual}}, \tag{11.17}
\end{aligned}
$$

so the calculation of the unnormalized Bayes factors is the same whether or not we "pad" θ, and we may then take

$$
\begin{aligned}
h_i(\theta) &= f(y \mid \theta, i) g(\theta \mid i) \\
&= f(y \mid \theta_{\text{actual}}, i) g_{\text{actual}}(\theta_{\text{actual}} \mid i) g_{\text{pad}}(\theta_{\text{pad}} \mid i)
\end{aligned}
$$

to be the unnormalized densities for the component distributions of the ST/US chain. It is clear that the normalizing constants for these distributions are just the unnormalized Bayes factors (Equation 11.17).

Thus these unknown normalizing constants are estimated the same way we estimate all unknown normalizing constants in ST/US. If preliminary trial and error has adjusted the pseudo-prior so that the ST/US chain frequently visits all components, then the unknown normalizing constants are approximately $d_i \approx 1/c_i$ (Section 11.2.4 above). Improved estimates with MCSEs of the unknown normalizing constants can be found by another run of the ST/US sampler. The normalizing constants (Equation 11.12), being simple averages over the run, have MCSEs straightforwardly estimated by the method of batch means.

One might say this method is "cheating" because it does not completely specify how the trial and error is done, and it is clear that the trial and error is crucial because ST/US does not "work" (by definition) until trial and error has successfully adjusted the pseudo-prior. Moreover, most of the work of estimation is done by the trial and error, which must adjust $c_i \approx 1/d_i$ to within a factor of 2–3. The final run only provides a little polishing and MCSE. Since Bayes factors may vary by 10^{10} or more, it is clear that trial and error does most of the work. It is now clear why any method that proposes to compute Bayes factors or other unknown normalizing constants from one run of one Markov chain cannot compete with this "cheating."

11.3.2 Practice

However, as shown by an example below, the trial and error can be simple and straightforward. Moreover, the trial and error does not complicate MCSE. In the context of the final run, the components c_i of the pseudo-prior are known constants and are treated as such in the computation of MCSE for the unknown normalizing constant estimates (Equation 11.12).

Let us see how this works. As stated above, we follow the example in the vignette `bfst.pdf` of the `mcmc` package. Simulated data for the problem are the same logistic regression data in the data frame `logit` in the `mcmc` package analyzed in Section 1.13. There are five variables in the data set, the response `y` and four predictors, `x1`, `x2`, `x3`, and `x4`. Here we assume the same Bayesian model as in Section 1.13, but now we wish to calculate Bayes factors for the $16 = 2^4$ possible submodels that include or exclude each of the predictors, `x1`, `x2`, `x3`, and `x4`.

11.3.2.1 Setup

We set up a matrix that indicates these models. In the R code shown below, `out` is the result of the frequentist analysis done by the `glm` function shown in Section 1.13:

```
varnam <- names(coefficients(out))
varnam <- varnam[varnam != "(Intercept)"]
nvar <- length(varnam)

models <- NULL
foo <- seq(0, 2^nvar - 1)
for (i in 1:nvar) {
    bar <- foo %/% 2^(i - 1)
    bar <- bar %% 2
    models <- cbind(bar, models, deparse.level = 0)
}
colnames(models) <- varnam
```

The slightly tricky code above essentially counts from 0 to 15 in binary, the ith row of the matrix `models` is $i - 1$ in binary. In each row, 1 indicates that the predictor is in the model and 0 indicates that it is out.

The function `temper` in the `mcmc` package that does tempering requires a notion of neighbors among models. It attempts jumps only between neighboring models. Here we choose models to be neighbors if they differ only by one predictor.

```
neighbors <- matrix(FALSE, nrow(models), nrow(models))
for (i in 1:nrow(neighbors)) {
    for (j in 1:ncol(neighbors)) {
        foo <- models[i, ]
        bar <- models[j, ]
        if (sum(foo != bar) == 1) neighbors[i, j] <- TRUE
    }
}
```

Now we specify the equilibrium distribution of the ST/US chain. Its state vector is (i, θ), where i is an integer between 1 and 16 and θ is the parameter vector "padded" to always be the same length, so we take it to be the length of the parameter vector of the full model which is `length(out$coefficients)`, or `ncol(models) + 1`, which makes the length of the state of the ST chain `ncol(models) + 2`. We take the within model priors for the "padded" components of the parameter vector to be the same as those for the "actual" components, normal with mean 0 and standard deviation 2 for all cases. As is seen in Equation 11.17, the priors for the "padded" components (parameters not in the model for the current state) do not matter because they drop out of the Bayes factor calculation. The choice does not matter much for this toy example; see the discussion below for more on this issue. It is important that we use normalized log priors, the term `dnorm(beta,0,2, log = TRUE)` in the function `ludfun` defined below. This is unlike when we are simulating only one model as in the function `lupost` defined in Section 1.13, where unnormalized log priors – `beta^2 / 8` were used.

The `temper` function wants the log unnormalized density of the equilibrium distribution. We include an additional argument, `log.pseudo.prior`, which is $\log(c_i)$ in our mathematical development, because this changes from run to run as we adjust it by trial and error. Other "arguments" are the model matrix of the full model `modmat`, the matrix `models` relating integer indices (the first component of the state vector of the ST chain) to which predictors are in or out of the model, and the data vector `y`, but these are not passed as arguments to our function and instead are found in the R global environment.

```
modmat <- out$x
y <- logit$y

ludfun <- function(state, log.pseudo.prior) {
    stopifnot(is.numeric(state))
    stopifnot(length(state) == ncol(models) + 2)
    icomp <- state[1]
    stopifnot(icomp == as.integer(icomp))
    stopifnot(1 <= icomp && icomp <= nrow(models))
    stopifnot(is.numeric(log.pseudo.prior))
    stopifnot(length(log.pseudo.prior) == nrow(models))
    beta <- state[-1]
    inies <- c(TRUE, as.logical(models[icomp, ]))
    beta.logl <- beta
    beta.logl[! inies] <- 0
```

```
        eta <- as.numeric(modmat %*% beta.logl)
        logp <- ifelse(eta < 0, eta - log1p(exp(eta)),
         - log1p(exp(- eta)))
        logq <- ifelse(eta < 0, - log1p(exp(eta)),
         - eta - log1p(exp(- eta)))
        logl <- sum(logp[y == 1]) + sum(logq[y == 0])
        logl + sum(dnorm(beta, 0, 2, log = TRUE))
         + log.pseudo.prior[icomp]
}
```

11.3.2.2   *Trial and Error*

With this setup we are ready for the trial-and-error process. We start with a flat log pseudo-prior (having no idea what it should be).

```
state.initial <- c(nrow(models), out$coefficients)
qux <- rep(0, nrow(models))
out <- temper(ludfun, initial = state.initial,
 neighbors = neighbors,
    nbatch = 1000, blen = 100, log.pseudo.prior = qux)
```

So what happens?

```
> ibar <- colMeans(out$ibatch)
> ibar
 [1]  0.00000 0.00000 0.00000 0.00000 0.00524 0.06489 0.00754
 [8]  0.06021 0.00033 0.00202 0.00008 0.00054 0.28473 0.31487
[15]  0.12478 0.13477
```

The ST/US chain did not mix well, several models not being visited even once. So we adjust the pseudo-priors to get uniform distribution.

```
> qux <- qux + pmin(log(max(ibar) / ibar), 10)
> qux <- qux - min(qux)
> qux
 [1] 10.0000000 10.0000000 10.0000000 10.0000000  4.0958384
 [6]  1.5794663  3.7319377  1.6543214  6.8608225  5.0490623
[11]  8.2778885  6.3683460  0.1006185  0.0000000  0.9256077
[16]  0.8485902
```

When a component of `ibar` is zero, the corresponding component of `ibar` is `Inf`, but the `pmin` function limits the increase to 10 (an arbitrarily chosen constant). The statement

```
qux <- qux - min(qux)
```

is unnecessary. An overall arbitrary constant can be added to the log pseudo-prior without changing the equilibrium distribution of the ST chain. We do this only to make `qux` more comparable from run to run.

Now we repeat this until the log pseudo-prior "converges" roughly.

```
qux.save <- qux
repeat {
    out <- temper(out, log.pseudo.prior = qux)
    ibar <- colMeans(out$ibatch)
    qux <- qux + pmin(log(max(ibar) / ibar), 10)
    qux <- qux - min(qux)
    qux.save <- rbind(qux.save, qux, deparse.level = 0)
    if (max(ibar) / min(ibar) < 2)
        break
}
```

The entire matrix qux.save is shown in the vignette. Here we just show a few columns:

```
> qux.save[ , 1:5]
          [,1]       [,2]      [,3]       [,4]      [,5]
[1,] 10.00000 10.000000 10.00000 10.000000 4.095838
[2,] 17.70751  9.999775 14.43037  9.714906 4.049512
[3,] 18.76818  9.325494 14.43382  9.014132 3.972982
[4,] 18.94703  9.733071 14.71371  9.478451 4.276229
```

We see we get fairly rapid (albeit sloppy) convergence to the log reciprocal normalizing constants.

Now that the pseudo-prior is adjusted well enough, we need to perhaps make other adjustments to get acceptance rates near 20%. The acceptance rates for jump updates and for within-component updates are shown in the vignette. Those for jump updates seemed OK, but those for within-component updates were too small (as low as 0.02) for some components. Hence the scaling for within-component updates was changed,

```
out <- temper(out, scale = 0.5, log.pseudo.prior = qux)
```

and this produces within-component acceptance rates that are acceptable (at least 0.15 for all components).

Inspection of autocorrelation functions for components of out$ibatch (not shown in the vignette or here) says that batch length needs to be at least 4 times longer. We make it 10 times longer for safety.

```
out <- temper(out, blen = 10 * out$blen, log.pseudo.prior = qux)
```

The total time for all runs of the temper function was 70 minutes on a slow old laptop and less than 7 minutes on a fast workstation.

11.3.2.3 *Monte Carlo Approximation*

Now we calculate log 10 Bayes factors relative to the model with the highest unnormalized Bayes factor:

```
> log.10.unnorm.bayes
 <- (qux - log(colMeans(out$ibatch)))/
 log(10)
> k <- seq(along = log.10.unnorm.bayes)[log.10.unnorm.bayes ==
```

```
+        min(log.10.unnorm.bayes)]
> models[k, ]
x1 x2 x3 x4
 1  1  0  1
> log.10.bayes <- log.10.unnorm.bayes - log.10.unnorm.bayes[k]
> log.10.bayes
 [1]  8.17814103  4.17098637  6.33069128  4.05292216  1.80254545
 [6]  0.67203156  1.40468558  0.70498671  2.58875400  1.93202268
[11]  2.82341431  2.37170521  0.08004553  0.00000000  0.37357715
[16]  0.35242443
```

These are the Monte Carlo approximations of the negatives of the base 10 logarithms of the unnormalized Bayes factors. Higher numbers mean lower posterior probability. The model with the highest Bayes factor (0.00000 in the vector shown above) is the model with predictors x1, x2, and x4 and intercept. The model with the lowest Bayes factor (8.17814 in the vector shown above) is the model with no predictors except the intercept. Thus there is a difference of more than eight orders of magnitude among the unnormalized Bayes factors.

Note that the trial-and-error process did most of the work. The log pseudo-prior for the model with the lowest Bayes factor was 18.94703 (shown above). Converted to base 10 logs, this is 8.22859, which is nearly the same as our final estimate 8.17814 (shown just above). The final run contributes only a final polishing to the work done by trial and error. However, it does do all the work for MCSE. The MCSE calculation is shown in the vignette. The MCSEs are about 0.02, so 95% nonsimultaneous confidence intervals have a margin of error of about 0.04. These are all relative to the model with highest estimated negative log 10 Bayes factor (0.00000 in the vector shown just above). Hence, we have only weak evidence (if we assume uniform prior on models) of the superiority of the model with the highest Monte Carlo Bayes factor to the model with the next closest (0.08005 in the vector shown just above). All of the other models have clearly lower posterior probability if we assume a uniform prior on models. Of course, the whole point of Bayes factors is that users are allowed to adopt a nonuniform prior on models, and would just subtract the base 10 logs of their prior on models from these numbers. The MCSE would stay the same.

11.3.3 Discussion

We hope that readers are impressed with the power of this method. We calculated the most extreme log 10 Bayes factor to be 8.17 ± 0.04. If we had simply sampled with uniform prior on models, we would have visited the no-intercept model with approximate probability $10^{-8.17}$—that is, never in any practical computer run. The key to the method is pseudo-prior adjustment by trial and error. The method could have been invented by any Bayesian who realized that the priors on models, pri(m) in our notation, do not affect the Bayes factors and hence are irrelevant to calculating Bayes factors. Thus the priors (or pseudo-priors in our terminology) should be chosen for reasons of computational convenience, as we have done, rather than to incorporate prior information.

The rest of the details of the method are unimportant. The `temper` function in R is convenient to use for this purpose, but there is no reason to believe that it provides optimal sampling. Samplers carefully designed for each particular application would undoubtedly do better. Our notion of "padding" so that the within-model parameters have the same dimension for all models follows Carlin and Chib (1995), but "reversible jump" samplers

(Green, 1995) would undoubtedly do better (see the Bayesian model selection example in Section 1.17.3). Unfortunately, there seems to be no way to code up a function like `temper` that uses "reversible jump" and requires no theoretical work from users that, if messed up, destroys the algorithm. The `temper` function is foolproof in the sense that if the log unnormalized density function written by the user (like our `ludfun`) is correct, then the ST Markov chain has the equilibrium distribution it is supposed to have. There is nothing the user can mess up except this user-written function. No analog of this for "reversible jump" chains is apparent (to your humble author).

Two issues remain, the first being about within-model priors for the "padding" components of within-model parameter vectors $g_{pad}(\theta_{pad} \mid m)$ in the notation in Equation 11.17. Rather than choose these so that they do not depend on the data (as we did), it would be better (if more trouble) to choose them differently for each "padding" component, centering $g_{pad}(\theta_{pad} \mid m)$ so the distribution of a component of θ_{pad} is near to the marginal distribution of the same component in neighboring models (according to the `neighbors` argument of the `temper` function).

The other remaining issue is adjusting acceptance rates for jumps. There is no way to adjust this other than by changing the number of models and their definitions. But the models we have cannot be changed; if we are to calculate Bayes factors for them, then we must sample them as they are. But we can insert new models between old models. For example, if the acceptance for swaps between model i and model j is too low, then we can insert distribution k between them that has unnormalized density

$$h_k(x) = \sqrt{h_i(x)h_j(x)}.$$

This idea is inherited from simulated tempering; Geyer and Thompson (1995) have much discussion of how to insert additional distributions into a tempering network. It is another key issue in using tempering to speed up sampling. It is less obvious in the Bayes factor context, but still an available technique if needed.

Acknowledgments

This chapter benefited from detailed comments by Christina Knudson, Leif Johnson, Galin Jones, and Brian Shea.

References

Carlin, B. P. and Chib, S. 1995. Bayesian model choice via Markov chain Monte Carlo methods. *Journal of the Royal Statistical Society, Series B*, 56:261–274.

Earl, D. J. and Deem, M. W. 2005. Parallel tempering: Theory, applications, and new perspectives. *Physical Chemistry Chemical Physics*, 7:3910–3916.

Gelfand, A. E. and Smith, A. F. M. 1990. Sampling-based approaches to calculating marginal densities. *Journal of the American Statistical Association*, 85:398–409.

Geweke, J. 1989. Bayesian inference in econometric models using Monte Carlo integration. *Econometrica*, 57:1317–1339.

Geyer, C. J. 1991. Markov chain Monte Carlo maximum likelihood for dependent data. In E. M. Keramidas (ed.), *Computing Science and Statistics: Proceedings of the 23rd Symposium on the Interface*, pp. 156–163. American Statistical Association, New York.

Geyer, C. J. 1994. On the convergence of Monte Carlo maximum likelihood calculations. *Journal of the Royal Statistical Society, Series B*, 56:261–274.

Geyer, C. J. 1999. Likelihood inference for spatial point processes. In O. E. Barndorff-Nielsen, W. S. Kendall, and M. N. M. van Lieshout (eds), *Stochastic Geometry: Likelihood and Computation*, pp. 79–140. Chapman & Hall/CRC, Boca Raton, FL.

Geyer, C. J. 2010. *mcmc: Markov Chain Monte Carlo*. R package version 0.8, available from CRAN.

Geyer, C. J. and Thompson, E. A. 1992. Constrained Monte Carlo maximum likelihood for dependent data (with discussion). *Journal of the Royal Statistical Society, Series B*, 54:657–699.

Geyer, C. J. and Thompson, E. A. 1995. Annealing Markov chain Monte Carlo with applications to ancestral inference. *Journal of the American Statistical Association*, 90:909–920.

Green, P. J. 1995. Reversible jump Markov chain Monte Carlo computation and Bayesian model determination. *Biometrika*, 82:711–732.

Insua, D. R. and Ruggeri, F. (eds) 2000. *Robust Bayesian Analysis*, Lecture Notes in Statistics, Vol. 152. Springer, New York.

Kirkpatrick, S., Gelatt, C. D. J., and Vecchi, M. P. 1983. Optimization by simulated annealing. *Science*, 220:671–680.

Marinari, E. and Parisi, G. 1992. Simulated tempering: A new Monte Carlo scheme. *Europhysics Letters*, 19:451–458.

Meyn, S. P. and Tweedie, R. L. 1993. *Markov Chains and Stochastic Stability*. Springer, London.

Torrie, G. M. and Valleau, J. P. 1977. Nonphysical sampling distributions in Monte Carlo free-energy estimation: Umbrella sampling. *Journal of Computational Physics*, 23:187–199.

Trotter, H. F. and Tukey, J. W. 1956. Conditional Monte Carlo for normal samples. In H. A. Meyer (ed.), *Symposium on Monte Carlo Methods*, pp. 64–79. Wiley, New York.

12

Likelihood-Free MCMC

Scott A. Sisson and Yanan Fan

12.1 Introduction

In Bayesian inference, the posterior distribution for parameters $\theta \in \Theta$ is given by $\pi(\theta \,|\, y) \propto \pi(y \,|\, \theta)\pi(\theta)$, where one's prior beliefs about the unknown parameters, as expressed through the prior distribution $\pi(\theta)$, are updated by the observed data $y \in \mathcal{Y}$ via the likelihood function $\pi(y \,|\, \theta)$. Inference for the parameters θ is then based on the posterior distribution. Except in simple cases, numerical simulation methods, such as Markov chain Monte Carlo (MCMC), are required to approximate the integrations needed to summarize features of the posterior distribution. Inevitably, increasing demands on statistical modeling and computation have resulted in the development of progressively more sophisticated algorithms.

Most recently there has been interest in performing Bayesian analyses for models which are sufficiently complex that the likelihood function $\pi(y \,|\, \theta)$ is either analytically unavailable or computationally prohibitive to evaluate. The classes of algorithms and methods developed to perform Bayesian inference in this setting have become known as *likelihood-free computation* or *approximate Bayesian computation* (Beaumont et al., 2002; Marjoram et al., 2003; Ratmann et al., 2009; Sisson et al., 2007; Tavaré et al., 1997). This name refers to the circumventing of explicit evaluation of the likelihood by a simulation-based approximation.

Likelihood-free methods are rapidly gaining popularity as a practical approach to fitting models under the Bayesian paradigm that would otherwise have been computationally impractical. To date they have found widespread usage in a diverse range of applications. These include wireless communications engineering (Nevat et al., 2008), quantile distributions (Drovandi and Pettitt, 2009), HIV contact tracing (Blum and Tran, 2010), the evolution of drug resistance in tuberculosis (Luciani et al., 2009), population genetics (Beaumont et al., 2002), protein networks (Ratmann et al., 2009, 2007), archeology (Wilkinson and Tavaré, 2009); ecology (Jabot and Chave, 2009), operational risk (Peters and Sisson, 2006), species migration (Hamilton et al., 2005), chain-ladder claims reserving (Peters et al., 2008), coalescent models (Tavaré et al., 1997), α-stable models (Peters et al., in press), models for extremes (Bortot et al., 2007), susceptible–infected–removed models (Toni et al., 2009), pathogen transmission (Tanaka et al., 2006), and human evolution (Fagundes et al., 2007).

ALGORITHM 12.1 LIKELIHOOD-FREE REJECTION SAMPLING ALGORITHM

```
1.   Generate θ' ~ π(θ) from the prior.
2.   Generate data set x from the model π(x|θ').
3.   Accept θ' if x ≈ y.
```

The underlying concept of likelihood-free methods may be simply encapsulated as shown in Algorithm 12.1, the likelihood-free rejection sampling algorithm (Tavaré et al., 1997). For a candidate parameter vector θ', a data set is generated from the model (i.e. the likelihood function) $x \sim \pi(x \mid \theta')$. If the simulated and observed data sets are similar (in some manner), so that $x \approx y$, then θ' is a good candidate to have generated the observed data from the given model, and so θ' is retained and forms as a part of the samples from the posterior distribution $\pi(\theta \mid y)$. Conversely, if x and y are dissimilar, then θ' is unlikely to have generated the observed data for this model, and so θ' is discarded. The parameter vectors accepted under this approach offer support for y under the model, and so may be considered to be drawn approximately from the posterior distribution $\pi(\theta \mid y)$. In this manner, the evaluation of the likelihood $\pi(y \mid \theta')$, essential to most Bayesian posterior simulation methods, is replaced by an estimate of the proximity of a simulated data set $x \sim \pi(x \mid \theta')$ to the observed data set y. While available in various forms, all likelihood-free methods and models apply this basic principle.

In this chapter, we aim to provide a tutorial-style exposition of likelihood-free modeling and computation using MCMC simulation. In Section 12.2 we provide an overview of the models underlying likelihood-free inference, and illustrate the conditions under which these models form an acceptable approximation to the true but intractable posterior $\pi(\theta \mid y)$. In Section 12.3 we examine how MCMC-based samplers are able to circumvent evaluation of the intractable likelihood function, while still targeting this approximate posterior model. We also discuss different forms of samplers that have been proposed in order to improve algorithm and inferential performance. Finally, in Section 12.4 we present a step-by-step examination of the various practical issues involved in performing an analysis using likelihood-free methods, before concluding with a discussion.

Throughout we assume a basic familiarity with Bayesian inference and the Metropolis–Hastings algorithm. For this relevant background information, the reader is referred to the many useful chapters in this volume.

12.2   Review of Likelihood-Free Theory and Methods

In this section we discuss the modeling principles underlying likelihood-free computation.

12.2.1   Likelihood-Free Basics

A common procedure to improve sampler efficiency in challenging settings is to embed the target posterior within an augmented model. In this setting, auxiliary parameters are introduced into the model whose sole purpose is to facilitate computations—see, for example, simulated tempering or annealing methods (Geyer and Thompson, 1995; Neal, 2003). Likelihood-free inference adopts a similar approach by augmenting the target posterior from $\pi(\theta \mid y) \propto \pi(y \mid \theta)\pi(\theta)$ to

$$\pi_{\mathrm{LF}}(\theta, x \mid y) \propto \pi(y \mid x, \theta)\pi(x \mid \theta)\pi(\theta) \tag{12.1}$$

where the auxiliary parameter x is a (simulated) data set from $\pi(x \mid \theta)$ (see Algorithm 12.1), on the same space as $y \in \mathcal{Y}$ (Reeves and Pettitt, 2005; Wilkinson, 2008). As discussed in more detail below (Section 12.2.2), the distribution $\pi(y \mid x, \theta)$ is chosen to weight the posterior

$\pi(\theta \mid x)$ with high density in regions where x and y are similar. The density $\pi(y \mid x, \theta)$ is assumed to be constant with respect to θ at the point $x = y$, so that $\pi(y \mid y, \theta) = c$, for some constant $c > 0$, with the result that the target posterior is recovered exactly at $x = y$. That is, $\pi_{\mathrm{LF}}(\theta, y \mid y) \propto \pi(y \mid \theta)\pi(\theta)$.

Ultimately interest is typically in the marginal posterior

$$\pi_{\mathrm{LF}}(\theta \mid y) \propto \pi(\theta) \int_y \pi(y \mid x, \theta)\pi(x \mid \theta)\, dx, \tag{12.2}$$

integrating out the auxiliary data set x. The distribution $\pi_{\mathrm{LF}}(\theta \mid y)$ then acts as an approximation to $\pi(\theta \mid y)$. In practice this integration is performed numerically by simply discarding the realizations of the auxiliary data sets from the output of any sampler targeting the joint posterior $\pi_{\mathrm{LF}}(\theta, x \mid y)$. Other samplers can target $\pi_{\mathrm{LF}}(\theta \mid y)$ directly (see Section 12.3.1).

12.2.2 The Nature of the Posterior Approximation

The likelihood-free posterior distribution $\pi_{\mathrm{LF}}(\theta \mid y)$ will only recover the target posterior $\pi(\theta \mid y)$ exactly when the density $\pi(y \mid x, \theta)$ is precisely a point mass at $y = x$ and zero elsewhere (Reeves and Pettitt, 2005). In this case

$$\pi_{\mathrm{LF}}(\theta \mid y) \propto \pi(\theta) \int_y \pi(y \mid x, \theta)\pi(x \mid \theta)\, dx = \pi(y \mid \theta)\pi(\theta).$$

However, as observed from Algorithm 12.1, this choice for $\pi(y \mid x, \theta)$ will result in a rejection sampler with an acceptance probability of zero unless the proposed auxiliary data set exactly equals the observed data $x = y$. This event will occur with probability zero for all but the simplest applications (involving very low-dimensional discrete data). In a similar manner, MCMC-based likelihood-free samplers (Section 12.3) will also suffer acceptance rates of zero.

In practice, two concessions are made on the form of $\pi(y \mid x, \theta)$, and each of these can induce some form of approximation into $\pi_{\mathrm{LF}}(\theta \mid y)$ (Marjoram et al., 2003). The first allows the density to be a standard kernel density function, K, centered at the point $x = y$ and with scale determined by a parameter vector ϵ, usually taken as a scalar. In this manner

$$\pi_\epsilon(y \mid x, \theta) = \frac{1}{\epsilon} K\left(\frac{|x - y|}{\epsilon}\right)$$

weights the intractable likelihood with high density in regions $x \approx y$ where the auxiliary and observed data sets are similar, and with low density in regions where they are not similar (Beaumont et al., 2002; Blum, 2010; Peters et al., 2008). The interpretation of likelihood-free models in the nonparametric framework is of current research interest (Blum, 2010).

The second concession on the form of $\pi_\epsilon(y \mid x, \theta)$ permits the comparison of the data sets, x and y, to occur through a low-dimensional vector of summary statistics $T(\cdot)$, where $\dim(T(\cdot)) \geq \dim(\theta)$. Accordingly, given the improbability of generating an auxiliary data set such that $x \approx y$, the density

$$\pi_\epsilon(y \mid x, \theta) = \frac{1}{\epsilon} K\left(\frac{|T(x) - T(y)|}{\epsilon}\right) \tag{12.3}$$

will provide regions of high density when $T(x) \approx T(y)$ and low density otherwise. If the vector of summary statistics is also sufficient for the parameters θ, then comparing the summary statistics of two data sets will be equivalent to comparing the data sets themselves. Hence there will be no loss of information in model fitting, and accordingly no further approximation will be introduced into $\pi_{LF}(\theta \,|\, y)$. However, the event $T(x) \approx T(y)$ will be substantially more likely than $x \approx y$, and so likelihood-free samplers based on summary statistics $T(\cdot)$ will in general be considerably more efficient in terms of acceptance rates than those based on full data sets (Pritchard et al., 1999; Tavaré et al., 1997). As noted by McKinley et al. (2009), the procedure of model fitting via summary statistics $T(\cdot)$ permits the application of likelihood-free inference in situations where the observed data y are incomplete.

Note that under the form (Equation 12.3), $\lim_{\epsilon \to 0} \pi_\epsilon(y \,|\, x, \theta)$ is a point mass on $T(x) = T(y)$. Hence, if $T(\cdot)$ are also sufficient statistics for θ, then $\lim_{\epsilon \to 0} \pi_{LF}(\theta \,|\, y) = \pi(\theta \,|\, y)$ exactly recovers the intractable posterior (Reeves and Pettitt, 2005). Otherwise, if $\epsilon > 0$ or if $T(\cdot)$ are not sufficient statistics, then the likelihood-free approximation to $\pi(\theta \,|\, y)$ is given by $\pi_{LF}(\theta \,|\, y)$ in Equation 12.2.

A frequently utilized weighting kernel $\pi_\epsilon(y \,|\, x, \theta)$ is the uniform kernel density (Marjoram et al., 2003; Tavaré et al., 1997), whereby $T(y)$ is uniformly distributed on the sphere centered at $T(x)$ with radius ϵ. This is commonly written as

$$\pi_\epsilon(y \,|\, x, \theta) \propto \begin{cases} 1, & \text{if } \rho(T(x), T(y)) \leq \epsilon, \\ 0, & \text{otherwise,} \end{cases} \qquad (12.4)$$

where ρ denotes a distance measure (e.g. Euclidean) between $T(x)$ and $T(y)$. In the form of Equation 12.3 this is expressed as $\pi_\epsilon(y \,|\, x, \theta) = \epsilon^{-1} K_u(\rho(T(x), T(y))/\epsilon)$, where K_u is the uniform kernel density. Alternative kernel densities that have been implemented include the Epanechnikov kernel (Beaumont et al., 2002), a nonparametric density estimate (Ratmann et al., 2009) (see Section 12.3.2 below), and the Gaussian kernel density (Peters et al., 2008), whereby $\pi_\epsilon(y \,|\, x, \theta)$ is centered at $T(x)$ and scaled by ϵ, so that $T(y) \sim N(T(x), \Sigma\epsilon^2)$ for some covariance matrix Σ.

12.2.3 A Simple Example

As an illustration, we examine the deviation of the likelihood-free approximation from the target posterior in a simple example. Consider the case where $\pi(\theta \,|\, y)$ is the univariate $N(0, 1)$ density. To realize this posterior in the likelihood-free setting, we specify the likelihood as $x \sim N(\theta, 1)$, define $T(x) = x$ as a sufficient statistic for θ (the sample mean), and set the observed data $y = 0$. With the prior $\pi(\theta) \propto 1$ for convenience, if the weighting kernel $\pi_\epsilon(y \,|\, x, \theta)$ is given by Equation 12.4, with $\rho(T(x), T(y)) = |x - y|$, or if $\pi_\epsilon(y \,|\, x, \theta)$ is a Gaussian density with $y \sim N(x, \epsilon^2/3)$, then

$$\pi_{LF}(\theta \,|\, y) \propto \frac{\Phi(\epsilon - \theta) - \Phi(-\epsilon - \theta)}{2\epsilon} \quad \text{and} \quad \pi_{LF}(\theta \,|\, y) = N\left(0, 1 + \frac{\epsilon^2}{3}\right),$$

respectively, where $\Phi(\cdot)$ denotes the standard Gaussian cumulative distribution function. The factor of 3 in the Gaussian kernel density ensures that both uniform and Gaussian kernels have the same standard deviation. In both cases $\pi_{LF}(\theta \,|\, y) \to N(0, 1)$ as $\epsilon \to 0$.

The two likelihood-free approximations are illustrated in Figure 12.1 which compares the target $\pi(\theta \,|\, y)$ to both forms of $\pi_{LF}(\theta \,|\, y)$ for different values of ϵ. Clearly, as ϵ gets smaller, $\pi_{LF}(\theta \,|\, y) \approx \pi(\theta \,|\, y)$ becomes a better approximation. Conversely, as ϵ increases, so does the

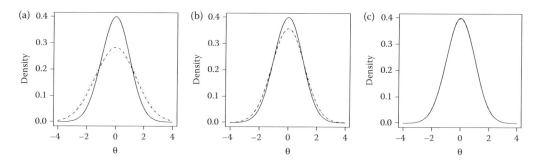

FIGURE 12.1
Comparison of likelihood-free approximations to the $N(0,1)$ target posterior (solid line) for ϵ values of (a) $\sqrt{3}$, (b) $\sqrt{3}/2$, (c) $\sqrt{3}/10$. Likelihood-free posteriors are constructed using uniform (dotted line) and Gaussian (dashed line) kernel weighting densities $\pi_\epsilon(y \mid x, \theta)$.

posterior variance in the likelihood-free approximation. There is only a small difference between using uniform and Gaussian weighting functions in this case.

Suppose now that an alternative vector of summary statistics $\tilde{T}(\cdot)$ also permits unbiased estimates of θ, but is less efficient than $T(\cdot)$, with a relative efficiency of $e \leq 1$. As noted by A. N. Pettitt (personal communication), for the above example with the Gaussian kernel density for $\pi_\epsilon(y \mid x, \theta)$, the likelihood-free approximation using $\tilde{T}(\cdot)$ becomes $\pi_{LF}(\theta \mid y) = N(0, 1/e + \epsilon^2/3)$. The $1/e$ term can easily be greater than the $\epsilon^2/3$ term, especially as practical interest is in small ϵ. This example illustrates that inefficient statistics can often determine the quality of the posterior approximation, and that this approximation can remain poor even for $\epsilon = 0$.

Accordingly, it is common in practice to aim to reduce ϵ as low as is computationally feasible. However, in certain circumstances, it is not clear that doing so will result in a better approximation to $\pi(\theta \mid y)$ than for a larger ϵ. This point is illustrated in Section 12.4.4.

12.3   Likelihood-Free MCMC Samplers

A Metropolis–Hastings sampler may be constructed to target the augmented likelihood-free posterior $\pi_{LF}(\theta, x \mid y)$ (given by Equation 12.1) without directly evaluating the intractable likelihood (Marjoram et al., 2003). Consider a proposal distribution for this sampler with the factorization

$$q[(\theta, x), (\theta', x')] = q(\theta, \theta')\pi(x' \mid \theta').$$

That is, when at a current algorithm state (θ, x), a new parameter vector θ' is drawn from a proposal distribution $q(\theta, \theta')$ and, conditionally on θ', a proposed data set x' is generated from the model $x' \sim \pi(x \mid \theta')$. Following standard arguments, to achieve a Markov chain with stationary distribution $\pi_{LF}(\theta, x \mid y)$, we enforce the detailed-balance (time-reversibility) condition

$$\pi_{LF}(\theta, x \mid y)P[(\theta, x), (\theta', x')] = \pi_{LF}(\theta', x' \mid y)P[(\theta', x'), (\theta, x)], \qquad (12.5)$$

where the Metropolis–Hastings transition probability is given by

$$P[(\theta, x), (\theta', x')] = q[(\theta, x), (\theta', x')]\alpha[(\theta, x), (\theta', x')].$$

The probability of accepting a move from (θ, x) to (θ', x') within the Metropolis–Hastings framework is then given by $\min\{1, \alpha[(\theta, x), (\theta', x')]\}$, where

$$\alpha[(\theta, x), (\theta', x')] = \frac{\pi_{LF}(\theta', x' \mid y)q[(\theta', x'), (\theta, x)]}{\pi_{LF}(\theta, x \mid y)q[(\theta, x), (\theta', x')]}$$

$$= \frac{\pi_\epsilon(y \mid x', \theta')\pi(x' \mid \theta')\pi(\theta')}{\pi_\epsilon(y \mid x, \theta)\pi(x \mid \theta)\pi(\theta)} \frac{q(\theta', \theta)\pi(x \mid \theta)}{q(\theta, \theta')\pi(x' \mid \theta')}$$

$$= \frac{\pi_\epsilon(y \mid x', \theta')\pi(\theta')q(\theta', \theta)}{\pi_\epsilon(y \mid x, \theta)\pi(\theta)q(\theta, \theta')}. \tag{12.6}$$

Note that the intractable likelihoods do not need to be evaluated in the acceptance probability calculation (Equation 12.6), leaving a computationally tractable expression which can now be evaluated. Without loss of generality we may assume that $\min\{1, \alpha[(\theta', x'), (\theta, x)]\} = 1$, and hence the detailed-balance condition (Equation 12.5), is satisfied since

$$\pi_{LF}(\theta, x \mid y)P[(\theta, x), (\theta', x')] = \pi_{LF}(\theta, x \mid y)q[(\theta, x), (\theta', x')]\alpha[(\theta, x), (\theta', x')]$$

$$= \frac{\pi_{LF}(\theta, x \mid y)q(\theta, \theta')\pi(x' \mid \theta')\pi_\epsilon(y \mid x', \theta')\pi(\theta')q(\theta', \theta)}{\pi_\epsilon(y \mid x, \theta)\pi(\theta)q(\theta, \theta')}$$

$$= \frac{\pi_\epsilon(y \mid x, \theta)\pi(x \mid \theta)\pi(\theta)q(\theta, \theta')\pi(x' \mid \theta')\pi_\epsilon(y \mid x', \theta')\pi(\theta')q(\theta', \theta)}{\pi_\epsilon(y \mid x, \theta)\pi(\theta)q(\theta, \theta')}$$

$$= \pi_\epsilon(y \mid x', \theta')\pi(x' \mid \theta')\pi(\theta')q(\theta', \theta)\pi(x \mid \theta)$$

$$= \pi_{LF}(\theta', x' \mid y)P[(\theta', x'), (\theta, x)].$$

ALGORITHM 12.2   LIKELIHOOD-FREE (LF)-MCMC ALGORITHM

1.   Initialize (θ_0, x_0) and ϵ. Set $t = 0$.

At step t:
2.   Generate $\theta' \sim q(\theta_t, \theta)$ from a proposal distribution.
3.   Generate $x' \sim \pi(x \mid \theta')$ from the model given θ'.
4.   With probability $\min\{1, \dfrac{\pi_\epsilon(y \mid x', \theta')\pi(\theta')q(\theta', \theta_t)}{\pi_\epsilon(y \mid x_t, \theta_t)\pi(\theta_t)q(\theta_t, \theta')}\}$ set $(\theta_{t+1}, x_{t+1}) = (\theta', x')$, otherwise set $(\theta_{t+1}, x_{t+1}) = (\theta_t, x_t)$.
5.   Increment $t = t + 1$ and go to 2.

The MCMC algorithm targeting $\pi_{LF}(\theta, x \mid y)$, adapted from Marjoram et al. (2003), is listed in Algorithm 12.2. The sampler generates the Markov chain sequence (θ_t, x_t) for $t \geq 0$, although in practice it is only necessary to store the vectors of summary statistics $T(x_t)$ and $T(x')$ at any stage in the algorithm. This is particularly useful when the auxiliary data sets x_t are large and complex.

An interesting feature of this sampler is that its acceptance rate is directly related to the value of the true likelihood function $\pi(y \mid \theta')$ at the proposed vector θ' (Sisson et al., 2007). This is most obviously seen when using the uniform kernel weighting density (Equation 12.4), as proposed moves to (θ', x') can only be accepted if $\rho(T(x'), T(y)) \leq \epsilon$, and this occurs with a probability in proportion to the likelihood. For low ϵ values this can result in very low acceptance rates, particularly in the tails of the distribution, thereby affecting chain mixing in regions of low posterior density (see Section 12.4.5 for an illustration). However, the LF-MCMC algorithm offers improved acceptance rates over rejection sampling-based likelihood-free algorithms (Marjoram et al., 2003).

We now examine a number of variations on the basic LF-MCMC algorithm which have been proposed either to improve sampler performance or to examine model goodness of fit.

12.3.1 Marginal Space Samplers

Given the definition of $\pi_{LF}(\theta \mid y)$ in Equation 12.2, an unbiased pointwise estimate of the marginal posterior distribution is available through Monte Carlo integration as

$$\pi_{LF}(\theta \mid y) \approx \frac{\pi(\theta)}{S} \sum_{s=1}^{S} \pi_\epsilon(y \mid x^s, \theta), \tag{12.7}$$

where $x^1, \ldots, x^S$ are independent draws from the model $\pi(x \mid \theta)$ (Marjoram et al., 2003; Peters et al., 2008; Ratmann et al., 2009; Reeves and Pettitt, 2005; Sisson et al., 2007; Toni et al., 2009; Wegmann et al., 2009). This then permits an MCMC sampler to be constructed directly targeting the likelihood-free marginal posterior $\pi_{LF}(\theta \mid y)$. In this setting, the probability of accepting a proposed move from θ to $\theta' \sim q(\theta, \theta')$ is given by $\min\{1, \alpha(\theta, \theta')\}$, where

$$\alpha(\theta, \theta') = \frac{\pi_{LF}(\theta' \mid y)q(\theta', \theta)}{\pi_{LF}(\theta \mid y)q(\theta, \theta')} \approx \frac{\frac{1}{S}\sum_{s} \pi_\epsilon(y \mid x'^s, \theta')\pi(\theta')q(\theta', \theta)}{\frac{1}{S}\sum_{s} \pi_\epsilon(y \mid x^s, \theta)\pi(\theta)q(\theta, \theta')} \tag{12.8}$$

and $x'^1, \ldots, x'^S \sim \pi(x \mid \theta')$. As the Monte Carlo approximation (Equation 12.7) becomes more accurate as S increases, the performance and acceptance rate of the marginal likelihood-free sampler will gradually approach that of the equivalent standard MCMC sampler.

However, the above ratio of two unbiased likelihood estimates is only unbiased as $S \to \infty$. Hence, the above sampler will only approximately target $\pi_{LF}(\theta \mid y)$ for large S, which makes it highly inefficient. However, note that estimating $\alpha(\theta, \theta')$ with $S = 1$ exactly recovers (Equation 12.6), the acceptance probability of the MCMC algorithm targeting $\pi_{LF}(\theta, x \mid y)$. That is, the marginal space likelihood-free sampler with $S = 1$ is precisely the likelihood-free MCMC sampler in Algorithm 12.2. As the sampler targeting $\pi_{LF}(\theta, x \mid y)$ also provides unbiased estimates of the marginal $\pi_{LF}(\theta \mid y)$, it follows that the likelihood-free sampler targeting $\pi_{LF}(\theta \mid y)$ directly is also unbiased in practice (Sisson et al., 2010). A similar argument for $S > 1$ can also be made, as outlined below.

An alternative augmented likelihood-free posterior distribution is given by

$$\pi_{LF}(\theta, x_{1:S} \mid y) \propto \pi_\epsilon(y \mid x_{1:S}, \theta)\pi(x_{1:S} \mid \theta)\pi(\theta)$$

$$:= \left[\frac{1}{S} \sum_{s=1}^{S} \pi_\epsilon(y \mid x^s, \theta) \right] \left[\prod_{s=1}^{S} \pi(x^s \mid \theta)] \right] \pi(\theta),$$

where $x_{1:S} = (x^1, \ldots, x^S)$ represents $s = 1, \ldots, S$ replicate auxiliary data sets $x^s \sim \pi(x \mid \theta)$. This posterior, generalized from Del Moral et al. (2008), is based on the more general expected auxiliary variable approach of Andrieu et al. (2008), where the summation form of $\pi_\epsilon(y \mid x_{1:S}, \theta)$ describes this expectation. The resulting marginal posterior $\pi_{\mathrm{LF}}^S(\theta \mid y) = \int_{y^S} \pi_{\mathrm{LF}}(\theta, x_{1:S}, \theta \mid y) dx_{1:S}$ is the same for all S, namely $\pi_{\mathrm{LF}}^S(\theta \mid y) = \pi_{\mathrm{LF}}(\theta \mid y)$.

The motivation for this form of posterior is that that a sampler targeting $\pi_{\mathrm{LF}}(\theta, x_{1:S} \mid y)$, for $S > 1$, will possess improved sampler performance compared to an equivalent sampler targeting $\pi_{\mathrm{LF}}(\theta, x \mid y)$, through a reduction in the variability of the Metropolis–Hastings acceptance probability. With the natural choice of proposal density given by

$$q[(\theta, x_{1:S}), (\theta', x'_{1:S})] = q(\theta, \theta') \prod_{s=1}^{S} \pi(x'^s \mid \theta'),$$

where $x'_{1:S} = (x'^1, \ldots, x'^S)$, the acceptance probability of a Metropolis–Hastings algorithm targeting $\pi_{\mathrm{LF}}(\theta, x_{1:S} \mid y)$ reduces to

$$\alpha[(\theta, x_{1:S}), (\theta', x'_{1:S})] = \frac{\dfrac{1}{S} \sum_s \pi_\epsilon(y \mid x'^s, \theta') \pi(\theta') q(\theta', \theta)}{\dfrac{1}{S} \sum_s \pi_\epsilon(y \mid x^s, \theta), \pi(\theta) q(\theta, \theta')}. \tag{12.9}$$

This is the same acceptance probability (Equation 12.8) as a marginal likelihood-free sampler targeting $\pi_{\mathrm{LF}}(\theta \mid y)$ directly, using S Monte Carlo draws to estimate $\pi_{\mathrm{LF}}(\theta \mid y)$ pointwise, via Equation 12.7. Hence, both marginal and augmented likelihood-free samplers possess identical mixing and efficiency properties. The difference between the two is that the marginal sampler acceptance probability (Equation 12.8) is approximate for finite S, whereas the augmented sampler acceptance probability (Equation 12.9) is exact. However, clearly the marginal likelihood-free sampler is, in practice, unbiased for all $S \geq 1$. See Sisson et al. (2010) for a more detailed analysis.

12.3.2 Error-Distribution Augmented Samplers

In all likelihood-free MCMC algorithms, low values of ϵ result in slowly mixing chains through low acceptance rates. However, this also provides a potentially more accurate posterior approximation $\pi_{\mathrm{LF}}(\theta \mid y) \approx \pi(\theta \mid y)$. Conversely, MCMC samplers with larger ϵ values may possess improved chain mixing and efficiency, although at the expense of a poorer posterior approximation (e.g. Figure 12.1). Motivated by a desire for improved sampler efficiency while realizing low ϵ values, Bortot et al. (2007) proposed augmenting the likelihood-free posterior approximation to include ϵ, so that

$$\pi_{\mathrm{LF}}(\theta, x, \epsilon \mid y) \propto \pi_\epsilon(y \mid x, \theta) \pi(x \mid \theta) \pi(\theta) \pi(\epsilon).$$

Accordingly, ϵ is treated as a tempering parameter in the manner of simulated tempering (Geyer and Thompson, 1995), with larger and smaller values respectively corresponding to "hot" and "cold" tempered posterior distributions. The density $\pi(\epsilon)$ is a pseudo-prior, which serves only to influence the mixing of the sampler through the tempered distributions. Bortot et al. (2007) suggested using a distribution which favors small ϵ values for accuracy, while permitting large values to improve chain acceptance rates. The approximation to the true posterior $\pi(\theta \,|\, y)$ is then given by

$$\pi^{\mathcal{E}}_{\mathrm{LF}}(\theta \,|\, y) = \int_{\mathcal{E}} \int_{y} \pi_{\mathrm{LF}}(\theta, x, \epsilon \,|\, y) \, dx \, d\epsilon$$

where $\epsilon \in \mathcal{E} \subseteq \mathbb{R}^{+}$. Sampler performance aside, this approach permits an *a posteriori* evaluation of an appropriate value $\epsilon = \epsilon^{*}$ such that $\pi^{\mathcal{E}}_{\mathrm{LF}}(\theta \,|\, y)$ with $\mathcal{E} = [0, \epsilon^{*}]$ provides an acceptable approximation to $\pi(\theta \,|\, y)$.

An alternative error-distribution augmented model was proposed by Ratmann et al. (2009) with the aim of diagnosing model misspecification for the observed data y. For the vector of summary statistics $T(x) = (T_1(x), \ldots, T_R(x))$, the discrepancy between the model $\pi(x \,|\, \theta)$ and the observed data is given by $\tau = (\tau_1, \ldots, \tau_R)$, where $\tau_r = T_r(x) - T_r(y)$, for $r = 1, \ldots, R$, is the error under the model in reproducing the rth element of $T(\cdot)$. The joint distribution of model parameters and model errors is defined as

$$\pi_{\mathrm{LF}}(\theta, x_{1:S}, \tau \,|\, y) \propto \pi_{\epsilon}(y \,|\, \tau, x_{1:S}, \theta) \pi(x_{1:S} \,|\, \theta) \pi(\theta) \pi(\tau)$$

$$:= \min_{r} \hat{\xi}_r(\tau_r \,|\, y, x_{1:S}, \theta) \pi(x_{1:S} \,|\, \theta) \pi(\theta) \pi(\tau), \qquad (12.10)$$

where the univariate error distributions

$$\hat{\xi}_r(\tau_r \,|\, y, x_{1:S}, \theta) = \frac{1}{S\epsilon_r} \sum_{s=1}^{S} K \left(\frac{\tau_r - \left[T_r(x^s) - T_r(y) \right]}{\epsilon_r} \right) \qquad (12.11)$$

are constructed from smoothed kernel density estimates of model errors, estimated from S auxiliary data sets $x^1, \ldots, x^S$, and where $\pi(\tau) = \prod_r \pi(\tau_r)$, the joint prior distribution for the model errors, is centered on zero, reflecting that the model is assumed plausible *a priori*. The terms $\min_r \hat{\xi}_r(\tau_r \,|\, y, x, \theta)$ and $\pi(\tau)$ take the place of the weighting density $\pi_{\epsilon}(y \,|\, \tau, x_{1:S}, \theta)$. The minimum of the univariate densities $\hat{\xi}_r(\tau_r \,|\, y, x, \theta)$ is taken over the R model errors to reflect the most conservative estimate of model adequacy, while also reducing the computation on the multivariate τ to its univariate component margins. The smoothing bandwidths ϵ_r of each summary statistic $T_r(\cdot)$ are dynamically estimated during sampler implementation as twice the interquartile range of $T_r(x^s) - T_r(y)$, given $x^1, \ldots, x^S$.

Assessment of model adequacy can then be based on

$$\pi_{\mathrm{LF}}(\tau \,|\, y) = \int_{\Theta} \int_{y^S} \pi_{\mathrm{LF}}(\theta, x_{1:S}, \tau \,|\, y) \, dx_{1:S} \, d\theta,$$

the posterior distribution of the model errors. If the model is adequately specified then $\pi_{\mathrm{LF}}(\tau \,|\, y)$ should be centered on the zero vector. If this is not the case then the model is misspecified. The nature of the departure of $\pi_{\mathrm{LF}}(\tau \,|\, y)$ from the origin, for example via one or more summary statistics $T_r(\cdot)$, may indicate the manner in which the model is deficient. See, for example, Wilkinson (2008) for further assessment of model errors in likelihood-free models.

12.3.3 Potential Alternative MCMC Samplers

Given the variety of MCMC techniques available for standard Bayesian inference, there are a number of currently unexplored ways in which these might be adapted to improve the performance of likelihood-free MCMC samplers.

For example, within the class of marginal space samplers (Section 12.3.1), the number of Monte Carlo draws S determines the quality of the estimate of $\pi_{LF}(\theta \mid y)$ (cf. Equation 12.7). A standard implementation of the delayed-rejection algorithm (Tierney and Mira, 1999) would permit rejected proposals based on poor but computationally cheap posterior estimates (i.e. using low to moderate S), to generate more accurate but computationally expensive second-stage proposals (using large S), thereby adapting the computational overheads of the sampler to the required performance.

Alternatively, coupling two or more Markov chains targeting $\pi_{LF}(\theta, x \mid y)$, each utilizing a different ϵ value, would achieve improved mixing in the "cold" distribution (i.e. the chain with the lowest ϵ) through the switching of states between neighboring (in an ϵ sense) chains (Pettitt, 2006). This could be particularly useful in multimodal posteriors. While this flexibility is already available with continuously varying ϵ in the augmented sampler targeting $\pi_{LF}(\theta, x, \epsilon \mid y)$ (Bortot et al., 2007; see also Section 12.3.2 above), there are benefits to constructing samplers from multiple chain sample-paths.

Finally, likelihood-free MCMC samplers have to date focused on tempering distributions based on varying ϵ. While not possible in all applications, there is clear scope for a class of algorithms based on tempering on the number of observed data points from which the summary statistics $T(\cdot)$ are calculated. Lower numbers of data points will produce greater variability in the summary statistics, in turn generating wider posteriors for the parameters θ, but with lower computational overheads required to generate the auxiliary data x.

12.4   A Practical Guide to Likelihood-Free MCMC

In this section we examine various practical aspects of likelihood-free computation under a simple worked analysis. For observed data $y = (y_1, \ldots, y_{20})$ consider two candidate models: $y_i \sim$ Exponential(λ) and $y_i \sim$ Gamma(k, ψ), where model equivalence is obtained under $k = 1, \psi = 1/\lambda$. Suppose that the sample mean and standard deviation of y are available as summary statistics $T(y) = (\bar{y}, s_y) = (4, 1)$, and that interest is in fitting each model and in establishing model adequacy. Note that the summary statistics $T(\cdot)$ are sufficient for λ but not for (k, ψ), where they form moment-based estimators. For the following we consider flat priors $\pi(\lambda) \propto 1$, $\pi(k, \psi) \propto 1$ for convenience. The true posterior distribution under the Exponential(λ) model is $\lambda \mid y \sim$ Gamma$(21, 80)$.

12.4.1   An Exploratory Analysis

An initial exploratory investigation of model adequacy is illustrated in Figure 12.2, which presents scatterplots of summary statistics versus summary statistics, and summary statistics versus parameter values under each model. Images are based on 2000 parameter realizations $\lambda, k, \psi \sim U(0, 20)$ followed by summary statistic generation under each model parameter. Horizontal and vertical lines denote the values of the observed summary statistics $T(y)$.

From the plots of sample means against standard deviations, $T(y)$ is clearly better represented by the gamma than the exponential model. The observed summary statistics (i.e. the intersection of horizontal and vertical lines) lie in regions of relatively lower prior predictive density under the exponential model, compared to the gamma. That is, *a priori*, the statistics $T(y)$ appear more probable under the more complex model.

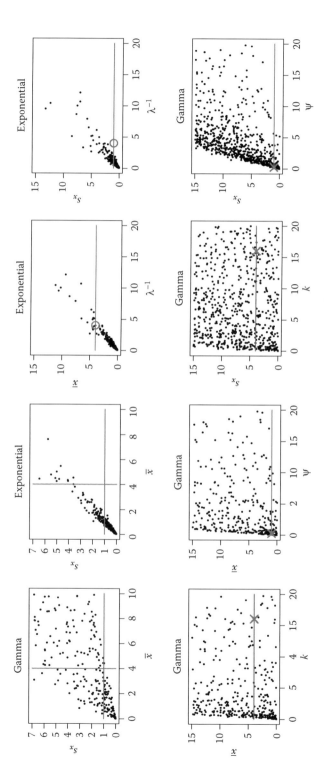

FIGURE 12.2

Scatterplots of summary statistics $T(x) = (\bar{x}, s_x)$ and parameter values λ, k, ψ under both Exponential(λ) and Gamma(k, ψ) models, based on 2000 realizations $\lambda, k, \psi \sim U(0, 20)$. Horizontal and vertical lines denote observed summary statistics $T(y) = (4, 1)$. Circles denote the maximum likelihood estimate of $\hat{\lambda} = 1/\bar{y} = 1/4$ under the exponential model. Crosses denote method of moments estimators $\hat{k} = \bar{y}^2/s_y^2 = 16$ and $\hat{\psi} = s_y^2/\bar{y} = 1/4$ under the gamma model.

Consider the plots of λ^{-1} versus $T(x)$ under the exponential model. The observed statistics $T(y)$ individually impose competing requirements on the exponential parameter. An observed sample mean of $\bar{y} = 4$ indicates that λ^{-1} is most likely in the approximate range $[3, 5]$ (indicated by those λ^{-1} values where the horizontal line intersects with the density). However, the sample standard deviation $s_y = 1$ independently suggests that λ^{-1} is most likely in the approximate range $[0.5, 1.5]$. If either $\bar{x}$ or s_x were the only summary statistic, then only one of these ranges is appropriate, and the observed data would be considerably more likely under the exponential model. However, the relative model fits and model adequacies of the exponential and gamma can only be evaluated by using the same summary statistics on each model. (Otherwise, the model with the smaller number of summary statistics will be considered the most likely model, simply because it is more probable to match fewer statistics.) As a result, the competing constraints on λ through the statistics $\bar{x}$ and s_y are so jointly improbable under the exponential model that simulated and observed data will rarely coincide, making $T(y)$ very unlikely under this model. This is a strong indicator of model inadequacy.

In contrast, the plots of k and ψ against $T(x)$ under the gamma model indicate no obvious restrictions on the parameters based on $T(y)$, suggesting that this model is flexible enough to have generated the observed data with relatively high probability. Note that from these marginal scatterplots it is not clear that these statistics are at all informative for the model parameters. This indicates the importance of parameterization for visualization, as alternatively considering method of moments estimators as summary statistics $(\hat{k}, \hat{\psi})$, where $\hat{k} = \bar{x}^2/s_x^2$ and $\hat{\psi} = s_x^2/\bar{x}$, will result in strong linear relationships between (k, ψ) and $(\hat{k}, \hat{\psi})$. Of course, in practice direct unbiased estimators are rarely known.

12.4.2   The Effect of ϵ

We now implement the LF-MCMC algorithm (Algorithm 12.2) targeting the Exponential(λ) model, with an interest in evaluating sampler performance for different ϵ values. Recall that small ϵ is required to obtain a good likelihood-free approximation to the intractable posterior $\pi_{\mathrm{LF}}(\theta \mid y) \approx \pi(\theta \mid y)$ (see Figure 12.1), where now $\theta = \lambda$. However, implementing the sampler with low ϵ can be problematic in terms of initializing the chain and in achieving convergence to the stationary distribution.

An initialization problem may occur when using weighting kernels $\pi_\epsilon(y \mid x, \theta)$ with compact support, such as the uniform kernel (Equation 12.4) defined on $[-\epsilon, \epsilon]$. Here, initial chain values (θ_0, x_0) are required such that $\pi_\epsilon(y \mid x_0, \theta_0) \neq 0$ in the denominator of the acceptance probability at time $t = 1$ (Algorithm 12.2). ϵ, this is unlikely to be the case for the first such parameter vector tried. Two naive strategies are to either repeatedly generate $x_0 \sim \pi(x \mid \theta_0)$, or similarly repeatedly generate $\theta_0 \sim \pi(\theta)$ and $x_0 \sim \pi(x \mid \theta_0)$, until $\pi_\epsilon(y \mid x_0, \theta_0) \neq 0$ is achieved. However, the former strategy may never terminate unless θ_0 is located within a region of high posterior density. The latter strategy may never terminate if the prior is diffuse with respect to the posterior. Relatedly, Markov chain convergence can be very slow for small ϵ when moving through regions of very low density, for which generating $x' \sim \pi(x \mid \theta')$ with $T(x') \approx T(y)$ is highly improbable.

One strategy to avoid these problems is to augment the target distribution from $\pi_{\mathrm{LF}}(\theta, x \mid y)$ to $\pi_{\mathrm{LF}}(\theta, x, \epsilon \mid y)$ (Bortot et al., 2007), permitting a time-variable ϵ to improve chain mixing (see Section 12.3 for discussion on this and other strategies to improve chain mixing). A simpler strategy is to implement a specified chain burn-in period, defined by a monotonic decreasing sequence $\epsilon_{t+1} \leq \epsilon_t$, initialized with large ϵ_0, for which $\epsilon_t = \epsilon$ remains constant at the desired level for $t \geq t^*$, beyond some (possibly random) time t^* (see Peters et al., 2010). For example, consider the linear sequence $\epsilon_t = \max\{\epsilon_0 - ct, \epsilon\}$ for some $c > 0$. However, the issue here is in determining the rate at which the sequence approaches the target ϵ: if c is too large, then $\epsilon_t = \epsilon$ before (θ_t, x_t) has reached a region of high density; if c is too small, then the chain mixes well but is computationally expensive through a slow burn-in.

One self-scaling option for the uniform weighting density (Equation 12.4) would be to define $\epsilon_0 = \rho(T(x_0), T(y))$ and, given the proposed pair (θ', x') at time t, propose a new ϵ value as

$$\epsilon'' = \max\{\epsilon, \min\{\epsilon', \epsilon_{t-1}\}\}, \tag{12.12}$$

where $\epsilon' = \rho(T(x'), T(y)) > 0$ is the distance between observed and simulated summary statistics. If the proposed pair (θ', x') are accepted then set $\epsilon_t = \epsilon''$, else set $\epsilon_t = \epsilon_{t-1}$. That is, the proposed ϵ'' is dynamically defined as the smallest possible value that results in a nonzero weighting function $\pi_{\epsilon_t}(y \mid x', \theta')$ in the numerator of the acceptance probability, without going below the target ϵ, and while decreasing monotonically. If the proposed move to (θ', x') is accepted, the value ϵ'' is accepted as the new state, else the previous value ϵ_{t-1} is retained. Similar approaches could be taken with nonuniform weighting densities $\pi_\epsilon(y \mid x, \theta)$.

Four trace plots of λ_t and ϵ_t for the Exponential(λ) model are illustrated in Figure 12.3a,b, using the above procedure. All Markov chains were initialized at $\lambda_0 = 10$ with target $\epsilon = 3$, proposals were generated via $\lambda' \sim N(\lambda_{t-1}, 1)$, and the distance measure

$$\rho(T(x), T(y)) = \left\{ [T(x) - T(y)]^\top \Sigma^{-1} [T(x) - T(y)] \right\}^{1/2} \tag{12.13}$$

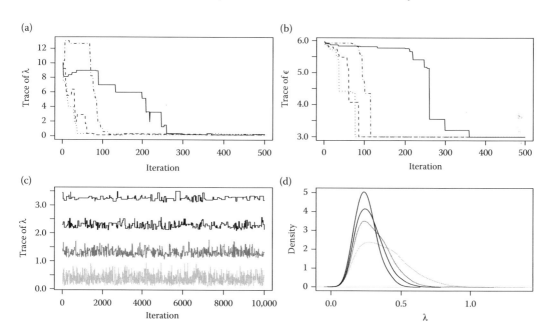

FIGURE 12.3
Performance of the LF-MCMC sampler for the Exponential(λ) model. Trace plots of (a) λ_t and (b) ϵ_t for four chains using the self-scaling $\{\epsilon_t\}$ sequence given by Equation 12.12. The maximum likelihood estimate of λ is 0.25 and the target ϵ is 3. (c) Jittered trace plots of λ_t with different target $\epsilon = 4.5$ (bottom), 4, 3.5, and 3 (top). (d) Posterior density estimates of λ for the same chains based on a chain length of 100,000 iterations.

is given by Mahalanobis distance. The covariance matrix $\Sigma = \text{cov}(T(y))$ is estimated by the sample covariance of 1000 summary vectors $T(x)$ generated from $\pi(x \mid \hat{\lambda})$ conditional on $\hat{\lambda} = 0.25$, the maximum likelihood estimate. All four chains converge to the high-density region at $\lambda = 0.25$ quickly, although at different speeds as the sampler takes different routes through parameter space. Mixing during burn-in is variable between chains, although overall convergence to $\epsilon_t = 3$ is rapid. The requirement of tuning the rate of convergence, beyond specifying the final tolerance ϵ, is clearly circumvented.

Figure 12.3c,d also illustrates the performance of the LF-MCMC sampler, post convergence, based on four chains of length 100,000, each with different target ϵ. As expected (see the discussion in Section 12.3), smaller ϵ results in lower acceptance rates. In Figure 12.3c, $\epsilon = 4.5$ (bottom trace), 4, 3.5, and 3 (top) result in post-convergence (of ϵ_t) mean acceptance rates of 12.2%, 6.1%, 2.9%, and 1.1%, respectively. Conversely, precision (and accuracy) of the posterior marginal distribution for λ increases with decreasing ϵ, as seen in Figure 12.3d.

In practice, a robust procedure to identify a suitable target ϵ for the likelihood-free MCMC sampler is not yet available. Wegmann et al. (2009) implement the LF-MCMC algorithm with a large ϵ value to enhance chain mixing, and then perform a regression-based adjustment (Beaumont et al., 2002; Blum and François, 2010) to improve the final posterior approximation. Bortot et al. (2007) implement the LF-MCMC algorithm targeting the augmented posterior $\pi_{\text{LF}}(\theta, x, \epsilon \mid y)$ (see Section 12.3.2), and examine the changes in $\pi_{\text{LF}}^{\mathcal{E}}(\theta \mid y) = \int_{\mathcal{E}} \int_{\mathcal{Y}} \pi_{\text{LF}}(\theta, x, \epsilon \mid y) \, dx \, d\epsilon$, with $\mathcal{E} = [0, \epsilon^*]$, for varying ϵ^*. The final choice of ϵ^* is the largest value for which reducing ϵ^* further produces no obvious improvement in the posterior approximation. This procedure may be repeated manually through repeated LF-MCMC sampler implementations at different fixed ϵ values (Tanaka et al., 2006). Nevertheless, in practice ϵ is often reduced as low as possible such that computation remains within acceptable limits.

12.4.3 The Effect of the Weighting Density

The optimal form of kernel weighting density $\pi_\epsilon(y \mid x, \theta)$ for a given analysis is unclear at present. While the uniform weighting kernel (Equation 12.4) is the most common in practice—indeed, many likelihood-free methods have this kernel written directly into the algorithm (sometimes implicitly)—it seems credible that alternative forms may offer improved posterior approximations for given computational overheads. Some support for this is available through recently observed links between the likelihood-free posterior approximation $\pi_{\text{LF}}(\theta \mid y)$ and nonparametric smoothing (Blum, 2010).

Here we evaluate the effect of the weighting density $\pi_\epsilon(y \mid x, \theta)$ on posterior accuracy under the Exponential (λ) model, as measured by the one-sample Kolmogorov–Smirnov distance between the likelihood-free posterior sample and the true Gamma(21, 80) posterior. To provide fair comparisons, we evaluate posterior accuracy as a function of computational overheads, measured by the mean post-convergence acceptance rate of the LF-MCMC sampler. The following results are based on posterior samples consisting of 1000 posterior realizations obtained by recording every 1000th chain state, following a burn-in period of 10,000 iterations. Figures are constructed by averaging the results of 25 sampler replications under identical conditions, for a range of ϵ values.

Figure 12.4a shows the effect of varying the form of the kernel weighting function based on the Mahalanobis distance (Equation 12.13). There appears little obvious difference in the accuracy of the posterior approximations in this example. However, it is credible to suspect that nonuniform weighting functions may be superior in general (Blum, 2010; Peters et al., 2008). This is more clearly demonstrated in Section 12.4.5 below. The slight worsening in the

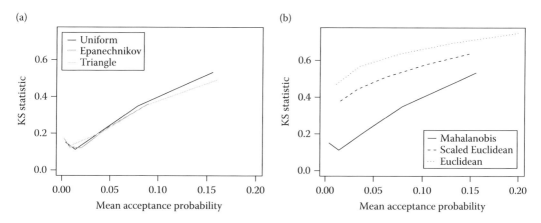

FIGURE 12.4
Performance of the LF-MCMC sampler for the Exponential(λ) model under varying kernel weighting densities: (a) Mahalanobis distance between $T(x)$ and $T(y)$ evaluated on uniform, Epanechnikov and triangle kernel functions; (b) Mahalanobis, scaled Euclidean and Euclidean distance between $T(x)$ and $T(y)$ evaluated on the uniform kernel function. Sampler performance is measured in terms of accuracy (y-axis: one-sample Kolmogorov–Smirnov test statistic evaluated between likelihood-free posterior sample and true posterior) versus computational overheads (x-axis: mean sampler acceptance probability).

accuracy of the posterior approximation, indicated by the upturn for low ϵ in Figure 12.4a, will be examined in more detail in Section 12.4.4.

Regardless of its actual form, the weighting density $\pi_\epsilon(y \mid x, \theta)$ should take the distribution of the summary statistics $T(\cdot)$ into consideration. Fan et al. (2010) note that using a Euclidean distance measure (given by Equation 12.13 with $\Sigma = I$, the identity matrix) within (say) the uniform weighting kernel (Equation 12.4), ignores the scale and dependence (correlation) structure of $T(\cdot)$, accepting sampler moves if $T(y)$ is within a circle of size ϵ centered on $T(x)$, rather than within an ellipse defined by $\Sigma = \text{cov}(T(y))$. In theory, the form of the distance measure does not matter as in the limit $\epsilon \to 0$ any effect of the distance measure ρ is removed from the posterior $\pi_{\text{LF}}(\theta \mid y)$, that is, $T(x) = T(y)$ regardless of the form of Σ. In practice, however, with $\epsilon > 0$, the distance measure can have a strong effect on the quality of the likelihood-free posterior approximation $\pi_{\text{LF}}(\theta \mid y) \approx \pi(\theta \mid y)$.

Using the uniform weighting density, Figure 12.4b demonstrates the effect of using Mahalanobis distance (Equation 12.13), with Σ given by estimates of $\text{cov}(T(y))$, $\text{diag}(\text{cov}(T(y)))$ (scaled Euclidean distance) and the identity matrix I (Euclidean distance). Clearly, for a fixed computational overhead (x-axis), greater accuracy is attainable by standardizing and orthogonalizing the summary statistics. In this sense, Mahalanobis distance represents an approximate standardization of the distribution of $T(y) \mid \tilde{\theta}$ at an appropriate point $\tilde{\theta}$ following indirect inference arguments (Jiang and Turnbull, 2004). As $\text{cov}(T(y))$ may vary with θ, Fan et al. (2010) suggest using an approximate maximum *a posteriori* estimate of θ, so that $\tilde{\theta}$ resides in a region of high posterior density. The assumption is then that $\text{cov}(T(y))$ varies little over the region of high posterior density.

12.4.4 The Choice of Summary Statistics

Likelihood-free computation is based on the reproduction of observed statistics $T(y)$ under the model. If the $T(y)$ are sufficient for θ, then the true posterior $\pi(\theta \mid y)$ can be recovered exactly as $\epsilon \to 0$. If $\dim(T(y))$ is large (Bortot et al., 2007), then likelihood-free algorithms become computationally inefficient through the need to reproduce large numbers of summary statistics (Blum, 2010). However, low-dimensional, nonsufficient summary vectors produce less efficient estimators of θ, and so generate wider posterior distributions $\pi_{LF}(\theta \mid y)$ than using sufficient statistics (see Section 12.2.3). Ideally, low-dimensional and near-sufficient $T(y)$ are the preferred option.

Unfortunately, it is usually difficult to know which statistics are near-sufficient in practice. A brute-force strategy to address this issue is to repeat the analysis while sequentially increasing the number of summary statistics each time (in order of their perceived importance), until no further changes to $\pi_{LF}(\theta \mid y)$ are observed (Marjoram et al., 2003; see also Joyce and Marjoram, 2008). If the extra statistics are *un*informative, the quality of approximation will remain the same, but the sampler will be less efficient. However, simply enlarging the number of informative summary statistics is not necessarily the best way to improve the likelihood-free approximation $\pi_{LF}(\theta \mid y) \approx \pi(\theta \mid y)$, and in fact may worsen the approximation in some cases.

An example of this is provided by the present Exponential(λ) model, where either of the two summary statistics $T(y) = (\bar{y}, s_y) = (4, 1)$ alone is informative for λ (and indeed, $\bar{y}$ is sufficient), as we expect that $\lambda \approx 1/\bar{y} \approx 1/s_y$ under any data generated from this model. In this respect, however, the observed values of the summary statistics provide conflicting information for the model parameter (see Section 12.4.1). Figure 12.5 examines the effect of this, by evaluating the accuracy of the likelihood-free posterior approximation $\pi_{LF}(\theta \mid y) \approx \pi(\theta \mid y)$ as a function of ϵ under different summary statistic combinations. As before, posterior accuracy is measured via the one-sample Kolmogorov–Smirnov test statistic with respect to the true Gamma$(21, 80)$ posterior.

With $T(y) = \bar{y}$, Figure 12.5a demonstrates that accuracy improves as ϵ decreases, as expected. For Figure 12.5b, with $T(y) = s_y$ (dots), the resulting $\pi_{LF}(\theta \mid y)$ posterior is clearly different from the true posterior for all ϵ. Of course, the limiting posterior as $\epsilon \to 0$ is (very) approximately Gamma$(21, 20)$, resulting from an exponential model with $\lambda = 1/s_y = 1$, rather than Gamma$(21, 80)$ resulting from an exponential model with $\lambda = 1/\bar{y} = 1/4$. The crosses in Figure 12.5b denote the Kolmogorov–Smirnov test statistic with respect to the Gamma$(21, 20)$ distribution, which indicates that $\pi_{LF}(\theta \mid y)$ is roughly consistent with this

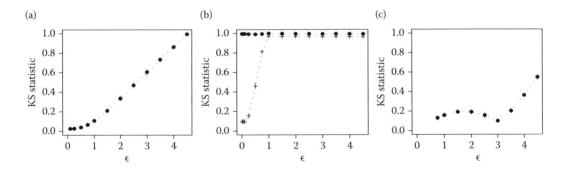

FIGURE 12.5
Likelihood-free posterior accuracy of the Exponential(λ) model as a function of ϵ for differing summary statistics: (a) $T(y) = \bar{y}$; (b) $T(y) = s_y$; (c) $T(y) = (\bar{y}, s_y)$. Posterior accuracy (y-axis) is measured by one-sample Kolmogorov–Smirnov (KS) test statistic evaluated between likelihood-free posterior sample and true posterior. Points and vertical lines represent KS statistic means and ranges based on 25 sampler replicates at fixed ϵ levels. Crosses in (b) denote KS statistic evaluated with respect to a Gamma$(21, 20)$ distribution.

distribution as ϵ decreases. That the Gamma$(21, 20)$ is not the exact limiting density (i.e. the KS statistic does not tend to zero as $\epsilon \to 0$) stems from the fact that s_y is not a sufficient statistic for λ, and is less than fully efficient.

In Figure 12.5c with $T(y) = (\bar{y}, s_y)$, which contains an exactly sufficient statistic $(\bar{y})$, the accuracy of $\pi_{\mathrm{LF}}(\theta \,|\, y)$ appears to improve with decreasing ϵ, and then actually worsens before improving again. This would appear to go against the generally accepted principle that, for sufficient statistics, decreasing ϵ will always improve the approximation $\pi_{\mathrm{LF}}(\theta \,|\, y) \approx \pi(\theta \,|\, y)$. Of course, the reality here is that both of these competing statistics are pulling the likelihood-free posterior in different directions, with the consequence that the limiting posterior as $\epsilon \to 0$ will be some combination of both gamma distributions, rather than the presumed (and desired) Gamma$(21, 80)$.

This observation leads to the uncomfortable conclusion that model comparison through likelihood-free posteriors with a fixed vector of summary statistics $T(y)$, will ultimately compare distortions of those models which are overly simplified with respect to the true data-generation process. This remains true even when using sufficient statistics and for $\epsilon \to 0$.

12.4.5 Improving Mixing

Recall that the acceptance rate of the LF-MCMC algorithm (Algorithm 12.2) is directly related to the value of the true likelihood $\pi(y \,|\, \theta')$ at the proposed vector θ' (Section 12.3). While this is a necessary consequence of likelihood-free computation, it does imply poor sampler performance in regions of low probability, as the Markov chain sample-path may persist in distributional tails for long periods of time due to low acceptance probabilities (Sisson et al., 2007). This is illustrated in Figure 12.6(a, b: lowest light gray lines), which displays the marginal sample paths of k and ψ under the Gamma(k, ψ) model, based on 5000 iterations of a sampler targeting $\pi(\theta, x \,|\, y)$ with $\epsilon = 2$ and using the uniform kernel density $\pi_\epsilon(y \,|\, x, \theta)$. At around 1400 iterations the sampler becomes stuck in the tail of the posterior for the following 700 iterations, with very little meaningful movement.

A simple strategy to improve sampler performance in this respect is to increase the number of auxiliary data sets S generated under the model, by targeting either the joint posterior $\pi_{\mathrm{LF}}(\theta, x_{1:S} \,|\, y)$ or the marginal posterior $\pi_{\mathrm{LF}}(\theta \,|\, y)$ with $S \geq 1$ Monte Carlo draws (see Section 12.3.1). This approach will reduce the variability of the acceptance probability (Equation 12.8), and allow the Markov chain acceptance rate to approach that of a sampler targeting the true posterior $\pi(\theta \,|\, y)$. The trace plots in Figure 12.6a,b (bottom to top) correspond to chains implementing $S = 1, 10, 20,$ and 50 auxiliary data set generations per likelihood evaluation. Visually, there is some suggestion that mixing is improved as S increases. Note, however, that for any fixed S, the LF-MCMC sampler may still become stuck if the sampler explores sufficiently far into the distributional tail.

Figure 12.6c,d investigates this idea from an alternative perspective. Based on 2 million sampler iterations, the lengths of sojourns that the k parameter spent above a fixed threshold κ were recorded. A sojourn length is defined as the consecutive number of iterations in which the parameter k remains above κ. Intuitively, if likelihood-free samplers tend to persist in distributional tails, the length of the sojourns will be much larger for the worse-performing samplers. Figure 12.6c,d shows the distributions of sojourn lengths for samplers with $S = 1,$ $10, 25,$ and 50 auxiliary data sets, with $\kappa = 45$ (c) and $\kappa = 50$ (d). Boxplot shading indicates use of the uniform (white) or Gaussian (gray) weighting kernel $\pi_\epsilon(y \,|\, x, \theta)$.

A number of points are immediately apparent. Firstly, chain mixing is poorer the further into the tails the sampler explores. This is illustrated by the increased scale of the sojourn

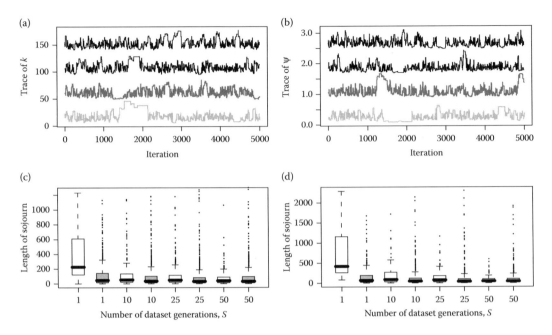

FIGURE 12.6

Aspects of LF-MCMC sampler performance. Trace plots of (a) k and (b) ψ parameters under the gamma model, for varying numbers of auxiliary data sets $S = 1$ (lower traces), $10, 25$ and 50 (upper traces) using $\epsilon = 2$ and the uniform kernel density $\pi_\epsilon(y \mid x, \theta)$. Distribution of sojourn lengths of parameter k above (c) $\kappa = 45$ and (d) $\kappa = 50$ for varying numbers of auxiliary data sets. Boxplot shading indicates uniform (white) or Gaussian (gray) kernel densities $\pi_\epsilon(y \mid x, \theta)$. The Gaussian kernel sampler used $\epsilon = 2/\sqrt{3}$ to ensure a comparable standard deviation with the uniform kernel sampler.

lengths for $\kappa = 50$ compared to $\kappa = 45$. Secondly, increasing S by a small amount substantially reduces chain tail persistence. As S increases further, the Markov chain performance approaches that of a sampler directly targeting the true posterior $\pi(\theta \mid y)$, and so less performance gains are observed by increasing S beyond a certain point. Finally, there is strong evidence to suggest that LF-MCMC algorithms using weighting kernel densities $\pi_\epsilon(y \mid x, \theta)$ that do not generate large numbers of zero-valued likelihoods will possess superior performance to those that do. Here use of the Gaussian weighting kernel clearly outperforms the uniform kernel in all cases. In summary, it would appear that the choice of kernel weighting function $\pi_\epsilon(\theta \mid y)$ has a larger impact on sampler performance than the number of auxiliary data sets S.

12.4.6   Evaluating Model Misspecification

In order to evaluate the adequacy of both exponential and gamma models in terms of their support for the observed data $T(y) = (\bar{y}, s_y)$, we fit the error-distribution augmented model (Equation 12.10) given by

$$\pi_{\mathrm{LF}}(\theta, x_{1:S}, \tau \mid y) := \min_r \hat{\xi}_r(\tau_r \mid y, x_{1:S}, \theta)\pi(x_{1:S} \mid \theta)\pi(\theta)\pi(\tau),$$

as described in Section 12.3.2 (Ratmann et al., 2009). The vector $\tau = (\tau_1, \tau_2)$, with $\tau_r = T_r(x) - T_r(y)$ for $r = 1, 2$, describes the error under the model in reproducing the observed summary statistics $T(y)$. The marginal likelihood-free posterior $\pi_{\mathrm{LF}}(\tau \,|\, y)$ should be centered on the zero vector for models which can adequately account for the observed data.

We follow Ratmann et al. (2009) in specifying K in Equation 12.11 as a biweight (quartic) kernel with an adaptive bandwidth ϵ_r determined by twice the interquartile range of $T_r(x^s) - T_r(y)$ given $x_{1:S} = (x^1, \ldots, x^S)$. The prior on the error τ is determined as $\pi(\tau) = \prod_r \pi(\tau_r)$, where $\pi(\tau_r) = \exp(-\,|\,\tau_r\,|\,/\delta_r)/(2\delta_r)$ with $\delta_1 = \delta_2 = 0.75$ for both exponential and gamma models.

Based on 50,000 sampler iterations using $S = 50$ auxiliary data sets, the resulting bivariate posterior $\pi_{\mathrm{LF}}(\tau \,|\, y)$ is illustrated in Figure 12.7 for both models. From these plots, the errors τ under the gamma model (bottom plots) are clearly centered on the origin, with 50% marginal high-density regions given by $\tau_1 \,|\, y \sim [-0.51, 0.53]$ and $\tau_2 \,|\, y \sim [-0.44, 0.22]$ (Ratmann et al., 2009). However for the exponential model (top plots), while the marginal 50% high density regions $\tau_1 \,|\, y \sim [-0.32, 1.35]$ and $\tau_2 \,|\, y \sim [-0.55, 0.27]$ also both contain zero, there is some indication of model misspecification as the joint posterior error distribution $\tau \,|\, y$ is not fully centered on the zero vector. Based on this assessment, and recalling the discussion on the exploratory analysis in Section 12.4.1, the gamma model would appear to provide a better overall fit to the observed data.

12.5 Discussion

In the early 1990s, the introduction of accessible MCMC samplers provided the catalyst for a rapid adoption of Bayesian methods and inference as credible tools in model-based research. Twenty years later, the demand for computational techniques capable of handling the types of models inspired by complex hypotheses has resulted in new classes of simulation-based inference, which are again expanding the applicability and relevance of the Bayesian paradigm to new levels.

While the focus of the present chapter is on Markov chain-based, likelihood-free simulation, alternative methods to obtain samples from $\pi_{\mathrm{LF}}(\theta \,|\, y)$ have been developed, each with their own benefits and drawbacks. While MCMC-based samplers can be more efficient than rejection sampling algorithms, the tendency of sampler performance to degrade in regions of low posterior density (see Section 12.4.5 above; see also Sisson et al., 2007) can be detrimental to sampler efficiency. One class of methods, based on the output of a rejection sampler with a high ϵ value (for efficiency), uses standard multivariate regression methods to estimate the relationship between the summary statistics $T(x)$ and parameter vectors θ (Beaumont et al., 2002; Blum and François, 2010; Marjoram and Tavaré, 2006). The idea is then to approximately transform the sampled observations from $(\theta, T(x))$ to $(\theta^*, T(y))$ so that the adjusted likelihood-free posterior $\pi_{\mathrm{LF}}(\theta, x \,|\, y) \to \pi_{\mathrm{LF}}(\theta^*, y \,|\, y) \approx \pi(\theta \,|\, y)$ is an improved approximation. Further attempts to improve sampler efficiency over MCMC-based methods have resulted in the development of likelihood-free sequential Monte Carlo and sequential importance sampling algorithms (Beaumont et al., 2009; Del Moral et al., 2008; Peters et al., 2008; Sisson et al., 2007; Toni et al., 2009). Several authors have reported that likelihood-free sequential Monte Carlo approaches can outperform their MCMC counterparts (McKinley et al., 2009; Sisson et al., 2007).

There remain many open research questions in likelihood-free Bayesian inference. These include how to select and incorporate the vectors of summary statistics $T(\cdot)$, how to perform posterior simulation in the most efficient manner, and which joint likelihood-free posterior

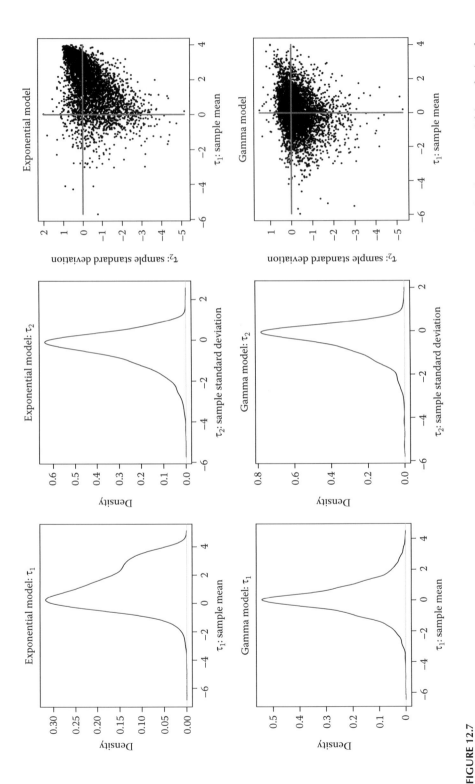

FIGURE 12.7
Marginal likelihood-free posterior distributions $\pi_{LF}(\tau \mid y)$ of the error-distribution augmented model (Equation 12.10), under the exponential (top plots) and gamma (bottom plots) models. Plots are based on 50,000 sampler iterations.

models and kernel weighting densities admit the most effective marginal approximation to the true posterior $\pi_{LF}(\theta \mid y) \approx \pi(\theta \mid y)$. Additionally, the links to existing bodies of research, including nonparametrics (Blum, 2010) and indirect inference (Jiang and Turnbull, 2004), are at best poorly understood.

Finally, there is an increasing trend toward using likelihood-free inference for model selection purposes (Grelaud et al., 2009; Toni et al., 2009). While this is a natural extension of inference for individual models, the analysis in Section 12.4.4 urges caution and suggests that further research is needed into the effect of the likelihood-free approximation both within models and on the marginal likelihoods $\pi_{LF}(y) = \int_{\mathcal{Y}} \pi_{LF}(\theta \mid y) \, d\theta$ upon which model comparison is based.

Acknowledgments

This work was supported by the Australian Research Council through the Discovery Project scheme (DP0664970 and DP1092805).

References

Andrieu, C., Berthelsen, K. K., Doucet, A., and Roberts, G. O. 2008. The expected auxiliary variable method for Monte Carlo simulation. Technical report, submitted for publication.

Beaumont, M. A., Cornuet, J.-M., Marin, J.-M., and Robert, C. P. 2009. Adaptive approximate Bayesian computation. *Biometrika*, 96(4):983–990.

Beaumont, M. A., Zhang, W., and Balding, D. J. 2002. Approximate Bayesian computation in population genetics. *Genetics*, 162:2025–2035.

Blum, M. G. B. 2010. Approximate Bayesian computation: A nonparametric perspective. *Journal of the American Statistical Association* 105:1178–1187.

Blum, M. G. B. and François, O. 2010. Non-linear regression models for approximate Bayesian computation. *Statistics and Computing*, 20(1):63–73.

Blum, M. G. B. and Tran, V. C. 2010. HIV with contact-tracing: A case study in approximate Bayesian computation. *Biostatistics*, 11(4):644–660.

Bortot, P., Coles, S. G., and Sisson, S. A. 2007. Inference for stereological extremes. *Journal of the American Statistical Association*, 102:84–92.

Del Moral, P., Doucet, A., and Jasra, A. 2008. Adaptive sequential Monte Carlo samplers. Technical report, University of Bordeaux.

Drovandi, C. C. and Pettitt, A. N. 2009. A note on Bayesian estimation of quantile distributions. Technical report, Queensland University of Technology.

Fagundes, N. J. R., Ray, N., Beaumont, M. A., Neuenschwander, S., Salzano, F. M., Bonatto, S. L., and Excoffier, L. 2007. Statistical evaluation of alternative models of human evolution. *Proceedings of the National Academy of Sciences of the USA*, 104:17614–17619.

Fan, Y., Peters, G. W., and Sisson, S. A. 2010. Improved efficiency in approximate Bayesian computation. Technical report, University of New South Wales.

Geyer, C. J. and Thompson, E. A. 1995. Annealing Markov chain Monte Carlo with applications to ancestral inference. *Journal of the American Statistical Association*, 90:909–920.

Grelaud, A., Robert, C. P., Marin, J.-M., Rodolphe, F., and Taly, J.-F. 2009. ABC likelihood-free methods for model choice in Gibbs random fields. *Bayesian Analysis*, 4:317–336.

Hamilton, G., Currat, M., Ray, N., Heckel, G., Beaumont, M. A., and Excoffier, L. 2005. Bayesian estimation of recent migration rates after a spatial expansion. *Genetics*, 170:409–417.

Jabot, F. and Chave, J. 2009. Inferring the parameters of the netural theory of biodiversity using phylogenetic information and implications for tropical forests. *Ecology Letters*, 12:239–248.

Jiang, W. and Turnbull, B. 2004. The indirect method: Inference based on intermediate statistics—A synthesis and examples. *Statistical Science*, 19:238–263.

Joyce, P. and Marjoram, P. 2008. Approximately sufficient statistics and Bayesian computation. *Statistical Applications in Genetics and Molecular Biology*, 7(1): no. 23.

Luciani, F., Sisson, S. A., Jiang, H., Francis, A., and Tanaka, M. M. 2009. The high fitness cost of drug resistance in *Mycobacterium tuberculosis*. *Proceedings of the National Academy of Sciences of the USA*, 106:14711–14715.

Marjoram, P. and Tavaré, S. 2006. Modern computational approaches for analysing molectular genetic variation data. *Nature Reviews Genetics*, 7:759–770.

Marjoram, P., Molitor, J., Plagnol, V., and Tavaré, S. 2003. Markov chain Monte Carlo without likelihoods. *Proceedings of the National Academy of Sciences of the USA*, 100:15324–15328.

McKinley, T., Cook, A. R., and Deardon, R. 2009. Inference in epidemic models without likelihoods. *International Journal of Biostatistics*, 5: article 24.

Neal, R. M. 2003. Slice sampling. *Annals of Statistics*, 31:705–767.

Nevat, I., Peters, G. W., and Yuan, J. 2008. Coherent detection for cooperative networks with arbitrary relay functions using likelihood-free inference. Technical report, University of New South Wales.

Peters, G. W. and Sisson, S. A. 2006. Bayesian inference, Monte Carlo sampling and operational risk. *Journal of Operational Risk*, 1(3).

Peters, G. W., Fan, Y., and Sisson, S. A. 2008. On sequential Monte Carlo, partial rejection control and approximate Bayesian computation. Technical report, University of New South Wales.

Peters, G. W., Nevat, I., Sisson, S. A., Fan, Y., and Yuan, J. 2010. Bayesian symbol detection in wireless relay networks via likelihood-free inference. *IEEE Transactions on Signal Processing*, 58:5206–5218.

Peters, G. W., Sisson, S. A., and Fan, Y. (in press). Likelihood-free Bayesian inference for alpha-stable models. *Computational Statistics and Data Analysis*, in press.

Pettitt, A. N. 2006. From doubly intractable distributions via auxiliary variables to likelihood free inference. Paper presented to Recent Advances in Monte Carlo Based Inference workshop, Isaac Newton Institute, Cambridge, UK, 30 October. http://www.newton.ac.uk/programmes/SCB/scbw01p.html

Pritchard, J. K., Seielstad, M. T., Perez-Lezaun, A., and Feldman, M. W. 1999. Population growth of human Y chromosomes: A study of Y chromosome microsatellites. *Molecular Biology and Evolution*, 16:1791–1798.

Ratmann, O., Andrieu, C., Hinkley, T., Wiuf, C., and Richardson, S. 2009. Model criticism based on likelihood-free inference, with an application to protein network evolution. *Proceedings of the National Academy of Sciences of the USA*, 106:10576–10581.

Ratmann, O., Jørgensen, O., Hinkley, T., Stumpf, M., Richardson, S., and Wiuf, C. 2007. Using likelihood-free inference to compare evolutionary dynamics of the protein networks of *H. pylori* and *P. falciparum*. *PLoS Computational Biology*, 3:e230.

Reeves, R. W. and Pettitt, A. N. 2005. A theoretical framework for approximate Bayesian computation. In A. R. Francis, K. M. Matawie, A. Oshlack, and G. K. Smyth (eds), *Statistical Solutions to Modern Problems: Proceedings of the 20th International Workshop on Statistical Modelling, Sydney, 10–15 July 2005*, pp. 393–396. University of Western Sydney Press, Sydney.

Sisson, S. A., Fan, Y., and Tanaka, M. M. 2007. Sequential Monte Carlo without likelihoods. *Proc. Natl. Acad. Sci.*, 104:1760–1765. Errata 2009, 106:16889.

Sisson, S. A., Peters, G. W., Briers, M., and Fan, Y. 2010. A note on target distribution ambiguity of likelihood-free samplers. arXiv:1005.5201v1.

Tanaka, M. M., Francis, A. R., Luciani, F., and Sisson, S. A. 2006. Using approximate Bayesian computation to estimate tuberculosis transmission parameters from genotype data. *Genetics*, 173:1511–1520.

Tavaré, S., Balding, D. J., Griffiths, R. C., and Donnelly, P. 1997. Inferring coalescence times from DNA sequence data. *Genetics*, 145:505–518.

Tierney, L. and Mira, A. 1999. Some adaptive Monte Carlo methods for Bayesian inference. *Statistics in Medicine*, 18:2507–2515.

Toni, T., Welch, D., Strelkowa, N., Ipsen, A., and Stumpf, M. P. H. 2009. Approximate Bayesian computation scheme for parameter inference and model selection in dynamical systems. *Journal of the Royal Society Interface*, 6:187–202.

Wegmann, D., Leuenberger, C., and Excoffier, L. 2009. Efficient approximate Bayesian computation coupled with Markov chain Monte Carlo without likelihood. *Genetics*, 182:1207–1218.

Wilkinson, R. D. 2008. Approximate Bayesian computation (ABC) gives exact results under the assumption of model error. Technical report, Department of Probability and Statistics, University of Sheffield.

Wilkinson, R. D. and Tavaré, S. 2009. Estimating primate divergence times by using conditioned birth-and-death processes. *Theoretical Population Biology*, 75:278–285.

Part II

Applications and Case Studies

13

MCMC in the Analysis of Genetic Data on Related Individuals

Elizabeth Thompson

13.1   Introduction

This chapter provides an overview of the use of Markov chain Monte Carlo (MCMC) methods in the analysis of data observed for multiple genetic loci on members of extended pedigrees in which there are many missing data. Rather than on the details of the MCMC sampling methods, our focus is first on the complex structure of these data that necessitates MCMC methods, and second on the use of Monte Carlo realizations of latent variables in statistical inference in this area.

MCMC should be a weapon of last resort, when exact computation and other Monte Carlo methods fail. When MCMC is needed, there are two prerequisites for its efficient use in complex stochastic systems. The first is a consideration of the conditional independence structure of the data observations and latent variables, and a choice of latent variable structure that will facilitate computation and sampling. While unnecessary augmentation of the latent variable space is clearly disadvantageous, there are classic cases where augmentation of the space greatly improves efficiency (Besag and Green, 1993). Second, and related, it is important to consider what parts of a computation may be performed exactly. Where a partial exact computation is feasible, this may be used to resample jointly subsets of the latent variables, and hence improve MCMC performance. Additionally, partial exact computation may permit the use of Rao-Blackwellized estimators (Gelfand and Smith, 1990), improving efficiency in the use of sampled realizations. Thus, in Section 13.3 we consider the structures and exact computational algorithms that will complement MCMC approaches.

As genetic marker data on observable individuals increase, and the traits requiring analysis become genetically more complex, the challenges both for exact computation and MCMC methods increase also. In Section 13.4, we describe MCMC samplers of genetic latent variables that have evolved from the single-site genotypic updating samplers of Sheehan (2000) to the most recent multiple-meiosis and locus sampling of inheritance patterns of Tong and Thompson (2008). The separation of the analysis of trait data from the MCMC sampling of latent variables conditional on genetic marker data was first proposed by Lange and Sobel (1991). With the increasing complexity of models for trait data, this becomes the approach of choice, and in Section 13.5 we discuss the sampling of latent inheritance patterns conditional only on dense marker data. In some cases, the model on which sampling is based is too simple to even approximate reality. Then, importance sampling reweighting becomes a key tool in improving the usefulness of this approach. Also in the arena of marker data based analyses is the question of genetic map estimation (Section 13.5.2).

Having developed the exact and Monte Carlo computational methods in Sections 13.3 and 13.4, in Section 13.6 we describe their use in the analysis of genetic data. In Section 13.6.1, we show how realizations of inheritance patterns can be used in the Monte Carlo estimation of multilocus linkage log-odds (lod) scores and other test statistics. In Section 13.6.2, we show how the variation in Monte Carlo realizations of latent variables can be used to measure uncertainty in inferences and test linkage detection, using the latent p-value or *fuzzy p-value* approach of Geyer and Meeden (2005). Finally, in Section 13.6.3 we discuss two approaches to localization of genes for complex traits, using the latent p-value approach. Overall, our thesis again is that a single set of realizations of latent genetic variables, made conditional on joint marker data on all individuals and over an entire genomic region, can be used in a broad variety of ways to analyze the genetic basis of complex traits.

While this chapter contains new material, particularly in relation to methods for approaching modern dense single nucleotide polymorphism (SNP) data using MCMC methods, much of the background information is based on earlier papers. These include a tutorial chapter on MCMC for genetic data (Thompson, 2005) and a chapter in the *Handbook of Statistical Genetics* on linkage analysis (Thompson, 2007). Many additional references may be found in these two previous papers.

13.2   Pedigrees, Genetic Variants, and the Inheritance of Genome

In this section, we introduce the specification of pedigrees and inheritance, and then discuss structure of genetic models. A pedigree is a specification of the genealogical relationships among a set of individuals. Each individual is given a unique identifier, and the two parents of each individual are specified. Individuals with unspecified parents are *founders*: the others are *nonfounders*. Graphically, males are traditionally represented by squares and females by circles. In the graphical representation of a pedigree known as a *marriage node graph*, a male and a female individual having shared offspring are connected to a *marriage node*, and the marriage node is connected to each offspring. An example pedigree we will use throughout this chapter is shown in Figure 13.1. For clarity, the marriage nodes are shown as bullets.

Each marriage node is connected upward to two parent individuals, and downward to at least one (and possibly many) offspring individual(s). Each nonfounder is connected upward to precisely one marriage node. A parent individual may be connected to multiple marriage nodes. In the example 28-member pedigree (Figure 13.1), the letters are the identifiers of the individuals (some not shown). There are 9 founders and 19 nonfounders, 12 males and 16 females, and 11 marriage nodes. One individual (H) has two marriages. One individual (C) is *inbred*, having related parents E and H. Note that it is not only inbreeding that causes loops in pedigrees. Even without C, the fact that E and H are double-first cousins creates a loop in the pedigree structure. Another loop is created by the fact that sibs D and F are double-first cousins to A, B, and J.

Human individuals are diploid: every cell nucleus contains two haploid copies of the DNA of the human genome, each of approximately 3×10^9 base pairs (bp). One of these copies derives from the DNA in the individual's mother (the maternal genome), and the other from the DNA in the individual's father (the paternal genome). Note that *all* DNA is double-stranded. The double-stranded nature of DNA has nothing to do with the haploid (single genome copy) or diploid (two-copy) genome content of a cell or organism. The biological process through which DNA in parent cells is copied and transmitted to offspring is known as *meiosis*, and *Mendel's first law* (1866) specifies this transmission marginally, at

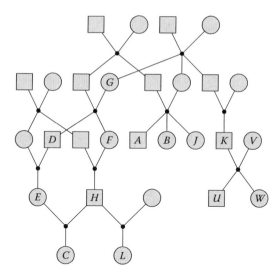

FIGURE 13.1
An example 28-member pedigree.

any location in the genome. A genome location is known as a *locus* (plural *loci*). In modern terminology, Mendel's first law states that the copy transmitted from parent to offspring is a randomly chosen one of the two parental copies, and that all meioses, whether to different offspring of a single parent or in different parental individuals, are independent.

At many loci of our genomes there is genetic variation. At a given locus, the possible variants are known as the alleles of that locus. The two (possibly identical) alleles at a locus carried by a diploid individual are the individual's *genotype* at that locus. The DNA of our cell nuclei is divided into 46 chromosomes (22 pairs and 2 sex chromosomes). The allelic types along a chromosome are known as the *haplotype*. In a given genomic region, the two haplotypes carried by an individual determine the individual's genotype at all loci in the region. The converse is not true; a set of single-locus genotypes of an individual may correspond to many different haplotype pairs. This is is problem of *phase* (Browning and Browning, 2007).

Inheritance is dependent among loci on the same chromosome pair. Specifically, DNA at nearby loci has a very high probability of being copied to an offspring from the same parental chromosome, and in fact chromosomes are inherited in chunks with length of order 10^8 bp. Mendel's first law implies only that, at each locus, an offspring will share an allele with each parent. The chromosomal dependence in inheritance resulting from the process of meiosis implies that, at least locally and with high probability, an offspring will share a haplotype with each parent.

13.3 Conditional Independence Structures of Genetic Data

The descent of DNA in a pedigree is not directly observable, and, even where individuals are available for observation, the DNA variants of their separate chromosomes (i.e. their *haplotypes*) are not normally observable. Thus, the framework for analyses of genetic data can be described through several complementary latent variable specifications. In a genetic

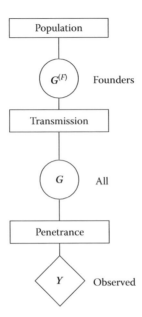

FIGURE 13.2
The structure of genetic models.

analysis the primary objective is often the computation of a likelihood, or the probability of observed data **Y**. The latent variables are the targets of MCMC approaches to Monte Carlo estimates of this likelihood.

13.3.1   Genotypic Structure of Pedigree Data

While the genetic data have become ever more complex, the basic structure outlined by Elston and Stewart (1971) remains. Figure 13.2 shows this structure, and in these and similar figures, models will be represented by boxes, latent variables by circles, and observable data by diamonds. For the founder members of the pedigree, the *population model* specifies the probabilities of the allelic types of DNA and hence also their genotypes. The *transmission model* specifies the probabilities of meiotic events, and hence the descent of DNA and thence the genotypes of all members of the pedigree. The *penetrance model* specifies the probability of data observations given the genotype. The data observation here may be qualitative or quantitative, and the penetrance probability may depend on other covariate information on the individual, such as age, sex, or geographic location. Given this classic structure of genetic models, it is natural to consider first the genotypes of individuals as defining the latent structure of genetic data.

The probability of data, **Y**, or likelihood of any model parameters Γ is given by

$$L(\Gamma) = P(\mathbf{Y}; \Gamma) = \sum_{\mathbf{G}} P(\mathbf{Y} \mid \mathbf{G}) P(\mathbf{G})$$

$$= \sum_{\mathbf{G}} \left(\prod_{\text{fou}} P(G_i) \right) \left(\prod_{\text{nonfou}} P(G_i \mid G_{M(i)}, G_{F(i)}) \right) \left(\prod_{\text{obs}} P(Y_i \mid G_i) \right). \qquad (13.1)$$

The probability structure here implies that data on offspring are conditionally independent given the genotypes of parents, or more generally that data on disjoint parts of the pedigree

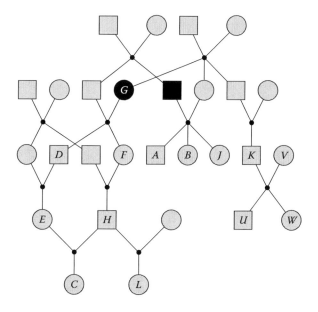

FIGURE 13.3
Genotypic peeling on pedigree structures. The two dark-shaded individuals are a cutset dividing the left- and right-hand parts of the pedigree.

are conditionally independent given the genotypes of individuals in a cutset dividing these parts (Figure 13.3). This led to the computational method of pedigree peeling proposed by Elston and Stewart (1971), and soon generalized to arbitrary pedigrees (Cannings et al., 1978) and more complex models (Cannings et al., 1980).

We will not labor the details here, since the approach is now well known through the generalization to other graphical models (Lauritzen, 1992; Lauritzen and Spiegelhalter, 1988). One point worth noting is that pedigrees are intrinsically directed, with DNA being copied from parents to offspring (Equation 13.1). Thus when the accumulated probability relates to individuals connected to a cutset member i via his offspring ("below i"), the natural probability to consider is

$$R^{\dagger}(g) = P(\text{data} \mid G_i = g).$$

However, if the accumulated probability is for data on individuals connected via the parents of i ("above i"), the natural probability is

$$R^*(g) = P(\text{data}, G_i = g).$$

On a complex pedigree, the accumulated probability may relate to a pedigree subset above some individuals but below others. For example, in Figure 13.3 the probability of the data in the right half of the pedigree could be expressed as

$$R_{\ell,r}^{(r)}(g_1, g_2) = P(\text{right data}, G_\ell = g_1 \mid G_r = g_2),$$

where ℓ and r denote the left and right member of the cutset pair in the middle of the pedigree. Equivalently, the data in the left half could be considered as

$$R_{\ell,r}^{(\ell)}(g_1, g_2) = P(\text{left data}, G_r = g_2 \mid G_\ell = g_1).$$

Since individuals ℓ and r are unrelated, multiplying these two functions and summing over (g_1, g_2) provides the overall likelihood $P(\mathbf{Y})$. Then also, the product of the two functions normalized by $P(\mathbf{Y})$ provides the conditional probability $P(G_\ell = g_1, G_r = g_2 \mid \mathbf{Y})$.

The methods of this section can be applied to data at several genetic loci along a chromosome, with the latent genotype being *phased*. That is, it is a specification of the collection of alleles on each of the two chromosomes of the individual: the two *haplotypes*. The model is then completely general, with the population model permitting any specified population haplotype frequencies, the meiosis model permitting any specified transmission of DNA from parents to offspring, and data observations being determined arbitrarily by the phased genotypes of individuals jointly over the loci. However, the number of potential phased genotypes of an individual increases exponentially with the number of genetic loci, and for more than a very few genetic loci computation becomes infeasible. An alternate structure of latent variables is then required.

13.3.2   Inheritance Structure of Genetic Data

One alternate structure of latent variables consists of a specification in all the meioses i (parent–offspring transmissions) of the pedigree of the inheritance of genome at any set of discrete loci j:

$$S_{i,j} = 0, \text{ if DNA at meiosis } i \text{ locus } j \text{ is parent's maternal DNA,}$$

$$= 1, \text{ if DNA at meiosis } i \text{ locus } j \text{ is parent's paternal DNA.}$$

For convenience, we define the two sets of vectors each of which makes up the array $\mathbf{S} = \{S_{i,j}\}$:

$$S_{\bullet,j} = \{S_{i,j}; i = 1, \ldots, m\}, j = 1, \ldots, l,$$

$$S_{i,\bullet} = \{S_{i,j}; j = 1, \ldots, l\}, i = 1, \ldots, m,$$

where m is the number of meioses in the pedigree (twice the number of nonfounders) and l the number of loci under consideration. In the literature, the vector $S_{\bullet,j}$ is known as the *inheritance vector* at locus j (Lander and Green, 1987).

According to Mendel's first law, the components of $S_{\bullet,j}$ are independent, and hence so also are the vectors $S_{i,\bullet}$. However, the components of $S_{i,\bullet}$ are dependent among loci j on the same chromosome pair. For any pair of loci j and j',

$$P(S_{i,j} = 0) = P(S_{i,j} = 1) = P(S_{i,j'} = 0) = P(S_{i,j'} = 1) = 1/2$$

by Mendel's first law. One additional parameter, $\rho(j,j') = P(S_{i,j} \neq S_{i,j'})$, suffices to specify the joint distribution. In reality, the value of the recombination parameter ρ depends on the meiosis i, most importantly on the sex of the parent in which the meiosis occurs (Kong et al., 2002). Sex-specific recombination parameters impose no computational burden, but for notational convenience we will ignore the dependence of ρ on i. For loci j and j' that are close in the genome, $\rho(j,j')$ is small and approximately equal to the genetic distance between the loci in morgans (Haldane, 1919). The relationship between genetic (meiotic) distance and physical (base-pair) distance is complex and variable across the genome, but a useful overall average is that 1 centimorgan (cM) corresponds to $\rho \approx 0.01$, and to 10^6 bp. The value $\rho(j,j') = \frac{1}{2}$ corresponds to independence of $S_{i,j}$ and $S_{i,j'}$, and under most models of meiosis $0 \le \rho \le \frac{1}{2}$.

For more than two loci, suppose that loci $j = 1, 2, \ldots, l$ are ordered along the chromosome. A convenient and adequately accurate assumption is that $S_{i,j}$ are Markov in j:

$$P(S_{i,j+1} = s \mid S_{i,1}, \ldots, S_{i,j}) = P(S_{i,j+1} = s \mid S_{i,j}) \propto \rho_j^{|s - S_{i,j}|} (1 - \rho_j)^{(1 - |s - S_{i,j}|)}, \tag{13.2}$$

where ρ_j now denotes the recombination parameter between successive loci j and $j + 1$. Then

$$P(S_{\bullet,j} \mid S_{\bullet,j-1}) = \prod_{i=1}^{m} P(S_{i,j} \mid S_{i,j-1})$$

$$\text{and} \quad P(\mathbf{S}) = P(S_{\bullet,1}) \left(\prod_{j=2}^{l} P(S_{\bullet,j} \mid S_{\bullet,j-1}) \right).$$

We now assume further that the data can be separated into components $Y_{\bullet,j}$ determined separately by genotypes at each locus j. These genotypes are a deterministic function of the allelic types at this locus of founder members of the pedigree, and of $S_{\bullet,j}$. Then

$$P(\mathbf{Y} \mid \mathbf{S}) = \prod_{j=1}^{l} P(Y_{\bullet,j} \mid S_{\bullet,j})$$

$$P(\mathbf{Y}) = \sum_{\mathbf{S}} P(\mathbf{Y} \mid \mathbf{S}) P(\mathbf{S}) \tag{13.3}$$

$$= \sum_{\mathbf{S}} \left(\prod_{j=1}^{l} P(Y_{\bullet,j} \mid S_{\bullet,j}) \right) P(S_{\bullet,1}) \left(\prod_{j=2}^{l} P(S_{\bullet,j} \mid S_{\bullet,j-1}) \right).$$

The data then have hidden Markov (HMM) structure as shown in Figure 13.4. The meiosis model provides the hidden layer of inheritance vectors $S_{\bullet,j}$ while population models determine the allelic types $\mathcal{A}_j^{(F)}$ of founders (F) at locus j. These latent variables determine the genotypes of all individuals at each locus j, and we shall find that computation of probabilities of data $Y_{\bullet,j}$ is straightforward for loci at which genotypes are observed (typically "marker loci"), but is more complex if single-locus genotypes are not observable (Section 13.3.3).

Given a method for computation of $P(Y_{\bullet,j} \mid S_{\bullet,j})$ (Section 13.3.3), standard HMM computational algorithms can be applied (Baum et al., 1970). Following standard notation, let

$$Y^{*(j)} = \{Y_{\bullet,1}, \ldots, Y_{\bullet,j}\} \quad \text{and} \quad Y^{\dagger(j)} = \{Y_{\bullet,j}, \ldots, Y_{\bullet,l}\},$$

$$R_j^*(s) = P(Y^{*(j)}, S_{\bullet,j} = s) \quad \text{and} \quad R_j^{\dagger}(s) = P(Y^{\dagger(j+1)} \mid S_{\bullet,j} = s).$$

Given $S_{\bullet,j}$, $Y^{*(j-1)}$, $Y_{\bullet,j}$, and $S_{\bullet,j+1}$ are mutually independent. Alternately, given $S_{\bullet,j}$, $Y^{\dagger(j+1)}$, $Y_{\bullet,j}$, and $S_{\bullet,j-1}$ are mutually independent. Unlike a pedigree, a chromosome has no direction, but we retain the conditional (†) and joint (*) forms for analogy with Section 13.3.1. As in that case, the likelihood $P(\mathbf{Y}) = P(Y^{*(l)})$ may be computed by successive elimination of each $S_{\bullet,j}$ (Baum, 1972), while

$$P(S_{\bullet,j} \mid \mathbf{Y}) = R^{*(j)}(s) R^{\dagger(j)}(s) / P(\mathbf{Y}). \tag{13.4}$$

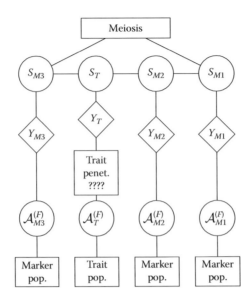

FIGURE 13.4
The HMM dependence structure of pedigree data. For details, see text.

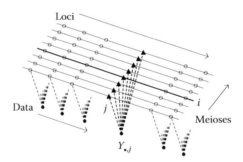

FIGURE 13.5
The dependence structure of pedigree data.

While the HMM approach permits the computation of likelihoods on pedigree data sets with data at multiple loci on a chromosome, it has limitations. First, not only is the meiosis model more restrictive than in the Elston–Stewart framework, but so also are the trait and marker penetrance models, with the data separating into components determined only by the inheritance vector at that locus. Second, we have replaced an algorithm exponential in the number of loci by one that is exponential in the size of the pedigree. If there are m meioses in the pedigree, each $S_{\bullet,j}$ can take 2^m values, and the basic HMM algorithm is of order $2^m \times 2^m \times l = l4^m$ (Lander and Green, 1987).

In fact, the situation is not so severe, due again to Mendel's first law. The (unconditional) independence of meioses provides a dependence structure of the form shown in Figure 13.5, which is a *factored* hidden Markov structure (Fishelson and Geiger, 2004). Although $S_{\bullet,j}$ still takes 2^m values, the forward computation of R_{j+1}^* from R_j^* may be accomplished for each of the m meioses in turn, providing an algorithm of order $l\,m2^m$. However, the approach remains exponential in m, so that exact computation of probabilities of genetic data observed at multiple dependent loci is limited to small pedigrees.

13.3.3 Identical by Descent Structure of Genetic Data

Segments of DNA in different genomes that are copies of the same genomic material in a recent common ancestor are said to be *identical by descent* (*ibd*). In the analysis of data on a fixed set of pedigree structures, *ibd* is defined relative to the founders of the pedigrees. By definition, the genomes of founders are nowhere *ibd*. An accurate model is that *ibd* segments of DNA carry the same allelic types; mutation has low probability and can be ignored. By definition, non-*ibd* segments carry independent allelic types. Thus, identity by descent underlies all similarity among relatives that results from the effects of their DNA.

At any given locus j, the pattern of *ibd* among pedigree members is a function of the inheritance vector $S_{\bullet,j}$. Consequently, given $S_{\bullet,j}$, we may define the *ibd*-graph among observed pedigree members as shown in Figure 13.6. In the pedigree on the left the individuals labeled A–L are assumed observed at the locus in question. In the *ibd*-graph on the right, the edges are the (data on) these observed individuals, and the nodes can be considered as (the allelic type of) the DNA shared *ibd* by the individuals. Edges join the nodes representing the two DNA segments carried by the individual at this locus. Thus, in this example, sibs A, B, and J all share DNA *ibd* from one of their parents, while B and J share also the DNA from their other parent, but A does not. An individual such as C connected to only one node is assumed to carry two *ibd* segments of DNA at the locus, one copied to him from each of his parents, who must necessarily then have a common ancestor within the pedigree.

For clarity the founder genomes are labeled in Figure 13.6, but it is important to recognize that the founder origins are irrelevant. Only *ibd* among the current observed individuals impacts the probabilities of data.

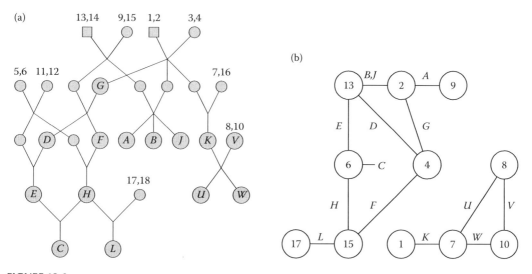

FIGURE 13.6

The *ibd*-graph as a function of inheritance. (a) Pedigree with observed individuals A–L and labeled founder genomes 1–18. (b) A possible *ibd*-graph resulting from descent of founder genomes in this pedigree. For details see text.

13.3.4   *ibd*-Graph Computations for Markers and Traits

Since *ibd* underlies all similarity among relatives that results from the effects of their DNA, at any locus j the *ibd*-graph on observed individuals is an equivalence class of all the inheritance vectors $S_{\bullet,j}$ that must give identical data probabilities $P(Y_{\bullet,j} \mid S_{\bullet,j})$.

For marker loci, at which, for observed individuals, it is assumed genotypes can be observed without error, Sobel and Lange (1996) and Kruglyak et al. (1996) independently provided efficient algorithms for computation of $P(Y_{\bullet,j} \mid S_{\bullet,j})$ using the *ibd*-graph. By assumption, the allelic types of *ibd*-nodes are independent, since each represents a non-*ibd* piece of genome. Suppose that each node g has allelic type a_k independently with probability q_k. Then

$$P(\mathcal{A}) = \prod_g q(\mathcal{A}(g)) = \prod_k q_k^{n(k)}, \tag{13.5}$$

where $n(k)$ is number of nodes g with type a_k, and

$$P(Y_{\bullet,j} \mid \mathbf{S}) = \sum P(\mathcal{A}_j),$$

where the sum is over all $\mathcal{A}_j$ consistent with $Y_{\bullet,j}$. There are always 2, 1, or 0 possible $\mathcal{A}_j$, and probabilities multiply over unconnected components of the *ibd*-graph.

As an example, consider the larger component of the *ibd*-graph in Figure 13.6. Suppose that A, B, J are all $a_1 a_4$, G is $a_1 a_6$, D is $a_4 a_6$, E is $a_4 a_2$, C is $a_2 a_2$, F is $a_3 a_6$, H is $a_2 a_3$, and L is $a_1 a_3$. It immediately follows that node 2 is a_1, nodes 9, 13 are a_4, 4 is a_6, 6 is a_2, 15 is a_3, and 17 is a_1. This is the only possible assignment, and results in two nodes of type a_1, two of type a_4, and one each of $a_2, a_3,$ and a_6. The probability is $q_1^2 q_2 q_3 q_4^2 q_6$.

Computation on the *ibd*-graph is not limited to error-free genotypic data. In 1997, S. Heath proposed and implemented computation for arbitrary two-allele single-locus penetrance models (Thompson and Heath, 1999). Again, the distinct nodes g_i have independent allelic types, (say) type $\mathcal{A}(g_i)$ with probabilities $q(\mathcal{A}(g_i))$. The allelic types, $\mathcal{A}_{n,1}$ and $\mathcal{A}_{n,2}$, of the two genome labels of an observed individual n determine his genotype, and hence probabilities of his observed data Y_n, $P(Y_n \mid \mathcal{A}_{n,1}, \mathcal{A}_{n,2})$, independently for each n:

$$P(\mathbf{Y} \mid ibd) = \sum_{\mathcal{A}(\mathbf{g})} \left(\prod_n P(Y_n \mid \mathcal{A}_{n,1}, \mathcal{A}_{n,2}) \right) \left(\prod_i q(\mathcal{A}(g_i)) \right). \tag{13.6}$$

The parallel between Equation 13.6 and Equations 13.1 and 13.3 is clear, and computation proceeds through the graph as for any graphical model. Since components of the distinct genome label graph are generally small, this computation is much easier than computing on the pedigree structure, where summation over unobserved types is required. Details are given by Thompson (2005).

13.4 MCMC Sampling of Latent Variables

13.4.1 Genotypes and Meioses

As in many areas of application of MCMC the earliest samplers involved single-site updates, either of the genotype of one individual (Sheehan, 2000), or of a single meiosis indicator (Thompson, 1994). However, many genotypic configurations on a pedigree have zero probability, since individuals must receive an allele from each parent. Likewise, given discrete data observations on individuals, inheritance vectors may be tightly constrained, and changes to a single component alone impossible. Thus, a single-site Gibbs update does not lead to an MCMC sampling process that is irreducible over the space of feasible latent variable configurations.

13.4.2 Some Block Gibbs Samplers

A natural extension is thus to consider block Gibbs samplers (Roberts and Sahu, 1997), and these have proven much more successful. The HMM structure of the data (Figures 13.4 and 13.5) suggests updating the latent genotypes or the inheritance vector $S_{\bullet,j}$ at a given locus, conditional on other latent variables and on the data observations (Heath, 1997). Note that

$$P(S_{\bullet,j} \mid \mathbf{Y}, \{S_{\bullet,k}, k \neq j\}) = P(S_{\bullet,j} \mid Y_{\bullet,j}, S_{\bullet,j-1}, S_{\bullet,j+1})$$

$$\propto P(Y_{\bullet,j} \mid S_{\bullet,j})P(S_{\bullet,j} \mid S_{\bullet,j-1}, S_{\bullet,j+1}). \qquad (13.7)$$

That is, the update involves only the data at locus j, and the inheritance vectors at the two neighboring loci. At a single locus, sampling from $P(S_{\bullet,j} \mid Y_{\bullet,j})$ is accomplished by *reverse peeling* (Ploughman and Boehnke, 1989). Only the transmission probabilities are changed by conditioning also on inheritance vectors at neighboring loci; these depend now on the recombination parameters. Analogously to Equation 13.2,

$$P(S_{i,j} = s \mid S_{i,j-1}, S_{i,j+1}) \propto \rho_{j-1}^{|s-S_{i,j-1}|}(1 - \rho_{j-1})^{(1-|s-S_{i,j-1}|)}\rho_j^{|s-S_{i,j+1}|}(1 - \rho_j)^{(1-|s-S_{i,j+1}|)}.$$

Reverse peeling samples directly from $P(S_{\bullet,j} \mid Y_{\bullet,j})$. Hence, any $S_{\bullet,j}$ consistent with data $Y_{\bullet,j}$ can be sampled. Provided recombination parameters ρ_{j-1} and ρ_j between locus j and its neighbors are strictly positive, conditioning on $S_{\bullet,j-1}$ and $S_{\bullet,j+1}$ does not affect the space of feasible realizations. Thus the full-locus Gibbs sampler (or L-sampler) is irreducible (Heath, 1997), although mixing may be poor if recombination parameters are small. Typically, an L-sampler scan is performed, resampling in a random order each $S_{\bullet,j}$ for each $j \in 1, 2, \ldots, l$ from its full conditional given $\{S_{\bullet,k}, k \neq j\}$ and $\mathbf{Y}$.

In contrast, the meiosis sampler (or M-sampler) updates jointly all components of the meiosis vector $S_{i,\bullet}$, which is easily accomplished by applying the Baum algorithm to the two-state HMM of $S_{i,j}, j = 1, \ldots, l$, keeping the remainder of $\mathbf{S}$ ($S_{k,\bullet}, k \neq i$) fixed (Thompson and Heath, 1999). Typically, an M-sampler scan is performed, resampling in a random order each $S_{i,\bullet}$ for each $i \in 1, 2, \ldots, m$ from its full conditional given $\{S_{k,\bullet}, k \neq i\}$ and $\mathbf{Y}$. Since resampling is jointly overall all loci, the M-sampler is not directly affected by small recombination parameters. However, the feasible space of meioses is often tightly constrained by data, and the M-sampler is often not irreducible.

The LM-sampler (Thompson and Heath, 1999) is a combination of L-sampler and M-sampler that performs markedly better than either. At each stage, a random choice is made to perform either an L-sampler update or an M-sampler update (see Figure 13.5). Again, typically updates are by scan, so the choice is of L-sampler updates for every locus in a random order, or of M-sampler updates of every meiosis in a random order. Since the L-sampler is irreducible, so is the LM-sampler, while incorporating M-sampler updates greatly improves mixing performance.

13.4.3 Gibbs Updates and Restricted Updates on Larger Blocks

In principle there is no reason why a block Gibbs MCMC update should be restricted to a single locus or to a single meiosis. When each locus has only two alleles (e.g. SNP markers), joint updating of the inheritance vectors $S_{\bullet,j}$ over several such loci j is feasible, although the practical benefits are unclear. More useful is the joint updating of several meiosis vectors $S_{i,\bullet}$. Here again the algorithm is a direct application of the forwards–backwards Baum algorithm (Baum et al., 1970) to some subset of meioses $i \in I$. As in Section 13.3.2, forward computation provides

$$R_j^*(s(I)) = P(Y^{*(j)}, S_{\bullet,j}(I) = s(I))$$

for each $j = 1, 2, \ldots, l$. Reversing the procedure, we have

$$P(S_{\bullet,j}(I) = s \mid \mathbf{Y}, S_{\bullet,k}, k = j+1, \ldots, l) = P(S_{\bullet,j}(I) = s \mid Y^{*j}, S_{\bullet,j+1}(I))$$

$$\propto R_j^*(s(I)) \prod_{i \in I} \rho_j^{|s_i - S_{i,j+1}|}(1 - \rho_j)^{1 - |s_i - S_{i,j+1}|}.$$

Hence, forward computation and backward resampling is feasible provided only the set of meioses in I is small enough for the HMM computations to be feasible in an MCMC framework. Tong and Thompson (2008) have implemented and tested a variety of multiple-meiosis Gibbs proposals. Updating jointly the meioses from a given individual, from a given parent couple, or within a small three-generation subset of a pedigree, can greatly improve mixing.

In some cases, the number of meioses in a nuclear family or three-generation subset of a pedigree can exceed practical bounds for a full Gibbs update. In this case, sampling within a restricted set of updates provides an alternative approach. For example, some updates considered by Thomas et al. (2000) include proposals to switch the binary indicators of all the maternal and/or paternal meioses of all offspring in a nuclear family. In this case, at a given locus, there are just four alternatives to be considered, including the current state. Formally, this is most easily considered as an auxiliary variable problem (Tong and Thompson, 2008). Given a current state $\mathbf{S} = \{S_{\bullet,j}, j = 1, \ldots, l\}$, define $X_j = 0, 1, 2, 3$ to indicate each of the possible alternative states at locus j. Since, given $\mathbf{S}$, X_j is a one-to-one function of $S_{\bullet,j}$, it retains the Markov structure of the inheritance vectors $S_{\bullet,j}$ over loci j. Since X_j has only four states, a full Gibbs update of $\mathbf{X} = (X_1, \ldots, X_l)$ is easily accomplished via the forwards–backwards Baum algorithm, and translates to an update of $\mathbf{S}$ within the restricted space of alternatives. Other restricted updates designed to improve mixing particularly in pedigrees with large sibships may be considered similarly (Tong and Thompson, 2008). While these proposals greatly improve mixing, it is important to recognize that only the L-sampler ensures irreducibility; all samplers must include some proportion of L-sampler steps.

13.5 MCMC Sampling of Inheritance Given Marker Data

13.5.1 Sampling Inheritance Conditional on Marker Data

We have seen in Section 13.3.4 that computation of data probabilities, and hence also MCMC sampling, is more straightforward for genetic markers than the analogous computations for trait data. As informative marker data become increasingly available, and trait models increasingly complex, basing MCMC realizations of latent inheritance only on marker data, and then using these realizations in multiple trait analyses and/or for multiple trait models becomes the approach of choice. In this section we therefore focus on this marker-based MCMC, and return to trait data analyses in Section 13.6.

13.5.2 Monte Carlo EM and Likelihood Ratio Estimation

While models for genetic marker data are generally more straightforward than are trait models, there are still unknown parameters of the meiosis and population marker models. Although MCMC-based EM algorithms have been more widely used in genetic analyses (Guo and Thompson, 1994), we will focus here on the estimation of genetic maps, or, equivalently, the recombination parameters ρ_j between marker j and $j+1, j = 1, \ldots, l-1$. Again, estimation for sex-specific recombination parameters is no more complex, but for notational convenience we restrict here to a single vector of recombination parameters.

In the EM framework, the *complete data* consist of latent variables $\mathbf{S}_M$ and observed marker data $\mathbf{Y}_M$, and (see Equation 13.3) the *complete-data log likelihood* is

$$\log P_\rho(\mathbf{Y}_M, \mathbf{S}_M) = \log P(\mathbf{Y}_M \mid \mathbf{S}_M) + \log P_\rho(\mathbf{S}_M)$$

$$= \sum_{j=1}^{l} \log P(Y_{\bullet,j} \mid S_{\bullet,j}) + \log P(S_{\bullet,1}) + \sum_{j=2}^{l} \log P(S_{\bullet,j} \mid S_{\bullet,j-1}).$$

Note that the recombination parameter ρ_j appears only in the term

$$\log P(S_{\bullet,j+1} \mid S_{\bullet,j}) = \sum_{i=1}^{m} \log P(S_{i,j+1} \mid S_{i,j})$$

$$= \left(\sum_{i=1}^{m} |S_{i,j+1} - S_{i,j}| \right) \log \rho_j + \left(m - \left(\sum_{i=1}^{m} |S_{i,j+1} - S_{i,j}| \right) \right) \log(1 - \rho_j)$$

(see Equation 13.2). In the m meioses, for each marker interval $j = 1, \ldots, m$, the E-step of the EM algorithm thus requires only computation of the expected number of recombinants, at the current recombination frequency vector ρ and conditionally on $\mathbf{Y}_M$:

$$E_\rho \left(| S_{i,j+1} - S_{i,j} | \mid \mathbf{Y}_M \right) = P_\rho(S_{i,j+1} \neq S_{i,j} \mid \mathbf{Y}_M).$$

The M-step then updates each ρ_j to $\sum_i P_\rho(S_{i,j+1} \neq S_{i,j} \mid \mathbf{Y}_M)/m$.

On very small pedigrees the E-step may be performed exactly. A slight generalization of Equation 13.4 gives

$$P(S_{\bullet,j}, S_{\bullet,j+1} \mid \mathbf{Y}) = R_{(j)}^*(s)P(Y_{\bullet,j+1} \mid S_{\bullet,j+1})P(S_{\bullet,j+1} \mid S_{\bullet,j})R_{(j+1)}^\dagger(s)/P(\mathbf{Y})$$

However, unless the number of meioses m is very small, use of this bivariate distribution of inheritance vectors at two loci is impractical. On the other hand, MCMC procedures provide realizations of $\mathbf{S}$ conditional on $\mathbf{Y}_M$ and hence straightforward Monte Carlo estimates of the conditional expected recombination counts. This Monte Carlo EM (MCEM) is very easily implemented, but a disadvantage is that MCMC sampling is repeated with each update of the vector of recombination parameters ρ. This can be very slow, particularly close to the maximum likelihood estimate, where very large sample sizes are required for accurate estimation of the recombination counts. Stewart and Thompson (2006) therefore propose a more general estimation and testing framework for genetic maps using MCEM only for the earlier iterates and stochastic approximation (Gu and Kong, 1998; Robbins and Monro, 1951) in the later stages.

In estimation of genetic maps, or more generally any genetic model Γ, it is important to be able to explore the local likelihood surface around a final estimate, or the variation of that local surface with changing values of nuisance parameters. The use of MCMC-based local likelihood ratio estimates (Thompson and Guo, 1991) provides a practical approach:

$$
\begin{aligned}
\frac{L(\Gamma^*)}{L(\Gamma)} = \frac{P(\mathbf{Y}; \Gamma^*)}{P(\mathbf{Y}; \Gamma)} &= E\left(\frac{P(\mathbf{Y}, \mathbf{S}; \Gamma^*)}{P(\mathbf{Y}, \mathbf{S}; \Gamma)} \middle| \mathbf{Y}; \Gamma \right) \\
&= E\left(\frac{P(\mathbf{Y} \mid \mathbf{S}; \Gamma^*) P(\mathbf{S}; \Gamma^*)}{P(\mathbf{Y} \mid \mathbf{S}; \Gamma) P(\mathbf{S}; \Gamma)} \middle| \mathbf{Y}, \Gamma \right),
\end{aligned}
\tag{13.8}
$$

where the expectation is over the distribution of latent variables $\mathbf{S}$ conditional on data $\mathbf{Y}$ under model Γ. That is, this single MCMC sample of $\mathbf{S}$ provides an estimate of the entire local likelihood surface $L(\Gamma^*)$ normalized by $L(\Gamma)$.

Again, for simplicity, we consider only the example of genetic marker data $\mathbf{Y}_M$, the related inheritance at marker loci $\mathbf{S}_M$ and genetic marker models Γ_M. For example, we might first wish to explore the likelihood surface for recombination frequencies in the neighborhood of an estimated map. In this case Γ and Γ^* differ only in these recombination frequencies and $P(\mathbf{Y} \mid \mathbf{S}; \Gamma^*) = P(\mathbf{Y} \mid \mathbf{S}; \Gamma)$. Furthermore, the ratio $P(\mathbf{S}; \Gamma^*)/P(\mathbf{S}; \Gamma)$ takes the simple form

$$
\begin{aligned}
\frac{P(\mathbf{S}; \Gamma^*)}{P(\mathbf{S}; \Gamma)} &= \prod_{j=2}^{l} \frac{P(S_{\bullet,j} \mid S_{\bullet,j-1}; \Gamma^*)}{P(S_{\bullet,j} \mid S_{\bullet,j-1}; \Gamma)} \\
&= \prod_{i=1}^{m} \prod_{j=1}^{l-1} \left(\frac{\rho_j^*}{\rho_j} \right)^{|S_{i,j+1} - S_{i,j}|} \left(\frac{1 - \rho_j^*}{1 - \rho_j} \right)^{1 - |S_{i,j+1} - S_{i,j}|}.
\end{aligned}
\tag{13.9}
$$

Thus computation of the likelihood ratio $L(\Gamma^*)/L(\Gamma)$ is very easily and efficiently accomplished.

Alternatively, we might wish to explore the sensitivity of the estimated likelihood to alternate assumptions about marker allele frequencies. In this case $P(\mathbf{S}; \Gamma^*) = P(\mathbf{S}; \Gamma)$, and the ratio $P(\mathbf{Y} \mid \mathbf{S}; \Gamma^*)/P(\mathbf{Y} \mid \mathbf{S}; \Gamma)$ is very easily computed, since the _ibd_-graph is the determined by $\mathbf{S}$ (Section 13.3.4). In fact, in many cases the ratio is simply a product of the ratio of allele frequencies under Γ^* and Γ (Equation 13.5), but can be slightly more complex where different allelic assignments are possible under a particular sampled $\mathbf{S}$.

13.5.3 Importance Sampling Reweighting

Importance sampling reweighting is a key tool in investigating the effect of alternative models on MCMC realizations, and for using these realizations under a variety of alternate model assumptions. Suppose that we have realizations of $\mathbf{S}_M$ conditional on marker data $\mathbf{Y}_M$ under a model Γ_M, but wish to consider an alternate model Γ_M^*:

$$
\begin{aligned}
\mathrm{E}(g(\mathbf{S}_M) \mid \mathbf{Y}_M; \Gamma_M^*) &= \sum_{\mathbf{S}_M} g(\mathbf{S}_M) \mathrm{P}(\mathbf{S}_M \mid \mathbf{Y}_M; \Gamma_M^*) \\
&= \sum_{\mathbf{S}_M} g(\mathbf{S}_M) \frac{\mathrm{P}(\mathbf{S}_M \mid \mathbf{Y}_M; \Gamma_M^*)}{\mathrm{P}(\mathbf{S}_M \mid \mathbf{Y}_M; \Gamma_M)} \mathrm{P}(\mathbf{S}_M \mid \mathbf{Y}_M; \Gamma_M) \\
&= \mathrm{E}(g(\mathbf{S}_M) \frac{\mathrm{P}(\mathbf{S}_M \mid \mathbf{Y}_M; \Gamma_M^*)}{\mathrm{P}(\mathbf{S}_M \mid \mathbf{Y}_M; \Gamma_M)} \mid \mathbf{Y}_M; \Gamma_M).
\end{aligned}
$$

That is, realizations $\mathbf{S}_M$ sampled conditional on $\mathbf{Y}_M$ under model Γ_M must be reweighted by a factor

$$
\frac{\mathrm{P}(\mathbf{S}_M \mid \mathbf{Y}_M; \Gamma_M^*)}{\mathrm{P}(\mathbf{S}_M \mid \mathbf{Y}_M; \Gamma_M)} = \frac{\mathrm{P}(\mathbf{Y}_M \mid \mathbf{S}_M; \Gamma_M^*)}{\mathrm{P}(\mathbf{Y}_M \mid \mathbf{S}_M; \Gamma_M)} \frac{\mathrm{P}(\mathbf{S}_M; \Gamma_M^*)}{\mathrm{P}(\mathbf{S}_M; \Gamma_M)} \frac{\mathrm{P}(\mathbf{Y}_M; \Gamma_M)}{\mathrm{P}(\mathbf{Y}_M; \Gamma_M^*)}.
$$

Note also, from Equation 13.8, that

$$
\frac{\mathrm{P}(\mathbf{Y}_M; \Gamma_M^*)}{\mathrm{P}(\mathbf{Y}_M; \Gamma_M)} = \mathrm{E}\left(\frac{\mathrm{P}(\mathbf{Y}_M \mid \mathbf{S}_M; \Gamma_M^*)}{\mathrm{P}(\mathbf{Y}_M \mid \mathbf{S}_M; \Gamma_M)} \frac{\mathrm{P}(\mathbf{S}_M; \Gamma_M^*)}{\mathrm{P}(\mathbf{S}_M; \Gamma_M)} \,\middle|\, \mathbf{Y}_M; \Gamma_M \right).
$$

That is, the relative weights

$$
\frac{\mathrm{P}(\mathbf{Y}_M \mid \mathbf{S}_M; \Gamma_M^*)}{\mathrm{P}(\mathbf{Y}_M \mid \mathbf{S}_M; \Gamma_M)} \frac{\mathrm{P}(\mathbf{S}_M; \Gamma_M^*)}{\mathrm{P}(\mathbf{S}_M; \Gamma_M)}
$$

may be simply normalized by their sum.

As in Section 13.5.2, we will consider just two examples, one relating to parameters of the distribution of $\mathbf{S}_M$ and the other to parameters of $\mathrm{P}(\mathbf{Y} \mid \mathbf{S})$. Both can provide substantial computational savings in genetic analyses involving multiple linked marker loci. In general, reweighting is a powerful tool in analyzing data under more complex models for which direct MCMC sampling is impractical, and Gibbs samplers infeasible.

For any model modification relating to the meiosis process,

$$
\mathrm{P}(\mathbf{Y}_M \mid \mathbf{S}_M; \Gamma_M^*) = \mathrm{P}(\mathbf{Y}_M \mid \mathbf{S}_M; \Gamma_M).
$$

Under Mendel's first law, meioses are independent, and only

$$
\frac{\mathrm{P}(\mathbf{S}; \Gamma_M^*)}{\mathrm{P}(\mathbf{S}; \Gamma)} = \prod_{i=1}^{m} \frac{\mathrm{P}(S_{i,\bullet}; \Gamma_M^*)}{\mathrm{P}(S_{i,\bullet}; \Gamma)}
$$

need be computed. As in Equation 13.9, clearly we can reweight to alternative genetic maps. For example, we may do the initial MCMC assuming equal recombination frequencies in male and female meioses, but then wish to investigate more carefully in a region of the genome where these recombination frequencies differ. Provided the probabilities of realized $\mathbf{S}_M$ do not differ too greatly, the previous MCMC samples may be reweighted, avoiding additional MCMC. As in any reweighting scheme, the distribution of the weights can be used to assess the increase in Monte Carlo variance due to reweighting. More generally, this approach may be used to assess deviations from the assumption of no genetic interference (Thompson, 2000); that is, $S_{i,j}$ are no longer Markov over loci j. While the Markov structure of the components of $S_{i,\bullet}$ greatly facilitates MCMC, reweighting permits analysis under any model for which $P(S_{i,\bullet})$ is computable.

Likewise, analogous to the example in Section 13.5.2, we may reweight to alternative allele frequencies. Then $P(\mathbf{Y}_M \mid \mathbf{S}_M; \Gamma_M^*) \neq P(\mathbf{Y}_M \mid \mathbf{S}_M; \Gamma_M)$, but the ratio of these probabilities is easily computed. However, as in the case of meiosis models, the approach is more general, permitting, at least to a limited degree, the incorporation of *allelic association* or linkage disequilibrium (LD) among loci. LD is normally the result of population history, and is maintained by very tight linkage $\rho \approx 0$. This results, at the population level, in association of allelic types at different loci along a founder haplotype. Just as the Markov structure of inheritance vectors $S_{\bullet,j}$ over loci j is essential to effective block Gibbs samplers, so also is the conditional independence of data $Y_{\bullet,j}$ given $S_{\bullet,j}$ (Equation 13.3). Thus LD cannot be directly incorporated into MCMC. However, since LD requires $\rho \approx 0$, realized $S_{\bullet,j}$ and hence *ibd*-graphs are almost always constant across loci in LD. Then reweighting is straightforward; the product of allele frequencies across loci that is assumed in the MCMC is adjusted to the haplotype frequencies that actually obtain in the population. Even where a realized recombination event does change the *ibd*-graph for observed pedigree members within a region of LD, reweighting is still possible, although it becomes less straightforward since then the assignment of haplotypes across both *ibd*-graphs must be considered.

13.6 Using MCMC Realizations for Complex Trait Inference

13.6.1 Estimating a Likelihood Ratio or lod Score

The MCMC sampling methods of Section 13.4 may be applied to genetic loci of any kind, whether genetic markers at which genotypes of individuals are available, or to trait loci where there is a more complex penetrance relationship between trait data and latent genotypes. Where the trait model and data together provide strong information on latent inheritance patterns, an MCMC sampling procedure that incorporates trait data can provide more accurate results with greater computational efficiency. One such set of procedures are those developed by George and Thompson (2003). For much modern data, however, marker genotype information proliferates, while the traits of interest are complex, and have no clear inheritance pattern. In such cases, MCMC-based samples of latent inheritance conditioned on marker data, with subsequent analyses of trait data conditional on these inheritances, is more computationally efficient and practically feasible.

We consider first the classic statistics used to detect genetic linkage of a trait to a given region of the genome in which data on genetic markers are available. The data consist of both marker data and trait data, $\mathbf{Y} = (\mathbf{Y}_M, \mathbf{Y}_T)$, and the full model is now indexed by parameter $\xi = (\beta, \gamma, \Gamma_M)$. Here Γ_M denotes all parameters relating to the markers, principally their allele frequencies, their order along the chromosome, and the recombination frequencies between adjacent marker loci. The parameter β denotes all parameters relating underlying inheritance at a causal locus to observable trait data, principally penetrance parameters and allele frequencies at loci contributing to the trait. The trait locus location γ is the parameter of interest: $\gamma = \infty$ implies absence of linkage of the trait to these markers. The statistical approach taken is then to compute a likelihood and hence a *location lod score*:

$$\text{lod}(\gamma) = \log_{10}\left(\frac{P(\mathbf{Y}; \Gamma_M, \beta, \gamma)}{P(\mathbf{Y}; \Gamma_M, \beta, \gamma = \infty)}\right). \tag{13.10}$$

Note that the lod score is simply a log-likelihood difference, although traditionally in this area logs to base 10 are used rather than natural logarithms. More importantly, note that the models in numerator and denominator differ only in γ. The likelihood of a particular location γ is compared to the likelihood of no linkage ($\gamma = \infty$), under the *same* trait model (β) and marker model (Γ_M).

Now

$$P(\mathbf{Y}; \Gamma_M, \beta, \gamma) = P(Y_T \mid \mathbf{Y}_M; \Gamma_M, \beta, \gamma)P(\mathbf{Y}_M; \Gamma_M)$$

and, when $\gamma = \infty$, Y_T and $\mathbf{Y}_M$ are independent. Thus Equation 13.10 reduces to

$$\text{lod}(\gamma) = \log_{10}\left(\frac{P(Y_T \mid \mathbf{Y}_M; \Gamma_M, \beta, \gamma)}{P(Y_T; \beta)}\right).$$

Finally,

$$P(Y_T \mid \mathbf{Y}_M; \Gamma_M, \beta, \gamma) = \sum_{\mathbf{S}_M} P(Y_T \mid \mathbf{S}_M; \beta, \gamma)P(\mathbf{S}_M \mid \mathbf{Y}_M; \Gamma_M)$$

$$= E_{\Gamma_M}\left(P_{\beta,\gamma}(Y_T \mid \mathbf{S}_M) \mid \mathbf{Y}_M\right), \tag{13.11}$$

where $\mathbf{S}_M$ denotes the inheritance vectors at all marker locations.

Equation 13.11 suggests the MCMC approach first proposed by Lange and Sobel (1991) to first sample $\mathbf{S}_M$ conditionally on marker data $\mathbf{Y}_M$ and then use exact computation to compute $P_{\beta,\gamma}(Y_T \mid \mathbf{S}_M)$ for choices of β and γ. As in Equation 13.7, this computation is directly analogous to the pedigree-peeling computation of the marginal probability $P_\beta(Y_T)$, with only the transmission probabilities being modified to condition on inheritance vectors $S_{\bullet j}$ and $S_{\bullet j'}$ at marker loci j and j' flanking the position(s) γ of hypothesized causal loci. A major advantage of this approach is that MCMC need be performed once only, to generate a large sample of $\mathbf{S}_M$ which can be used to estimate $\text{lod}(\gamma)$ for a variety of γ and under a variety of trait models β.

With increasing density of marker data over the genome, this approach becomes even more effective. In part, this is because of the large amounts of marker data that must be incorporated into the MCMC, and the long MCMC runs required for adequate mixing of the MCMC with dense marker data. In addition, with inheritance sampled at locations dense in the genome, normally only these locations need be considered as potential causal trait-locus locations, and trait analyses may therefore be done directly on the MCMC sample of *ibd*-graphs at these locations (Section 13.3.4).

Due to the conditional independence structure of inheritance patterns $\mathbf{S} = \{S_{i,j}\}$, MCMC conditional on marker data can only be efficiently performed at the level of $\mathbf{S}_M$. Define an individual to be "observed" (O_M) if there are any marker or trait data for that individual. Since the *ibd*-graph at any location j is a deterministic function of the inheritance vector

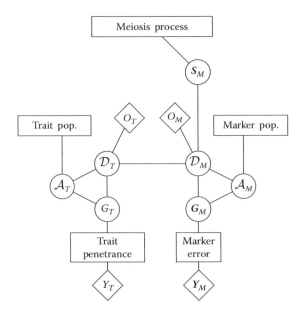

FIGURE 13.7
The structure of genetic data with modern dense SNP data. For details, see text.

$S_{\bullet,j}$ and the set of observed individuals, this MCMC sampling will provide a set of distinct *ibd*-graphs $\mathcal{D}_M$ and their sampling counts (Figure 13.7). This is a substantial reduction in the relevant output, first since many different inheritance patterns give rise to the same *ibd*-graph, but also since *ibd*-graphs remain constant across regions encompassing many marker locations. At the location of any marker, the *ibd*-graph for trait analysis, $\mathcal{D}_T$, is a subgraph of $\mathcal{D}_M$ at that location, since now only individuals $\mathcal{O}_T$ observed for the trait need be considered. Since, provided γ is a marker location, $P_{\beta,\gamma}(Y_T \mid \mathbf{S}_M) = P_\beta(Y_T \mid \mathcal{D}_\gamma)$, lod score estimation (Equation 13.11), can be carried out entirely in the *ibd*-graph framework (Section 13.3.4).

13.6.2 Uncertainty in Inheritance and Tests for Linkage Detection

The lod score is estimated via the log of the average of contributions $W_\gamma(\mathbf{S}_M) = P_{\beta,\gamma}(Y_T \mid \mathbf{S}_M)$, where $\mathbf{S}_M$ are realized conditional on $\mathbf{Y}_M$ (Equation 13.11). Typically, the lod score is plotted as a function of the hypothesized trait locus location γ. However, there is information also in the distribution of the contributions, $W(\mathbf{S}_M)$. They provide estimates of the Monte Carlo standard error, for example by using batch means (Glynn and Whitt, 1991). More simply, a plot of the 10th and 90th quantiles of the lod-score contributions, along with the lod-score curve, provides a clear visualization of positions γ of high uncertainty. For example, where the estimate is dominated by a few extreme values the estimate can lie well above the 90th quantile.

This leads naturally to the assessment of uncertainty via latent p-values (Thompson and Geyer, 2007), which is based on the earlier proposal of fuzzy p-values by Geyer and Meeden (2005). Tests for detection and localization of causal trait loci typically condition on the trait data Y_T. Since our tests will condition on trait data Y_T, we will drop explicit use of Y_T and the marker subscript M on the latent variables. Then, given any test statistic $W(\mathbf{S})$, and realizations $\mathbf{S}_\infty$ under the null hypothesis $\gamma = \infty$,

$$\pi(\mathbf{S}) = \mathrm{P}(W(\mathbf{S}_\infty) \geq W(\mathbf{S}) \mid \mathbf{S}) = \mathrm{P}(W(\mathbf{S}_\infty) \geq W(\mathbf{S}) \mid \mathbf{S}, \mathbf{Y}_M)$$

is a latent p-value, having a uniform $U(0,1)$ distribution under the null hypothesis (Thompson and Geyer, 2007). Let $\mathbf{S}_\infty^{(h)}$, $h = 1, \ldots, m \sim \mathrm{P}_\infty$, and $\mathbf{S}^{(k)}$, $k = 1, \ldots, n \sim \mathrm{P}(\cdot \mid \mathbf{Y}_M)$. Then, for each $k = 1, \ldots, n$,

$$\eta(\mathbf{S}^{(k)}, \mathbf{Y}_M) = \mathrm{P}(W(\mathbf{S}_\infty) \geq W(\mathbf{S}^{(k)}) \mid \mathbf{S}^{(k)}, \mathbf{Y}_M),$$

estimated by

$$m^{-1} \sum_{h=1}^{m} I(W(\mathbf{S}_\infty^{(h)}) \geq W(\mathbf{S}^{(k)})),$$

is a realization from the latent p-value distribution. Strictly, this is only so if the random variables $W(\mathbf{S})$ are continuous. Thompson and Geyer (2007) show how discreteness can be simply dealt with to provide, under H_∞, an exact $U(0,1)$ distribution over data sets $\mathbf{Y}_M$.

The approach of the previous paragraph leads to tests for linkage detection using the (latent) lod score $\log(\mathrm{P}_{\beta,\gamma}(Y_T \mid \mathbf{S}) / \mathrm{P}_\beta(Y_T))$, at any particular position γ. If γ is the position of marker locus j, then this test statistic is a function only of $S_{\bullet j}$. More importantly, the procedure can be applied equally easily to any omnibus test statistics such as the maximum lod score. While the maximum lod score has been used for over 50 years as a test statistic providing evidence of linkage (Smith, 1953), formal p-value evaluation is seldom performed, due to the lack of distributional theory for general pedigrees and hence need for extensive resimulation of marker data $\mathbf{Y}_M$. The use of a latent p-value requires only realization of latent $\mathbf{S}$ conditional on the observed $\mathbf{Y}_M$.

The latent p-value approach can be equally applied to any linkage detection test statistic that is a function of $\mathbf{S}$. The null hypothesis is again $\gamma = \infty$, or independence of trait data Y_T and inheritance $\mathbf{S}$ and marker loci. There are a wide variety of such statistics, $W(\mathbf{S})$; see, for example, McPeek (1999). Typically the approach has been to compute $\mathrm{E}(W(\mathbf{S}) \mid \mathbf{Y}_M)$ and to compare the value with the null distribution (Whittemore and Halpern, 1994). As in the lod-score case, use of the latent p-value approach avoids the need for distributional approximations or extensive simulation. Recently, Di and Thompson (2009) have used this approach to provide "marginal" tests for linkage detection, testing the null hypothesis $\gamma = \infty$ of no causal locus on a chromosome.

13.6.3 Localization of Causal Loci Using Latent *p*-Values

Although test statistics such as the lod score are formally tests of linkage detection, with null hypothesis $\gamma = \infty$, they are often used as evidence for localization of trait loci, for example by selecting the position γ where the test statistic is most extreme. However, using the latent p-value approach, confidence intervals and formal localization tests are possible.

We consider first the classical confidence interval approach. That is, a value γ_0 is in a $(1 - \alpha)$-level confidence set if a test size α fails to reject the hypothesis $\gamma = \gamma_0$. The latent p-value approach can be used to provide randomized confidence sets for the location of a causal locus (Thompson, 2008). A lod-score-based latent test statistic $W(\mathbf{S}; \gamma_0)$ for testing $\gamma = \gamma_0$ may be constructed for any γ_0, and the resulting randomized p-value inverted to provide a randomized confidence interval for γ (Geyer and Meeden, 2005). We obtain the latent p-value $\pi^*_{\gamma^*}(\mathbf{S})$ by realizing $\mathbf{S}$ under $\gamma = \gamma^*$ conditional on Y_T, and obtain realizations from the conditional distribution of $\pi^*_{\gamma^*}(\mathbf{S})$ by MCMC realization of $\mathbf{S}$ under $\gamma = \gamma^*$ given $\mathbf{Y}_M$ and Y_T. This procedure is also computationally intensive, since MCMC at each γ^* is required. Further, since sampling is conditional on trait data Y_T, the nuisance parameters β enter into the procedure, and values must be assumed.

An alternative approach to localization has been taken by Di and Thompson (2009). The null hypothesis considered is that there is no causal locus in an interval (γ_ℓ, γ_r). Rejection of the null hypothesis localizes some causal gene(s) to the interval. Unlike the confidence interval approach, the conditional test does not require specification of the number or effects of causal genes, and the validity of the test does not rely on such model assumptions. The test rests on the result that, if there no causal locus in (γ_ℓ, γ_r), then at any point j within the interval the inheritance vector $S_{\bullet,j}$ given $S_{\bullet,\ell}$ and $S_{\bullet,r}$ follows the conditional null inheritance distribution regardless of data Y_T. As before, test statistics $W(S_{\bullet,j})$ are functions of the inheritance vector at j and the trait data, and the null Markov chain distribution of $S_{\bullet,j}$ given $S_{\bullet,\ell}$ and $S_{\bullet,r}$ provides the null distribution of $W(S_{\bullet,j})$. If these variables were observable, a p-value $\pi(\mathbf{S})$ would result. In fact, as before, $\mathbf{S}$ is latent, but can be sampled by MCMC, conditionally on $\mathbf{Y}_M$, providing a probability distribution for $\pi(\mathbf{S})$ given $\mathbf{Y}_M$. This probability distribution is again a latent p-value, and a randomized test follows. Although a single MCMC run can provide a sample of realizations of $\mathbf{S}$ and hence of all required pairs (γ_ℓ, γ_r), the conditional testing procedure is computationally intensive, requiring Monte Carlo realization of $S_{\bullet,j}$ conditional on each MCMC realization.

13.7 Summary

In this chapter we have focused less on the mechanics of MCMC samplers, and more on the need for MCMC in the analyses of genetic data on related individuals, and on the uses that can be made of the realizations in making inferences from these data.

We first explore the structure of genetic data on pedigrees, leading to three complementary sets of latent variables at each genetic locus j. These are the genotypes of individuals, $G_{\bullet,j}$, the inheritance vector at the locus, $S_{\bullet,j}$, and the allelic types of founder DNA, $\mathcal{A}_j$, with $G_{\bullet,j}$ being a function of $S_{\bullet,j}$ and $\mathcal{A}_j$. We explore the limits of exact computation of likelihoods, and then show how these latent variables are the useful targets of MCMC simulation when exact computation is infeasible.

With modern genetic data, marker data are plentiful and informative, while the traits requiring analysis are often complex. It has therefore become the paradigm of choice to sample latent variables conditionally only on marker data, and then to use these realizations in multiple trait and/or trait-model analyses. In Section 13.5 we therefore focus first on questions involving only marker-based sampling, such as the estimation of genetic maps. This section emphasizes the key role of importance sampling reweighting of MCMC realizations. This enables a variety of marker models to be explored based on a single set of MCMC realizations, and to use marker models for which the full conditionals are not available.

Finally, in Section 13.6, we turn to the use of MCMC realizations of latent variables in the analysis of trait data, considering both linkage detection and linkage localization. We describe recent work in this area, including the use of the probability distribution of latent p-values to express both the significance of a result and the degree of uncertainty about that significance, due to uncertainty in the latent variables. This work is recent, and it remains to be seen which methods will stand and which will be superseded.

Acknowledgment

This work was supported in part by National Institutes of Health Grant GM 46255.

References

Baum, L. E. 1972. An inequality and associated maximization technique in statistical estimation for probabilistic functions on Markov processes. In O. Shisha (ed.), *Inequalities-III: Proceedings of the Third Symposium on Inequalities. University of California Los Angeles, 1969*, pp. 1–8. Academic Press, New York.

Baum, L. E., Petrie, T., Soules, G., and Weiss, N. 1970. A maximization technique occurring in the statistical analysis of probabilistic functions on Markov chains. *Annals of Mathematical Statistics*, 41:164–171.

Besag, J. and Green, P. J. 1993. Spatial statistics and Bayesian computation. *Journal of the Royal Statistical Society, Series B*, 55:25–37.

Browning, S. R. and Browning, B. L. 2007. Rapid and accurate haplotype phasing and missing-data inference for whole-genome association studies by use of localized haplotype clustering. *American Journal of Human Genetics*, 81:1084–1097.

Cannings, C., Thompson, E. A., and Skolnick, M. H. 1978. Probability functions on complex pedigrees. *Advances of Applied Probability*, 10:26–61.

Cannings, C., Thompson, E. A., and Skolnick, M. H. 1980. Pedigree analysis of complex models. In J. Mielke and M. Crawford (eds), *Current Developments in Anthropological Genetics*, pp. 251–298. Plenum Press, New York.

Di, Y. and Thompson, E. A. 2009. Conditional tests for linkage localization. *Human Heredity*, 68:139–150.

Elston, R. C. and Stewart, J. 1971. A general model for the analysis of pedigree data. *Human Heredity*, 21:523–542.

Fishelson, M. and Geiger, D. 2004. Optimizing exact linkage computations. *Journal of Computational Biology*, 11:263–275.

Gelfand, A. E. and Smith, A. F. M. 1990. Sampling based approaches to calculating marginal densities. *Journal of the American Statistical Association*, 46:193–227.

George, A. W. and Thompson, E. A. 2003. Multipoint linkage analyses for disease mapping in extended pedigrees: A Markov chain Monte Carlo approach. *Statistical Science*, 18:515–531.

Geyer, C. J. and Meeden, G. D. 2005. Fuzzy and randomized confidence intervals and *p*-values (with discussion). *Statistical Science*, 20:358–387.

Glynn, P. W. and Whitt, W. 1991. Estimating the asymptotic variance with batch means. *Operations Research Letters*, 10:431–435.

Gu, M. G. and Kong, F. H. 1998. A stochastic approximation algorithm with Markov chain Monte Carlo method for incomplete data estimation problems. *Proceedings of the National Academy of Sciences of the USA*, 95:7270–7274.

Guo, S. W. and Thompson, E. A. 1994. Monte Carlo estimation of mixed models for large complex pedigrees. *Biometrics*, 50:417–432.

Haldane, J. B. S. 1919. The combination of linkage values and the calculation of distances between the loci of linked factors. *Journal of Genetics*, 8:229–309.

Heath, S. C. 1997. Markov chain Monte Carlo segregation and linkage analysis for oligogenic models. *American Journal of Human Genetics*, 61(3):748–760.

Kong, A., Gudbjartsson, D. F., Sainz, J., Jonsdottir, G. M., Gudjonsson, S. A., Richardsson, B., Sigurdardottir, S., et al. 2002. A high-resolution recombination map of the human genome. *Nature Genetics*, 31:241–247.

Kruglyak, L., Daly, M. J., Reeve-Daly, M. P., and Lander, E. S. 1996. Parametric and nonparametric linkage analysis: A unified multipoint approach. *American Journal of Human Genetics*, 58(6):1347–1363.

Lander, E. S. and Green, P. 1987. Construction of multilocus genetic linkage maps in humans. *Proceedings of the National Academy of Sciences of the USA*, 84(8):2363–2367.

Lange, K. and Sobel, E. 1991. A random walk method for computing genetic location scores. *American Journal of Human Genetics*, 49:1320–1334.

Lauritzen, S. J. 1992. Propagation of probabilities, means and variances in mixed graphical association models. *Journal of the American Statistical Association*, 87:1098–1108.

Lauritzen, S. L. and Spiegelhalter, D. J. 1988. Local computations with probabilities on graphical structures and their application to expert systems. *Journal of the Royal Statistical Society, Series B*, 50:157–224.

McPeek, M. S. 1999. Optimal allele-sharing statistics for genetic mapping using affected relatives. *Genetic Epidemiology*, 16:225–249.

Ploughman, L. M. and Boehnke, M. 1989. Estimating the power of a proposed linkage study for a complex genetic trait. *American Journal of Human Genetics*, 44:543–551.

Robbins, H. and Monro, S. 1951. A stochastic approximation method. *Annals of Mathematical Statistics*, 22:400–407.

Roberts, G. O. and Sahu, S. K. 1997. Updating schemes, correlation structure, blocking and parameterization for the Gibbs sampler. *Journal of the Royal Statistical Society, Series B*, 59:291–317.

Sheehan, N. A. 2000. On the application of Markov chain Monte Carlo methods to genetic analyses on complex pedigrees. *International Statistical Review*, 68:83–110.

Smith, C. A. B. 1953. Detection of linkage in human genetics. *Journal of the Royal Statistical Society, Series B*, 15:153–192.

Sobel, E. and Lange, K. 1996. Descent graphs in pedigree analysis: Applications to haplotyping, location scores, and marker-sharing statistics. *American Journal of Human Genetics*, 58:1323–1337.

Stewart, W. C. L. and Thompson, E. A. 2006. Improving estimates of genetic maps: A maximum likelihood approach. *Biometrics*, 62:728–734.

Thomas, A., Gutin, A., and Abkevich, V. 2000. Multilocus linkage analysis by blocked Gibbs sampling. *Statistics and Computing*, 10:259–269.

Thompson, E. A. 1994. Monte Carlo estimation of multilocus autozygosity probabilities. In J. Sall and A. Lehman (eds), *Computing Science and Statistics: Proceedings of the 26th Symposium on the Interface*, pp. 498–506. Interface Foundation of North America, Fairfax Station, VA.

Thompson, E. A. 2000. MCMC estimation of multi-locus genome sharing and multipoint gene location scores. *International Statistical Review*, 68:53–73.

Thompson, E. A. 2005. MCMC in the analysis of genetic data on pedigrees. In F. Liang, J.-S. Wang, and W. Kendall (eds), *Markov Chain Monte Carlo: Innovations and Applications*, pp. 183–216. World Scientific, Singapore.

Thompson, E. A. 2007. Linkage analysis. In D. J. Balding, M. Bishop, and C. Cannings (eds), *Handbook of Statistical Genetics* 3rd edn, pp. 1141–1167. Wiley, Chichester.

Thompson, E. A. 2008. Uncertainty in inheritance: assessing linkage evidence. In *Joint Statistical Meetings Proceedings, Salt Lake City 2007*, pp. 3751–3758.

Thompson, E. A. and Geyer, C. J. 2007. Fuzzy p-values in latent variable problems. *Biometrika*, 90:49–60.

Thompson, E. A. and Guo, S. W. 1991. Evaluation of likelihood ratios for complex genetic models. *IMA Journal of Mathematics Applied in Medicine and Biology*, 8(3):149–169.

Thompson, E. A. and Heath, S. C. 1999. Estimation of conditional multilocus gene identity among relatives. In F. Seillier-Moiseiwitsch (ed.), *Statistics in Molecular Biology and Genetics: Selected Proceedings of a 1997 Joint AMS-IMS-SIAM Summer Conference on Statistics in Molecular Biology,*

IMS Lecture Note–Monograph Series Vol. 33, pp. 95–113. Institute of Mathematical Statistics, Hayward, CA.

Tong, L. and Thompson, E. A. 2008. Multilocus lod scores in large pedigrees: Combination of exact and approximate calculations. *Human Heredity*, 65:142–153.

Whittemore, A. and Halpern, J. 1994. A class of tests for linkage using affected pedigree members. *Biometrics*, 50:118–127.

14

An MCMC-Based Analysis of a Multilevel Model for Functional MRI Data

Brian Caffo, DuBois Bowman, Lynn Eberly, and Susan Spear Bassett

14.1 Introduction

Functional neuroimaging technologies have inspired a revolution in the study of brain function and its correlation with behavior, disease, and environment. In these techniques, temporal three-dimensional images of the brain are analyzed to produce a quantitative description of brain function. Such techniques can, among other things, present evidence of localization of brain function within and between subjects. For example, when subjects perform motor tasks in a functional magnetic resonance imaging (fMRI) scanner, such as finger tapping, the analysis typically will present increased regional cerebral blood flow (activation) in the motor cortex. Moreover, these techniques can also provide some information about how areas of the brain connect and communicate. In this chapter we further investigate a novel Markov chain Monte Carlo (MCMC) based analysis of a model from Bowman et al. (2008) that combines activation studies with the study of brain connectivity in a single unified approach.

This idea of localization of brain function underlying fMRI activation studies has a long history, with early attempts from the debunked science of phrenology in the early nineteenth century and later breakthroughs by such luminaries as Broca, Wernicke, and Brodman (see Gazzaniga et al., 2002, for an accessible brief history). Prior to new measurement techniques, studies of brain function and localization were limited to animal studies, or post-mortem evaluation of patients with stroke damage or injuries. However, new measurement techniques, such as fMRI, positron emission tomography (PET), and electroencephalography, allow modern researchers noninvasively to study brain function in human subjects.

In contrast with the study of functional localization, the companion idea of connectivity has a shorter history. Functional connectivity is defined as correlation between remote neurophysiological events (Friston et al., 2007). This idea is based on the principal of functional integration of geographically separated areas of the brain. Such integration is supported by the existence of anatomical connections between cortical areas as well as ones within the cortical sheet (see the discussion in Friston et al., 2007). This neuroanatomical model suggests a hierarchical structure of connectivity that includes correlations within and between areas of functional specialization. Therefore, we use this hierarchical biological model of brain function to explore a multilevel statistical model that simultaneously considers potentially long-range correlations as well as shorter-range ones.

We focus on analyzing fMRI data in particular, though the statistical and computational techniques apply more broadly to other functional neuroimaging modalities. Functional

MRI has its roots in the late nineteenth-century discovery that neuronal activity was accompanied by a localized increase in blood flow (Roy and Sherrington, 1890). More specifically, neuronal activity requires energy, which is supplied by chemical reactions from oxygenated hemoglobin. Therefore, provided a cognitive task is localized, a temporal comparison of blood oxygenation levels when the task is being executed versus when it is not would reveal areas of the brain where neurons are active. This is the principle of blood oxygenation level dependent (BOLD) fMRI (Ogawa et al., 1990). In this technique, a subject in an MRI scanner is asked to perform a task at specific timings while images targeting the BOLD signal are taken in rapid succession, usually one image every 2–3 seconds. Examples of tasks are motor tasks, pressing a button after a visual stimulus, mentally rotating figures and so on. The development of a well-controlled task, or paradigm, that isolates the particular cognitive function of interest is not covered in this chapter.

We focus on using Bayesian multilevel models via MCMC for the analysis of functional neuroimaging studies. We emphasize the analysis of so-called group-level fMRI data. In such studies one is interested in the commonality of activation and connectivity within groups and differences between groups, such as comparing diseased and control subjects.

In the following two subsections, we provide an overview of existing related fMRI research and introduce the data used to illustrate the methods. In Section 14.2, we give details on the processing and first-stage analysis of the data. In Section 14.3, we introduce the multilevel model used for analysis and outline the details of the MCMC procedure. In Section 14.4, we propose novel methods for analyzing and visualizing the output from the Markov chain, including the analysis of voxel means, regional means, and intra- and inter-regional connectivity. We conclude with a discussion.

14.1.1   Literature Review

Traditional inter-group analyses of fMRI data employ a two-stage procedure, where a first stage relates the paradigm to the images and a second stage compares contrast estimates from the first stage across subjects groups. This two-stage process is motivated by classical two-stage procedures for linear mixed effects models (see Verbeke and Molenberghs, 2000) and has the benefit of greatly reducing the amount of data to be considered in the second stage. Two-stage analysis of fMRI is proposed and considered in Beckmann et al. (2003), Friston et al. (1999, 2005), and Holmes and Friston (1998), among others. See also Worsley et al. (2002) for a more formal discussion of two-stage random effects approaches for fMRI data.

Standard methods for analyzing two-stage data ignore spatial dependence and connectivity at the modeling level and instead incorporate the spatial dependence into the analysis of statistical maps created from the models. The map of statistics is assumed to possess a conditionally independent neighborhood structure, typically a Markov random field. Further descriptions of the use of Markov random fields in neuroimaging analysis can be found in Worsley (1994), Cao and Worsley (2001), Worsley and Friston (1995), Worsley et al. (1996), and Friston et al. (2007). These approaches are notable for their speed and general applicability. However, they are also characterized by several issues of concern. First, the random field assumptions are somewhat restrictive, as they do not allow for long-range functional correlations and impose a rigid distributional structure. Moreover, many desirable summary statistics have unknown distributions when the statistical map follows a random field. Therefore, indirect inference is typically used by considering a fairly narrow class of statistics with tractable distributions. Conceptually, this could be combated via simulation from the random field under the null hypothesis. However, a more popular approach uses resampling methods, mostly focused on permutation testing. At the expense of computational complexity, these methods can flexibly handle any test statistic, and make few assumptions on the underlying distribution and correlation structure of the data. Permutation methods, as applied to neuroimaging data, are reviewed in Nichols and Holmes (2002). An example for factorial experiments is given in Suckling and Bullmore (2004). Comparisons between cluster-size permutation tests and random-field intensity tests are given in Hayasaka and Nichols (2003). Despite their numerous benefits, permutation methods are focused on testing and do not offer generative models for the data. That is, unlike model-based methods, permutation testing lacks a formal mechanism for connecting the data to a population.

Multilevel models for inter-group analysis of fMRI data have become increasingly popular. Bowman et al. (2008) gave a Bayesian approach and applied it to both Alzheimer's disease and substance abuse disorder data sets; the model from that article motivates the analysis in this chapter. Bowman and Kilts (2003) give a multilevel model applied to the related area of functional PET imaging. The theory and application of Bayesian models is discussed in Friston et al. (2002a,b). Woolrich et al. (2004) use reference priors for inter-group Bayesian fMRI analysis.

Functional MRI connectivity studies have largely focused on the analysis of resting state data, based on the hypothesis of a default-mode brain network (Biswal et al., 1995; Greicius, 2003). These networks represent functional correlations in brain activity between voxels while resting in the scanner. Xiong et al. (1999) considered such resting-state connectivity between regions and compared results to those motivated by other techniques. Greicius (2004) used independent component analysis to explore resting state connectivity (see Calhoun et al., 2003). Arfanakis et al. (2000) considered connectivity via regional correlations and independent components analysis. However, unlike the previous references, they considered active-state data collected along with an experimental fMRI paradigm, though focused on connectivity results in areas unassociated with the paradigm. Our approach differs drastically from these references, both in terms of the methodology considered and the goal. With regard to methodology, we consider a model-based approach to connectivity and decompose connectivity into both short-range connections and longer-range connections. Moreover, our focus is on connectivity associated with a paradigm, and how this connectivity varies across experimental groups. That is, we consider areas of the brain that act in concert to perform the paradigm, rather than considering a default-mode brain network.

14.1.2 Example Data

The data used in our examples come from a study of subjects at high familial risk for Alzheimer's disease and controls with little familial risk. Alzheimer's disease is a degenerative memory disorder affecting millions of adults in the United States alone (Brookmeyer et al., 1998). Typically, Alzheimer's disease affects adults older than 65 years, though early onset cases do occur. The disease causes dementia, with the most common early symptom being short-term memory loss. Because precursors of the disease, such as mild cognitive impairment, occur well before clinical diagnoses, the study of at-risk individuals yields important information about early disease pathology (Bassett et al., 2006; Fox et al., 2001). Because some of the at-risk subjects will not become eventual cases, and some of the controls may become cases, larger sample sizes are necessary for a prospective or cross-sectional study of familiar risk. In this study, the at-risk subjects had at least one parent with autopsy-confirmed Alzheimer's disease and at least one additional affected first-degree relative as per a clinical diagnosis of probable Alzheimer's disease. However, the subjects themselves

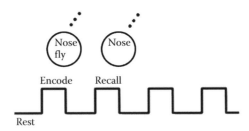

FIGURE 14.1
Illustration of the paradigm.

had no clinical Alzheimer's disease symptoms. Control subjects were also asymptomatic and had no affected first-degree relatives. The study is impressive in its scope, with over 80 subjects in each group, which is atypically large for an fMRI study.

The fMRI paradigm, an auditory word-pair-associate task developed by Bookheimer et al. (2000), was chosen because its primary locus of activation is in the medial temporal lobe, a site of early neuropathological changes associated with Alzheimer's disease. Moreover, loss of verbal memory is an important component of Alzheimer's disease symptoms (Martin et al., 1985). The task consisted of encoding and recall blocks, where subjects heard an unrelated pair of words in the encoding phase and were asked to remember the second word when prompted with the first in the recall phase. The paradigm included two six-minute sessions, each consisting of seven unique word-pairs. A pictorial description of the task is given in Figure 14.1, while further technical information is given in Bassett et al. (2006). A sagital profile of the image acquisition area for a specific subject is given in Figure 14.2.

Known anatomically-derived (Tzourio-Mazoyer et al., 2002) regions of interest (ROIs) are overlaid onto the single-subject maps. This parcelation allows for the study of inter- and intra-regional connectivity. Only those voxels in the image acquisition area with a substantial (greater than 10 voxels) intersection with an ROI are retained. Specifically, let v be a voxel, Im be the collection of voxels in the image acquisition area and ROI_i be the collection of voxels in region of interest i. Voxel v is retained if $v \in ROI_i$ for some i and $I \cap ROI_i$ contains more than 10 voxels. This drastically reduces the number of voxels under

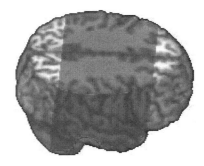

FIGURE 14.2
Image acquisition area (darker gray region) overlaid on template brain. The image is displayed such that anterior is to the right and posterior to the left. The superior portion has been cropped to display an axial slice. (See online supplement for color figures.)

(a)

(b)

(c)

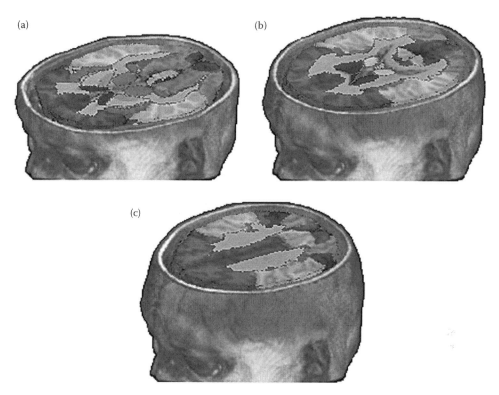

FIGURE 14.3
Example region of interest definitions for three axial slices. Each shade of gray represents a different region. (See online supplement for color figures.)

consideration and further limits the imaging area available for study. Figure 14.3 depicts the ROI definitions for three axial slices.

14.2 Data Preprocessing and First-Level Analysis

Following Friston et al. (2005), Holmes and Friston (1998), and others, we approach our analysis using a two-stage procedure. In the first stage, the data are preprocessed and contrast estimates obtained by linear regression over the time series, such as those comparing active states to rest, are retained for a second-stage analysis. The results is an image of contrast estimates, one per subject, that are then compared across subjects. This approach approximates random effect modeling and has several notable benefits. A principal one, however, is the issue of data reduction, as the contrast maps retained for the second stage are much smaller than the raw fMRI time series of images. However, we emphasize that this approach can have limiting assumptions, such as the inability to incorporate within-session temporal effects into group-level analyses. Such criticisms can be addressed by retaining some of the within-session temporal information for the second stage, such as by fitting separate effects for each block of the paradigm and analyzing them jointly in the second-stage analysis (Bowman and Kilts, 2003). However, this strategy results in less data reduction for the second stage. Hence, in this chapter, we perform a second-stage analysis using only the contrast maps from the first while stipulating this potential limiting assumption.

Our preprocessing strategy was similar to those discussed in Friston et al. (1995, 2007) and Frackowiak et al. (2004). First, the within-subject images were spatially renormalized to the first image via rigid-body transformations. Secondly, these images were transformed into Montreal Neurological Institute (MNI) template space, so that valid comparisons could be made across subjects. Contrary to standard practice, we did not then smooth the images, as the second-stage model contains random effects that shrink voxel-level means within regions.

Next, a canonical hemodynamic response function (HRF) was convolved with indicator functions for the task sequence. This step is necessary as the BOLD signal is only a proxy for neuronal activity. In fact, initially after the onset of the task, blood oxygenation levels will be slightly decreased before being replenished. Therefore, task-related increases in the BOLD signal are expected after a short lag subsequent to the onset of the task. The use of a canonical HRF is another somewhat limiting assumption for this and many other studies, as it varies both across subjects and spatially within subjects due to, among other processes, kinetics of the vasculature.

Let $y_i(v) = \{y_{i1}(v), \ldots, y_{iT}(v)\}^t$ be the temporal response vector for subject $i = 1, \ldots, I$, voxel $v = 1, \ldots, V$, and time index $t = 1, \ldots, T$. Let X be a $T \times p$ matrix of an intercept and hemodynamically convolved indicator functions associated with the paradigm. Let H be a $T \times q$ design matrix of slowly varying trend terms, such as a linear trend and low frequency trigonometric terms, to serve as a high-pass filter. This can account for slowly varying systematic effects, such as scanner drift, where the signal can steadily increase or decrease over a session. Then the general linear model fit for subject i at voxel v is given by

$$y_i(v) = X\beta_i(v) + H\gamma_i(v) + \epsilon_i(v), \tag{14.1}$$

where $\epsilon_i(v) = \{\epsilon_{i1}(v), \ldots, \epsilon_{iT}(v)\}^t$ are assumed to be an AR(1) process, $\text{corr}\{\epsilon_{it}(v), \epsilon_{i,t-k}(v)\} = \rho(v)^k$, with an innovation variance $\sigma^2(v)$. Our two-stage procedure retains a contrast estimate from Equation 14.1 for the inter-subject model considered in the next section. For example, we consider the comparison of encoding blocks versus rest, taking the contrast estimate at each voxel, hence creating a contrast map for each subject.

14.3 A Multilevel Model for Incorporating Regional Connectivity

14.3.1 Model

We consider decomposing the template brain into G regions, as depicted in Figure 14.3 (Tzourio-Mazoyer et al., 2002). As in Bowman et al. (2008), let $\hat{\beta}_{igj}(v)$ denote the contrast estimate from model (Equation 14.1) for subject $i = 1, \ldots, I_j$ having condition $j = 1, \ldots, J$, in region $g = 1 \ldots G$ and voxel $v = 1, \ldots, V_g$, where V_g is the number of voxels contained in region g. In our application $J = 2$, differentiating at-risk and control subjects, $G = 46$, $I_1 = 71$ and $I_2 = 83$. Throughout we adopt the convention that omitting the voxel-level parentheses refers to a vector over voxels, such as $\hat{\beta}_{igj} = \{\hat{\beta}_{igj}(1), \ldots, \hat{\beta}_{igj}(V_g)\}^t$.

The multilevel model that we explore is the following:

$$\hat{\beta}_{igj}(v) \mid \mu_{gj}(v), \alpha_{igj}, \sigma_{gj}^{-2} \sim N\{\mu_{gj}(v) + \alpha_{igj}, \sigma_{gj}^2\}$$

$$\mu_{gj}(v) \mid \lambda_{gj}^2 \sim N\{\mu_{0gj}, \lambda_{gj}^2\}$$

$$\sigma_{gj}^{-2} \sim \Gamma(a_0, b_0) \tag{14.2}$$

$$\alpha_{ij} \mid \Gamma_j \sim MVN(0, \Gamma_j)$$

$$\lambda_{gj}^{-2} \sim \Gamma(c_0, d_0)$$

$$\Gamma_j^{-1} \sim \text{Wishart}\left\{(h_0 H_{0j})^{-1}, h_0\right\},$$

where $\alpha_{ij} = (\alpha_{i1j}, \ldots, \alpha_{iGj})'$. Here, $\mu_{gj}(v)$ is the mean contrast across subjects but within groups. The term α_{igj} represents subject- and region-specific deviations from the mean. This term forces a conditional exchangeable correlation structure within regions of $\rho_{gj} = \frac{\gamma_{gj}}{\gamma_{gj} + \sigma_{gj}^2}$, where γ_{gj} is diagonal element g from Γ_j. Thus, ρ_{gj} measures the correlation of contrast estimates within a region and group. We refer to it as a measure of *intra*-regional paradigm-related connectivity. In contrast, Γ_j is the variance–covariance matrix of the random effect terms between the G ROIs for condition j. Hence, we view the corresponding correlation matrix, say R_j, as a measure of *inter*-regional paradigm-related connectivity.

The residual variance, σ_{gj}^2, is constant within regions, unlike many models for fMRI that presume separate voxel-specific variances. That is, instead of smoothing voxel-specific variances with further hierarchies, our model uses anatomical information to smooth variances within regions. Note that separate variances are assumed for each of the groups. The other variance term, λ_{gj}, measures variation in the voxel-level means around the prior mean, μ_{0gj}.

14.3.2 Simulating the Markov Chain

The block full conditionals associated with model (Equation 14.2) (see Bowman et al., 2008, for more discussion) are given below:

$$\mu_{gj} \sim N\left[\left(\lambda_{gj}^{-2} + I_j \sigma_{gj}^{-2}\right)\left\{\lambda_{gj}^{-2}\mu_{0gj} + I_j\sigma_{gj}^{-2}\left(\bar{\beta}_{gj} - \mathbf{1}\bar{\alpha}_{gj}\right)\right\}, \text{Diag}\left(\lambda_{gj}^{-2} + I_j\sigma_{gj}^{-2}\right)\right]$$

$$\sigma_{gj}^{-2} \sim \Gamma\left\{a_0 + I_j V_g/2, \left(\frac{1}{b_0} + \frac{1}{2}\sum_{i=1}^{I_j} \|\hat{\beta}_{igj} - \mu_{gj} - \mathbf{1}\alpha_{igj}\|^2\right)^{-1}\right\}$$

$$\alpha_{ij} \sim N\left[\left(\Gamma_j^{-1} + D_j^{-1}\right)^{-1}\left\{D_j^{-1}\left(\bar{\beta}_{ij} - \bar{\mu}_j\right)\right\}, \left(\Gamma_j^{-1} + D_j^{-1}\right)\right] \tag{14.3}$$

$$\lambda_{gj}^{-2} \sim \Gamma\left\{c_0 + V_g/2, \left(\frac{1}{d_0} + \frac{(\mu_{gj} - \mu_{0gj})'(\mu_{gj} - \mu_{0gj})}{2}\right)^{-1}\right\}$$

$$\Gamma_j^{-1} \sim \text{Wishart}\left\{\left(h_0 H_0 + \sum_{i=1}^{I_j} \alpha_{ij}\alpha_{ij}'\right)^{-1}, h_0 + I_j\right\},$$

where $\mu_{gj} = \{\mu_{gj}(1),\ldots,\mu_{gj}(V_G)\}^t$, $\bar{\beta}_{gj}$ is $\frac{1}{I_j}\sum_{i=1}^{I_j}\{\hat{\beta}_{igj}(1),\ldots,\hat{\beta}_{igj}(V_G)\}^t$, $\bar{\alpha}_{gj} = \frac{1}{I_j}\sum_{i=1}^{I_j}\alpha_{igj}$,
$D_j = \mathrm{Diag}(V_1\sigma_{1j}^{-2},\ldots,V_G\sigma_{Gj}^{-2})$ and $\bar{\mu}_j = \left\{\frac{1}{V_1}\sum_{v=1}^{V_1}\mu_{1j}(v),\ldots,\frac{1}{V_G}\sum_{v=1}^{V_G}\mu_{Gj}(v)\right\}^t$. The update
order proceeded with Γ_j first, then a loop over g for μ_{gj}, σ_{gi}, λ_{gj}, and then the update for α_{ij}.

The full conditionals display the benefit of the use of the linear mixed effects model and "Gibbs-friendly" prior distributions. That is, the full conditionals are based on simple matrix summaries that can be executed quickly. We discuss simple extensions with less restrictive priors in Section 14.6. Also note that none of the block updates have dimension larger than $\max_g V_g = 1784$ and, more importantly, no matrix inversions are required for matrices with dimension larger than $\max_g V_g \times \max_g V_g$. This is a primary strength of the model, as any approach that requires matrix manipulations over all of the voxels would not allow the fast block updates. The code was written in MATLAB® (Mathworks version 2006b) and is available from the first author's website. Ten thousand iterations were run for the results presented; however, later runs of 100,000 iterations confirm the conclusions. Note that, with over 60,000 variables updated in each iteration (roughly 30,000 per group), this resulted in over 100 million basic operations. Regardless, the sampler was run on a standard laptop in under an hour (2.16 GHz dual core Intel processor and 2 GB of RAM).

With dimension in excess of 60,000, the posterior raises numerous issues regarding simulating the chain and analyzing the output. First, we note that storage of the output is itself a challenge. Memory allocation limits were reached if the entire chain of voxelwise results was stored for any reasonably long chain. We adopted the following strategy to combat this issue. The complete chain was stored for all of the values that have only tens of measurements per region. That is, the complete chain was stored for the σ_{gj}, α_{ij}, λ_{gj}, and Γ_j. For the μ_{gj}, the complete posterior mean was updated each iteration and stored. In addition, a batch means estimate (Jones et al., 2006) of the variance of this posterior mean was also stored. To utilize simple update rules, adaptive batch sizes were not employed and, instead, fixed batch sizes of size 100 were used. Moreover, the complete chains for several regional summaries, such as the mean and quantiles of the μ_{gj} within regions, were also stored. Finally, the complete value of μ_{gj} was stored for every 20th iteration, resulting in 500 total stored iterations. However, it should be noted that considerable loss of information is incurred if the chain is subsampled (MacEachern and Berliner, 1994). We do not recommend combining the values, subsampled or not, into a matrix or other single data structure. Instead, we recommend that the value for each saved iteration be stored in a separate file, with the filename indicating the iteration number.

For starting values, we used empirical moments and cumulants. Specifically, we let μ_{gj} be the empirical vector mean over the I_j subjects within region g and group j. We let σ_{gj}^2 be the average (across voxels) of the inter-subject variances within region g and group j. We let λ_{gj} be the between-region variance of the region- and group-specific means of the $\hat{\beta}_{igj}(v)$ ($\bar{\beta}_{gj}$ from above). We further let α_{igj} be the mean of the $\hat{\beta}_{igj}(v)$ within subject, region, and group. Then Γ_j was set to be the variance–covariance matrix of the starting value for α_{ij}, calculated by taking variances and covariances over subjects. The least accurate of these starting values are those for the α_{igj} and Γ_j, as the starting value for α_{igj} has mean $\frac{1}{V_g}\sum_{v=1}^{V_g}\mu_{gj}(v) + \alpha_{igj}$ (see Equation 14.2). However, recall that the outcomes are contrasts estimates and ideally the task should specialize to only a small portion of the brain, and therefore the term $\frac{1}{V_g}\sum_{v=1}^{V_g}\mu_{gj}(v)$ should be small in absolute value. Hence, for starting values, ignoring the fixed effects contribution in the moment estimates is not problematic.

To empirically evaluate the results of the chain, posterior mean estimates were compared with these starting values. In all cases they agreed well, though we stipulate that this only adds to the face validity of the chain and is not a formal method of convergence assessment. To further evaluate properties of the chain, trace plots of the parameters were investigated. However, the volume of parameters precludes investigation of all plots for the $\mu_{gj}(v)$. Instead, a random sample of voxels was selected and investigated in greater detail. To investigate sensitivity to hyperparameter settings, several chains were run, varying these parameters.

14.4 Analyzing the Chain

14.4.1 Activation Results

We consider the distribution of voxel-level contrast means to answer the question of whether the μ_{gj} are systematically larger than zero in any area and whether they differ across groups. The former question is of greater initial importance, as there is less interest in assessing inter-group differences when there is little evidence of localized within-group activation. We use a novel adaptation of a supra-threshold clustering technique widely used in the frequentist analysis of statistical maps. To evaluate a unit-free statistic, consider the map of voxel-level signal-to-noise statistics, $\{|\mu_{gj}(v)|/\sigma_{gj}\}_v$. Following traditional analysis, we consider clusters of contiguous connected voxels above a threshold (Cao and Worsley, 2001; Friston et al., 1993; Nichols and Holmes, 2002; Worsley, 1994; Worsley et al., 1996). Here, voxels are connected if they share a face, edge or corner. Figure 14.4 illustrates with fictitious one-dimensional data. We refer to the number of contiguous voxels in a cluster as the *extent*; in Figure 14.4 this is the width of the cluster above the chosen threshold. We also considered the center of mass of the cluster, the area of the cluster, and the peak value within the cluster. Here the center of mass is simply the average of the X, Y, and Z coordinates of each cluster surviving the threshold. The area of the cluster is proportional to taking the product of the voxels in the cluster and the associated heights of the statistics, and summing the results. Two cutoffs were considered for these supra-threshold statistics, 0.1 and 0.2. These were obtained empirically, by considering the inter-voxel distribution of the posterior means and

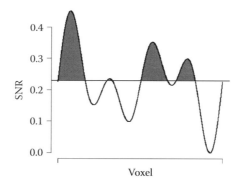

FIGURE 14.4
Illustration of supra-threshold cluster level statistics.

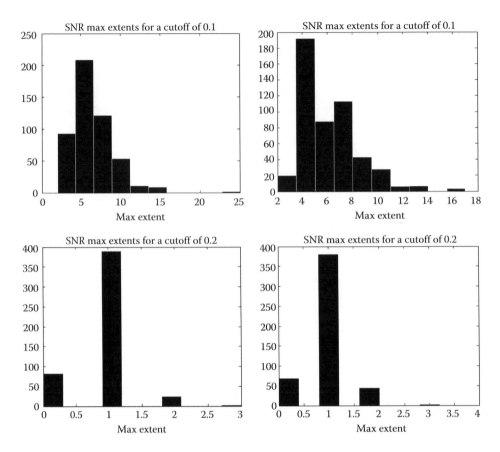

FIGURE 14.5
Estimated posterior distributions for the maximum supra-threshold cluster extents. The top and bottom images use an SNR cutoff of 0.1 and 0.2 respectively, and the left and right images correspond to at-risk and control groups respectively.

inspection of related data sets. Finally, to reduce cluster results to single numbers, we took the maximal statistic, such as considering the maximum cluster extent per MCMC iteration.

We considered thresholding the signal-to-noise statistic map generated at each MCMC iteration. As previously mentioned, a complication arose in that it is most convenient to have the chain of $\mu_{gj}(v)$ maps saved, as different thresholding values and statistics need to be evaluated interactively. As the complete chain is generally too large to save, we subsampled the chain and saved 500 equally spaced iterations. For each saved map, we determined the cluster size with the largest supra-threshold extent. Histogram estimates of the posterior distributions for the two groups and for the two cutoffs are given in Figure 14.5. Here the maximal extents appear to be quite small, suggesting little voxel-level activation across subjects within groups. Figure 14.6 displays the centroids for the clusters surviving the threshold (of 0.1) for the control group for the 500 saved iterations. The color and width of the points are related to the extent of the cluster. There is little evidence to suggest voxel-level localization of the clusters. In addition to considering the maximal extents, we also considered the areas under the clusters as well as the maximal peak value of the clusters. In each case, there was little suggestion of interesting voxel-level results. These conclusions are consistent with those of more standard analyses.

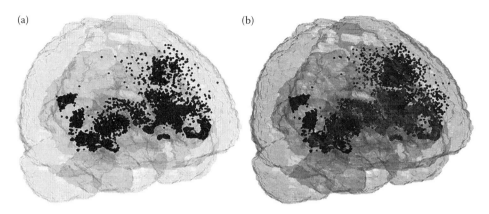

FIGURE 14.6

Plot of supra-threshold (threshold of 0.1) cluster centroids across 500 saved iterations for the control (a) and at-risk (b) groups. The shading and size of the points are proportional to the cluster extent. Transparent template brain overlays are provided for reference. (See online supplement for color figures.)

We also considered regional mean level effects. That is, let $\theta_{gj} = \sum_{v=1}^{V_g} \mu_{gj}(v)/V_g$, where V_g is the number of voxels in region g. As with the voxel-level results, we consider both within- and between-group effects. We first calculated the MCMC estimate of the minimum of the posterior tail probabilities of θ_{gj} being larger or smaller than 0, for each g and j:

$$\min\{P(\theta_{gj} < 0 \mid \text{Data}), P(\theta_{gj} > 0 \mid \text{Data})\}.$$

This quantity combines the information from the two one-sided tail probabilities similar to taking the smaller of two p-values from one-sided tests to perform a two-sided test. This is useful to answer whether or not the regional mean appears either much larger or smaller than zero.

In the control group, this quantity was the smallest in the right superior temporal lobe, with a posterior mean for θ_{gj} of -0.0021 and minimum tail probability of 0.0057. This was followed by the left supplementary motor area (0.0532) and the right mid frontal area (0.0586). In the at-risk group, there were only modestly small minimum tail probabilities in the left supplementary motor area (0.0560) and the right mid frontal area (0.0563).

To compare the two groups, we again evaluated a minimum of posterior tail probabilities. Specifically, we considered the minimum posterior probability of one mean being smaller or larger than the other. That is, for each g, we considered

$$\min\{P(\theta_{g1} < \theta_{g2} \mid \text{Data}), P(\theta_{g1} > \theta_{g2} \mid \text{Data})\},$$

where $j = 2$ refers to the at-risk group and $j = 1$ refers to the controls. This quantity was the smallest in the right superior temporal pole, with a minimum tail probability of 0.032, and in the right superior temporal lobe (0.0648). Histogram estimates of $\theta_{g1} - \theta_{g2}$ for these two regions are given in Figure 14.7, showing that activation related to this contrast is (largely speaking) slightly higher in the at-risk group compared to the controls.

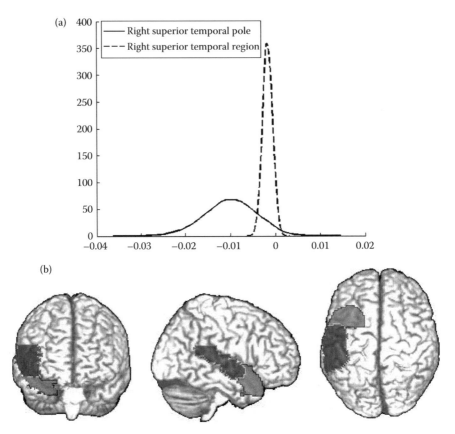

FIGURE 14.7

Posterior distributions for $\theta_{g1} - \theta_{g2}$ for the two regions with the smallest posterior tail probabilities. Projection plots of the two regions overlaid on a template brain are given for reference. (See online supplement for color figures.)

14.5 Connectivity Results

14.5.1 Intra-Regional Connectivity

We first consider results for intra-regional connectivity; that is, we consider the posterior distributions for the $\{\rho_{gj}\}$. Figure 14.8 displays posterior credible intervals and posterior medians for the $\{\rho_{gj}\}$ for the control and at-risk subjects for the 46 regions of interest, based on all 10,000 iterations. It is perhaps surprising that the correlations are as high as they are, especially given that no spatial smoothing was performed. We note that some of the variation in the intra-regional correlation arises from the size of the region in consideration, with, as expected, the smaller regions tending to demonstrate greater connectivity.

Figure 14.9 shows 95% equi-tail posterior credible intervals and posterior medians for the ratio ρ_1/ρ_2, with a gray vertical reference line drawn at one. The data exhibit regions both with greater and lesser intra-regional connectivity. Figure 14.9 also displays projection maps of the regions with higher intra-regional connectivity among the controls (shown in the upper plots, as determined by a credible interval entirely above one) and lower intra-regional connectivity among the at risk (shown in the lower plots). Lower intra-regional

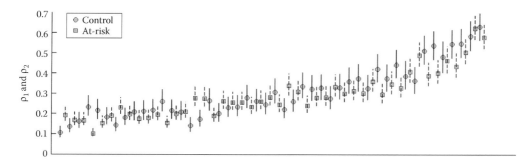

FIGURE 14.8
Posterior credible intervals and medians for ρ_1 (controls, circles) and ρ_2 (at-risk, squares). Intervals are ordered by the average of the two groups' medians.

connectivity among the control subjects was most apparent in the areas near the anterior cingulum. Higher intra-regional connectivity among the controls was most apparent in more frontal areas and was diffusely spread out.

14.5.2 Inter-Regional Connectivity

The proposed hierarchical model also allows for the study of inter-regional connectivity associated with the paradigm. Figure 14.10 connects the centroids of regions whose posterior mean correlation (from the off-diagonal entries of R_j) was above 0.6 for control (left) and at-risk (right) subjects, with estimates obtained using all 10,000 iterations. Visually, the picture suggests a denser network of connectivity for the at-risk subjects, perhaps suggesting that this population has to attend to the task more rigorously to complete it or that more regions are called upon to complete the task to compensate for weaknesses in a few regions. Below we explore more formal methods for comparing the chain of variance and covariance matrices. We consider the eigenvalue decomposition of the variance matrices, Γ_j, and the correlation matrices, R_j. This is analogous to a principal component analysis of the α_{ij}. In particular, we focus on the eigenvalue decomposition of the R_j, as the region-specific variances are of less interest.

The posterior mean of the percentage of variation explained by each component was 29, 13, 10, 8, and 6 for the control group and 29, 12, 9, 7, and 6 for the at-risk group. Figure 14.11 displays the posterior distribution for the largest eigenvalues for the control (solid) and at-risk subjects, respectively. We also looked at the distribution of the eigenvectors corresponding to the maximum eigenvalue. The at-risk group had larger loadings (across the board) for the first eigenvalue. The control group loaded most heavily (in absolute value) on the left precentral gyrus, the right mid cingulum, the right supplementary motor area, and the left postcentral gyrus. The at-risk group loaded more heavily on the left insula, the left precentral gyrus, the left caudate and left mid cingulum.

The equi-tail 95% posterior credible interval for the ratio of the largest eigenvalues (control over at-risk) was [0.58, 0.89], with a posterior distribution shown in the middle plot of Figure 14.11. To consider the variances, we considered the largest eigenvalue of $(\Gamma_1 + \Gamma_2)^{-1}\Gamma_1$, the greatest root statistic (Mardia et al., 1979). The equi-tail 95% credible interval for the greatest root statistic was [0.981, 0.989], with a posterior given in the rightmost plot of Figure 14.11.

Overall, the results suggest much greater connectivity in the at-risk group. We found this particular result to be the most intriguing for this data set and believe that it is suggestive of the idea that at-risk subjects have to engage more cognitive resources to attend to the task.

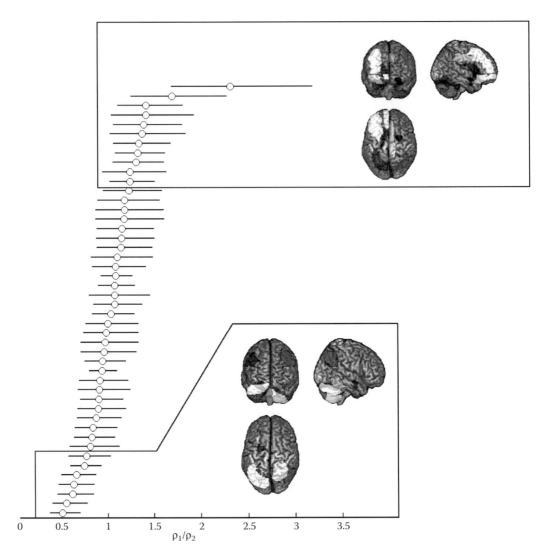

FIGURE 14.9
Credible intervals and posterior medians for ratios of ρ_1/ρ_2 accompanied by projection maps of the regions with credible intervals showing higher intra-regional connectivity among the at-risk (top plots, 11 regions) and higher connectivity among the controls (bottom plots, seven regions), respectively. Note that one of the top 11 and one of the bottom seven have credible intervals that overlap zero. (See online supplement for color figures.)

14.6   Discussion

In this chapter, we investigated a model from Bowman et al. (2008) and a data set from Bassett et al. (2006) and introduced some novel methods for analyzing, interpreting, and visualizing the output. The data are suggestive of some interesting findings on functional differences between a group of subjects at high risk for the development of Alzheimer's disease and a group of controls. First, the voxel-level contrast map results suggest little difference between the groups in terms of activation, while the regional mean results suggest

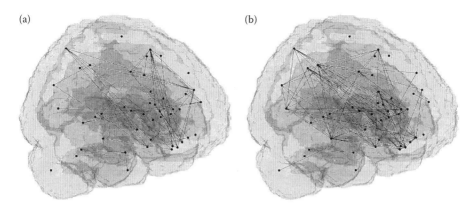

FIGURE 14.10
Posterior mean connectivity estimates exceeding a threshold of 0.6 for control (a) and at-risk (b) subjects. Lines connect the centers of regions with posterior mean connectivity estimates exceeding the threshold. (See online supplement for color figures.)

a modest decrease in activation for the controls in two regions of the temporal lobe. These results, which have been confirmed with more traditional parametric modeling (not shown), differ from those of Bassett et al. (2006), based on the same subjects at an earlier visit. The discrepancy could be due to a variety of factors, such as a learning effect, differences in the sample (as there was dropout for the subsequent visit), or actual physiological longitudinal changes. We defer a full longitudinal analysis of this data to future research.

The connectivity results are perhaps more interesting in demonstrating greater differences between the two groups. The inter-regional results suggest greater connectivity among the at-risk groups. This result potentially suggests that the at-risk group are calling on greater cognitive reserves to perform the tasks. The intra-regional results suggest important differences in areas of intrinsic connectivity for the two groups.

The at-risk Alzheimer's disease data set is uniquely suited to this model. First, the smaller imaging acquisition area limits the number of regions of interest to consider. Secondly, the large number of subjects also allows for the estimation of a finer regional parcelation of the connectivity matrix. For example, if the study had typical group sizes of 15 or 20 per group, estimation of 46×46 covariance matrix would not be feasible, and hence regions would have to be aggregated to employ the model.

With regard to the model, its weakest point is the reliance on Gibbs-friendly priors for the variance components. In particular, the use of inverted gamma priors (with small rates and scales) and the inverse Wishart distribution for the variance components has been widely discussed and criticized (Daniels, 1999; Daniels and Kass, 1999; Daniels and Pourahmadi, 2002; He and Hodges, 2008; Yang and Berger, 1994). The previous references provide several alternative priors and approaches, including placing the priors on the eigenspace rather than the natural units. Such approaches are appealing in this setting, because principal component analysis of the region-specific random effects is of interest. However, a very practical solution would simply use a mixture of two or three gammas for the precisions and a mixture of two or three Wishart distributions for the inverse variance matrices. These solutions may add enough hyperparameters to allow for needed flexibility for the prior distributions, while still retaining a simple structure.

We discuss possible methods for further computational acceleration, though, as previously mentioned, the chain ran adequately fast for our application. However, for whole

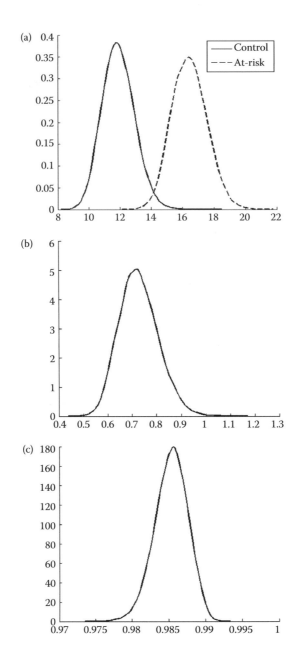

FIGURE 14.11
(a) Density estimates for the posterior distribution of the largest eigenvalue for R_j for the control (solid) and at-risk (dashed) groups. (b) Density estimates for the posterior distribution of the ratio of the largest eigenvalues for R_1 divided by that of R_2. (c) Density estimate for the posterior distribution for the greatest root statistic.

brain results and next generation scanners, the extent of the computations will increase dramatically. A possible acceleration could be obtained with parallel processing. To be specific, the region- and group-specific parameters, μ_{gj}, σ_{gj} and λ_{gj}, are all conditionally independent given the inter-regional parameters, α_{ij} and Γ_j. Hence, they could be updated in parallel, hopefully speeding up calculations by an order of magnitude. We have successfully applied this approach in unrelated areas with good success, though it was not applied here.

Perhaps the greatest challenge in this setting, and most germane to the topic of this book, is the question of the overall validity of the use of MCMC as a mechanism for analysis. While our application possessed only tens of thousands of parameters, current MRI and genomic technology puts the relevant number closer to millions. To our knowledge, convergence, implementation, diagnostic, and inferential issues for such large chains have had little discussion in the MCMC literature and represent a great challenge for future MCMC research.

References

Arfanakis, K., Cordes, D., Haughton, V., Moritz, C., Quigley, M., and Meyerand, M. 2000. Combining independent component analysis and correlation analysis to probe interregional connectivity in fMRI task activation datasets. *Magnetic Resonance Imaging*, 18(8):921–930.

Bassett, S., Yousem, D., Cristinzio, C., Kusevic, I., Yassa, M., Caffo, B., and Zeger, S. 2006. Familial risk for Alzheimer's disease alters fMRI activation patterns. *Brain*, 129(5):1229.

Beckmann, C., Jenkinson, M., and Smith, S. 2003. General multilevel linear modeling for group analysis in fMRI. *NeuroImage*, 20(2):1052–1063.

Biswal, B., Yetkin, F., Haughton, V., and Hyde, J. 1995. Functional connectivity in the motor cortex of resting human brain using echo-planar MRI. *Magnetic Resonance Medicine*, 34(4): 537–541.

Bookheimer, S., Strojwas, M., Cohen, M., Saunders, A., Pericak-Vance, M., Mazziotta, J., and Small, G. 2000. Patterns of brain activation in people at risk for Alzheimer's disease. *New England Journal of Medicine*, 343:450–456.

Bowman, F., Caffo, B., Bassett, S., and Kilts, C. 2008. A Bayesian hierarchical framework for spatial modeling of fMRI data. *NeuroImage*, 39(1):146–156.

Bowman, F. and Kilts, C. 2003. Modeling intra-subject correlation among repeated scans in positron emission tomography (PET) neuroimaging data. *Human Brain Mapping*, 20:59–70.

Brookmeyer, R., Gray, S., and Kawas, C. 1998. Projections of Alzheimer's disease in the United States and the public health impact of delaying disease onset. *American Journal of Public Health*, 88:1337–1342.

Calhoun, V., Adali, T., Hansen, L., Larsen, J., and Pekar, J. 2003. ICA of functional MRI data: An overview. *Fourth International Symposium on Independent Component Analysis and Blind Source Separation*, pp. 281–288.

Cao, J. and Worsley, K. 2001. Applications of random fields in human brain mapping. *Spatial Statistics: Methodological Aspects and Applications*, 159:170–182.

Daniels, M. 1999. A prior for the variance in hierarchical models. *Canadian Journal of Statistics*, 27(3):567–578.

Daniels, M. and Kass, R. 1999. Nonconjugate Bayesian estimation of covariance matrices and its use in hierarchical models. *Journal of the American Statistical Association*, 94(448):1254–1263.

Daniels, M. and Pourahmadi, M. 2002. Bayesian analysis of covariance matrices and dynamic models for longitudinal data. *Biometrika*, 89(3):553–566.

Fox, N., Crum, W., Scahill, R., Stevens, J., Janssen, J., and Rossor, M. 2001. Imaging of onset and progression of Alzheimer's disease with voxel-compression mapping of serial magnetic resonance images. *Lancet*, 358(9277):201–205.

Frackowiak, R., Friston, K., Firth, C., Dolan, R., Price, C., Zeki, S., Ashburner, J., and Penny, W. (eds) 2004. *Human Brain Function*, 2nd edn. Elsevier Academic Press, Amsterdam.

Friston, K., Ashburner, J., Stefan, K., Nichols, T., and Penny, W. (eds) 2007. *Statistical Parametric Mapping: The Analysis of Functional Brain Images*. Elsevier Academic Press, Amsterdam.

Friston, K., Glaser, D., Henson, R., Kiebel, S., Phillips, C., and Ashburner, J. 2002. Classical and Bayesian inference in neuroimaging: Applications. *NeuroImage*, 16:484–512.

Friston, K., Holmes, A., Price, C., Büchel, C., and Worsley, K. 1999. Multisubject fMRI studies and conjunction analyses. *NeuroImage*, 10(4):385–396.

Friston, K., Holmes, A., Worsley, K., Poline, J., Frith, C., Frackowiak, R. et al. 1995. Statistical parametric maps in functional imaging: A general linear approach. *Human Brain Mapping*, 2(4):189–210.

Friston, K., Penny, W., Phillips, C., Kiebel, S., Hinton, G., and Ashburner, J. 2002. Classical and Bayesian inference in neuroimaging: Theory. *NeuroImage*, 16(2):465–483.

Friston, K., Stephan, K., Lund, T., Morcom, A., and Kiebel, S. 2005. Mixed-effects and fMRI studies. *NeuroImage*, 24(1):244–252.

Friston, K., Worsley, K., Frackowiak, R., Mazziotta, J., and Evans, A. 1993. Assessing the significance of focal activations using their spatial extent. *Human Brain Mapping*, 1(3):210–220.

Gazzaniga, M., Ivry, R., and Mangun, G. 2002. *Cognitive Neuroscience: The Biology of the Mind*, 2nd edn. Norton, New York.

Greicius, M. 2003. Functional connectivity in the resting brain: A network analysis of the default mode hypothesis. *Proceedings of the National Academy of Sciences of the USA*, 100(1): 253–258.

Greicius, M. 2004. Default-mode network activity distinguishes Alzheimer's disease from healthy aging: Evidence from functional MRI. *Proceedings of the National Academy of Sciences of the USA*, 101(13):4637–4642.

Hayasaka, S. and Nichols, T. 2003. Validating cluster size inference: Random field and permutation methods. *NeuroImage*, 20(4):2343–2356.

He, Y. and Hodges, J. 2008. Point estimates for variance-structure parameters in Bayesian analysis of hierarchical models. *Computational Statisics and Data Analysis*, 52:2560–2577.

Holmes, A. and Friston, K. 1998. Generalisability, random effects and population inference. *NeuroImage*, 7(4):754.

Jones, G., Haran, M., Caffo, B., and Neath, R. 2006. Fixed-width output analysis for Markov chain Monte Carlo. *Journal of the American Statistical Association*, 101:1537–1547.

MacEachern, S. and Berliner, L. 1994. Subsampling the Gibbs sampler. *American Statistician*, 48:188–190.

Mardia, K., Kent, J., and Bibby, J. 1979. *Multivariate Analysis*. Academic Press, San Diego.

Martin, A., Brouwers, P., Cox, C., and Fedio, P. 1985. On the nature of the verbal memory deficit in Alzheimer's disease. *Brain and Language*, 25(2):323–341.

Nichols, T. and Holmes, A. 2002. Nonparametric permutation tests for functional neuroimaging: A primer with examples. *Human Brain Mapping*, 15(1):1–25.

Ogawa, S., Lee, T., Kay, A., and Tank, D. 1990. Brain magnetic resonance imaging with contrast dependent on blood oxygenation. *Proceedings of the National Academy of Sciences of the USA*, 87(24):9868–9872.

Roy, C. and Sherrington, C. 1890. On the regulation of the blood supply of the brain. *Journal of Physiology*, 11(85):108.

Suckling, J. and Bullmore, E. 2004. Permutation tests for factorially designed neuroimaging experiments. *Human Brain Mapping*, 22(3):193–205.

Tzourio-Mazoyer, N., Landeau, B., Papathanassiou, D., Crivello, F., Etard, O., Delcroix, N., Mazoyer, B., and M. J. 2002. Automated anatomical labeling of activations in SPM using a macroscopic anatomical parcellation of the MNI MRI single-subject brain. *NeuroImage*, 15:273–289.

Verbeke, G. and Molenberghs, G. 2000. *Linear Mixed Models for Longitudinal Data*. Springer, New York.

Woolrich, M., Behrens, T., Beckmann, C., Jenkinson, M., and Smith, S. 2004. Multilevel linear modelling for fMRI group analysis using Bayesian inference. *NeuroImage*, 21(4):1732–1747.

Worsley, K. 1994. Local maxima and the expected Euler characteristic of excursion sets of χ^2, F and t fields. *Advances in Applied Probability*, 26(1):13–42.

Worsley, K. and Friston, K. 1995. Analysis of fMRI time-series revisited, again. *NeuroImage*, 2(3):173–181.

Worsley, K., Liao, C., Aston, J., Petre, V., Duncan, G., Morales, F., and Evans, A. 2002. A general statistical analysis for fMRI data. *NeuroImage*, 15(1):1–15.

Worsley, K., Marrett, S., Neelin, P., Vandal, A., Friston, K., and Evans, A. 1996. A unified statistical approach for determining significant signals in images of cerebral activation. *Human Brain Mapping*, 458:73.

Xiong, J., Parsons, L., Gao, J., and Fox, P. 1999. Interregional connectivity to primary motor cortex revealed using MRI resting state images. *Human Brain Mapping*, 8(2–3):151–156.

Yang, R. and Berger, J. 1994. Estimation of a covariance matrix using the reference prior. *Annals of Statistics*, 22(3):1195–1211.

15

Partially Collapsed Gibbs Sampling and Path-Adaptive Metropolis–Hastings in High-Energy Astrophysics

David A. van Dyk and Taeyoung Park

15.1 Introduction

As the many examples in this book illustrate, Markov chain Monte Carlo (MCMC) methods have revolutionized Bayesian statistical analyses. Rather than using off-the-shelf models and methods, we can use MCMC to fit application-specific models that are designed to account for the particular complexities of a problem. These complex multilevel models are becoming more prevalent throughout the natural, social, and engineering sciences largely because of the ease of using standard MCMC methods such as the Gibbs and Metropolis–Hastings (MH) samplers. Indeed, the ability to easily fit statistical models that directly represent the complexity of a data-generation mechanism has arguably lead to the increased popularity of Bayesian methods in many scientific disciplines.

Although simple standard methods work surprisingly well in many problems, neither the Gibbs nor the MH sampler can directly handle problems with very high posterior correlations among the parameters. The marginal distribution of a given parameter is much more variable than the corresponding full conditional distribution in this case, causing the Gibbs sampler to take small steps. With MH a proposal distribution that does not account for the posterior correlation either has far too much mass in regions of low posterior probability or has such small marginal variances that only small steps are proposed, causing high rejection rates and/or high autocorrelations in the resulting Markov chains. Unfortunately, accounting for the posterior correlation requires more information about the posterior distribution than is typically available when the proposal distribution is constructed.

Much work has been devoted to developing computational methods that extend the usefulness of these standard tools in the presence of high correlations. For Gibbs sampling, for example, it is now well known that blocking or grouping steps (Liu et al., 1994), nesting steps (van Dyk, 2000), collapsing or marginalizing parameters (Liu et al., 1994; Meng and van Dyk, 1999), incorporating auxiliary variables (Besag and Green, 1993), certain parameter transformations (Gelfand et al., 1995; Yu and Meng, 2011), and parameter expansion (Liu and Wu, 1999) can all be used to improve the convergence of certain samplers. By embedding an MH sampler within the Gibbs sampler and updating one parameter at a time (i.e. the well-known Metropolis-within-Gibbs sampler in the terminology of Gilks et al., 1995), the same strategies can be used to improve MH samplers.

In this chapter, we describe two newer methods that are designed to improve the performance of Gibbs and Metropolis-within-Gibbs samplers. The partially collapsed Gibbs (PCG) sampler (van Dyk and Park, 2008; Park and van Dyk, 2009) takes advantage of the fact that we expect reducing conditioning to increase the variance of the complete conditional distributions of a Gibbs sampler. Thus, by replacing a subset of the complete conditional distributions by distributions that condition on fewer of the unknown quantities, that is, conditional distributions of some marginal distributions of the target posterior distribution, we expect the sampler to take larger steps and its overall convergence characteristics to improve. This strategy must be used with care, however, since the resulting set of conditional distributions may not be functionally compatible and changing the order of the draws can alter the stationary distribution of the chain. The second strategy involves updating the Metropolis proposal distribution to take account of what is known about the target distribution given an initial set of draws.

Although these are both general strategies with many potential applications, they were both motivated by a particular model fitting task in high-energy astrophysics. In recent years, technological advances have dramatically increased the quality and quantity of data available to astronomers. Multilevel statistical models are used to account for these complex data-generation mechanisms, which can include both the physical data sources and sophisticated instrumentation. Bayesian methods and MCMC techniques both find numerous applications among the many resulting statistical problems and are becoming ever more popular among astronomers. Examples include the search for planets orbiting distant stars (Gregory, 2005), the analysis of stellar evolution using sophisticated physics-based computer models (DeGennaro et al., 2008; van Dyk et al., 2009), the analysis of the composition and temperature distribution of stellar coronae (Kashyap and Drake, 1998), and the search for multi-scale structure in X-ray images (Esch et al., 2004; Connors and van Dyk, 2007), to name just a few. In this chapter, we describe the PCG sampler and the path-adaptive MH sampler and show how they can dramatically improve the computational performance of MCMC samplers designed to search for narrow emission lines in high-energy astronomical spectral analysis.

15.2 Partially Collapsed Gibbs Sampler

Collapsing in a Gibbs sampler involves integrating a joint posterior distribution over a subset of unknown quantities to construct a marginal or *collapsed* posterior distribution under which a new collapsed Gibbs sampler is built (Liu et al., 1994). This strategy is similar to the efficient data augmentation strategy used to improve the rate of convergence of the EM algorithm (van Dyk and Meng, 1997). Efficient data augmentation aims to construct an EM algorithm using as little missing data as possible. That is, a portion of the missing data is collapsed out of the distribution of unknown quantities. Just as collapsing is known to improve the convergence of a Gibbs sampler, it is known that reducing the missing data in this way can only improve the rate of convergence of the EM algorithm (Meng and van Dyk, 1997). Generally speaking, there is a strong relationship between the rate of convergence of EM-type algorithms and Gibbs samplers constructed with the same set of conditional distributions (see Tanner and Wong, 1987; Liu, 1994; Liu and Wu, 1999; van Dyk and Meng, 2001). Strong correlations slow both types of algorithms.

Although these collapsing or marginalizing strategies typically improve convergence, they may not be easy to implement. For example, the complete conditional distributions of the collapsed posterior distribution may be harder to work with than the conditional distributions of the original posterior distribution. The PCG sampler aims to take partial computational advantage of the collapsing strategy while maintaining simple implementation by mixing conditional distributions from the original posterior distribution with those of one or more collapsed posterior distributions. Thus, we use collapsing only in those conditional distributions where it does not complicate parameter updating. This strategy is analogous to the ECME and AECM algorithms which generalize EM by allowing different amounts of missing data when updating different model parameters (Liu and Rubin, 1994; Meng and van Dyk, 1997); see Park and van Dyk (2009) and van Dyk and Meng (2010) for discussion.

To see both the potential advantages and the potential pitfalls of partially collapsing a Gibbs sampler, consider a simple example where a three-step Gibbs sampler is constructed to simulate the trivariate Gaussian distribution, $(X, Y, Z) \sim N_3(\mathbf{0}, \mathbf{\Sigma})$ with

$$\mathbf{\Sigma} = \begin{pmatrix} 1 & \rho & 0.5 \\ \rho & 1 & 0.5 \\ 0.5 & 0.5 & 1 \end{pmatrix},$$

where ρ is a known constant that controls the convergence rate of the Gibbs sampler. The three-step Gibbs sampler iterates among the following steps:

Step 1. Draw X from $p(X \mid Y, Z)$.   (Sampler 1)

Step 2. Draw Y from $p(Y \mid X, Z)$.

Step 3. Draw Z from $p(Z \mid X, Y)$.

The convergence rate of the Gibbs sampler is equal to the spectral radius of the corresponding forward operator (Liu, 2001). Letting $\mathbf{Q} = \mathbf{\Sigma}^{-1}$ with $\mathbf{Q} = \{q_{ij}\}$, Amit (1991) showed that the spectral radius of the forward operator for Sampler 1 is the largest norm of the eigenvalues of $\prod_{i=1}^{3}(\mathbf{I} - \mathbf{D}_i\mathbf{Q})$, where $\mathbf{I}$ is the 3×3 identity matrix and $\mathbf{D}_i$ is the 3×3 matrix of zeros except that the ith diagonal entry is q_{ii}^{-1}. For example, with $\rho = 0.99$, the spectral radius is 0.98, indicating slow convergence. The convergence characteristics of the Gibbs sampler and the sampled correlation structure between X and Y are shown in the first row of Figure 15.1, which illustrates slow convergence and strong correlation. In this simple example, we can easily reduce the conditioning in any of the steps in the hope of improving convergence. In particular, the marginal distribution of (Y, Z) is a bivariate Gaussian distribution and we can eliminate the conditioning on X in Step 2:

Step 1. Draw X from $p(X \mid Y, Z)$.   (Sampler 2)

Step 2. Draw Y from $p(Y \mid Z)$.

Step 3. Draw Z from $p(Z \mid X, Y)$.

This is advantageous because the draws of Y are independent of the draws of X in Sampler 2, eliminating a high correlation in Sampler 1. Unfortunately, however, the three conditional distributions in Sampler 2, $p(X \mid Y, Z)$, $p(Y \mid Z)$, and $p(Z \mid X, Y)$, are functionally incompatible and imply inconsistent dependence structure. Sampling Y from $p(Y \mid Z)$ suggests that X and Y are conditionally independent given Z, whereas sampling X from $p(X \mid Y, Z)$ suggests conditional dependence. The result is that the stationary distribution of Sampler 2 does not correspond to the target distribution $p(X, Y, Z)$; information on the correlation between X

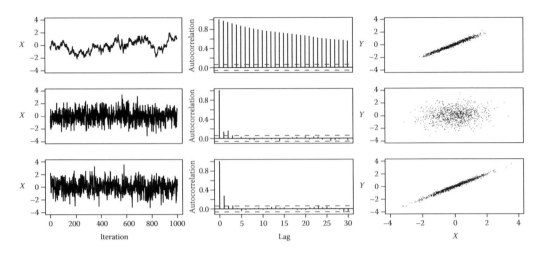

FIGURE 15.1
Comparing three MCMC samplers for a simple Gaussian example. The three rows correspond to the Gibbs sampler (Sampler 1), the Gibbs sampler with the conditioning on X eliminated in Step 2 (Sampler 2), and the partially collapsed Gibbs sampler (Sampler 3). The first two columns show the mixing and autocorrelation of X, and the last column presents the sampled correlation structure between X and Y, based on 1000 iterations. The PCG sampler dramatically improves convergence while maintaining the target stationary distribution. In this simple case, the PCG sampler is simply a blocked Gibbs sampler.

and Y is lost (see the second row of Figure 15.1). Of course, there is an obvious solution. If we simply change the order of the draws in Sampler 2, that is, first sample Y from its conditional distribution given Z and then X from its conditional distribution given (Y, Z), we obtain a correct joint draw from $p(X, Y \mid Z)$. This results in the following sampler:

> **Step 1.** Draw Y from $p(Y \mid Z)$.   (Sampler 3)
>
> **Step 2.** Draw X from $p(X \mid Y, Z)$.
>
> **Step 3.** Draw Z from $p(Z \mid X, Y)$.

Although the conditional distributions remain incompatible, the third row of Figure 15.1 shows the fast convergence of the subchain for X and the correctly sampled correlation between X and Y. In this case, Sampler 3 is simply a blocked version of Sampler 1: sampling $p(Y \mid Z)$ and then $p(X \mid Y, Z)$ combines into a single draw from $p(X, Y \mid Z)$. As we shall illustrate, however, partial collapse is a more general technique than blocking. Liu et al. (1994) showed that the spectral radius of the forward operator for the blocked Gibbs sampler iterating between $p(X, Y \mid Z)$ and $p(Z \mid X, Y)$ is the square of the maximal correlation between (X, Y) and Z, that is, $1/\{2(1 + \rho)\}$. Thus, with $\rho = 0.99$, the spectral radius is 0.25, confirming the faster convergence of Sampler 3 than Sampler 1.

This simple three-step sampler illustrates an important point: care must be taken if we are to maintain the target stationary distribution when reducing the conditioning in some but not all of the steps of a Gibbs sampler. Van Dyk and Park (2008) describe three basic tools that can be used to transform a Gibbs sampler into a PCG sampler that maintains the target stationary distribution. The first tool is *marginalization*, which involves moving a group of unknowns from being conditioned upon to being sampled in one or more steps of a Gibbs sampler; the marginalized group can differ among the steps. In Sampler 1 this involves replacing the sampling of $p(Y \mid X, Z)$ with the sampling of $p(X, Y \mid Z)$ in Step 2;

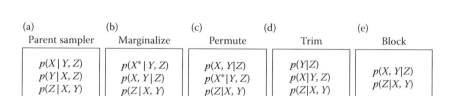

$$p(Z|X, Y)$$

(a) Parent sampler	(b) Marginalize	(c) Permute	(d) Trim	(e) Block
$p(X \mid Y, Z)$ $p(Y \mid X, Z)$ $p(Z \mid X, Y)$	$p(X^* \mid Y, Z)$ $p(X, Y \mid Z)$ $p(Z \mid X, Y)$	$p(X, Y \mid Z)$ $p(X^* \mid Y, Z)$ $p(Z \mid X, Y)$	$p(Y \mid Z)$ $p(X \mid Y, Z)$ $p(Z \mid X, Y)$	$p(X, Y \mid Z)$ $p(Z \mid X, Y)$

FIGURE 15.2

Transforming the Gibbs sampler (Sampler 1) into the partially collapsed Gibbs sampler (Sampler 3) sequentially using marginalization, permutation, and trimming. The sampler in (e) is a blocked version of Sampler 1.

see Figure 15.2a,b. Notice that rather than simply reducing the conditioning by eliminating X, we are moving X from being conditioned upon to being sampled. This can be done by combining a distribution that conditions on less with a conditional distribution available from the parent sampler, that is, $p(X, Y \mid Z)$ can be sampled by first sampling the reduced conditional distribution, $p(Y \mid Z)$, and then sampling the conditional distribution from the original sampler, $p(X \mid Y, Z)$. This preserves the stationary distribution of the Markov chain. The second tool is *permutation* of the steps. We may need to permute steps in order to use the third tool, which is to *trim* sampled components from steps if the components can be removed from the sampler without altering its Markov transition kernel. In Figure 15.2c we permute the steps so that we can trim X^* from the sampler in (d). Here and elsewhere we use a superscript "$\star$" to designate an *intermediate quantity* that is sampled but is not part of the output of an iteration. Finally, we block the first two steps in (e).

Both marginalization and permutation clearly maintain the stationary distribution of the chain and both can affect its convergence properties; marginalization can dramatically improve convergence, while the effect of permutation is typically small. Reducing conditioning (i.e. marginalization) increases variance and hence the sizes of the sampling jumps; see van Dyk and Park (2008) for a technical treatment. Trimming is explicitly designed to maintain the kernel of the chain. The primary advantage of trimming is to reduce the complexity of the individual steps. In doing so, trimming may introduce incompatibility into a sampler.

To illustrate how the three tools are used in a more realistic setting we use the simple four-step example given in Figure 15.3a, where the target distribution is $p(\mathbf{W}, \mathbf{X}, \mathbf{Y}, \mathbf{Z})$. Suppose it is possible to directly sample from $p(\mathbf{Y} \mid \mathbf{X}, \mathbf{Z})$ and $p(\mathbf{Z} \mid \mathbf{X}, \mathbf{Y})$, which are both conditional distributions of $\int p(\mathbf{W}, \mathbf{X}, \mathbf{Y}, \mathbf{Z}) d\mathbf{W}$, with $p(\mathbf{Y} \mid \mathbf{X}, \mathbf{Z}) \neq p(\mathbf{Y} \mid \mathbf{W}, \mathbf{X}, \mathbf{Z})$ and $p(\mathbf{Z} \mid \mathbf{X}, \mathbf{Y}) \neq p(\mathbf{Z} \mid \mathbf{W}, \mathbf{X}, \mathbf{Y})$. If we were to simply replace the third and fourth draws in

(a) Parent sampler	(b) Marginalize	(c) Permute	(d) Trim	(e) Block
$p(W \mid X, Y, Z)$ $p(X \mid W, Y, Z)$ $p(Y \mid W, X, Z)$ $p(Z \mid W, X, Y)$	$p(W^* \mid X, Y, Z)$ $p(X \mid W, Y, Z)$ $p(W^*, Y \mid X, Z)$ $p(W, Z \mid X, Y)$	$p(W^*, Y \mid X, Z)$ $p(W^*, Z \mid X, Y)$ $p(W \mid X, Y, Z)$ $p(X \mid W, Y, Z)$	$p(Y \mid X, Z)$ $p(Z \mid X, Y)$ $p(W \mid X, Y, Z)$ $p(X \mid W, Y, Z)$	$p(Y \mid X, Z)$ $p(W, Z \mid X, Y)$ $p(X \mid W, Y, Z)$

FIGURE 15.3

Transforming a four-step Gibbs sampler into a partially collapsed Gibbs sampler. The sampler in (e) is composed of incompatible conditional distributions, is not a blocked version of the sampler in (a), and is therefore not a Gibbs sampler *per se*.

Figure 15.3a with draws from $p(\mathbf{Y} \mid \mathbf{X}, \mathbf{Z})$ and $p(\mathbf{Z} \mid \mathbf{X}, \mathbf{Y})$, we would have no direct way of verifying that the stationary distribution of the resulting chain is the target joint distribution. Instead, we use the three basic tools to derive a PCG sampler. This allows us to reap the benefits of partial collapse while ensuring that the stationary distribution of the chain is the target distribution.

In Figure 15.3b, we use marginalization to move $\mathbf{W}$ from being conditioned upon to being sampled in the last two steps. In each step we condition on the most recently sampled value of each quantity that is not sampled in that step. The output of the iteration consists of the most recently sampled value of each quantity at the end of the iteration: $\mathbf{X}$ sampled in the second step, $\mathbf{Y}$ sampled in the third step, and $(\mathbf{W}, \mathbf{Z})$ sampled in the last step. Although sampling $\mathbf{W}$ three times in each iteration may be inefficient, removing any two of the three draws affects the transition kernel of the chain: the draw in the first step is conditioned upon in the second step and the draw in the last step is part of the output of the iteration. In order to preserve the stationary distribution, we only remove intermediate quantities whose values are not conditioned upon subsequently. Permuting the steps of a Gibbs sampler does not alter its stationary distribution but can enable certain intermediate quantities to meet the criterion for removal. In Figure 15.3c we permute the steps so that two of the draws of $\mathbf{W}$ can be trimmed in (d). The intermediate draws of $\mathbf{W}$ sampled in the first and second steps of Figure 15.3c are not used subsequently and both can be removed from the sampler. Finally, the middle two steps of Figure 15.3d can be combined to derive the final sampler given in (e). After blocking, the set of conditional distributions in Figure 15.3e remains incompatible, illustrating that partial collapse is a more general technique than blocking.

The samplers in Figure 15.3c, d have the same stationary distribution because removing the intermediate quantities does not affect the transition kernel. Thus, we know the stationary distribution of Figure 15.3d is the target joint distribution. This illustrates how careful use of the three basic tools can lead to PCG samplers with the target stationary distribution. Notice that the samplers in Figure 15.3d,e are not Gibbs samplers *per se*. The conditional distributions that are sampled in each are incompatible and permuting their order may alter the stationary distribution of the chain.

15.3   Path-Adaptive Metropolis–Hastings Sampler

The second computational method aims to improve the convergence of the MH sampler by updating the proposal distribution using information about the target distribution obtained from an initial run of the chain. Suppose a target distribution of interest has density $\pi(\mathbf{X})$. Given a current state $\mathbf{X}^{(t)}$, the MH sampler proposes a state $\mathbf{X}'$ using a proposal distribution $p_1(\mathbf{X}' \mid \mathbf{X}^{(t)})$; we use a one in the subscript because we update this proposal distribution below. The move from $\mathbf{X}^{(t)}$ to $\mathbf{X}'$ is accepted with probability

$$q_1(\mathbf{X}' \mid \mathbf{X}^{(t)}) = \min \left\{ 1, \frac{\pi(\mathbf{X}')/p_1(\mathbf{X}' \mid \mathbf{X}^{(t)})}{\pi(\mathbf{X}^{(t)})/p_1(\mathbf{X}^{(t)} \mid \mathbf{X}')} \right\}.$$

That is, $\mathbf{X}^{(t+1)}$ is set to $\mathbf{X}'$ with probability $q_1(\mathbf{X}' \mid \mathbf{X}^{(t)})$ and to $\mathbf{X}^{(t)}$ otherwise. Thus, for any $\mathbf{X}^{(t+1)} \neq \mathbf{X}^{(t)}$, the transition kernel of the MH sampler is

$$\mathcal{K}_1(\mathbf{X}^{(t+1)} \mid \mathbf{X}^{(t)}) = p_1(\mathbf{X}^{(t+1)} \mid \mathbf{X}^{(t)}) q_1(\mathbf{X}^{(t+1)} \mid \mathbf{X}^{(t)}).$$

The path-adaptive Metropolis–Hastings (PAMH) sampler is an efficient MH sampler that uses an empirical distribution generated from an initial run of the chain (i.e. the path samples of the chain) as a second proposal distribution. This is used to construct a second transition kernel that is mixed with the original transition kernel in subsequent draws. In this way, we use the sample generated by MH to construct a proposal distribution that more closely resembles the target distribution. This can dramatically improve performance if the original MH sampler is either slow mixing or computationally demanding.

The strategy of mixing transition kernels for MCMC methods had been suggested in the literature as one of the basic forms of hybrid strategies (Tierney, 1994). When different Markov transition kernels with a common stationary distribution are available, we can combine them in a mixture by specifying positive probabilities to the kernels and selecting one of the kernels according to the probabilities in each iteration. The random-scan Gibbs sampler is a common example of such a hybrid sampler (Roberts and Rosenthal, 1997). The PAMH sampler is a mixture of two MH samplers: with probability α, a proposal state $\mathbf{X}'$ is generated from $p_1(\mathbf{X}' \mid \mathbf{X}^{(t)})$ and accepted with probability $q_1(\mathbf{X}' \mid \mathbf{X}^{(t)})$; and with probability $1 - \alpha$, a proposal state $\mathbf{X}'$ is generated from an empirical distribution $\hat{\pi}(\mathbf{X})$ and accepted with probability

$$q_2(\mathbf{X}' \mid \mathbf{X}^{(t)}) = \min \left\{ 1, \frac{\pi(\mathbf{X}')/\hat{\pi}(\mathbf{X}')}{\pi(\mathbf{X}^{(t)})/\hat{\pi}(\mathbf{X}^{(t)})} \right\}.$$

Thus, for any $\mathbf{X}^{(t+1)} \neq \mathbf{X}^{(t)}$, the transition kernel of the PAMH sampler is given by

$$\mathcal{K}_+(\mathbf{X}^{(t+1)} \mid \mathbf{X}^{(t)}) = \alpha \mathcal{K}_1(\mathbf{X}^{(t+1)} \mid \mathbf{X}^{(t)}) + (1 - \alpha)\mathcal{K}_2(\mathbf{X}^{(t+1)} \mid \mathbf{X}^{(t)}), \qquad (15.1)$$

where $\mathcal{K}_2(\mathbf{X}^{(t+1)} \mid \mathbf{X}^{(t)}) = \hat{\pi}(\mathbf{X}^{(t+1)})q_2(\mathbf{X}^{(t+1)} \mid \mathbf{X}^{(t)})$.

An adaptive MCMC sampler (Roberts and Rosenthal, 2009) attempts to learn about a target distribution using information available from MCMC draws while they run. Thus, the PAMH sampler can be viewed as an adaptive MCMC sampler in that it mixes the original transition kernel with a transition kernel learned from an initial run of an MCMC sampler. It does not, however, continually adapt the empirical transition kernel. The mixture proportion α in Equation 15.1 is a tuning parameter that is set in advance. In effect the value of α is set to one during the initial run that uses only the original proposal distribution to generate samples from $\pi(\mathbf{X})$, and the path samples from the initial run are then used to compute an approximation to the target distribution, $\hat{\pi}(\mathbf{X})$. After the initial run, α is fixed at some value between 0 and 1, and the mixture kernel in Equation 15.1 is used. In other words, the original MH sampler is run for the first N_1 iterations, and the PAMH sampler that mixes the original proposal distribution with an approximation to the target distribution is run for an additional N_2 iterations. The number of iterations for the initial run, N_1, is usually set to a reasonably small number.

If the dimension of $\mathbf{X}$ is small, the empirical distribution $\hat{\pi}(\mathbf{X})$ can be computed by discretizing the space into sufficiently small pixels and calculating the proportion of the initial N_1 draws that fall into each pixel. In some cases the approximation can be improved by discarding an initial burn-in from the N_1 draws. In this way, we approximate $\hat{\pi}(\mathbf{X})$ with a step function that is sampled by first selecting a pixel according to the empirical pixel probabilities and then sampling uniformly within the pixel. To get a more precise approximation to the target distribution, we can use a more sophisticated nonparametric density estimation method, such as kernel density estimation. This strategy is more efficient in higher dimensions and can improve the empirical approximation even in lower dimensions. Of course, if the target distribution is discrete, no pixeling or smoothing is necessary.

Detailed balance is satisfied by the mixture transition kernel in Equation 15.1 because

$$\pi(\mathbf{X}^{(t)})\mathcal{K}_+(\mathbf{X}^{(t+1)} \mid \mathbf{X}^{(t)}) = \alpha \min \left\{ \pi(\mathbf{X}^{(t)})p_1(\mathbf{X}^{(t+1)} \mid \mathbf{X}^{(t)}), \pi(\mathbf{X}^{(t+1)})p_1(\mathbf{X}^{(t)} \mid \mathbf{X}^{(t+1)}) \right\}$$

$$+ (1 - \alpha) \min \left\{ \pi(\mathbf{X}^{(t)})\hat{\pi}(\mathbf{X}^{(t+1)}), \pi(\mathbf{X}^{(t+1)})\hat{\pi}(\mathbf{X}^{(t)}) \right\}$$

is a symmetric function of $\mathbf{X}^{(t)}$ and $\mathbf{X}^{(t+1)}$. Thus the resulting Markov chain is reversible with respect to $\pi(\mathbf{X})$. The PAMH sampler uses a mixture of two MH samplers rather than a single MH sampler with the mixture of two proposal distributions because the mixture of two MH samplers requires only the computation of one proposal distribution at each iteration. Thus, the PAMH sampler reduces the number of evaluations of $\pi(\mathbf{X})$. This significantly improves the overall computation in the example of Section 15.4 where this evaluation is computationally costly. See Tierney (1998) for a comparison of the asymptotic efficiency of these two strategies.

To illustrate the advantage of the PAMH sampling strategy, we introduce a simple example where both the Gibbs sampler and the MH sampler exhibit slow convergence. Consider the following bivariate distribution which has Gaussian conditional distributions but is not a bivariate Gaussian distribution:

$$p(X, Y) \propto \exp \left\{ -\frac{1}{2} \left(8X^2 Y^2 + X^2 + Y^2 - 8X - 8Y \right) \right\}. \tag{15.2}$$

This is a bimodal special case of a parameterized family of distributions derived by Gelman and Meng (1991).

A Gibbs sampler can easily be constructed to simulate from Equation 15.2:

Step 1. Draw X from $p(X \mid Y)$, where $X \mid Y \sim N(4/(8Y^2 + 1), 1/(8Y^2 + 1))$.

Step 2. Draw Y from $p(Y \mid X)$, where $Y \mid X \sim N(4/(8X^2 + 1), 1/(8X^2 + 1))$.

An MH sampler can also be used to simulate the target distribution in Equation 15.2. We use an independent bivariate Gaussian distribution for the proposal distribution. That is, given the current state $(X^{(t)}, Y^{(t)})$, we generate a proposal state $(X', Y') = (X^{(t)} + \epsilon_1, Y^{(t)} + \epsilon_2)$, where $\epsilon_i \overset{\text{i.i.d.}}{\sim} N(0, \tau^2)$ for $i = 1, 2$, and accept the proposal state with probability $p(X', Y')/p(X^{(t)}, Y^{(t)})$. In this case, τ is a tuning parameter that is chosen in advance and affects the convergence of the resulting sampler (Roberts and Rosenthal, 2001). A value of τ that is too small produces small jumps which are often accepted but lead to a Markov chain that moves slowly. On the other hand, when τ is too large, the sampler will propose large jumps that are too often rejected. Thus, it is important to find a reasonable choice of τ between these two extremes. For illustration, we use three MH samplers, run with $\tau = 0.5$, 1, and 2.

We ran the Gibbs sampler and the MH sampler with three different values of τ for 20,000 iterations each. Convergence of the four samplers is described in the first four rows of Figure 15.4. The first two columns of Figure 15.4 show the trace plot of the last 5000 iterations and autocorrelations computed using the last 10,000 iterations of each subchain of X. The last column compares each simulated marginal distribution of X based on the last 10,000 draws (histogram) with the target distribution (solid line). The first four rows of Figure 15.4 illustrate the slow mixing and high autocorrelations of all four samplers; the simulated marginal distributions do not approximate the target distribution as well as we

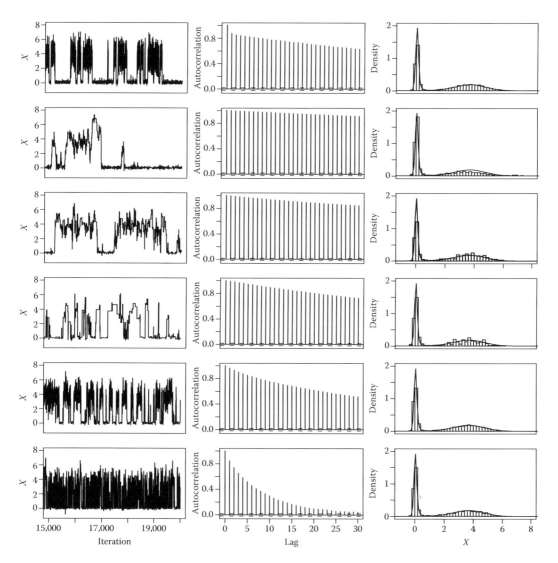

FIGURE 15.4

Comparing six MCMC samplers constructed for simulating a bivariate distribution that has Gaussian conditional distributions but is not a bivariate Gaussian distribution. The rows correspond to the Gibbs sampler, the MH sampler with $\tau = 0.5$, the MH sampler with $\tau = 1$, the MH sampler with $\tau = 2$, the MH-within-PCG sampler run with $\tau = 1$, and the PAMH-within-PCG sampler run with $\tau = 1$ and $\alpha = 0.5$. The first column shows trace plots of the last 5000 iterations of each chain; the second column contains autocorrelations of the last 10,000 draws; and the last column compares each simulated marginal distribution of X based on the last 10,000 draws (histogram) with the true marginal distribution (solid line).

might hope. Among the three MH samplers, the choice of $\tau = 1$ (row 3) results in the best convergence.

We can use a PCG sampler to improve convergence as described in Section 15.2. In particular, if we could eliminate the conditioning on Y in Step 1 of the Gibbs sampler, we could generate independent draws by iterating between the following steps:

Step 1. Draw X from $p(X)$, where

$$p(X) \propto \frac{1}{\sqrt{8X^2 + 1}} \exp\left\{ -\frac{1}{2}\left(X^2 - 8X - \frac{16}{8X^2 + 1}\right)\right\}.$$

Step 2. Draw Y from $p(Y \mid X)$.

This PCG sampler would be a blocked one-step sampler if we could simulate $p(X)$ directly. Because we cannot, we consider indirect sampling in Step 1 using MH with a Gaussian proposal distribution, $X' \mid X^{(t)} \sim N(X^{(t)}, \tau^2)$. This results in an MH-within-PCG sampler that we implement with $\tau = 1$.* To further improve convergence, we use PAMH sampling in Step 1. This results in a PAMH-within-PCG sampler, that we also implement with $\tau = 1$ and with $\alpha = 1$ for the first 1000 iterations and $\alpha = \frac{1}{2}$ for the next 19,000 iterations. We discretize the space into 200 bins equally spaced between -1 and 8, and approximate $\hat{\pi}(X)$ using the bin proportions from the first 1000 iterations. The last two rows of Figure 15.4 illustrate the convergence of the MH and PAMH-within-PCG samplers, respectively. The PAMH-within-PCG sampler exhibits a clear improvement over the other five MCMC samplers.

15.4 Spectral Analysis in High-Energy Astrophysics

We now turn to the illustration of PCG and PAMH in spectral analysis in high-energy astrophysics. In recent years technological advances have dramatically increased the quality and quantity of data available to astronomers. Instrumentation is tailored to data-collection challenges associated with specific scientific goals. These instruments provide massive new surveys resulting in new catalogs containing terabytes of data, high resolution spectrography and imaging across the electromagnetic spectrum, and incredibly detailed movies of dynamic and explosive processes in the solar atmosphere. The spectrum of new instruments is helping make impressive strides in our understanding of the universe, but at the same time generating massive data-analytic and data-mining challenges for scientists who study the data.

High-energy astrophysics is concerned with ultraviolet rays, X-rays, and γ-rays, that is, photons with energies of a few electron-volts (eV), a few kiloelectron-volts (keV), or more than a megaelectron-volt, respectively. Roughly speaking, the production of high-energy electromagnetic waves requires temperatures of millions of degrees and signals the release of deep wells of stored energy such as those in very strong magnetic fields, extreme gravity, explosive nuclear forces, and shock waves in hot plasmas. Thus, X-ray and γ-ray telescopes can map nearby stars with active magnetic fields, the remnants of exploding stars, areas of star formation, regions near the event horizon of a black hole, very distant turbulent galaxies, or even the glowing gas embedding a cosmic cluster of galaxies. The distribution of the energy of the electromagnetic emissions is called the spectrum and gives insight into these deep energy wells: the composition, density, and temperature/energy distribution of the emitting material; any chaotic or turbulent flows; and the strengths of the magnetic, electrical, or gravitational fields.

---

* Verifying the stationary distribution of a MH-within PCG sampler or a PHMH-within-PCG sampler considers somewhat subtle considerations that we do not discuss here. See van Dyk and Park (2011) for details.

In this chapter we focus on X-ray spectral analysis. A typical spectrum can be formulated as a finite mixture distribution composed of one or more continuum terms, which are smooth functions across a wide range of energies, and one or more emission lines, which are local features highly focused on a narrow band of energies. For simplicity we focus on a case where there is one continuum term and one emission line. Because of instrumental constraints, photons are counted in a number of energy bins. These photon counts are modeled as an inhomogeneous Poisson process with expectation in energy bin j modeled as

$$\Lambda_j(\boldsymbol{\theta}) = f_j\left(\boldsymbol{\theta}^C\right) + \lambda \pi_j\left(\mu, \sigma^2\right), \tag{15.3}$$

where $\boldsymbol{\theta}$ is the set of model parameters, $f_j(\boldsymbol{\theta}^C)$ is the expected continuum count in bin j, $\boldsymbol{\theta}^C$ is the set of free parameters in the continuum model, λ is the total expected line count, and $\pi_j(\mu, \sigma^2)$ is the proportion of an emission line with location μ and width σ^2 falling into bin j. Various emission line profiles such as Gaussian distributions, t distributions, and delta functions can be used to derive the emission line bin proportions as a function of μ and σ^2. We focus on the use of a delta function which is parameterized only in terms of μ.

Due to instrumental constraints, the photon counts are subject to blurring of the individual photon energies, stochastic censoring with energy dependent rates, and background contamination. To account for these processes, we embed the scientific model in Equation 15.3 within a more complex observed-data model. In particular, the observed photon counts in detector channel l are modeled with a Poisson distribution,

$$Y_{\text{obs} l} \sim \text{Poisson}\left(\sum_j M_{lj} \Lambda_j(\boldsymbol{\theta}) u_j(\boldsymbol{\theta}^A) + \theta_l^B\right), \tag{15.4}$$

where M_{lj} is the probability that a photon that arrives with energy corresponding to bin j is recorded in channel l, $u_j(\boldsymbol{\theta}^A)$ is the probability that a photon with energy corresponding to bin j is observed, and θ_l^B is the expected background counts in channel l. A multilevel model can be constructed to incorporate both the finite mixture distribution of the spectral model and the complexity of the data-generation mechanism. Using a missing-data/latent-variable setup, a standard Gibbs sampler can be constructed to fit the model (van Dyk et al., 2001; van Dyk and Kang, 2004).

15.5 Efficient MCMC in Spectral Analysis

As a specific example, we consider data collected using the Chandra X-ray Observatory in an observation of the quasar PG1634+706 (Park et al., 2008). Quasars are extremely distant astronomical objects that are believed to contain supermassive black holes with masses exceeding that of our Sun by a factor of a million. Because quasars are very distant, the universe was a fraction of its current age when the light we now see as a quasar was emitted. They are also very luminous and therefore give us a way to study the "young" universe. Thus, the study of quasars is important for cosmological theory and their spectra can give insight into their composition, temperature, distance, and velocity.

We are particularly interested in an emission feature of the quasar's spectrum, which is a narrow Fe-K-alpha emission line whose location indicates the ionization state of iron in the emitting plasma. To fit the location of a narrow emission line, we model the emission line with a delta function, so that the entire line falls within one data bin.

Unfortunately, the standard Gibbs sampler described in van Dyk et al. (2001) breaks down when delta functions are used to model emission lines. Using the method of data augmentation, the standard Gibbs sampler is constructed in terms of missing data that include unobserved Poisson photon counts with expectation given in Equation 15.3 and unobserved mixture indicator variables for the mixture given in Equation 15.3. To see why the standard Gibbs sampler fails, we examine how the mixture indicator variables and line location are updated. The components of the mixture indicator variable are updated for each photon within each bin as a Bernoulli variable with probability of being from an emission line,

$$\frac{\lambda \pi_j(\mu)}{f_j(\boldsymbol{\theta}^C) + \lambda \pi_j(\mu)} \tag{15.5}$$

in energy bin j. (We suppress the width, σ^2, of the emission line $\pi_j(\mu, \sigma^2)$, because a delta function has no width.) Because the delta function is contained in a single bin, $\pi_j(\mu) = 1$ if μ is within bin j, and 0 otherwise. This means that the probability in Equation 15.5 is zero for all energy bins except the one containing the current line location, μ. Thus, in each iteration of the standard Gibbs sampler, the only bin that can have photons attributed to the emission line is the bin that contains the current iterate of the line location. When the line location is updated using the photons attributed to the emission line, it is necessarily set to the same value as the current iterate. Thus, unless there are no photons attributed to the line, its location is fixed. As a result, unless the line is very weak, the standard Gibbs sampler is in effect not positive recurrent and does not converge to the target distribution. Although this sampler works fine with emission lines of appreciable width, it fails for delta functions (Park and van Dyk, 2009; van Dyk and Park, 2004).

To understand the computational challenges of fitting this model, we must go into some of the technical details of the Gibbs sampler. Let $\mathbf{Y} = \{Y_{\text{obs}\,l}\}$ be the observed data modeled in Equation 15.4, $\mathbf{Y}_{\text{mis}} = (\mathbf{Y}_{\text{mis}\,1}, \mathbf{Y}_{\text{mis}\,2})$ be a collection of missing data, where $\mathbf{Y}_{\text{mis}\,1}$ denotes the unobserved Poisson photon counts with expectation given in Equation 15.3 and $\mathbf{Y}_{\text{mis}\,2}$ the unobserved mixture indicator variable for each photon under the finite mixture model given in Equation 15.3, μ be the delta function line location, and $\boldsymbol{\psi}$ be the model parameters other than μ. To sample from the target distribution $p(\mathbf{Y}_{\text{mis}}, \boldsymbol{\psi}, \mu \mid \mathbf{Y}_{\text{obs}})$, the parent Gibbs sampler is constructed by iteratively sampling from its conditional distributions, as shown in Figure 15.5a. This is a special case of the "standard" Gibbs sampler discussed above and derived by van Dyk et al. (2001). We devise a PCG sampler to improve the convergence of the parent Gibbs sampler. To construct a PCG sampler, we eliminate the conditioning on all or some of $\mathbf{Y}_{\text{mis}}$ in the step that updates μ. In combination with PAMH, this results in three new efficient samplers.

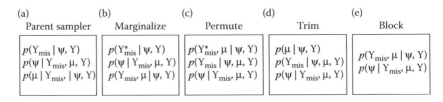

FIGURE 15.5
Transforming the parent Gibbs sampler into PCG I. The PCG I sampler in (e) is constructed by partially collapsing out the missing data and corresponds to a blocked version of its parent sampler in (a).

First, PCG I is constructed by eliminating the conditioning on all of $\mathbf{Y}_{mis}$ in the step that updates μ. Figure 15.5 shows how the parent Gibbs sampler shown in (a) is transformed into PCG I shown in (e) by partially collapsing $\mathbf{Y}_{mis}$ out of the sampler. In Figure 15.5b, $\mathbf{Y}_{mis}$ is moved from being conditioned upon to being sampled in the step that updates μ. The steps are then permuted in Figure 15.5c in order to make one of the two updates of $\mathbf{Y}_{mis}$ redundant. This allows us to trim the unused sample of $\mathbf{Y}_{mis}^{\star}$ from the first step in Figure 15.5d. Finally, we can combine the first two steps into a single sampling of $p(\mathbf{Y}_{mis}, \mu \mid \psi, \mathbf{Y})$. The resulting PCG sampler in Figure 15.5e is a blocked version of the parent Gibbs sampler.

Because the likelihood function is flat within each bin as a function of μ, we can treat μ as a discrete parameter. Its distribution given ψ and $\mathbf{Y}_{obs}$ is multinomial with values corresponding to the midpoints of data bins and probability vector proportional to the product of the Poisson distributions given in Equation 15.4. (We use a flat prior distribution on μ.) This probability vector must be computed at each iteration of the sampler, which is computationally expensive owing to the large blurring matrix $\mathbf{M} = \{M_{lj}\}$ and the large number of energy bins. Because sampling from $p(\mu \mid \psi, \mathbf{Y}_{obs})$ is so expensive, we consider a second PCG sampler that avoids this update. In particular, we consider eliminating only the mixture indicator variables, $\mathbf{Y}_{mis\,2}$, from the step that updates μ in the derivation of PCG II. Because the resulting update for μ conditions on $\mathbf{Y}_{mis\,1}$, ψ, and $\mathbf{Y}_{obs}$, its distribution is multinomial with probability vector proportional to the product of the Poisson distributions given in Equation 15.3. This distribution does not involve the large dimension of the blurring matrix and is much quicker to compute.

Figure 15.6 illustrates the construction of PCG II which is identical to that of PCG I except that only $\mathbf{Y}_{mis\,2}$ is moved from being conditioned upon to being sampled in the step that updates μ. Unlike PCG I, however, PCG II consists of a set of incompatible conditional distributions and does not correspond to a blocked version of the parent Gibbs sampler; see Figure 15.6d. Due to the greater degree of collapsing, PCG I is expected to have better convergence characteristics than PCG II (see van Dyk and Park, 2008). The tradeoff is, however, that an iteration of PCG II is much faster to compute than one of PCG I. A numerical comparison of the two samplers appears below.

In order to further improve computational speed, we consider using an MH step to update μ in an effort to avoid the expense of computing a lengthy probability vector at each iteration. This requires us to evaluate only two components of the multinomial probability vector, the components corresponding to the current value and the proposed value of the line location. Although this can significantly reduce computation time per iteration, it is difficult to find a good proposal distribution because the posterior distribution of the line location can be highly multimodal. A proposal distribution with relatively high variance is required to allow jumping among the modes, but this leads to many rejected proposals in

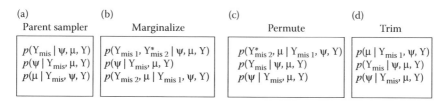

FIGURE 15.6
Transforming the parent Gibbs sampler into PCG II. This PCG sampler is constructed by partially collapsing out part of the missing data, is composed of a set of incompatible conditional distributions, and is not a blocked version of the sampler in (a).

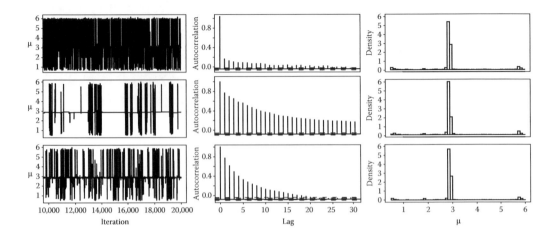

FIGURE 15.7
Comparing three efficient MCMC samplers constructed for spectral analysis. The rows correspond to the PCG I, PCG II, and the PAMH-within-PCG I samplers, respectively. The PAMH sampler was run with $\tau = 2$ and $\alpha = 0.5$. The first column shows trace plots of the last 10,000 iterations of each chain; the second column contains autocorrelation plots of the last 10,000 draws; and the last column presents a simulated marginal distributions of the line location μ based on the last 10,000 draws. Although PCG I mixes more quickly than PAMH-within-PCG I, it requires about 50 times more computing time for the 20,000 draws.

regions between the modes. To improve the convergence of the MH sampler, we consider the PAMH sampling strategy to update μ. We use a Gaussian proposal distribution with standard deviation $\tau = 2$ for the initial MH within Gibbs sampler, which allows jumps across the range of energies (5.5 keV). After 1000 iterations we use an estimate of the discrete marginal posterior distribution of μ in a second MH transition kernel that we mix with the original kernel in a fifty–fifty mixture for the remaining draws of the PAMH-within-PCG I sampler.

The convergence characteristics of the three samplers are compared in Figure 15.7. Each of the three samplers is run for 20,000 iterations. The rows correspond to the PCG I, PCG II, and PAMH-within-PCG I samplers, respectively; the columns correspond to the trace and autocorrelation plots of the last 10,000 iterations, and the simulated marginal posterior distribution of the line location based on the last 10,000 iterations, respectively. Comparing the first two columns of Figure 15.7 illustrates that PCG I has the quickest convergence among the three PCG samplers, but the other two PCG samplers also have fairly fast convergence. When sampling μ from its multinomial conditional distribution, however, PCG I requires significantly more computing time than PCG II, which in turn takes significantly more time than PAMH-within-PCG I. The total computing time for 20,000 iterations of the PCG I, PCG II, and PAMH-within-PCG I samplers on a UNIX machine was 15 hours 35 minutes, 1 hour 55 minutes, and 19 minutes, respectively.

To further compare the three PCG samplers, we compute their effective sample sizes (ESS), defined by

$$\text{ESS} = \frac{n}{1 + 2\sum_{k=1}^{\infty} \rho_k(\theta)},$$

where n is the total sample size and $\rho_k(\theta)$ is the lag-k autocorrelation for θ; see Kass et al. (1998) and Liu (2001). The infinite sum in effective sample size is truncated at lag k when $\rho_k(\theta) < 0.05$. For n equal to 10,000, the effective sample sizes of the PCG I, PCG II, and PAMH-within-PCG I samplers are 2436, 304, and 894, respectively. When computation time is accounted for, the PCG I and PCG II samplers have similar ESS per second: 0.087 and 0.088, respectively. The PAMH-within-PCG I sampler, on the other hand, has ESS per second of 1.568, which is about 18 times larger. Thus, PAMH-within-PCG I offers a dramatic improvement in computation time with very good mixing.

15.6 Conclusion

In this chapter we illustrate the use of two computational techniques to improve the performance of MCMC samplers in a particular example from high-energy astrophysics. These techniques are of course useful in other settings as well. For example, PCG samplers are generally useful when eliminating conditioning on some unobserved quantities in a step of a Gibbs sampler does not complicate the draw. Reducing the conditioning in this way can only improve convergence, but must be implemented carefully to be sure the target stationary distribution is maintained (see Section 15.2). Even when the resulting draw is more complicated, PCG samplers may be worth pursuing if the conditional variance is increased substantially by reducing the conditioning. In this case, the extra effort in obtaining the draw may be offset by an improvement in the overall convergence of the chain. PAMH, on the other hand, is most useful when the initial MH sampler mixes poorly but visits all important areas of the parameter space. In this case, the initial draws can easily be used to construct an improved proposal distribution. Even if mixing is acceptable, PAMH may be useful if the MH proposal distribution is expensive to evaluate. In this case, the initial draws can be used to construct a proposal distribution that provides similar mixing, but with faster evaluation.

There are many other computational techniques and variants on MCMC samplers that can be applied to the myriad of complex model fitting challenges in astronomy. Puzzling together the appropriate computational and statistical methods for the numerous outstanding data-analytic problems offers a gold mine for methodological researchers. We invite all interested readers to join us in this seemingly endless but ever enjoyable endeavor!

Acknowledgments

The authors gratefully acknowledge funding for this project partially provided by NSF grants DMS-04-06085, SES-05-50980, and DMS-09-07522 and by NASA Contract NAS8-39073 and NAS8-03060 (CXC). The spectral analysis example stems from work of the California Harvard Astrostatistics Collaboration (www.ics.uci.edu/~dvd/astrostat.html).

References

Amit, Y. 1991. On rates of convergence of stochastic relaxation for Gaussian and non-Gaussian distributions. *Journal of Multivariate Analysis*, 38:82–89.

Besag, J. and Green, P. J. 1993. Spatial statistics and Bayesian computation. *Journal of the Royal Statistical Society, Series B*, 55:25–37.

Connors, A. and van Dyk, D. A. 2007. How to win with non-Gaussian data: Poisson goodness-of-fit. In E. Feigelson and G. Babu (eds), *Statistical Challenges in Modern Astronomy IV*, Astronomical Society of the Pacific Conference Series, Vol. 371, pp. 101–117. Astronomical Society of the Pacific, San Francisco.

DeGennaro, S., von Hippel, T., Jefferys, W. H., Stein, N., van Dyk, D. A., and Jeffery, E. 2008. Inverting color-magnitude diagrams to access precise cluster parameters: A new white dwarf age for the Hyades. *Astrophysical Journal*, 696:12–23.

van Dyk, D. A. and Park, T. 2004. Efficient EM-type algorithms for fitting spectral lines in high-energy astrophysics. In A. Gelman and X.-L. Meng (eds), *Applied Bayesian Modeling and Causal Inference from Incomplete-Data Perspectives: Contributions by Donald Rubin's Statistical Family*. Wiley, Chichester.

van Dyk, D. A. and Park, T. 2008. Partially collapsed Gibbs samplers: Theory and methods. *Journal of the American Statistical Association*, 103:790–796.

van Dyk, D. A. 2000. Nesting EM algorithms for computational efficiency. *Statistical Sinica*, 10: 203–225.

van Dyk, D. A., Connors, A., Kashyap, V., and Siemiginowska, A. 2001. Analysis of energy spectra with low photon counts via Bayesian posterior simulation. *Astrophysical Journal*, 548:224–243.

van Dyk, D. A., DeGennaro, S., Stein, N., Jefferys, W. H., and von Hippel, T. 2009. Statistical analysis of stellar evolution. *Annals of Applied Statistics*, 3(1): 117–143.

van Dyk, D. A. and Kang, H. 2004. Highly structured models for spectral analysis in high-energy astrophysics. *Statistical Science*, 19:275–293.

van Dyk, D. A. and Meng, X.-L. 1997. Some findings on the orderings and groupings of conditional maximizations within ECM-type algorithms. *Journal of Computational and Graphical Statistics*, 6:202–223.

van Dyk, D. A. and Meng, X.-L. 2001. The art of data augmentation. *Journal of Computational and Graphical Statistics*, 10:1–111.

van Dyk, D. A. and Meng, X.-L. 2010. Cross-fertilizing strategies for better EM mountain climbing and DA field exploration: A graphical guide book. *Statistical Science*, in press.

van Dyk, D. A. and Park, T. 2011. The metropolis within partially collapsed Gibbs sampler. In progress.

Esch, D. N., Connors, A., Karovska, M., and van Dyk, D. A. 2004. An image reconstruction technique with error estimates. *Astrophysical Journal*, 610:1213–1227.

Gelfand, A. E., Sahu, S. K., and Carlin, B. P. 1995. Efficient parameterization for normal linear mixed models. *Biometrika*, 82:479–488.

Gelman, A. and Meng, X.-L. 1991. A note on bivariate distributions that are conditionally normal. *American Statistician*, 45:125–126.

Gilks, W. R., Best, N. G., and Tan, K. K. C. 1995. Adaptive rejection Metropolis sampling within Gibbs sampling. *Applied Statistics*, 44:455–472.

Gregory, P. C. 2005. A Bayesian analysis of extrasolar planet data for HD 73526. *Astrophysical Journal*, 631:1198–1214.

Kashyap, V. and Drake, J. J. 1998. Markov-chain Monte Carlo reconstruction of emission measure distributions: Application to solar extreme-ultraviolet spectra. *Astrophysical Journal*, 503:450–466.

Kass, R. E., Carlin, B. P., Gelman, A., and Neal, R. M. 1998. Markov chain Monte Carlo in practice: A roundtable discussion. *American Statistician*, 52:93–100.

Liu, J. S. 1994. Fraction of missing information and convergence rate of data augmentation. In J. Sall and A. Lehman (eds), *Computing Science and Statistics: Proceedings of the 26th Symposium on the Interface*, pp. 490–496. Interface Foundation of North America, Fairfax Station, VA.

Liu, J. S. 2001. *Monte Carlo Strategies in Scientific Computing*. Springer, New York.

Liu, C. and Rubin, D. B. 1994. The ECME algorithm: A simple extension of EM and ECM with faster monotone convergence. *Biometrika*, 81:633–648.

Liu, J. S. and Wu, Y. N. 1999. Parameter expansion for data augmentation. *Journal of the American Statistical Association*, 94:1264–1274.

Liu, J. S., Wong, W. H., and Kong, A. 1994. Covariance structure of the Gibbs sampler with applications to comparisons of estimators and augmentation schemes. *Biometrika*, 81:27–40.

Meng, X.-L. and van Dyk, D. A. 1997. The EM algorithm—an old folk song sung to a fast new tune (with discussion). *Journal of the Royal Statistical Society, Series B*, 59:511–567.

Meng, X.-L. and van Dyk, D. A. 1999. Seeking efficient data augmentation schemes via conditional and marginal augmentation. *Biometrika*, 86:301–320.

Park, T. and van Dyk, D. A. 2009. Partially collapsed Gibbs samplers: Illustrations and applications. *Journal of Computational and Graphical Statistics*, 18:283–305.

Park, T., van Dyk, D. A., and Siemiginowska, A. 2008. Searching for narrow emission lines in X-ray spectra: Computation and methods. *Astrophysical Journal*, 688:807–825.

Roberts, G. O. and Rosenthal, J. S. 1997. Geometric ergodicity and hybrid Markov chains. *Electronic Communications in Probability*, 2:13–25.

Roberts, G. O. and Rosenthal, J. S. 2001. Optimal scaling for various Metropolis-Hastings algorithms. *Statistical Science*, 16:351–367.

Roberts, G. O. and Rosenthal, J. S. 2009. Examples of adaptive MCMC. *Journal of Computational and Graphical Statistics*, 18:349–367.

Tanner, M. A. and Wong, W. H. 1987. The calculation of posterior distributions by data augmentation (with discussion). *Journal of the American Statistical Association*, 82:528–550.

Tierney, L. 1994. Markov chains for exploring posterior distributions. *Annals of Statistics*, 22:1701–1762.

Tierney, L. 1998. A note on Metropolis-Hastings kernels for general state spaces. *Annals of Applied Probability*, 8:1–9.

Yu, Y. 2005. Three contributions to statistical computing. PhD thesis, Department of Statistics, Harvard University.

Yu, Y. and Meng, X.-L. 2011. To center or not to center, that is not the question: A sufficiency-ancillarity interweaving strategy for boosting MCMC efficiency (with discussion). *Journal of Computational and Graphical Statistics*, in press.

16

Posterior Exploration for Computationally Intensive Forward Models

David Higdon, C. Shane Reese, J. David Moulton, Jasper A. Vrugt, and Colin Fox

16.1 Introduction

In a common inverse problem, we wish to infer about an unknown spatial field $x = (x_1, \ldots, x_m)^T$, given indirect observations $y = (y_1, \ldots, y_n)^T$. The observations, or data, are linked to the unknown field x through some physical system

$$y = \zeta(x) + \epsilon,$$

where $\zeta(x)$ denotes the physical system and ϵ is an n-vector of observation errors. Examples of such problems include medical imaging (Kaipio and Somersalo, 2004), geologic and hydrologic inversion (Stenerud et al., 2008), and cosmology (Jimenez et al. 2004). When a forward model, or simulator, of the physical process $\eta(x)$ is available, one can model the data using the simulator

$$y = \eta(x) + e,$$

where e includes observation error as well as error due to the fact that the simulator $\eta(x)$ may be systematically different from reality $\zeta(x)$ for input condition x. Our goal is to use the observed data y to make inference about the spatial input parameters x—predict x and characterize the uncertainty in the prediction for x.

The likelihood $L(y|x)$ is then specified to account for both mismatch and sampling error. We will assume zero-mean Gaussian errors so that

$$L(y|x) \propto \exp\left\{-\frac{1}{2}(y - \eta(x))^T \Sigma_e^{-1}(y - \eta(x))\right\}, \tag{16.1}$$

with Σ_e known. It is worth noting that the data often come from only a single experiment. So while it is possible to quantify numerical errors, such as those due to discretization (see Kaipio and Somersalo, 2004; Nissinen et al., 2008), there is no opportunity to obtain data from additional experiments for which some controllable inputs have been varied. Because of this limitation, there is little hope of determining the sources of error in e due to model inadequacy. Therefore, the likelihood specification will often need to be done with some care, incorporating the modeler's judgment about the appropriate size and nature of the mismatch term.

In many inverse problems we wish to reconstruct x, an unknown process over a regular two-dimensional lattice. We consider systems for which the model input parameters x denote a spatial field or image. The spatial prior is specified for x, $\pi(x)$, which typically takes into account modeling, and possibly computational considerations.

The resulting posterior is then given by

$$\pi(x|y) \propto L(y|\eta(x)) \times \pi(x).$$

This posterior can, in principle, be explored via Markov chain Monte Carlo (MCMC). However the combined effects of the high dimensionality of x and the computational demands of the simulator make implementation difficult, and often impossible, in practice. By itself, the high dimensionality of x is not necessarily a problem. MCMC has been carried out with relative ease in large image applications (Rue, 2001; Weir, 1997). However, in these examples, the forward model was either trivial or non-existent. Unfortunately, even a mildly demanding forward simulation model can greatly affect the feasibility of doing MCMC to solve the inverse problem.

In this chapter we apply a standard single-site updating scheme that dates back to Metropolis et al. (1953) to sample from this posterior. While this approach has proven effective in a variety of applications, it has the drawback of requiring hundreds of thousands of calls to the simulation model. In Section 16.3 we consider two MCMC schemes that use highly multivariate updates to sample from $\pi(x|y)$: the multivariate random-walk Metropolis algorithm (Gelman et al., 1996) and the *differential evolution* MCMC (DE-MCMC) sampler of ter Braak (2006). Such multivariate updating schemes are alluring for computationally demanding inverse problems since they have the potential to update many (or all) components of x at once, while requiring only a single evaluation of the simulator. Next, in Section 16.4, we consider augmenting the basic posterior formulation with additional formulations based on faster, approximate simulators. The faster, approximate simulators are created by altering the multigrid solver used to compute $\eta(x)$. These approximate simulators can be used in a delayed acceptance scheme (Christen and Fox, 2005; Fox and Nicholls, 1997), as well as in an augmented formulation (Higdon et al., 2002). Both of these recipes can be utilized with any of the above MCMC schemes, often leading to substantial improvements in efficiency. In each section we illustrate the updating schemes with an electrical impedance tomography (EIT) application described in the next section, where the values of x denote electrical conductivity of a two-dimensional object. The chapter concludes with a discussion and some general recommendations.

16.2 An Inverse Problem in Electrical Impedance Tomography

Bayesian methods for EIT applications have been described in Fox and Nicholls (1997), Kaipio et al. (2000), and Andersen et al. (2003). A notional inverse problem is depicted in Figure 16.1; this setup was given previously in Moulton et al. (2008). Here a two-dimensional object composed of regions with differing electrical conductivity is interrogated by 16 electrodes. From each electrode, in turn, a current I is injected into the object and taken out at a rate of $I/(16 - 1)$ at the remaining 15 electrodes. The voltage is then measured at each of the 16 electrodes. These 16 experimental configurations result in $n = 16 \times 16$ voltage observations which are denoted by the n-vector y. The measurement error is simulated by adding independent and identically distributed mean-zero Gaussian noise to each of the voltage

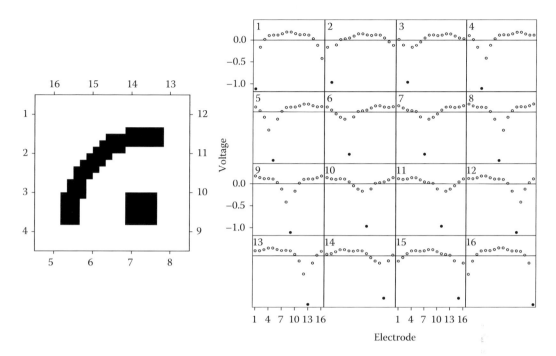

FIGURE 16.1

A synthetic EIT application. A two-dimensional object is surrounded by electrodes at 16 evenly spaced locations around its edge. The conductivity of the object is 3 in the white regions, and 4 in the black regions (the units are arbitrary since the data are invariant to scalings of the conductivity). First, a current of I is injected at electrode 1, and extracted evenly at the other 15 electrodes. The voltage is measured at each electrode. This data is shown in the plot labeled 1 on the right. Similar experiments are carried out with each electrode taking a turn as the injector. The resulting voltages are shown in the remaining 15 plots. In each plot, the voltage corresponding to the injector electrode is given by a black plotting symbol.

measurements. The standard deviation σ of this noise is chosen so that the signal to noise ratio is about 1000 : 3, which is typical of actual EIT measurements. The resulting simulated data is shown on the right in Figure 16.1—one plot for each of the 16 circuit configurations. In each of those plots, the injector electrode is denoted by the black plotting symbol.

We take s to denote spatial locations within the object $\Omega = [0, 1] \times [0, 1]$, and take $x(s)$ to denote the electrical conductivity at site s. We also take $v(s)$ to be the potential at location s, and $j(s)$ to be the current at boundary location s. A mathematical model for the measurements is then the Neumann boundary-value problem

$$-\nabla \cdot x(s)\nabla v(s) = 0, \quad s \in \Omega,$$

$$x(s)\frac{\partial v(s)}{\partial n(s)} = j(s), \quad s \in \partial\Omega,$$

where $\partial\Omega$ denotes the boundary of the object Ω and $n(s)$ is the unit normal vector at the boundary location $s \in \partial\Omega$. The conservation of current requires that the sum of the currents at each of the 16 electrodes be 0.

In order to numerically solve this problem for a given set of currents at the electrodes and a given conductivity field, $x(s)$, the conductivity field is discretized into an $m = 24 \times 24$ lattice. We use a robust multigrid solver called Black Box MG (Dendy, 1987). In addition to being rather general and fast, we can also exploit the multigrid nature of the algorithm to develop fast approximations using the MCMC scheme described in Section 16.4.

Now, for any specified conductivity configuration x and current configuration, the multigrid solver produces 16 voltages. For all 16 current configurations, 16 forward solves produces an $n = 256$-vector of resulting voltages $\eta(x)$. Hence, the sampling model for the data y given the conductivity field x is given by Equation 16.1, where $\Sigma_e = \sigma^2 I_n$.

For the conductivity image prior, we adapt a Markov random field (MRF) prior from Geman and McClure (1987). This prior has the form

$$\pi(x) \propto \exp\left\{\beta \sum_{i\sim j} u(x_i - x_j)\right\}, \quad x \in [2.5, 4.5]^m, \tag{16.2}$$

where β and s control the regularity of the field, and $u(\cdot)$ is the tricube function of Cleveland (1979):

$$u(d) = \begin{cases} \dfrac{1}{s}(1 - [d/s]^3)^3, & \text{if } -s < d < s, \\ 0, & \text{if } |d| \geq s. \end{cases}$$

The sum is over all horizontal and vertical nearest neighbors, denoted by $i \sim j$, and given by the edges in the Markov random field graph in Figure 16.2. Hence, this prior encourages neighboring x_i to have similar values, but once x_i and x_j are more than s apart, the penalty does not grow. This allows occasional large shifts between neighboring x_i. For this chapter, we fix $(\beta, s) = (0.5, 0.3)$. A realization from this prior is shown on the right in Figure 16.2. A typical prior realization shows patches of homogeneous values, along with abrupt changes in intensity at patch boundaries. This prior also allows an occasional, isolated, extreme single pixel value.

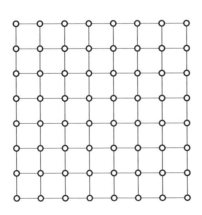 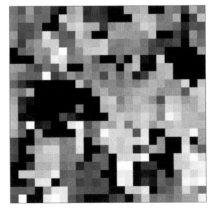

FIGURE 16.2
(Left) First-order neighborhood MRF graph corresponding to the prior in Equation 16.2; each term in the sum corresponds to an edge in the MRF graph. (Right) A realization from this gray level prior.

The resulting posterior density has the form

$$\pi(x|y) \propto \exp\left\{-\frac{1}{2\sigma^2}(y - \eta(x))^T(y - \eta(x))\right\} \times \exp\left\{\beta \sum_{i \sim j} u(x_i - x_j)\right\}, \quad x \in [2.5, 4.5]^m.$$

(16.3)

The patchiness and speckle allowed by this prior, and the rather global nature of the likelihood, make posterior exploration for this inverse problem rather challenging, and a good test case for various MCMC schemes that have been developed over the years. We note that the nature of the posterior can be dramatically altered by changing the prior specification for x. This is discussed later in this section.

This chapter considers a number of MCMC approaches for sampling from this posterior distribution. We start at the beginning.

16.2.1 Posterior Exploration via Single-Site Metropolis Updates

A robust and straightforward method for computing samples from the posterior $\pi(x|y)$ is the single-site Metropolis scheme, originally carried out in Metropolis et al. (1953) on the world's first computer with addressable memory, the MANIAC. A common formulation of this scheme is summarized in Algorithm 1 using pseudocode. This scheme is engineered to maintain detailed balance—so that the relative movement between any two states x and x^* is done in proportion to the posterior density at these two points. The width of the proposal distribution σ_z should be adjusted so that inequality in line 5 is satisfied roughly half the time (Gelman et al., 1996), but an acceptance rate between 70% and 30% does nearly as well for single-site updates. After scanning through each of the parameter elements (for loop, steps 3–7), one typically records the current value of x. We do so every 10 scans through the parameter vector.

ALGORITHM 1   SINGLE-SITE METROPOLIS

```
1: initialize x
2: for  k = 1 : niter  do
3:     for  i = 1 : m  do
4:         x'_i = x_i + z,  where  z ~ N(0, σ_z)
5:         if   u < π(x'|y)/π(x|y),  where  u ~ U(0, 1)  then
6:             set  x_i = x'_i
7:         end if
8:     end for
9: end for
```

This single-site scheme was originally intended for distributions with very local dependencies within the elements of x so that the ratio in line 5 simplifies dramatically. In general, this simplification depends on the full conditional density of x_i,

$$\pi(x_i|x_{-i}, y), \quad \text{where } x_{-i} = (x_1, \ldots, x_{i-1}, x_{i+1}, \ldots, x_n)^T.$$

This density is determined by keeping all of the product terms in $\pi(x|y)$ that contain x_i, and ignoring the terms that do not. Hence the ratio in line 5 can be rewritten as

$$\frac{\pi(x'|y)}{\pi(x|y)} = \frac{\pi(x_i'|x_{-i}, y)}{\pi(x_i|x_{-i}, y)}.$$

In many cases this ratio becomes trivial to compute. However, in the case of this particular inverse problem, we must still evaluate the simulator to compute this ratio. This is exactly what makes the MCMC computation costly for this problem.

Nonetheless, this straightforward sampling approach does adequately sample the posterior, given sufficient computational effort. Figure 16.3 shows realizations produced by the single-site Metropolis algorithm, separated by 1000 scans through each element of x. Inspection of these realizations makes it clear that posterior realizations yield a crisp distinction between the high- and low-conductivity regions, as was intended by the MRF prior for x. Around the boundary of the high conductivity region, there is a fair amount of uncertainty as to whether or not a given pixel has high or low conductivity.

Figure 16.4 shows the resulting posterior mean for x and the history of three pixel values over the course of the single-site updating scheme. The sampler was run until 40,000 $\times m$ forward simulations were carried out. An evenly spaced sample of 6000 values for three of the m pixels is shown on the left in Figure 16.4. Note that for the middle pixel (blue circle), the marginal posterior distribution is bimodal—some realizations have the conductivity value near 3, others near 4. Being able to move between these modes is crucial for a well-mixing chain. Getting this pixel to move between modes is not simply a matter of getting that one pixel to move by itself; the movement of that pixel is accomplished by getting the entire image x to move between local modes of the posterior.

This local multimodality is largely induced by our choice of prior. For example, if we alter the prior model in Equation 16.2 so that

$$u(d) = -d^2, \tag{16.4}$$

we have a standard Gaussian Markov random field (GMRF) prior for x. If, in addition, the simulator is a linear mapping from inputs x to ouputs $\eta(x)$, the resulting posterior is necessarily Gaussian, and hence unimodal. While this is not true for nonlinear forward models/simulators, the GMRF prior still has substantial influence on the nature of the posterior. Figure 16.5 shows two realizations and the posterior mean resulting from such a prior with $\beta = 2$. Here posterior realizations are locally more variable—the difference between neighboring pixels is generally larger. However, the global nature of the posterior realizations is far more controlled than those in Figure 16.3 since the GMRF prior suppresses

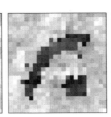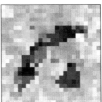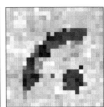

FIGURE 16.3
Five realizations from the single-site Metropolis scheme. Realizations are separated by 1000 scans through the m-dimensional image parameter x.

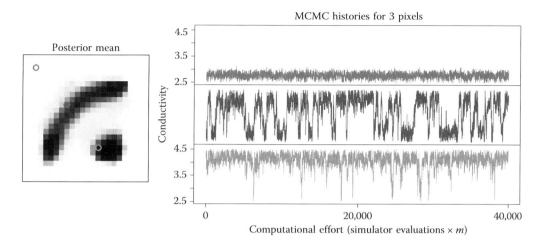

FIGURE 16.4

Posterior mean image for x and MCMC traces of three pixels: one which is predominantly light (small conductivity); one which is predominantly dark (high conductivity) and one which is on the edge of the object. This MCMC run carries out 40,000 $\times m$ forward simulator evaluations. The value of x_i is given every 10th iteration (i.e. every $10 \times m$ single-site updates).

local modes that appear under the previous formulation. This resulting formulation is also far easier to sample, requiring about one tenth of the effort needed for formulation in Equation 16.3. An alternate, controlling prior formulation uses a process convolution prior for x is given in the Appendix to this chapter. In addition to yielding a more easily sampled posterior, the prior also represents the image x with far fewer parameters than the m used in the MRF specifications.

While these alternative specifications lead to simpler posterior distributions, they do so while giving overly smooth posterior realizations. Still, such realizations may be useful for exploratory purposes, and for initializing other samplers; we do not further pursue such formulations here. Instead, we focus on comparison of various MCMC schemes to sample the original gray level posterior in Equation 16.3. We use the sample traces from the three pixels circled in Figure 16.4 to make comparisons between a variety of samplers which are discussed in the next sections—the movement of these three pixels is representative of all the image pixels. In particular, we focus on the frequency of movement between high and low conductivity at these sites.

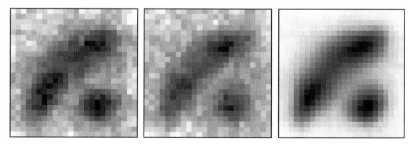

FIGURE 16.5

Two realizations and the posterior mean from the single-site Metropolis scheme run on the posterior resulting from the GMRF prior. Realizations are separated by 1000 scans through the m-dimensional image x.

16.3   Multivariate Updating Schemes

Schemes that propose to update more than just a single component of x at a time have the potential to reduce the computational burden of producing an MCMC sample from $\pi(x|y)$. The single-site scheme above is also applicable when the proposal for x' changes some or all of the components of x. However, producing a multivariate candidate x' that has an appreciable chance of being accepted (i.e. satisfying the inequality in line 5 of Algorithm 1) while allowing appreciable movement, is very difficult. This highlights a very appealing aspect of the single-site Metropolis scheme: even fairly thoughtless one-dimentional proposals have an appreciable chance of being accepted while adequately exploring the posterior.

There are clustering MCMC algorithms from statistical physics that allow for many pixels in x to be updated at once (Edwards and Sokal, 1988). Such methods can be adapted to this particular problem as in Higdon (1998); however, such methods typically show decreased efficiency relative to single-site updating when the likelihood is strong relative to the prior. This is certainly the case with our attempts on this application whose results are not worth discussing here. Instead, we look to multivariate random-walk Metropolis updating and the DE-MCMC scheme of ter Braak (2006) as competitors to the costly single-site Metropolis updating for our EIT application.

16.3.1   Random-Walk Metropolis

The multivariate random-walk Metropolis scheme (RWM) has been the focus of a number of theoretical investigations (Gelman et al., 1996; Tierney, 1994). But to date this scheme has not been widely used in applications, and has proven advantageous only in simple, unimodal settings. The preference for single-site, or limited multivariate updates in practice may be attributed to how the full conditionals often simplify computation, or may be due to the difficulty in tuning highly multivariate proposals. In our EIT application, the univariate full conditionals do not lead to any computational advantages. If there is ever an application for which RWM may be preferable, this is it. Single-site updating is very costly, and may be inefficient relative to multivariate updating schemes for this multimodal posterior.

A multivariate Gaussian RMW scheme for the m-vector x is summarized in Algorithm 2 using pseudocode.

ALGORITHM 2   RANDOM-WALK METROPOLIS

```
1: initialize x
2: for k = 1 : niter do
3:     x' = x + z, where z ~ N_m(0, Σ_z)
4:         if  u < π(x'|y)/π(x|y),  where u ~ U(0, 1)   then
5:             set x = x'
6:         end if
7: end for
```

We consider three different proposals for this scheme:

$$\Sigma_z \propto \Sigma_1 = I_m,$$
$$\Sigma_z \propto \Sigma_2 = \mathrm{diag}(s_1^2, \dots, s_m^2),$$
$$\Sigma_z \propto \Sigma_3 = S^2,$$

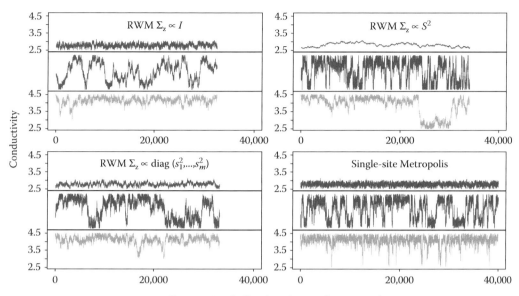

FIGURE 16.6
MCMC traces of three pixels circled in Figure 16.4 under three multivariate random-walk Metropolis schemes, and single-site Metropolis. For each run, $40,000 \times m$ forward simulator evaluations are carried out. While the RWM scheme with $\Sigma_z \propto S^2$ results in good movement for the central pixel, the movement of the top and bottom pixels is clearly inferior to that of single-site Metropolis.

where s_i^2 is the posterior marginal sample variance for the conductivity x_i, and S^2 is the $m \times m$ sample covariance matrix—both estimated from the previously obtained single-site MCMC run. In each case we set $\Sigma_z = \alpha_i \Sigma_i$, where the scalar α_i is chosen so that the candidate x' is accepted 30% of the time, which is close to optimal in a Gaussian setting.

MCMC traces for these three implementations of RWM are shown in Figure 16.6. The traces from the single-site Metropolis scheme are also given for comparison. Interestingly, the behavior of the traces varies with the choice of Σ_z. The scheme with $\Sigma_z \propto S^2$ shows the most movement for the central pixel, which moves between high and low conductivity over the run. However, its performance for the top, low-conductivity pixel is noticeably worse. None of the RWM schemes do as well as single-site Metropolis when looking at the bottom, high conductivity pixel. These results suggest that a scheme that utilizes both single-site and RWM updates with $\Sigma_z \propto S^2$ might give slightly better posterior exploration than single-site Metropolis alone.

16.3.2 Differential Evolution and Variants

In ter Braak's DE-MCMC algorithm, a collection of independent chains $\{x^1, \ldots, x^P\}$ are constructed, each sampling from the posterior. Chain x^p is updated according to a multivariate Metropolis step where the candidate is a perturbation of x^p based on the difference between two randomly chosen chains in the collection. This DE-MCMC scheme is described in Algorithm 3. In the original implementation, σ^2 is chosen to be quite small so that a proposal of the form $x' = x + e$ would nearly always be accepted; γ is chosen so that the proposal is accepted about 30% of the time. Hence, it is the $\gamma(x^q - x^r)$ part of the proposal that accounts for nearly all of the movement from the current location x.

ALGORITHM 3   DIFFERENTIAL EVOLUTION

1: initialize P copies $\{x^1, \ldots, x^P\}$
2: **for** $k = 1 :$ niter **do**
3:     **for** $p = 1 : P$ **do**
4:         chose indices q and r without replacement from $\{1, \ldots,$
            $p-1, p+1, \ldots, P\}$
5:         $x^{p\prime} = x^p + \gamma(x^q - x^r) + e$, where $e \sim N(0, \sigma^2 I_n)$
6:         **if** $u < \frac{\pi(x^{p\prime}|y)}{\pi(x^p|y)}$, where $u \sim U(0, 1)$ **then**
7:             $x^p = x^{p\prime}$
8:         **end if**
9:     **end for**
10: **end for**

One interpretation of the DE-MCMC algorithm is as an empirical version of the RWM algorithm. The proposal distribution is a scaled difference between random draws from $\pi(x|y)$; the dependence between the parallel chains means that these draws are not independent. Theoretical considerations make $2.38/\sqrt{m}$ a useful starting choice for γ (Gelman et al., 1996). However, some tuning of γ is usually appropriate. An obvious appeal of this DE scheme is that it avoids the difficult task of determining the appropriate Σ_z used in the Gaussian RWM implementation from earlier in this section. By carrying P copies of the chain, fruitful multivariate candidates can be generated on the fly. Such schemes have proven useful in difficult, low-dimensional posterior distributions, but the utility of such an approach has yet to be demonstrated on highly multivariate posteriors resulting from applications such as this.

As a first step in illustrating DE-MCMC on the EIT application, we initialized the $P = 400$ chains by taking equally spaced realizations from the first $6000 \times m$ iterations from the single-site Metropolis scheme described earlier. Then each of the 400 chains were updated in turn according to the DE-MCMC algorithm. The sampler continued until $40{,}000 \times m$ simulator evaluations were carried out. Thus each chain was updated $100 \times m$ times. The resulting MCMC traces for the three representative pixels are shown in Figure 16.7 for three of the 400 chains used in our implementation. For comparison, the trace from $100 \times m$ single-site Metropolis is also given on the bottom right of the figure. Also, the mean and (marginal) standard deviation for the central pixel (marked by the blue circle in Figure 16.4) are shown in Figure 16.8 for each of the 400 chains. Within a given chain, the pixels show very little movement; the final value of the 400 chains is not far from the starting point, as is clear from Figure 16.8.

We also consider an alteration to the basic formulation of ter Braak in which the scalar γ is drawn from a $U(-a, a)$ distribution. We set $a = 0.02$ so that the proposal is accepted about 30% of the time. For this alteration, we set $e = 0$ since these small steps had very little impact on the sampler. While this alteration leads to noticeably better movement than our standard DE-MCMC implementation, the movement of this chain is still clearly inferior to single-site Metropolis. Given the less than stellar performance of the multivariate RWM scheme, the lack of success here is not a big surprise since both schemes make use of highly multivariate updates based on $\pi(x|y)$. The larger surprise is that the general failure of these multivariate updating schemes to provide any improvement over single-site Metropolis updating, even when there are no computational savings to be had by considering univariate full conditionals. We note that the poor performance of these multivariate updating schemes does not preclude the existence of some modification that will eventually prove beneficial for this application; we simply did not find one.

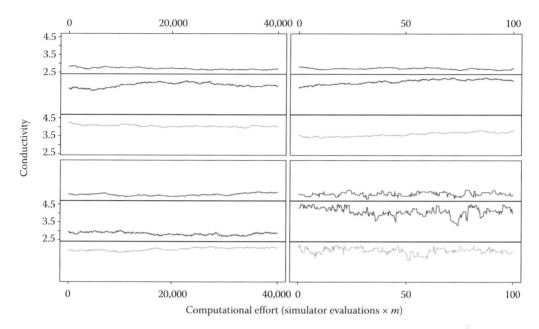

FIGURE 16.7
MCMC traces of the same three pixels shown in Figure 16.4 for three of the 400 chains used in the DE-MCMC scheme. The algorithm ran until 40,000 $\times m$ forward simulator evaluations so that the computational effort matches the other MCMC schemes. For each of the 400 chains, $100 \times m$ updates are carried out. The bottom right plot shows movement from $100 \times m$ single-site Metropolis iterations for comparison. The resulting movement is clearly inferior to that of the standard single-site scheme when normalized by computational effort.

16.4 Augmenting with Fast, Approximate Simulators

In many applications, a faster, approximate simulator is available for improving the MCMC sampling. There are a limited number of rigorous approaches for utilizing fast, approximate simulators: delayed acceptance schemes that limit the number of calls to the expensive,

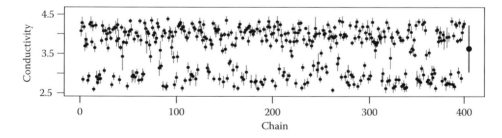

FIGURE 16.8
The marginal posterior mean and lines extending ± 1 standard deviation for the central pixel circled in Figure 16.4. The mean and standard deviation is estimated separately from each of the 400 copies in the DE-MCMC scheme. The chains were initialized from evenly spaced realizations taken from the single-site Metropolis scheme. The spread of the means relative to the estimated standard deviation indicates a very poor mixing, or posterior movement for the DE-MCMC scheme. For comparison, the marginal mean ± 1 standard deviation of single-site Metropolis is shown at the rightmost edge of the plot.

exact simulator (Christen and Fox, 2005; Efendiev et al., 2006; Fox and Nicholls, 1997); and augmented or tempering schemes (Higdon et al., 2002, 2003). For this chapter, we consider simple implementations of both of these approaches and discuss their implementation in context of the EIT application.

For the multigrid EIT simulator $\eta(x)$, an approximate simulator can be created by altering how the multigrid steps are carried out during the solve. Specifically, multigrid algorithms achieve their efficiency through the recursive use of a local smoothing iteration and successively coarser discrete problems; see (Briggs et al., 2000) for an overview of multigrid principles. The most common multigrid cycle, the V-cycle, is shown in Figure 16.9. Here a local smoothing iteration, such as a Gauss–Seidel relaxation, smooths the error of the current iterate. In turn, the smoothed error is represented on a coarser grid through the restriction (weighting) of the current residual. The coarser grid provides a means to find an inexpensive correction to the current iterate; however, this grid may still be too large for a direct solve. In this case the process is repeated until the coarsest grid is reached and a direct solve may be performed. The correction is then interpolated and smoothed, repeatedly, until the finest grid is reached. If a single smoothing iteration is applied at each grid level of the coarsening and refining phases, then the multigrid cycle is denoted as $V(1,1)$.

The complementarity of the smoothing and coarse-grid correction processes leads to multigrid's optimal algorithmic scaling (i.e. solution cost grows only linearly with number of unknowns), and to a uniform reduction in the error with each cycle. It is this latter property that creates the opportunity to develop efficient approximate solvers using elements of robust variational multigrid algorithms. For example, MacLachlan and Moulton (2006) developed the Multilevel Upscaling (MLUPS) algorithm to efficiently model flow through highly heterogeneous porous media. MLUPS leveraged the hierarchy of discrete operators provided by the operator-induced variational coarsening of the Black Box Multigrid (BoxMG) algorithm (see Dendy, 1982), and eliminated the smoothing iterations from the finest few levels.

In this work, we produce approximate solvers by limiting the number of $V(1,1)$ cycles carried out. Starting with a fixed initial solution, the approximate solvers will produce

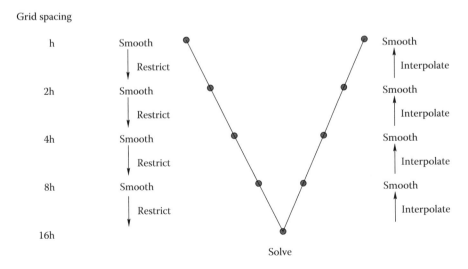

FIGURE 16.9
Schematic of the V-cycle multigrid iterative algorithm.

ALGORITHM 4   DIFFERENTIAL ACCEPTANCE METROPOLOIS

```
 1: initialize x
 2: for k = 1 : niter do
 3:     for i = 1 : m do
 4:         x'ᵢ = xᵢ + z, where z ~ N(0, σ_z) )
 5:         if  u₁ < π₁(x'|y)/π₁(x|y),  where u₁ ~ U(0, 1)   then
 6:             if u₂ < π₀(x'|y)π₁(x|y)/π₀(x|y)π₁(x'|y),  where u₂ ~ U(0, 1)   then
 7:                 set xᵢ = x'ᵢ
 8:             end if
 9:         end if
10:     end for
11: end for
```

solutions of fitted voltages which depend on the conductivity field x. These approximate solutions are obtained more quickly, but do not match the voltages obtained from the "exact" solve. Here we consider two approximate solvers: one that stops after two $V(1, 1)$ cycles; and another that stops after a single $V(1, 1)$ cycle. The resulting approximate simulators $\eta_1(x)$ (two V cycles) and $\eta_2(x)$ (a single V cycle) are less accurate overall, but faster. In this case, $\eta_1(x)$ typically takes a third of the computing time to evaluate relative to the exact solver $\eta_0(x)$, while $\eta_2(x)$ typically takes a quarter of the time.

16.4.1 Delayed Acceptance Metropolis

The delayed acceptance approach of Fox and Nicholls (1997) and Christen and Fox (2005) uses a fast, approximate simulator to "pretest" a proposal. This approach adapts ideas from the surrogate transition method of Liu (2001) for dealing with complex forward models. For now, we define different posterior formulations, depending on which simulator is used:

$$\pi_v(x|y) \propto L_v(y|x) \times \pi(x)$$

$$\propto \exp\left\{-\frac{1}{2\sigma^2}(y - \eta_v(x))^T(y - \eta_v(x))\right\} \times \exp\left\{\beta \sum_{i \sim j} u(x_i - x_j)\right\} \times I[x \in [2.5, 4.5]^m].$$

We note that one could alter the sampling model for the approximate formulations, though it is not done here.

A simple Metropolis-based formulation of this scheme is given in Algorithm 4, where $\pi_0(x|y)$ and $\pi_1(x|y)$ denote the posteriors using the exact and approximate simulators, respectively. Notice that the exact simulator need only be run if the pretest condition ($u_1 < \frac{\pi_1(x'|y)}{\pi_1(x|y)}$) involving the faster, approximate simulator is satisfied. Hence, if the proposal width is chosen so that the pretest condition is satisfied only a third of the time, the exact simulator is only run for a third of the MCMC iterations. If we use the first approximate simulator $\eta_1(x)$, then the $40,000 \times m$ iterations required for our original single-site Metropolis scheme take about 66% of the computational effort using this delayed acceptance approach.

If $\eta_1(x)$ is a very good approximation to the exact simulator $\eta_0(x)$, then this delayed acceptance sampler is equally efficient if one normalizes by iteration. This is the case for the first approximate simulator in this example–the difference in log likelihood is typically no more than ±0.2 over the range of posterior samples. However, if the approximate simulator

poorly matches the exact one, any savings obtained by reducing the number of exact simulator evaluations will be more than offset by reduced efficiency in the delayed acceptance
sampler. In our application here, $|\eta_2(x) - \eta_0(x)|$ can be as large as 2 for some realizations
x from the posterior. In using $\eta_2(x)$ in the delayed acceptance scheme, we detect a slight
increase in autocorrelation which is more than offset by gains in computational efficiency
from evaluating $\eta_2(x)$ in place of $\eta_0(x)$.

Apparently, this potential loss of efficiency is not present in this application since $\eta_2(x)$ is
still an adequate approximation to the exact simulator $\eta_0(x)$. This loss of efficiency due to
poor approximation is readily apparent if one takes a univariate example in which $\pi_0(x|y)$ is
the standard normal density, and $\pi_1(x|y)$ is the normal density with mean one and standard
deviation 0.5. In this case, the delayed acceptance sampler must take occasional, slow-
moving excursions in the negative numbers to offset the lack of support in $\pi_1(x|y)$ in that
region, reducing the efficiency of the sampler.

Finally, we note that Christen and Fox (2005) give a more general formulation for the
delayed acceptance sampler for which the approximate simulator can depend on the
current state x of the chain. While a bit more demanding computationally, the more
general algorithm can make use of local approximations which are available in some
applications.

16.4.2   An Augmented Sampler

By augmenting the posterior of interest with auxiliary distributions one can use Metropolis
coupling (Geyer, 1991), simulated tempering (Marinari and Parisi, 1992), or related schemes
(Liu and Sabatti, 1999). Here we augment our posterior with additional posteriors based
on the two approximate simulators. We introduce the auxiliary variable $v \in \{0, 1, 2\}$ to our
formulation, which indexes the simulator to be used, and treat v as an additional parameter in a larger formulation. We specify a uniform prior for v over $\{0, 1, 2\}$ resulting in the
augmented formulation

$$\pi(x, v|y) \propto L(y|x, v) \times \pi(x) \times \pi(v)$$

$$\propto \exp\left\{-\frac{1}{2\sigma^2}(y - \eta_v(x))^T(y - \eta_v(x))\right\} \times \exp\left\{\beta \sum_{i \sim j} u(x_i - x_j)\right\}$$

$$\times I[x \in [2.5, 4.5]^m] \times I[v \in \{0, 1, 2\}].$$

This augmented formulation can be sampled as before, except that after scanning through
the elements of x to carry out single-site Metropolis updates, a simple Metropolis update
is then carried out for v by making a uniform proposal over $\{0, 1, 2\} \setminus v$. Ideally, this chain
should move somewhat often between the states of v.

A small subsequence from this chain is shown in Figure 16.10. As this sampler runs, the
draws for which $v = 0$ are from the posterior of interest. While $v = 1$ or 2, the chain is
using one of the faster, approximate simulators. Hence, it can more quickly carry out the
single-site Metropolis updates, so that the chain moves more rapidly through this auxiliary
posterior. By the time the chain returns to $v = 0$, the realizations of x will generally show
more movement than a sampler based solely on the exact simulator $\eta_0(x)$.

Marginally, the augmented sampler spends about 20% of its iterations at $v = 2$, 42% at
$v = 1$, and 38% at $v = 0$. This augmented formulation allows about twice the number of
single-site Metropolis updates as compared to the the standard single-site Metropolis chain

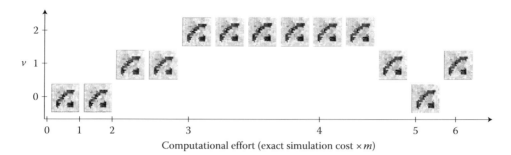

FIGURE 16.10
A sequence from the augmented sampler chain. A scan of m of single-site Metropolis updates is followed by a Metropolis update of the formulation index variable v. Here the sequence of images x starts with $v = 0$, then moves up to $v = 2$, and then back down to $v = 0$. While the chain is not using the exact simulator ($v > 0$), the computational cost of carrying out the m single-site Metropolis updates for each x_i is substantially reduced.

on the exact formulation alone. In all this nearly doubles the efficiency when normalized by computing effort. The efficiency of the sampler could be improved slightly by altering the prior for v so that the chain spends more time at $v = 2$ and less at $v = 0$.

A feature of both the delayed acceptance algorithm and the augmented formulation is that they utilize the most efficient MCMC scheme available. Both of these methods could be used with an alternative to single-site Metropolis if it is found to be more efficient. For the augmented example above, we could improve the computational efficiency by employing delayed acceptance, using $\eta_1(x)$ when carrying out the m single-site Metropolis updates for x_i when $v = 0$. There is no practical benefit in using delayed acceptance using $\eta_2(x)$ when $v = 1$ since the relative speed of the two simulators is not that different.

16.5 Discussion

For the EIT example, single-site Metropolis requires about 2 million simulator evaluations to effectively sample this posterior distribution. Multivariate updating schemes such as random-walk Metropolis or DE-MCMC—as we implemented them here—do not offer any real relief. Utilizing fast approximations through delayed acceptance and/or tempering schemes may reduce the computational burden by a factor of 4 or so, more if a very fast and accurate approximation is available. This means that sampling this $m = 576$-dimensional posterior is going to require at least a half a million simulator evaluations. This number will most certainly increase as the dimensionality m increases. Hence, a very fast simulator is required if one hopes to use such an image-based MCMC approach for a three-dimensional inverse problem.

One challenging feature of this application is the multimodal nature of the posterior which is largely induced by our choice of prior. By specifying a more regularizing prior, such as the GMRF (Equation 16.4) or the process convolution (Equation A.1), the resulting posterior will more likely be unimodal, so that standard MCMC schemes will be more efficient. Of course, the sacrifice is that one is now less able to recover small-scale structure that may be present in the inverse problem.

In some applications the simulator is sufficiently complicated that one can only hope to run it a few hundred times. In such cases, there is no possibility of reconstructing an entire image of unknown pixel intensities. However, one can construct a very fast surrogate by replacing the simulator by a response surface model built from the limited number of simulations that have been carried out. Craig et al. (2001) and Kennedy and O'Hagan (2001) are two examples of applications which utilize a response surface to aid the resulting simulation-based inference. Of course, this requires a low-dimensional representation of the unknown parameters to be input to the simulator. It also requires that the simulation output be amenable to a response surface model.

Finally, we note that the traditional way to speed up the computation required to solve an inverse problem is to speed up the simulator $\eta(x)$. A substantial amount of progress has been made in creating simulators that run on highly distributed computing machines. Comparatively little progress has been made in utilizing modern computing architectures to speed up MCMC-based posterior exploration in difficult inverse problems. Clearly schemes such as Metropolis coupling chains and DE-MCMC are quite amenable to distributed implementations. The integration of modern computing architecture with MCMC methods will certainly extend the reach of MCMC based solutions to inverse problems.

Appendix: Formulation Based on a Process Convolution Prior

An alternative to treating each pixel in the image as a parameter to be estimated is to use a lower-dimensional representation for the prior. Here we describe a process convolution (Higdon, 2002) prior for the underlying image x.

We define $x(s)$, $s \in \Omega$, to be a mean-zero Gaussian process. But rather than specify $x(s)$ through its covariance function, it is determined by a latent process u and a smoothing kernel $k(s)$. The latent process $u = (u_1, \ldots, u_p)^T$ is located at the spatial sites $\omega_1, \ldots, \omega_p$, also in Ω (shown in Figure 16.11). The u_j are then modeled as independent draws from a $N(0, \sigma_u^2)$ distribution. The resulting continuous Gaussian process model for $x(s)$ is then

$$x(s) = \sum_{j=1}^{p} u_j k(s - \omega_j), \tag{A.1}$$

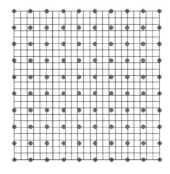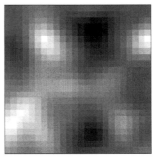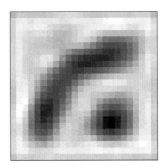

FIGURE 16.11
(Left) A 10×10 lattice of locations $\omega_1, \ldots, \omega_p$, for the u_j of the process convolution prior; the 24×24 image pixels are shown for reference. (Middle) A realization from the process convolution prior for $x(s)$. (Right) Posterior mean from the single-site Metropolis scheme run on the u vector that controls the image x.

where $k(\cdot - \omega_j)$ is a kernel centered at ω_j. For the EIT application, we define the smoothing kernel $k(\cdot)$ to be a radially symmetric bivariate Gaussian density, with standard deviation $\sigma_u = 0.11$. Figure 16.11 shows a prior draw from this model over the 24×24 pixel sites in Ω. Under this formulation, the image x is controlled by $p = 100$ parameters in u. Thus a single-site Metropolis scan of u takes less than 20% of the computational effort required to update each pixel in x. In addition, this prior enforces very smooth realizations for x. This makes the posterior distribution better behaved, but may make posterior realizations of x unreasonably smooth. The resulting posterior mean for x is shown in Figure 16.11. For a more detailed look at process convolution models, see Higdon (2002); Paciorek and Schervish (2004) give non-stationary extensions of these spatial models.

Acknowledgments

This work was partially supported by the Los Alamos National Laboratory LDRD/ER research grant 20080300ER, *Multilevel Adaptive Sampling for Multiscale Inverse Problems*, and the DOE Office of Science Advanced Computing Research (ASCR) program in Applied Mathematical Sciences.

References

Andersen, K., Brooks, S., and Hansen, M. 2003. Bayesian inversion of geoelectrical resistivity data. *Journal of the Royal Statistical Society, Series B*, 65(3):619–642.

Briggs, W. L., Henson, V. E., and McCormick, S. F. 2000. *A Multigrid Tutorial*, 2nd edn. SIAM Books, Philadelphia.

Christen, J. and Fox, C. 2005. Markov chain Monte Carlo using an approximation. *Journal of Computational and Graphical Statistics*, 14(4):795–810.

Cleveland, W. S. 1979. Robust locally weighted regression and smoothing scatterplots. *Journal of the American Statistical Association*, 74:829–836.

Craig, P. S., Goldstein, M., Rougier, J. C., and Seheult, A. H. 2001. Bayesian forecasting using large computer models. *Journal of the American Statistical Association*, 96:717–729.

Dendy, J. E. 1982. Black box multigrid. *Journal of Computational Physics*, 48:366–386.

Dendy Jr, J. 1987. Two multigrid methods for three-dimensional problems with discontinuous and anisotropic coefficients. *SIAM Journal on Scientific and Statistical Computing*, 8:673.

Edwards, R. G. and Sokal, A. D. 1988. Generalization of the Fortuin-Kasteleyn-Swendsen-Wang representation and Monte Carlo algorithm. *Physical Review Letters*, 38:2009–2012.

Efendiev, Y., Hou, T., and Luo, W. 2006. Preconditioning Markov chain Monte Carlo simulations using coarse-scale models. *SIAM Journal on Scientific Computing*, 28(2):776–803.

Fox, C. and Nicholls, G. 1997. Sampling conductivity images via MCMC. In K. V. Mardia, C. A. Gill and R. G. Aykroyd (eds), *The Art and Science of Bayesian Image Analysis. Proceedings of the Leeds Annual Statistics Research Workshop*, pp. 91–100. Leeds University Press, Leeds.

Gelman, A., Roberts, G., and Gilks, W. 1996. Efficient Metropolis jumping rules. In J. M. Bernardo, J. O. Berger, A. P. Dawid, and A. F. M. Smith (eds), *Bayesian Statistics 5: Proceedings of the Fifth Valencia International Meeting*, pp. 599–607. Oxford University Press, Oxford.

Geman, S. and McClure, D. 1987. Statistical methods for tomographic image reconstruction. *Bulletin of the International Statistical Institute*, 52(4):5–21.

Geyer, C. J. 1991. Monte Carlo maximum likelihood for dependent data. In E. M. Keramidas (ed.), *Computing Science and Statistics: Proceedings of the 23rd Symposium on the Interface*, pp. 156–163. American Statistical Association, New York.

Higdon, D. 2002. Space and space-time modeling using process convolutions. In C. Anderson, V. Barnett, P. C. Chatwin, and A. H. El-Shaarawi (eds), *Quantitative Methods for Current Environmental Issues*, pp. 37–56. Springer, London.

Higdon, D., Lee, H., and Bi, Z. 2002. A bayesian approach to characterizing uncertainty in inverse problems using coarse and fine scale information. *IEEE Transactions in Signal Processing*, 50:389–399.

Higdon, D. M. 1998. Axiliary variable methods for Markov chain Monte Carlo with applications. *Journal of the American Statistical Association*, 93:585–595.

Higdon, D. M., Lee, H., and Holloman, C. 2003. Markov chain Monte Carlo-based approaches for inference in computationally intensive inverse problems. In J. M. Bernardo, M. J. Bayarri, J. O. Berger, A. P. Dawid, D. Heckerman, A. F. M. Smith, and M. West (eds), *Bayesian Statistics 7. Proceedings of the Seventh Valencia International Meeting*, pp. 181–197. Oxford University Press, Oxford.

Jimenez, R., Verde, L., Peiris, H., and Kosowsky, A. 2004. Fast cosmological parameter estimation from microwave background temperature and polarization power spectra. *Physical Review D*, 70(2):23005.

Kaipio, J., Kolehmainen, V., Somersalo, E., and Vauhkonen, M. 2000. Statistical inversion and Monte Carlo sampling methods in electrical impedance tomography. *Inverse Problems*, 16(5):1487–1522.

Kaipio, J. P. and Somersalo, E. 2004. *Statistical and Computational Inverse Problems*. Springer, New York.

Kennedy, M. and O'Hagan, A. 2001. Bayesian calibration of computer models (with discussion). *Journal of the Royal Statistical Society, Series B*, 68:425–464.

Liu, J. 2001. *Monte Carlo Strategies in Scientific Computing*. Springer, New York.

Liu, J. and Sabatti, C. 1999. Simulated sintering: Markov chain Monte Carlo with spaces of varying dimensions. In J. M. Bernardo, J. O. Berger, A. P. Dawid, and A. F. M. Smith (eds), *Bayesian Statistics 6: Proceedings of the Sixth Valencia International Meeting*. Oxford University Press, Oxford.

MacLachlan, S. P. and Moulton, J. D. 2006. Multilevel upscaling through variational coarsening. *Water Resources Research*, 42.

Marinari, E. and Parisi, G. 1992. Simulated tempering: a new Monte Carlo scheme. *Europhysics Letters*, 19:451–458.

Metropolis, N., Rosenbluth, A., Rosenbluth, M., Teller, A., and Teller, E. 1953. Equation of state calculations by fast computing machines. *Journal of Chemical Physics*, 21:1087–1092.

Moulton, J. D., Fox, C., and Svyatskiy, D. 2008. Multilevel approximations in sample-based inversion from the Dirichlet-to-Neumann map. *Journal of Physics: Conference Series*, 124:012035 (10 pp.).

Nissinen, A., Heikkinen, L. M., and Kaipio, J. P. 2008. The Bayesian approximation error approach for electrical impedance tomography—experimental results. *Measurement Science and Technology*, 19(1):015501 (9 pp.).

Paciorek, C. J. and Schervish, M. J. 2004. Nonstationary covariance functions for Gaussian process regression. In S. Thrun, L. Saul, and B. Schölkopf (eds), *Advances in Neural Information Processing Systems 16*. MIT Press, Cambridge, MA.

Rue, H. 2001. Fast sampling of Gaussian Markov random fields. *Journal of the Royal Statistical Society, Series B*, 63(2):325–338.

Stenerud, V., Kippe, V., Lie, K., and Datta-Gupta, A. 2008. to appear. Adaptive multiscale streamline simulation and inversion for high-resolution geomodels. *SPE J*.

ter Braak, C. F. J. 2006. A Markov chain Monte Carlo version of the genetic algorithm Differential Evolution: Easy Bayesian computing for real parameter spaces. *Statistics and Computing*, 16(3):239–249.

Tierney, L. 1994. Markov chains for exploring posterior distributions (with discussion). *Annals of Statistics*, 21:1701–1762.

Weir, I. 1997. Fully Bayesian reconstructions from single photon emission computed tomography. *Journal of the American Statistical Association*, 92:49–60.

17

Statistical Ecology

Ruth King

17.1 Introduction

In recent years there has been an explosion in the application of Bayesian methods within the field of statistical ecology. This is evidenced by the huge increase in the number of publications that use (and develop) Bayesian methods for analyzing ecological data in both statistical and ecological journals. In addition, in recent years there have been a number of books published that focus solely on the use of Bayesian methods within statistical ecology (King et al., 2009; Link and Barker, 2009; McCarthy, 2007; Royle and Dorazio, 2008). One reason why the Bayesian approach has enjoyed a significant increase in its application to ecology is the particularly complex data that are collected on a typical population under study. This can make standard frequentist analyses difficult to implement, often resulting in simplifying assumptions being made. For example, typical issues that can arise relate to complex distributional assumptions for the observed data; intractable likelihood expressions; and large numbers of biologically plausible models. Within a Bayesian framework it is possible to make use of standard Markov chain Monte Carlo (MCMC) tools, such as data augmentation techniques, so that these simplifications do not need to be made.

The analysis of ecological data is often motivated by the aim of understanding the given system and/or of conservation management. This is of particular interest in recent years with the potential impact of climate change. Identifying relationships between demographic parameters (such as survival probabilities, productivity rates, and migrational behavior) and environmental conditions may provide significant insight into the potential impact of changing climate on a given system. In particular, the different biological processes are often separated into individual components (Buckland et al., 2004), allowing a direct interpretation of the processes and explicit relationships with different factors to be expressed. An area of particular recent interest in statistical ecology relates to the use of hidden Markov models (or state–space models) to separate the different underlying processes. For example, Newman et al. (2006) describe how these models can be applied to data relating to animal populations within a Bayesian framework. Royle (2008) uses a state–space formulation to separate the life history of the individuals (i.e. survival process) with the observation of the individuals (recapture process) in the presence of individual heterogeneity. The appeal of this type of approach is its conceptual simplicity, along with the readily available computational tools for fitting these models. The typical linear and normal assumptions can also be relaxed within this framework, permitting more realistic population models to be fitted. These methods have been applied to a number of areas, including fisheries models (Millar and Meyer, 2000), species richness (Dorazio et al., 2006), abundance data (Reynolds et al., 2009), animal telemetry data (Jonsen et al., 2005), and occupancy models (Royle and Kéry, 2007). It is anticipated that the use of these methods will continue to increase within these, and additional, areas as a result of the more complex statistical analyses that can be performed and an increase in the number of relatively easy-to-use programs particularly using WinBUGS (or OpenBUGS). For example, Brooks et al. (2000, 2002, 2004), Gimenez et al. (2009), King et al. (2009), O'Hara et al. (2009), Royle (2008), and Royle et al. (2007) all provide WinBUGS code for different ecological examples. Within this chapter we focus on two forms of common ecological data: ring-recovery data and count data. We consider a number of issues related that typically arise when analysing such data, including mixed effects models, model selection, efficient MCMC algorithms, and integrated data analyses, extending the models previously fitted to the data considered by Besbeas et al. (2002), Brooks et al. (2004), and King et al. (2008b). The individual models described can be fitted in WinBUGS; however, the length of the computer simulations makes the analysis for the count data prohibitive in this case.

17.2 Analysis of Ring-Recovery Data

We consider in detail an application of Bayesian inference, using MCMC, to a common form of ecological data (particularly for avian populations), namely ring-recovery data. These data are collected by biologists or volunteers over a number of time periods (typically years). At the beginning of each time period, $i = 1, \ldots, I$, individuals are marked (e.g. a ring or tag applied) and released. The number of individuals recovered dead in each subsequent time period is then recorded. For simplicity we assume that individuals are ringed and recovered on an annual basis. Furthermore, we assume that for an individual that dies, any subsequent recovery is immediate.

The data are typically presented in the form of an array. The first column details the number of ringed individuals in each year of release (denoted by $R_i, i = 1, \ldots, I$); and each following column provides the number of individuals recovered dead within each subsequent year (denoted by $m_{i,j}$ for $i = 1, \ldots, I$ and $j = 1, \ldots, J$, where $J \geq I$). Clearly $m_{i,j} = 0$ for $j < i$, since an individual cannot be recovered dead before it is marked and released! Table 17.1 provides sample ring-recovery data on lapwings ringed from 1963 to 1973; all lapwings are ringed as chicks at the beginning of each year, and the year corresponds to a "bird year" lasting from April to March. Note that we will consider the UK ring-recovery data for lapwings released from 1963 to 1997 and recovered up to 1998 (so that $I = J = 35$). Finally, we let $m_{i,J+1}$ denote the number of individuals released in year i that are not seen again (either because they survive to the end of the study, or die and are not recovered). In other words,

$$m_{i,J+1} = R_i - \sum_{j=1}^{J} m_{i,j}.$$

The array $\mathbf{m} = \{m_{i,j} : i = 1, \ldots, I, j = 1, \ldots, J+1\}$ is typically referred to as an *m*-array, and is a sufficient statistic for ring-recovery data.

The corresponding likelihood of the data is straightforward to calculate, and for this data set is a function of three parameters:

TABLE 17.1

Ring-Recovery Data for UK Lapwings for the Years 1963–1973

Year of Ringing	Number Ringed	Year of Recovery										
		1964	1965	1966	1967	1968	1969	1970	1971	1972	1973	1974
1963	1147	14	4	1	2	1	0	1	1	0	0	0
1964	1285		20	3	4	0	1	1	0	0	0	0
1965	1106			10	1	2	2	0	2	2	1	1
1966	1615				9	7	4	2	1	1	0	0
1967	1618					12	1	6	2	0	0	1
1968	2120						9	6	4	0	2	2
1969	2003							10	8	5	3	1
1970	1963								8	3	2	0
1971	2463									4	1	1
1972	3092										7	2
1973	3442											15

- $\phi_{1,j} = P$ (an individual in their first year survives until year $j+1$ | alive in year j);
- $\phi_{a,j} = P$ (an individual adult (i.e. age ≥ 1 year) survives until year $j+1$ | alive in year j);
- $\lambda_j = P$ (an individual is recovered dead in the interval $[j, j+1)$ | dies in the interval $[j, j+1)$).

We use standard vector notation, $\boldsymbol{\phi}_1 = \{\phi_{1,j} : j = 1, \ldots, J\}$, and similarly for $\boldsymbol{\phi}_a$ and $\boldsymbol{\lambda}$.
For each row of the m-array, the data have a multinomial distribution,

$$\mathbf{m}_i \sim \text{Multinomial}(R_i, \mathbf{q}_i),$$

where $\mathbf{m}_i = \{m_{i,j} : j = 1, \ldots, J+1\}$ denotes the ith row of the m-array, and $\mathbf{q}_i$ the corresponding multinomial cell probabilities. In particular, we have for $i = 1, \ldots, I$ and $j = 1, \ldots, J$,

$$q_{i,j} = \begin{cases} 0, & i > j, \\ (1 - \phi_{1,i})\lambda_i, & i = j, \\ \phi_{1,i}\lambda_j(1 - \phi_{a,j}) \prod_{k=i+1}^{j-1} \phi_{a,k}, & i < j, \end{cases}$$

where we use the standard notation that if $j - 1 < i + 1$, the product is the null product and simply equal to one. Finally, to complete the specification, we need to calculate the probability, $q_{i,J+1}$, that an individual is not seen again. We do this by simply noting that the multinomial cell probabilities must sum to unity, so that for $i = 1, \ldots, I$,

$$q_{i,J+1} = 1 - \sum_{j=1}^{J} q_{i,j}.$$

The likelihood is the product over each row of the corresponding multinomial probabilities,

$$f(\mathbf{m} \mid \boldsymbol{\phi}_1, \boldsymbol{\phi}_a, \boldsymbol{\lambda}) \propto \prod_{i=1}^{I} \prod_{j=1}^{J+1} q_{i,j}^{m_{i,j}}.$$

17.2.1 Covariate Analysis

The likelihood given above is a function of three demographic parameters: first-year survival probabilities, adult survival probabilities, and recovery probabilities. The number of parameters is typically fairly large, when considering fully time-dependent models. In addition, arbitrary time-dependent parameters do not provide any understanding as to the potential factors of temporal variability. One approach that can reduce the number of parameters to be estimated (typically providing a greater precision of the estimated parameters) and potentially provide a greater understanding of the factors driving the temporal variability is the use of covariates. In particular, we begin by considering the model proposed by Besbeas et al. (2002). This model specifies a relationship between the survival probabilities and winter severity. To represent this environmental covariate we use the number of days that the minimum temperature falls below freezing in Central England over the winter months, which we denote by *fdays*, and regard as a surrogate for the harshness of the winter. We use a logit link function between the survival probabilities and the environmental covariate, to ensure that the survival probabilities are constrained to the interval [0, 1], so that

$$\text{logit } \phi_{1,j} = \log\left(\frac{\phi_{1,j}}{1 - \phi_{1,j}}\right) = \alpha_1 + \beta_1 f_j,$$

$$\text{logit } \phi_{a,j} = \log\left(\frac{\phi_{a,j}}{1 - \phi_{a,j}}\right) = \alpha_a + \beta_a f_j,$$

where f_j denotes the normalized value of *fdays* in year j (so that f_j has mean 0 and variance 1 over values $j = 1, \ldots, J$). Normalized covariate values are used in order to improve the mixing of the Markov chain and for interpretability of the corresponding intercept and slope parameters of the logistic regression. Alternatively, for recovery probabilities, there is some evidence from other studies that recovery probabilities have been decreasing with time (Baillie and Green, 1987). Thus, we specify a linear temporal dependence on the logit scale (once more ensuring $\lambda_j \in [0, 1]$),

$$\text{logit } \lambda_j = \log\left(\frac{\lambda_j}{1 - \lambda_j}\right) = \alpha_\lambda + \gamma_\lambda t_j,$$

where t_j denotes the normalized value for year $j = 1, \ldots, J$. We note that Brooks et al. (2004) fail to normalize the years in their logistic regression for the recovery probabilities.

The parameters in the model are α_1, β_1, α_a, β_a, α_λ, and γ_λ (i.e. a total of six), significantly reducing the number of parameters in the model compared to the arbitrary time-dependence model (where there are a total of $I \times 2J$ parameters, although for the lapwing data since only chicks are ringed we cannot estimate $\phi_{a,1}$ i.e. adult survival probability in year 1, resulting in $I \times 2J - 1$ parameters). In addition, when using covariates to explain temporal heterogeneity, we note that increasing the length of the study (i.e. increasing J) does not result in an increase in the number of parameters to be estimated (simply the number of covariate values of which the demographic parameters are a function). We need to specify priors on each of the parameters. In particular, without any prior information, we specify an independent normal prior on each parameter with mean zero and variance 10, which can considered to be vague in this context. Note that this does not induce a flat prior on the corresponding demographic parameter. See, for example, Newman (2003) and King et al. (2009) for further discussion.

17.2.1.1 *Posterior Conditional Distributions*

Using Bayes' theorem, we combine the likelihood of the data with the priors specified on each of the parameters. To explore and summarize the posterior distribution we use MCMC. We begin by calculating the posterior conditional distribution for each of the parameters in the model. For notational convenience, we let $\alpha = \{\alpha_1, \alpha_a, \alpha_\lambda\}$ and $\beta = \{\beta_1, \beta_a\}$. We let $\alpha_{(1)}$ ($\beta_{(1)}$) denote the set of parameters, excluding α_1 (β_1). The posterior conditional distribution for α_1 is given by

$$\pi(\alpha_1 \mid \mathbf{m}, \alpha_{(1)}, \beta, \gamma_\lambda) \propto p(\alpha_1) \prod_{i=1}^{I} \prod_{j=1}^{J+1} q_{i,j}^{m_{i,j}}$$

$$\propto \exp\left(-\frac{\alpha_1^2}{20}\right) \prod_{i=1}^{I}\left[\Delta_i^{m_{i,J+1}} (1 - \phi_{1,i})^{m_{i,i}} \prod_{j=i+1}^{J} \phi_{1,i}^{m_{i,j}} \right]$$

$$\propto \exp\left(-\frac{\alpha_1^2}{20}\right) \prod_{i=1}^{I}\left[\Delta_i^{m_{i,J+1}} \exp\left(\alpha_1 \sum_{j=i+1}^{J} m_{i,j}\right) \right.$$

$$\left. \times \prod_{j=i}^{J} \left(\frac{1}{1 + \exp(\alpha_1 + \beta_1 f_i)}\right)^{m_{i,j}} \right],$$

where $\Delta_i \equiv q_{i,J+1}$ corresponds to the probability of not being observed again within the study, given by

$$\Delta_i = 1 - (1 - \phi_{1,i})\lambda_i - \sum_{j=i+1}^{J}\left[\phi_{1,i}\lambda_j(1 - \phi_{a,j}) \prod_{k=i+1}^{j-1} \phi_{a,k} \right]$$

$$= 1 - \left(\frac{1}{1 + \exp(\alpha_1 + \beta_1 f_i)}\right)\lambda_i - \sum_{j=i+1}^{J}\left[\left(\frac{\exp(\alpha_1 + \beta_1 f_i)}{1 + \exp(\alpha_1 + \beta_1 f_i)}\right)\lambda_j(1 - \phi_{a,j-1}) \prod_{k=i}^{j-2} \phi_{a,k} \right].$$

This posterior conditional distribution is clearly a nonstandard distribution. Similar posterior conditional distributions exist for all the other regression coefficient parameters—for example,

$$\pi(\beta_1 \mid \mathbf{m}, \alpha, \beta_{(1)}, \gamma_\lambda) \propto \exp\left(-\frac{\beta_1^2}{20}\right) \prod_{i=1}^{I}\left[\Delta_i^{m_{i,J+1}} \exp\left(\beta_1 f_i \sum_{j=i+1}^{J} m_{i,j}\right) \right.$$

$$\left. \times \prod_{j=i}^{J} \left(\frac{1}{1 + \exp(\alpha_1 + \beta_1 f_i)}\right)^{m_{i,j}} \right].$$

Thus we use a Metropolis–Hastings random-walk single-update algorithm for updating each of the parameters $\alpha_1, \beta_1, \alpha_a, \beta_a, \alpha_\lambda,$ and γ_λ within the Markov chain. For example, suppose that we propose to update parameter α_1. We propose the new candidate value,

$$\alpha_1' \sim U[\alpha_1 - \delta, \alpha_1 + \delta],$$

where δ is chosen arbitrarily. We accept this proposed parameter value with probability

$$\min\left(\frac{\pi(\alpha_1' \mid \mathbf{m}, \boldsymbol{\alpha}_{(1)}, \boldsymbol{\beta}, \gamma_\lambda)}{\pi(\alpha_1 \mid \mathbf{m}, \boldsymbol{\alpha}_{(1)}, \boldsymbol{\beta}, \gamma_\lambda)}\right),$$

since the proposal distribution is symmetric.

Note that δ is chosen via pilot tuning. For example, we calculate the mean acceptance rate over 1000 iterations of the Markov chain, and increase or decrease the proposal variance if this is deemed to be too high or too low, respectively. Gelman et al. (1996) and Roberts and Rosenthal (2001) suggest an optimal mean acceptance rate of (approximately) 0.234, but more generally a mean acceptance rate of 20–40% for well-performing chains. For the above example, Table 17.2 provides the value of δ used for each regression parameter for the pilot tuning steps performed and the corresponding mean acceptance rate, all lying in the interval 20–40% for the final proposal values. Note that there is clearly a tradeoff between the length of time used for pilot tuning and the computation time in performing the MCMC iterations. For this example, simulations are very quick to perform, so that relatively minimal pilot tuning is required.

17.2.1.2 Results

Implementing the above MCMC algorithm, the convergence to the stationary distribution appears to be very fast (i.e. within 1000 iterations from reasonable starting points). We run multiple chains starting from over-dispersed starting points for 100,000 iterations, using a conservative burn-in of 10,000 iterations. Independent replications provided essentially identical posterior results (to 3 decimal places for the summary statistics for each parameter) so that we assume that convergence has been achieved. More formally, the Brooks–Gelman–Rubin (BGR) statistic (Brooks and Gelman, 1998) also did not indicate any lack of convergence. The corresponding posterior means and standard deviations for each of the regression parameters are presented in column (a) of Table 17.3.

Clearly there is a negative relationship between both first-year and adult survival probabilities with the covariate *fdays*. Note that within the Markov chain, only negative values for the survival slope parameters (i.e. β_1 and β_a) are visited within the Markov chain following the burn-in period (even with initial positive starting values), demonstrating the

TABLE 17.2

The Values of the Proposal Parameter δ Used in the Iterative Pilot Tuning Procedure for Each Regression Coefficient for the Random-Walk Single-Update Metropolis–Hastings Algorithm with Uniform Proposal Distribution Within $\pm\delta$ of the Current Value and Corresponding Mean Acceptance Rate for 1000 Iterations (Ignoring the First 100 Iterations)

	Initial Value		Attempt 2		Attempt 3	
Parameter	δ	Mean Acceptance Probability (%)	δ	Mean Acceptance Probability (%)	δ	Mean Acceptance Probability (%)
α_1	0.1	73.4	0.6	16.4	0.4	25.4
β_1	0.1	67.4	0.5	19.3	0.3	33.0
α_a	0.1	70.6	0.6	17.5	0.4	24.6
β_a	0.1	60.9	0.5	13.4	0.3	24.5
α_λ	0.1	48.0	0.2	26.9	0.2	25.5
β_λ	0.1	54.2	0.2	27.6	0.2	30.5

TABLE 17.3

The Posterior Mean and Standard Deviation (SD) of each Regression Parameter for the Lapwing Data Set Using Different Independent Normal Priors Specified on the Regression Coefficients

Parameter	(a) Posterior Mean (SD)	(b) Posterior Mean (SD)	(c) Posterior Mean (SD)
α_1	0.536 (0.069)	0.533 (0.069)	0.536 (0.069)
β_1	−0.208 (0.062)	−0.207 (0.062)	−0.208 (0.062)
α_a	1.531 (0.070)	1.526 (0.069)	1.532 (0.070)
β_a	−0.310 (0.044)	−0.310 (0.044)	−0.311 (0.044)
α_λ	−4.567 (0.035)	−4.563 (0.035)	−4.567 (0.035)
γ_λ	−0.346 (0.039)	−0.345 (0.039)	−0.346 (0.039)

The priors used are: (a) $N(0, 10)$; (b) $N(0, 1)$; and (c) $N(0, 100)$.

strength of the negative association. This result is unsurprising since *fdays* is a surrogate for the harshness of the winter, when the majority of mortalities occur. In addition, there appears to be a decrease in the recovery probabilities with time (this is not unusual for ring-recovery studies).

Finally, we consider a prior sensitivity analysis. Columns (b) and (c) of Table 17.3 provide the corresponding posterior mean and standard deviations for the regression parameters assuming independent $N(0, 1)$ and $N(0, 100)$ priors on the regression parameters (i.e. changing the prior variances by a factor of 10). Clearly the posterior is data-driven with very little sensitivity on the posterior distributions of the parameters with the different prior specifications. The results here differ slightly with respect to those obtained by Brooks et al. (2004), (who use a $N(0, 100)$ prior specification, independently on each regression parameter) due to the fact that we normalize the time covariate for the recovery probabilities, while they logistically regress the recovery probability on the (raw) times, $2, \ldots, T$. Thus, the interpretation of the logistic regression parameters for the recovery probabilities differs between analyses (and hence so do the posterior estimates for these parameters). However, the posterior distributions for the other parameters are essentially identical.

17.2.2   Mixed Effects Model

The covariate model above assumes a deterministic relationship between the demographic parameters and covariates of interest. However, we now relax this assumption and allow additional temporal dependence not explained by the covariates considered, extending the models previously fitted to these data and considered by Besbeas et al. (2002), Brooks et al. (2004), and King et al. (2008b). We consider a mixed effects model (Pinheiro and Bates, 2000), with both fixed effects (covariate dependence) and additional random effects (on an annual level). In particular, we specify the mixed model on the first-year survival probabilities to be of the form

$$\text{logit } \phi_{1,j} = \alpha_1 + \beta_1 f_j + \epsilon_{1,j},$$

where $\epsilon_{1,j} \sim N(0, \sigma_1^2)$. This essentially changes the deterministic relationship between the survival probability and the covariate to be a stochastic relationship. An alternative specification of this model is

$$\text{logit } \phi_{1,j} \sim N(\alpha_1 + \beta_1 f_j, \sigma_1^2).$$

We once more use the standard vector notation $\epsilon_1 = \{\epsilon_{1,i} : i = 1, \ldots, I\}$. We consider analogous models for ϕ_a (regressed on $fdays$) and λ (regressed on year), with additional random effect terms ϵ_a and ϵ_λ and corresponding random effect variance terms, σ_a^2 and σ_λ^2, respectively. For notational convenience, we set $\epsilon = \{\epsilon_1, \epsilon_a, \epsilon_\lambda\}$ and $\sigma^2 = \{\sigma_1^2, \sigma_a^2, \sigma_\lambda^2\}$. Finally, note that we consider the same priors for the regression coefficients as before, and for the random effect variances specify $\sigma_k^2 \sim \Gamma^{-1}(a_k, b_k)$ for $k \in \{1, a, \lambda\}$. The parameters of interest in the model are typically the hyperparameters $\alpha, \beta, \gamma_\lambda$ and σ^2, as for standard mixed models, although we can also estimate the ϵ terms and hence ϕ_1, ϕ_a and λ. Note that, within the Bayesian framework, we could consider the random effects components of the model as simply specifying a hierarchical prior on the ϵ_1, ϵ_a and ϵ_λ terms.

17.2.2.1 Obtaining Posterior Inference

We wish to calculate the posterior distribution $\pi(\alpha, \beta, \gamma_\lambda, \sigma^2 \mid \mathbf{m})$. In order to do this, we need to specify the corresponding likelihood for the data, given the parameters. For a mixed model, the corresponding likelihood is expressed as an integral over the ϵ values. In particular, we can express the likelihood in the form

$$f(\mathbf{m} \mid \alpha, \beta, \gamma_\lambda, \sigma^2) = \int f(\mathbf{m} \mid \alpha, \beta, \gamma_\lambda, \epsilon) p(\epsilon \mid \sigma^2) \, d\epsilon.$$

However, this integral is analytically intractable. Thus, we consider a computationally intensive method for performing the integration, using MCMC. In particular, we regard the ϵ as parameters (or auxiliary variables) to be estimated. We then form the joint posterior distribution over both the auxiliary variables and model parameters,

$$\pi(\alpha, \beta, \gamma_\lambda, \sigma^2, \epsilon \mid \mathbf{m}) \propto f(\mathbf{m} \mid \alpha, \beta, \gamma_\lambda, \sigma^2, \epsilon) p(\alpha, \beta, \gamma_\lambda, \sigma^2, \epsilon)$$

$$= f(\mathbf{m} \mid \alpha, \beta, \gamma_\lambda, \epsilon) p(\alpha) p(\beta) p(\gamma_\lambda) p(\epsilon \mid \sigma^2) p(\sigma^2),$$

taking into account the conditional independence of the different parameters, and where $f(\mathbf{m} \mid \alpha, \beta, \gamma_\lambda, \epsilon)$ can once more be easily calculated using the standard likelihood for ring-recovery data, since the demographic parameters ϕ_1, ϕ_a and λ are a deterministic function of the parameters $\alpha, \beta, \gamma_\lambda$ and ϵ. The required posterior distribution is simply the marginal distribution,

$$\pi(\alpha, \beta, \gamma_\lambda, \sigma^2 \mid \mathbf{m}) = \int \pi(\alpha, \beta, \gamma_\lambda, \sigma^2, \epsilon \mid \mathbf{m}) \, d\epsilon.$$

To obtain a sample from this marginal distribution, we use an MCMC algorithm to obtain a sample from the full posterior distribution of all the model parameters and auxiliary variables and simply consider the sampled values of the parameters of interest (i.e. irrespective of the values for the auxiliary variable (ϵ) terms). We note that the posterior distribution over both the model parameters and auxiliary variables is sampled from within the MCMC algorithm, so that we can also obtain posterior summary statistics of the auxiliary variables if they are of interest—see Section 17.3.1 for a particular example. In addition, since we impute the ϵ values, we also impute the demographic parameters ϕ_1, ϕ_a and λ. Thus we can once more easily obtain posterior estimates of these demographic rates from the MCMC algorithm. Finally, we note that mixed models often take significantly longer to run than fixed effects models, due to the increase in the number of parameters that need to be updated at each iteration of the Markov chain. For example, for the lapwing data set, the mixed model takes approximately 15 times longer to run than the fixed effects model (with only six parameters), due to the large number of random effects terms (ϵ_1, ϵ_a, and ϵ_λ) that need to be imputed within the MCMC algorithm.

17.2.2.2 *Posterior Conditional Distributions*

We implement a single-update Metropolis–Hastings algorithm. In particular, we implement the same proposal distributions for the regression coefficients, α and β, as for the fixed effects model above. In addition, for each random effect term, we again use a random-walk Metropolis–Hastings step, using a uniform proposal distribution. Finally, for the random effect variances, we use a Gibbs step, since the posterior conditional distributions are of standard form. For example, for σ_1^2, we have the posterior conditional distribution,

$$\sigma_1^2 \mid \alpha, \beta, \gamma_\lambda, \epsilon, \mathbf{m} \sim \Gamma^{-1}\left(a_1 + \frac{I}{2}, b_1 + \frac{1}{2}\sum_{i=1}^{I}\epsilon_{1,i}^2\right).$$

With similar results for σ_a^2 and σ_λ^2. Without any prior information we set $a_k = b_k = 0.001$ for $k \in \{1, a, \lambda\}$. Note that Gelman (2006) suggests an alternative prior specification for the random effect variance terms, when there is no prior information, where the standard deviation (rather than variance) is an (improper) uniform distribution on the positive real line. This induces a prior on the variance of the form, $p(\sigma_1^2) \propto \sigma_1^{-1}$. The corresponding posterior conditional distribution is again of standard form with

$$\sigma_1^2 \mid \alpha, \beta, \gamma_\lambda, \epsilon, \mathbf{m} \sim \Gamma^{-1}\left(\frac{I-1}{2}, \frac{1}{2}\sum_{i=1}^{I}\epsilon_{1,i}^2\right).$$

We initially retain the inverse gamma prior, but do consider a prior sensitivity analysis using this alternative prior.

17.2.2.3 *Results*

The simulations are run for 100,000 iterations, with the first 10,000 simulations discarded as burn-in. Note that, typically, for random effects models longer simulations are necessary (since more parameter space needs to be explored). However, convergence appears to be rather swift yet again. Column (a) of Table 17.4 provides the corresponding posterior summary statistics for the regression coefficients and random effects variance terms. Independent replications from over-dispersed starting points differed only slightly (typically in the third decimal place), so that we assume that convergence has been achieved (with BGR statistics approximately equal to one for all parameters). Comparing the posterior estimates for the regression coefficients with Table 17.3 for the corresponding fixed effects model (i.e. no random effects present), it is clear that these parameter estimates are very similar, as we would typically expect. The effect of the additional random effects terms is most easily seen by considering the corresponding demographic parameter estimates, since the magnitude of the random effect variance needs to be interpreted with respect to regression coefficient

TABLE 17.4

The Posterior Mean and Standard Deviation (SD) of Each Parameter for the Lapwing Data Set for the Mixed Effects Model for (a) $\Gamma^{-1}(0.001, 0.001)$ Prior on the Random Effect Variance Terms; and (b) Gelman's Prior ($\propto \sigma^{-1}$) on the Random Effect Variance Terms

	(a)	(b)
Parameter	Posterior Mean (SD)	Posterior Mean (SD)
α_1	0.535 (0.081)	0.538 (0.086)
β_1	−0.216 (0.079)	−0.214 (0.082)
α_a	1.525 (0.073)	1.533 (0.073)
β_a	−0.315 (0.050)	−0.317 (0.052)
α_λ	−4.567 (0.039)	−4.567 (0.040)
γ_λ	−0.350 (0.044)	−0.350 (0.044)
σ_1^2	0.061 (0.059)	0.090 (0.071)
σ_a^2	0.009 (0.011)	0.012 (0.016)
σ_λ^2	0.008 (0.009)	0.010 (0.012)

parameters and link function. Figure 17.1 provides the posterior mean and 95% highest posterior density interval (HPDI) for the survival and recovery probabilities for both the fixed effects and mixed effects models for comparison. Note that producing posterior estimates of the demographic parameters are straightforward, as they are calculated within the MCMC algorithm at each iteration (since the ϵ terms are imputed).

The random effects variance terms appear to be very small for the adult survival probabilities and recovery probabilities. This is demonstrated in Figure 17.1 with the very similar posterior estimates for the fixed effects and mixed effects models. This would suggest that the covariates largely explain the temporal variability within the demographic parameters. However, for the first-year survival probabilities, the addition of a random effect in the model appears to significantly increase the posterior uncertainty, suggesting that the covariate *fdays* may not adequately model the temporal variability for the parameter. Thus, this result could in itself prompt further investigation for the first-year survival probabilities, for example, the consideration of further environmental covariates or the addition of an individual heterogeneity component to the model.

We conduct a prior sensitivity analysis, using the prior suggested by Gelman (2006). The corresponding posterior summary statistics of the regression parameters and random effect variance terms are given in column (b) of Table 17.4. There is typically very little difference between the posterior results obtained using the different priors. The largest difference observed is in the posterior mean (and standard deviation) for σ_1^2. This results in a very slight increase in the posterior variance of the first-year survival probabilities, but the difference is minimal.

17.2.3 Model Uncertainty

Previously, we have assumed a known covariate structure for each of the demographic parameters—a dependence on *fdays* for the survival probabilities or time for the recovery probabilities. Typically the models are developed from biological understanding; however, there will generally be a level of model uncertainty regarding the presence of absence of the covariates within the model. For example, for the lapwing data, we assumed that

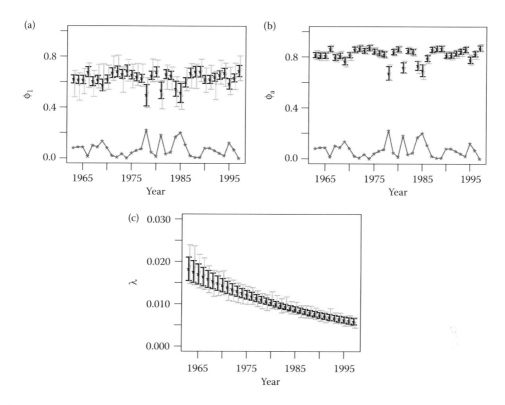

FIGURE 17.1
The posterior mean (*) and 95% HPDI for the fixed effects model (in black) and mixed model (in gray) (a) ϕ_1; (b) ϕ_a; and (c) λ. In (a) and (b) the bottom line denotes the covariate *fdays*.

the survival probabilities were a function of only *fdays* and the recovery probabilities of only time. However, more generally we may wish to consider additional models, allowing for different combinations of *fdays* and time dependence on the demographic parameters. Discriminating between these different competing models is often of particular biological interest, in order to understand the underlying dynamics of the biological system.

Within the Bayesian framework model uncertainty is conceptually easy to introduce, by simply considering the model itself to be an unknown parameter to be estimated. The joint posterior distribution over both parameter and model space is given by

$$\pi(\boldsymbol{\theta}_n, n \mid \mathbf{m}) \propto f(\mathbf{m} \mid \boldsymbol{\theta}_n, n) p(\boldsymbol{\theta}_n \mid n) p(n),$$

where $f(\mathbf{m} \mid \boldsymbol{\theta}_n, n)$ denotes the likelihood of the data given model n with corresponding parameters $\boldsymbol{\theta}_n$. We quantitatively discriminate between competing models by calculating the corresponding posterior model probability given by the marginal distribution,

$$\pi(n \mid \mathbf{m}) \propto \int \pi(\boldsymbol{\theta}_n, n \mid \mathbf{m}) \, d\boldsymbol{\theta}_n. \tag{17.1}$$

To obtain a sample from the joint posterior distribution, $\pi(\boldsymbol{\theta}_n, n \mid \mathbf{m})$, we use the reversible jump (RJ)MCMC algorithm (Green, 1995), since the posterior distribution is multidimensional with the number of parameters (i.e. regression coefficients) differing between models.

17.2.3.1   Model Specification

We need to define the set of models that we wish to consider and discriminate between. In order to do this we define the saturated model of the form:

$$\text{logit } \phi_{1,j} = \alpha_1 + \beta_1 f_j + \gamma_1 t_j + \epsilon_{1,j}, \quad \epsilon_{1,j} \sim N(0, \sigma_1^2),$$

$$\text{logit } \phi_{a,j} = \alpha_a + \beta_a f_j + \gamma_a t_j,$$

$$\text{logit } \lambda_j = \alpha_\lambda + \beta_\lambda f_j + \gamma_\lambda t_j.$$

We adopt the standard notation $\boldsymbol{\alpha} = \{\alpha_1, \alpha_a, \alpha_\lambda\}$, and similarly for $\boldsymbol{\beta}$ and $\boldsymbol{\gamma}$. We consider a mixed effects model for the first-year survival probabilities and a fixed effects model for the adult survival probabilities and recovery probabilities, motivated by the results in Section 17.2.2.

Alternative (sub)models are then obtained by specifying restrictions on the regression coefficients. For example, setting $\beta_a = 0$ implies that the adult survival probabilities are not related to *fdays*. We consider all four possible combinations of covariate dependence for each demographic parameter, allowing for the inclusion or exclusion of each covariate. Taking all possible combinations of covariate dependence for each demographic parameter gives a total of $4^3 = 64$ models. Within the Bayesian framework, we need to specify a prior over the model space, that is, define the prior probability of each model. In this instance, with no prior information, we specify an equal prior probability on each model, which also corresponds to a marginal prior probability of 0.5 that a demographic parameter is dependent on a given covariate. We note that, more generally, placing a flat prior over the full model space may not be the most sensible course of action. See King et al. (2006) who discuss this in further detail in the presence of additional age dependence on the demographic parameters, and King and Brooks (2003) for the case of multi-site data.

17.2.3.2   Reversible Jump Algorithm

Within each iteration of the Markov chain we consider the following steps:

1. Update each parameter, conditional on it being present in the model; for each (nonzero) parameter $\boldsymbol{\alpha}, \boldsymbol{\beta}, \boldsymbol{\gamma}$ and ϵ_1 we use a single-update uniform random-walk Metropolis–Hastings update and for σ_1^2 a Gibbs step.
2. Update the covariate dependence in the model using a reversible jump step.

We consider only Step 2 in detail, since the previous Metropolis–Hastings steps are implemented in the same way as before. For the reversible jump step of the MCMC algorithm, we cycle through each demographic parameter in turn and propose to add or remove a single covariate dependence. Without loss of generality, suppose that we are considering the adult survival probabilities. We randomly select one of the covariates (*fdays* or time). Suppose that we select *fdays*. If the covariate dependence is present ($\beta_a \neq 0$), we propose to remove the dependence; else if the covariate dependence is absent ($\beta_a = 0$), we propose to add the dependence. We initially consider the case where we propose to add the dependence.

We denote the current model by n, with parameters $\boldsymbol{\alpha}, \boldsymbol{\beta}, \boldsymbol{\gamma}, \epsilon_1$, and σ_1^2 and with $\beta_a = 0$. We propose to move to model n' which has the additional covariate dependence on *fdays* for adult survival probabilities. We propose a candidate value β_a' from some proposal distribution q. All other parameter values remain the same. Finally, we let $\boldsymbol{\beta}' = \{\beta_1, \beta_a', \beta_\lambda\}$

denote the regression coefficients for *fdays* in the proposed model. We accept the proposed move with probability $\min(1, A)$, where

$$A = \frac{\pi(\boldsymbol{\alpha}, \boldsymbol{\beta}', \boldsymbol{\gamma}, \boldsymbol{\epsilon}_1, \sigma_1^2, n' \mid \mathbf{m})}{\pi(\boldsymbol{\alpha}, \boldsymbol{\beta}, \boldsymbol{\gamma}, \boldsymbol{\epsilon}_1, \sigma_1^2, n \mid \mathbf{m}) q(\beta_a')}.$$

Note that the Jacobian term in the standard reversible jump acceptance probability is simply equal to one in this case.

Now, to consider the reverse move, suppose that the chain is in model n' with current parameter values $\boldsymbol{\alpha}, \boldsymbol{\beta}', \boldsymbol{\gamma}, \boldsymbol{\epsilon}_1$, and σ_1^2, such that $\boldsymbol{\beta}' = \{\beta_1, \beta_a', \beta_\lambda\}$. We propose the new model n, such that $\beta_a = 0$, and set $\boldsymbol{\beta} = \{\beta_1, \beta_a, \beta_\lambda\}$. We accept this move with probability $\min(1, A^{-1})$ for A given above.

17.2.3.3 *Proposal Distribution*

In order to improve the efficiency of the reversible jump updating step we consider the proposal distribution, q, in more detail. Without loss of generality, suppose that we propose to add in the parameter δ to the model, so that $\delta \in \{\beta_1, \beta_a, \beta_\lambda, \gamma_1, \gamma_a, \gamma_\lambda\}$. We consider a normal proposal distribution, namely,

$$\delta \sim N(\mu_\delta, \sigma_\delta^2).$$

The values of μ_δ and σ_δ^2 are chosen via a pilot tuning exercise. In particular, we run the saturated model for 10,000 iterations (discarding the first 1000 iterations as burn-in). We then set μ_δ and σ_δ^2 to be the posterior mean and variance of the corresponding regression coefficient from this pilot run. In other words,

$$\mu_\delta = E_\pi(\delta); \qquad \sigma_\delta^2 = \text{var}_\pi(\delta),$$

where we take the expectation and variance with respect to the posterior distribution of the parameters in the saturated model. See King and Brooks (2002) and King et al. (2009) for further discussion of proposal distributions of this form.

17.2.3.4 *Results*

We run the simulations for a total of 100,000 iterations, discarding the first 10,000 iterations as burn-in. Trace plots suggest that again the burn-in is very conservative. Independent replications from over-dispersed starting points obtain essentially identical results, so that we assume the estimates have converged. Table 17.5a provides the corresponding (marginal) posterior probability that each covariate is present in the model for each demographic parameter. Clearly there is very strong evidence that the adult survival probability is dependent on *fdays*, with a marginal posterior probability of the dependence equal to 0.959 (equivalent to a Bayes factor of 23). Similarly, the recovery probability appears to be strongly dependent on time, with a posterior probability of 1.000 to three decimal places. There is evidence that the survival probabilities are not dependent on time for this data (Bayes factor of 33 for first-years and 10 for adults). Finally, there is uncertainty as to whether the first-year survival probabilities are dependent on *fdays*, with a posterior probability of 0.567 of no dependence (or a Bayes factor of 1.3).

TABLE 17.5

The Marginal Posterior Probability of the Dependence of Each Demographic Parameter on the Combination of Covariates *fdays* and Time, Assuming Fixed Effects Models for ϕ_a and λ and a Mixed Effects Model for ϕ_1, Specifying Independent Normal Distributions on the Regression Parameters With Mean Zero and Variance of (a) 10, (b) 1, and (c) 100, and (d) Gelman's Prior on the Random Effect Variance Component With $N(0, 10)$ Priors on the Regression Parameters

Covariate Dependence	(a) $N(0, 10)$ Priors			(b) $N(0, 1)$ Priors			(c) $N(0, 100)$ Priors			(d) Gelman's Prior		
	ϕ_1	ϕ_a	λ	ϕ_1	ϕ_a	λ	ϕ_1	ϕ_a	λ	ϕ_1	ϕ_a	λ
fdays and time	0.012	0.084	0.125	0.055	0.222	0.171	0.002	0.029	0.085	0.012	0.084	0.111
fdays only	0.420	0.875	0.000	0.578	0.751	0.000	0.227	0.922	0.000	0.368	0.880	0.000
time only	0.017	0.007	0.870	0.033	0.018	0.829	0.008	0.002	0.915	0.020	0.005	0.889
none	0.550	0.039	0.000	0.334	0.019	0.010	0.764	0.047	0.000	0.600	0.031	0.000

We once more consider a prior sensitivity analysis, changing the prior variance on each of the parameters. In particular, we consider independent $N(0, 1)$ and $N(0, 100)$ priors on each of the regression parameters (i.e. increasing and decreasing the variance by a factor of 10). Alternatively, we consider Gelman's prior on the random effect variance term (with independent $N(0, 10)$ priors on the regression coefficients). The corresponding posterior model probabilities obtained are given in Table 17.5b–d. Recall that previously, when considering only a single model, changing the prior specification had virtually no impact on the corresponding posterior distributions of the parameters (see Section 17.2.1 and Tables 17.3 and 17.4). For the adult survival probabilities and recovery probabilities, the posterior model probabilities (and corresponding interpretation of the results) are generally fairly insensitive to the choice of prior. However, we can clearly see that the prior specification does influence the posterior model probabilities for the first-year survival probabilities, where there is the greatest uncertainty regarding the presence or absence of the covariates in the model. Increasing (decreasing) the prior variance results in a decrease (increase) in the corresponding posterior probability of the covariate being present in the model. This can be explained by considering the form of the posterior model probability in equation 17.1, which involves integrating out the parameters within the joint posterior distribution over both parameter and model space. Specifying a larger prior variance increases the area being integrated over with small posterior mass, decreasing the corresponding integral value (and hence posterior probability), and is often referred to as Lindley's paradox (Lindley, 1957). This has a greater effect (as for first-year survival probabilities) when there is relatively little information contained within the data relating to the dependence structure. Alternatively, specifying Gelman's prior on the random effect variance term for the first-year survival probability has relatively little impact. The posterior mean of the random effect variance term is slightly greater (0.123 for Gelman's prior compared to 0.093 for the $\Gamma^{-1}(0.001, 0.001)$ prior), which has the impact of slightly decreasing the posterior probability that the first-year survival probability is dependent on *fdays*, with the random effect term essentially explaining a greater amount of the temporal variability (compared to *fdays*).

17.2.3.5 Comments

RJMCMC algorithms typically require longer simulations than standard MCMC algorithms, since the additional model space needs to be explored as well as parameter space. However, tuning the reversible jump step is typically more difficult, and the corresponding mean acceptance probabilities are lower than for standard Metropolis–Hastings steps. For example, in the reversible jump step above, we obtain a mean acceptance probability of 15.4% for changing the dependence of the first-year survival probability on *fdays* (with a proposal mean of -0.377 and a proposal standard deviation of 0.165). However, if we had considered a "plain vanilla" reversible jump step, and used the prior distribution as the proposal distribution, we would have obtained a mean acceptance probability of 3.8%. In this instance, the prior distribution is rather diffuse, so we could consider an alternative pilot tuning exercise for the proposal variance (keeping the proposal mean equal to zero). For example, setting the proposal variance to be equal to unity provided a mean acceptance rate of 8.0%. However, taking the posterior mean and variance of the parameters in the saturated model as the proposal mean and variance involves only a single simulation, and also updates the location of the proposal distribution, although it does assume that the parameter values are generally similar across models. Note that additional pilot tuning could also be performed here. In this instance, changing the proposal variance did not make any significant improvement in the acceptance probabilities.

We performed model selection on the covariates present in the model. However, we could also consider the analogous model selection technique in relation to the random effects, in terms of their presence or absence. This is generally more difficult since we need to specify a "sensible" prior on the random effect variance term. As we have seen for the covariate terms, posterior model probabilities are typically more sensitive to the priors specified on the parameters than the posterior distribution of the parameter. This is as a result of integrating out over the parameter space in the derivation of the posterior model probability, as described above. Typically we specify a vague prior on the random effect variance term. In order to use the RJMCMC algorithm, the prior needs to be a proper prior so that the acceptance probability can be calculated (since we need to evaluate the prior in the acceptance probability, and the constant of proportionality is infinite for improper priors). For proper vague priors (such as the $\Gamma^{-1}(0.001, 0.001)$ distribution), the prior distribution is very diffuse, so that the simpler fixed effects model will often be chosen (i.e. Lindley's paradox occurs). Thus, for this data set, we consider the posterior distributions of the random effects variances and corresponding demographic parameters (assuming a mixed model for each demographic parameter) to see whether it appears that random effects are important or not. In particular, we conclude that random effects are only present for the first-year survival probabilities. A more rigorous approach would be to consider the form of the distribution of the random effect variance term in more detail, such that the induced prior on the corresponding random effect terms (and hence demographic parameter) are "sensible."

17.3 Analysis of Count Data

For the UK lapwing population, there are additional, independent, count data. We consider data from 1965 to 1998. These data correspond to estimates of the number of breeding female lapwings at a number of sites throughout the United Kingdom and can be regarded as an index for the total population. However, these counts are only an estimate of the index (or population size). In order to account for this uncertainty regarding these estimates we consider a state–space approach, following the ideas and model suggested by Besbeas et al. (2002).

17.3.1 State–Space Model

State–space models consider two separate processes: a system process, describing how the population size changes over time; and the observation process, which takes into account the uncertainty in the observed count data. We consider each of these processes in turn.

17.3.1.1 System Process

For the UK lapwing data, the individuals are described as either first-years or adults. Thus, we consider these population sizes separately. Let $N_{1,i}$ and $N_{a,i}$ denote the true number of first-year (female) birds and adult (breeding female) birds at time i. We assume that all adult birds breed; and that no first-year birds breed. A natural model for the number of first-year birds at time i would be

$$N_{1,i} \sim \text{Poisson}(N_{a,i-1}\rho_{i-1}\phi_{1,i-1}),$$

where ρ_i denotes the productivity rate of females per female in year i. In addition, for the number of female adults we assume

$$N_{a,i} \sim \text{Binomial}(N_{1,i-1} + N_{a,i-1}, \phi_{a,i-1}).$$

For further discussion of this model, see, for example, Besbeas et al. (2002). We note that $\rho_i\phi_{1,i}$ are confounded in this model, since the terms only appear as a product. In order to separate these two processes (first-year survival and productivity), additional data is necessary. For example, specifying a logistic regression on the first-year survival probability on a given covariate can remove this confounding. Alternatively, the use of additional data (such as the ring-recovery data considered previously) can also remove this confounding issue, since the parameter ϕ_1 is estimable from the ring-recovery data.

  The system process is defined for $i = 1, \ldots, T$ (with $i = 1$ corresponding to the year 1965 and $T = 34$). However, for $i = 1$, $N_{1,i}$ and $N_{a,i}$ are a function of $N_{1,0}$ and $N_{a,0}$. To allow for this (without truncating the likelihood), we consider $N_{1,0}$ and $N_{a,0}$ as parameters, and place vague uniform priors on them. Note that this essentially induces a prior on all other population sizes, $N_{1,i}$ and $N_{a,i}$ for $i = 1, \ldots, T$, by the relationship expressed above in the system process.

17.3.1.2 Observation Process

We do not observe the true population sizes, $\mathbf{N}_1 = \{N_{1,1}, \ldots, N_{1,T}\}$ and $\mathbf{N}_a = \{N_{a,1}, \ldots, N_{a,T}\}$, but only an estimate of some of the population sizes. For the UK lapwings, we have an estimate of only the number of breeding females, denoted by $\mathbf{y} = \{y_1, \ldots, y_T\}$, that is, $\mathbf{y}$ is an estimate of $\mathbf{N}_a$. In order to model the observation uncertainty, we assume

$$y_i \sim N(N_{a,i}, \sigma_y^2),$$

for $i = 1, \ldots, T$, where σ_y^2 is to be estimated. Clearly many other models are possible, such as a lognormal distribution (King et al., 2008b), or where the observation error variance is proportional to the true population size. We retain the simplest normal observation model. We place a conjugate inverse gamma prior on the observation variance, $\sigma_y^2 \sim \Gamma^{-1}(a_y, b_y)$, with $a_y = b_y = 0.001$.

17.3.1.3 Model

In addition to the survival and recovery parameters, there are the additional productivity rates. We specify a logarithmic regression for the productivity rate of the form

$$\log \rho_i = \alpha_\rho + \gamma_\rho t_i,$$

where t_i denotes the (normalized) variable corresponding to time (i.e. year). We extend our vector notation, so that $\boldsymbol{\alpha} = \{\alpha_1, \alpha_a, \alpha_\lambda, \alpha_\rho\}$ and $\boldsymbol{\gamma} = \{\gamma_\lambda, \gamma_\rho\}$.

The overall model has a random effects component for the first-year survival probabilities, also dependent on *fdays*, a fixed effects model for adult survival probabilities dependent on *fdays*, and fixed effects models for the recovery probabilities and productivity rate, both dependent on time. Thus, this model considers additional random effects for the first-year survival probabilities, not considered in the previous analyses by Besbeas et al. (2002), Brooks et al. (2004), and King et al. (2008b). Notationally, we specify this model in the form

$$\phi_1(fdays, \sigma_1^2)/\phi_a(fdays)/\lambda(t)/\rho(t).$$

This model is motivated by the model identified from the ring-recovery data considered previously with a logistic regression specified on $\boldsymbol{\phi}_1$, $\boldsymbol{\phi}_a$ and $\boldsymbol{\lambda}$ and the analysis by Besbeas et al. (2002) (for the productivity rates).

17.3.1.4 Obtaining Inference

The likelihood of the count data is analytically intractable. We consider an auxiliary variable approach (analogous to the approach used for the random effects model). We treat the true population sizes $\mathbf{N}_1$ and $\mathbf{N}_a$ as parameters (or auxiliary variables). The corresponding joint likelihood of the count data y and true population sizes (given the demographic parameters) can be written in the form

$$f(y, \mathbf{N}_1, \mathbf{N}_a \mid N_{1,0}, N_{a,0}, \boldsymbol{\phi}_1, \boldsymbol{\phi}_a, \boldsymbol{\lambda}, \boldsymbol{\rho}, \sigma_y^2) = f_{obs}(y \mid \mathbf{N}_a, \sigma_y^2) f_{sys}(\mathbf{N}_1, \mathbf{N}_a \mid N_{1,0}, N_{a,0}, \boldsymbol{\phi}_1, \boldsymbol{\phi}_a, \boldsymbol{\lambda}, \boldsymbol{\rho}),$$

where f_{obs} and f_{sys} denote the likelihood functions associated with the observation and system processes. Thus f_{obs} is a product over normal distributions, and f_{sys} a product over Poisson and binomial distributions. For notational simplicity, we specify the likelihood as a function of the demographic parameters. Equivalently, we can express the likelihood given the regression parameters (and random effect terms if present). This likelihood can be combined with the priors to form the joint posterior distribution of the parameters and auxiliary variables (true population sizes and random effects terms). However, note that there is typically very little information relating to the first-year parameters (population size and survival probabilities), with only estimates of the adult population sizes. In addition, the posterior precision of the other parameters can also be poor (see, e.g. Brooks et al., 2004). However, there is the additional, independent, ring-recovery data that can be combined with the count data within a single integrated analysis, using all the available information.

17.3.2 Integrated Analysis

Following the approach of Besbeas et al. (2002), we combine the independent ring-recovery data and count data within an integrated analysis, but consider a different underlying model (as we assume a mixed effects model for the first-year survival probabilities). Since the data sources are independent of each other, we can write the joint likelihood of the ring-recovery data and count data as the product of the individual likelihoods. Thus, we can express the joint posterior distribution of the regression parameters and auxiliary variables (population sizes and random effect terms) in the form

$$\pi(\boldsymbol{\alpha}, \boldsymbol{\beta}, \boldsymbol{\gamma}, \sigma_1^2, \boldsymbol{\epsilon}_1, \mathbf{N}_1, \mathbf{N}_a, \sigma_y^2 \mid \boldsymbol{m}, \boldsymbol{y}) \propto f_{\text{obs}}(\boldsymbol{y} \mid \mathbf{N}_a, \sigma_y^2) f_{\text{sys}}(\mathbf{N}_1, \mathbf{N}_a \mid \boldsymbol{\alpha}, \boldsymbol{\beta}, \boldsymbol{\gamma}, \boldsymbol{\epsilon}, N_{1,0}, N_{a,0})$$

$$\times f(\mathbf{m} \mid \boldsymbol{\alpha}, \boldsymbol{\beta}, \boldsymbol{\gamma}, \boldsymbol{\epsilon})$$

$$\times p(\boldsymbol{\alpha}) p(\boldsymbol{\beta}) p(\boldsymbol{\gamma}) p(\boldsymbol{\epsilon}_1 \mid \sigma_1^2) p(\sigma_1^2) p(\sigma_y^2) p(N_{0,1}) p(N_{0,a}).$$

The first line corresponds to the likelihood for the count data, the second line to the likelihood for the ring-recovery data, and the final terms to the priors for all the parameters in the model.

17.3.2.1 MCMC Algorithm

We consider a single-update Metropolis–Hastings algorithm. We implement a uniform random-walk algorithm for the demographic regression parameters, $\boldsymbol{\alpha}, \boldsymbol{\beta}$, and $\boldsymbol{\gamma}$, and the population sizes, $N_{1,0}$ and $N_{a,0}$. Alternatively, for σ_y^2, we use a Gibbs update, since the conditional distribution is of standard form,

$$\sigma_y^2 \mid \mathbf{N}_a \sim \Gamma^{-1}\left(a_y + \frac{T}{2}, b_y + \frac{1}{2}\sum_{i=1}^{T}(y_i - N_{a,i})^2\right).$$

Finally, we consider two different updating schemes for $\mathbf{N}_1$ and $\mathbf{N}_a$.

> **Algorithm 1**. Uniform random-walk single-update Metropolis–Hastings algorithm (as implemented by King et al., 2008b).
>
> **Algorithm 2.** Single-update Metropolis–Hastings algorithm using the system process (i.e. binomial or Poisson distribution) as the proposal distribution for times $i = 1, \ldots, T$.

For algorithm 1, we initially performed a pilot tuning exercise to obtain the lower and upper bound of the uniform proposal. We set the proposal distribution such that the candidate value is within ± 75 of the current value, providing a mean acceptance probability of 18–39% for first-years and adults over all years. Algorithm 2 does not typically require any pilot tuning, and results in mean acceptance probabilities of 56–75% for first-years and adults for years $i = 1, \ldots, T - 1$ (for year T the mean acceptance probabilities are 100% and 98% for first-years and adults). Figure 17.2 provides a trace plot of the number of first-years and adults for a typical year, while Figure 17.3 provides the corresponding autocorrelation function (ACF) plot, for both algorithms. The trace plots suggest that algorithm 2 may have better mixing properties (particularly for the adults). This is supported by the ACF plots, with a reduction of approximately 20% in the autocorrelation (by lag 50) of algorithm 2 compared to algorithm 1 (for this particular year). Thus, we retain the use of algorithm 2.

One problem with the single-update approach is that population sizes are highly correlated from one year to the next, due to the underlying system process. This means that the algorithm can exhibit poor mixing and high autocorrelation (as demonstrated above). Block

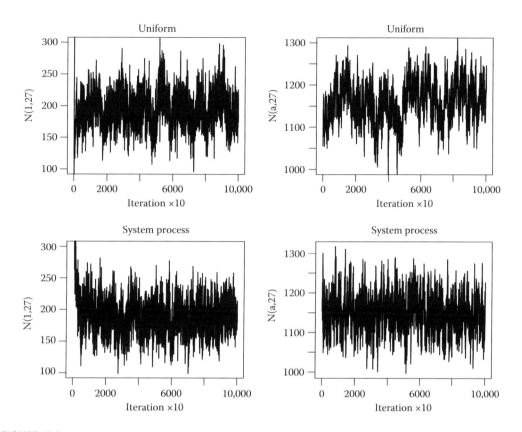

FIGURE 17.2

Trace plots for the number of first-years (top left) and adults (top right) for algorithm 1 (uniform proposal); and first-years (bottom left) and adults (bottom right) for algorithm 2 (using system process) in the year 1989.

updating can improve the mixing, and was considered for algorithm 2 (e.g. simultaneously updating $N_{1,i}$ and $N_{a,i}$ within a single update), but did not appear to improve the mixing in this case.

17.3.2.2 Results

The simulations are run for 1 million iterations, with the first 10% discarded as burn-in, to be conservative. Table 17.6 provides the corresponding posterior mean and standard deviation of each parameter in the model. Note that the posterior distributions for the regression parameters are very similar to those obtained for the ring-recovery alone, given in Table 17.3. This is a result of there being relatively little direct information in the count data on these parameters, so that the ring-recovery data dominate with respect to these parameters. For further discussion of similar issues and for a comparison of results for ring-recovery data only, count data only, and integrated data for the analogous fixed effects model, see Brooks et al. (2004) and King et al. (2009). Recall that within the MCMC algorithm we also impute the true population sizes. Thus, we can also draw inference on the true population sizes, which is typically of particular interest. Figure 17.4 provides the posterior mean and 95% HPDI of the estimates of the true population size for first-years and adults, along with the corresponding observed count data, y.

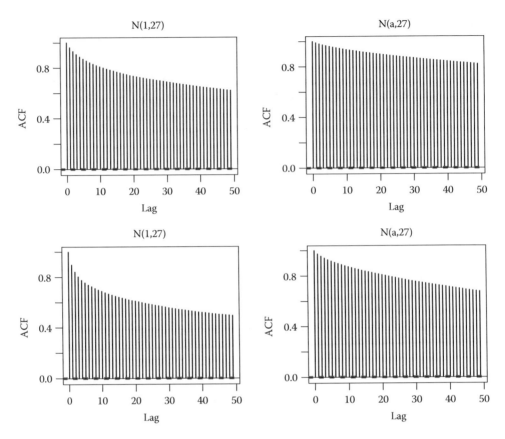

FIGURE 17.3
ACF plots for the number of first-years (top left) and adults (top right) for algorithm 1 (uniform proposal); and first-years (bottom left) and adults (bottom right) for algorithm 2 (using system process) in the year 1989. Note that the ACF at lag 50 is equal to 0.619 and 0.494 for the number of first-years and 0.820 and 0.677 for the number of adults for algorithms 1 and 2, respectively.

TABLE 17.6

The Posterior Mean and Standard Deviation (SD) of Each Parameter for the Integrated Analysis of Ring-Recovery Data and Count Data

Parameter	Posterior Mean (SD)
α_1	0.545 (0.082)
β_1	−0.202 (0.075)
α_a	1.545 (0.071)
β_a	−0.245 (0.039)
α_λ	−4.563 (0.035)
γ_λ	−0.351 (0.039)
α_ρ	−1.142 (0.090)
γ_ρ	−0.253 (0.053)
σ_1^2	0.063 (0.056)

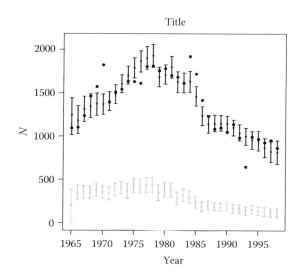

FIGURE 17.4

The posterior mean (*) and 95% HPDI for the population size relating to adults (in black) and first-years (in gray). The solid dots give the corresponding data, y, relating to the estimates of the adult population size.

In addition, within the MCMC algorithm we can obtain a sample from the posterior distribution of functions of the population sizes (and/or other parameters). For example, one particular quantity of interest is the (log) adult population growth rate over time. For generality, we let $\log r_{i,j}$ denote the log change in adult growth to year i at lag j, defined to be

$$\log r_{i,j} = \log N_{a,i} - \log N_{a,i-j}.$$

We can estimate additional quantities of interest, such as the posterior probability that the log adult population growth rate at lag j is positive, corresponding to an increase in the adult population between times $i - j$ and i. Consider, for example, $r_{i,25}$ (i.e. the change in population size over 25 years), used in the determination of species of conservation concern (Gregory et al., 2002). The posterior distribution for $r_{i,25}$ is plotted in Figure 17.5. An estimate from the observed count data, ignoring the uncertainty in relation to these estimates, is plotted for comparison (although no associated uncertainty intervals can be calculated). Clearly, there appears to be a significant change in the adult lapwing population over the 25-year period 1973–1998. For example, in 1998 we obtain a 42% posterior probability that the population has declined by more than 50%. See King et al. (2008b) and Brooks et al. (2008) for further discussion of assessing changing population sizes and their relation to conservation concern.

17.3.3 Model Selection

We once more consider the issue of the underlying covariate dependence for each of the demographic parameters. Allowing each parameter (i.e. survival probabilities, recovery probabilities, and productivity rates) to be dependent on *fdays* and/or time, there are a total of $4^4 = 256$ possible models. Once more, we assume that the first-year survival probabilities have a random effects component, whereas all other parameters are fixed effects models, thus extending the set of models considered by King et al. (2008b). We assume an equal

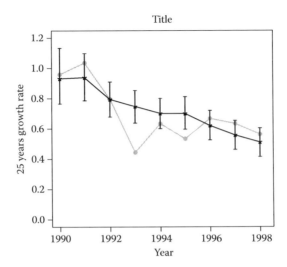

FIGURE 17.5
The posterior mean (*) and 95% HPDI for the 25-year adult growth rate, $r_{i,25}$ (in black) and corresponding 25-year adult growth rate from raw data, y (in gray).

prior probability on each of the possible models and use the analogous RJMCMC algorithm as before, extending the method to updating the dependence of the productivity rate on *fdays* and time, so we omit the details for brevity.

17.3.3.1 Results

The simulations are run for a total of 10 million iterations, discarding the first 10% as burn-in. Independent replications did provide some mild variation with respect to the posterior model probabilities (within the second decimal place), but the interpretation of the results remained consistent between simulations. We note that the poor mixing over model space appears to be a result of bimodality in the model space between nonneighboring models (see further discussion below). Tables 17.7 and 17.8 provide the corresponding posterior (model-averaged) mean and standard deviation of the model parameters and corresponding marginal posterior model probabilities for each of the demographic parameters in terms of the dependence on *fdays* and time. It is interesting to compare Table 17.8 with Table 17.5a, the analogous marginal posterior model probabilities for only the ring-recovery data. One of the main differences is that in the integrated analysis, the adult survival probability has a significantly higher probability of being dependent on time (0.679 compared to 0.091). From Table 17.7 we see that if the time dependence is present, then the adult survival probability is declining with time (i.e. the posterior distribution for γ_a is clearly negative). This is demonstrated in Figure 17.6, which provides the (model-averaged) estimates of all the demographic parameters for both the integrated analysis and for the ring-recovery data only, for comparison. Clearly, for the first-year survival probabilities and recovery probabilities there is very little difference in the posterior estimates of these parameters for the integrated and ring-recovery only analyses (although there are some differences in the posterior model probabilities for these parameters). We note that the count data do not contain any information on the recovery probabilities and there is no direct information on the first-year survival probabilities (only indirect information contained in the estimates of the

TABLE 17.7

The Marginal (Model-Averaged) Posterior Mean and Standard Deviation (SD) of Each Regression Parameter (Conditional on Being Present in the Model) and the Corresponding Posterior Probability that the Parameter is Present in the Model for the Integrated Analysis of Ring-Recovery Data and Count Data. Note that the Logistic Intercept Terms (α) and σ_1^2 are Always Assumed To Be Present

Parameter	Posterior Mean (SD)	Posterior Probability
α_1	0.503 (0.087)	1.000
β_1	−0.192 (0.086)	0.243
γ_1	−0.045 (0.093)	0.033
σ_1^2	0.078 (0.065)	1.000
α_a	1.488 (0.076)	1.000
β_a	−0.209 (0.063)	0.835
γ_a	−0.222 (0.049)	0.679
α_λ	−4.594 (0.040)	1.000
β_λ	0.148 (0.054)	0.511
γ_λ	−0.378 (0.045)	1.000
α_ρ	−1.061 (0.098)	1.000
β_ρ	−0.175 (0.122)	0.101
γ_ρ	−0.260 (0.086)	0.371

number of adults). There is a slight discrepancy between the estimates of the adult-survival probabilities; this appears to be most likely a result of the differing posterior probabilities of being dependent on time between the two analyses.

Finally, we note a couple of differences between these results and those obtained by King et al. (2008b) who consider a similar approach, but using only fixed effects models with a different observation process. In particular, the (marginal) posterior model probabilities for the first-year survival probabilities differ substantially, with 0.644 posterior probability for $\phi_1(f)$ and 0.323 posterior probability for ϕ_1. This is compared to the posterior probabilities of 0.234 for model $\phi(f)$ and 0.732 for model ϕ_1 within our analysis using a mixed effects model for first-year survival probabilities. Thus, allowing for additional random effects significantly reduces the posterior probability that the first-year survival probabilities are dependent on *fdays*. This is often the case, since the additional random effects account for

TABLE 17.8

The Marginal Posterior Probability of the Dependence of Each Demographic Parameter on the Combination of Covariates *fdays* and Time

Covariates	ϕ_1	ϕ_a	λ	ρ
fdays and time	0.009	0.588	0.511	0.033
fdays only	0.234	0.247	0.000	0.068
time only	0.025	0.090	0.489	0.322
no dependence	0.732	0.074	0.000	0.577

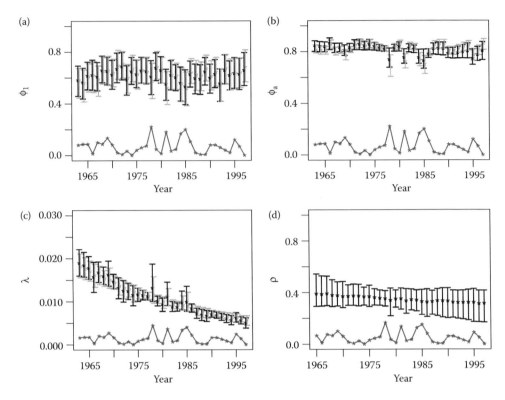

FIGURE 17.6
The posterior model-averaged mean (*) and 95% HPDI for each parameter for the integrated analysis (in black)
and for the ring-recovery data only (in gray), assuming a mixed effects model for ϕ_1 and fixed effects models for
ϕ_a, λ and ρ, for (a) ϕ_1; (b) ϕ_a; (c) λ; and (d) ρ. The bottom line in each plot denotes the values for the covariate *fdays*.

some of the temporal heterogeneity; whereas if we assume a fixed effects model, the covariate (*fdays*) may not explain the temporal variability very well, but is better than assuming a constant survival probability. In addition, we note that in the analysis by King et al. (2008b), although the same two (marginal) models for the recovery probabilities dominate the posterior distribution, the corresponding posterior model probability for $\lambda(t)$ is 0.748 and for $\lambda(f, t)$ is 0.252. This corresponds to a Bayes factor of 2.97—bordering on "positive evidence" (Kass and Raftery, 1995) for only time dependence for the recovery probabilities compared to the additional dependence on *fdays*. This is in contrast to our results, with posterior probabilities of 0.489 for $\lambda(t)$ and 0.511 for $\lambda(f, t)$ (and a Bayes factor of ≈ 1), so that there is greater posterior uncertainty as to the presence/absence of the dependence of *fdays* for the recovery probability.

17.3.3.2 Comments

Care needs to be taken when considering marginal posterior distributions, of parameters and/or models. Intricate detail (and interesting interpretations) can be missed if parameters and/or models are highly correlated, and this detail can be difficult to identify, particularly over large model spaces. For example, in this analysis the marginal posterior model

TABLE 17.9

The Posterior Probabilities for the Models With Largest
Posterior Support

Model	Posterior Probability
$\phi_1(\sigma_1^2)/\phi_a(fdays,t)/\lambda(t)/\rho$	0.183
$\phi_1(\sigma_1^2)/\phi_a(fdays,t)/\lambda(fdays,t)/\rho$	0.164
$\phi_1(fdays,\sigma_1^2)/\phi_a(fdays,t)/\lambda(t)/\rho$	0.144
$\phi_1(\sigma_1^2)/\phi_a(fdays)/\lambda(fdays,t)/\rho(t)$	0.109
$\phi_1(\sigma_1^2)/\phi_a(fdays)/\lambda(t)/\rho(t)$	0.059
$\phi_1(\sigma_1^2)/\phi_a/\lambda(fdays,t)/\rho(t)$	0.056
$\phi_1(fdays,\sigma_1^2)/\phi_a(fdays)/\lambda(t)/\rho(t)$	0.052

The terms in brackets correspond to the parameter dependence: *fdays*
for the covariate *fdays*; *t* for time; and σ_1^2 corresponding to the random
effect component for first-year survival probabilities.

probabilities given in Table 17.8 cannot provide any information relating to the correlation
between the models. Thus in Table 17.9 we present the overall models with largest posterior
support.

Careful consideration of these models suggests that the marginal models for the adult survival probability and productivity rates may be correlated (when ϕ_a is dependent on time,
ρ is not, and vice versa). We investigate this further. In particular, we calculate the posterior
probability that productivity is dependent on time, given that adult survival probability
is not time-dependent, to be equal to 0.999; whereas the posterior probability that productivity is dependent on time, given that the adult survival probability is time-dependent, is
only 0.049. Conversely, the posterior probability that the adult survival probability is time-dependent, given the productivity rate is time-dependent (not time-dependent), is 0.094
(0.999). Clearly there is a strong negative posterior correlation between the time dependence
of the adult survival probability and productivity rate. Overall, the posterior probability
that either the adult survival probability *or* the productivity rate is time-dependent, but not
both, is equal to 0.966 (i.e. a Bayes factor of 28). This corresponds to strong evidence that
only the adult survival probabilities or productivity rates are time-dependent, but with
slightly larger posterior support that it is the adult survival probabilities (a Bayes factor
of 1.5 that it is the adult survival probability rather than productivity rate, conditional on
only one being time-dependent). Similar results were obtained by King et al. (2008b) when
considering only fixed effects models. Interestingly, additional (independent) studies have
identified declining chick numbers, which would support the model with time-dependent
productivity rates; see Besbeas et al. (2002) and Wilson et al. (2001) for further discussion.
Finally, we note that when parameter dependencies are highly correlated, moving between
the different models can be difficult (essentially we have a bimodal distribution over model
space). This can result in poor mixing within the RJMCMC algorithm. To improve the mixing between models, "block" updates can be performed with respect to model updates,
essentially proposing to update the model for the different parameters simultaneously. For
example, in this case, we could add an additional move type that proposes to update both
the covariate dependence on the adult survival probabilities and productivity rates within
a single model move.

17.4 Discussion

The use of MCMC algorithms can greatly simplify analyses of complex data that arise in eco-logical problems, and allows more realistic models to be fitted to the data. The use of these algorithms in statistical ecology is likely to continue due to an increased awareness of the methods and freely available published codes. One area where Bayesian approaches have had a significant impact in statistical ecology is in relation to the inclusion of heterogene-ity within models. Individual heterogeneity is an example which has received particular attention due not only to its perceived importance and relevance within ecological mod-els but also to associated problems. The use of MCMC greatly aids the fitting of (even nonnormal) random effects models, allowing more complex and realistic models to be fit-ted to data (Barry et al., 2003; Brooks et al., 2002; King and Brooks, 2008; Royle and Link, 2002). Alternatively, covariate models are often used to describe the relationship between demographic parameters and factors of interest (on an individual or temporal scale), and so are often of particular interest to biologists. However, missing values often arise in these circumstances, adding an additional level of complexity to an analysis. Once more, a data augmentation approach can be implemented, with the missing values treated as auxiliary variables and imputed within the MCMC algorithm (Bonner and Schwarz, 2006; Dupuis, 1995; King et al., 2006, 2008a). Discriminating between competing covariate models pro-vides information relating to the underlying dynamics of the system and is typically of real biological interest. Two issues often arise in such analyses: constructing efficient RJMCMC algorithms; and the prior specification on the parameters. These continue to be active areas of research. Additionally, within covariate analyses, a parametric relationship is typically assumed between the covariate(s) and demographic parameters. However, this is generally a very restrictive assumption, and often not tested within analyses. An alternative approach has been presented by Gimenez et al. (2006) who consider the use of spline functions to describe the relationship between the demographic parameter and the covariate of interest. This flexible modeling approach is a real step in developing more complex and realistic models, to link observed data with potential factors.

The MCMC algorithm typically allows more complex models to be fitted to the data within a Bayesian framework. However, there is typically a tradeoff between fitting increasingly complex models, using advanced techniques, and the corresponding computation time needed to obtain posterior estimates of interest. Thus, one area of particular interest (and more generally) is the development of efficient MCMC algorithms. For example, the gen-eral implementation of the Metropolis–Hastings algorithm is typically straightforward; however, problems can still arise, such as poor mixing, so that more "intelligent" algo-rithms need to be developed. An example of this problem, as a result of high correlation between parameters, is given in Section 17.3.1. Although the "plain vanilla" algorithm was improved, the algorithm implemented still suffered from high autocorrelation. More generally, Link and Barker (2008) have considered efficient Metropolis–Hastings updates for the recapture and survival probabilities relating to capture–recapture data. The leap from MCMC to RJMCMC, in the presence of model uncertainty, often brings additional mixing problems, in terms of proposed moves between different models having very low acceptance probabilities, so that the development of alternative model updating algorithms would be of particular interest. The complexity of the likelihood expression in many eco-logical applications can make it difficult to implement some efficient RJMCMC algorithms, such as the method proposed by Brooks et al. (2003), so that alternative algorithms need to be developed. For example, Gramacy et al. (2010) apply the method of "importance tempering" to mark–recapture–recovery data in the presence of model uncertainty.

The explosion in the application of Bayesian methods within statistical ecology shows no sign of slowing. On the contrary, application of the methods continues to increase. The publication of freely available computer codes (in WinBUGS/OpenBUGS and R) and books devoted to the area will no doubt fuel the further expansion of the use of MCMC within the field of statistical ecology. The complexity of data collected will help to drive statistical advances within the field, and more generally, such as the development of efficient (RJ)MCMC algorithms.

References

Baillie, S. and Green, R. E. 1987. The importance of variation in recovery rates when estimating survival rates from ringing recoveries. *Acta Ornithologica*, 23:41–60.

Barry, S. C., Brooks, S. P., Catchpole, E. A., and Morgan, B. J. T. 2003. The analysis of ring-recovery data using random effects. *Biometrics*, 59:54–65.

Besbeas, P., Freeman, S. N., Morgan, B. J. T., and Catchpole, E. A. 2002. Integrating mark-recapture-recovery and census data to estimate animal abundance and demographic parameters. *Biometrics*, 58:540–547.

Bonner, S. J. and Schwarz, C. J. 2006. An extension of the Cormack-Jolly-Seber model for continuous covariates with application to *Microtus pennsylvanicus*. *Biometrics*, 62:142–149.

Brooks, S. P., Catchpole, E. A., and Morgan, B. J. T. 2000. Bayesian animal survival estimation. *Statistical Science*, 15:357–376.

Brooks, S. P., Catchpole, E. A., Morgan, B. J. T., and Harris, M. 2002. Bayesian methods for analysing ringing data. *Journal of Applied Statistics*, 29:187–206.

Brooks, S. P., Freeman, S. N., Greenwood, J. J. D., King, R., and Mazzetta, C. 2008. Quantifying conservation concern—Bayesian statistics and the red data lists. *Biological Conservation*, 141:1436–1441.

Brooks, S. P. and Gelman, A. 1998. Alternative methods for monitoring convergence of iterative simulations. *Journal of Computational and Graphical Statistics*, 7:434–455.

Brooks, S. P., Giudici, P., and Roberts, G. O. 2003. Efficient construction of reversible jump MCMC proposal distributions (with discussion). *Journal of the Royal Statistical Society, Series B*, 65: 3–55.

Brooks, S. P., King, R., and Morgan, B. J. T. 2004. A Bayesian approach to combining animal abundance and demographic data. *Animal Biodiversity and Conservation*, 27:515–529.

Buckland, S. T., Newman, K. N., Thomas, L., and Koesters, N. B. 2004. State–space models for the dynamics of wild animal populations. *Ecological Modelling*, 171:157–175.

Dorazio, R. M., Royle, J. A., Söderström, B., and Glimskär 2006. Estimating species richness and accumulation by modeling species occurrence and detectability. *Ecology*, 87:842–854.

Dupuis, J. A. 1995. Bayesian estimation of movement and survival probabilities from capture-recapture data. *Biometrika*, 82:761–772.

Gelman, A. 2006. Prior distributions for variance parameters in hierarchical models. *Bayesian Analysis*, 1:515–534.

Gelman, A., Roberts, G. O., and Gilks, W. R. 1996. Efficient Metropolis jumping rules. In J. M. Bernardo, J. O. Berger, A. P. Dawid, and A. F. M. Smith (eds), *Bayesian Statistics 5: Proceedings of the Fifth Valencia International Meeting*, pp. 599–607. Oxford University Press, Oxford.

Gimenez, O., Bonner, S. J., King, R., Parker, R. A., Brooks, S. P., Jamieson, L. E., Grosbois, V., Morgan, B. J. T., and Thomas, L. 2009. WinBUGS for population ecologists: Bayesian modelling using

Markov chain Monte Carlo. In D. L. Thomson, E. G. Cooch, and M. J. Conroy (eds), *Modeling Demographic Processes in Marked Populations*, Environmental and Ecological Statistics Series, Volume 3, pp. 883–915. Springer, New York.

Gimenez, O., Crainiceanu, C., Barbraud, C., Jenouvrier, S., and Morgan, B. J. T. 2006. Semiparametric regression in capture–recapture modeling. *Biometrics*, 62:691–698.

Gramacy, R. B., Samworth, R. J., and King, R. 2010. Importance tempering. *Statistics and Computing*, 20:1–7.

Green, P. J. 1995. Reversible jump Markov chain Monte Carlo computation and Bayesian model determination. *Biometrika*, 82:711–732.

Gregory, R. D., Wilkinson, N. I., Noble, D. G., Robinson, J. A., Brown, A. F., Hughes, J., Procter, D., Gibbons, D. W., and Galbraith, C. A. 2002. The population status of birds in the United Kingdom, Channel Islands and the Isle of Man: An analysis of conservation concern 2002-2007. *British Birds*, 95:410–448.

Jonsen, I. D., Flemming, J. M., and Myers, R. A. 2005. Robust state–space modelling of animal movement data. *Ecology*, 86:2874–2880.

Kass, R. E. and Raftery, A. E. 1995. Bayes factors. *Journal of the American Statistical Association*, 90: 773–793.

King, R. and Brooks, S. P. 2002. Model selection for integrated recovery/recapture data. *Biometrics*, 58:841–851.

King, R. and Brooks, S. P. 2003. Survival and spatial fidelity of mouflon: the effect of location, age and sex. *Journal of Agricultural, Biological, and Environmental Statistics*, 8:486–531.

King, R. and Brooks, S. P. 2008. On the Bayesian estimation of a closed population size in the presence of heterogeneity and model uncertainty. *Biometrics*, 64:816–824.

King, R., Brooks, S. P., and Coulson, T. 2008a. Analysing complex capture-recapture data in the presence of individual and temporal covariates and model uncertainty. *Biometrics*, 64:1187–1195.

King, R., Brooks, S. P., Mazzetta, C., Freeman, S. N., and Morgan, B. J. T. 2008b. Identifying and diagnosing population declines: A Bayesian assessment of lapwings in the UK. *Applied Statistics*, 57:609–632.

King, R., Brooks, S. P., Morgan, B. J. T., and Coulson, T. N. 2006. Factor influencing Soay sheep survival: A Bayesian analysis. *Biometrics*, 62:211–220.

King, R., Morgan, B. J. T., Gimenez, O., and Brooks, S. P. 2009. *Bayesian Analysis for Population Ecology*. Chapman & Hall/CRC, Boca Raton, FL.

Lindley, D. V. 1957. A statistical paradox. *Biometrika*, 44:187–192.

Link, W. A. and Barker, R. J. 2008. Efficient implementation of the Metropolis-Hastings algorithm, with application to the Cormack-Jolly-Seber model. *Environmental and Ecological Statistics*, 15:79–87.

Link, W. A. and Barker, R. J. 2009. *Bayesian Inference: With Ecological Applications*. Academic Press, London.

McCarthy, M. A. 2007. *Bayesian Methods for Ecology*. Cambridge University Press, Cambridge.

Millar, R. B. and Meyer, R. 2000. Non-linear state space modelling of fisheries biomass dynamics by using Metropolis-Hastings within-Gibbs sampling. *Applied Statistics*, 49:327–342.

Newman, K. B. 2003. Modelling paired release-recovery data in the presence of survival and capture heterogeneity with application to marked juvenile salmon. *Statistical Modelling*, 3:157–177.

Newman, K. N., Buckland, S. T., Lindley, S. T., Thomas, L., and Fernández, C. 2006. Hidden process models for animal population dynamics. *Ecological Applications*, 16:74–86.

O'Hara, R. B., Lampila, S., and Orell, M. 2009. Estimation of rates of births, deaths, and immigration from mark-recapture data. *Biometrics*, 65:275–281.

Pinheiro, J. C. and Bates, D. M. 2000. *Mixed-Effects Models in S and S-PLUS*. Springer, New York.

Reynolds, T., King, R., Harwood, J., Frediksen, M., Wanless, S., and Harris, M. 2009. Integrated data analyses in the presence of emigration and tag-loss. *Journal of Agricultural, Biological, and Environmental Statistics*, 14:411–431.

Roberts, G. O. and Rosenthal, J. S. 2001. Optimal scaling for various Metropolis-Hastings algorithms. *Statistical Science*, 16:351–367.

Royle, J. A. 2008. Modeling individual effects in the Cormack-Jolly-Seber model: A state–space formulation. *Biometrics*, 64:364–370.

Royle, J. A. and Dorazio, R. M. 2008. *Hierarchical Modeling and Inference in Ecology*. Academic Press, London.

Royle, J. A., Dorazio, R. M., and Link, W. A. 2007. Analysis of multinomial models with unknown index using data augmentation. *Journal of Computational and Graphical Statistics*, 16:67–85.

Royle, J. A. and Kéry, M. 2007. A Bayesian state–space formulation of dynamic occupancy models. *Ecology*, 88:1813–1823.

Royle, J. A. and Link, W. A. 2002. Random effects and shrinkage estimation in capture-recapture models. *Journal of Applied Statistics*, 29:329–351.

Wilson, A. M., Vickery, J. A., and Browne, S. J. 2001. Numbers and distribution of northern lapwings *Vanellus vanellus* breeding in England and Wales in 1998. *Bird Study*, 48:2–17.

18

Gaussian Random Field Models for Spatial Data

Murali Haran

18.1 Introduction

Spatial data contain information about both the attribute of interest and its location. Examples can be found in a large number of disciplines, including ecology, geology, epidemiology, geography, image analysis, meteorology, forestry, and geosciences. The location may be a set of coordinates, such as the latitude and longitude associated with an observed pollutant level, or it may be a small region such as a county associated with an observed disease rate. Following Cressie (1993), we categorize spatial data into three distinct types: (i) *geostatistical or point-level data*, as in the pollutant levels observed at several monitors across a region; (ii) *lattice or "areal" (regionally aggregated) data*, for example, US disease rates provided by county; and (iii) *point process data*, where the locations themselves are random variables and of interest, as in the set of locations where a rare animal species was observed. Point processes where random variables associated with the random locations are also of interest are referred to as *marked point processes*. In this chapter, we only consider spatial data that fall into categories (i) and (ii).

We will use the following notation throughout. Denote a real-valued spatial process in d dimensions by $\{Z(\mathbf{s}) : \mathbf{s} \in D \subset \mathbb{R}^d\}$, where $\mathbf{s}$ is the location of the process $Z(\mathbf{s})$ and $\mathbf{s}$ varies over the index set D, resulting in a multivariate random process. For point-level data D is a continuous, fixed set, while for lattice or areal data D is discrete and fixed. For spatial point processes, D is stochastic and usually continuous. The distinctions among the above categories may not always be apparent in any given context, so determining a category is part of the modeling process.

The purpose of this chapter is to discuss the use of Gaussian random fields for modeling a variety of point-level and areal spatial data, and to point out the flexibility in model choices afforded by Markov chain Monte Carlo (MCMC) algorithms. Details on theory, algorithms and advanced spatial modeling can be found in Cressie (1993), Stein (1999), Banerjee et al. (2004), and other standard texts. The reader is referred to the excellent monograph by Møller and Waagepetersen (2004) for details on modeling and computation for spatial point processes.

18.1.1 Some Motivation for Spatial Modeling

Spatial modeling can provide a statistically sound approach for performing interpolations for point-level data, which is at the heart of "kriging", a body of work originating from mineral exploration (see Matheron, 1971). Even when interpolation is not the primary goal, accounting for spatial dependence can lead to better inference, superior predictions, and more accurate estimates of the variability of estimates. We describe toy examples to illustrate two general scenarios where modeling spatial dependence can be beneficial: when there is dependence in the data, and when we need to adjust for an unknown spatially varying mean. Learning about spatial dependence from observed data may also be of interest in its own right, for example, in research questions where detecting spatial clusters is of interest.

Example 18.1   Accounting Appropriately for Dependence

Let $Z(s) = 6s + \epsilon(s)$ be a random variable indexed by its location $s \in (0, 1)$, with dependent errors $\epsilon(s)$ generated via a simple autoregressive model: $\epsilon(s_1) = 7$, $\epsilon(s_i) \sim N(0.9\epsilon(s_{i-1}), 0.1)$, $i = 2, \ldots, 100$, for equally spaced locations $x_1, \ldots, x_{100}$ in $(0, 1)$. Figure 18.1a shows how a model that assumes the errors are dependent, such as a linear Gaussian process (GP) model (solid curves) described later in Section 18.2.1, provides a much better fit than a regression model with independent errors (dotted lines). Note that for spatial data, s is usually in two- or three-dimensional space; we are only considering one-dimensional space here in order to better illustrate the ideas.

Example 18.2   Adjusting for an Unknown Spatially Varying Mean

Let $Z(s) = \sin(s) + \epsilon(s)$ where, for any set of locations $s_1, \ldots, s_k \in (0, 1)$, and $\epsilon(s_1), \ldots, \epsilon(s_k)$ are independent and identically distributed normal random variables with mean 0 and variance σ^2. Suppose that $Z(s)$ is observed at ten locations. From Figure 18.1b the dependent error model (solid curves) is superior to an independent error model (dotted lines), even though there was no dependence in the generating process. Adding dependence can thus act as a form of protection against a poorly specified model.

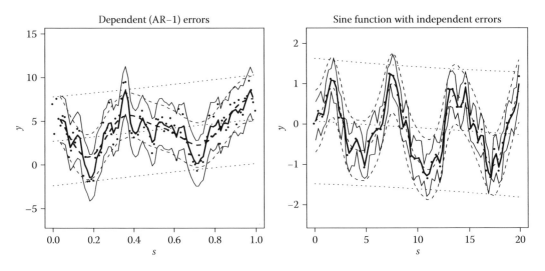

FIGURE 18.1
Black dots: simulated data. Solid curves: Gaussian process with exponential covariance. Dashed curves: Gaussian process with gaussian covariance. Dotted lines: independent error model. In all cases, the mean and 95% prediction intervals are provided.

Example 18.2 shows how accounting for spatial dependence can adjust for a misspecified mean, thereby accounting for important missing spatially varying covariate information (for instance, the $\sin(x)$ function above). As pointed out in Cressie (1993, p. 25), "What is one person's (spatial) covariance structure may be another person's mean structure." In other words, an interpolation based on assuming dependence (a certain covariance structure) can be similar to an interpolation that utilizes a particular mean structure ($\sin(x)$ above). Example 18.2 also shows the utility of GPs for modeling the relationship between "inputs" ($s_1, \ldots, s_n$) and "outputs" ($Z(s_1), \ldots, Z(s_n)$) when little is known about the parametric form of the relationship. In fact, this flexibility of GPs has been exploited for modeling relationships between inputs and outputs from complex computer experiments (see Currin et al., 1991; Sacks et al., 1989). For more discussion on motivations for spatial modeling see, for instance, Cressie (1993, p. 13) and Schabenberger and Gotway (2005, p. 31).

18.1.2 MCMC and Spatial Models: A Shared History

Most algorithms related to MCMC originated in statistical physics problems concerned with lattice systems of particles, including the original Metropolis et al. (1953) paper. The Hammersley–Clifford theorem (Besag, 1974; Clifford, 1990) provides an equivalence between the local specification via the conditional distribution of each particle given its neighboring particles, and the global specification of the joint distribution of all the particles. The specification of the joint distribution via local specification of the conditional distributions of the individual variables is the Markov random field specification, which has found extensive applications in spatial statistics and image analysis, as outlined in a series of papers by Besag and co-authors (see Besag, 1974, 1989; Besag and Kempton, 1986; Besag et al., 1995), and several papers on Bayesian image analysis (Amit et al., 1991; Geman and Geman, 1984; Grenander and Keenan, 1989). It is also the basis for variable-at-a-time Metropolis–Hastings and Gibbs samplers for simulating these systems. Thus, spatial statistics was among the earliest fields to recognize the power and generality of MCMC. A historical perspective on the connection between spatial statistics and MCMC, along with related references, can be found in Besag and Green (1993).

While these original connections between MCMC and spatial modeling are associated with Markov random field models, this discussion of Gaussian random field models includes both GP and Gaussian Markov random field (GMRF) models in Section 18.2. In Section 18.3, we describe the generalized versions of both linear models, followed by a discussion of non-Gaussian Markov random field models in Section 18.4 and a brief discussion of more flexible models in Section 18.5.

18.2 Linear Spatial Models

In this section we discuss linear Gaussian random field models for both geostatistical and areal (lattice) data. Although a wide array of alternative approaches exist (see Cressie, 1993), we model the spatial dependence via a parametric covariance function or, as is common for lattice data, via a parameterized precision (inverse covariance) matrix, and consider Bayesian inference and prediction.

18.2.1 Linear Gaussian Process Models

We first consider geostatistical data. Let the spatial process at location $\mathbf{s} \in D$ be defined as

$$Z(\mathbf{s}) = \mathbf{X}(\mathbf{s})\boldsymbol{\beta} + w(\mathbf{s}), \quad \text{for } \mathbf{s} \in D, \tag{18.1}$$

where $\mathbf{X}(\mathbf{s})$ is a set of p covariates associated with each site $\mathbf{s}$, and $\boldsymbol{\beta}$ is a p-dimensional vector of coefficients. Spatial dependence can be imposed by modeling $\{w(\mathbf{s}) : \mathbf{s} \in D\}$ as a zero-mean stationary GP. Distributionally, this implies that for any $\mathbf{s}_1, \ldots, \mathbf{s}_n \in D$, if we let $\mathbf{w} = (w(\mathbf{s}_1), \ldots, w(\mathbf{s}_n))^T$ and Θ be the parameters of the model, then

$$\mathbf{w} \mid \Theta \sim N(0, \Sigma(\Theta)), \tag{18.2}$$

where $\Sigma(\Theta)$ is the covariance matrix of the n-dimensional normal density. We need $\Sigma(\Theta)$ to be symmetric and positive definite for this distribution to be proper. If we specify $\Sigma(\Theta)$ by a positive definite parametric covariance function, we can ensure that these conditions are satisfied. For example, consider the exponential covariance with parameters $\Theta = (\psi, \kappa, \phi)$, with $\psi, \kappa, \phi > 0$. The exponential covariance $\Sigma(\Theta)$ has the form $\Sigma(\Theta) = \psi I + \kappa H(\phi)$, where I is the identity matrix, the (i,j)th element of $H(\phi)$ is $\exp(-\|\mathbf{s}_i - \mathbf{s}_j\|/\phi)$, and $\|\mathbf{s}_i - \mathbf{s}_j\|$ is the Euclidean distance between locations $\mathbf{s}_i, \mathbf{s}_j \in D$. Alternatives to Euclidean distance may be useful—for instance, geodesic distances are often appropriate for spatial data over large regions (Banerjee, 2005). This model is interpreted as follows: the "nugget" ψ is the variance of the nonspatial error, say from measurement error or from a micro-scale stochastic source associated with each location, and κ and ϕ dictate the scale and range of the spatial dependence, respectively. Clearly, this assumes that the covariance and hence dependence between two locations decreases as the distance between them increases.

The exponential covariance function is important for applications, but is a special case of the more flexible Matérn family (Handcock and Stein, 1993). The Matérn covariance between $Z(\mathbf{s}_i)$ and $Z(\mathbf{s}_j)$ with parameters $\psi, \kappa, \phi, \nu > 0$ is based only on the distance x between $\mathbf{s}_i$ and $\mathbf{s}_j$,

$$\text{cov}(x; \psi, \kappa, \phi, \nu) = \begin{cases} \dfrac{\kappa}{2^{\nu-1}\Gamma(\nu)}(2\nu^{1/2}x/\phi)^\nu K_\nu(2\nu^{1/2}x/\phi), & \text{if } x > 0, \\ \psi + \kappa, & \text{if } x = 0, \end{cases} \tag{18.3}$$

where $K_\nu(x)$ is a modified Bessel function of order ν (Abramowitz and Stegun, 1964), and ν determines the smoothness of the process. As ν increases, the process becomes increasingly smooth. As an illustration, Figure 18.1 compares prediction (interpolation) using GPs with exponential ($\nu = 0.5$) and gaussian ($\nu \to \infty$) covariance functions (we use lower case for "gaussian" as suggested in Schabenberger and Gotway, 2005, since the covariance is not related to the Gaussian distribution). Notice how the gaussian covariance function produces a much smoother interpolator (dashed curves) than the more "wiggly" interpolation produced by the exponential covariance (solid curves). Stein (1999) recommends the Matérn since it is flexible enough to allow the smoothness of the process to also be estimated. He cautions against GPs with gaussian correlations since they are overly smooth (they are infinitely differentiable). In general the smoothness ν may be hard to estimate from data; hence, a popular default is to use the exponential covariance for spatial data where the physical process producing the realizations is unlikely to be smooth, and a gaussian covariance for modeling output from computer experiments or other data where the associated smoothness assumption may be reasonable.

Let $\mathbf{Z} = (Z(\mathbf{s}_1), \ldots, Z(\mathbf{s}_n))^T$. From Equations 18.1 and 18.2, once a covariance function is chosen (say, according to Equation 18.3), $\mathbf{Z}$ has a multivariate normal distribution with unknown parameters $\Theta, \boldsymbol{\beta}$. Maximum likelihood inference for the parameters is then simple in principle, though strong dependence among the parameters and expensive matrix operations may sometimes make it more difficult. A Bayesian model specification is completed with prior distributions placed on $\Theta, \boldsymbol{\beta}$. "Objective" priors (perhaps more appropriately referred to as "default" priors) for the linear GP model have been derived by several authors (Berger et al., 2001; De Oliveira, 2007; Paulo, 2005). These default priors are very useful since it is often challenging to quantify prior information about these parameters in a subjective manner. However, they can be complicated and computationally expensive, and proving posterior propriety often necessitates analytical work. To avoid posterior impropriety when building more complicated models, it is common to use proper priors and rely on approaches based on exploratory data analysis to determine prior settings. For example, one could use a uniform density that allows for a reasonable range of values for the range parameter ϕ, and inverse gamma densities with an infinite variance and mean set to a reasonable guess for κ and ψ (see, e.g. Finley et al., 2007), where the guess may again depend on some rough exploratory data analysis such as looking at variograms. For a careful analysis, it is critical to study sensitivity to prior settings.

18.2.1.1 MCMC for Linear GPs

Inference for the linear GP model is based on the posterior distribution $\pi(\Theta, \boldsymbol{\beta} \mid \mathbf{Z})$ that results from Equations 18.1 and 18.2 and a suitable prior for $\Theta, \boldsymbol{\beta}$. Although π is of fairly low dimensions as long as the number of covariates is not too large, MCMC sampling for this model can be complicated by two issues: (i) the strong dependence among the covariance parameters, which leads to autocorrelations in the sampler; (ii) the fact that matrix operations involved at each iteration of the algorithm are of order N^3, where N is the number of data points. Reparameterization-based MCMC approaches, such as those proposed in Yan et al. (2007) and Cowles et al. (2009), or block updating schemes, where multiple covariance parameters are updated at once in a single Metropolis–Hastings step (*cf.* Tibbits et al., 2010), may help with the dependence. Also, there are existing software implementations of MCMC algorithms for linear GP models (Finley et al., 2007; Smith et al., 2008). A number of approaches can be used to speed up the matrix operations, including changing the covariance function in order to induce sparseness or other special matrix structures that are amenable to fast matrix algorithms; we discuss this further in Section 18.5.

Predictions of the process, $\mathbf{Z}^* = (Z(\mathbf{s}_1^*), \ldots, Z(\mathbf{s}_m^*))^T$, where $\mathbf{s}_1^*, \ldots, \mathbf{s}_m^*$ are new locations in D, are obtained via the posterior predictive distribution,

$$\pi(\mathbf{Z}^* \mid \mathbf{Z}) = \int \pi(\mathbf{Z}^* \mid \mathbf{Z}, \Theta, \boldsymbol{\beta}) \pi(\Theta, \boldsymbol{\beta} \mid \mathbf{Z}) \, d\Theta \, d\boldsymbol{\beta}. \tag{18.4}$$

Under the GP assumption the joint distribution of $\mathbf{Z}, \mathbf{Z}^*$ given $\Theta, \boldsymbol{\beta}$ is

$$\begin{bmatrix} \mathbf{Z} \\ \mathbf{Z}^* \end{bmatrix} \mid \Theta, \boldsymbol{\beta} \sim N\left(\begin{bmatrix} \boldsymbol{\mu}_1 \\ \boldsymbol{\mu}_2 \end{bmatrix}, \begin{bmatrix} \Sigma_{11} & \Sigma_{12} \\ \Sigma_{21} & \Sigma_{22} \end{bmatrix} \right),$$

where $\boldsymbol{\mu}_1$ and $\boldsymbol{\mu}_2$ are the linear regression means of $\mathbf{Z}$ and $\mathbf{Z}^*$ (functions of covariates and $\boldsymbol{\beta}$), and $\Sigma_{11}, \Sigma_{12}, \Sigma_{21}, \Sigma_{22}$ are block partitions of the covariance matrix $\Sigma(\Theta)$ (functions of covariance parameters Θ). By basic normal theory (e.g. Anderson, 2003), $\mathbf{Z}^* \mid \mathbf{Z}, \boldsymbol{\beta}, \Theta,$

corresponding to the first term in the integrand in (18.4), is normal with mean and covariance

$$E(\mathbf{Z}^* \mid \mathbf{Z}, \boldsymbol{\beta}, \Theta) = \boldsymbol{\mu}_2 + \Sigma_{21}\Sigma_{11}^{-1}(\mathbf{Z} - \boldsymbol{\mu}_1), \quad \mathrm{var}(\mathbf{Z}^* \mid \mathbf{Z}, \boldsymbol{\beta}, \Theta) = \Sigma_{22} - \Sigma_{21}\Sigma_{11}^{-1}\Sigma_{12}. \quad (18.5)$$

Note, in particular, that the prediction for $\mathbf{Z}^*$ given $\mathbf{Z}$ has expectation obtained by adding two components: (i) the mean $\boldsymbol{\mu}_2$ which, in the simple linear case, is $\boldsymbol{\beta}\mathbf{X}^*$, where $\mathbf{X}^*$ are the covariates at the new locations; (ii) a product of the residual from the simple linear regression on the observations $(\mathbf{Z} - \boldsymbol{\mu}_1)$ weighted by $\Sigma_{21}\Sigma_{11}^{-1}$. If there is no dependence, the second term is close to 0, but if there is a strong dependence, the second term pulls the expected value at a new location closer to the values at nearby locations. Draws from the posterior predictive distribution (Equation 18.4) are obtained in two steps: (i) simulate $\Theta', \boldsymbol{\beta}' \sim \pi(\Theta, \boldsymbol{\beta} \mid \mathbf{Z})$ by the Metropolis–Hastings algorithm; (ii) simulate $\mathbf{Z}^* \mid \Theta', \boldsymbol{\beta}', \mathbf{Z}$ from a multivariate normal density with conditional mean and covariance from Equation 18.5 using the $\Theta', \boldsymbol{\beta}'$ draws from step (i).

Example 18.3

Haran et al. (2010) interpolate flowering dates for wheat crops across North Dakota as part of a model to estimate crop epidemic risks. The flowering dates are only available at a few locations across the state, but using a linear GP model with a Matérn covariance, it is possible to obtain distributions for interpolated flowering dates at sites where other information (weather predictors) is available for the epidemic model, as shown in Figure 18.2. Although only point estimates are displayed here, the full distribution of the interpolated flowering dates is used when estimating crop epidemic risks.

18.2.2 Linear Gaussian Markov Random Field Models

A direct specification of spatial dependence via $\Sigma(\Theta)$, while intuitively appealing, relies on measuring spatial proximity in terms of distances between the locations. When modeling areal data, it is possible to use measures such as inter-centroid distances to serve this purpose, but this can be awkward due to irregularities in the shape of the regions. Also, since the data are aggregates, assuming a single location corresponding to multiple random variables may be inappropriate. An alternative approach is a conditional specification, by assuming that a random variable associated with a region depends primarily on its neighbors. A simple neighborhood could consist of adjacent regions, but more complicated neighborhood structures are possible depending on the specifics of the problem. Let the spatial process at location $\mathbf{s} \in D$ be defined as in Equation 18.1 so $Z(\mathbf{s}) = X(\mathbf{s})\boldsymbol{\beta} + w(\mathbf{s})$, but now assume that the spatial random variables ("random effects") $\mathbf{w}$ are modeled *conditionally*. Let $\mathbf{w}_{-i}$ denote the vector $\mathbf{w}$ excluding $w(\mathbf{s}_i)$. For each $\mathbf{s}_i$ we model $w(\mathbf{s}_i)$ in terms of its *full conditional distribution*, that is, its distribution given the remaining random variables, $\mathbf{w}_{-i}$:

$$w(\mathbf{s}_i) \mid \mathbf{w}_{-i}, \Theta \sim N\left(\sum_{j=1}^{n} c_{ij}w(\mathbf{s}_j), \kappa_i^{-1}\right), \quad i = 1, \ldots, n, \quad (18.6)$$

where c_{ij} describes the neighborhood structure. c_{ij} is nonzero only if i and j are neighbors, while the κ_i are the precision (inverse variance) parameters. To make the connection to the linear GP model (Equation 18.2) apparent, we let Θ denote the precision parameters. Each

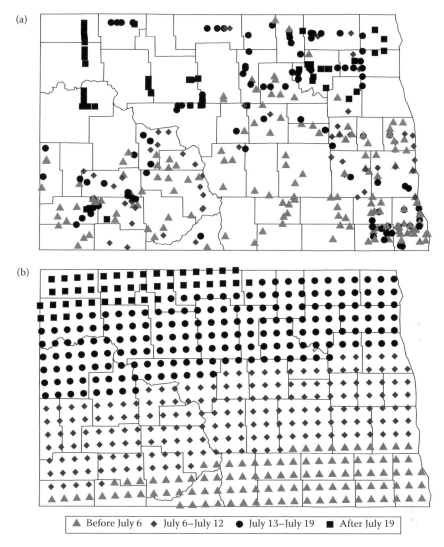

FIGURE 18.2
(a) Raw flowering date. (b) Interpolated flowering dates at desired grid locations, using means from posterior predictive distribution from linear Gaussian process model.

$w(\mathbf{s}_i)$ is therefore a normal random variate with mean based on neighboring values of $w(\mathbf{s}_i)$. Just as we need to ensure that the covariance is positive definite for a valid GP, we need to ensure that the set of conditional specifications result in a valid joint distribution. Let Q be an $n \times n$ matrix with ith diagonal element κ_i and (i, j)th off-diagonal element $-\kappa_i c_{ij}$. Besag (1974) proved that if Q is symmetric and positive definite (Equation 18.6) specifies a valid joint distribution,

$$\mathbf{w} \mid \Theta \sim N(0, Q^{-1}), \tag{18.7}$$

with Θ the set of precision parameters (note that the c_{ij} and κ_i depend on Θ). Usually a common precision parameter, say τ, is assumed so $\kappa_i = \tau$ for all i, and hence $Q(\tau) = \tau(I + C)$ where C is a matrix which has 0 on its diagonals and (i, j)th off-diagonal element

$-c_{ij}$, though a more attractive smoother may be obtained by using weights in a GMRF model motivated by a connection to thin-plate splines (Yue and Speckman, 2009). To add flexibility to the above GMRF model, some authors have included an extra parameter in the matrix C (see Ferreira and De Oliveira, 2007). Inference for the linear GMRF model specified by Equations 18.1 and 18.7 can therefore proceed after assuming a prior distribution for τ, β, often an inverse gamma and flat prior respectively. An alternative formulation is an improper version of the GMRF prior, the so-called "intrinsic Gaussian Markov random field" (Besag and Kooperberg, 1995):

$$f(\mathbf{w} \mid \Theta) \propto \tau^{(N-1)/2} \exp\{-\mathbf{w}^T Q(\tau)\mathbf{w}\}, \tag{18.8}$$

where Q has $-\tau c_{ij}$ on its off-diagonals (as above) and ith diagonal element $\tau \sum_j c_{ij}$. The notation $j \sim i$ implies that i and j are neighbors. In the special case where $c_{ij} = 1$ if $j \sim i$ and 0 otherwise, Equation 18.8 simplifies to the "pairwise-difference form,"

$$f(\mathbf{w} \mid \Theta) \propto \tau^{(N-1)/2} \exp\left(-\frac{1}{2}\sum_{i \sim j}\{w(\mathbf{s}_i) - w(\mathbf{s}_j)\}^2\right),$$

which is convenient for constructing MCMC algorithms with univariate updates since the full conditionals are easy to evaluate. Q is rank deficient so the above density is improper. This form is a very popular prior for the underlying spatial field of interest. For instance, denote noisy observations by $\mathbf{y} = (y(\mathbf{s}_1), \ldots, y(\mathbf{s}_n))^T$, so $y(\mathbf{s}_i) = w(\mathbf{s}_i) + \epsilon_i$ where $\epsilon_i \sim N(0, \sigma^2)$ is independent error. Then an estimate of the smoothed underlying spatial process $\mathbf{w}$ can be obtained from the posterior distribution of $\mathbf{w} \mid \mathbf{y}$ as specified by Equation 18.8. If the parameters, say τ and σ^2, are also to be estimated and have priors placed on them, inference is based on the posterior $\mathbf{w}, \tau, \sigma^2 \mid \mathbf{y}$. The impropriety of the intrinsic GMRF is not an issue as long as the posterior is proper. If $c_{ij} = 1$ when i and j are neighbors and 0 otherwise, this corresponds to an intuitive conditional specification:

$$f(w_j \mid \mathbf{w}_{-i}, \tau) \sim N\left(\frac{\sum_{j \in N(i)}^{n} w(\mathbf{s}_j)}{n}, \frac{1}{n_i \tau}\right),$$

where n_i is the number of neighbors for the ith region, and $N(i)$ is the set of neighbors of the ith region. Hence, the distribution of $w(\mathbf{s}_i)$ is normal with mean given by the average of its neighbors and its variance decreases as the number of neighbors increases. See Rue and Held (2005) for a discussion of related theory for GMRF models, and Sun et al. (1999) for conditions under which posterior propriety is guaranteed for various GMRF models.

Although GMRF-based models are very popular in statistics and numerous other fields, particularly computer science and image analysis, there is some concern about whether they are reasonable models even for areal or lattice data (McCullagh, 2002). The marginal dependence induced can be complicated and counterintuitive (Besag and Kooperberg, 1995; Wall, 2004). In addition, a GMRF model on a lattice is known to be inconsistent with the corresponding GMRF model on a subset of the lattice, that is, the corresponding marginal distributions are not the same. However, quoting Besag (2002), this is not a major issue if "The main purpose of having the spatial dependence is to absorb spatial variation (dependence) rather than produce a spatial model with scientifically interpretable parameters." GMRF models can help produce much better individual estimates by "borrowing strength" from the neighbors of each individual (region). This is of particular importance in small area estimation problems (see Ghosh and Rao, 1994), where many observations are based on small populations, for instance disease rate estimates in sparsely populated counties. Spatial dependence allows the model to borrow information from neighboring counties which may collectively have larger populations, thereby reducing the variability of the estimates. Similar considerations apply in disease mapping models (Mollié, 1996) where small regions and the rarity of diseases have led to the popularity of variants of the GMRF-based Bayesian image restoration model due to Besag et al. (1991). More sophisticated extensions of such models in the context of environmental science and public health are described in several recent books (see, e.g. Lawson, 2008; Le and Zidek, 2006; Waller and Gotway, 2004). Several of these models fall under the category of spatial generalized linear models, as discussed in Section 18.3.

18.2.2.1 MCMC for Linear GMRFs

The conditional independence structure of a GMRF makes it natural to write and compute the full conditional distributions of each $w(\mathbf{s}_i)$, without any matrix computations. Hence MCMC algorithms which update a single variable at a time are easy to construct. When this algorithm is efficient, it is preferable due to its simplicity. Unfortunately, such univariate algorithms may often result in slow mixing Markov chains. In the linear GMRF model posterior distribution, it is possible to analytically integrate out all the spatial random effects ($\mathbf{w}$), that is, it is easy to integrate the posterior distribution $\pi(\mathbf{w}, \Theta, \boldsymbol{\beta} \mid \mathbf{Z})$ with respect to $\mathbf{w}$ to obtain the marginal $\pi(\Theta, \boldsymbol{\beta} \mid \mathbf{Z})$ in closed form. This is a fairly low-dimensional distribution, similar to the linear GP model posterior, and similar strategies as described for sampling from the linear GP model posterior may be helpful here. However, unlike the linear GP model posterior, all matrices involved in linear GMRF models are sparse. A reordering of the nodes corresponding to the graph can exploit the sparsity of the precision matrices of GMRFs, thereby reducing the matrix operations from $O(n^3)$ to $O(nb^2)$ where b^2 is the bandwidth of the sparse matrix; see Rue (2001) and Golub and Van Loan (1996, p. 155). For instance, Example 18.3.2.2 (see Section 18.3.2) involves $n = 454$ data points, but the reordered precision matrix has a bandwidth of just 24. The matrix computations are therefore speeded up by a factor of 357 each, and the ensuing increase in computational speed is even larger.

18.2.3 Summary

Linear Gaussian random fields are a simple and flexible approach to modeling dependent data. When the data are point-level, GPs are convenient since the covariance can be specified as a function of the distance between any two locations. When the data are aggregated or on a lattice, GMRFs are convenient as dependence can be specified in terms of adjacencies and neighborhoods. MCMC allows for easy simulation from the posterior distribution for both categories of models, especially since the low-dimensional posterior distribution of the covariance (or precision) parameters and regression coefficients may be obtained in closed form. Relatively simple univariate Metropolis–Hastings algorithms may work well, and existing software packages can implement reasonably efficient MCMC algorithms. When the simple approaches produce slow mixing Markov chains, reparameterizations or block updating algorithms may be helpful. Many strategies are available for reducing the considerable computational burden posed by matrix operations for linear GP models, including the use of covariance functions that result in special matrix structures amenable to fast computations. GMRFs have significant computational advantages over GPs due to the conditional independence structure which naturally results in sparse matrices and greatly reduced computations for each update of the MCMC algorithm.

18.3 Spatial Generalized Linear Models

Linear GP and GMRF models are very flexible, and work surprisingly well in a variety of situations, including many where the process is quite non-Gaussian and discrete, such as some kinds of spatial count data. When the linear Gaussian assumption provides a poor fit to data, transforming the data via the Box–Cox family of transformations, say, and modeling the transformed response via a linear GP or GMRF may be adequate (see "trans-Gaussian kriging," for instance, in Cressie, 1993, with the use of delta method approximations to estimate the variance and perform bias correction). However, when it is important to model the known sampling mechanism for the data, and this mechanism is non-Gaussian, spatial generalized linear models (SGLMs) may be very useful. SGLMs are generalized linear models (McCullagh and Nelder, 1983) for spatially associated data. The spatial dependence (the error structure) for SGLMs can be modeled via GPs for point-level ("geostatistical") data as described in the seminal paper by Diggle et al. (1998). Here, we also include the use of GMRF models for the errors, as commonly used for lattice or areal data. Note that the SGLMs here may also be referred to as "spatial generalized linear mixed models" since the specification of spatial dependence via a generalized linear model framework always involves random effects.

18.3.1 Generalized Linear Model Framework

We begin with a brief description of SGLMs using GP models. Let $\{Z(\mathbf{s}) : \mathbf{s} \in D\}$ and $\{w(\mathbf{s}) : \mathbf{s} \in D\}$ be two spatial processes on $D \subset \mathbb{R}^d$ ($d \in \mathbb{Z}^+$.) Assume that the $Z(\mathbf{s}_i)$ are conditionally independent given $w(\mathbf{s}_1), \ldots, w(\mathbf{s}_n)$, where $\mathbf{s}_1, \ldots, \mathbf{s}_n \in D$, the $Z(\mathbf{s}_i)$ conditionally follow some common distributional form, for example, Poisson for count data or Bernoulli for binary data, and

$$\mathrm{E}(Z(\mathbf{s}_i) \mid \mathbf{w}) = \mu(\mathbf{s}_i), \quad \text{for } i = 1, \ldots, n. \tag{18.9}$$

Let $\eta(\mathbf{s}) = h\{\mu(\mathbf{s})\}$ for some known link function $h(\cdot)$ (e.g. the logit link, $h(x) = \log\left(\frac{x}{1-x}\right)$, or log link, $h(x) = \log(x)$). Furthermore, assume that

$$\eta(\mathbf{s}) = \mathbf{X}(\mathbf{s})\boldsymbol{\beta} + w(\mathbf{s}), \tag{18.10}$$

where $\mathbf{X}(\mathbf{s})$ is a set of p covariates associated with each site $\mathbf{s}$, and $\boldsymbol{\beta}$ is a p-dimensional vector of coefficients. Spatial dependence is imposed on this process by modeling $\{w(\mathbf{s}) : \mathbf{s} \in D\}$ as a stationary GP so $\mathbf{w} = (w(\mathbf{s}_1), \ldots, w(\mathbf{s}_n))^T$ is distributed as

$$\mathbf{w} \mid \Theta \sim N(0, \Sigma(\Theta)). \tag{18.11}$$

$\Sigma(\Theta)$ is a symmetric, positive definite covariance matrix usually defined via a parametric covariance such as a Matérn covariance function (Handcock and Stein, 1993), where Θ is a vector of parameters used to specify the covariance function. Note that with the identity link function and Gaussian distributions for the conditional distribution of the $Z(\mathbf{s}_i)$, we can obtain the linear GP model as a special case. The model specification is completed with prior distributions placed on Θ, β, where proper priors are typically chosen to avoid issues with posterior impropriety. There has been little work on prior settings for SGLMs, with researchers relying on a mix of heuristics and experience to derive suitable priors. Prior sensitivity analyses are, again, crucial, as also discussed in Section 18.6. It is important to carefully interpret the regression parameters in SGLMs *conditional* on the underlying spatial random effects, rather than as the usual marginal regression coefficients (Diggle et al., 1998, p. 302).

The GMRF version of SGLMs is formulated in similar fashion, so Equations 18.9 and 18.10 stay the same but Equation 18.11 is replaced by Equation 18.7. Inference for the SGLM model is based on the posterior distribution $\pi(\Theta, \beta, \mathbf{w} \mid \mathbf{Z})$. Predictions can then be obtained easily via the posterior predictive distribution. In principle, the solution to virtually any scientific question related to these models is easily obtained via sample-based inference. Examples of such questions include finding maxima (see the example in Diggle et al., 1998), spatial cumulative distribution functions when finding the proportion of area where $Z(\mathbf{s})$ is above some limit (Short et al., 2005), and integrating over subregions in the case of Gaussian process SGLMs when inference is required over a subregion.

18.3.2 Examples

18.3.2.1 Binary Data

Spatial binary data occur frequently in environmental and ecological research, for instance when the data correspond to presence or absence of a certain invasive plant species at a location, or when the data happen to fall into one of two categories, say two soil types. Interpolation in point-level data and smoothing in areal/lattice data may be of interest. Often, researchers may be interested in learning about relationships between the observations and predictors while adjusting appropriately for spatial dependence, and in some cases learning about spatial dependence may itself be of interest.

Example 18.4

The coastal marshes of the mid-Atlantic are an extremely important aquatic resource. An invasive plant species called *Phragmites australis* or "phrag" is a major threat to this aquatic ecosystem (see Saltonstall, 2002), and its rapid expansion may be the result of human activities causing habitat disturbance (Marks et al., 1994). Data from the Atlantic Slopes Consortium (Brooks et al., 2006) provide information on presence or absence of phrag in the Chesapeake Bay area, along with predictors of phrag presence such as land use characteristics. Accounting for spatial dependence when studying phrag presence is important since areas near a phrag-dominated region are more likely to have phrag. Of interest is estimating both the smoothed probability surface associated with phrag over the entire region as well as the most important predictors of phrag presence. Because the response (phrag presence/absence) is binary and spatial dependence is a critical component of the model, there is a need for a spatial regression model for binary data. This can be easily constructed via an SGLM, as discussed below.

An SGLM for binary data may be specified following Equations 18.9 and 18.10:

$$Z(\mathbf{s}) \mid p(\mathbf{s}) \sim \text{Bernoulli}(p(\mathbf{s})),$$

$$\Phi^{-1}\{p(\mathbf{s})\} = \beta \mathbf{X}(\mathbf{s}) + w(\mathbf{s}), \tag{18.12}$$

where $\Phi^{-1}\{p(\mathbf{s})\}$ is the inverse cumulative density function of a standard normal density, so $p(\mathbf{s}) = \Phi\{\boldsymbol{\beta}X + w(\mathbf{s})\}$. $\mathbf{X}(\mathbf{s})$, as before, is a set of p covariates associated with each site $\mathbf{s}$, and $\boldsymbol{\beta}$ is a p-dimensional vector of coefficients. $\mathbf{w}$ is modeled as a dependent process via a GP or GMRF as discussed in Section 18.3.1. The model described by Equation 18.12 is the clipped Gaussian random field (De Oliveira, 2000) since it can equivalently be specified as:

$$Z(\mathbf{s}) \mid Z^*(\mathbf{s}) = \begin{cases} 1, & \text{if } Z^*(\mathbf{s}) > 0, \\ 0, & \text{if } Z^*(\mathbf{s}) \leq 0. \end{cases}$$

$Z^*(\mathbf{s})$ is then modeled as a linear GP or GMRF as in Section 18.2. This is an intuitive approach to modeling spatial binary data since the underlying latent process may correspond to a physical process that was converted to a binary value due to the detection limits of the measuring device. It may also just be considered a modeling device to help smooth the binary field, when there is reason to assume that the binary field will be smooth. Alternatively, a logit model may be used instead of the probit in the second stage in Equation 18.12, so $\log\left\{\frac{p(\mathbf{s})}{1-p(\mathbf{s})}\right\} = \boldsymbol{\beta}\mathbf{X}(\mathbf{s}) + w(\mathbf{s})$.

Several of the covariance function parameters are not identifiable. Hence, for a GP model the scale and smoothness parameters are fixed at appropriate values. These identifiability issues are common in SGLMs, but are made even worse in SGLMs for binary data since they contain less information about the magnitude of dependence. A potential advantage of GMRF-based models over GP-based models for binary data is that they can aggregate pieces of binary information from neighboring regions to better estimate spatial dependence.

18.3.2.2 Count Data

SGLMs are well suited to modeling count data. For example, consider the model

$$Z(\mathbf{s}) \mid \mu(\mathbf{s}) \sim \text{Poisson}(E(\mathbf{s})\mu(\mathbf{s})),$$

$$\log(\mu(\mathbf{s})) = \boldsymbol{\beta}X + w(\mathbf{s}),$$

where $E(\mathbf{s})$ is a known expected count at $\mathbf{s}$ based on other information or by assuming uniform rates across the region, say by multiplying the overall rate by the population at $\mathbf{s}$.

Example 18.5

Yang et al. (2009) study infant mortality rates by county in the southern US states of Alabama, Georgia, Mississippi, North Carolina, and South Carolina (Health Resources and Services Administration, 2003) between 1998 and 2000. Of interest is finding regions with unusually elevated levels in order to study possible socio-economic contributing factors. Since no interpolation is required here, the purpose of introducing spatial dependence via a GMRF model is to improve individual county-level estimates using spatial smoothing by "borrowing information" from neighboring counties. The raw and smoothed posterior means for the maps are displayed in Figure 18.3. Based on the posterior distribution, it is possible to make inferences about questions of interest, such as the probability that the rate exceeds some threshold, and the importance of different socio-economic factors.

The two main examples in Diggle et al. (1998) involve count data, utilizing a Poisson and binomial model respectively. SGLMs for count data are also explored in Christensen

(a)

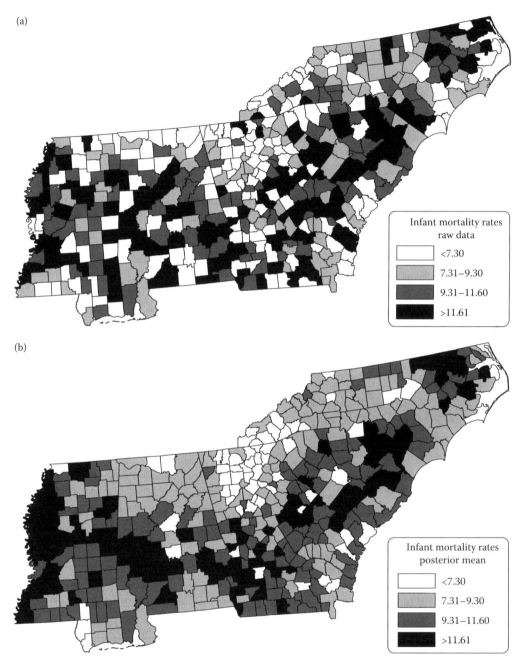

FIGURE 18.3
(a) Raw infant mortality rates. (b) Posterior mean infant mortality rates.

and Waagepetersen (2002), where a Langevin–Hastings MCMC approach is also developed for simulating from the posterior distribution. Note that count data with reasonably large counts may be modeled well by linear GP models. Given the added complexity of implementing SGLMs, it may therefore be advisable to first try a linear GP model before using an SGLM. However, when there is scientific interest in modeling a known sampling mechanism, SGLMs may be a better option.

18.3.2.3 Zero-Inflated Data

In many disciplines, particularly ecology and environmental sciences, observations are often in the form of spatial counts with an excess of zeros (see Welsh et al., 1996). SGLMs provide a nice framework for modeling such processes. For instance, Rathbun and Fei (2006) describe a model for oak trees which determines the species range by a spatial probit model which depends on a set of covariates thought to determine the species' range. Within that range (corresponding to suitable habitat), species counts are assumed to follow an independent Poisson distribution depending on a set of environmental covariates. The model for isopod nest burrows in Agarwal et al. (2002) generates a zero with probability p and a draw from a Poisson with probability $1 - p$. The excess zeros are modeled via a logistic regression and the Poisson mean follows a log-linear model. Spatial dependence is imposed via a GMRF model.

Example 18.6

Recta et al. (2011) study the spatial distribution of Colorado potato beetle populations in potato fields where a substantial proportion of observations were zeros. From the point of view of population studies, it is important to identify the within-field factors that predispose the presence of an adult. The distribution may be seen as a manifestation of two biological processes: incidence, as shown by presence or absence; and severity, as shown by the mean of positive counts. The observation at location $\mathbf{s}$, $Z(\mathbf{s})$, is decomposed into two variables: incidence (binary) variable $U(\mathbf{s}) = 1$ if $Z(\mathbf{s}) > 0$, else $U(\mathbf{s}) = 0$, and severity (count) variable $V(\mathbf{s}) = Z(\mathbf{s})$ if $Z(\mathbf{s}) > 0$ (irrelevant otherwise). Separate linear GP models can be specified for the $U(\mathbf{s})$ and $V(\mathbf{s})$ processes, with different covariance structures and means. This formulation allows a great deal of flexibility, including the ability to study spatial dependence in severity, and spatial dependence between severity and incidence, and the potential to relate predictors specifically to severity and independence.

For convenience, we order the data so that the incidences, observations where $U(\mathbf{s}) = 1$, are the first n_1 observations. Hence, there are n_1 observations for V, corresponding to the first n_1 observations of U, and $n_1 \leq n$. Our observation vectors are therefore $\mathbf{U} = (U(\mathbf{s}_1), \ldots, U(\mathbf{s}_n))^T$ and $\mathbf{V} = (V(\mathbf{s}_1), \ldots, V(\mathbf{s}_{n_1}))^T$. Placing this model in an SGLM framework:

$$U(\mathbf{s}) = \text{Bernoulli}(A(\mathbf{s})), \quad \text{so } \Pr(U(\mathbf{s}) = 1 \mid w_U(\mathbf{s}), \boldsymbol{\alpha}) = A(\mathbf{s}),$$

$$V(\mathbf{s}) = \text{TruncPoisson}(B(\mathbf{s})), \quad \text{so } \mathrm{E}(V(\mathbf{s}) \mid w_V(\mathbf{s}), \boldsymbol{\beta}) = \frac{B(\mathbf{s})}{1 - e^{-B(\mathbf{s})}}.$$

TruncPoisson is a truncated Poisson random variable (cf. David and Johnson, 1952) with $\Pr(V(\mathbf{s}) = r \mid B(\mathbf{s})) = \frac{B(\mathbf{s})^r e^{-B(\mathbf{s})}}{r!(1 - e^{-B(\mathbf{s})})}$, $r = 1, 2, \ldots$, and $w_U(\mathbf{s}), w_V(\mathbf{s}), \boldsymbol{\alpha}, \boldsymbol{\beta}$ are described below. Furthermore, using canonical link functions,

$$\log\left(\frac{A(\mathbf{s})}{1 - A(\mathbf{s})}\right) = X_U(\mathbf{s})\boldsymbol{\alpha} + w_u(\mathbf{s}),$$

$$\log(B(\mathbf{s})) = X_V(\mathbf{s})\boldsymbol{\beta} + w_V(\mathbf{s}),$$

where $X_U(\mathbf{s}), X_V(\mathbf{s})$, are vectors of explanatory variables, $\boldsymbol{\alpha}, \boldsymbol{\beta}$ are regression coefficients, and $\mathbf{w}_U = (w_U(\mathbf{s}_1), \ldots, w_U(\mathbf{s}_n))^T$ and $\mathbf{w}_V = (w_V(\mathbf{s}_1), \ldots, w_V(\mathbf{s}_{n_1}))^T$ are modeled via GPs. The resulting covariance matrices for $\mathbf{w}_U, \mathbf{w}_V$ are Σ_U, Σ_V respectively, specified by exponential covariance functions as described in Section 18.2.1. The parameters of our model are therefore $(\boldsymbol{\alpha}, \boldsymbol{\beta}, \Theta)$, with Θ representing covariance function parameters. Note that the model structure also allows for a flexible cross-covariance relating these two latent processes (Recta et al., 2011), though we do not discuss this here. Priors for $\boldsymbol{\alpha}, \boldsymbol{\beta}, \Theta$ are specified in standard fashion, with a flat prior for the regression parameters and log-uniform priors for the covariance function parameters. Inference and prediction for this model are based on the posterior $\pi(\mathbf{w}_U, \mathbf{w}_V, \boldsymbol{\alpha}, \boldsymbol{\beta}, \Theta \mid \mathbf{U}, \mathbf{V})$. An algorithm for sampling from this distribution is discussed in the next section.

18.3.3 MCMC for SGLMs

For SGLMs, unlike spatial linear models, marginal distributions are not available in closed form for any of the parameters. In other words, for linear spatial models it is possible to study $\pi(\Theta, \boldsymbol{\beta} \mid \mathbf{Z})$ ("marginalizing out" $\mathbf{w}$), while for SGLMs, inference is based on the joint distribution $\pi(\Theta, \boldsymbol{\beta}, \mathbf{w} \mid \mathbf{Z})$. Hence, the dimension of the distribution of interest is typically of the same order as the number of data points. While one can easily construct variable-at-a-time Metropolis–Hastings samplers for such distributions, the strong dependence among the spatial random effects (w_i) and the covariance/precision parameters (Θ) typically results in heavily autocorrelated MCMC samplers, which makes sample-based inference a challenge in practice. In addition, expensive matrix operations involved in each iteration of the algorithm continue to be a major challenge for GP-based models, though some recent approaches have been proposed to resolve this (see Section 18.5). Hence not only are standard MCMC algorithms for SGLMs slow mixing, but also each update can be computationally expensive, leading to very inefficient samplers.

A general approach to improve mixing in the Markov chain is to update parameters jointly in large blocks. This is a well-known approach for improving mixing (see, e.g. Liu et al., 1994) and is particularly useful in SGLMs due to the strong dependence among the components of the posterior distribution. However, constructing joint updates effectively can be a challenge, especially in high dimensions. Approaches proposed for constructing joint updates for such models often involve deriving a multivariate normal approximation to the joint conditional distribution of the random effects, $\pi(\mathbf{w} \mid \Theta, \boldsymbol{\beta}, \mathbf{Z})$. We now briefly discuss two general approaches for constructing MCMC algorithms for SGLMs.

18.3.3.1 Langevin–Hastings MCMC

Langevin–Hastings updating schemes (Roberts and Tweedie, 1996) and efficient reparameterizations for GP-based SGLMs are investigated in Christensen et al. (2006) and Christensen and Waagepetersen (2002). The Langevin–Hastings algorithm is a variant of the Metropolis–Hastings algorithm inspired by considering continuous-time Markov processes that have as their stationary distributions the target distribution. Since it is only possible to simulate from the discrete-time approximation to this process and the discrete-time approximation does not result in a Markov chain with desirable properties (it is not even recurrent), Langevin–Hastings works by utilizing the discrete-time approximation as a proposal for a standard Metropolis–Hastings algorithm (Roberts and Tweedie, 1996). Hence, Langevin–Hastings, like most algorithms for block updating of parameters, is an approach for constructing a multivariate normal approximation that can be used as a proposal for a block update. The significant potential improvement offered by Langevin–Hastings over simple random-walk type Metropolis algorithms is due to the fact that the local property of the distribution, specifically the gradient of the target distribution, is utilized. This can help move the Markov chain in the direction of modes. To illustrate the use of MCMC block sampling approaches based on Langevin–Hastings MCMC, we return to Example 18.3.2.3.

Example 18.7   Langevin–Hastings MCMC for a Two-Stage Spatial Zero-Inflated Poisson Model

This description follows closely the more detailed discussion in Recta et al. (2011). Simple univariate Metropolis–Hastings updates worked poorly and numerous Metropolis random-walk block update schemes for the random effects resulted in very slow mixing Markov chains as well. Hence, Langevin–Hastings updates were applied to the random effects. Borrowing from the notation and description in Christensen et al. (2006) and Diggle and Ribeiro (2007), let

$$\nabla(\gamma) = \frac{\partial}{\partial \gamma} \log \pi(\gamma \mid \ldots) = -\gamma + (\Sigma^{1/2})^T \left[\begin{array}{c} \left\{ (U(\mathbf{s}_i) - A(\mathbf{s}_i)) \dfrac{h_c'(A(\mathbf{s}_i))}{h'(A(\mathbf{s}_i))} \right\}_{i=1}^{n} \\[3ex] \left\{ (V(\mathbf{s}_j) - B(\mathbf{s}_j)) \dfrac{g_c'(B(\mathbf{s}_j))}{g'(B(\mathbf{s}_j))} \right\}_{j=n+1}^{n+n_1} \end{array} \right]$$

denote the gradient of the log target density evaluated at γ (denoted by $\pi(\gamma \mid \ldots)$) where h_c' and g_c' are the partial derivatives of the canonical link functions for the Bernoulli and truncated Poisson distributions respectively, h' and g' are partial derivatives of the actual link functions used, and $\Sigma^{1/2}$ is the Choleski factor of the joint covariance matrix for $\mathbf{w}_U, \mathbf{w}_V$. Since we used canonical links in both cases, $\dfrac{h_c'\{A(\mathbf{s}_i)\}}{h'\{A(\mathbf{s}_i)\}} = \dfrac{g_c'\{B(\mathbf{s}_i)\}}{g'\{B(\mathbf{s}_i)\}} = 1$ for each i. However, since the Langevin–Hastings algorithm above is not geometrically ergodic (Christensen et al., 2001), we use a truncated version where the gradient is

$$\nabla^{\text{trunc}}(\gamma) = \frac{\partial}{\partial \gamma} \log \pi(\gamma \mid \ldots) = -\gamma + (\Sigma^{1/2})^T \left[\begin{array}{c} \{U(\mathbf{s}_i) - A(\mathbf{s}_i)\}_{i=1}^{n} \\[2ex] \left\{ V(\mathbf{s}_j) - (B(\mathbf{s}_j) \wedge H) \right\}_{j=n+1}^{n+n_1} \end{array} \right],$$

where $H \in (0, \infty)$ is a truncation constant. This results in a geometrically ergodic algorithm (Christensen et al., 2001) so a central limit theorem holds for the estimated expectations and a consistent estimate for standard errors can be used to provide a theoretically justified stopping rule for the algorithm (Jones et al., 2006). The binomial part of the gradient does not need to be truncated because the expectation, $A(\mathbf{s})$, is bounded. Given that the current value of the random effects vector is γ, the Langevin–Hastings update for the entire vector of spatial random effects involves using a multivariate normal proposal, $N(\gamma + \frac{h}{2} \nabla(\gamma)^{\text{trunc}}, hI)$, $h > 0$. The tuning parameter h may be selected based on some initial runs, say by adapting it so that acceptance rates are similar to optimal rates given in Roberts and Rosenthal (1998).

Unfortunately, the above MCMC algorithm still mixes slowly in practice, which may be due to the fact that Langevin–Hastings works poorly when different components have different variances (Roberts and Rosenthal, 2001); this is certainly the case for the random effects $\mathbf{w}_U, \mathbf{w}_V$. To improve the mixing of the Markov chain we follow Christensen et al. (2006) and transform the vector of random effects into approximately (*a posteriori*) uncorrelated components with homogeneous variance. For convenience, let $\mathbf{w} = (\mathbf{w}_U^T, \mathbf{w}_V^T)^T$ be the vector of spatial random effects and $\mathbf{Y} = (\mathbf{U}^T, \mathbf{V}^T)^T$. The covariance matrix for $\mathbf{w} \mid \mathbf{Y}$ is approximately $\tilde{\Sigma} = (\Sigma^{-1} + \Lambda(\hat{\mathbf{w}}))^{-1}$, where $\Lambda(\hat{\mathbf{w}})$ is a diagonal matrix with entries $\frac{\partial^2}{(\partial w_j)^2} \log \pi(Y_j \mid w_j)$, and w_j and Y_j are the jth elements of $\mathbf{w}$ and $\mathbf{Y}$ respectively. Also, $\hat{\mathbf{w}}$ is assumed to be a typical value of $\mathbf{w}$, such as the posterior mode of $\mathbf{w}$. Let $\tilde{\mathbf{w}}$ be such that $\mathbf{w} = \tilde{\Sigma}^{1/2}\tilde{\mathbf{w}}$. Christensen et al. (2006) suggest updating $\tilde{\mathbf{w}}$ instead of $\mathbf{w}$, since $\tilde{\mathbf{w}}$ has approximately uncorrelated components with homogeneous variance, simplifying the construction of an efficient MCMC algorithm. For our application, setting $\Lambda(w(x)) = 0$ for all x is a convenient choice and appears to be adequate, though there are alternative approaches (see Christensen et al., 2006). An efficient Metropolis–Hastings algorithm is obtained by updating the transformed parameter vector $\tilde{\mathbf{w}}$ via Langevin–Hastings. The remaining parameters α, β, Θ may then be updated using simple Metropolis random-walk updates. This worked reasonably well in our examples, but certain reparameterizations may also be helpful in cases where mixing is poor.

The above Langevin–Hastings algorithm was found in Recta et al. (2011) to be efficient in a number of real data and simulated examples. Similar efficiencies were seen for SGLMs for count data in Christensen et al. (2006). For prediction at a new set of locations, say $\mathbf{s}_1^*, \dots, \mathbf{s}_m^*$, we would first predict the spatial random effect vectors $\mathbf{w}_U^* = (w_U(\mathbf{s}_1^*), \dots, w_U(\mathbf{s}_m^*))^T$ and $\mathbf{w}_V^* = (w_V(\mathbf{s}_1^*), \dots, w_V(\mathbf{s}_n^*))^T$. Again, sample-based inference provides a simple and effective way to obtain these predictions. Given a sampled vector of $(\mathbf{w}_U, \mathbf{w}_V, \alpha, \beta, \Theta)$ from above, we can easily sample the vectors $\mathbf{w}_U^*, \mathbf{w}_V^* \mid \mathbf{w}_U, \mathbf{w}_V, \alpha, \beta, \Theta$ from the posterior predictive distribution as it is a multivariate normal, similar in form to Equation 18.5. Once these vectors are obtained, the corresponding predictions for the incidence and prevalence (U and V) processes at the new locations are produced by simulating from the corresponding Bernoulli and truncated Poisson distributions. Many scientific question related to prediction or inference may be easily answered based on the samples produced from the posterior distribution of the regression parameters and spatial dependence parameters, along with the samples from the posterior predictive distribution.

18.3.3.2 Approximating an SGLM by a Linear Spatial Model

Another approach for constructing efficient MCMC algorithms involves approximating an SGLM by a linear spatial model. This can be done by using an appropriate normal approximation to the non-Gaussian model. Consider an SGLM of the form described in Section 18.3.1. A linear spatial model approximation may be obtained as follows:

$$M(\mathbf{s}_i) \mid \beta, w(\mathbf{s}_i) \sim N(X(\mathbf{s}_i)\beta + w(\mathbf{s}_i), c(\mathbf{s}_i)), \quad i = 1, \dots, n,$$
$$\mathbf{w} \mid \Theta \sim N(0, \Sigma(\Theta)),$$

$$(18.13)$$

with $M(\mathbf{s}_i)$ representing the observation or some transformation of the observation at location $\mathbf{s}_i$, and $c(\mathbf{s}_i)$ an approximation to the variance of $M(\mathbf{s}_i)$. It is clear that an approximation of the above form results in a joint normal specification for the model. Hence, the approximate model is a linear spatial model of the form described in Section 18.2.1 and the resulting full conditional distribution for the spatial random effects, $\pi(\mathbf{w} \mid \Theta, \beta, \mathbf{Z})$, is multivariate normal. Note that an approximation of this form can also be obtained for SGLMs that have underlying GMRFs. We consider, as an example, the following version of the well-known Poisson-GMRF model (Besag et al., 1991):

$$Z(\mathbf{s}_i) \mid w(\mathbf{s}_i) \sim \text{Poisson}(E_i \exp\{w(\mathbf{s}_i)\}), \quad i = 1, \dots, n,$$
$$f(\mathbf{w} \mid \tau) \propto \tau^{(N-1)/2} \exp\{-\mathbf{w}^T Q(\tau)\mathbf{w}\},$$

with a proper inverse gamma prior for τ. An approximation to the Poisson likelihood above may be obtained by following Haran (2003) (see also Haran and Tierney, 2011; Haran et al., 2003). By using the transformation $M(\mathbf{s}_i) = \log(Y_i/E_i)$, and a delta method approximation to obtain $c(\mathbf{s}_i) = \min(1/Y_i, 1/0.5)$, we derive the approximation $M(\mathbf{s}_i) \stackrel{\cdot}{\sim} N(w(\mathbf{s}_i), c(\mathbf{s}_i))$. Other accurate approximations of the form Equation 18.13, including versions of the Laplace approximation, have also been studied (*cf.* Knorr-Held and Rue, 2002; Rue and Held, 2005).

The linear spatial model approximation to an SGLM has been pursued in constructing block MCMC algorithms where the approximate conditional distribution of the spatial random effects can be used as a proposal for a block Metropolis–Hastings update (Haran et al., 2003; Knorr-Held and Rue, 2002). The spatial linear model approximation above also allows for the random effects to be integrated out analytically, resulting in low-dimensional approximate marginal distributions for the remaining parameters of the model. The approximate marginal and conditional may be obtained as follows:

1. The linear spatial model approximation of the form Equation 18.13 results in a posterior distribution $\hat{\pi}(\Theta, \boldsymbol{\beta}, \mathbf{w} \mid \mathbf{Z})$. This can be used as an approximation to the posterior distribution of the SGLM ($\pi(\Theta, \boldsymbol{\beta}, \mathbf{w} \mid \mathbf{Z})$).

2. The approximate distribution $\hat{\pi}(\Theta, \boldsymbol{\beta}, \mathbf{w} \mid \mathbf{Z})$ can be analytically integrated with respect to $\mathbf{w}$ to obtain a low-dimensional approximate marginal posterior $\hat{\pi}(\Theta, \boldsymbol{\beta} \mid \mathbf{Z})$. The approximate conditional distribution of $\mathbf{w}$ is also easily obtained in closed form, $\hat{\pi}(\mathbf{w} \mid \Theta, \boldsymbol{\beta}, \mathbf{Z})$.

The general approach above has been explored in the development of heavy-tailed proposal distributions for rejection samplers, perfect samplers and efficient MCMC block samplers (Haran, 2003; Haran and Tierney, 2011). Separate heavy-tailed approximations to the marginal $\hat{\pi}(\Theta, \boldsymbol{\beta} \mid \mathbf{Z})$ and the conditional $\hat{\pi}(\mathbf{w} \mid \Theta, \boldsymbol{\beta}, \mathbf{Z})$ can be used to obtain a joint distribution, which may then be used as a proposal for an independence Metropolis–Hastings algorithm that proposes from the approximation at every iteration (see Haran and Tierney, 2011, for details). This algorithm is uniformly ergodic in some cases (Haran and Tierney, 2011) so rigorous ways to determine MCMC standard errors and the length of the Markov chain (Flegal et al., 2008; Jones et al., 2006) are available. The general framework described above for obtaining a linear spatial model approximation and integrating out the random effects (Haran, 2003; Haran and Tierney, 2011) has been extended in order to obtain fast, fully analytical approximations for SGLMs and related latent Gaussian models (Rue et al., 2009). While their fully analytical approximation may not have the same degree of flexibility offered by Monte Carlo-based inference, the approach in Rue et al. (2009) completely avoids MCMC and is therefore a promising approach for routine, efficient fitting of such models especially when model comparisons are of interest or large data sets are involved. However, it is worth noting that with Monte Carlo approaches, unlike with purely analytical approaches, it is possible in principle to obtain arbitrarily precise estimates. That is, as the Monte Carlo sample size gets large, the standard error reduces to zero. Sample-based inference is also an extremely useful tool for appropriately propagating uncertainty; this is increasingly important in complex scientific problems where multiple models are used—output from one model often acts as input to another model. In the spatial modeling context, when joint distributions are of particular interest, sample-based inference provides a convenient approach for propagating uncertainty while preserving properties of the joint distribution.

Both the Langevin–Hastings algorithm and MCMC based on "linearizing" an SGLM, along with their variants, result in efficient MCMC algorithms in many cases. An advantage of algorithms that use proposals that depend on the current state of the Markov chain (like the Langevin–Hastings algorithm or other block sampling algorithms discussed here), over fully-blocked independence Metropolis–Hastings approaches is that they take into account local properties of the target distribution when proposing updates. This may result in a better algorithm when a single approximation to the entire distribution is inaccurate. However, if the approximation is reasonably accurate, the independence

Metropolis–Hastings algorithm using this approximation can explore the posterior distribution very quickly, does not get stuck in local modes, and can be easily parallelized as all the proposals for the algorithm can be generated independently of each other. Since massively parallel computing is becoming increasingly affordable, this may be a useful feature.

We note that computing and inference for GP-based models and GP-based SGLMs for binary data, particularly for large data sets, have been studied extensively in the machine learning literature as well. While the context of the problems may not always be obviously spatial in nature, the models used are very similar. Therefore, several of the associated computational approaches may also be very useful and practical in the context of the spatial models discussed here. Both analytical approaches and sophisticated MCMC-based approaches (see Neal, 1999) have been developed. An excellent review of this literature may be found in Chapters 3 and 8 of Rasmussen and Williams (2006).

18.3.4 Maximum Likelihood Inference for SGLMs

It is important to note that even in a non-Bayesian framework, computation for SGLMs is nontrivial. For the SGLMs described in the previous section, the maximum likelihood estimator (MLE) maximizes the integrated likelihood. Hence, the MLE for Θ, β maximizes

$$\int \mathcal{L}(\Theta, \beta, \mathbf{w}; \mathbf{Z}) \, d\mathbf{w},$$

with respect to Θ, β. Evaluating the likelihood requires high-dimensional integration and the most rigorous approach to solving this problem uses MCMC maximum likelihood (Geyer, 1996; Geyer and Thompson, 1992). Alternatives include Monte Carlo expectation-maximization (MCEM) (Wei and Tanner, 1990), as explored by Zhang (2002) for SGLMs, although in some cases fast approximate approaches such as composite likelihood may be useful, as discussed for binary data by Heagerty and Lele (1998). In general, computation for maximum likelihood-based inference for SGLMs may often be at least as demanding as in the Bayesian formulation. On the other hand, the Bayesian approach also provides a natural way to incorporate the uncertainties (variability) in each of the parameter estimates when obtaining predictions and estimates of other parameters in the model.

18.3.5 Summary

SGLMs provide a very flexible approach for modeling dependent data when there is a known non-Gaussian sampling mechanism at work. Either GP or GMRF models can be used to specify the dependence in a hierarchical framework. Constructing MCMC algorithms for SGLMs can be challenging due to the high-dimensional posterior distributions, strong dependence among the parameters, and expensive matrix operations involved in the updates. Recent work suggests that MCMC algorithms that involve block updates of the spatial random effects can result in improved mixing in the resulting Markov chains. Constructing efficient block updating algorithms can be challenging, but finding accurate linear spatial model approximations to SGLMs or using Langevin–Hastings-based approaches may improve the efficiency of the MCMC algorithm in many situations. Matrix operations can be greatly speeded up by using similar tools to those used for linear spatial models. These will be discussed again in Section 18.5 in the context of spatial modeling for large data sets. Appropriate SGLMs and efficient sample-based inference allow for statistical inference for a very wide range of interesting scientific problems.

18.4 Non-Gaussian Markov Random Field Models

Non-Gaussian Markov random field (NMRF) models provide an alternative to SGLM approaches for modeling non-Gaussian lattice/areal data. These models were first proposed as "auto-models" in Besag (1974) and involve specifying dependence among spatial random variables conditionally, rather than jointly. NMRFs may be useful alternatives to SGLMs, especially when used to build space-time models, since they can model some interactions in a more direct and interpretable fashion, for example, when modeling the spread of contagious diseases from one region to its neighbor, thereby capturing some of the dynamics of a process. GMRFs, as described by Equation 18.6, are special cases of Markov random field models. A more general formulation is provided as follows:

$$p(Z_i \mid \mathbf{Z}_{-i}) \propto \exp\left(X_i \boldsymbol{\beta} + \psi \sum_{j \neq i} c_{ij} Z_j \right),$$

with $\psi > 0$. This conditional specification results in a valid joint specification (Besag, 1974; Cressie, 1993) and belongs to the exponential family. Consider a specific example of this for binary data, the autologistic model (Besag, 1974; Heikkinen and Högmander, 1994):

$$\log \frac{p(Z_i = 1)}{p(Z_i = 0)} = X_i \boldsymbol{\beta} + \psi \sum_{j \neq i} w_{ij} Z_j,$$

where $w_{ij} = 1$ if i and j are neighbors, and $w_{ij} = 0$ otherwise. For a fixed value of the parameters $\boldsymbol{\beta}$ and ψ, the conditional specification above leads to an obvious univariate Metropolis–Hastings algorithm that cycles through all the full conditional distributions in turn. However, when inference for the parameters is of interest, as is often the case, the joint distribution can be derived via Brook's lemma (Brook, 1964; see also Cressie, 1993, Chapter 6), to obtain

$$p(Z_1, \ldots, Z_n) = c(\psi, \boldsymbol{\beta})^{-1} \exp\left(\boldsymbol{\beta} \sum_i X_i Z_i + \psi \sum_{i,j} w_{ij} Z_i Z_j \right),$$

where $c(\psi, \boldsymbol{\beta})$ is the intractable normalizing constant, which is actually a normalizing *function* of the parameters $\psi, \boldsymbol{\beta}$. Other autoexponential models can be specified in similar fashion to the autologistic above, for example, the auto-Poisson model for count data (see Ferrándiz et al., 1995), or the centered autologistic model (Caragea and Kaiser, 2009). Specifying conditionals such that they lead to a valid joint specification involves satisfying mathematical constraints like the positivity condition (see, e.g. Besag, 1974; Kaiser and Cressie, 2000) and deriving the joint distribution for sound likelihood-based analysis can be challenging. Also, the resulting dependence can be nonintuitive. For example, Kaiser and Cressie (1997) propose a "Winsorized" Poisson automodel since it is not possible to model positive dependence with a regular Poisson auto-model. In addition, it is nontrivial to extend these models to other scenarios, say to accommodate other sources of information or data types such as zero-inflated data. These challenges, along with the considerable computational burden involved with full likelihood-based inference for such models (as we will see below), have made non-Gaussian Markov random fields more difficult to use routinely than SGLMs.

Since the joint distributions for NMRFs contain intractable normalizing functions involving the parameters of interest, Besag (1975) proposed the "pseudolikelihood" approximation to the likelihood, which involves multiplying the full conditional distributions together. The pseudolikelihood is maximized to provide an approximation to the MLE. This approximation works well in spatial models when the dependence is weak, but under strong dependence the maximum pseudolikelihood estimate may be a very poor approximation to the maximum likelihood estimate (see Gumpertz et al., 1997). MCMC maximum likelihood (Geyer, 1996; Geyer and Thompson, 1992) provides a sound methodology for estimating the maximum likelihood via a combination of MCMC and importance sampling. This is yet another instance of the enormous flexibility in model specification due to the availability of MCMC-based algorithms. MCMC maximum likelihood is a very general approach for maximum likelihoods involving intractable normalizing functions that also automatically provides sample-based estimates of standard errors for the parameters, though the choice of importance function plays a critical role in determining the quality of the estimates.

Now consider a Bayesian model obtained by placing a prior on the parameters (say, (β, ψ) for the autologistic model). Since the normalizing function is intractable, the Metropolis–Hastings acceptance ratio cannot be evaluated and constructing an MCMC algorithm for the model is therefore nontrivial. Approximate algorithms replacing the likelihood by pseudolikelihood (Heikkinen and Högmander, 1994) or by using estimated ratios of normalizing functions have been proposed, but these do not have a sound theoretical basis, though recent work by Atchadé et al. (2008) is a first attempt at providing some theory for the latter algorithm. A recent auxiliary variables approach (Møller et al., 2006) has opened up possibilities for constructing a Markov chain with the desired stationary distribution, though it requires samples from the *exact distribution* of the auto-model at a fixed value of the parameter, which is typically very difficult. Perfect sampling algorithms that produce samples from the stationary distribution of the Markov chains do exist for some models such as the autologistic model (Møller, 1999; Propp and Wilson, 1996). Perfect samplers are attractive alternatives to regular MCMC algorithms but are typically computationally very expensive relative to MCMC algorithms; Bayesian inference for non-Gaussian Markov random fields, however, is one area where perfect sampling has potential to be useful. Zheng and Zhu (2008) describe how the Møller et al. (2006) approach can be used to construct an MCMC algorithm for Bayesian inference for a space-time autologistic model. While there has been some recent activity in this area (*cf.* Hughes et al., 2011), Bayesian inference and computation for auto-models is still a relatively open area for research.

To summarize, non-Gaussian Markov random fields are an alternative to modeling dependent non-Gaussian data. The specification of dependence does not involve link functions and can therefore provide a more direct or intuitive model than SGLMs for some problems. Unfortunately, the mathematical constraints that allow conditional specifications to lead to valid joint specifications of non-Gaussian Markov random fields can be complicated and nonintuitive. Also, such models are not easily extended to more complicated scenarios (as discussed in Section 18.5). Sound inference for NMRFs has been a major hurdle due to intractable normalizing functions that appear in the likelihood. Maximum likelihood based inference for such models can be done via MCMC maximum likelihood; Bayesian inference for such models has been an even greater challenge, but recent research in MCMC methods has opened up some promising possibilities. Potential advantages and disadvantages of NMRFs over SGLMs are yet to be fully explored.

18.5   Extensions

The classes of models described in the previous three sections, while very rich, are relatively simple and only have two or three (hierarchical) levels each. Using random field models as building blocks, a very large number of more flexible models can be developed for tackling an array of important scientific problems. In particular, when there is interest in incorporating mechanistic models and physical constraints, spatial models can be specified via a series of conditional models, capturing the physical characteristics while still accounting for spatial and temporal dependence and various sources of error. For instance, Wikle et al. (2001) describe a series of conditionally specified models to obtain a very flexible space-time model for tropical ocean surface winds.

In similar fashion, joint models for data that are both point-level and areal can be easily specified in a hierarchical framework, providing a model-based approach to dealing with data available at different levels of aggregation (see the discussion of spatial misalignment in Banerjee et al., 2004, Chapter 6). Methods for spatial processes that either occur or are observed at multiple scales (Ferreira and Lee, 2007) take advantage of much of the same basic machinery described here. Models for spatiotemporal processes are particularly important since many problems involve space-time data, and it is critical to jointly model both sources of dependence. Spatiotemporal processes are particularly useful when mechanistic models are of interest and when there are interesting dynamics to be captured. Assuming space-time "separability"—that is, that the dependencies across time and space do not interact (mathematically, the covariance is multiplicative in the spatial and temporal dimensions)—allows the use of Kronecker products and dramatic increases in the speed of matrix computations. However, nonseparability is often not a tenable assumption. Classes of computationally tractable spatial models with stationary, nonseparable covariances have been proposed (Cressie and Huang, 1999; Gneiting, 2002) to address this issue, but in many cases computation can quickly become very challenging with increase in space-time data, particularly for more flexible models. While MCMC-based approaches are feasible in some cases (see Wikle et al., 1998), approximate approaches based on dimension reduction and empirical Bayes estimation combined with Kalman filtering (Wikle and Cressie, 1999) or sequential Monte Carlo-based approaches (Doucet et al., 2001) may be more computationally efficient, for example, in fitting multiresolution space-time models (Johannesson et al., 2007). It is also often of interest to model dependencies among multiple space-time variables, along with various sources of missing data and covariate information; multiple variables, covariates and missing data are very common in many studies, particularly in public health and social science related research. These variables and missing information can just be treated as additional random variables in the model, and added into the sampler, thereby accounting for any uncertainties associated with them. Multivariate models can be explored via both multivariate GMRFs (Carlin and Banerjee, 2003; Gelfand and Vounatsou, 2003) and multivariate or hierarchical GP models (Royle and Berliner, 1999).

As with many other areas of statistics, a major challenge for spatial modelers is dealing with massive data sets. This is particularly problematic for GP-based models since matrix operations involving very large matrices can be computationally prohibitive. One set of approaches centers around fast matrix computations that exploit the sparsity of matrices in GMRF models. Rue and Tjelmeland (2002) attempt to approximate GPs by GMRFs to exploit these computational advantages in the GP model case as well, but discover that the approximation does not always work as desired, particularly when the dependence is strong. However, utilizing sparsity does seem to be among the more promising general strategies, as shown in recent work by Cornford et al. (2005) who describe a framework to first impose sparsity and then exploit it in order to speed up computations for large data sets. Furrer et al. (2006) and Kaufman et al. (2008) use covariance tapering while Cressie and Johannesson (2008) use a fixed number of basis functions to construct a nonstationary covariance and exploit the special structure in the resulting covariance matrix to drastically reduce computations. Other approaches that have been investigated in recent years include using a Fourier basis representation of a GP (see Fuentes, 2007; Paciorek, 2007), and fast likelihood approximations for a GP model based on products of conditionals (Caragea, 2003; Stein et al., 2004; Vecchia, 1988). Wikle (2002) presents an approach for modeling large-scale spatial count data using an SGLM where he uses a spectral-domain representation of the spatial random effects to model dependence. Higdon (1998) describes a kernel mixing approach by utilizing the fact that a dependent process can be created by convolving a continuous white noise process with a convolution kernel. By using a discrete version of this process, for instance with a relatively small set of independent normal random variates, it is possible to model a very large spatial data set. The recently developed "predictive process" approach due to Banerjee et al. (2008) involves working with a low-dimensional projection of the original process, thereby greatly reducing the computational burden.

Stationarity and isotropy may be restrictive assumptions for the spatial process, particularly when there are strong reasons to suspect that dependence may be different in different directions or regions. Including anisotropy in GP models is fairly standard (Cressie, 1993); however, it is more difficult to obtain valid nonstationary processes that are also computationally tractable. Several such models for nonstationary processes have been proposed, including spatially-varying kernel based approaches for GPs (Higdon et al., 1999; Paciorek and Schervish, 2006) and GMRFs, or by convolving a fixed kernel over independent spatial processes with different kernels (Fuentes and Smith, 2001). MCMC plays a central role in fitting these flexible models.

To summarize, linear Gaussian random field models and SGLMs provide nice building blocks for constructing much more complicated models in a hierarchical Bayesian framework. Much of the benefit of Bayesian modeling and sample-based inference is realized in spatial modeling in situations where a flexible and potentially complicated model is desired, but is much more easily specified via a series of relatively simple conditional models, where one or more of the conditional models are spatial models.

18.6   Conclusion

The linear Gaussian random fields discussed in Section 18.2 are enormously powerful and flexible modeling tools when viewed in a maximum likelihood or Bayesian perspective, as are some of the non-Gaussian random fields discussed briefly in Section 18.4. Although the discussion here has focused on spatial data, the methods are useful in a very wide array of nonspatial problems, including machine learning and classification (primarily using GP models; see Rasmussen and Williams, 2006), time series, analysis of longitudinal data, and image analysis (primarily GMRF models; see references in Rue and Held, 2005). The major advantage of using a Bayesian approach accrues from the ability to easily specify coherent joint models for a variety of complicated scientific questions and data sets with multiple sources of variability. Bayesian approaches are particularly useful for the broad classes of models and problems discussed in Sections 18.3 and 18.5.

MCMC and sample-based inference in general greatly expands the set of questions that can be answered when studying spatial data, where for complicated conditionally specified models, asymptotic approximations may be difficult to obtain and hard to justify. Sample-based inference allows for easy assessment of the variability of all estimates, joint and marginal distributions of any subset of parameters, incorporation of nonlinearities, multiple sources of variability, allowing for missing data, and the use of scientific information in the model as well as via prior distributions. It is also typically much harder to study likelihood surfaces in maximum likelihood inference, but one can routinely study posterior surfaces—this provides a much more detailed picture regarding parameters in complicated models, which is important when likelihood or posterior surfaces are multimodal, flat or highly skewed. In principle, once the computational problem is solved (a good sampler is implemented), essentially all such questions can be based on the estimated posterior distribution. Since this is a very concise overview of Gaussian random field models for spatial data, we have neither discussed the theory underlying Gaussian random fields nor important principles of exploratory data analysis and model checking, but these can be found in many texts (Banerjee et al., 2004; Cressie, 1993; Diggle and Ribeiro, 2007; Schabenberger and Gotway, 2005).

It is perhaps best to end with some words of caution. While MCMC algorithms and powerful computing have made it possible to fit increasingly flexible spatial models, it is not always clear that there is enough information to learn about the parameters in such models. Zhang (2004) shows that not all parameters are consistently estimable in maximum likelihood-based inference for Gaussian process SGLMs; however, one quantity is consistently estimable (the ratio of the scale and range parameters in a Matérn covariance), and this is the quantity that drives prediction. It has been noted that the likelihood surface for the covariance parameters in a linear GP model is relatively flat, which accounts for the large standard errors of the estimates in a maximum likelihood setting (see Handcock and Stein, 1993; Li and Sudjianto, 2005). In our experience, this can also lead to large posterior standard deviations for the parameters in a Bayesian framework, both in GMRF and GP-based models. Remarkably, prediction based on these models often works extremely well—Stein (1999) provides a theoretical discussion of this in the context of linear GP models. This can be viewed as both a positive and a negative: a positive because often researchers are most concerned with prediction, and a negative because inference about the parameters is unreliable, and model validation techniques such as cross-validation cannot detect problems with inference about these parameters, as noted by Zhang (2002). The large uncertainties about parameter values can also be a problem in other situations. For example, the inferred dependence (range parameter) may not differ significantly even when differences in dependence exist, say at different time points or across different subregions. While the discussion above has centered on GP-based models and there has been less theoretical work on studying properties of GMRF-based models, many of the same practical concerns appear to exist, including identifiability issues, especially for the variance components (see, e.g. Bernardinelli et al., 1995). Prior sensitivity analysis should be a critical component of a careful spatial analysis.

As discussed at length here, Gaussian random field models are very useful but present considerable computational challenges. Software to meet this challenge includes the R (Ihaka and Gentleman, 1996) packages `geoR` (Ribeiro and Diggle, 2001), `spBayes` (Finley et al., 2007) for various linear GP models, `geoRglm` (Christensen and Ribeiro, 2002) for GP models for count data, `ramps` (Smith et al., 2008) for joint linear models for point-level and areal data, and the `GeoBugs` module in WinBUGS (Lunn et al., 2000) for various GMRF models. However, although these software packages are a great advance, there still remains a lot of room for development of new algorithms and software for the classes of spatial models discussed here.

Acknowledgments

The author thanks Tse-Chuan Yang and Sham Bhat for help with drawing maps and Sudipto Banerjee, Brad Carlin, Victor de Oliveira, Don Richards, Margaret Short, Lance Waller, and Tse-Chuan Yang for helpful discussions. Comments from the editor and an anonymous referee also helped improving this chapter.

References

Abramowitz, M. and Stegun, I. A. 1964. *Handbook of Mathematical Functions*. Dover, New York.

Agarwal, D. K., Gelfand, A. E., and Citron-Pousty, S. 2002. Zero-inflated models with application to spatial count data. *Environmental and Ecological Statistics*, 9(4):341–355.

Amit, Y., Grenander, U., and Piccioni, M. 1991. Structural image restoration through deformable templates. *Journal of the American Statistical Association*, 86:376–387.

Anderson, T. 2003. *An Introduction to Multivariate Statistical Analysis*, 3rd edn. Wiley, Hoboken, NJ.

Atchadé, Y., Lartillot, N., and Robert, C. 2008. Bayesian computation for statistical models with intractable normalizing constants. Technical report, Department of Statistics, University of Michigan.

Banerjee, S. 2005. On geodetic distance computations in spatial modeling. *Biometrics*, 61(2):617–625.

Banerjee, S., Carlin, B., and Gelfand, A. 2004. *Hierarchical Modeling and Analysis for Spatial Data*. Chapman & Hall/CRC, Boca Raton, FL.

Banerjee, S., Gelfand, A., Finley, A. O., and Sang, H. 2008. Gaussian predictive process models for large spatial datasets. *Journal of the Royal Statistical Society*, Series B, 70:825–848.

Berger, J. O., De Oliveira, V., and Sansó, B. 2001. Objective Bayesian analysis of spatially correlated data. *Journal of the American Statistical Association*, 96(456):1361–1374.

Bernardinelli, L., Clayton, D., and Montomoli, C. 1995. Bayesian estimates of disease maps: How important are priors? *Statistics in Medicine*, 14:2411–2431.

Besag, J. 1974. Spatial interaction and the statistical analysis of lattice systems (with discussion). *Journal of the Royal Statistical Society*, Series B, 36:192–236.

Besag, J. 1975. Statistical analysis of non-lattice data. *The Statistician*, 24:179–196.

Besag, J. 1989. Towards Bayesian image analysis. *Journal of Applied Statistics*, 16:395–407.

Besag, J. 2002. Discussion of "What is a Statistical Model?" by Peter McCullagh. *Annals of Statistics*, 30:1267–1277.

Besag, J. and Green, P. 1993. Spatial statistics and Bayesian computation (with discussion). *Journal of the Royal Statistical Society*, Series B, 55:25–37, 53–102.

Besag, J. and Kempton, R. 1986. Statistical analysis of field experiments using neighbouring plots. *Biometrics*, 42:231–251.

Besag, J. and Kooperberg, C. 1995. On conditional and intrinsic autoregressions. *Biometrika*, 82: 733–746.

Besag, J., Green, P., Higdon, D., and Mengersen, K. 1995. Bayesian computation and stochastic systems (with discussion). *Statistical Science*, 10:3–66.

Besag, J., York, J., and Mollié, A. 1991. Bayesian image restoration, with two applications in spatial statistics (with discussion). *Annals of the Institute of Statistical Mathematics*, 43:1–59.

Brook, D. 1964. On the distinction between the conditional probability and the joint probability approaches in the specification of nearest-neighbour systems. *Biometrika*, 51:481–483.

Brooks, R., Wardrop, D., Thornton, K., Whigham, D., Hershner, C., Brinson, M., and Shortle, J. 2006. Integration of ecological and socioeconomic indicators for estuaries and watersheds of the Atlantic slope. Final Report to US EPA STAR Program, Agreement R-82868401, Washington, DC. Prepared by the Atlantic Slope Consortium, University Park, PA. Technical report, Pennsylvania State University, http://www.asc.psu.edu.

Caragea, P. C. 2003. Approximate likelihoods for spatial processes. PhD dissertation. Technical report, Department of Statistics, University of North Carolina, Chapel Hill.

Caragea, P. C. and Kaiser, M. 2009. Autologistic models with interpretable parameters. *Journal of Agricultural, Biological, and Environmental Statistics*, 14:281–300.

Carlin, B. P. and Banerjee, S. 2003. Hierarchical multivariate CAR models for spatio-temporally correlated survival data. In J. M. Bernardo, M. J. Bayarri, J. O. Berger, A. P. Dawid, D. Heckerman, A. F. M. Smith, and M. West (eds), *Bayesian Statistics 7. Proceedings of the Seventh Valencia International Meeting*, pp. 45–64. Oxford University Press, Oxford.

Christensen, O., Møller, J., and Waagepetersen, R. 2001. Geometric ergodicity of Metropolis-Hastings algorithms for conditional simulation in generalized linear mixed models. *Methodology and Computing in Applied Probability*, 3(3):309–327.

Christensen, O. F. and Ribeiro, P. J. 2002. GeoRglm: A package for generalised linear spatial models. *R News*, 2(2):26–28.

Christensen, O. F. and Waagepetersen, R. 2002. Bayesian prediction of spatial count data using generalized linear mixed models. *Biometrics*, 58(2):280–286.

Christensen, O. F., Roberts, G. O., and Sköld, M. 2006. Robust Markov chain Monte Carlo methods for spatial generalized linear mixed models. *Journal of Computational and Graphical Statistics*, 15(1):1–17.

Clifford, P. 1990. Markov random fields in statistics. In G. R. Grimmett and D. J. A. Welsh (eds), *Disorder in Physical Systems: A Volume in Honour of John M. Hammersley on His 70th Birthday*, pp. 19–32. Clarendon Press, Oxford.

Cornford, D., Csato, L., and Opper, M. 2005. Sequential, Bayesian geostatistics: a principled method for large data sets. *Geographical Analysis*, 37(2):183–199.

Cowles, M., Yan, J., and Smith, B. 2009. Reparameterized and marginalized posterior and predictive sampling for complex Bayesian geostatistical models. *Journal of Computational and Graphical Statistics*, 18:262–282.

Cressie, N. and Huang, H.-C. 1999. Classes of nonseparable, spatio-temporal stationary covariance functions. *Journal of the American Statistical Association*, 94:1330–1340.

Cressie, N. and Johannesson, G. 2008. Fixed rank kriging for very large spatial data sets. *Journal of the Royal Statistical Society, Series B*, 70(1):209–226.

Cressie, N. A. 1993. *Statistics for Spatial Data*, 2nd edn. Wiley, New York.

Currin, C., Mitchell, T., Morris, M., and Ylvisaker, D. 1991. Bayesian prediction of deterministic functions, with applications to the design and analysis of computer experiments. *Journal of the American Statistical Association*, 86:953–963.

David, F. and Johnson, N. 1952. The truncated Poisson. *Biometrics*, 8(4):275–285.

De Oliveira, V. 2000. Bayesian prediction of clipped Gaussian random fields. *Computational Statistics and Data Analysis*, 34(3):299–314.

De Oliveira, V. 2007. Objective Bayesian analysis of spatial data with measurement error. *Canadian Journal of Statistics*, 35(2):283–301.

Diggle, P. and Ribeiro, P. 2007. *Model-Based Geostatistics*. Springer, New York.

Diggle, P. J., Tawn, J. A., and Moyeed, R. A. 1998. Model-based geostatistics (with discussion). *Applied Statistics*, 47:299–350.

Doucet, A., de Freitas, N., and Gordon, N. 2001. *Sequential Monte Carlo Methods in Practice*. Springer, New York.

Ferrándiz, J., López, A., Llopis, A., Morales, M., and Tejerizo, M. L. 1995. Spatial interaction between neighbouring counties: Cancer mortality data in Valencia (Spain). *Biometrics*, 51:665–678.

Ferreira, M. A. R. and De Oliveira, V. 2007. Bayesian reference analysis for Gaussian Markov random fields. *Journal of Multivariate Analysis*, 98(4):789–812.

Ferreira, M. A. R. and Lee, H. 2007. *Multiscale Modeling: A Bayesian Perspective*. Springer, New York.

Finley, A., Banerjee, S., and Carlin, B. 2007. spBayes: an R package for univariate and multivariate hierarchical point-referenced spatial models. *Journal of Statistical Software*, 19:1–20.

Flegal, J., Haran, M., and Jones, G. 2008. Markov chain Monte Carlo: Can we trust the third significant figure? *Statistical Science*, 23:250–260.

Fuentes, M. 2007. Approximate likelihood for large irregularly spaced spatial data. *Journal of the American Statistical Association*, 102:321–331.

Fuentes, M. and Smith, R. 2001. A new class of nonstationary spatial models. North Carolina State University Institute of Statistics Mimeo Series, 2534.

Furrer, R., Genton, M., and Nychka, D. 2006. Covariance tapering for interpolation of large spatial datasets. *Journal of Computational and Graphical Statistics*, 5:502–523.

Gelfand, A. E. and Vounatsou, P. 2003. Proper multivariate conditional autoregressive models for spatial data analysis. *Biostatistics*, 4(1):11–15.

Geman, S. and Geman, D. 1984. Stochastic relaxation, Gibbs distributions, and the Bayesian restoration of images. *IEEE Transcations in Pattern Analysis and Machine Intelligence*, 6:721–741.

Geyer, C. 1996. Estimation and optimization of functions. In W.R. Gilks, S. Richardson, and D.J. Spiegelhalter (eds), *Markov Chain Monte Carlo in Practice*, pp. 241–258. Chapman & Hall, London.

Geyer, C. J. and Thompson, E. A. 1992. Constrained Monte Carlo maximum likelihood for dependent data (with discussion). *Journal of the Royal Statistical Society, Series B*, 54:657–683.

Ghosh, M. and Rao, J. N. K. 1994. Small area estimation: An appraisal (with discussion). *Statistical Science*, 9:55–93.

Gneiting, T. 2002. Nonseparable, stationary covariance functions for space-time data. *Journal of the American Statistical Association*, 97(458):590–601.

Golub, G. and Van Loan, C. 1996. *Matrix Computation*, 3rd edn. Johns Hopkins University Press, Baltimore, MD.

Grenander, U. and Keenan, D. M. 1989. Towards automated image understanding. *Journal of Applied Statistics*, 16:207–221.

Gumpertz, M. L., Graham, J. M., and Ristaino, J. B. 1997. Autologistic model of spatial pattern of Phytophthora epidemic in bell pepper: Effects of soil variables on disease presence. *Journal of Agricultural, Biological, and Environmental Statistics*, 2:131–156.

Handcock, M. S. and Stein, M. L. 1993. A Bayesian analysis of kriging. *Technometrics*, 35:403–410.

Haran, M. 2003. Efficient perfect and MCMC sampling methods for Bayesian spatial and components of variance models, PhD dissertation. Technical report, Department of Statistics, University of Minnesota, Minneapolis.

Haran, M. and Tierney, L. 2011. On automating Markov chain Monte Carlo for a class of spatial models. To appear in *Bayesian Analysis*.

Haran, M., Bhat, K., Molineros, J., and De Wolf, E. 2010. Estimating the risk of a crop epidemic from coincident spatio-temporal processes. *Journal of Agricultural, Biological, and Environmental Statistics*, 15(2):158–175.

Haran, M., Hodges, J. S., and Carlin, B. P. 2003. Accelerating computation in Markov random field models for spatial data via structured MCMC. *Journal of Computational and Graphical Statistics*, 12:249–264.

Heagerty, P. J. and Lele, S. R. 1998. A composite likelihood approach to binary spatial data. *Journal of the American Statistical Association*, 93:1099–1111.

Health Resources and Services Administration 2003. Health professions, area resource file (ARF) system. Technical report, Quality Resource Systems, Inc., Fairfax, VA.

Heikkinen, J. and Högmander, H. 1994. Fully Bayesian approach to image restoration with an application in biogeography. *Applied Statistics*, 43:569–582.

Higdon, D. 1998. A process-convolution approach to modelling temperatures in the North Atlantic Ocean (Disc: P191-192). *Environmental and Ecological Statistics*, 5:173–190.

Higdon, D., Swall, J., and Kern, J. 1999. Non-stationary spatial modeling. In J. M. Bernardo, J. O. Berger, A. P. Dawid, and A. F. M. Smith (eds), *Bayesian Statistics 6: Proceedings of the Sixth Valencia International Meeting*, pp. 761–768. Oxford University Press, Oxford.

Hughes, J., Haran, M., and Caragea, P.C. 2011. Autologistic models for binary data on a lattice. To appear in *Environmetrics*.

Ihaka, R. and Gentleman, R. 1996. R: A language for data analysis and graphics. *Journal of Computational and Graphical Statistics*, 5:299–314.

Johannesson, G., Cressie, N., and Huang, H.-C. 2007. Dynamic multi-resolution spatial models. *Environmental and Ecological Statistics*, 14(1):5–25.

Jones, G. L., Haran, M., Caffo, B. S., and Neath, R. 2006. Fixed-width output analysis for Markov chain Monte Carlo. *Journal of the American Statistical Association*, 101(476):1537–1547.

Kaiser, M. S. and Cressie, N. 1997. Modeling Poisson variables with positive spatial dependence. *Statistics and Probability Letters*, 35:423–432.

Kaiser, M. S. and Cressie, N. 2000. The construction of multivariate distributions from Markov random fields. *Journal of Multivariate Analysis*, 73(2):199–220.

Kaufman, C., Schervish, M., and Nychka, D. 2008. Covariance tapering for likelihood-based estimation in large spatial data sets. *Journal of the American Statistical Association*, 103(484): 1545–1555.

Knorr-Held, L. and Rue, H. 2002. On block updating in Markov random field models for disease mapping. *Scandinavian Journal of Statistics*, 29(4):597–614.

Lawson, A. 2008. *Bayesian Disease Mapping: Hierarchical Modeling in Spatial Epidemiology*. Chapman & Hall/CRC, Boca Raton, FL.

Le, N. and Zidek, J. 2006. *Statistical Analysis of Environmental Space-Time Processes*. Springer, New York.

Li, R. and Sudjianto, A. 2005. Analysis of computer experiments using penalized likelihood in Gaussian kriging models. *Technometrics*, 47(2):111–120.

Liu, J. S., Wong, W. H., and Kong, A. 1994. Covariance structure of the Gibbs sampler with applications to the comparisons of estimators and augmentation schemes. *Biometrika*, 81: 27–40.

Lunn, D. J., Thomas, A., Best, N., and Spiegelhalter, D. 2000. WinBUGS—A Bayesian modelling framework: Concepts, structure, and extensibility. *Statistics and Computing*, 10:325–337.

Marks, M., Lapin, B., and Randall, J. 1994. *Phragmites australis* (*P. communis*): Threats, management, and monitoring. *Natural Areas Journal*, 14(4):285–294.

Matheron, G. 1971. The theory of regionalized variables and its applications. *Cahiers du Centre de Morphologie Mathématique de Fontainebleau*, 5:211.

McCullagh, P. 2002. What is a statistical model? *Annals of Statistics*, 30(5):1225–1310.

McCullagh, P. and Nelder, J. A. 1983. *Generalized Linear Models*. Chapman & Hall, London.

Metropolis, N., Rosenbluth, M., Rosenbluth, A., Teller, A., and Teller, E. 1953. Equation of state calculations by fast computing machines. *Chemical Physics*, 21:1087–1092.

Møller, J. 1999. Perfect simulation of conditionally specified models. *Journal of the Royal Statistical Society, Series B*, 61:251–264.

Møller, J. and Waagepetersen, R. 2004. *Statistical Inference and Simulation for Spatial Point Processes*. Chapman and Hall/CRC, Boca Raton, FL.

Møller, J., Pettitt, A., Berthelsen, K., and Reeves, R. 2006. An efficient Markov chain Monte Carlo method for distributions with intractable normalising constants. *Biometrika*, 93(2):451–458.

Mollié, A. 1996. Bayesian mapping of disease. In W.R. Gilks, S. Richardson, and D.J. Spiegelhalter (eds), *Markov Chain Monte Carlo in Practice*, pp. 359–379. Chapman & Hall, London.

Neal, R. M. 1999. Regression and classification using Gaussian process priors. In J. M. Bernardo, J. O. Berger, A. P. Dawid, and A. F. M. Smith (eds), *Bayesian Statistics 6: Proceedings of the Sixth Valencia International Meeting*, pp 475–501. Oxford University Press, Oxford.

Paciorek, C. 2007. Bayesian smoothing with Gaussian processes using Fourier basis functions in the spectralGP package. *Journal of Statistical Software*, 19(2).

Paciorek, C. and Schervish, M. 2006. Spatial modelling using a new class of nonstationary covariance functions. *Environmetrics*, 17:483–506.

Paulo, R. 2005. Default priors for Gaussian processes. *Annals of Statistics*, 33(2):556–582.

Propp, J. G. and Wilson, D. B. 1996. Exact sampling with coupled Markov chains and applications to statistical mechanics. *Random Structures and Algorithms*, 9(1-2):223–252.

Rasmussen, C. and Williams, C. 2006. *Gaussian Processes for Machine Learning*. MIT Press, Cambridge, MA.

Rathbun, S. L. and Fei, S. 2006. A spatial zero-inflated Poisson regression model for oak regeneration. *Environmental and Ecological Statistics*, 13(4):409–426.

Recta, V., Haran, M., and Rosenberger, J. L. 2011. A two-stage model for incidence and prevalence in point-level spatial count data. To appear in Environmetrics.

Ribeiro, P. and Diggle, P. 2001. geoR: A package for geostatistical analysis. *R News*, 1(2):14–18.

Roberts, G. and Rosenthal, J. 2001. Optimal scaling for various Metropolis-Hastings algorithms. *Statistical Science*, 16(4):351–367.

Roberts, G. O. and Rosenthal, J. S. 1998. Optimal scaling of discrete approximations to Langevin diffusions. *Journal of the Royal Statistical Society*, Series B, 60:255–268.

Roberts, G. O. and Tweedie, R. L. 1996. Exponential convergence of Langevin distributions and their discrete approximations. *Bernoulli*, 2:341–363.

Royle, J. and Berliner, L. 1999. A hierarchical approach to multivariate spatial modeling and prediction. *Journal of Agricultural, Biological, and Environmental Statistics*, 4(1):29–56.

Rue, H. 2001. Fast sampling of Gaussian Markov random fields. *Journal of the Royal Statistical Society*, Series B, 63(2):325–338.

Rue, H. and Held, L. 2005. *Gaussian Markov Random Fields: Theory and Applications*. Chapman & Hall/CRC, Boca Raton, FL.

Rue, H. and Tjelmeland, H. 2002. Fitting Gaussian Markov random fields to Gaussian fields. *Scandinavian Journal of Statistics*, 29(1):31–49.

Rue, H., Martino, S., and Chopin, N. 2009. Approximate Bayesian inference for latent Gaussian models by using integrated nested Laplace approximations. *Journal of the Royal Statistical Society*, Series B, 71(2):319–392.

Sacks, J., Welch, W. J., Mitchell, T. J., and Wynn, H. P. 1989. Design and analysis of computer experiments (with discussion). *Statistical Science*, 4:409–435.

Saltonstall, K. 2002. Cryptic invasion by a non-native genotype of the common reed, *Phragmites australis*, into North America. *Proceedings of the National Academy of Sciences of the USA*, 99(4):2445.

Schabenberger, O. and Gotway, C. 2005. *Statistical Methods for Spatial Data Analysis*. Chapman & Hall/CRC, Boca Raton, FL.

Short, M., Carlin, B., and Gelfand, A. 2005. Bivariate spatial process modeling for constructing indicator or intensity weighted spatial CDFs. *Journal of Agricultural, Biological, and Environmental Statistics*, 10(3):259–275.

Smith, B. J., Yan, J., and Cowles, M. K. 2008. Unified geostatistical modeling for data fusion and spatial heteroskedasticity with R package RAMPS. *Journal of Statistical Software*, 25:91–110.

Stein, M. L. 1999. *Interpolation of Spatial Data: Some Theory for Kriging*. Springer, New York.

Stein, M. L., Chi, Z., and Welty, L. J. 2004. Approximating likelihoods for large spatial data sets. *Journal of the Royal Statistical Society*, Series B, 66(2):275–296.

Sun, D., Tsutakawa, R. K., and Speckman, P. L. 1999. Posterior distribution of hierarchical models using CAR(1) distributions. *Biometrika*, 86:341–350.

Tibbits, M. M., Haran, M., and Liechty, J. C. 2010. Parallel multivariate slice sampling. *Statistics and Computing*, to appear.

Vecchia, A. V. 1988. Estimation and model identification for continuous spatial processes. *Journal of the Royal Statistical Society*, Series B, 50:297–312.

Wall, M. 2004. A close look at the spatial structure implied by the CAR and SAR models. *Journal of Statistical Planning and Inference*, 121(2):311–324.

Waller, L. and Gotway, C. 2004. *Applied Spatial Statistics for Public Health Data*. Wiley, Hoboken, NJ.

Wei, G. C. G. and Tanner, M. A. 1990. A Monte Carlo implementation of the EM algorithm and the poor man's data augmentation algorithms. *Journal of the American Statistical Association*, 85:699–704.

Welsh, A., Cunningham, R., Donnelly, C., and Lindenmayer, D. 1996. Modelling the abundance of rare species: Statistical models for counts with extra zeros. *Ecological Modelling*, 88(1–3):297–308.

Wikle, C. 2002. Spatial modeling of count data: A case study in modelling breeding bird survey data on large spatial domains. In A. B. Lawson and D. G. T. Denison (eds), *Spatial Cluster Modelling*, pp. 199–209. Chapman & Hall/CRC, Boca Raton, FL.

Wikle, C. K. and Cressie, N. 1999. A dimension-reduced approach to space-time Kalman filtering. *Biometrika*, 86:815–829.

Wikle, C. K., Berliner, L. M., and Cressie, N. 1998. Hierarchical Bayesian space-time models. *Environmental and Ecological Statistics*, 5:117–154.

Wikle, C. K., Milliff, R. F., Nychka, D., and Berliner, L. M. 2001. Spatiotemporal hierarchical Bayesian modeling: Tropical ocean surface winds. *Journal of the American Statistical Association*, 96(454):382–397.

Yan, J., Cowles, M. K., Wang, S., and Armstrong, M. P. 2007. Parallelizing MCMC for Bayesian spatiotemporal geostatistical models. *Statistics and Computing*, 17(4):323–335.

Yang, T., Teng, H., and Haran, M. 2009. The impacts of social capital on infant mortalities in the U.S.: A spatial investigation. *Applied Spatial Analysis and Policy*, 2(3): 211–227.

Yue, Y. and Speckman, P. L. 2009. Nonstationary spatial Gaussian Markov random. Technical report, Department of Statistics, University of Missouri-Columbia.

Zhang, H. 2002. On estimation and prediction for spatial generalized linear mixed models. *Biometrics*, 58(1):129–136.

Zhang, H. 2004. Inconsistent estimation and asymptotically equal interpolations in model-based geostatistics. *Journal of the American Statistical Association*, 99(465):250–261.

Zheng, Y. and Zhu, J. 2008. Markov chain Monte Carlo for a spatial-temporal autologistic regression model. *Journal of Computational and Graphical Statistics*, 17(1):123–137.

19

Modeling Preference Changes via a Hidden Markov Item Response Theory Model

Jong Hee Park

19.1 Introduction

Over the past two decades, political scientists have made great advances in the empirical estimation of ideal points (Bafumi et al., 2005; Bailey and Chang, 2001; Clinton et al., 2000; Heckman and Snyder, 1997; Jackman, 2001; Londregan, 2000; Martin and Quinn, 2002; Poole and Rosenthal, 1997). An ideal point, or preference, is a foundational theoretical concept for explaining the choices a political actor makes. For example, in simple unidimensional spatial models of voting, a legislator's vote choice is modeled as a rational decision based on a (Euclidean geometric) calculation of differences in utility values between the legislator's ideal point, a proposed bill, and the status quo.

Although an ideal point is often assumed to be static for theoretical convenience, dynamics in ideal points pose an important theoretical and empirical puzzle to researchers. For example, examining the judicial opinion writing of 16 US Supreme Court justices, Epstein et al. (1998) conclude that there is enough evidence to invalidate the assumption of preference stability over time.* They also go on to claim that any inference about a justice's "revealed preference" that is based on the stable preference assumption can be misleading if the justice actually underwent several preference changes over a lifetime. However, the development of statistical methods for dynamic ideal point estimation has been limited to a few published works (Martin and Quinn, 2002; McCarty et al., 1997). Also, the existing methods for dynamic ideal point estimation fail to distinguish fundamental changes from random drifts. In this paper I propose a method to detect *sharp, discontinuous changes in ideal points*.

The approach I take in this paper is to combine Chib's (1998) hidden Markov model (HMM) with the two-parameter item response theory (IRT) model. In this model, the dynamics in ideal points are modeled as agent-specific hidden regime changes. I demonstrate the utility of the hidden Markov IRT model by analyzing changes in ideal points among the 43 US Supreme Court justices serving between 1937 and 2006, and conclude that the model provides an effective benchmark for making probabilistic inferences about the timing of preference changes.

---

* The study of linkages between judges' opinions and their ideological leanings has become an important area of research in the last two decades, as political scientists have rejected "apolitical" legal understandings of judicial opinions in favor of attitudinal and rational models that introduce "political" factors into the decision-making process (Epstein and Knight, 1998; Segal and Spaeth, 1993).

19.2 Dynamic Ideal Point Estimation

Assuming a quadratic utility loss function, the utility of voting for item i by legislator j at time t is

$$U_{jt}(Y_i) = -(\theta_{jt} - Y_i)^2 + \delta_{ijt}^{(Y)},$$

where θ_{jt} is legislator j's ideal point at t, Y_i is the location of Yay, and $\delta^{(Y)}$ is a stochastic error drawn from a Gaussian distribution. For simple notation, I assume that θ_{jt} and Y_i are scalar, which means that the underlying political space is one-dimensional. The utility of voting against item i is defined similarly:

$$U_{jt}(N_i) = -(\theta_{jt} - N_i)^2 + \delta_{ijt}^{(N)}.$$

In this random utility model, a legislator votes for a bill i when $U_{jt}(Y_i) - U_{jt}(N_i) > 0$. If the utility difference between two vote choices is treated as a latent variable, the process can be modeled as a Bernoulli trial in which the probability of a yes vote is a function of a legislator's ideal point and the proposed bill's location:

$$y_{ijt} = \begin{cases} 1, & \text{if } z_{ijt} = U_{jt}(Y_i) - U_{jt}(N_i) > 0, \\ 0, & \text{if } z_{ijt} = U_{jt}(Y_i) - U_{jt}(N_i) \leq 0. \end{cases}$$

Then, as shown by Jackman (2001), some simple algebra shows the connection between the random utility voting model and the two-parameter IRT model:

$$\begin{aligned} z_{ijt} = U_{jt}(Y_i) - U_{jt}(N_i) &= -(\theta_{jt} - Y_i)^2 + \delta_{ijt}^{(Y)} + (\theta_{jt} - N_i)^2 - \delta_{ijt}^{(N)} \\ &= -2(N_i - Y_i)\theta_{jt} - (Y_i^2 - N_i^2) + \delta_{ijt}^{(Y)} - \delta_{ijt}^{(N)} \\ &= \beta_i \theta_{jt} - \alpha_i + \varepsilon_{ijt}. \end{aligned}$$

Note that the t subscript in ideal points is carried through the equation to denote the dynamics in ideal points.

If a political actor only makes a few decisions or is active for a short period of time, ignoring ideal point temporal dynamics is unlikely to pose a problem. However, for someone such as a legislator who serves multiple terms in office sessions, the conventional IRT model with constant ideal points very likely fails to capture any political evolution. As the legislator ages, exposure to exogenous shocks in the form of economic shifts, social upheavals, and new political environments is likely to affect voting decisions. It would be unrealistic to attribute all time-varying patterns in voting behavior to bill characteristics (α and β).[*]

Since the constant IRT model itself is highly parameterized with $2I + J$ parameters, where I is the number of items and J is the number of legislators, letting ideal points (θ_{jt}) vary over time is not a trivial modification. Two methods have been proposed so far. The first method, which does not rely on the IRT framework, is to specify ideal points as a polynomial function of time (McCarty et al., 1997). The other is to model the transition of ideal points as a first-order Markov process while the observed voting data are generated from the IRT model

---

[*] Note that the variance parameter in the IRT model is not identified as in the binary response models.

(Martin and Quinn, 2002). One major difference between two methods is the source of the dyanmics. McCarty et al.'s (1997) method assumes that the effect of time on ideal points is deterministic. By contrast, Martin and Quinn's dynamic ideal point method decomposes the source of changes in ideal points into a deterministic part and a stochastic part, and estimates the variance of legislator-specific transitions. These two methods are successfully applied to developing dynamic measurements of ideal points in US legislators and US Supreme Court justices with DW-NOMINATE (McCarty et al., 1997) and the Martin–Quinn score (Martin and Quinn, 2002), respectively.

However, while both methods are effective in uncovering transitions in ideal points, neither is specifically designed to detect the timing of changes in ideal points. In other words, the existing dynamic ideal point estimation methods are not optimal for modeling sharp, discontinuous changes in ideal points. This is an important issue since theoretical discussions on changes in ideal points pit continuous transitions of ideal points against discontinuous transitions. To put it differently, researchers who are more interested in the existence and timing of ideal point shifts rather than with smooth evolutions of ideal points over time would not find the existing methods helpful. This is why I have introduced a dynamic IRT model specifically designed to capture sharp, discontinuous ideal point shifts.

19.3 Hidden Markov Item Response Theory Model

The approach I take combines a HMM with the standard two parameter IRT model. Specifically, I use Chib's (1998) model to capture hidden regime changes in a legislator's ideal point. Note that in Chib's model the regime transition is constrained so that a Markov chain only moves forward to the terminal state. This constraint generates a nonergodic Markov chain, which turns out to be computationally efficient and as flexible as HMMs with ergodic Markov chains.*

Let s_{jt} be an indicator of hidden regimes for legislator j's ideal point at t, and P_j be a transition matrix for the hidden regimes. Due to the nonergodic constraint, it is trivial to compute the initial probability: $\pi_0 = (1, \dots, 0)$. The latent propensity of voting for an item i can be expressed as a function of item characteristics and ideal points, which are subject to agent-specific regime changes:

$$z_{ijk} = \beta_i \theta_{j,s_{jt}} - \alpha_i + \varepsilon_{ijt}, \quad \varepsilon_{ijt} \sim N(0,1), \tag{19.1}$$

$$s_{jt} \mid s_{j,t-1} \sim \text{Markov}(\pi_0, P_j). \tag{19.2}$$

---

* Let $p_{ij} = \Pr(s_t = j \mid s_{t-1} = i)$ be the probability of moving to state j from state i at time t when the state at $t-1$ is i. Then, the transition matrix of Chib (1998) is

$$\mathbf{P} = \begin{pmatrix} p_{11} & p_{12} & 0 & \cdots & 0 \\ 0 & p_{22} & p_{23} & \cdots & 0 \\ \vdots & \vdots & \vdots & \vdots & \vdots \\ 0 & 0 & 0 & p_{M-1,M-1} & p_{M-1,M} \\ 0 & 0 & & 0 & 1 \end{pmatrix}.$$

The value of $\theta_{j,s_{jt}}$ can take M different values at each time point subject to the first-order Markov process. In other words, s_{jt} indicates the preference regime associated with a legislator's ideal point at t.

It should be stressed that the hidden Markov IRT model can be considered as a special type of the dynamic IRT model developed by Martin and Quinn (2002). While the dynamic IRT model assumes that ideal points change at each time point due to random shocks, the hidden Markov IRT model assumes that ideal points change only when the underlying regime changes. When there is no detected change point, $s_{jt} = 1$ for $t = 1, \ldots, T_j$, the hidden Markov IRT model reduces to the constant IRT model.

Albert (1992) and Johnson and Albert (1999) provide an efficient Gibbs sampling algorithm for the constant IRT model. Once hidden state variables s_{jt} are sampled, the rest of the sampling scheme is similar to the constant IRT model.

Normal distributions are used as prior distributions for ideal points and item parameters. For identification, I use the standard normal distribution as a prior distribution of ideal points:*

$$\lambda_i \sim N(\mu_0, \mathbf{V}_0),$$

$$\theta_{j,s_t} \sim N(0, 1),$$

$$p_{ii} \sim \text{Beta}(a, b),$$

where $\lambda_i = (\alpha_i, \beta_i)'$.

The MCMC sampling algorithm for the hidden Markov IRT model consists of five steps, including two steps for augmented variables. We have

$$p(\boldsymbol{\alpha}, \boldsymbol{\beta}, \boldsymbol{\theta}, \mathbf{P} \mid \mathbf{y}) = \int p(\boldsymbol{\alpha}, \boldsymbol{\beta}, \boldsymbol{\theta}, \mathbf{P}, \mathbf{s}, \mathbf{z} \mid \mathbf{y}) \, d\mathbf{s} \, d\mathbf{z}$$

$$= \int p(\boldsymbol{\alpha}, \boldsymbol{\beta} \mid \boldsymbol{\theta}, \mathbf{P}, \mathbf{s}, \mathbf{z}) p(\boldsymbol{\theta} \mid \mathbf{P}, \mathbf{s}, \mathbf{z}) p(\mathbf{P} \mid \mathbf{s}, \mathbf{z}) p(\mathbf{s} \mid \mathbf{z}, \mathbf{y}) p(\mathbf{z} \mid \mathbf{y}) \, d\mathbf{s} \, d\mathbf{z}.$$

Step 1. Simulation of latent utilities. Following Albert and Chib (1993), the latent variable (z_{ijk}) in Equation 19.1 is sampled from two truncated normal distributions, the support of which changes depending on realized binary outcomes:

$$z_{ijk} \sim \begin{cases} N_{(-\infty, 0]}(\alpha_i + \beta_i \theta_{j,s_{jt}}, 1), & \text{if } y_{ijt} = 0, \\ N_{(0, \infty)}(\alpha_i + \beta_i \theta_{j,s_{jt}}, 1), & \text{if } y_{ijt} = 1. \end{cases}$$

Step 2. Simulation of item parameters. A vectorized notation is used to explain the simulation of item parameters. Latent utilities are formed as a $J \times 1$ vector $(\mathbf{z}_{it})$, and ideal point estimates are transformed into a $J \times 2$ matrix $\boldsymbol{\Theta}_t = (\mathbf{1}, \boldsymbol{\theta}_{j,s_{jt}})$ where $\boldsymbol{\theta}_{j,s_{jt}}$ is a vector of ideal points for all legislators at time t: $\boldsymbol{\theta}_{j,s_{jt}} = (\theta_{1,s_{1,t}}, \ldots, \theta_{J,s_{J,t}})'$. Finally, item parameters are stacked as a 2×1 matrix $\lambda_i = (\alpha_i, \beta_i)'$. Then we have a multivariate linear regression model

$$\mathbf{z}_{it} = \boldsymbol{\Theta}_t \lambda_i + \boldsymbol{\varepsilon}_{it}.$$

---

* See Clinton et al. (2000) and Jackman (2001) for the identification of the IRT model in Bayesian estimation.

Note that $\mathbf{\Theta}_t$ serves as a design matrix and $\mathbf{\lambda}_i$ serves as a parameter vector at this sampling step. From this,

$$\mathbf{\lambda}_i \mid \mathbf{\theta}, \mathbf{P}, \mathbf{z}, \mathbf{y} \sim N(\mathbf{\mu_\lambda}, \mathbf{V_\lambda}),$$

$$\mathbf{V_\lambda} = \left(\sum_{t=1}^{T} \mathbf{\Theta}_t' \mathbf{\Theta}_t + \mathbf{V}_0^{-1} \right)^{-1},$$

$$\mathbf{\mu_\lambda} = \mathbf{S_\lambda} \left(\sum_{t=1}^{T} \mathbf{\Theta}_t' \mathbf{z}_{it} + \mathbf{V}_0^{-1} \mathbf{\mu}_0 \right).$$

Step 3. Simulation of a latent state vector. For the simulation of ideal points, I transform Equation 19.1 into a multivariate time series model by subtracting the difficulty parameter $\mathbf{\alpha}_t^j$ from latent utilities ($\mathbf{z}_{jt}^* = \mathbf{z}_{jt} - \mathbf{\alpha}_t^j$) and stacking them as an $I_t^j \times 1$ vector. I_t^j indicates the number of items considered by legislator j at time t and varies across legislators. $\mathbf{\alpha}_t^j$ indicates difficulty parameters for all items considered by legislator j at time t. The dimension of $\mathbf{\alpha}_t^j$ also changes by legislators and time. Similarly, let $\mathbf{\beta}_t^j$ denote discrimination parameters for all items considered by legislator j at time t. The new equation can take the form of a linear regression model with $\mathbf{\beta}_t^j$ as a $I_t^j \times 1$ design matrix and θ_{j,s_t} as a 1×1 parameter vector as follows:

$$\mathbf{z}_{jt}^* = \mathbf{\beta}_t^j \theta_{j,s_t} + \mathbf{\varepsilon}_{jt}. \tag{19.3}$$

Sampling a latent state vector for each legislator is done using Chib's (1998) recursive sampling algorithm. The algorithm is identical for all legislators and needs to be repeated J times. Thus, I drop subscript j for notational simplicity. $\mathbf{\beta}, \mathbf{z}^*, \mathbf{\theta}$ and P should be read as $\mathbf{\beta}^j, \mathbf{z}_j^*, \mathbf{\theta}_j$ and P_j in the following. Note that I suppress time subscripts of $\mathbf{\beta}, \mathbf{z}^*, \mathbf{\theta}$ to denote them as matrices containing all observations. The joint sampling of latent states can be decomposed as follows:

$$p(s_1, \ldots, s_T \mid \mathbf{\beta}, \mathbf{z}^*, \mathbf{\theta}, P) = p(s_T \mid \mathbf{\beta}, \mathbf{z}^*, \mathbf{\theta}, P) p(s_{T-1}, s_{T-2}, \ldots, s_1 \mid s_T, \mathbf{\beta}, \mathbf{z}^*, \mathbf{\theta}, P)$$

$$= p(s_T \mid \mathbf{\beta}, \mathbf{z}^*, \mathbf{\theta}, P) \ldots p(s_t \mid S^{t+1}, \mathbf{\beta}, \mathbf{z}^*, \mathbf{\theta}, P) \ldots$$

$$p(s_1 \mid \mathbf{\beta}, \mathbf{z}^*, S^2, \mathbf{\theta}, P), \tag{19.4}$$

where $\mathbf{S}^{t+1}$ indicates the history of the state from $t+1$ to T. Using Bayes' theorem, a typical form of Equation 19.4 can be decomposed as follows:

$$p(s_t \mid S^{t+1}, \mathbf{\beta}, \mathbf{z}^*, \mathbf{\theta}, P) \propto p(s_{t+1} \mid s_t, P) p(s_t \mid \mathbf{\beta}, \mathbf{z}_t^*, \mathbf{\theta}, P).$$

The first part of the right-hand side is a transition probability from t to $t+1$, which is obtained from a transition matrix (P). The second part of the right-hand side should

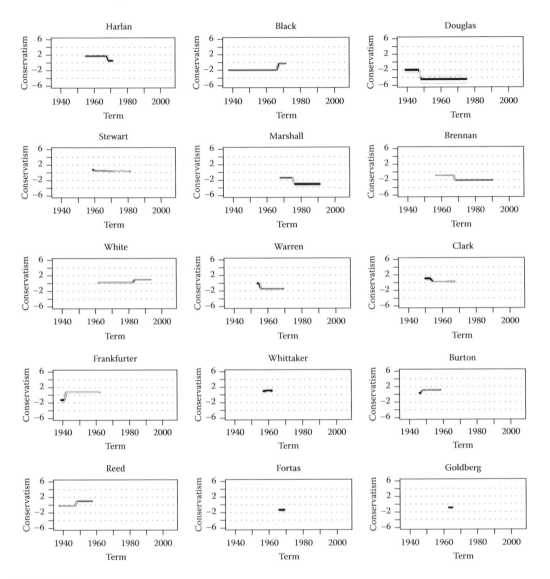

FIGURE 19.1
Estimated ideal points with one break. The chain was run for 20,000 draws after throwing out the first 10,000.
Every 10th draw was stored for the analysis. Thick lines indicate posterior means and light lines are 95% Bayesian
credible intervals.

be obtained via recursive calculation. Let $\mathbf{Z}_t^*$ denote all $\mathbf{z}^*$ up to t. Then

$$p(s_t \mid \mathbf{Z}_{t-1}^*, \boldsymbol{\beta}, \boldsymbol{\theta}, P) = \int_{\mathbf{S}} p(s_t \mid s_{t-1}) p(s_{t-1} \mid \mathbf{Z}_{t-1}^*, \boldsymbol{\beta}, \boldsymbol{\theta}, P) ds_t$$

$$= \sum_{m=1}^{M} p(s_t \mid s_{t-1} = m) p(s_{t-1} = m \mid \mathbf{Z}_{t-1}^*, \boldsymbol{\beta}, \boldsymbol{\theta}, P),$$

$$p(s_t \mid \mathbf{Z}_t^*, \boldsymbol{\beta}, \boldsymbol{\theta}, P) = \frac{p(s_t \mid \mathbf{Z}_{t-1}^*, \boldsymbol{\beta}, \boldsymbol{\theta}, P) p(\mathbf{z}_t^* \mid \mathbf{Z}_{t-1}^*, \theta_{s_t})}{\sum_{m=1}^{M} p(s_t = m \mid \mathbf{Z}_{t-1}^*, \boldsymbol{\beta}, \boldsymbol{\theta}, P) p(\mathbf{z}_t^* \mid \mathbf{Z}_{t-1}^*, \theta_{s_t=m})}.$$

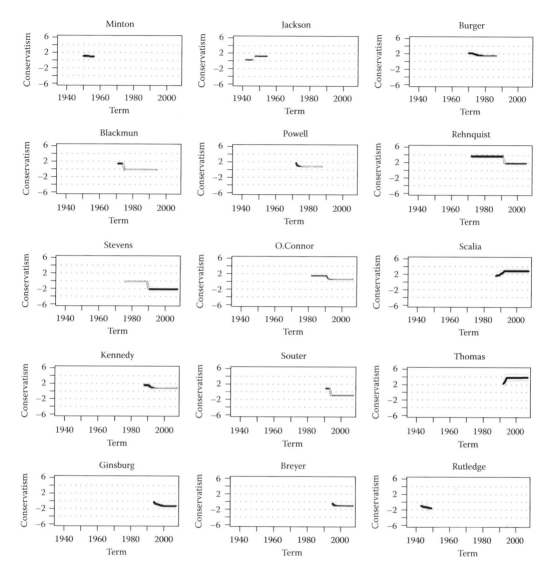

FIGURE 19.1
Continued.

Step 4. Simulation of transition probabilities. Simulating transition probabilities given sampled state variables is a standard beta update from binary outcomes. For each legislator, let n_{ii} be the number of one-step transitions from state i to i, and n_{ij} be the number of one-step transitions from state i to j. Then for the posterior distributions of legislator-specific transition probabilities we have

$$p(p_{ii} \mid \mathbf{s}) \propto p(\mathbf{s} \mid p_{ii})\text{Beta}(a, b)$$

$$\propto p_{ii}^{n_{ii}}(1 - p_{ii})^1 p_{ii}^{a-1}(1 - p_{ii})^{b-1},$$

$$p_{ii} \mid \mathbf{s} \sim \text{Beta}(a + n_{ii}, b + 1).$$

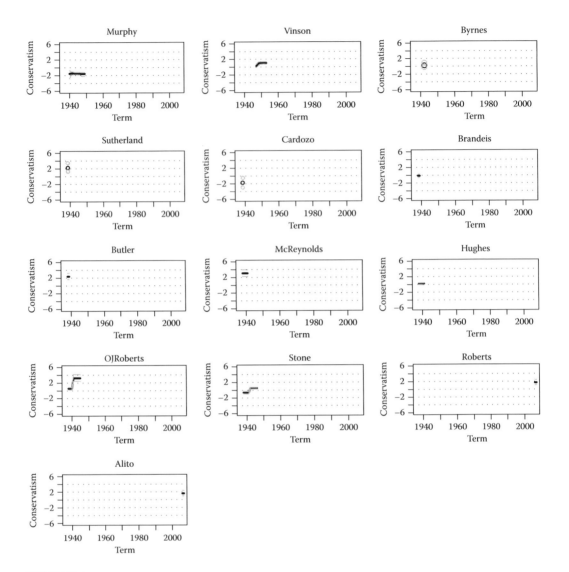

FIGURE 19.1
Continued.

Step 5. Simulation of ideal points. Ideal points are sampled using the transformation
   shown in Equation 19.3. Based on the sampled transition matrices, state variables,
   item characteristic parameters, and latent variables, θ_{j,s_t} is updated across all legisla-
   tors by treating $\boldsymbol{\beta}^j$ as a design matrix and $\mathbf{z}^*_{j,t}$ as response variables. Let $\boldsymbol{\beta}_{j,m}$ and $\mathbf{z}^*_{j,m}$
   denote sampled parameters for legislator j's mth state. Then

$$\theta_{j,m} \mid \mathbf{z}^*_{j,m} \sim N(\mu^{j,m}_\theta, \mathbf{V}^{j,m}_\theta),$$

$$\mathbf{V}^{j,m}_\theta = (\boldsymbol{\beta}'_{j,m}\boldsymbol{\beta}_{j,m} + 1)^{-1},$$

$$\mu^{j,m}_\theta = \mathbf{V}^{j,m}_\theta (\boldsymbol{\beta}'_{j,m}\mathbf{z}^*_{j,m}).$$

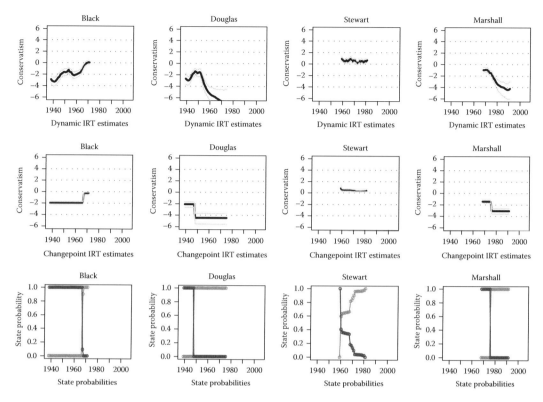

FIGURE 19.2

Comparison of ideal point estimates from the dynamic IRT model by Martin and Quinn (2002) and the hidden Markov IRT model: Black, Douglas, Stewart, and Marshall. The chain was run for 20,000 draws after throwing out the first 10,000. Thick lines on the plots in the top two rows indicate posterior means and light lines are 95% Bayesian credible intervals. The plots in the bottom row show posterior probabilities of being in state 1 (dark lines) and in state 2 (light lines).

19.4 Preference Changes in US Supreme Court Justices

Using the hidden Markov IRT model, I analyze ideal point changes of 43 U.S. Supreme Court justices who served between 1937 and 2006.* The 43 justices considered 4868 cases during the period, and on average each justice considers 113 cases throughout their terms on the bench.

In this analysis, I drop six justices with 2 years of service or less: Sutherland, Cardozo, Brandeis, Butler, Roberts, and Alito. Following Martin and Quinn (2007), I use informative priors for three justices, the liberal Hugo Black, the moderate Potter Stewart, and the conservative William Rehnquist, in order to interpret results in such a way that positive ideal point estimates indicate the conservative position and negative ideal point estimates indicate the liberal position:

$$\theta_{Black} \sim N(-2, 0.1),$$

---

* I thank Martin and Quinn for providing data. For details, see Martin and Quinn (2007).

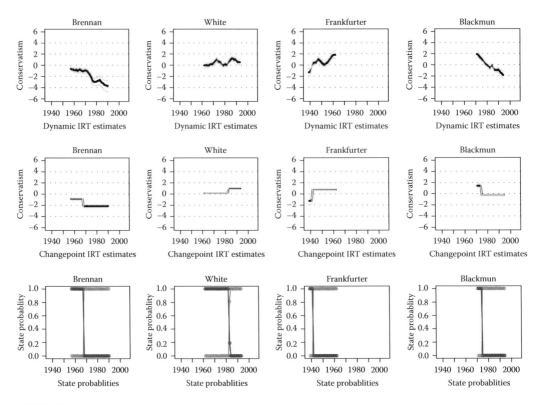

FIGURE 19.3
Comparison of ideal point estimates from the dynamic IRT model by Martin and Quinn (2002) and the hidden Markov IRT model: Brennan, White, Frakfurter, and Blackmun. The chain was run for 20,000 draws after throwing out the first 10,000. Thick lines on the plots in the top two rows indicate posterior means and light lines are 95% Bayesian credible intervals. The plots in the bottom row show posterior probabilities of being in state 1 (dark lines) and in state 2 (light lines).

$$\theta_{\text{Stewart}} \sim N(1, 0.1),$$

$$\theta_{\text{Rehnquist}} \sim N(3, 0.1).$$

Also, these informative priors serve to limit the bounds of ideal point estimates; ideal point estimates near -2 and 3 are highly extreme values in this scale.

Figure 19.1 shows the results of the hidden Markov analysis of the 43 US Supreme Court justices. The fitted hidden Markov IRT model finds a break in ideal points for each justice. By checking the size of the break, we can tell whether a justice's preferences have actually changed.

Sixteen justices exhibit dramatic ideal points changes over their careers in the Court. Harlan, Black, Douglas, Marshall, Brennan, Warren, Frankfurter, Reed, Jackson, Blackmun, Rehnquist, Stevens, Souter, Thomas, O. J. Roberts, and Stone have dramatic shifts in their terms. However, significant preference changes are not found in Stewart, White, Whittacker, Burger, Kennedy, Scalia, Ginsburg, Breyer, Murphy, and O'Connor.

The results of the hidden Markov analysis identify substantively important issues about the timing and grouping of ideal point changes that may be indicative of broader social and political contextual factors. First, ideal points of the justices who served early in the sample period—Frankfurter, McReynolds, O. J. Roberts, and Stone—changed dramatically

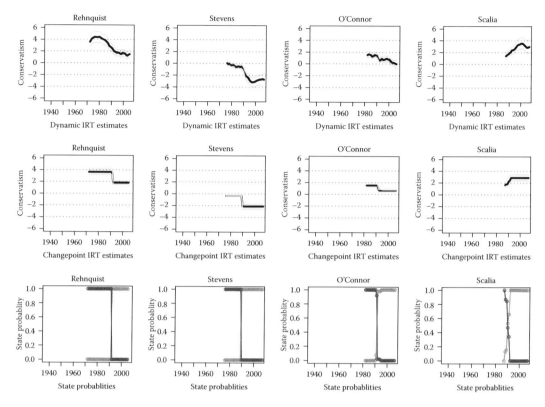

FIGURE 19.4

Comparison of ideal point estimates from the dynamic IRT model by Martin and Quinn (2002) and the hidden Markov IRT model: Rehnquist, Stevens, O'Connor, and Scalia. The chain was run for 20,000 draws after throwing out the first 10,000. Thick lines on the plots in the top two rows indicate posterior means and light lines are 95% Bayesian credible intervals. The plots in the bottom row show posterior probabilities of being in state 1 (dark lines) and in state 2 (light lines).

between the late 1930s and the early 1940s. Given the timing of the breaks, these preference changes seem likely to be related to what is known commonly as "the switch in time that saved nine," when Justice O. J. Roberts shifted his alignment to the liberal bloc of justices on a key 1937 case, a move that is often viewed as a means to protect the Court's independence from President Franklin Roosevelt's attempts to reorganize it through expansion (Epstein and Walker, 2007; Ho and Quinn, 2010).

Another interesting finding is timing of David Souter's preference shift. Souter, who was nominated by George H. W. Bush in 1990, has drawn the ire of conservatives for voting with liberal justices on many important cases including *Planned Parenthood* v. *Casey* and *Bush* v. *Gore*. When George H. W. Bush's son, George W. Bush, sought to fill two openings during his presidential term, conservatives fretted over whether his conservative picks would exhibit a similar leftward drift over their careers. The hidden Markov IRT model detects Souter's movement to the left in the early 1990s, very shortly after his confirmation. This movement was not so much a slow evolution, but a quick about-face followed by a long, consistent liberal preference.

Figures 19.2 through 19.4 compare the estimates from the hidden Markov IRT model with the estimates from Martin and Quinn (2002)'s dynamic IRT model. To save space, I select 12 justices with more than 20 years of service in the Court. In each figure, the top row replicates Martin and Quinn (2002)'s dynamic ideal point estimates, the middle row shows the hidden Markov ideal point estimates, and the bottom row shows the posterior regime probabilities from the hidden Markov IRT model.

The difference is clear. While Martin and Quinn's (2002) dynamic IRT model tracks trends in ideal points over time, the hidden Markov IRT model provides a sharp estimate of preference changes. For example, in the case of Douglas in Figure 19.2, it is hard to pinpoint the timing of Douglas's change from the Martin and Quinn (2002) estimate. By contrast, the hidden Markov IRT model clearly shows the timing of the shift. However, when justices' ideal points change slowly as in the case of Blackmun, the timing of the break uncovered by the hidden Markov IRT model is not as informative as the estimate from Martin and Quinn's (2002) dynamic IRT model.

19.5   Conclusions

In this chapter, I present a statistical model for dynamic ideal point estimation. The model combines the hidden Markov model with the standard two-parameter IRT model. The application of the the model to the US Supreme Court data demonstrates that the hidden Markov IRT model is an effective method to detect preference changes from longitudinal voting data.

Acknowledgments

The author would like to thank Andrew D. Martin and Kevin M. Quinn for providing the data and C++ codes for their dynamic IRT model; as well as John Balz, Andrew D. Martin, and Michael Peress for providing comments on the manuscript.

References

Albert, J. H. 1992. Bayesian estimation of normal ogive item response curves using Gibbs sampling. *Journal of Educational Statistics*, 17:251–269.

Albert, J. H. and Chib, S. 1993. Bayesian analysis of binary and polychotomous response data. *Journal of the American Statistical Association*, 88(422):669–679.

Bafumi, J., Gelman, A., Park, D. K., and Kaplan, N. 2005. Practical issues in implementing and understanding Bayesian ideal point estimation. *Political Analysis*, 13(2):171–187.

Bailey, M. and Chang, K. 2001. Comparing presidents, senators, and justices: Interinstitutional preference estimation. *Journal of Law, Economics, and Organization*, 17(2):477–506.

Chib, S. 1998. Estimation and comparison of multiple change-point models. *Journal of Econometrics*, 86(2):221–241.

Clinton, J., Jackman, S., and Rivers, D. 2000. The statistical analysis of legislative behavior: A unified approach. Paper presented at the Annual Meeting of the Political Methodology Society.

Epstein, L. and Knight, J. 1998. *The Choices Justices Make*. CQ Press, Washington, D.C.

Epstein, L. and Walker, T. G. 2007. *Constitutional Law for a Changing America: Institutional Powers and Constraints*. CQ Press, Washington, DC.

Epstein, L., Hoekstra, V., Segal, J. A., and Spaeth, H. J. 1998. Do political preferences change? A longitudinal study of U.S. Supreme Court justices. *Journal of Politics*, 60(3):801–818.

Heckman, J. J. and Snyder, J. M. 1997. Linear probability models of the demand for attributes with an empirical application to estimating the preferences of legislators. *Rand Journal of Economics*, 28:142–189.

Ho, D. E. and Quinn, K. M. 2010. Did a switch in time save nine? *Journal of Legal Analysis*, 2(1):69–113.

Jackman, S. 2001. Multidimensional analysis of roll call data via Bayesian simulation: Identification, estimation, inference and model checking. *Political Analysis*, 9:227–241.

Johnson, V. E. and Albert, J. H. 1999. *Ordinal Data Modeling*. Springer, New York.

Londregan, J. 2000. Estimating legislators' preferred points. *Political Analysis*, 8(1):35–56.

Martin, A. D. and Quinn, K. M. 2002. Dynamic ideal point estimation via Markov chain Monte Carlo for the U.S. Supreme Court, 1953–1999. *Political Analysis*, 10(2):134–153.

Martin, A. D. and Quinn, K. M. 2007. Assessing preference change on the US Supreme Court. *Journal of Law, Economics, and Organization*, 23(2):365–385.

McCarty, N. M., Poole, K. T., and Rosenthal, H. 1997. *Income Redistribution and the Realignment of American Politics*. AEI Press, Washington, DC.

Poole, K. T. and Rosenthal, H. 1997. *Congress: A Political-Economic History of Roll-Call Voting*. Oxford University Press, Oxford.

Segal, J. A. and Spaeth, H. J. 1993. *The Supreme Court and the Attitudinal Model*. Cambridge University Press, Cambridge.

20

Parallel Bayesian MCMC Imputation for Multiple Distributed Lag Models: A Case Study in Environmental Epidemiology

Brian Caffo, Roger Peng, Francesca Dominici, Thomas A. Louis, and Scott Zeger

20.1 Introduction

Patterned missing covariate data is a challenging issue in environmental epidemiology. For example, particulate matter measures of air pollution are often collected only every third day or every sixth day, while morbidity and mortality outcomes are collected daily. In this setting, many desirable models cannot be directly fit. We investigate such a setting in so-called "distributed lag" models when the lagged predictor is collected on a cruder time scale than the response. In multi-site studies with complete predictor data at some sites, multilevel models can be used to inform imputation for the sites with missing data.

We focus on the implementation of such multilevel models, in terms of both model development and computational implementation of the sampler. Specifically, we parallelize single chain runs of sampler. This is of note, since the Markovian structure of Markov chain Monte Carlo (MCMC) samplers typically makes effective parallelization of single chains difficult. However, the conditional independence relationships of our developed model allow us to exploit parallel computing to run the chain. As a first attempt at using parallel MCMC for Bayesian imputation on such data, this chapter largely represents a proof of principle, though we demonstrate some promising potential for the methodology. Specifically, the methodology results in proportional decreases in run-time over the nonparallelized version near one over the number of available nodes.

In addition, we describe a novel software implementation of parallelization that is uniquely suited to disk-based shared memory systems. We use a "blackboard" parallel computing scheme where shared network storage is a used as a blackboard to tally currently completed and queued tasks. This strategy allows for easy addition and subtraction of compute nodes and control of load balancing. Moreover, it builds in automatic checkpointing.

Our investigation is motivated by multi-site time series studies of the short-term effects of air pollution on disease or death rates. A common measure of air pollution used for such studies is the amount in micrograms per cubic meter of particulate matter of a specified maximum aerodynamic diameter. We focus on $PM_{2.5}$ (see Samet et al., 2000). Unfortunately, the definitive source of particulate matter data in the United States, the Environmental Protection Agency's air pollution network of monitoring stations, collects data only a few times per week at some locations. One of the most frequent observed data patterns for

$PM_{2.5}$ is data being recorded every third day. However, the disease rates that we consider are collected daily.

In this setting, directly fitting a model that includes several lags of $PM_{2.5}$ simultaneously is not possible. Such models are useful, for example, to investigate a cumulative weekly effect of air pollution on health. They are also useful to more finely investigate the dynamics of the relationship between the exposure and response. As an example, one might postulate that after an increase in air pollution, high air pollution levels on later days may have a smaller impact, as the risk set has been depleted from the initial increase (Dominici et al., 2002; Schwartz, 2000; Zeger et al., 1999).

We focus on distributed lag models that relate the current-day disease rate to particulate matter levels over the past week. That is, our model includes the current day's $PM_{2.5}$ levels as well as the previous six days. While direct estimation of the effect for any particular lag is possible, joint estimation of the distributed lag model is not possible (see Section 20.3). Moreover, missing-data imputation for counties with patterned missing data is difficult. We consider a situation where several independent time series are observed at different geographical regions, some with complete $PM_{2.5}$ data. We use multilevel models to borrow information across series to fill in the missing data via Bayesian imputation. The hierarchical model is also used to combine county-specific distributed lag effects into national estimates.

The rest of the chapter is organized as follows. In Section 20.2 we outline the data set used for analysis and follow in Section 20.3 with a discussion of Bayesian imputation. In Section 20.4 we describe the distributed lag models of interest, and in Section 20.5 we illustrate a multiple imputation strategy. Section 20.6 uses the imputation algorithm to analyze hospitalization rates of chronic obstructive pulmnonary disease (COPD). Finally, Section 20.7 gives some conclusions, discussion and proposals for future work.

20.2 The Data Set

The Johns Hopkins Environmental Biostatistics and Epidemiology Group has assembled a national database comprising time series data on daily hospital admission rates for respiratory outcomes, fine particles ($PM_{2.5}$), and weather variables for the 206 largest US counties having a population larger than 200,000 and with at least one full year of $PM_{2.5}$ data available. The study population, derived from Medicare claims, includes 21 million adults older than 65 with a place of residence in one of the 206 counties included in the study.

Daily counts of hospital admissions and daily number of people enrolled in the cohort are constructed from the Medicare National Claims History Files. These counts are obtained from billing claims of Medicare enrollees residing in the 206 counties. Each billing claim contains the following information: date of service, treatment, disease (ICD 9 codes), age, gender, race and place of residence (zip and county).

Air pollution data for fine particles are collected and posted by the United States Environmental Protection Agency Aerometric Information Retrieval Service (AIRS, now called the Air Quality System, AQS). To protect against outlying observations, a 10% trimmed mean is used to average across monitors after correction for yearly averages for each monitor. Specifically, after removing a smoothly varying annual trend from each monitor time series, the trimmed mean was computed using the deviations from this smooth trend. Weather data is

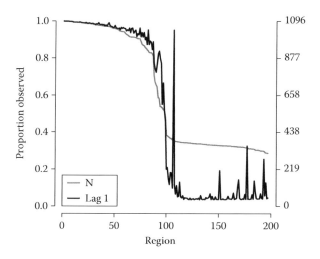

FIGURE 20.1

Summary of the missing-data pattern. The gray line displays the proportion of the total days in the study with observed $PM_{2.5}$ data for each county, with the actual number of days displayed on the right axis. The black line shows the proportion and count of the days with observed air pollution data where the lag-1 day is also observed.

obtained from the National Weather Monitoring Network which comprises daily temperature and daily dew points temperature for approximately 8000 monitoring stations in the USA. We aggregate data across monitors to obtain temperature time series data for each of the 206 counties, of which 196 were used in analysis. Details about aggregation algorithms for the air pollution and weather are posted at http://www.biostat.jhsph.edu/MCAPS and further information about data collection is given in Dominici et al. (2006).

Figure 20.1 illustrates the salient features of the missing-data pattern for $PM_{2.5}$ in this database. This study considered 1096 monitoring days. Figure 20.1 displays the proportion of the 1096 days with observed $PM_{2.5}$ data for each county (dark gray line). The associated number of observed days is displayed on the right scale. This figure also displays the proportion of 1096 days with observed $PM_{2.5}$ data where the lag-1 day was also observed (black line).

The plots show that nearly half of the 196 counties have measurements on roughly one third of the total possible days. Ninety-six of these counties have over 40% of the air pollution data observed and enough instances of seven consecutive observed $PM_{2.5}$ days to estimate the desired distributed lag model (see Section 20.4). For these counties, any missing data is often due to a large contiguous block, for example, several weeks where the monitor malfunctioned. Such uninformative missing data leaves ample daily measurements to estimate distributed lag models, so is ignored in our model. The remaining counties have $PM_{2.5}$ data collected every third day and possibly also have blocks of missing data, hence have data on less than 33% of the days under study. Because of the systematically missing $PM_{2.5}$ data in these counties, there is little hope of fitting a distributed lag model without borrowing information on the exposure-response curve from daily time series data from other counties. The plot further highlights this by showing that direct estimates of the lag-1 autocovariances are not available for roughly half of the counties. However, because of the missing-data pattern, all of the counties have direct estimates of the lag-3 autocovariances.

20.3 Bayesian Imputation

In this section, we discuss the relative merits of Bayesian imputation. We focus on our particular missing-data problem, and refer the reader to Carlin and Louis (2009) and Little and Rubin (2002) for general introductions to Bayesian statistics, missing data, and computation. We argue that imputation for systematic missingness in the predictor time series is relevant for distributed lag models, and particularly for the data set in question, while it is less relevant for single-lag models. In this section, we restrict our discussion to the consideration of a single outcome time series, say Y_t, and single predictor time series, X_t. For context, consider the outcome to be the natural log of the county-specific Medicare emergency admissions rate for COPD, and the predictor to be $PM_{2.5}$ levels for that county. To make this thought experiment more realistic, let Y_t and X_t be the residual time series obtained after having regressed out relevant confounding variables. We assume that the $\{Y_t\}$ are completely observed and the $\{X_t\}$ are observed only every third day, so that $X_0, X_3, X_6, \ldots$ are recorded; and we evaluate whether or not to impute the missing predictors. In our subsequent analysis of the data, we will treat this problem more formally using Poisson regression.

20.3.1 Single-Lag Models

A single-lag model relates the Y_t to X_{t-u} for some $u = 0, 1, 2, \ldots$ via the mean model $E[Y_t] = \theta_u X_{t-u}$, when an identity link function is used. We argue that, for any such single-lag model, implementing imputation strategies for the missing predictor values is unnecessary. Consider that direct evidence regarding any single-lag model is available in the form of simple lagged cross-correlations. For example, the pairs $(Y_0, X_0), (Y_3, X_3), (Y_6, X_6), \ldots$ provide direct evidence for $u = 0$; the pairs $(Y_1, X_0), (Y_4, X_3), (Y_7, X_6), \ldots$ provide direct evidence for $u = 1$ and so on. Imputing the missing predictors only serves to inject unneeded assumptions. Furthermore, there is a tradeoff where more variation in the predictor series benefits the model's ability to estimate the associated parameter, yet hampers the ability to impute informatively. Hence, in the typical cases where the natural variation in the predictor series is large enough to be of interest, we suggest that imputing systematically missing predictor data for single-lag models is not worth the trouble. In less desirable situations with low variation in the predictor series, imputation for single-lag models may be of use.

20.3.2 Distributed Lag Models

Now consider a distributed lag model, such as $E[Y_t] = \sum_{u=0}^{d} \theta_u X_{t-u}$. Here, if a county has predictor data recorded every third day, there is no direct information to estimate this relationship. Specifically, let $t = 0, \ldots, T-1$ and $\mathbf{D}$ be the design matrix associated with the distributed lag model and Y be the vector of responses. Then the least squares estimates of the coefficients are $(\frac{1}{T}\mathbf{D}^t\mathbf{D})^{-1}\frac{1}{T}\mathbf{D}^tY$. The off-diagonal terms of $\frac{1}{T}\mathbf{D}^t\mathbf{D}$ contain the lagged autocovariances in the $\{X_t\}$ series; $\frac{1}{T}\mathbf{D}^tY$ contains the lagged cross-covariances between the $\{Y_t\}$ and $\{X_t\}$. As was previously noted, these lagged cross-covariances are directly estimable, even with patterned missing data in the predictor series. In contrast, the autocovariances in the predictor series are only directly estimable for lags that are multiples of 3. Thus, without addressing the missing predictor data, the distributed lag model cannot be fit. For our data, this would eliminate information from nearly 50% of the counties studied. Hence, a study of the utility of predicting the missing data is warranted.

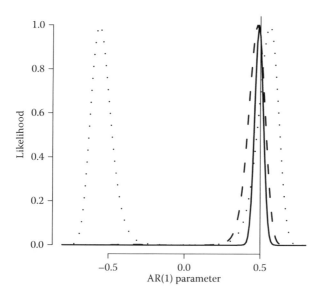

FIGURE 20.2

Likelihood for the AR(1) coefficient for AR(1) simulated data with an assumed correctly known innovation variance of 1 and coefficient of 0.5, for three missing-data patterns: completely observed (solid), observed only every three days (dashed) and observed only every six days (dotted). A solid vertical line denotes the actual coefficient value of 0.2.

One might consider using a model, such as an AR(p), to extrapolate the missing autocorrelations. However, a single time series with this degree of systematic missingness may not have enough information to estimate the parameters. Consider an AR(1) process. To illustrate, Figure 20.2 shows the likelihood for the AR(1) coefficient for data simulated under an AR(1) model with a correctly known innovation variance of 1 and data observed every day (solid), every third day (dashed) and every sixth day (dotted). Any inference for the AR(1) parameter (at 0.5, depicted with a horizontal line) would be imprecise with the systematically missing data. For the data observed every sixth day, notice that the likelihood is multimodal and symmetric about zero. This is because the likelihood only depends on the AR coefficient raised to even powers. This poses a problem even for our every-third-day data, because additional missing observations create patterns of data collected only every sixth day. Such multimodal likelihoods for AR models are described in Wallin and Isaksson (2002) (also see Broersen et al., 2004). In addition, here we assume that the correct model is known exactly, which is unlikely to be true in practice.

In our data set, there is important information in the counties with completely observed data that can be used to help choose models for the predictor time series and estimate parameters. Figure 20.3 demonstrates such model fits with an AR(4) model applied to a detrended version of the log of the $PM_{2.5}$ process. This plot is informative because when the data are available only once every third day, it is not possible to estimate the autocorrelation function using data only from that county. However, it also illustrates that the population distribution of autoregressive parameters appears to be well defined by the counties with mostly observed data. Figure 20.4 shows the estimated residual standard deviations from these model fits, suggesting that these are well estimated even if the autoregressive parameters are not. The data suggest that an AR(1) process is perhaps sufficient, though we continue to focus on AR(4) models to highlight salient points regarding imputation.

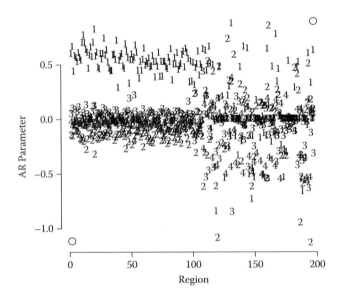

FIGURE 20.3
Estimated AR(4) coefficients, labeled "1" to "4," by counties ordered by decreasing percentage of observed data
from left to right. Roughly the first 100 counties have substantial consecutively observed data to estimate the AR
parameters while the remaining 100 do not.

20.4 Model and Notation

In this section, we present notation and modeling assumptions. A summary of the most
important parameters and hyperparameters is given in Table 20.1. Let Y_{ct}, for county
$c = 0, \ldots, C - 1$ and day $t = 0, \ldots, T - 1$, denote a response time series of counts, such

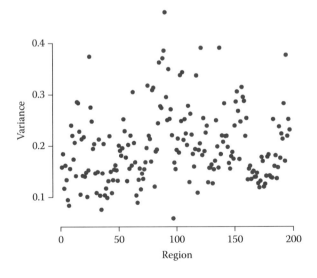

FIGURE 20.4
Variances from individual autoregressive time series by counties ordered by decreasing percentage of observed
data from left to right.

TABLE 20.1

Parameters and their Definitions

Y_{ct}	Response count for county c at time t
R_{ct}	Size of the risk set for county c at time t
λ_{ct}	County-specific expected rate for Y_{ct}
X_{ct}	Log $PM_{2.5}$ level in micrograms per cubic meter for county c at time t
θ_{cu}	Distributed lag parameters for county c
θ_{cu}^*	Constrained distributed lag parameters for county c
$\theta_{cu}^\dagger$	Constrained and reparameterized distributed lag parameters for county c
$\tilde{\ell}(\boldsymbol{\theta}_c)$	Profile log likelihood for $\boldsymbol{\theta}_c$
$\mathbf{W}_{ct}, \boldsymbol{\psi}_c$	Slowly varying trend model on the log $PM_{2.5}$ series $\log(X_{ct}) = \mathbf{W}_{ct}\boldsymbol{\psi}_c + \epsilon_{ct}$
ϵ_{ct}	Residuals from above model, where we presume $\boldsymbol{\psi}_c$ is known and fixed at the estimated value; these terms are the actually imputed terms
$\boldsymbol{\mu}$	Inter-county mean of the $\boldsymbol{\theta}_c$
$\boldsymbol{\Sigma}_\theta$	Inter-county variance–covariance matrix of the $\boldsymbol{\theta}_c$
$\boldsymbol{\alpha}_c$	County-specific autoregressive parameters on the ϵ_{ct}
$\boldsymbol{\zeta}$	Inter-county mean of the $\boldsymbol{\alpha}_c$
$\boldsymbol{\Sigma}_\alpha$	Inter-county variance–covariance matrix of the $\boldsymbol{\alpha}_c$

as the daily incident counts of COPD. Let R_{ct} denote the number of persons in county c at risk for disease on day t. We assume that both of these processes are completely observed. Let $\lambda_{ct} = E[Y_{ct} \mid R_{ct}, X_{ct}, \boldsymbol{\theta}_c, \boldsymbol{\beta}_c]$ be the daily mean, X_{ct} denote $PM_{2.5}$, and Z_{ctj}, for $j = 0, \ldots, J-1$, be other covariates of interest, which we assume are completely observed. We assume that the response process follows a Poisson law with means satisfying:

$$\log(\lambda_{ct}/R_{ct}) = \sum_{u=0}^{d} \theta_{cu} X_{c,t-u} + \sum_{j=0}^{J-1} Z_{ctj}\beta_{cj}.$$

For the hospital admissions data, we consider $d = 6$. That is, the model relates the prior week's air pollution to the current day's mean disease rate.

The sum of the lagged air pollution parameters, labeled the distributed lag "total" or "cumulative" effect, $\sum_{u=0}^{d} \theta_{cu}$, is a parameter of primary interest. The total effect is the change in the log mean response rate given a one-unit across-the-board increase in the $PM_{2.5}$ over the current and prior d days.

To mitigate the variance inflation incurred by including many obviously collinear lagged covariates, sometimes a functional form is placed on the θ_{cu}, especially if d is large (see Zanobetti et al., 2000). A particularly effective approach is to assume that $\theta_{cu} = \mathbf{A}_u \boldsymbol{\theta}_c^*$, where $\mathbf{A}_u$ is column u from a smoothing design matrix, say $\mathbf{A}$, on the time points $0, \ldots, d$. When $d = 6$, choosing $\mathbf{A}$ to be a bin smoother with bins for the current day, days lag 1 to 2, and days lag 3 to 6, has been shown to be a useful approach in the air pollution time series literature (Bell et al., 2004). Such a smoother requires three parameters, $\boldsymbol{\theta}_c^* = (\theta_{c1}^*, \theta_{c2}^*, \theta_{c3}^*)$, so that the model becomes

$$\log(\lambda_{ctr}/R_{ctr}) = \theta_{c1}^* x_{ct} + \theta_{c2}^* \sum_{u=1}^{2} X_{c,t-u} + \theta_{c3}^* \sum_{u=3}^{6} X_{c,t-u} + \sum_{j=0}^{k} z_{ctj}\beta_{cjr}.$$

This restriction on the distibuted lag parameters is equivalent to a convenient form for the rate model that considers the seven-day average air pollution and its deviation from the three-day average and current day:

$$\log(\lambda_{ctr}/R_{ctr}) = \theta_{c1}^{\dagger}\bar{x}_{ct}^{(6)} + \theta_{c2}^{\dagger}(\bar{x}_{ct}^{(2)} - \bar{x}_{ct}^{(6)}) + \theta_{c3}^{\dagger}(x_{ct} - \bar{x}_{ct}^{(2)}) + \sum_{j=0}^{k} z_{ctj}\beta_{cjr}. \qquad (20.1)$$

Here $\bar{x}_{ct}^{(k)}$ is the average of the current-day and k previous days' $PM_{2.5}$ values. These parameters are related to the θ_i^* via the equalities

$$\theta_{c1}^* = \frac{1}{7}\theta_{c1}^{\dagger} + \frac{4}{21}\theta_{c2}^{\dagger} + \frac{6}{7}\theta_{c3}^{\dagger},$$

$$\theta_{c2}^* = \frac{1}{7}\theta_{c1}^{\dagger} + \frac{4}{21}\theta_{c2}^{\dagger} - \frac{1}{7}\theta_{c3}^{\dagger},$$

$$\theta_{c3}^* = \frac{1}{7}\theta_{c1}^{\dagger} - \frac{1}{7}\theta_{c2}^{\dagger} - \frac{1}{7}\theta_{c3}^{\dagger}.$$

In this constrained model the total effect is $\theta_{c1}^{\dagger} = \theta_{c1}^* + 2\theta_{c2}^* + 4\theta_{c3}^*$. We use the constrained and reparameterized specification from Equation 20.1 for analysis. For convenience, we have dropped the superscript $*$ or $\dagger$ from θ when generically discussing the likelihood or MCMC sampler.

We denote the Poisson log likelihood for county c by $\ell_c(\boldsymbol{\theta}_c, \boldsymbol{\beta}_c)$, where bold face represents a vector of the relevant parameters, such as $\boldsymbol{\theta}_c = (\theta_{c1}, \ldots, \theta_{cd})^t$. Our approach uses Bayesian methodology to explore the joint likelihood by smoothing parameters across counties. However, the number of nuisance parameters makes implementation and prior specification unwieldy. Therefore, we replace the county-specific log likelihoods with the associated profile log likelihoods:

$$\tilde{\ell}_c(\boldsymbol{\theta}_c) = \ell_c\{\boldsymbol{\theta}_c, \hat{\boldsymbol{\beta}}_c(\boldsymbol{\theta}_c)\}, \quad \text{where } \hat{\boldsymbol{\beta}}_c(\boldsymbol{\theta}_c) = \operatorname{argmax}_{\boldsymbol{\beta}_c} \ell_c(\boldsymbol{\theta}_c, \boldsymbol{\beta}_c).$$

This step greatly reduces the complexity of the MCMC fitting algorithm. However, it does so at the cost of theoretical unity, as the profile likelihood used for Bayesian inference is not a proper likelihood (Monahan and Boos, 1992), as well as computing time. We stipulate that this choice may impact the validity of the sampler and inference. Currently, we assess validity by comparing results with maximum likelihood results for counties with complete data.

The model for the air pollution time series contains trend variables and AR(p) distributed errors. We assume that

$$\log(X_{ct}) = \mathbf{W}_{ct}\boldsymbol{\psi}_c + \epsilon_{ct}, \qquad (20.2)$$

where the ϵ_{ct} are a stationary autoregressive process of order p with conditional means and variances

$$E[\epsilon_{ct}|\epsilon_{c,t-1}, \ldots, \epsilon_{c,t-p}] = \sum_{j=1}^{p} \alpha_{cj}\epsilon_{c,t-j}, \quad \operatorname{var}(\epsilon_{ct}|\epsilon_{c,t-1}, \ldots, \epsilon_{c,t-p}) = \sigma_c^2.$$

Here the trend term, $\mathbf{W}_{ct}\boldsymbol{\psi}_c$, represents the slowly varying correlation between air pollution and seasonality. Specifically, we set $\mathbf{W}_{ct}$ to be a natural cubic spline with 24 degrees of freedom per year. Throughout, we set $p = 4$.

20.4.1 Prior and Hierarchical Model Specification

We place a $N(\boldsymbol{\mu}, \boldsymbol{\Sigma}_\theta)$ prior on the distributed lag parameters, and a diffuse normal prior for $\boldsymbol{\mu}$ and an inverted Wishart prior with an identity matrix scale on $\boldsymbol{\Sigma}_\theta$ with 4 degrees of freedom. Here, $\boldsymbol{\mu}$ is a parameter of central interest, estimating the between-county mean distributed lag parameters.

We do not place a prior on the missing-data trend term $\boldsymbol{\psi}_c$, instead fixing it from the onset at the least squares estimated value. For the autoregressive parameters, α_{cj}, we place the prior on the lagged partial autocorrelations (Barnett et al., 1996; Monahan, 1983). We refer the reader to Diggle (1990) for a definition of partial autocorrelations and Huerta and West (1999) for a different perspective for placing priors on autoregressive parameters.

We use a recursive formula of Durbin (1960), to transform the autoregressive parameters to and from the partial autocorrelations. Let $\tilde{\alpha}_{cj}$ represent the p partial autocorrelations for county c; we specify that

$$0.5 \log\{(1 + \tilde{\boldsymbol{\alpha}}_c)/(1 - \tilde{\boldsymbol{\alpha}}_c)\} \sim N(\boldsymbol{\zeta}, \boldsymbol{\Sigma}_\alpha),$$

where the Fisher's Z transformation, $\log\{(1 + a)/(1 - a)\}$, is assumed to operate componentwise on vectors. Here, taking Fisher's Z transformation is useful as the partial correlations are bounded by 1 in absolute value for stationary series.

We use a diffuse normal prior for $\boldsymbol{\zeta}$ and an inverse Wishart distribution centered at an identity matrix with 10 degrees of freedom. The prior on σ_c^{-2} is gamma with a mean set at the county-specific method of moments estimates and a coefficient of variation of 10. Note that we chose not to shrink variance estimates across counties, as they appear to be well estimated from the data.

20.5 Bayesian Imputation

20.5.1 Sampler

Here we give an overview of the Bayesian imputation algorithm. Let brackets generically denote a density, and let $\mathbf{X}_{c,\text{obs}}$ and $\mathbf{X}_{c,\text{miss}}$ be the collection of X_{tc} observed and missing components for county c respectively, $\mathbf{Y}_c$ be the collection of Y_{tc}, $\mathbf{P}_c = \{\psi_c, \alpha_{1c}, \ldots, \alpha_{pc}, \sigma_c\}$, $\mathbf{P}$ be the collection of between-county parameters and $\mathbf{H}$ denote hyperparameters. Then, the full join posterior is

$$[\mathbf{X}_{0,\text{miss}}, \ldots, \mathbf{X}_{C-1,\text{miss}}, \boldsymbol{\theta}_0, \ldots, \boldsymbol{\theta}_{C-1}, \mathbf{P}_0, \ldots, \mathbf{P}_{C-1}, \mathbf{P} \mid \mathbf{Y}_0, \ldots, \mathbf{Y}_{C-1}, \mathbf{X}_{0,\text{obs}}, \mathbf{X}_{C-1,\text{obs}}, \mathbf{H}]$$

$$\propto \left\{ \prod_c [\mathbf{Y}_c \mid \mathbf{X}_{c,\text{miss}}, \mathbf{X}_{c,\text{obs}}, \boldsymbol{\theta}_c][\mathbf{X}_{c,\text{miss}}, \mathbf{X}_{c,\text{obs}} \mid \mathbf{P}_c][\mathbf{P}_c \mid \mathbf{P}, \mathbf{H}] \right\} [\mathbf{P} \mid \mathbf{H}].$$

Here, recall that $[\mathbf{Y}_c \mid \mathbf{X}_{c,\text{miss}}, \mathbf{X}_{c,\text{obs}}, \boldsymbol{\theta}_c]$ uses the profile likelihood, rather than the actual likelihood. Our sampler proceeds as follows (where EE is "everything else"):

$$[\mathbf{X}_{0,\text{miss}} \mid EE] \propto [\mathbf{Y}_0 \mid \mathbf{X}_{0,\text{miss}}, \mathbf{X}_{0,\text{obs}}, \boldsymbol{\theta}_0][\mathbf{X}_{0,\text{miss}}, \mathbf{X}_{0,\text{obs}} \mid \mathbf{P}_0],$$

$$[\mathbf{X}_{1,\text{miss}} \mid EE] \propto [\mathbf{Y}_1 \mid \mathbf{X}_{0,\text{miss}}, \mathbf{X}_{0,\text{obs}}, \boldsymbol{\theta}_1][\mathbf{X}_{0,\text{miss}}, \mathbf{X}_{0,\text{obs}} \mid \mathbf{P}_1],$$

$$\vdots$$

$$[\mathbf{X}_{C-1,\text{miss}} \mid EE] \propto [\mathbf{Y}_C \mid \mathbf{X}_{C-1,\text{miss}}, \mathbf{X}_{C-1,\text{obs}}, \boldsymbol{\theta}_{C-1}][\mathbf{X}_{C-1,\text{miss}}, \mathbf{X}_{C-1,\text{obs}} \mid \mathbf{P}_{C-1}],$$

$$[\mathbf{P}_0 \mid EE] \propto [\mathbf{Y}_0 \mid \mathbf{X}_{c,\text{miss}}, \mathbf{X}_{c,\text{obs}}, \boldsymbol{\theta}_0][\mathbf{X}_{0,\text{miss}}, \mathbf{X}_{0,\text{obs}} \mid \mathbf{P}_0],$$

$$[\mathbf{P}_1 \mid EE] \propto [\mathbf{Y}_1 \mid \mathbf{X}_{c,\text{miss}}, \mathbf{X}_{c,\text{obs}}, \boldsymbol{\theta}_1][\mathbf{X}_{1,\text{miss}}, \mathbf{X}_{1,\text{obs}} \mid \mathbf{P}_1],$$

$$\vdots$$

$$[\mathbf{P}_{C-1} \mid EE] \propto [\mathbf{Y}_{C-1} \mid \mathbf{X}_{C-1,\text{miss}}, \mathbf{X}_{C-1,\text{obs}}, \boldsymbol{\theta}_C][\mathbf{X}_{C-1,\text{miss}}, \mathbf{X}_{C-1,\text{obs}} \mid \mathbf{P}_{C-1}],$$

$$[\mathbf{P} \mid EE] \propto [\mathbf{P}_c \mid \mathbf{P}, \mathbf{H}][\mathbf{P} \mid \mathbf{H}].$$

Because of the Gibbs-friendly priors, $\boldsymbol{\mu}$ and $\boldsymbol{\zeta}$ have multivariate normal full conditionals. Moreover, $\boldsymbol{\Sigma}_\theta$ and $\boldsymbol{\Sigma}_\alpha$ have inverse Wishart full conditionals, while the $\{\sigma_c^2\}$ have an inverted gamma. The county-specific distributed lag parameters and AR parameters, $\{\boldsymbol{\theta}_c\}$ and $\boldsymbol{\alpha}_c$, require a Metropolis step. We use a variable-at-a-time, random-walk update. Further details on the full conditionals are given in the Appendix to this chapter.

The update of the missing data deserves special attention. We use a variable-at-a-time Metropolis step to impute ϵ_{tc} for each missing day conditional on the remaining. Consider $p = 4$ and let ϵ_{c5} be a missing day to be imputed. We use the autoregressive prior for the day under consideration given all of the remaining days as the proposal. For example, the distribution of ϵ_{c5} given $\{\epsilon_{c1}, \ldots, \epsilon_{c4}, \epsilon_6, \ldots, \epsilon_9\}$ is used to generate the proposal for ϵ_{c5}, that is, the four neighboring days before and after the day under consideration. Because of the AR(4) assumption, this is equivalent to the distribution of ϵ_5 given all of the days. After imputation, $X_{c5} = \exp(\mathbf{W}_{c5}\boldsymbol{\psi}_c + \epsilon_{c5})$, is calculated. By simulating from the prior distribution of the current missing day given the remainder, only the contribution of X_{c5} to the profile likelihood remains in the Metropolis ratio. To summarize, the distribution of the current day given the remainder, disregarding the profile likelihood, is used to generate proposals; the profile likelihood is then used in a Metropolis correction. Of course, since $PM_{2.5}$ has a relatively weak relationship with the response, the acceptance rate is high.

20.5.2 A Parallel Imputation Algorithm

Given the large number of days that need to be imputed for the counties with missing data, and the difficult calculation of the county-specific profile likelihoods, the time for running such a sampler is quite long. In this section we propose a parallel computing approach that can greatly speed up computations.

Notice that the conditional-independence structure from Section 20.5.1 illustrates *that all of the county-specific full conditionals are conditionally independent*. Thus, the imputation of the missing predictor data, the simulation of the county-specific parameters, and the calculation of the profile likelihoods can be performed simultaneously. Hence, it represents an ideal instance where we can increase the efficiency of the simulation of a single Markov chain with parallel computation.

To elaborate, let f be the time required to update $[\mathbf{P} \mid EE]$, $g_0, g_1, \ldots, g_{C-1}$ be the time required to transfer the relevant information to and from the separate nodes for parallel processing, and $h_0, h_1, \ldots, h_{C-1}$ be the time required to perform the processing, as depicted in Figure 20.5. Suppose that C nodes are available for computation. Then, conceptually, the

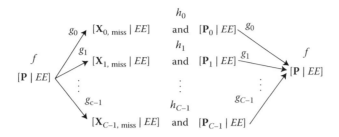

FIGURE 20.5
Parallel computing diagram.

run-time per parallel MCMC iteration is $f + \max_c(g_c + h_c)$. In contrast, the single processor run-time would be $f + \sum_c(h_c)$. Clearly, if the transfer times, $\{g_c\}$ are small relative to the county-specific processing times, $\{h_c\}$, then substantial savings can be made by parallelizing the process, with the gains scaling proportional to the number of conditionally independent full conditionals. This is exactly the setting of the Medicare claims data, where the profile likelihood and imputation county-specific calculations are very time-consuming. Of course, this simple schematic is extremely optimistic in not accounting for several factors, such as variability in the number of available nodes, node-specific run-times and the added time for the software to manage the parallelization. However, it does suggest that substantial gains can be made with parallelization.

While we know of few implementations of parallel MCMC of this scope, this approach to parallelizing Markov chains and its generalizations has been discussed previously (Kontoghiorghes, 2006; Rosenthal, 2000; Winkler, 2003). Moreover, other approaches could be used for parallelizing Markov chains. When applicable, perfect sampling (Fill, 1998; Fill et al., 2000; Propp and Wilson, 1998) could be easily parallelized. Specifically, each perfect sample is an independent and identically distributed draw from the stationary distribution and hence can be generated independently. Also regeneration times (Hobert et al., 2002; Jones et al., 2006; Mykland et al., 1995) create independent tours of the chain from other regeneration times. Therefore, given a starting value at a regeneration time, the tours could be generated in parallel. These two techniques have the drawback that a substantial amount of mathematics needs to be addressed to simply implement the sampler prior to any discussion of parallel implementation. A less theoretically justified, yet computationally simple, approach parallelizes and combines multiple independent chains (Gelman and Rubin, 1992; Geyer, 1992).

Most work on statistical parallel computing algorithms depends on existing network-based parallel computing algorithms, such as Parallel Virtual Machines (Beguelin et al., 1995) or Message Passing Interface (Snir et al., 1995), such as implemented in the R package SNOW (Rossini et al., 2007). These programs are not optimized for particular statistical problems or computational infrastructures and, furthermore, require direct computer-to-computer communication. While such parallel computing architectures are used, large computing clusters that employ queuing management software often cannot take advantage of these approaches.

In contrast, our approach uses a disk-based shared memory blackboard system which required building the parallelization software. Specifically in our approach, a collection of tokens, one for each county, are used to represent which counties currently need processing. A collection of identical programs, which we refer to as spiders, randomly select a token from the bin and move it to a bin of tokens representing counties currently being operated on. We have adopted several strategies to avoid race conditions, where two spiders simultaneously attempt to grab the same token, including: using file system locks and creating small random delays before the spider grabs the token. The spider then performs the county-specific update and moves its token to another bin of counties with finished calculations. The spider then goes back to the original bin and repeats the process. If there are no tokens remaining, the first spider to discover this fact then performs the national update while the remaining sit idle. It then moves the tokens back to the original bin to restart the process. Disk-based shared memory is used for all of the data transfer.

The benefits of this strategy for parallel MCMC are many. Notably, nodes or spiders can be dynamically added or subtracted. Moreover, load balancing can be accomplished easily. In addition, the system allowed us to use a storage area network (SAN) as the shared memory resource (blackboard). While having much slower data transfer than direct computer-to-computer based solutions, this approach allowed us to implement a parallel programming in spite of scheduling software that precludes more direct parallelization. As an added benefit, using the SAN for data transfer builds in automatic checkpointing for the algorithm. We've also found that this approach facilitates good MCMC practice, such as using the ending value from initial runs as the starting value for final runs. Of course, an overwhelming negative property of this approach is the need to create the custom, setting-specific, parallelization software.

20.6   Analysis of the Medicare Data

We analyzed the Medicare data using our parallel MCMC algorithm. We used 30 processors, resulting in a run-time of 5–10 seconds per MCMC iteration. In contrast, the run-time for a single processor was over 2 minutes. That is, there is a 90% decrease in run-time due to parallelization.

The sampler was run for 13,000 iterations. This number was used as simply the largest feasible in the time given. Final values from testing-iterations were used as starting values. Trace plots were investigated to evaluate the behavior of the chains, and were also used to change the step size of the random-walk samplers.

Figure 20.6 displays an example imputation for 1000 monitoring days for a county. The black lines connects observed days while the gray line depicts the estimated trend. The points depict the imputed data set. Figure 20.7 depicts a few days for a county where pollution data is observed every three days; the separate lines are iterations of the MCMC process. These figures illustrate the reasonableness of the imputed data. A possible concern is that the imputed data are slightly less variable than the actual data. Moreover, the data is more regular, without extremely high air pollution days. However, this produces conservatively wider credible intervals for the distributed lag estimates.

Figure 20.8 shows the estimated posterior medians for the exponential of the cumulative effect for a 10-unit increase in air pollution, along with 95% credible intervals. The estimates range from 0.746 to 1.423. The cumulative effect is interpreted as the relative increase or decrease in the rate of COPD corresponding to an across-the-board 10-unit increase in $PM_{2.5}$ over the prior six days. Therefore, for example, 1.007 (the national average) represents a 0.7% increase in the rate of COPD per microgram per 10 cubic meter increase in fine particulate matter over six days.

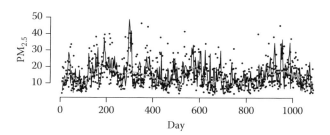

FIGURE 20.6
Example imputation from the MCMC sampler for a specific county. The black line connects observed points while the gray line shows the estimated trend. The points are from a specific iteration of the MCMC sampler.

The national estimate, μ_1, represents the variance-weighted average of the county-specific cumulative effects. The 95% equi-tail credible interval ranges from an estimated 2.6% decrease to a 4.0% increase in the rate of COPD. The posterior median was a 0.7% increase. In contrast, a meta-analysis model using the maximum likelihood fits and variances for only those counties with adequate data for fitting the distributed lag model results in a confidence interval for the national cumulative effect ranging from a 5.1% decrease to a 7.5% increase, while the mean is a 1.0% increase. That is, adding the data from the counties with systematic missing data does not appear to introduce a bias, but does greatly reduce the width of the interval.

The shape of the distributed lag function is of interest to environmental health researchers, as different diseases can have very different profiles, such as rates of hospitalization, recurrence, and complications. Examining the shape of the distributed lag function can shed light on the potential relationship of air pollution and the disease. For example, a decline over time could be evidence of the "harvesting" hypothesis, whereby a large air pollution effect for early lags would deplete the risk set of its frailest members, through hospitalization or mortality. Hence, the latter days would have lower effects. Figure 20.9 shows the exponent of 10 times the distributed lag parameters' posterior medians, θ_{c1}^*, θ_{c2}^*, and θ_{c3}^*, by county. Here θ_{c1}^* is the current-day estimate, while θ_{c2}^* and θ_{c3}^* are cumulative effect for days lag 1 to 2 and 3 to 6, respectively. The current-day effect tends to be much larger, and more variable, by county. The comparatively smaller values for the later lags are supportive of the harvesting hypothesis, though we emphasize that other mechanisms could be in place. Further, this model is not ideal for studying such phenomena, as the bin smoothing of the distributed lag parameters may be too crude to explore the distributed lag function's shape.

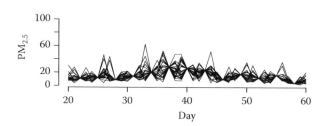

FIGURE 20.7
Several example imputations for a subset of the days for a county. The lines converge on observed days.

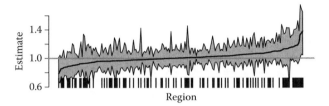

FIGURE 20.8
Estimates and credible intervals for the exponential of the distributed lag cumulative effect by county for a 10-unit increase in air pollution, $10\theta_{c1}$. The solid middle line shows the posterior medians; the gray area shows the estimated 95% equi-tail credible intervals. A horizontal reference line is drawn at 1. Hash marks denote counties with systematic missing data, where the distributed lag model could not be fit without imputation.

   Figure 20.10 shows 95% credible intervals for the AR parameters across counties. The counties are organized so that the rightmost 97 counties have the systematic missing data. Notice that, in these counties, their estimate for the AR(1) parameter is attenuated toward zero over the counties with complete data. The estimated posterior median of ζ is $(0.370, -0.010, -0.011, -0.025)^t$. The primary AR(1) parameter is slightly below the 0.5, because of the contribution of those counties with missing data. Regardless, we note that

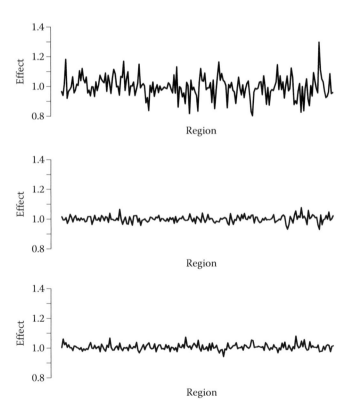

FIGURE 20.9
Exponent of ten times the distributed lag parameters' posterior medians: θ_{c1}^* (top), θ_{c2}^* (middle), and θ_{c3}^* (bottom), by county.

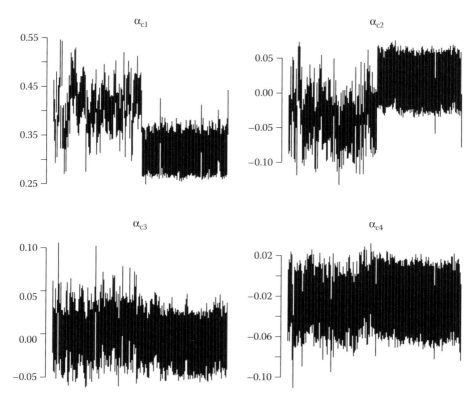

FIGURE 20.10
Posterior credible intervals for the AR parameters by county.

this shrinkage estimation greatly improves on county-specific estimation (see Figure 20.3) for those cites with incomplete data.

20.7 Summary

In this chapter, we propose an MCMC algorithm for fitting distributed lag models for time series with systematically missing predictor data. We emphasize that our analysis only scratches the surface for the analysis of the Medicare claims data. One practical issue is the effect of varying degrees of confounder adjustment, in terms of both the degrees of freedom employed in nonlinear fits and the confounders included. Moreover, a more thorough analysis would consider other health outcomes and different numbers of lags included. Also, commonly air pollution effects are interacted with age or age categories, because of the plausibility of different physical responses to air pollution with aging. In addition, the PM measurements are aggregates of several chemical pollutants. Determining the effects of individual component parts may help explain some of the county variation in the effect of air pollution on health.

The bin smoother on the distributed lag parameters allows for a simpler algorithm and interpretation. However, more reasonable smoothers, such as setting **A** to the design matrix for a regression-spline model, should be considered. This would allow much more accurate exploration of the shape of the distributed lag model, as well as variations in its shape by counties.

The use of the profile likelihood instead of the actual likelihood raises numerous issues and concerns. Foremost is the propriety of the posterior and hence the validity of the sampler and inference. The theoretical consequences of this approach should be evaluated. Moreover, comparisons with other strategies, such as placing independent diffuse priors on the nuisance parameters, are of interest.

Also of interest is to eliminate the attenuation of the estimates of the autoregressive parameters for the counties with missing data. To highlight this problem more clearly, suppose that instead of 97 counties with missing data, we had 9700 with data recorded every other day. Then the accurate information regarding ζ and Σ_α contained in the counties with complete data would be swamped by the noisy bimodal likelihoods from the counties with systematic missingness. More elaborate hierarchies on this component of the model may allow for the counties with observed data to have control over estimation of these parameters.

In addition, the potential informativeness of counties having missing data (see Little and Rubin, 2002) should be investigated. To elaborate, clearly the pattern of missing data is uninformative for any given county; however, whether or not a county collected data every day or every third day may be informative. For example, counties with air pollutions levels well below or above standards may be less likely to collect data every day. Such missingness may impact national estimates.

We also did not use external variables to impute the missing predictor data. Ideally, a completely observed instrumental variable that is causally associated with the predictor, yet not with the response, would be observed. Such variables could be used to impute the predictor, but would not confound its relationship with the response. However, such variables are rarely available. More often variables that are potentially causally associated with the predictor are also potentially causally associated with response. For example, seasonality and temperature are thought to be causally associated with *PM* levels and many health outcomes. Hence, using those variables to impute the missing predictor data would immediately raise the criticism that any association found was due to residual confounding between the response and the variables used for imputation.

These points notwithstanding, this work suggests potential for the ability to impute missing data for distributed lag models. The Bayesian model produces a marked decrease in the width of the inter-county estimate of the cumulative effect. Moreover, the imputed data sets are consistent with the daily observed data, though perhaps being more regular and less variable. However, we note the bias incurred by lower variability in the imputed air pollution is conservative, and would attenuate the distribute lag effects.

A second accomplishment of this chapter is the parallelization algorithm and software development. The computational overhead for the parallelization software was small, and hence the decrease in run-time was nearly linear with the number of nodes.

Race conditions, times when multiple spiders attempted to access the same token, represent a difficult implementation problem. For example, after the county-specific updates finish, all spiders attempt to obtain the token representing the national update simultaneously. A colleague implementing a similar system proposed a potential solution (Fernando Pineda, personal communication). Specifically, he uses atomic operations on a lock file to only allow access to the bin of tokens to one program at a time. As an analogy, this approach has a queue of programs waiting for access to the bins to obtain a token. In contrast, our approach allows simultaneous access to the bins, thus increasing speed, though also increasing the likelihood of race conditions. A fundamental problem we have yet to solve is the need for truly atomic operations on networked file systems to prevent these race conditions. Our use of file system locks when moving files as the proposed atomic operation made the system very fragile, given the complex nature and inherent lag of an NFS-mounted SAN.

Our current solutions to these problems are inelegant. First, as described earlier, random waiting times were added. Secondly, spiders grabbed tokens in a random order. Finally, a worker program was created that searched for and cleaned up lost tokens and ensured that the appropriate number of spiders were operational. We are currently experimenting with the use of an SQL database, with database queries, rather than file manipulations to manage the tokens.

Appendix: Full Conditionals

We have

$$\boldsymbol{\theta}_c \propto \exp\left(\tilde{\ell}_c(\boldsymbol{\theta}_c) - \frac{1}{2}(\boldsymbol{\theta}_c - \boldsymbol{\mu})^t \boldsymbol{\Sigma}_\theta (\boldsymbol{\theta}_c - \boldsymbol{\mu})\right),$$

$$\boldsymbol{\mu} \sim N\left\{\left(C\boldsymbol{\Sigma}_\theta^{-1} + G_1^{-1}\right)^{-1} \boldsymbol{\Sigma}_\theta^{-1} \sum_{c=0}^{C-1} \boldsymbol{\theta}_c, \left(C\boldsymbol{\Sigma}_\theta^{-1} + G_1^{-1}\right)^{-1}\right\},$$

$$\boldsymbol{\Sigma}_\theta^{-1} \sim \text{Wishart}\left(G_2 + \sum_{c=0}^{C-1}(\boldsymbol{\theta}_c - \boldsymbol{\mu})(\boldsymbol{\theta}_c - \boldsymbol{\mu})^t, df_1 + C\right),$$

$$\boldsymbol{\alpha}_c \sim N\left\{(\mathbf{E}_c^t \mathbf{E}_c/\sigma_c^2 + \boldsymbol{\Sigma}_\alpha^{-1})^{-1}\mathbf{E}_c^t \boldsymbol{\epsilon}_c, (\mathbf{E}_c^t \mathbf{E}_c/\sigma_c^2 + \boldsymbol{\Sigma}_\alpha^{-1})\right\},$$

$$\boldsymbol{\zeta} \sim N\left\{\left(c\boldsymbol{\Sigma}_\alpha^{-1} + G_4^{-1}\right)^{-1} \boldsymbol{\Sigma}_\alpha^{-1} \sum_{c=0}^{C-1} \boldsymbol{\alpha}_c, \left(c\boldsymbol{\Sigma}_\alpha^{-1} + G_4^{-1}\right)^{-1} \boldsymbol{\Sigma}_\alpha^{-1}\right\},$$

$$\boldsymbol{\Sigma}_\alpha \sim \text{Wishart}\left(G_5 + \sum_{c=0}^{C-1}(\boldsymbol{\alpha}_c - \boldsymbol{\zeta})(\boldsymbol{\alpha}_c - \boldsymbol{\zeta})^t), df_2 + C\right),$$

$$\sigma_c^{-2} \sim \Gamma\left\{C/2 + G_6, \sum_{c=0}^{C-1}\left(\epsilon_{ct} - \sum_{u=1}^{p} \epsilon_{c,t-u}\alpha_u\right)^2 + G_7\right\}.$$

Here $\boldsymbol{\epsilon}_c = (\epsilon_{c1}, \ldots, \epsilon_{cT_c})^t$, where ϵ_{ct} are the residuals after fitting model (Equation 20.2). The matrix $\mathbf{E}_c$ denotes the lagged values of $\boldsymbol{\epsilon}_c$. $G_1, G_2, \ldots$ denote generic hyperparameters whose values are described in the chapter, while df_1 and df_2 correspond to prior Wishart degrees of freedom. That is, G_1 is the prior variance on $\boldsymbol{\mu}$; G_2 represents the Wishart scale matrix for $\boldsymbol{\Sigma}_\theta$; (G_3, G_4) represent the prior means and variance on $\boldsymbol{\zeta}$; G_5 represents the Wishart scale matrix for $\boldsymbol{\Sigma}_\alpha$; and G_6 and G_7 are the gamma shape and rate on σ_c^{-2}.

Acknowledgment

The authors would like to thank Dr. Fernando Pineda for helpful discussions on parallel computing.

References

Barnett, G., Kohn, R., and Sheather, S. 1996. Bayesian estimation of an autoregressive model using Markov chain Monte Carlo. *Journal of Econometrics*, 74:237–254.

Beguelin, A., Dongarra, J., Jiang, W., Manchek, R., and Sunderam, V. 1995. *PVM: Parallel Virtual Machine: A Users' Guide and Tutorial for Networked Parallel Computing*. MIT Press, Cambridge, MA.

Bell, M., McDermott, A., Zeger, S., Samet, J., and Dominici, F. 2004. Ozone and short-term mortality in 95 urban communities, 1987–2000. *Journal of the American Medical Association*, 292(19):2372–2378.

Broersen, P., de Waele, S., and Bos, R. 2004. Autoregressive spectral analysis when observations are missing. *Automatica*, 40(9):1495–1504.

Carlin, B. and Louis, T. 2009. *Bayesian Methods for Data Analysis*, 3rd edn. Chapman & Hall/CRC Press, Boca Raton, FL.

Diggle, P. 1990. *Times Series: A Biostatistical Introduction*. Oxford University Press, Oxford.

Dominici, F., McDermott, A., Zeger, S., and Samet, J. 2002. On the use of generalized additive models in time-series studies of air pollution and health. *American Journal of Epidemiology*, 156(3):193.

Dominici, F., Peng, R., Bell, M., Pham, L., McDermott, A., Zeger, S., and Samet, J. 2006. Fine particulate air pollution and hospital admission for cardiovascular and respiratory diseases. *Journal of the American Medical Association*, 295(10):1127–1134.

Durbin, J. 1960. The fitting of time series models. *International Statistical Review*, 28:233–244.

Fill, J. 1998. An interruptible algorithm for perfect sampling via Markov chains. *Annals of Applied Probability*, 8(1):131–162.

Fill, J., Machida, M., Murdoch, D., and Rosenthal, J. 2000. Extension of Fill's perfect rejection sampling algorithm to general chains. *Random Structures and Algorithms*, 17(3–4):290–316.

Gelman, A. and Rubin, D. 1992. Inference from iterative simulation using multiple sequences. *Statistical Science*, 7:457–511.

Geyer, C. 1992. Practical Markov chain Monte Carlo. *Statistical Science*, 7:473–511.

Hobert, J., Jones, G., Presnell, B., and Rosenthal, J. 2002. On the applicability of regenerative simulation in Markov chain Monte Carlo. *Biometrika*, 89(4):731–743.

Huerta, G. and West, M. 1999. Priors and component structures in autoregressive time series models. *Journal of the Royal Statistical Society, Series B*, 61(4):881–899.

Jones, G., Haran, M., Caffo, B., and Neath, R. 2006. Fixed-width output analysis for Markov chain Monte Carlo. *Journal of the American Statistical Association*, 101(476):1537–1547.

Kontoghiorghes, E. 2006. *Handbook of Parallel Computing and Statistics*. Chapman & Hall/CRC, Boca Raton, FL.

Little, R. and Rubin, D. 2002. *Statistical Analysis with Missing Data*, 2nd edn. Wiley, Hoboken, NJ.

Monahan, J. 1983. Fully Bayesian analysis of ARMA time series models. *Journal of Econometrics*, 21:307–331.

Monahan, J. and Boos, D. 1992. Proper likelihoods for Bayesian analysis. *Biometrika*, 79(2):271–278.

Mykland, P., Tierney, L., and Yu, B. 1995. Regeneration in Markov chain samplers. *Journal of the American Statistical Association*, 90(429):233–241.

Propp, J. and Wilson, D. 1998. How to get a perfectly random sample from a generic Markov chain and generate a random spanning tree of a directed graph. *Journal of Algorithms*, 27(2):170–217.

Rosenthal, J. 2000. Parallel computing and Monte Carlo algorithms. *Far East Journal of Theoretical Statistics*, 4(2):207–236.

Rossini, A., Tierney, L., and Li, N. 2007. Simple parallel statistical computing in R. *Journal of Computational and Graphical Statistics*, 16(2):399.

Samet, J., Dominici, F., Currieo, F., Coursac, I., and Zeger, S. 2000. Fine particulate air pollution and mortality in 20 U.S. cities 1987-1994. *New England Journal of Medicine*, 343(24):1742–1749.

Schwartz, J. 2000. Harvesting and long term exposure effects in the relation between air pollution and mortality. *American Journal of Epidemiology*, 151(5):440.

Snir, M., Otto, S., Walker, D., Dongarra, J., and Huss-Lederman, S. 1995. *MPI: The Complete Reference*. MIT Press, Cambridge, MA.

Wallin, R. and Isaksson, A. 2002. Multiple optima in identification of ARX models subject to missing data. *EURASIP Journal on Applied Signal Processing*, 1:30–37.

Winkler, G. 2003. *Image Analysis, Random Fields, and Dynamic Monte Carlo Methods*, 2nd edn. Springer, Berlin.

Zanobetti, A., Wand, M. P., Schwartz, J., and Ryan, L. M. 2000. Generalized additive distributed lag models: Quantifying mortality displacement. *Biostatistics*, 1(3):279–292.

Zeger, S., Dominici, F., and Samet, J. 1999. Harvesting-resistant estimates of air pollution effects on mortality. *Epidemiology*, 10(2):171.

21

MCMC for State–Space Models

Paul Fearnhead

21.1   Introduction: State–Space Models

In this chapter, we look at Markov chain Monte Carlo (MCMC) methods for a class of time series models called *state–space models*. The idea of state–space models is that there is an unobserved state of interest the evolves through time, and that partial observations of the state are made at successive time points. We will denote the state by X and observations by Y, and assume that our state–space model has the following structure:

$$X_t \mid \{x_{1:t-1}, y_{1:t-1}\} \sim p(x_t \mid x_{t-1}, \theta), \tag{21.1}$$

$$Y_t \mid \{x_{1:t}, y_{1:t-1}\} \sim p(y_t \mid x_t, \theta). \tag{21.2}$$

Here, and throughout, we use the notation $x_{1:t} = (x_1, \ldots, x_t)$, and write $p(\cdot \mid \cdot)$ for a generic conditional probability density or mass function (with the arguments making it clear which conditional distribution it relates to). To fully define the distribution of the hidden state we further specify an initial distribution $p(x_1 \mid \theta)$. We have made explicit the dependence of the model on an unknown parameter θ, which may be multidimensional. The assumptions in this model are that, conditional on the parameter θ, the state model is Markov, and that we have a conditional independence property for the observations: observation Y_t only depends on the state at that time, X_t.

For concreteness we give three examples of state–space models:

Example 21.1:   Stochastic Volatility

The following simple stochastic volatility (SV) model has been used for modeling the time-varying variance of log-returns on assets; for fuller details, see Hull and White (1987) and Shephard (1996). The state–space model is

$$X_t \mid \{x_{1:t-1}, y_{1:t-1}\} \sim N(\phi x_{t-1}, \sigma^2),$$

where $|\phi| < 1$, with initial distribution $X_1 \sim N(0, \sigma^2/(1 - \phi^2))$, and

$$Y_t \mid \{x_{1:t}, y_{1:t-1}\} \sim N(0, \beta^2 \exp\{x_t\}).$$

The parameters of the model are $\theta = (\beta, \phi, \sigma)$. The idea of the model is that the variance of the observations depends on the unobserved state, and the unobserved state is modeled by an AR(1) process.

Example 21.2: Discrete Hidden Markov Model

A general class of models occurs when the underlying state is a discrete-valued Markov model, with a finite state space. Thus we can assume without loss of generality that $X_t \in \{1, 2, \ldots, K\}$ and that the model for the dynamics of the state (Equation 21.1) is defined by a $K \times K$ transition matrix P. Thus, for all $i, j \in \{1, \ldots, K\}$,

$$\Pr(X_t = j \mid X_{t-1} = i, x_{1:t-2}, y_{1:t-1}) = P_{ij}.$$

Usually it is assumed that the distribution for X_1 is given by the stationary distribution of this Markov chain. The observation Equation 21.2 will depend on the application, but there will be K observation regimes (depending on the value of the state). Thus we can write

$$Y_t \mid \{x_t = k, x_{1:t-1}, y_{1:t-1}\} \sim f_k(y_t \mid \theta). \tag{21.3}$$

The parameters of this model will be the parameters of Equation 21.3 and the parameters of the transition matrix P.

Examples of such models include models of ion channels (Ball and Rice, 1992; Hodgson, 1999), DNA sequences (Boys et al., 2000), and speech (Juang and Rabiner, 1991).

Example 21.3: Change-Point Model

Change-point models partition the data into homogeneous regions. The model for the data is the same within each region, but differs across regions. Change-point models have been used for modeling stock prices (Chen and Gupta, 1997), climatic time series (Beaulieu et al., 2007; Lund and Reeves, 2002), DNA sequences (Didelot et al., 2007; Fearnhead, 2008), and neuronal activity in the brain (Ritov et al., 2002), among many other applications.

A simple change-point model can be described as a state–space model with the following state equation:

$$X_t \mid \{x_{1:t-1}, y_{1:t-1}\} = \begin{cases} x_{t-1}, & \text{with probability } 1 - p, \\ Z_t, & \text{otherwise,} \end{cases}$$

where the Z_ts are independent and identically distributed random variables with density function $p_Z(\cdot \mid \phi)$. Initially $X_1 = Z_1$, and the observation equation is given by

$$Y_t \mid \{x_{1:t}, y_{1:t-1}\} \sim p(y_t \mid x_t).$$

The parameters of this model are $\theta = (p, \phi)$, where p governs the expected number of change points in the model, and ϕ the marginal distribution for the state at any time.

We will focus on models for which we can calculate, for any $t < s$,

$$Q(t, s) = \int \left(\prod_{i=t}^{s} p(y_i \mid x) \right) p_Z(x \mid \phi) \, dx. \tag{21.4}$$

This is the marginal likelihood of the observations $y_{t:s}$, given that the observations come from a single segment. The functions $Q(t, s)$ depend on ϕ, but for notational convenience we have suppressed this.

21.2 Bayesian Analysis and MCMC Framework

Our aim is to perform Bayesian inference for a state–space model given data $y_{1:n}$. We assume a prior for the parameters, $p(\theta)$, has been specified, and we wish to obtain the posterior of the parameters $p(\theta \mid y_{1:n})$, or in some cases we may be interested in the joint distribution of the state and the parameters $p(\theta, x_{1:n} \mid y_{1:n})$.

How can we design an MCMC algorithm to sample from either of these posterior distributions? In both cases, this can be achieved using data augmentation (see Chapter 10, this volume). That is, we design a Markov chain whose state is $(\theta, X_{1:n})$, and whose stationary distribution is $p(\theta, x_{1:n} \mid y_{1:n})$ (samples from the marginal posterior $p(\theta \mid y_{1:n})$ can be obtained from samples from $p(\theta, x_{1:n} \mid y_{1:n})$ just by discarding the $x_{1:n}$ component of each sample). The reason for designing an MCMC algorithm on this state space is that, for state–space models of the form Equations 21.1 through 21.2, we can write down the stationary distribution of the MCMC algorithm up to proportionality:

$$
p(\theta, x_{1:n} \mid y_{1:n}) \propto p(\theta) p(x_1 \mid \theta) \left(\prod_{t=2}^{n} p(x_t \mid x_{t-1}, \theta) \right) \left(\prod_{t=1}^{n} p(y_t \mid x_t, \theta) \right). \tag{21.5}
$$

Hence, it is straightforward to use standard moves within our MCMC algorithm.

In most applications it is straightforward to implement an MCMC algorithm with Equation 21.5 as its stationary distribution. A common approach is to design moves that update θ conditional on the current values of $X_{1:n}$ and then update $X_{1:n}$ conditional on θ. We will describe various approaches within this framework. We first focus on the problem of updating the state; and to evaluate different methods we will consider models where θ is known. Then we will consider moves to update the parameters.

21.3 Updating the State

The simplest approach to updating the state $X_{1:n}$ is to update its components one at a time. Such a move is called a *single-site update*. While easy to implement, this move can lead to slow mixing if there is strong temporal dependence in the state process. In these cases it is better to update blocks of state components, $X_{t:s}$, or the whole state process $X_{1:n}$ in a single move. (As we will see, in some cases it is possible to update the whole process $X_{1:n}$ directly from its full conditional distribution $p(x_{1:n} \mid y_{1:n}, \theta)$, in which case these moves are particularly effective.)

We will give examples of single-site moves, and investigate when they do and do not work well, before looking at designing efficient block updates. For convenience we drop the conditioning on θ in the notation that we use within this section.

21.3.1 Single-Site Updates of the State

The idea of single-site updates is to design MCMC moves that update a single value of the state, x_t, conditional on all other values of the state process (and on θ). Repeated application of this move for $t = 1, \ldots, n$ will enable the whole state process to be updated.

Let $x_{-t} = (x_1, \ldots, x_{t-1}, x_{t+1}, \ldots, x_n)$ denote the whole state process excluding x_t. So a single-site update will update x_t for fixed x_{-t}, θ. The target distribution of such a move is the full conditional distribution $p(x_t \mid x_{-t}, \theta, y_{1:t})$, which as mentioned above we will write as $p(x_t \mid x_{-t}, y_{1:t})$, dropping the conditioning on θ in the notation that we use, as we are considering moves for fixed θ. Due to the Markov structure of our model, this simplifies to $p(x_t \mid x_{t-1}, x_{t+1}, y_t)$ for $t = 2, \ldots, n - 1$, $p(x_1 \mid x_2, y_1)$ for $t = 1$, and $p(x_n \mid x_{n-1}, y_n)$ for $t = n$. Sometimes we can simulate directly from these full conditional distributions, and such (*Gibbs*) moves will always be accepted. Where this is not possible, then if x_t is low-dimensional we can often implement an efficient *independence sampler* (see below).

We now give details of single-site update for Examples 21.2 (Gibbs move) and 21.1 (independence sample), and in both cases we investigate the mixing properties of the move in updating $X_{1:n}$.

Example 21.4: Single-Site Gibbs Move

For the hidden Markov model (HMM) of Example 21.2, with state transition matrix, P, we have for $t = 2, \ldots, n - 1$ that

$$\Pr(X_t = k \mid X_{t-1} = i, X_{t+1} = j, y_t) \propto \Pr(X_t = k \mid X_{t-1} = i)\Pr(X_{t+1} = j \mid x_t = k)p(y_t \mid X_t = k)$$

$$= P_{ik}P_{kj}f_k(y_t),$$

for $k = 1, \ldots, K$. Now as X_t has a finite state space, we can calculate the normalizing constant of this conditional distribution, and we get

$$\Pr(X_t = k \mid X_{t-1} = i, X_{t+1} = j, y_t) = \frac{P_{ik}P_{kj}p_k(y_t)}{\sum_{l=1}^{K} P_{il}P_{lj}f_l(y_t)}.$$

Similarly, we obtain $\Pr(X_1 = k \mid X_2 = j, y_1) \propto \Pr(X_1 = k)P_{kj}f_k(y_1)$ and $\Pr(X_n = k \mid X_{n-1} = i, y_n) \propto P_{ik}f_k(y_n)$. In both cases the normalizing constants of these conditional distributions can be obtained.

Thus for this model we can simulate from the full conditionals directly, which is the optimal proposal for x_t for fixed x_{-t}. Note that the computational cost of simulation is $O(K)$, due to calculation of the normalizing constants. For large K it may be more computationally efficient to use other proposals (such as an independence sample) whose computational cost does not depend on K.

We examine the efficiency of this MCMC move to update the state $X_{1:n}$ by focusing on an HMM for DNA sequences (see, e.g. Boys et al., 2000). The data consists of a sequence of DNA, so $y_t \in \{A,C,G,T\}$ for all t. For simplicity we consider a two-state HMM, with the likelihood function for $k = 1, 2$ being

$$\Pr(Y_t = y \mid X_t = k) = \pi_y^{(k)}, \quad \text{for } y \in \{A,C,G,T\}.$$

We denote the parameter associated with $X_t = k$ as $\pi^{(k)} = (\pi_A^{(k)}, \pi_C^{(k)}, \pi_G^{(k)}, \pi_T^{(k)})$.

We will consider the effect of both the dependence in the state dynamics, and the information in the observations, on the mixing rate of the MCMC move. To do this we will assume that state transition matrix satisfies $P_{12} = P_{21} = \alpha$, and

$$\pi^{(1)} = (1, 1, 1, 1)/4 + \beta(1, 1, -1, -1), \quad \pi^{(2)} = (1, 1, 1, 1)/4 - \beta(1, 1, -1, -1),$$

for $0 < \alpha < 1$ and $0 < \beta < 1/4$. Small values of α correspond to large dependence in the state dynamics, and small values of β correspond to less informative observations.

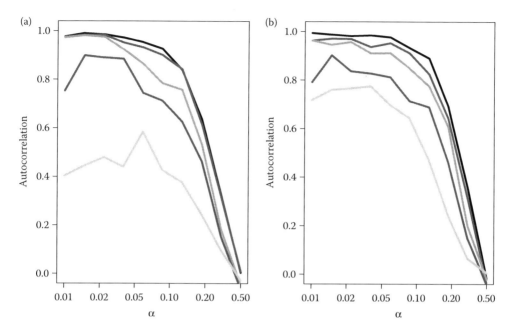

FIGURE 21.1
Lag-1 autocorrelation values for differing α for a two-state hidden Markov model: (a) $n = 200$; (b) $n = 500$. In each plot, different lines refer to different values of β; from top to bottom: $\beta = 0.02$; $\beta = 0.065$; $\beta = 0.11$; $\beta = 0.155$; and $\beta = 0.2$.

To measure the mixing properties of the single-site MCMC update we (i) simulated data for a given value of (α, β); (ii) ran an MCMC algorithm with single-site updates; and (iii) calculated an autocorrelation function for the MCMC output after discarding a suitable burn-in. For simplicity, we summarized the output based on the autocorrelation at lag-1 (all MCMC runs suggested auto-correlations that decayed approximately exponentially). We calculated the autocorrelation for the number of differences between the true value of the hidden state and the inferred value of the state.

Results are shown in Figure 21.1, where we see that the value of α is the main determinant of the mixing of the MCMC algorithm. Small values of α, which correspond to large dependence, result in poor mixing. Similarly, as β decreases, which relates to less informative observations, the mixing gets worse—though the dependence on β is less than on α. Qualitatively similar results are observed for the two values of n, but for smaller n we see that the value of β has more impact on the mixing properties.

Example 21.5: Single-Site Independence Sampler

Now consider the SV model of Example 21.1. We describe an independence sampler that was derived by Shephard and Pitt (1997). With this model we obtain, for $t = 2, \ldots, n - 1$,

$$p(x_t \mid x_{t-1}, x_{t+1}, y_t) \propto p(x_t \mid x_{t-1}) p(x_{t+1} \mid x_t) p(y_t \mid x_t) \tag{21.6}$$

$$\propto \exp\left\{-\frac{1}{2\sigma^2}((x_t - \phi x_{t-1}) + (x_{t+1} - \phi x_t)^2)\right\} \exp\left\{-\frac{x_t}{2}\right\} \exp\left\{-\frac{\exp\{-x_t\}y_t^2}{2\beta^2}\right\},$$

where we have removed any constants of proportionality that do not depend on x_t; the first term of the final expression corresponds to the two state transition densities, and the final two terms come from the likelihood.

Simulating directly from this conditional distribution is not possible, so we resort to approximation. Our approximation is based on a Taylor expansion of $\log p(x_t \mid x_{t-1}, x_{t+1}, y_t)$ about an estimate of x_t, which we call $\hat{x}_t$. Now if we define $\mu_t = \phi(x_{t-1} + x_{t+1})/(1 + \phi^2)$ and $\tau^2 = \sigma^2/(1 + \phi^2)$, then the first term in Equation 21.7 can be rewritten, up to a constant of proportionality, as $\exp\{-(x_t - \mu_t)^2/(2\tau^2)\}$. Thus without any observation, our conditional distribution of x_t would have a mean μ_t, and this appears a sensible value about which to take a Taylor expansion. Doing this leads to

$$\log p(x_t \mid x_{t-1}, x_{t+1}, y_t) \approx -\frac{(x_t - \mu_t)^2}{2\tau^2} - \frac{x_t}{2} - \frac{y_t^2}{2\beta^2} \exp\{-\mu_t\}\left(1 - (x_t - \mu_t) + \frac{1}{2}(x_t - \mu_t)^2\right).$$

As this approximation to the log-density is quadratic, this gives us a normal approximation to the conditional distribution, which we denote by $q(x_t \mid x_{t-1}, x_{t+1}, y_t)$. (For full details of the mean and variance of the approximation, see Shephard and Pitt, 1997.) Thus we can implement an MCMC move of X_t by using an independence sampler with proposal $q(x_t \mid x_{t-1}, x_{t+1}, y_t)$.

Similar normal approximations can be obtained for $p(x_1 \mid x_2, y_1)$ and $p(x_n \mid x_{n-1}, y_n)$, the only difference being in the values of μ_t and τ. Note that better estimates of $\hat{x}_t$ can be found, for example, by numerically finding the mode of $p(x_t \mid x_{t-1}, x_{t+1}, y_t)$ (Smith and Santos, 2006), but for single-site updates any increase in acceptance rate is unlikely to be worth the extra computation involved.

We investigate the efficiency of single-site updates for the SV model via simulation. We fix $\beta = 1$ and consider how mixing of the MCMC algorithm depends on the time dependence of the state process, ϕ, and marginal variance of the state process, $\tau^2 = \sigma^2/(1 - \phi^2)$. As above, we evaluate mixing by looking at the lag-1 autocorrelation of the mean square error in the estimate of the state process. Results are shown in Figure 21.2, where we see that ϕ has a sizeable effect on mixing—with $\phi \approx 1$, which corresponds to strong correlation in the state process, resulting in poor mixing. By comparison both n and τ^2 have little effect. For all MCMC runs the acceptance rate of the MCMC move was greater than 99%.

21.3.2 Block Updates for the State

While the single-site updates of Section 21.3.1 are easy to implement, we have seen that the resulting MCMC algorithms can mix slowly if there is strong dependence in the state process. This leads to the idea of *block updates*—updating the state at more than one time point in a single move. Ideally we would update the whole state process in one move, and in some cases it turns out that this is possible to do from the full conditional, so that moves are always accepted. These include the linear Gaussian models, where we can use the Kalman filter (see, e.g. Carter and Kohn, 1994; Harvey, 1989), as well as the HMM of Example 21.2 and the change-point model of Example 21.3. We give details of the methods used for the latter two below.

In situations where it is not possible to update the whole state process from its full conditional, one possibility is to use an independence proposal to update jointly a block of state values. We will describe such an approach for the SV model of Example 21.1, and then discuss alternative approaches for block updates for models where it is not possible to draw from the full conditional distribution of the state.

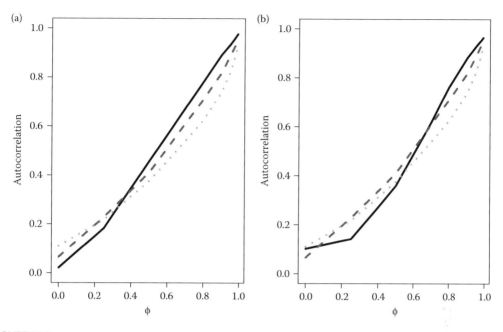

FIGURE 21.2
Lag-1 autocorrelation values for differing ϕ for the stochastic volatility model: (a) $n = 200$; (b) $n = 500$. In each plot, different lines refer to different values of τ^2: $\tau^2 = 0.5$ (full lines); $\tau^2 = 1$ (dashed lines); $\tau^2 = 2.0$ (dotted lines).

Example 21.6: Updating the State from its Full Conditional

The forward–backward algorithm is a method for sampling from the full conditional of the state process for discrete HMMs. See Rabiner and Juang (1986) for a review of this method, and Scott (2002) for further examples of its use within Bayesian inference. Here we describe its implementation for the model of Example 21.2.

The algorithm is based upon a forward recursion which calculates the filtering densities $\Pr(X_t \mid y_{1:t})$ for $t = 1, \ldots, n$; followed by a backward simulation step that simulates from $\Pr(X_n \mid y_{1:n})$ and then $\Pr(X_t \mid y_{1:n}, x_{t+1})$ for $t = n - 1, \ldots, 1$. The forward recursion is initialized with

$$\Pr(X_1 = k \mid y_1) \propto \Pr(X_1 = k)f_k(y_1), \quad \text{for } k = 1, \ldots, K,$$

where the normalizing constant is $p(y_1) = \sum_{l=1}^{K} \Pr(X_1 = l)f_l(y_1)$. Then for $t = 2, \ldots, n$ we have

$$\Pr(X_t = k \mid y_{1:t}) \propto f_k(y_t) \sum_{l=1}^{K} \Pr(X_{t-1} = l \mid y_{1:t-1})P_{lk}, \quad \text{for } k = 1, \ldots, K,$$

where the normalizing constant is $p(y_t \mid y_{1:t-1})$. (Note that a byproduct of the forward recursions is that we obtain the likelihood as a product of these normalizing constants, as $p(y_{1:n}) = p(y_1)\prod_{t=2}^{n} p(y_t \mid y_{1:t-1})$).

Once these filtering densities have been calculated and stored, we then simulate backwards. First, we simulate X_n from the filtering density $\Pr(X_n \mid y_{1:n})$; then, for $t = n - 1, \ldots, 1$, we iteratively simulate X_t given our simulated value for X_{t+1}, from

$$\Pr(X_t = l \mid y_{1:n}, X_{t+1} = k) = \Pr(X_t = l \mid y_{1:t}, X_{t+1} = k) \propto \Pr(X_t = l \mid y_{1:t})P_{lk}.$$

The computational complexity of the forward–backward algorithm is $O(nK^2)$ for the forward recursion, and $O(nK)$ for the backward simulation. This compares with $O(nK)$ for applying the single-site update to all state values. Thus, particularly for values large K, it may be computationally more efficient to use single-site updates. As seen above, whether this is the case will depend on the amount of dependence in the state model.

In the above description, we suppressed the dependence on the unknown parameter θ. Standard MCMC algorithms will update $X_{1:n}$ given θ and then θ given $X_{1:n}$ in one iteration. Thus each iteration will (potentially) have a new θ value, and will require the reapplication of the forward–backward algorithm to simulate $X_{1:n}$. One approach to reducing the computational cost of using the forward–backward algorithm within MCMC, suggested by Fearnhead (2006), is to (i) obtain a good point estimate of the parameters, $\hat{\theta}$; (ii) apply the forward recursion for this value of the parameter; and (iii) use $\Pr(X_{1:n} \mid y_{1:n}, \hat{\theta})$ as an independence proposal for updating the state. The advantage of this is that the costly forward recursion is only required once, as opposed to at every iteration of the MCMC algorithm. Furthermore, Fearnhead (2006) describes an efficient algorithm for simulating large samples of $X_{1:n}$ from the backward simulation step. In applications, providing a good estimate is obtained in (i), this approach has shown to produce efficient MCMC updates. Note that estimation in (i) could be performed in an adaptive manner during the burn-in period of the MCMC algorithm.

Our forward–backward description has focused on discrete-time processes. It is possible to extend the idea to continous-time (though still discrete-valued) HMMs; see, for example, Fearnhead and Meligkotsidou (2004) and Fearnhead and Sherlock (2006).

Example 21.7: Updating the State from its Full Conditional

We now show how the forward–backward algorithm can be applied to the change-point model of Example 21.3. The idea behind this application dates back to Yao (1984), but see also Barry and Hartigan (1992), Liu and Lawrence (1999), and Fearnhead (2006).

We introduce a new state variable, C_t, which we define to be the time of the most recent change point prior to t. Mathematically this is a function of $x_{1:t}$, with

$$C_t = \max\{s : x_s \neq x_{s+1} \text{ for } s < t\},$$

and $C_t = 0$ if there has been no change point prior to t (i.e. the set on the right-hand side is empty). Note that $C_t \in \{0, \ldots, t-1\}$, and C_t is a Markov process with

$$\Pr(C_t = j \mid C_{t-1} = i) = \begin{cases} p, & \text{if } j = t-1, \\ 1-p, & \text{if } i = j, \end{cases}$$

with all other transitions being impossible. Note that these two transitions correspond to there either being or not being a change point at time $t-1$.

We can now derive the forward–backward algorithm. The forward recursion is initialized with $\Pr(C_1 = 0 \mid y_1) = 1$, and for $t = 2, \ldots, n$ we have

$$\Pr(C_t = j \mid y_{1:t}) \propto (1-p)\frac{Q(j+1, t)}{Q(j+1, t-1)}\Pr(C_{t-1} = j \mid y_{1:t-1}), \quad \text{for } j = 0, \ldots, t-2,$$

$$\Pr(C_t = t-1 \mid y_{1:t}) \propto pQ(t, t).$$

The first equation corresponds to there not being a change point at time $t-1$. This happens with probability $1-p$, and in this case $C_t = C_{t-1}$. The second corresponds to there being a change point, which happens with probability p. The $Q(\cdot, \cdot)$ are defined in Equation 21.4. In both equations, the term involving $Q(\cdot, \cdot)$ is the likelihood of the observation y_t given C_t and $y_{1:t-1}$.

Once the filtering recursions have been solved, backward simulation proceeds using the conditional distributions

$$\Pr(C_t = j \mid C_{t+1} = t, y_{1:n}) = \Pr(C_t = j \mid y_{1:t}),$$

where conditioning on $C_{t+1} = t$ is equivalent to conditioning on a change point at t. Thus we can simulate the time of the last change point from $\Pr(C_n \mid y_{1:n})$, and then recursively, given a change point at t, simulate the next most recent change point from $\Pr(C_t \mid y_{1:t})$. This simulation continues until we simulate $C_t = 0$, which corresponds to no more change points.

The computational complexity of this algorithm is $O(n^2)$. The main cost is in solving the recursions, and one approach to reduce computational cost is to solve these for a specific value of the parameters, and then use the resulting conditional distribution for $X_{1:n}$ as an independence proposal (see Fearnhead, 2006, and the discussion for Example 21.2 above). Note that this forward–backward algorithm can be generalized to allow for different distributions of time between successive change points (see Fearnhead, 2008), and for HMM dependence in the state value for neighbouring segments (Fearnhead and Vasileiou, 2009).

Example 21.8: Block Independence Sampler

For the SV model of Example 21.1, we cannot sample directly from the full conditional distribution $p(x_{1:n} \mid y_{1:n})$. Instead we follow Shephard and Pitt (1997) and consider an independence sampler for block updating. The proposal distribution for the independence sampler is based on a natural extension of the independence sampler for singe-site updates.

Consider an update for $X_{t:s}$ for $s > t$. For an efficient independence proposal we require a good approximation to $p(x_{t:s} \mid x_{t-1}, x_{s+1}, y_{t:s})$. (If $t = 1$ we would drop the conditioning on x_{t-1}, and if $s = n$ we would drop the conditioning on x_{s+1} here and in the following.) Now we can write

$$p(x_{t:s} \mid x_{t-1}, x_{s+1}, y_{t:s}) \propto p(x_{t:s} \mid x_{t-1}, x_{s+1}) \prod_{j=t}^{s} p(y_j \mid x_j),$$

where the first term on the right-hand side is a multivariate Gaussin density. Thus if, for all $j = t, \ldots, s$, we approximate $p(y_j \mid x_j)$ by a Gaussian likelihood, we obtain a Gaussian approximation to $p(x_{t:s} \mid x_{t-1}, x_{s+1}, y_{t:s})$ which can be used as an independence proposal. We can obtain a Gaussian approximation to $p(y_j \mid x_j)$ by using a quadratic (in x_j) approximation to $\log p(y_j \mid x_j)$ via a Taylor expansion about a suitable estimate $\hat{x}_j$. The details of this quadratic approximation are the same as for the single-step update described above. Further details can be found in Shephard and Pitt (1997). The resulting quadratic approximation to $p(x_{t:s} \mid x_{t-1}, x_{s+1}, y_{t:s})$ can be calculated efficiently using the Kalman filter (Kalman and Bucy, 1961), or efficient methods for Gaussian Markov random field models (Rue and Held, 2005), and its complexity is $O(s - t)$.

Implementation of this method requires a suitable set of estimates $\hat{x}_{t:s} = (\hat{x}_t, \ldots, \hat{x}_s)$. If we denote by $q(x_{t:s} \mid \hat{x}_{t:s})$ the Gaussian approximation to $p(x_{t:s} \mid x_{t-1}, x_{s+1}, y_{t:s})$ obtained by using the estimate $\hat{x}_{t:s}$, then one approach is to: (i) choose an initial estimate $\hat{x}_{t:s}^{(0)}$; and (ii) for $i = 1, \ldots, I$, set $\hat{x}_{t:s}^{(i)}$ to be the mean of $q(x_{t:s} \mid \hat{x}_{t:s}^{(i-1)})$. In practice, choosing $\hat{x}_{t:s}^{(0)}$ to be the mean of $p(x_{t:s} \mid x_{t-1}, x_{s+1})$ and using small values of I appears to work well.

This approach to designing independence proposals can be extended to other models where the model of the state is linear Gaussian (see Jungbacker and Koopman, 2007). Using the resulting independence sampler within an MCMC algorithm is straightforward if it is efficient to update the complete state path $X_{1:n}$. If not, we must update the state in smaller blocks. A simplistic approach would be to split the data into blocks of (approximately) equal size, τ say, and then update in turn $X_{1:\tau}$, $X_{(\tau+1):2\tau}$, and so on. However, this approach will mean that state values toward the boundaries of each block will mix slowly due to the conditioning on the state values immediately outside the boundary of the blocks. To avoid this, Shephard and Pitt (1997) suggest randomly choosing the blocks to be updated for each application of the independence proposal. Another popular alternative is to choose overlapping blocks, for example, $X_{1:2\tau}$, $X_{(\tau+1):3\tau}$, $X_{(2\tau+1):4\tau}$ and so on.

A further important consideration in implementation is the choice of block size. Too small and we will obtain poor mixing due to the strong dependence of $X_{t:s}$ on X_{t-1} and X_{s+1}; too large and we will have poor mixing due to low acceptance rates. (One approach is to use adaptive MCMC methods to choose appropriate block sizes; see Roberts and Rosenthal, 2009.) Here we will look at the effect that block size has on acceptance probabilities for the SV model.

Plots of average acceptance rates for different block sizes and different data sets are shown in Figure 21.3. Two features are striking. The first is that efficiency varies substantially with ϕ, with values of $\phi \approx 1$ producing higher average acceptance rates. This is because for $\phi \approx 1$ there is stronger dependence in the state process, and thus the (Gaussian) $p(x_{t:s} \mid x_{t-1}, x_{s+1})$ dominates the (non-Gaussian) likelihood $p(y_{t:s} \mid x_{t:s})$. The second is that there is great variability in acceptance rates across different runs: thus choice of too large block sizes can lead to the chain becoming easily stuck (e.g. acceptance probabilities of 10^{-8} or less were observed for blocks of 2000 or more observations when $\phi = 0.8$). This variability suggests that a sensible strategy is either to randomly choose block sizes, or to adaptively choose block sizes for a given data set.

However, overall we see that the block updates are particularly efficient for the SV model. For block updates, acceptance rates greater than 0.01 are reasonable, and the average acceptance rate was greater than this for all combinations of ϕ and block size that we considered. Even looking at the worse-case acceptance rates across all runs, we have acceptances rates greater than 0.01 for blocks of size 400 when $\phi = 0.8$, and for blocks of size 2500 when $\phi = 0.99$.

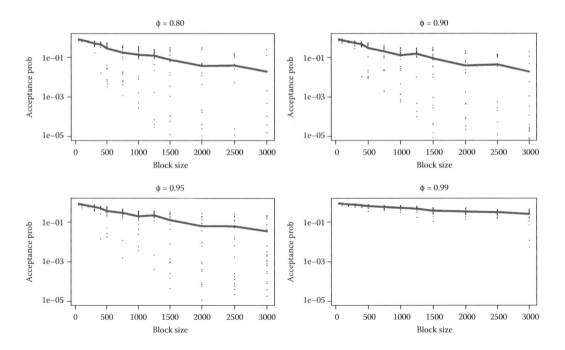

FIGURE 21.3

Average acceptance rates for different block sizes, and different ϕ values. Dots show mean acceptance rates for 20 different data-sets for each block size. Lines show mean acceptance rates for each block size. All runs had $\tau^2 = \sigma^2/(1 - \phi^2) = 0.2$. (Some MCMC runs had acceptance rates that are too small to appear on the plot.)

21.3.3 Other Approaches

Our examples have shown how to simulate directly from the full conditional of the state, or how to approximate the full conditional for use within an independence proposal. However, the former method can only be applied to a limited class of models, and the latter used the linear Gaussian nature of the state model. It is possible to obtain good independence proposals for more general state models, but this can become challenging, particularly for high-dimensional states and models with strong nonlinearities.

One general approach to block updates of the states has recently been proposed in Andrieu et al. (2010), which is based upon using sequential Monte Carlo methods (see Liu and Chen, 1998) within MCMC. Sequential Monte Carlo methods can be efficient for analysing state–space models where parameters are known, and the idea is that these are used to generate a proposal distribution for the path of the state within an MCMC algorithm.

21.4 Updating the Parameters

We now consider how to update the parameter, θ, within the MCMC algorithm. The natural approach is to update θ conditional on the current value of the state path $x_{1:n}$. Often this is simple to implement as either conjugate priors for θ can be chosen so that we can sample directly from $p(\theta \mid x_{1:n}, y_{1:n})$, or θ is of sufficiently low dimension that we can use efficient independence proposals. In some cases we need to update components or blocks of θ at a time, rather than the updating the whole parameter vector in one go.

However, even if we can sample from the full conditional $p(\theta \mid x_{1:n}, y_{1:n})$, the overall efficiency of the MCMC algorithm can still be poor if there is strong correlation between θ and $x_{1:n}$. The rate of convergence of an algorithm that alternates between sampling from $p(x_{1:n} \mid \theta, y_{1:n})$ and $p(\theta \mid x_{1:n}, y_{1:n})$ is given by Liu (1994) and Roberts and Sahu (1997). If, for a square-integrable function f of the parameters, we define the *Bayesian fraction of missing information*,

$$\gamma_f = 1 - \frac{\mathrm{E}\left(\mathrm{var}\left(f(\theta) \mid X_{1:n}, y_{1:n}\right) \mid y_{1:n}\right)}{\mathrm{var}\left(f(\theta) \mid y_{1:n}\right)}, \qquad (21.7)$$

then the geometric rate of convergence of the MCMC algorithm is $\gamma = \sup_f \gamma_f$. Values of $\gamma \approx 1$ suggest a poorly mixing MCMC algorithm. This will occur when, after conditioning on the data, there are functions f for which most of the variation in $f(\theta)$ is explained by the value of the state, $X_{1:n}$.

When there is strong dependence between θ and $X_{1:n}$, there are two techniques for improving mixing. The first is to consider a different parameterization, with the hope that for this new parameterization there will be less dependence between the state and the parameter. The second is to use moves that jointly update θ and $X_{1:n}$. We will describe and evaluate approaches for updating θ given $X_{1:n}$, and then consider these two approaches for improving mixing in turn.

21.4.1 Conditional Updates of the Parameters

Here we focus on Examples 21.1 and 21.2, and give outlines of how parameter updates can be made with these models. We will also investigate the mixing properties of the resulting MCMC algorithms.

Example 21.9: Conditional Parameter Updates

Following Shephard and Pitt (1997), we will consider independent priors for β, σ^2, and ϕ. As β is a scale parameter, we choose the canonical uninformative prior, $p(\beta) \propto 1/\beta$. For σ^2 our prior is $S_0 \chi_p^{-2}$. As it is normal to restrict $|\phi| < 1$, we choose a Beta(a, b) prior for $(\phi + 1)/2$. For these choices we have that, conditional on $\{x_{1:n}, y_{1:n}\}$, β is independent of ϕ, σ^2, and has distribution

$$\beta^2 \mid \{x_{1:n}, y_{1:n}\} \sim \chi_n^{-2} \sum_{t=1}^{n} y_t^2 \exp\{-x_t\}. \tag{21.8}$$

To update ϕ and σ it is simplest to use their conditional distributions,

$$\sigma^2 \mid \{x_{1:n}, y_{1:n}, \phi\} \sim \chi_{n+p}^{-2} \left\{ S_0 + x_1^2(1 - \phi^2) + \sum_{t=2}^{n}(x_t - \phi x_{t-1})^2 \right\},$$

$$p(\phi \mid x_{1:n}, y_{1:n}, \sigma) \propto (1 + \phi)^{a-1/2}(1 - \phi)^{b+1/2} \exp \left\{ -\frac{(1 - \phi^2)x_1^2}{2\sigma^2} - \frac{1}{2\sigma^2} \sum_{t=2}^{n}(x_t - \phi x_{t-1})^2 \right\}.$$

The distribution for σ^2 can be sampled from directly. For ϕ, a simple procedure is an independence sampler with Gaussian proposal. The Gaussian proposal is chosen proportional to

$$\exp \left\{ -\frac{(1 - \phi^2)x_1^2}{2\sigma^2} - \frac{1}{2\sigma^2} \sum_{t=2}^{n}(x_t - \phi x_{t-1})^2 \right\},$$

which corresponds to a mean of $\sum_{t=2}^{n} x_t x_{t-1} / \sum_{t=2}^{n-1} x_t^2$ and a variance of $\sigma^2 / \sum_{t=2}^{n-1} x_t^2$. (Note that this distribution can propose values outside $(-1, 1)$, and such values will always be rejected.)

An example of how the mixing of the MCMC algorithm is affected by the dependence within the state model is shown in the top row of Table 21.1 (labeled noncentered parameterization). We notice that as ϕ increases—that is, the dependence in the state model increases—so the mixing deteriorates. This is because in this limit the amount of information about β contained in the state path remains roughly constant as ϕ increases, but the amount of information about β contained just in the observations is decreasing. This means that the Bayesian fraction of missing information is increasing, and thus the MCMC algorithm mixes more poorly.

Example 21.10: Conditional Parameter Updates

Let P_k denote the kth row of the transition matrix, P. Consider the case where the parameter vector can be written as $\theta = (P, \phi_1, \ldots, \phi_K)$, with the likelihood function given $X_t = k$ is of the

TABLE 21.1

Lag-1 Autocorrelation for β for Both Noncentered and Centered Parameterizations

ϕ	0.8	0.9	0.95	9.75	0.99
Noncentered	0.11	0.21	0.37	0.62	0.98
Centered	0.89	0.79	0.64	0.43	0.29

Results are for $\sigma^2 = 0.02^2$, $\beta = 1$ and $n = 200$, and different values of ϕ.

form $f_k(y \mid \theta) = f_k(y \mid \phi_k)$. That is, we have a disjoint set of parameters for each of the K likelihood models. Further assume first that the distribution of X_1 is independent of θ. In this case, if our priors for the P_k and ϕ_k are independent, then the full conditional $p(\theta \mid x_{1:n}, y_{1:n})$ simplifies. Conditional on $\{x_{1:n}, y_{1:n}\}$, we have independence of $P_1, \ldots, P_K, \phi_1, \ldots, \phi_K$. Thus we can perform independent updates of each of these $2K$ parameters in turn. (If the distribution of X_1 depends on P, then this will introduce weak dependence in the posterior distribution of the P_k.)

If we choose a Dirichlet prior for the entries of P_k, then the $p(P_k \mid x_{1:n}, y_{1:n})$ will be a Dirichlet distribution. Updating of ϕ_k will depend on the specific likelihood model and priors used. However, for the DNA model introduced in Section 21.3.1, we have $\phi_k = \pi^{(k)} = (\pi_A^{(k)}, \pi_C^{(k)}, \pi_G^{(k)}, \pi_T^{(k)})$, and if we have a Dirichlet prior then $p(\phi_k \mid x_{1:n}, y_{1:n})$ will again be Dirichlet.

21.4.2 Reparameterization of the Model

We have seen that dependence between $X_{1:n}$ and θ can result in an MCMC algorithm for $(X_{1:n}, \theta)$ that mixes poorly. One approach to alleviate this is to consider alternative parameterizations.

Papaspiliopoulos et al. (2007) describe two possible general parameterizations for hierarchical models (see also Gelfand et al., 1995; Papaspiliopoulos et al., 2003), and these can be used for state–space models. These are *centered parameterizations*, which in our setup are defined by a model where $p(\theta \mid x_{1:n}, y_{1:n}) = p(\theta \mid x_{1:n})$, and *noncentered parameterizations*, where *a priori* θ and $X_{1:n}$ are independent. If we consider Examples 21.9 and 21.10 above, then for the SV model of Example 21.9 our parameterization for β is noncentered—as our model for $X_{1:n}$ does not depend on β. By comparison, for Example 21.10 our parameterization for P is a centered parameterization.

While it is nontrivial to introduce a noncentered parameterization for Example 21.10—though Papaspiliopoulos (2003) and Roberts et al. (2004) propose approaches that could be used—it is straightforward to introduce a centered parameterization for Example 21.9. We define $\mu = 2\log\beta$ and a new state model $X'_{1:n}$ where

$$X'_t \mid \{x'_{1:t-1}, y_{1:t-1}\} \sim N(\mu + \phi(x'_{t-1} - \mu), \sigma^2),$$

with $X'_1 \sim N(\mu, \sigma^2/(1 - \phi^2))$, and

$$Y_t \mid \{x'_{1:t}, y_{1:t-1}\} \sim N(0, \exp\{x'_t\}).$$

For this parameterization we have (Pitt and Shephard, 1999)

$$\mu \mid \{x'_{1:n}, y_{1:n}\} \sim N(b/a, \sigma^2/a),$$

where $a = (n-1)(1-\phi)^2 + (1-\phi^2)$ and $b = (1-\phi)\{\sum_{t=2}^n (x'_t - \phi x'_{t-1})\} + x'_1(1-\phi^2)$.

For large n we can compare γ_f (Equation 21.7) for $f(\theta) = \mu$ for both centered and noncentered parameterizations. If we conjecture that $\gamma \approx \gamma_f$, then these values will inform us about the relative efficiency of the two parameterizations. To compare γ_f for the two parameterizations we need only compare $E(\text{var}(2\log\beta \mid X_{1:n}, y_{1:n}) \mid y_1 : n)$ and $E(\text{var}(\mu \mid X'_{1:n}, y_{1:n}) \mid y_{1:n})$. If the former is larger, than the centered parameterization will have a smaller value for γ_f, and we may conjecture will have a better rate of convergence. Otherwise γ_f will be smaller for the noncentered parameterization.

Now for the noncentered parameterization we have $\mathrm{var}(\mu \mid X'_{1:n}, y_{1:n}) = 1/a \approx \sigma^2/(n(1 - \phi)^2)$. Thus, as this does not depend on $X'_{1:n}$, we have

$$\mathrm{E}(\mathrm{var}(\mu \mid X'_{1:n}, y_{1:n}) \mid y_{1:n}) \approx \frac{\sigma^2}{n(1 - \phi)^2}.$$

For the centered parameterization, from Equation 21.8, we have that $\mathrm{E}(\mathrm{var}(2 \log \beta \mid X_{1:n}, y_{1:n}) \mid y_{1:n}) = \mathrm{var}(\log \chi_n^2)$, thus for large n,

$$\mathrm{E}(\mathrm{var}(2 \log \beta \mid X_{1:n}, y_{1:n}) \mid y_{1:n}) \approx \frac{2}{n}.$$

Thus γ_f is smaller for the centered parameterization if $2/n > \sigma^2/(n(1 - \phi)^2)$ or

$$\phi > 1 - \frac{\sigma}{\sqrt{2}}.$$

This suggests that as $\phi \to 1$ we should prefer using the centered parameterization, but for small ϕ the noncentered parameterization would be preferred. This is confirmed by simulation (see Table 21.1). Similarly, when σ is small we should prefer the centered parameterization.

For the specific model we consider in Example 21.1, we have centered parameterizations for σ and ϕ. It is possible to extend the noncentered parameterizations for β to one for (β, σ) and even (β, σ, ϕ). For (β, σ) we introduce a state $X'_{1:n}$ where

$$X'_t \mid \{x'_{1:t-1}, y_{1:t-1}\} \sim N(\phi x'_{t-1}, 1),$$

with $X'_1 \sim N(0, 1/(1 - \phi^2))$, and

$$Y_t \mid \{x'_{1:t}, y_{1:t-1}\} \sim N(0, \beta^2 \exp\{\sigma x'_t\}).$$

For (β, σ, ϕ) we can parameterize the state in terms of the standardized residuals in the AR model, $(X_t - \phi X_{t-1})/\sigma$, and $X_1\sqrt{1 - \phi^2}$, which are independent standard normal random variables. Together with related ideas, this idea has been used extensively within continuous-time SV models (see Golightly and Wilkinson, 2008; Roberts and Stramer, 2001).

21.4.3 Joint Updates of the Parameters and State

One way of thinking about why strong correlation between θ and $X_{1:n}$ produces poor mixing, is that large moves of θ are likely to be rejected as they will be inconsistent with the current value of the state. This will happen even if the proposed new value for θ is consistent with the data. This motivates jointly updating θ and $X_{1:n}$ from a proposal $q(\theta', x'_{1:n} \mid \theta, x_{1:n}) = q(\theta' \mid \theta)q(x'_{1:n} \mid \theta')$. Thus $q(\theta' \mid \theta)$ could propose large moves, and then values of the state process consistent with θ' will be simulated from $q(x'_{1:n} \mid \theta')$.

This is most easily and commonly implemented for models where we can simulate directly from $p(x_{1:n} \mid \theta, y_{1:n})$, in which case we choose $q(x'_{1:n} \mid \theta') = p(x'_{1:n} \mid \theta', y_{1:n})$. The resulting acceptance ratio then simplifies to

$$\min\left\{1, \frac{q(\theta \mid \theta')p(\theta' \mid y_{1:n})}{q(\theta' \mid \theta)p(\theta \mid y_{1:n})}\right\}.$$

This acceptance ratio does not depend on $x_{1:n}$ or $x'_{1:n}$. The marginal chain for θ is equivalent to an MCMC chain for $p(\theta \mid y_{1:n})$ with proposal distribution $q(\theta' \mid \theta)$.

Providing an efficient proposal $q(\theta' \mid \theta)$ can be found, such an MCMC algorithm will always be more efficient than one that updates θ and $X_{1:n}$ independently. However, the difficulty with implementing this idea is how to choose $q(\theta' \mid \theta)$. For Markov modulated Poisson processes, Sherlock et al. (2008) found that a Gibbs sampler that updated $X_{1:n}$ given θ and θ given $X_{1:n}$ performed better than this joint update where $q(\theta' \mid \theta)$ was chosen to be a symmetric random walk. A further advantage of the Gibbs sampler is that it avoids tuning $q(\theta' \mid \theta)$, though this problem can be alleviated by using adaptive MCMC schemes (Andrieu and Thoms, 2008; Sherlock et al., 2008).

A simple extension of this joint updating idea is possible if we have an efficient independence proposal for $x_{1:n}$ given θ—as this proposal could be used as $q(x'_{1:n} \mid \theta')$. Here the efficiency of the resulting algorithm will depend on both the efficiency of $q(\theta' \mid \theta)$ as a proposal for an MCMC algorithm that explores $p(\theta \mid y_{1:n})$, and also the closeness of $q(x'_{1:n} \mid \theta')$ to $p(x'_{1:n} \mid \theta', y_{1:n})$.

21.5 Discussion

This chapter has given an introduction to MCMC methods for state–space models. Two main issues have been covered. Firstly, if there is strong, or long-range, dependence in the state–space model, then an efficient MCMC algorithm will need to update blocks of the state process in a single move. Secondly, strong correlation between the parameters and the state process can lead to slow mixing of the MCMC algorithm (even if there are efficient methods for updating the state process). To improve mixing, either reparameterization of the model, or joint updates of the state and the parameters will be needed.

While we have looked at examples where it is possible to construct efficient moves for updating the state, in many applications this can be difficult to achieve. Recent research has looked at the use of sequential Monte Carlo methods within MCMC (Andrieu et al., 2010), and these ideas show promise for providing a general-purpose approach for updating the whole state process (or large batches of it) in a single MCMC move. Related methods are able to allow for efficient joint updates of the state and parameter process (Andrieu et al., 2010) or for methods that mix over the marginal posterior of the parameters (Andrieu and Roberts, 2009).

While we have focused on discrete-time state processes, many of the issues extend naturally to continuous-time processes. For example, the issue of model parameterization for diffusion models is discussed in Roberts and Stramer (2001), and for these models certain parameterizations can lead to MCMC algorithms which are reducible. Extensions of the forward–backward algorithm to continuous-time models are considered in Fearnhead and Meligkotsidou (2004), and independence sampler updates for the state process in diffusion models are developed in Golightly and Wilkinson (2008).

References

Andrieu, C. and Roberts, G. O. 2009. The pseudo-marginal approach for efficient computations. *Annals of Statistics*, 37:697–725.

Andrieu, C. and Thoms, J. 2008. An overview of controlled MCMC. *Statistics and Computing*, 18:343–373.

Andrieu, C., Doucet, A., and Holenstein, R. 2010. Particle Markov chain Monte Carlo methods (with discussion). *Journal of the Royal Statistical Society, Series B*, 72(3):269–342.

Ball, F. G. and Rice, J. A. 1992. Stochastic models for ion channels: Introduction and bibliography. *Mathematical Biosciences*, 112:189–206.

Barry, D. and Hartigan, J. A. 1992. Product partition models for change point problems. *Annals of Statistics*, 20:260–279.

Beaulieu, C., Ouarda, T. B. M. J., and Seidou, O. 2007. A review of homogenization techniques for climate data and their applicability to precipitation series. *Hydrological Sciences Journal*, 52:18–37.

Boys, R. J., Henderson, D. A., and Wilkinson, D. J. 2000. Detecting homogeneous segments in DNA sequences by using hidden Markov models. *Applied Statistics*, 49:269–285.

Carter, C. K. and Kohn, R. 1994. On Gibbs sampling for state space models. *Biometrika*, 81(3):541–553.

Chen, J. and Gupta, A. K. 1997. Testing and locating changepoints with application to stock prices. *Journal of the American Statistical Association*, 92:739–747.

Didelot, X., Achtman, M., Parkhill, J., Thomson, N. R., and Falush, D. 2007. A bimodal pattern of relatedness between the *Salmonella* Paratyphi A and Typhi genomes: Convergence or divergence by homologous recombination? *Genome Research*, 17:61–68.

Fearnhead, P. 2006. Exact and efficient inference for multiple changepoint problems. *Statistics and Computing*, 16:203–213.

Fearnhead, P. 2008. Computational methods for complex stochastic systems: A review of some alternatives to MCMC. *Statistics and Computing*, 18:151–171.

Fearnhead, P. and Meligkotsidou, L. 2004. Exact filtering for partially-observed continuous-time Markov models. *Journal of the Royal Statistical Society, Series B*, 66:771–789.

Fearnhead, P. and Sherlock, C. 2006. Bayesian analysis of Markov modulated Poisson processes. *Journal of the Royal Statistical Society, Series B*, 68:767–784.

Fearnhead, P. and Vasileiou, D. 2009. Bayesian analysis of isochores. *Journal of the American Statistical Association*, 104(485):132–141.

Gelfand, A. E., Sahu, S., and Carlin, B. P. 1995. Efficient parameterisations for normal linear mixed models. *Biometrika*, 82:479–488.

Golightly, A. and Wilkinson, D. J. 2008. Bayesian inference for nonlinear multivariate diffusion models observed with error. *Computational Statistics and Data Analysis*, 52:1674–1693.

Harvey, A. C. 1989. *Forecasting, Stuctural Time Series and the Kalman Filter*. Cambridge University Press, Cambridge.

Hodgson, M. E. A. 1999. A Bayesian restoration of an ion channel signal. *Journal of the Royal Statistical Society, Series B*, 61:95–114.

Hull, J. and White, A. 1987. The pricing of options on assets with stochastic volatilities. *Journal of Finance*, 42:281–300.

Juang, B. H. and Rabiner, L. R. 1991. Hidden Markov models for speech recognition. *Technometrics*, 33:251–272.

Jungbacker, B. and Koopman, S. J. 2007. Monte Carlo estimation for nonlinear non-Gaussian state space models. *Biometrika*, 94:827–839.

Kalman, R. and Bucy, R. 1961. New results in linear filtering and prediction theory. *Journal of Basic Engineering, Transactions ASME Series D*, 83:95–108.

Liu, J. S. 1994. Fraction of missing information and convergence rate of data augmentation. In J. Sall and A. Lehman (eds), *Computing Science and Statistics: Proceedings of the 26th Symposium on the Interface*, pp. 490–496. Interface Foundation of North America, Fairfax Station, VA.

Liu, J. S. and Chen, R. 1998. Sequential Monte Carlo methods for dynamic systems. *Journal of the American Statistical Association*, 93:1032–1044.

Liu, J. S. and Lawrence, C. E. 1999. Bayesian inference on biopolymer models. *Bioinformatics*, 15:38–52.

Lund, R. and Reeves, J. 2002. Detection of undocumented changepoints: A revision of the two-phase regression model. *Journal of Climate*, 15:2547–2554.

Papaspiliopoulos, O. 2003. Non-centered parameterizations for hierarchical models and data augmentation. PhD thesis, Department of Mathematics and Statistics, Lancaster University.

Papaspiliopoulos, O., Roberts, G. O., and Sköld, M. 2003. Non-centred parameterizations for hierarchical models and data augmentation (with discussion). In J. M. Bernardo, M. J. Bayarri, J. O. Berger, A. P. Dawid, D. Heckerman, A. F. M. Smith, and M. West (eds), *Bayesian Statistics 7*, pp. 307–326. Clarendon Press, Oxford.

Papaspiliopoulos, O., Roberts, G. O., and Sköld, M. 2007. A general framework for the parameterization of hierarchical models. *Statistical Science*, 22:59–73.

Pitt, M. K. and Shephard, N. 1999. Filtering via simulation: auxiliary particle filters. *Journal of the American Statistical Association*, 94:590–599.

Rabiner, L. R. and Juang, B. H. 1986. An introduction to hidden Markov models. *IEEE ASSP Magazine*, pp. 4–15.

Ritov, Y., Raz, A., and Bergman, H. 2002. Detection of onset of neuronal activity by allowing for heterogeneity in the change points. *Journal of Neuroscience Methods*, 122:25–42.

Roberts, G. O. and Rosenthal, J. 2009. Examples of adaptive MCMC. *Journal of Computational and Graphical Statistics* 18(2):349–367.

Roberts, G. O. and Sahu, S. K. 1997. Updating schemes, correlation structure, blocking and parameterization for the Gibbs sampler. *Journal of the Royal Statistical Society, Series B*, 59:291–317.

Roberts, G. O. and Stramer, O. 2001. On inference for partially observed nonlinear diffusion models using the Metropolis-Hastings algorithm. *Biometrika*, 88:603–621.

Roberts, G. O., Papaspiliopoulos, O., and Dellaportas, P. 2004. Bayesian inference for non-Gaussian Ornstein–Uhlenbeck stochastic volatility processes. *Journal of the Royal Statistical Society, Series B*, 66:369–393.

Rue, H. and Held, L. 2005. *Gaussian Markov Random Fields: Theory and Applications*. Chapman & Hall/CRC, Boca Raton, FL.

Scott, S. L. 2002. Bayesian methods for hidden Markov models: Recursive computing in the 21st century. *Journal of the American Statistical Association*, 97:337–351.

Shephard, N. 1996. Statistical aspects of ARCH and stochastic volatility. In D. R. Cox, D. V. Hinkley, and O. E. Barndorff-Nielsen (eds), *Time Series Models in Econometrics, Finance and Other Fields*, pp. 1–67. Chapman & Hall, London.

Shephard, N. and Pitt, M. K. 1997. Likelihood analysis of non-Gaussian measurement time series. *Biometrika*, 84:653–667.

Sherlock, C., Fearnhead, P., and Roberts, G. O. 2008. The random walk Metropolis: linking theory and practice through a case study. *Statistical Science*, to appear.

Smith, J. Q. and Santos, A. F. 2006. Second order filter distribution approximations for financial time series with extreme outlier. *Journal of Business and Economic Statistics*, 24:329–337.

Yao, Y. 1984. Estimation of a noisy discrete-time step function: Bayes and empirical Bayes approaches. *Annals of Statistics*, 12:1434–1447.

22

MCMC *in Educational Research*

Roy Levy, Robert J. Mislevy, and John T. Behrens

22.1 Introduction

Quantitative educational research has traditionally relied on a broad range of statistical models that have evolved in relative isolation to address different facets of its subject matter. Experiments on instructional interventions employ Fisherian designs and analyses of variance; observational studies use regression techniques; and longitudinal studies use growth models in the manner of economists. The social organization of schooling—of students within classrooms, sometimes nested within teachers, of classrooms within schools, schools within districts, districts within states, and states within nations—necessitates hierarchical analyses. Large-scale assessments employ the complex sampling methodologies of survey research. Missing data abound across levels. And most characteristically, measurement error and latent variable models from psychometrics address the fundamental fact that what is ultimately of most interest, namely what students know and can do, cannot be directly observed: a student's performance on an assessment may be an *indicator* of proficiency but, no matter how well the assessment is constructed, it is *not the same thing* as proficiency. This measurement complexity exacerbates computational complexity when researchers attempt to combine models for measurement error with models addressing the aforementioned structures. Further difficulties arise from an extreme reliance on frequentist interpretations of statistical methods that limit the computational and interpretive machinery available (Behrens and Smith, 1996). In sum, most applied educational research has been marked by interpretive limitations inherent in the frequentist approach to testing, estimation, and model building, a plethora of independently created and applied conceptual models, and computational limitations in estimating models that would capture the complexity of this applied domain.

This chapter discusses how a Markov chain Monte Carlo (MCMC) approach to model estimation and associated Bayesian underpinnings address these issues in three ways. First, the Bayesian conceptualization and the form of results avoid a number of interpretive problems in the frequentist approach while providing probabilistic information of great value to applied researchers. Second, the flexibility of the MCMC models allows a conceptual unification of previously disparate modeling approaches. Third, the MCMC approach allows for the estimation of the more complex and complete models mentioned above, thereby providing conceptual and computational unification.

Because MCMC estimation is a method for obtaining empirical approximations of posterior distributions, its impact as calculation *per se* is joint with an emerging Bayesian revolution in reasoning about uncertainty—a statistical mindset quite different from that of hypothesis testing, parameter estimation, and model-fitting in the classical paradigm that has characterized educational research. A fertile groundwork was laid in this field from the 1960s through the 1980s by Melvin Novick. Two lines of Novick's work are particularly relevant to the subject of this chapter. First is the subjectivist Bayesian approach to model-based reasoning about real-world problems—building models in terms what one knows and does not know, from experience and theory, and what is important to the inferential problem at hand (see, e.g. Lindley and Novick, 1981, on exchangeability). His application of these ideas to prediction across multiple groups (Novick and Jackson, 1974) foreshadows the modular model-building to suit the complexities of real-world problems that MCMC enables. In particular, the ability to "borrow" information across groups to a degree determined by the data, rather than pooling the observations or estimating groups separately, was a major breakthrough of the time—natural from a Bayesian perspective, but difficult to frame and interpret under the classical paradigm. Second is the realization that broad use of the approach would require computing frameworks to handle the mathematics, so the analyst could concentrate on the substance of the problem. His Computer-Assisted Data Analysis (CADA; Libby et al., 1981) pioneered Bayesian reasoning about posteriors in ways that are today reflected in the output of MCMC programs such as WinBUGS (Spiegelhalter et al., 2007).

22.2 Statistical Models in Education Research

Hierarchical or multilevel models extend more basic statistical models to model dependencies between subjects or measures that have a hierarchical structure (e.g. test scores over time nested within students, students within classrooms, classrooms within schools, schools within districts/states, etc.). Regression-like models are formulated at the lowest level (level 1) of analysis. Parameters from this level of analysis are in turn modeled, frequently by regression-like models, to specify level-2 parameters that capture the effects of covariates at that level, such as school policies. This may be extended to any number of levels. Applications in education typically employ linear regression-like models at each level, frequently assuming normality in each case. The effects of primary interest differ from one study to another, but properly modeling the structure better captures patterns of shared influence and appropriately models the levels at which effects occur.

Within models for educational effects, the lowest level of modeling often addresses students' responses conditional on unobservable or latent variables that characterize students. These psychometric models facilitate inference from observations of behaviors made by subjects to more broadly conceived statements about the subjects and/or the domain of interest. Though surface features vary, modern psychometric modeling paradigms are characterized by the use of probabilistic reasoning in the form of statistical models to facilitate such inferences in light of uncertainty (Mislevy and Levy, 2007).

Table 22.1 summarizes several of the more popular psychometric models in terms of assumptions about the latent variables capturing subject proficiency and the observables serving as indicators of the latent variables. Factor analysis (FA; Bollen, 1989; Gorsuch, 1983) posits that both the observables and latent variables are continuous and frequently additionally assumes the observables to be normally distributed. Structural equation modeling (SEM; Bollen, 1989) can be historically viewed as extending the factor-analytic tradition

TABLE 22.1

Taxonomy of Popular Psychometric Models

Observable Variables	Latent Variable(s)			
	Continuous		Discrete	
	Univariate	Multivariate	Univariate	Multivariate
Dichotomous	Item response theory	Multidimensional item response theory	Latent class analysis	Bayesian networks Cognitive diagnosis models
Polytomous unordered	Item response theory		Latent class analysis	Bayesian networks Cognitive diagnosis models
Polytomous ordered	Item response theory	Multidimensional item response theory Factor analysis	Latent class analysis	Bayesian networks Cognitive diagnosis models
Normal	Factor analysis Structural equation modeling	Factor analysis Structural equation modeling		

with regression-like structures that relate latent variables to one anther. Item response theory (IRT; Lord, 1980) assumes the observables to be discrete and, when polytomous, usually ordered. Latent class analysis (LCA; Lazarsfeld and Henry, 1968) and related models (discussed in more detail below) assume that both the observables and latent variables are discrete.

The nomenclature in Table 22.1 reflects historical traditions, with associated purposes, assumptions, and estimation frameworks. As each of the modeling frameworks have expanded the historical lines have become blurred. For example, multidimensional latent variable models for discrete data may be framed as either a multidimensional extension of (unidimensional) IRT models or the application of common factor models to discrete data (Takane and de Leeuw, 1987). Moreover, the models can be combined in nuanced ways, such as the recently developed mixtures of IRT models that synthesize hitherto separate streams of work in IRT and latent class modeling (Rost, 1990).

This treatment is far from exhaustive and beyond the intent of this chapter (though in later sections we will discuss these and other models in use in education research, some of which can be viewed as extensions or combinations of those already mentioned). For the focus of this chapter, it is important to recognize that these modeling paradigms grew out of their own independent traditions, with at best only partially overlapping foci, literatures, notational schemes, and—principally related to the current discussion—estimation frameworks and routines. For example, FA and SEM have historically been employed to model relationships among constructs, rather than features of subjects. Estimation typically involves least squares or maximum likelihood using first- and second-order moments from sample data, with an emphasis on the estimation of structural parameters, that is, parameters for the conditional distributions of observed scores given latent variables, here interpreted as factor loadings and factor covariances, but not on the values of the latent variables for individual persons, here factors (Bollen, 1989). In contrast, IRT models are commonly employed to scale test items and examinees. Estimation usually involves the analysis of individual-level data or frequencies of response patterns and assumptions regarding the distribution of the latent variables for individuals, here interpreted as student proficiencies. Once again the estimation of structural parameters, now interpreted as item parameters, is important, but here, estimation of students' proficiencies is important to guide desired inferences about individuals (Lord, 1980). Disparate estimation approaches that optimized these different target of inferences evolved in IRT and in FA and SEM, with the unfortunate consequence of obscuring fundamental similarities among the models.

As a consequence of these fragmented strands of development, analysts faced choosing from a palette of partial and incomplete solutions as to both models and computer programs. The most sophisticated techniques available for IRT and LCA, for example, assumed simple random sampling, while the most widely used programs for hierarchical analysis and for complex-sampled data offered at best simple error models for student proficiencies. Each of the features addressed in the various models, however, represented a recurring structure in educational research settings, often at the same time. It is to solving this problem that the Bayesian inferential approach and MCMC estimation make their greatest contribution.

22.3 Historical and Current Research Activity

This section traces key developments and current applications of MCMC in educational research. The focus is on research settings where the power and flexibility of MCMC are leveraged to conduct modeling that, without MCMC, would prove difficult computationally or in terms of desired inferences. It is no coincidence that this collection of work is mainly Bayesian in nature, though we note that MCMC estimation has been employed in frequentist applications as well (Song and Lee, 2003). Aside from the natural linkage between MCMC estimation and Bayesian inference, a Bayesian approach in which models are formulated hierarchically, prior information can be easily incorporated, and uncertainty in unknown parameters is propagated offers advantages regardless of the estimation routine.*

22.3.1 Multilevel Models

Applications of multilevel models commonly assume linearity and normality within a level, either of the random effects themselves or of residuals given covariates at that level. A thorough overview of Bayesian and Gibbs sampling approaches to hierarchal models of this type as they are used in education is given by Seltzer et al. (1996). Anticipating the growth in popularity of MCMC, Draper (1995) points out that marginal likelihoods of variance parameters at level-2 units may be considerably skewed with few level-2 units (the logic of which may extended for hierarchies with more levels). The implication is that maximum likelihood (ML) estimation of point estimates fails to account for such skewness and will poorly account for the heterogeneity implied by higher-level variance components. Raudenbush et al. (1999) echoed this concern, and employed Gibbs sampling in an analysis of data on the Trial State Assessment from 41 states. They further noted a concern with a classical approach, as the 41 states were a nonrandom

---

* See Lindley and Smith (1972) and Mislevy (1986) for illustrative applications not involving MCMC.

sample and were better treated as strata; however, the traditional approach to treating strata as fixed effects contradicts the goal of modeling between-state variation. Kasim and Raudenbush (1998) noted the importance of properly accounting for uncertainty in higher levels of multilevel models, and turned to Gibbs sampling for estimation of variance components.

Assuming linearity and normality permits estimation via Gibbs sampling, in a straight-forward way of applying MCMC. A key strength of MCMC, however, lies in its flexibility to be applied to models that pose challenges for other estimation strategies, such as those with nonnormal distributional assumptions. In the context of multilevel models, MCMC has proven useful in conducting sensitivity analysis for distributional assumptions at various levels of hierarchical models (Seltzer, 1993; Seltzer et al., 2002) and in fitting hierarchical structures in multilevel logistic regression models (Schulz et al., 2004).

A related use of hierarchical modeling ideas appears in meta-analysis (Glass, 1976; Glass et al., 1981), which was originated to synthesize evidence across studies in educational, medical, and social science research. The most common statistical procedures (see Hedges and Olkin, 1985) rely on a fixed effects model and a series of binary decisions regarding which set of studies constitutes a homogeneous set of sub-studies from which to estimate effects. Unfortunately this approach uses χ^2 tests that suffer from the sample size sensitivity that meta-analysis was designed to solve in the first place: large studies lead to conclusions of separateness regardless of effect size and small studies lead to the opposite conclusion. The Bayesian approach (supported by MCMC estimation) takes a random effects view that models the degree of homogeneity of effects, thereby sidestepping the bifurcations required in the classical approach (Smith et al., 1995; Sutton and Abrams, 2001).

In meta-analytic hierarchical linear modeling, as in other contexts, a Bayesian MCMC approach provides probabilistic information through the posterior distributions that are of great interest to researchers and consumers of research alike; for example, "what is the probability that an effect will be negative?" or "what is the probability that school X had an effect greater than school Y?" These are common-sense questions that are unaddressed in the non-Bayesian framework (Gelman et al., 1995), but flow naturally from a Bayesian inferential framework and MCMC estimation (Gelman and Hill, 2007).

22.3.2 Psychometric Modeling

Table 22.1 classifies several popular psychometric models in terms of their assumptions regarding the latent and observable variables and serves to guide the current discussion.

22.3.2.1 Continuous Latent and Observable Variables

Standard factor-analytic and structural equation models, characterized by linear equations relating the latent and observed variables and (conditional) normality assumptions, do not pose challenges for traditional estimation routines. However, in detailing the use of Gibbs sampling for such models, Scheines et al. (1999) pointed out many advantages of Gibbs sampling over normal-theory ML estimation: Gibbs sampling does not rely on asymptotic arguments for estimation or model checking, inequality constraints may be easily imposed, information about multimodality—undetectable by standard ML estimation—may be seen in marginal posterior densities, and information for underidentified parameters may be supplied via informative priors.

A key advantage of MCMC for SEM lies in its power to estimate nonstandard models that pose considerable challenges for ML and least-squares estimation. Examples of such applications include models with quadratic, interaction, and other nonlinear relationships among latent variables (Arminger and Muthén, 1998; Lee et al., 2007), covariates (Lee et al., 2007), finite (latent) mixtures of structural equation models (Lee and Song, 2003; Zhu and Lee, 2001), heterogeneous factor analysis (Ansari et al., 2002), complex growth curve models (Zhang et al., 2007), and nonignorable missingness (Lee and Tang, 2006). Contrary to a common belief, the computation and programming necessary to implement a Bayesian solution via MCMC in such complex models is less intense than that necessary to conduct ML estimation (Ansari et al., 2002; Zhang et al., 2007).

22.3.2.2 *Continuous Latent Variables and Discrete Observable Variables*

Models in which a set of discrete, possibly ordinal observables (say, scored task or item responses) are modeled via continuous latent variables are widely used in assessment settings. In this section, we survey applications of MCMC to these models from both an IRT and FA perspective, highlighting aspects in which existing estimation traditions limit our modeling potential.

Working in an IRT framework, Albert's (1992) seminal work considered a model for dichotomous observables based on the normal distribution function (i.e. a probit model) and showed how posterior distributions for person and item parameters could be estimated via a Gibbs sampler. The algorithm was extended to handle polytomous data by Albert and Chib (1993); Sahu (2002) described a similar Gibbs sampling approach to modeling dichotomous item responses allowing for examinee guessing in assessment contexts.

A turning point in the application of MCMC for IRT and psychometric modeling more generally arrived with the work of Patz and Junker (1999a), who considered a model for dichotomous observables based on the logistic distribution function and offered a Metropolis–Hastings-within-Gibbs sampling approach, in which a Metropolis–Hastings step is employed to sample from the full conditional distributions. A particularly noteworthy aspect of the Metropolis(–Hastings)-within-Gibbs approach is its applicability to situations in which it is not possible to sample directly from the full conditional distributions. This flexibility has produced an explosion in the use of MCMC for complex, IRT-based models. Examples include models for polytomous data (Patz and Junker, 1999b), nominal data (Wollack et al., 2002), missing data (Patz and Junker, 1999b), rater effects (Patz and Junker, 1999b), testlets (Bradlow et al., 1999), multilevel models (Fox and Glas, 2001), and hierarchical models for mastery classification (Janssen et al., 2000).

An alternative perspective on continuous latent variable models for discrete data stems from the FA tradition, which views the observables as discretized versions of unobservable continuous data. Following the normality assumptions of FA for continuous variables, this approach is akin to a probit model. In surveying estimation approaches to such models, Wirth and Edwards (2007) concluded that traditional factor-analytic methods, even with corrections to estimates and standard errors for discrete data, can fail to capture the true fit. The underlying problem is that the traditional factor-analytic estimation based on minimizing some function of residual covariances or correlations was developed for continuous data, not discrete data. This illustrates the restrictions and limitations imposed by remaining within an estimation paradigm when trying to fit models beyond the scope of those originally intended for the estimation routine. What is needed is an estimation framework flexible enough to handle a variety of assumptions about the distributional features of the data and the data-generating process, not to mention the all too real potential for missingness or sparseness. MCMC provides such a framework.

An arena where the intersection of different modeling paradigms and their associated traditional estimation routines poses unnecessary limits involves multidimensional models for discrete observables, which may be viewed from an IRT perspective as an increase in the number the latent variables (over traditional unidimensional models) or from a (multidimensional) factor-analytic perspective as the factor analysis of discrete data. Interestingly, Wirth and Edwards (2007) tied common misconceptions associated with each perspective to the historical traditions of estimation within each paradigm. Traditional factor-analytic estimation approaches to discrete data have relied on integration over the distribution of the *observable* variables. As the number of observables increases, this integration becomes increasingly difficult, and the applicability of ML and weighted least squares routines requiring large sample sizes relative to the number of observables becomes suspect. Hence, an FA perspective prefers (relatively) few observables in the model, without regard to the number of latent variables. In contrast, traditional estimation approaches in IRT focus on the integration over the *latent* variable(s), which becomes increasingly difficult as the number of latent variables increases. Hence, an IRT perspective prefers (relatively) few latent variables in the model, but is silent with respect to the number of observables. Thus the particulars of the estimation paradigms restrict the scope of the models to be employed. MCMC may be seen as a unifying framework for estimation, freeing the analyst from these restrictive (and conflicting) biases. Examples of the use of MCMC in multidimensional modeling from both IRT and FA perspectives include the consideration of dichotomous data (Béguin and Glas, 2001; Bolt and Lall, 2003; Lee and Song, 2003), polytomous data (Yao and Boughton, 2007), combinations of continuous, dichotomous, and polytomous data (Lee and Zhu, 2000; Shi and Lee, 1998), multiple group models (Song and Lee, 2001), missing data (Song and Lee, 2002), and nonlinear relationships among latent variables (Lee and Zhu, 2000).

22.3.2.3 Discrete Latent Variables and Discrete Observable Variables

Traditional, unrestricted latent class models that model discrete observables as dependent on discrete latent variables are commonly estimated via ML. MCMC may still be advantageous for such models in handling missingness by design, large data sets with outliers, and constructing credibility intervals for inference when an assumption of multivariate normality (of ML estimates or posterior distributions) is unwarranted (Hoijtink, 1998; Hoijtink and Notenboom, 2004).

Turning to more complex models, MCMC has been shown to be useful in the estimation of models with covariates (Chung et al., 2006) and with ordinal and inequality constraints (van Onna, 2002). In assessment, cognitive diagnostic models involve modeling discrete observables (i.e. scored item responses) as dependent on different combinations of the latent, typically binary, attributes characterizing mastery of componential skills necessary to complete the various tasks. The models frequently involve conjunctive or disjunctive effects among parameters to model the probabilistic nature of student responses. These models pose estimation difficulties for traditional routines but can be handled by MCMC (de la Torre and Douglas, 2004; Hartz, 2002).

Such models may be also be cast in a graph-theoretic light as Bayesian networks, which allow for the estimation of a wide variety of complex effects via MCMC. Examples include compensatory, conjunctive, disjunctive, and inhibitor relationships for dichotomous and polytomous data assuming dichotomous or ordered latent student skills or attributes (Almond et al., 2007; Levy and Mislevy, 2004).

We note that the import of cognitive theory is receiving an increasing amount of attention here and across psychometric modeling more generally. Other examples include the use of multidimensional IRT models that posit conjunctive relationships among the latent variables in attempt to align the models with cognitive theories of student processing in solving tasks. Applications of these models have been limited due to difficulties associated with traditional estimation. Bolt and Lall (2003) fit a conjunctive multidimensional model via MCMC, illustrating how the flexibility of MCMC opens the door for the application of complex statistical models aligned with substantive theories regarding students.

22.3.2.4   *Combinations of Models*

The preceding discussions have been couched in terms of traditional divisions between models (Table 22.1), highlighting applications that pose difficulties for estimation routines typically employed. Expanding on that theme, an advanced approach to model construction takes a modular approach in which the statistical model is constructed in a piecewise manner, interweaving and overlaying features from the traditional paradigms. Simple examples include the models that bridge the FA and IRT divide by modeling discrete and continuous observables simultaneously (Lee and Zhu, 2000; Shi and Lee, 1998). More complex examples embed IRT and FA models in latent classes to construct finite mixtures of IRT or FA models (Bolt et al., 2001; Cohen and Bolt, 2005; Lee and Song, 2003; Zhu and Lee, 2001). Table 22.1 no longer reflects choices that must be made about models and associated estimation procedures, but rather modules of recurring relationships that can be adapted and assembled to suit the substantive problem at hand, then fit to data using the overarching framework of Bayesian inference and MCMC estimation.

Missing data are dealt with naturally under MCMC in such models when they are missing at random (see, e.g. Chung et al., 2006, in LCA). Indeed, there is no distinction conceptually between latent variables and missing data (Bollen, 2002), and under MCMC no new impediments are introduced.

Furthering this theme, recent work has sought to simultaneously address two key hallmarks of educational research, namely hierarchical structures of data and the presence of measurement error. Examples of the use of MCMC for multilevel psychometric models can be found in Ansari et al. (2002), Fox and Glas (2001), and Mariano and Junker (2007). Traditional estimation strategies have not been established for these models. Prior to MCMC, overlaying hierarchical structures on latent variable models in a single analysis was intractable. In the following section, we extend these ideas further and consider a model that interweaves hierarchical structures, regression models, and IRT models.

22.4   NAEP Example

This section describes a practical application that combines several of the prototypical structures of educational research discussed above. Johnson and Jenkins (2005) model the distribution of latent proficiencies of student populations, from clustered student-sampling designs, with matrix-sampled item presentation (a version with a multivariate model is given in Johnson, 2002). After reviewing the problem context, we note how previous analyses addressed some aspects of the complex whole while simplifying or ignoring others. We then describe Johnson and Jenkins's solution: a unified Bayesian model, made viable through the use of MCMC.

Large-scale educational assessments such as the National Assessment of Educational Progress (NAEP), the National Adult Literacy Survey (NALS), and the Trends in International Mathematics and Science Study (TIMSS) are developed to collect information on the knowledge and skills of targeted populations, and report how that knowledge varies across different groups in the population and may be related to demographic and educational background variables (Johnson and Jenkins, 2005). These projects simultaneously exhibit several of the recurring structural features that are common to educational research:

- Hierarchical organization of the focal groups. In design, analysis, and interpretation, large-scale educational surveys must address the fact that schooling is organized in terms of students within classes, sometimes crossed with teachers, nested within schools, typically within districts, within states, and, in international surveys, within countries.

- Complex sampling designs for students. Related to the structure of education is the necessity of stratified and cluster sampling designs for schooling. In NAEP, primary sampling units (PSUs) are standard metropolitan sampling areas or similar regions, from which schools are sampled, from which in turn students are sampled.

- Complex sampling designs for tasks. Students' knowledge is better represented by a broad sampling of tasks than by a small single sample of tasks. To reduce respondent burden, many overlapping blocks of tasks are presented to different students in order to better cover content domains at the level of populations. Furthermore, item samples frequently differ across time points and age or grade populations.

- Latent variables. In order to synthesize data across the different samples of tasks that different students take, many projects use latent variables models, notably IRT models. Key inferences are thus based on variables that are not observed from any respondent.

- Regression models. Covariates related to educational outcomes are available, and have effects, at all levels in the hierarchy. NAEP includes student background questionnaires on demographic and educational history, teacher surveys on pedagogical practices, and school-level data on socioeconomic variables.

The history of large-scale educational surveys exhibits continual efforts to incorporate these complexities in analysis. Limitations of special-purpose software would allow analysts to address some features, at the expense of simplifying or ignoring others. Until the 1980s, for example, NAEP accounted for cluster sampling with balanced half replicate designs and employed matrix sampled booklets of tasks, but reported results only in terms of single items or total scores in small sets (Chromy et al., 2004). Longford (1995) and Raudenbush et al. (1999) provide superpopulation-based analyses for educational surveys with hierarchical structures, but consider only error-free dependent variables. The multiple-imputation NAEP analyses introduced in 1984 (Beaton, 1987) accounted for the sampling design with jackknife procedures and used IRT to combine information across booklets, but the point estimates of the IRT item parameters and latent regression models were treated as known. Scott and Ip (2002) demonstrated a Bayesian framework for the multivariate IRT model that NAEP employs, but did not address the complex sampling design.

In contrast, the Johnson and Jenkins (2005) model allows the analyst to simultaneously estimate parameters for a joint model that addresses all of these design features. Its components are as follows. The item responses X_{ij} of each sampled student i are modeled as conditionally independent given the latent proficiency θ_i through an IRT model:

$$p\left(X_i \mid \theta_i, \beta\right) = \prod p\left(X_{ij} \mid \theta_i, \beta_j\right),$$

where β_j are parameters for item j with independent prior distributions $p(\beta_j)$. The forms and parameterizations of the IRT models depend on item types, with the three-parameter logistic IRT model for multiple-choice items and the partial credit graded response model for open-ended tasks with ordered rating scales. (Johnson and Jenkins did not model effects for individual raters and multiple ratings, but could have done so using the aforementioned Patz and Junker approach.)

A regression structure is employed to model the relationship between θ and student-level covariates y_i, with school-level and PSU-level clustering accounted for a linear mixed effects model (Laird and Ware, 1982). Letting student i attend school $s(i)$ and school s belong to PSU $p(s)$, Johnson and Jenkins posit

$$\theta \mid \left(y_i, \gamma, \sigma, \upsilon_{s(i)}\right) \sim N\left(\upsilon_{s(i)} + \gamma' y_i, \sigma_{s(i)}^2\right),$$

$$\upsilon_s \mid \tau \sim N\left(\eta_{p(s)}, \tau^2\right),$$

$$\eta_p \mid \omega \sim N\left(0, \omega^2\right),$$

again with independent prior distributions on the vector of regression coefficients γ, residual variances σ^2, and school and PSU effects and their variances. To estimate any function $G(\Theta)$ of the finite population, Johnson and Jenkins calculate the appropriately weighted mean of that function calculated with the MCMC draws of sampled students in each cycle, and monitor its distribution in the Gibbs chain. Figure 22.1 provides a "plate diagram" of the model (with covariates for schools added to illustrate where they would appear in the hierarchy).

Johnson and Jenkins compared the results from this unified model to the standard NAEP analysis with its piecewise approximations, using both simulations and data from operational NAEP assessments. They found that both the standard analysis and their unified model provided consistent estimates of subpopulation features, but the unified model more appropriately captured the variance of those estimates; the standard analysis, by treating IRT item parameters and population variances as known, tended to underestimate posterior uncertainty by about 10%. Furthermore, the unified model and MCMC estimation provided more stable estimates of sampling variance than the standard jackknife procedures. In sum, the use of MCMC estimation supported an analytic model that at once better captured significant features of the design and provided better-calibrated inferences for population and subpopulation characteristics of interest.

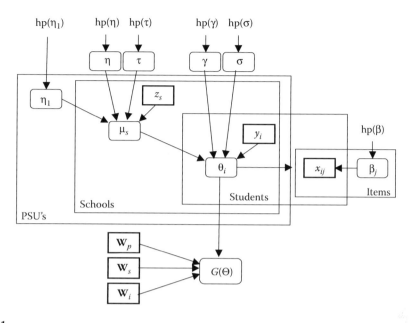

FIGURE 22.1

Plate diagram based on Johnson and Jenkins (2005), with school-level covariates z_s added. Solid rectangles represent observed data, rounded rectangles represent variables, and hp(.) represent highest-level prior distributions for indicated variables. Plates indicate replication over structurally similar relationships for PSUs, schools, students, and items. $\mathbf{W}_p$, $\mathbf{W}_s$, and $\mathbf{W}_i$ are known weights for PSUs, schools, and students respectively, that are used in the calculation of the target statistic $G(\Theta)$.

22.5 Discussion: Advantages of MCMC

While many of the above examples highlight complex statistical models that cannot be practically estimated by conventional means, MCMC can be gainfully employed in settings where alternative estimation routines already exist. For example, traditional estimation of IRT and LCA models proceed first with the estimation of conditional probability parameters for the observables which are then treated as known in estimating values of the latent variables. This divide-and-conquer approach understates the uncertainty in estimation, whereas a fully Bayesian analysis (facilitated by MCMC) propagates uncertainty appropriately.

A common criticism critique of MCMC is that it is difficult, both computationally in terms of computing resources and conceptually in terms of constructing the chains. As to the former, the availability of general use software such as WinBUGS (Spiegelhalter et al., 2007) and the publishing of code for various models represent considerable steps forward. As to the latter, there is no debate but that a certain level of technical sophistication is required to properly conduct an MCMC analysis. However, the criticism that MCMC is conceptually difficult is somewhat ironic, given that—for substantively motivated, statistically complex models—it is actually easier to set up an MCMC estimation routine than it is to proceed through the necessary steps (e.g. solving for first- and second- order derivatives) in ML and least squares estimation routines. A number of historically reoccurring features and assumptions of models in educational research (e.g. linear relationships, independence and normality of errors, few latent variables in IRT, few discrete observables in FA) have evolved in part from limitations on estimation routines. The flexibility of MCMC frees the analyst from the bonds associated with other estimation approaches, to construct models based on substantive theory. The lasting impact of the work by Patz and Junker (1999a, 1999b) was not only that MCMC could be employed to estimate existing models of varying complexity, but also that MCMC was a general approach to estimation flexible enough to handle any model that *could* be constructed. The explosion of MCMC in education research in the past decade serves as a testament to this new state of affairs. Applications of MCMC estimation for models such as cognitive diagnosis and mixtures of latent growth curves illustrate an interplay among statistical advances, more encompassing substantive models, and increasingly ambitious applications.

22.6   Conclusion

The subject matter of educational research is inherently complex. Schooling is hierarchically organized, with distinct covariates and resulting effects at each level. Concomitant variables for students must be included to address preexisting sources of variation. The dependent variables of interest, namely aspects students' knowledge and skill, are not directly observable and are often multivariate. Advances in the learning psychology ground increasingly complex within-student models, which interact with characteristics of tasks (as addressed in cognitive diagnosis). The tradition of disparate models that could each address only a few of these features, and did so in terms of hypothesis tests and point estimates of effects, was clearly inadequate to the substantive challenges of the field.

The way forward, as Novick realized nearly half a century ago, was the Bayesian inferential paradigm: a conceptual way to create models that address questions of substantive importance, built up from model fragments that addressed recurring structures and problems, that would enable researchers to understand patterns of variation in complex situations and properly account for both what could be learned and what remained uncertain. The advent of MCMC estimation makes this vision eminently achievable. Continued progress in user-friendly analytic frameworks, further examples of the superiority of the inferential approach, and infusion of this way of reasoning into the training of the next generation of educational researchers will complete the task.

References

Albert, J. H. 1992. Bayesian estimation of normal ogive item response curves using Gibbs sampling. *Journal of Educational Statistics*, 17:251–269.

Albert, J. H. and Chib, S. 1993. Bayesian analysis of binary and polychotomous response data. *Journal of the American Statistical Association*, 88:669–679.

Almond, R. G., DiBello, L. V., Moulder, B., and Zapata-Rivera, J. D. 2007. Modeling diagnostic assessments with Bayesian networks. *Journal of Educational Measurement*, 44:341–359.

Ansari, A., Jedidi, K., and Dube, L. 2002. Heterogeneous factor analysis models: A Bayesian approach. *Psychometrika*, 67:49–78.

Arminger, G. and Muthén, B. O. 1998. A Bayesian approach to nonlinear latent variable models using the Gibbs sampler and the Metropolis-Hastings algorithm. *Psychometrika*, 63:271–300.

Beaton, A. 1987. *The NAEP 1983–1984 Technical Report*. ETS, Princeton, NJ.

Béguin, A. A. and Glas, C. A. W. 2001. MCMC estimation and some model-fit analysis of multidimensional IRT models. *Psychometrika*, 66:541–562.

Behrens, J. T. and Smith, M. L. 1996. Data and data analysis. In D. C. Berliner and R. C. Calfee (eds), *Handbook of Educational Psychology*, pp. 949–989. Macmillan, New York.

Bollen, K. A. 1989. Structural Equations with Latent Variables. Wiley, New York.

Bollen, K. A. 2002. Latent variables in psychology and the social sciences. *Annual Review of Psychology*, 53:605–634.

Bolt, D. M., Cohen, A. S., and Wollack, J. A. 2001. A mixture item response model for multiple-choice data. *Journal of Educational and Behavioral Statistics*, 26:381–409.

Bolt, D. M. and Lall, V. F. 2003. Estimation of compensatory and noncompensatory multidimensional item response models using Markov chain Monte Carlo. *Applied Psychological Measurement*, 27:395–414.

Bradlow, E. T., Wainer, H., and Wang, X. 1999. A Bayesian random effects model for testlets. *Psychometrika*, 64:153–168.

Chromy, J. R., Finkner, A. L., and Horvitz, D. G. 2004. Survey design issues. In L. V. Jones and I. Olkin (eds), *The Nation's Report Card: Evolution and Perspectives*, pp. 383–427. Phi Delta Kappa Educational Foundation, Bloomington, IN.

Chung, H., Flaherty, B. P., and Schafer, J. L. 2006. Latent class logistic regression: Application to marijuana use and attitudes among high school seniors. *Journal of the Royal Statistical Society, Series A*, 169:723–743.

Cohen, A. S. and Bolt, D. M. 2005. A mixture model analysis of differential item functioning. *Journal of Educational Measurement*, 42:133–148.

de la Torre, J. and Douglas, J. 2004. Higher-order latent trait models for cognitive diagnosis. *Psychometrika*, 69:333–353.

Draper, D. 1995. Inference and hierarchical modeling in the social sciences. *Journal of Educational and Behavioral Statistics*, 20:115–147.

Fox, J. P. and Glas, C. A. W. 2001. Bayesian estimation of a multilevel IRT model using Gibbs sampling. *Psychometrika*, 66:271–288.

Gelman, A., Carlin, J. B., Stern, H. S., and Rubin, D. B. 1995. *Bayesian Data Analysis*. Chapman & Hall, London.

Gelman, A. and Hill, J. 2007. *Data Analysis Using Regression and Multilevel/Hierarchical Models*. Cambridge University Press, New York.

Glass, G. V. 1976. Primary, secondary, and meta-analysis of research. *Educational Researcher*, 5:3–8.

Glass, G. V., McGaw, B., and Smith, M. L. 1981. *Meta-analysis in Social Research*. Sage, Beverly Hills, CA.

Gorsuch, R. L. 1983. *Factor Analysis*, 2nd edn. Lawrence Erlbaum Associates, Hillsdale, NJ.

Hartz, S. M. 2002. A Bayesian framework for the unified model for assessing cognitive abilities: Blending theory with practicality. Doctoral dissertation, Department of Statistics, University of Illinois Champaign-Urbana.

Hedges, L. V. and Olkin, I. 1985. *Statistical Methods for Meta-analysis*. Academic Press, New York.

Hoijtink, H. 1998. Constrained latent class analysis using the Gibbs sampler and posterior predictive p-values: Applications to educational testing. *Statistica Sinica*, 8:691–711.

Hoijtink, H. and Notenboom, A. 2004. Model based clustering of large data sets: Tracing the development of spelling ability. *Psychometrika*, 69:481–498.

Janssen, R., Tuerlinckx, F., Meulders, M., and De Boeck, P. 2000. A hierarchical IRT model for criterion-referenced measurement. *Journal of Educational and Behavioral Statistics*, 25: 285–306.

Johnson, M. S. 2002. A Bayesian hierarchical model for multidimensional performance assessments. Paper presented at the annual meeting of the National Council on Measurement in Education, New Orleans, LA.

Johnson, M. S. and Jenkins, F. 2005. A Bayesian hierarchical model for large-scale educational surveys: An application to the National Assessment of Educational Progress. Research Report RR-04-38. ETS, Princeton, NJ.

Kasim, R. M. and Raudenbush, S. W. 1998. Application of Gibbs sampling to nested variance components models with heterogeneous within-group variance. *Journal of Educational and Behavioral Statistics*, 23:93–116.

Laird, N. and Ware, J. 1982. Random-effects models for longitudinal data. *Biometrics*, 38:963–974.

Lazarsfeld, P. F. and Henry, N. W. 1968. *Latent Structure Analysis*. Houghton Mifflin, Boston.

Lee, S. Y. and Song, X. Y. 2003. Bayesian model selection for mixtures of structural equation models with an unknown number of components. *British Journal of Mathematical and Statistical Psychology*, 56:145–165.

Lee, S. Y. and Tang, N. S. 2006. Bayesian analysis of nonlinear structural equation models with nonignorable missing data. *Psychometrika*, 71:541–564.

Lee, S. Y. and Zhu, H. T. 2000. Statistical analysis of nonlinear structural equation models with continuous and polytomous data. *British Journal of Mathematical and Statistical Psychology*, 53: 209–232.

Lee, S. Y., Song, X. Y. and Tang, N. S. 2007. Bayesian methods for analyzing structural equation models with covariates, interaction, and quadratic latent variables. *Structural Equation Modeling*, 14:404–434.

Levy, R. and Mislevy, R. J. 2004. Specifying and refining a measurement model for a simulation-based assessment. *International Journal of Measurement*, 4:333–369.

Libby, D. L., Novick, M. R., Chen, J. J., Woodworth, G. G., and Hamer, R. M. 1981. The Computer-Assisted Data Analysis (CADA) monitor. *American Statistician*, 35:165–166.

Lindley, D. V. and Novick, M. R. 1981. The role of exchangeability in inference. *Annals of Statistics*, 9:45–58.

Lindley, D. V. and Smith, A. F. M. 1972. Bayes estimates for the linear model. *Journal of the Royal Statistical Society, Series B*, 34:1–41.

Lord, F. M. 1980. *Applications of Item Response Theory to Practical Testing Problems*. Erlbaum, Hillsdale, NJ.

Longford, N. T. 1995. Model-based methods for analysis of data from the 1990 NAEP trial state assessment. Research and Development Report 95-696, National Center for Education Statistics, Washington, DC.

Mariano, L. T. and Junker, B. W. 2007. Covariates of the rating process in hierarchical models for multiple ratings of test items. *Journal of Educational and Behavioral Statistics*, 32:287–314.

Mislevy, R. J. 1986. Bayes modal estimation in item response models. *Psychometrika*, 51:177–196.

Mislevy, R. J. and Levy, R. 2007. Bayesian psychometric modeling from an evidence-centered design perspective. In C. R. Rao and S. Sinharay (eds), *Handbook of Statistics, Volume 26*, pp. 839–865. North-Holland, Amsterdam.

Novick, M. R. and Jackson, P. H. 1974. Further cross-validation analysis of the Bayesian m-group regression method. *American Educational Research Journal*, 11:77–85.

Patz, R. J. and Junker B. W. 1999a. A straightforward approach to Markov chain Monte Carlo methods for item response models. *Journal of Educational and Behavioral Statistics*, 24:146–178.

Patz, R. J. and Junker, B. W. 1999b. Applications and extensions of MCMC in IRT: Multiple item types, missing data, and rated responses. *Journal of Educational and Behavioral Statistics*, 24:342–366.

Raudenbush, S., Fotiu, R., and Cheong, Y. F. 1999. Synthesizing results from the Trial State Assessment. *Journal of Educational and Behavioral Statistics*, 24:413–438.

Rost, J. 1990. Rasch models in latent classes—an integration of two approaches to item analysis. *Applied Psychological Measurement*, 14:271–282.

Sahu, S. K. 2002. Bayesian estimation and model choice in item response models. *Journal of Statistical Computation and Simulation*, 72:217–232.

Scott, S. L. and Ip, E. H. 2002. Empirical Bayes and item-clustering in a latent variable hierarchical model: A case study from the National Assessment of Educational Progress. *Journal of the American Statistical Association*, 97:409–419.

Scheines, R., Hoijtink, H., and Boomsma, A. 1999. Bayesian estimation and testing of structural equation models. *Psychometrika*, 64:37–52.

Schulz, E. M., Betebenner, D., and Ahn, M. 2004. Hierarchical logistic regression in course placement. *Journal of Educational Measurement*, 41:271–286.

Seltzer, M. H. 1993. Sensitivity analysis for fixed effects in the hierarchical model: A Gibbs sampling approach. *Journal of Educational Statistics*, 18:207–235.

Seltzer, M., Novak, J., Choi, K., and Lim, N. 2002. Sensitivity analysis for hierarchical models employing t level-1 assumptions. *Journal of Educational and Behavioral Statistics*, 27: 181–222.

Seltzer, M. H., Wong, W. H., and Bryk, A. S. 1996. Bayesian analysis in applications of hierarchical models: Issues and methods. *Journal of Educational and Behavioral Statistics*, 21:131–167.

Shi, J. Q. and Lee, S. Y. 1998. Bayesian sampling-based approach for factor analysis model with continuous and polytomous data. *British Journal of Mathematical and Statistical Psychology*, 51: 233–252.

Smith, T. C., Spiegelhalter, D. J., and Thomas, A. 1995. Bayesian graphical modelling applied to random effects meta-analysis. *Statistics in Medicine*, 14:2685–2699.

Song, X. Y. and Lee, S. Y. 2001. Bayesian estimation and test for factor analysis model with continuous and polytomous data in several populations. *British Journal of Mathematical and Statistical Psychology*, 54:237–263.

Song, X. Y. and Lee, S. Y. 2002. Analysis of structural equation model with ignorable missing continuous and polytomous data. *Psychometrika*, 67:261–288.

Song, X. Y. and Lee, S. Y. 2003. Full maximum likelihood estimation of polychoric and polyserial correlations with missing data. *Multivariate Behavioral Research*, 38:57–79.

Spiegelhalter, D. J., Thomas, A., Best, N. G., and Lunn, D. 2007. *WinBUGS User Manual: Version 1.4.3*. MRC Biostatistics Unit, Cambridge. http://www.mrc-bsu.cam.ac.uk/bugs/winbugs/contents.shtml

Sutton, A. J. and Abrams, K. R. 2001. Bayesian methods in meta-analysis and evidence synthesis. *Statistical Methods in Medical Research*, 10:277–303.

Takane, Y. and de Leeuw, J. 1987. On the relationship between item response theory and factor analysis of discretized variables. *Psychometrika*, 52:393–408.

van Onna, M. J. H. 2002. Bayesian estimation and model selection in ordered latent class models for polytomous items. *Psychometrika*, 67:519–538.

Wirth, R. J. and Edwards, M. C. 2007. Item factor analysis: Current approaches and future directions. *Psychological Methods*, 12:58–79.

Wollack, J. A., Bolt, D. M., Cohen, A. S., and Lee, Y. S. 2002. Recovery of item parameters in the nominal response model: A comparison of marginal likelihood estimation and Markov chain Monte Carlo estimation. *Applied Psychological Measurement*, 26:339–352.

Yao, L. and Boughton, K. A. 2007. A multidimensional item response modeling approach for improving subscale proficiency estimation and classification. *Applied Psychological Measurement*, 31:83–105.

Zhang, Z., Hamagami, F., Wang, L., and Nesselroade, J. R. 2007. Bayesian analysis of longitudinal data using growth curve models. *International Journal of Behavioral Development*, 31:374–383.

Zhu, H. T. and Lee, S. Y. 2001. A Bayesian analysis of finite mixtures in the LISREL model. *Psychometrika*, 66:133–152.

23

Applications of MCMC in Fisheries Science

Russell B. Millar

23.1 Background

It has been said that counting fish is like counting trees, except that you can't see them, and they move. In a sea of uncertainty and variability, fisheries science has the daunting task of providing advice to fisheries managers who are charged with exploiting fisheries for maximum sustainable social and economic benefit.

There has been some progress toward ecosystem-based models of fisheries (see Browman and Stergiou, 2004, for a perspective), but typically, scientific advice about fisheries is provided to managers on a case-by-case basis. A fishery for a particular species is often spatially partitioned into separate management units, particularly if there is little movement of the species from one management unit to another (as determined by tagging studies, say). These management units are called stocks, and each stock of sufficient importance will be the subject of a stock assessment. Stock assessments are as varied as the species they assess, but in a nutshell, they seek to predict the consequences of exploiting the fishery under alternative regulations on the harvest. These regulations could include specification of total allowable catch, minimum (and/or maximum) legal size, duration of fishing season, area open for fishing, size and type of fishing gear, maximum size or horsepower of fishing vessel, and so on.

The amount of effort and expense invested in a stock assessment is typically commensurate with the perceived social and economic value of the stock, and the will of relevant stakeholders to fund the work. The latter can be particularly problematic for a stock that straddles or traverses geopolitical boundaries. In simpler cases, the stock assessment may utilize only the annual commercial catch rate (Section 23.4.1). At the other extreme, a high-value stock may be surveyed annually by a dedicated research vessel. A subset of the commercial and/or research catch may be measured for length, and where appropriate also for weight, sex, sexual maturity and age. Age can be determined by counting annual rings deposited in hard body parts – in fish this is typically the otolith (ear bone). Aging is more challenging for crustaceans because they molt their exoskeleton, and also for animals in the tropics because annual growth rings will not be formed if there is little seasonal variability. Larval surveys may also be conducted regularly, by research fishing with a small trawl with a very fine mesh (finer than 1 mm, for example). In addition, further research may be undertaken to investigate other features of the dynamics of the fishery, such as the effects of environmental change and variability, the relationship between recruitment (the number of young fish entering the fishery) and the size of the stock that spawned them (Section 23.4.3), behavior of fish to the fishing gear (Section 23.4.2), impact of recreational fishers, or the amount of (often unaccounted) wastage from the discard of fish that are not of legal or commercially viable size.

Unfortunately, even in the most data-rich situations, stock assessment models are often ill-conditioned and vastly different models can achieve similar fits to the available data. It is commonly the case that a stock will have been fished for many years prior to the establishment of a formal stock assessment and associated data collection. In this case, the two following scenarios may fit the data about equally well: first, that the stock is potentially highly productive but has been overfished to the extent that recruitment and productivity have been severely reduced; second, that the stock is of relatively low productivity and is being fished at an optimal level. The difficulty of distinguishing between scenarios arises because, even if the stock assessment is able to estimate overall mortality reasonably well, it cannot easily separate natural mortality from fishing-induced mortality. In the first case, a temporary reduction in fishing mortality should allow the biomass of the stock to increase, with a corresponding increase in recruitment and therefore long-term productivity. Failure to do so may push the stock to commercial extinction. In the second case, the prevailing management strategies for the stock are appropriate and any temporary reduction in fishing mortality would merely result in a temporary loss of economic benefit from the resource.

As a consequence of the ill-conditioning of many stock assessment models, it has been traditional to take certain key parameters to be known. For example, the value of 0.2 has ubiquitously been used as the rate of instantaneous natural mortality of cod* and many other ground-fish species (Myers and Cadigan, 1995). A typical stock assessment would be implemented by fitting a baseline model using fixed values of key parameters, and a sensitivity analysis would subsequently be performed using alternative values of those fixed parameters. However, there would generally be no cohesive framework for producing clear expressions of risk and uncertainty to managers. Moreover, the collection of stock assessments produced from the sensitivity analysis presented management with the opportunity to emphasize the particular model that best suited political objectives, or to reject the stock assessments outright due to their perceived unreliability. These were contributing factors in the demise of the Grand Banks cod fishery where, in particular, overweighting of commercial data (relative to research data) in the 1989 assessment produced considerably less pessimistic estimates of the fishery. Even so, there was a strong reluctance to make the reduction in total allowable catch that was indicated under even the least pessimistic assessment (Shelton, 2005), and the reductions that were made were insufficient to prevent the end of this thousand-year-old fishery in 1992 (Kurlansky, 1997).

A second major feature of traditional stock assessment models was their lack of realism and, in particular, their inability to include sources of variability in addition to observation error in the measured data. Until recently, a typical stock assessment model assumed that the population dynamics of a stock were deterministic. In effect, given the model parameters, these models provide a perfect prediction of the status of the stock, past and present. This deterministic ideology gave rise to nebulous practice. For example, maximum sustainable yield (MSY) is defined to be "The largest average catch or yield that can continuously be taken from a stock under existing environmental conditions" (Ricker, 1975). Mangel et al. (2002) argue that this definition of MSY gives a useful management concept and note that it implicitly allows for variability. The failure of deterministic models to include this variability gave the false impression that MSY was effectively "the yield that could continuously be taken." Fisheries managers were aware of the danger of interpreting MSY

---

* In the absence of fishing, $\frac{\partial N_t}{\partial t} = -mN_t$, where m is instantaneous natural mortality and N_t is the number of fish in any given cohort (i.e. from the same spawning year) at time t. Wizened fisheries scientists tell the story of a fractious round-table meeting where agreement on a suitable value for the natural mortality of cod could not be achieved. The minute-keeper of the meeting recorded this lack of consensus by writing $m =?$ on the handwritten minutes. This was later mistyped by a secretary, as $m = .2$.

in this naive way, and it was usual practice to incorporate a small safety factor factor by setting a target yield slightly lower than MSY. However, this was not based on any formal assessment of risk.

The major shortcomings of the early stock assessment models were well known, and were the subject of regular attention in the fisheries literature (see, for example, Hilden, 1988; Walters and Ludwig, 1981). Bayesian approaches appearing in the fisheries literature during the 1980s and early 1990s (see Hilborn et al., 1994, for a list) made some progress in this regard, but were necessarily confined to relatively simple models, often employed *ad hoc* or approximate methods of calculation, and were not applicable to mainstream stock assessments. These works had little impact on the implementation of stock assessments.

It was not until later in the 1990s that methodological and computational advances provided the opportunity for more realistic representation of uncertainty and variability in stock assessment models. Of particular note, Sullivan (1992) presented a linear-normal state–space model for incorporating uncertainty in the dynamics of a length-structured stock assessment, and demonstrated a maximum likelihood implementation via the classical Kalman filter. Schnute (1994) presented the general matrix recursion formulas for the Kalman filter and included the extended Kalman filter for nonlinear-normal state–space models. He also noted the natural Bayesian interpretation of the state–space framework. At about the same time, the first fully Bayesian stock assessment models appeared in the primary literature. McAllister et al. (1994) and McAllister and Ianelli (1997) used the sampling-importance resampling (SIR) algorithm to fit age-structured models that used deterministic stock dynamics, but did incorporate random variability in initial conditions. Raftery et al. (1995) also used the SIR algorithm in a deterministic population dynamics model for bowhead whales. Formal use of Bayesian hierarchical models for a fisheries meta-analysis was presented by Liermann and Hilborn (1997), notwithstanding that the posterior was approximated using profile likelihood to eliminate nuisance parameters.

The first mention of Markov chain Monte Carlo (MCMC) in the primary fisheries literature appears to be a brief comment in McAllister et al. (1994). Subsequently, in their discussion of methodologies for Monte Carlo sampling from the posterior, McAllister and Ianelli (1997) reported that they had also fitted their age-structured model using the Metropolis–Hastings algorithm, but preferred the SIR algorithm. Punt and Hilborn (1997) provided a description of several approaches for approximating or sampling from the posterior. This included the Metropolis–Hastings algorithm, but did not provide an example of its implementation. The first fully described implementation of MCMC in the primary fisheries literature appears to be Meyer and Millar (1999a), who fitted a nonlinear state–space model using Metropolis-within-Gibbs sampling with the aid of adaptive rejection sampling routines provided by Gilks et al. (1995). Patterson (1999) also used this algorithm, to fit an age-structured model which included model uncertainty in the choice of the error distribution and shape of the assumed stock–recruitment curve. Later that same year, Meyer and Millar (1999b) introduced fisheries scientists to the BUGS language, in the context of a state–space surplus production model, and Millar and Meyer (2000) provided more detail concerning the evaluation of this model. This is the example presented in Section 23.4.1.

23.2 The Current Situation

Bayesian stock assessments are now routinely used by fisheries agencies around the globe. For example, at present, of the ten most commercially important species assessed by the

New Zealand Ministry of Fisheries, seven use formal models of stock dynamics and these seven utilize Bayesian inference in some or all of the assessment.* At fisheries agencies where classical models are still used, it is often due to inertia rather than deliberate rejection of Bayesian principles. For example, the 2006 Workshop on Advanced Fish Stock Assessment of the International Council for Exploration of the Sea[†] regarded Bayesian methodology as a specialist area (ICES, 2006).

23.2.1 Software

WinBUGS is frequently used for simpler fisheries models, but the ADMB (Automatic Differentiation Model Builder, freely available from http://admb-project.org/) software has made by far the biggest contribution to the widespread use of Bayesian methodology and MCMC in fisheries. Automatic differentiation gives ADMB great ability to find the posterior mode and to evaluate the Hessian of the log-posterior to high precision, even for models that may contain thousands of parameters. This software is aided by additional features to improve stability of the optimization, and to enhance the efficiency of its implementation of the Metropolis–Hastings algorithm (Section 23.3.2). Command-line options enable an ADMB executable to switch from Bayesian mode to classical (penalized) maximum likelihood, with the priors either ignored or treated as penalty terms. It therefore also has wide acceptance by fisheries agencies which continue to use classical methods. Moreover, classical mode can be used as a preliminary model selection tool, leaving full Bayesian analysis and risk assessment to a smaller subset of models.

ADMB was first made available as commercial proprietary software (Otter Research, 2007). Its extensive modeling capabilities earned it a loyal and proactive following among the fisheries modeling community, but its cost and proprietary restrictions limited its use by a broader audience. The nonprofit ADMB Foundation (http://admb-foundation.org/) was incorporated in 2007, with one objective being to coordinate development of ADMB and promote its use amongst the wider scientific community. Through generous grants, the ADMB Foundation was able to purchase the rights to the ADMB software, and it was made freely available in late 2008 from the ADMB Project website. ADMB was made open source a few months later.

The ADMB Project has greatly improved the experience of installing and using ADMB. In particular, there are now utilities for running ADMB from within the R language, and for input and output of data files and model results between ADMB and R. Nonetheless, there is a steep learning curve to using ADMB because the model must be explicitly coded in a C++ like ADMB template language.

Programming in the ADMB template language can be prohibitively complicated to the vast majority of fisheries scientists, especially given the required complexity of many types of fisheries models. Consequently, several freely available stock assessment packages have been created using an ADMB executable (or dynamic link library) as the computational engine behind a user-friendly interface. Three such packages are ASAP (Age Structured Assessment Program), Stock Synthesis, and Coleraine (named for a New Zealand-made Cabernet Merlot). The first two can be downloaded from the US National Oceanic and Atmospheric Administration (NOAA) Fisheries Toolbox (http://nft.nefsc.noaa.gov/). The

---

* In decreasing order of 2006 commercial value: hoki, lobster, paua (abalone), arrow squid, orange roughy, snapper, lingcod, hake, scampi and tarakihi. Arrow squid (a reoccurring annual stock), scampi and tarakihi assessments currently do not utilize stock dynamic models. A Bayesian model for scampi is currently under development.
† The ICES organization has 20 member countries, and is responsible for coordinating marine research in the North Atlantic. See www.ices.dk.

ASAP technical manual also includes an appendix containing the ADMB template program. Coleraine is available from the School of Aquatic and Fisheries Sciences at the University of Washington, Seattle (www.fish.washington.edu/research/coleraine/). As the name would suggest, Coleraine has the strongest Bayesian flavor of these three packages.

23.2.2 Perception of MCMC in Fisheries

In fisheries modeling, MCMC is often touted as an alternative to bootstrapping for the purpose of including uncertainty. This is so much the case that many users of MCMC are unaware that they are operating within the Bayesian paradigm. For example, because it uses an ADMB computational engine, the ASAP software offers an MCMC option to "estimate uncertainty in the model solution." However, searches of the ASAP documentation for the words "Bayes," "Bayesian," "prior," and "posterior" all drew blanks. A good proportion of fisheries stock assessment reports exhibit the same characteristics. That is, some fisheries modelers are performing stock assessments using the convenient "MCMC option," with no conceptual understanding that they are employing a Bayesian model and hence with no notion of the priors that are implicitly being assumed.

23.3 ADMB

In its base form, ADMB is a sophisticated tool for general-purpose optimization. It includes many features for coping with high-dimensional problems. These include implicit transformation of bounded parameters, centering of parameter vectors, and the ability to fit a model in phases (Section 23.3.2). However, the raw optimization power of ADMB is derived from its use of automatic differentiation, giving it the ability to perform quasi-Newton optimization using accurate and computationally efficient calculation of derivatives.

23.3.1 Automatic Differentiation

When using ADMB in Bayesian mode, the negative log joint density function, $-\log f(\mathbf{y}, \boldsymbol{\theta})$, is specified within an ADMB template file using operator-overloaded C++ code. Automatic differentiation facilitates exact algebraic calculation (to within machine precision) of the derivative of the joint density with respect to all elements of the parameter vector $\boldsymbol{\theta}$. The calculation is efficient, and the Jacobian vector is typically obtained in less than three times the number of operations (Griewank, 2003) required to evaluate $\log f(\mathbf{y}, \boldsymbol{\theta})$.

In crude form, automatic differentiation can be considered an application of the chain rule of differentiation. By way of example, the hierarchical model in Section 23.4.2 uses the assumption $q_i \sim N(\mu_q, \sigma_q^2)$, where q_i is the log catchability of stock i. This contributes a term of the form

$$-\log \sigma_q - \frac{(q_i - \mu_q)^2}{2\sigma_q^2} \tag{23.1}$$

to the log joint density. The Jacobian of Equation 23.1 with respect to model parameters is zero except for partial derivatives with respect to q_i, μ_q, and σ_q, and attention will be restricted to these three partial derivatives only. In ADMB, the objects $f_1 = \log \sigma_q$ and

$f_2 = \frac{(q_i - \mu_q)^2}{2\sigma_q^2}$ also contain derivative information. That is,

$$f_1 \equiv \left(\log \sigma_q, \left[\frac{\partial \log \sigma_q}{\partial q_i}, \frac{\partial \log \sigma_q}{\partial \mu_q}, \frac{\partial \log \sigma_q}{\partial \sigma_q} \right] \right)$$

$$= \left(\log \sigma_q, \left[0, 0, \frac{1}{\sigma_q} \right] \right) \tag{23.2}$$

and

$$f_2 \equiv \left(\frac{(q_i - \mu_q)^2}{2\sigma_q^2}, \left[\frac{\partial \frac{(q_i - \mu_q)^2}{2\sigma_q^2}}{\partial q_i}, \frac{\partial \frac{(q_i - \mu_q)^2}{2\sigma_q^2}}{\partial \mu_q}, \frac{\partial \frac{(q_i - \mu_q)^2}{2\sigma_q^2}}{\partial \sigma_q} \right] \right). \tag{23.3}$$

In Equation 23.2, the derivative (with respect to σ_q) is obtained directly from overloading of the log operator, so that it evaluates its derivative in addition to its value. In Equation 23.3, the derivatives are obtained by overloading of the unary power operator and the binary multiplication/division and addition/subtraction operators, and successive application of the chain rule. For example, the partial derivative with respect to σ_q is obtained as

$$\frac{\partial \frac{(q_i - \mu_q)^2}{2\sigma_q^2}}{\partial \sigma_q} = \frac{\partial \frac{(q_i - \mu_q)^2}{2\sigma_q^2}}{\partial 2\sigma_q^2} \times \frac{\partial 2\sigma_q^2}{\partial \sigma_q^2} \times \frac{\partial \sigma_q^2}{\partial \sigma_q}$$

$$= -\frac{(q_i - \mu_q)^2}{4\sigma_q^4} \times 2 \times 2\sigma_q$$

$$= -\frac{(q_i - \mu_q)^2}{\sigma_q^3}.$$

Griewank (2003) notes that obtaining the derivative is far from "automatic" and recommends "algorithmic" differentiation as a more apt name for this methodology.

23.3.2 Metropolis–Hastings Implementation

The ADMB template program uses overloaded C++ code to specify an objective function to be minimized. In the Bayesian context this is the (negative of the) log joint density function $\log f(\mathbf{y}, \boldsymbol{\theta})$, and hence the optimization finds the mode of the posterior density $f(\boldsymbol{\theta} \mid \mathbf{y})$. Moreover, the Jacobian of $-\log f(\boldsymbol{\theta}, \mathbf{y})$ is efficiently obtained to machine precision, and so the Hessian can quickly be obtained from first-order differences. In Bayesian mode, ADMB uses the Metropolis–Hastings algorithm with the default initial proposal density being multivariate normal with covariance matrix, Σ, obtained as the inverse of this Hessian. The posterior mode is the default initial parameter value.

Gelman et al. (2003, Section 11.9) state that they found the above simple form of Metropolis–Hastings algorithm to be useful for problems with up to 50 parameters. ADMB provides several enhancements to this simple form of implementation and has successfully been deployed for Bayesian fisheries models containing at least several hundred parameters. For example, the 2007 Gulf of Alaska walleye pollock assessment contained 308 parameters and the posterior was sampled using a chain of length 1 million with a thinning factor of 200.

The ADMB implementation of the Metropolis–Hastings follows the recommendation in Gelman et al. (2003, Section 11.9) to enhance the algorithm by adjusting the rejection rate through scaling the covariance matrix of the multivariate normal proposal density. Notable additional features of ADMB include

- Automatic methods (e.g. smooth transformations) to cope with bounded parameters.
- The ability to fit the model in phases. That is, in optimization mode, it can introduce model parameters in steps (see the next item).
- Automatic centering of blocks of parameters. This is particularly useful when fitting random effects because the (centered) random effects can be introduced into the optimization at a later stage, having allowed ADMB to first fit the mean effects. In Bayesian mode this feature is likely to be highly beneficial to mixing of the Metropolis–Hastings algorithm, due to reduced correlations in Σ.
- Command-line options for the ADMB executable are used to specify MCMC options. In addition to specifying standard options (such as length of the chain, degree of thinning, input–output options) there is an option to allow it to use a mixture proposal density to fatten the tails relative to the multivariate normal, and another to reduce the extreme correlations in Σ.

23.4 Bayesian Applications to Fisheries

The examples below have been chosen to give a taste of the variety of modeling challenges that have been met by application of MCMC in fisheries. However, they do not include a formal stock assessment. A complete assessment of a high-value stock can be very lengthy, and the interested reader will find that numerous Bayesian assessments are publicly available online. For instance, the 118-page Gulf of Alaska walleye pollock assessment for 2007 can be found at www.afsc.noaa.gov/REFM/docs/2007/GOApollock.pdf.

The first example presents a state–space formulation of a surplus production model, and it is employed to assess a stock of albacore tuna where only annual catch and catch rate information is available. This example is demonstrative only—it uses historical data taken from Polacheck et al. (1993), but these data have since been substantially revised and extended. Moreover, length-disaggregated data are now measured on this species and the Highly Migratory Species Division of the US National Marine Fisheries Service is currently implementing a length-structured model for this stock using the MULTIFAN-CL software (another ADMB-engined stock assessment tool, freely available from www.multifan-cl.org/). The second and third examples demonstrate two different meta-analyses that have been applied to North-East Pacific (i.e. West Coast of USA and Canada) rockfish stocks.

23.4.1 Capturing Uncertainty

23.4.1.1 State–Space Model of South Atlantic Albacore Tuna Biomass

Surplus production models are widely used in fisheries stock assessment and are appropriate when the measurements on the fishery consist of just the annual catches and a measure of relative abundance. The deterministic version of these models takes the form

$$B_t = B_{t-1} + s(B_{t-1}) - C_{t-1},$$ (23.4)

where B_t is the fishable biomass at the start of year t and C_t is the catch during year t (for simplicity, assumed known). The surplus production function $s(B)$ denotes the overall change in biomass due to fish growth, recruitment of fish reaching legal size, and natural mortality.

The simplest plausible form for $s(B)$ is the quadratic Schaefer (1954) surplus production function, $s(B) = rB(1 - B/K)$, where r is the intrinsic growth rate of the population and K is virgin biomass. The Schaefer surplus production function takes its maximum value of $rK/4$ when biomass is half of virgin, $B = K/2$. This maximum value of surplus production is often regarded by management as the maximum sustainable yield of the fishery, and is the unknown quantity of primary interest.

Surplus production models are fitted to an annual index of abundance, $\mathbf{y} = (y_1, \dots, y_n)$. These could be obtained from research surveys, but most often catch-per-unit-effort (CPUE) data are used. CPUE is simply the catch divided by the fishing effort expended. For the example herein, the tuna are caught by longline, and CPUE was calculated as the catch weight (in kilograms) per 100 hooks deployed (Figure 23.1). The index of abundance is commonly assumed to be proportional to the biomass (but see Harley et al., 2001, who investigated a power relationship between CPUE and biomass) and the assumption of lognormal error is most commonly used. That is,

$$y_t = QB_t e^{v_t},$$ (23.5)

where v_t are independent and identically distributed (i.i.d.) $N(0, \tau^2)$. The parameter Q is the so-called "catchability coefficient."

Previously, Schaefer surplus production models had traditionally been fitted using non-linear least squares (Hilborn and Walters, 1992; Polacheck et al., 1993). If year $t = 1$ denotes the year in which fishing commenced, the nonlinear least squares model sets $\hat{B}_1 = K$ and uses the deterministic process (Equation 23.4) to obtain the predicted values $\hat{B}_t, t = 2, \dots, n$. The primary deficiency of this classical approach is that subsequent risk assessments fail to incorporate variability in the process equation. For example, recruitment of fish can vary an order of magnitude due to environmental variability.

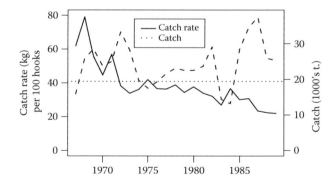

FIGURE 23.1
Catch rate and catch of South Atlantic albacore tuna, 1968–1989. The Bayes estimate of maximum surplus production (19,400 tonnes) is shown with a dotted line.

Meyer and Millar (1999b) applied a Bayesian state–space implementation of the Schaefer surplus production model to the albacore tuna data. They also chose to use the terminology "maximum surplus production" instead of the potentially misleading "maximum sustainable yield." The state–space implementation modeled variability in the process equation and the Bayesian framework permitted existing prior knowledge to be utilized. Under this model the biomass in year t is given by

$$B_1 = Ke^{u_1},$$

$$B_t = (B_{t-1} + rB_{t-1}(1 - B_{t-1}/K) - C_{t-1}) e^{u_t}, \quad t > 1,$$

where u_t are i.i.d. $N(0, \sigma^2)$.

Meyer and Millar (1999b) specified lognormal priors for model parameters K and r. The prior on r was obtained from formal hierarchical modeling of six other albacore tuna stocks. Punt et al. (1995) specified that the virgin biomass of this tuna stock was between 80 and 300 (1000 t) and the lognormal prior on K was derived by setting 80 and 300 as its 5th and 95th percentiles, respectively. Model parameters σ^2 and τ^2 were assigned inverse gamma priors. The hyperparameters of these priors were based on knowledge from other tuna stocks and indices of biomass. The catchability parameter, Q, was given a reference prior (uniform on the log scale). All model parameters were assumed *a priori* independent, that is, $\pi(K, r, Q, \sigma^2, \tau^2) = \pi(K)\pi(r)\pi(Q)\pi(\sigma^2)\pi(\tau^2)$.

23.4.1.2 Implementation

The state–space Schaefer surplus production model was applied to the albacore tuna data using WinBUGS, and the program code is available from www.stat.auckland.ac.nz/~millar/Bayesian/BayesIndex.html. The implementation reparameterized the model using $P_t = B_t/K$ because this was found to greatly reduce autocorrelation of the samples from the joint posterior. Note that P_t gives the biomass in year t as a proportion of virgin biomass. The process equation then becomes

$$P_1 = e^{u_1},$$

$$P_t = (P_{t-1} + rP_{t-1}(1 - P_{t-1}) - C_{t-1}/K) e^{u_t}, \quad t > 1.$$

A few additional lines of WinBUGS code enable the biomass trajectory to be extended beyond the last year of available data under different harvest scenarios (Figure 23.2). Such presentation of biomass uncertainty is instantly meaningful to fisheries managers. It is quick and convenient to produce (WinBUGS can draw a plot very much like Figure 23.2 with a couple of mouse clicks using the `Inference > Compare` menu), yet is also formally rigorous, subject to validity of the model.

23.4.2 Hierarchical Modeling of Research Trawl Catchability

The *absolute* catchability of a fishing gear can loosely be defined as the proportion of fish contacting the gear that are caught by the gear. For example, a small-meshed research trawl has absolute catchability of unity if it catches all fish in its path. For bottom dwelling fish, this gives rise to the swept-area estimate of biomass. This is applicable if the sea-bed habitat is sufficiently homogeneous (within strata) that it can be assumed that fish are evenly distributed, so that the biomass can be estimated by scaling up the catch using the fraction of habitat "swept" by the research trawl.

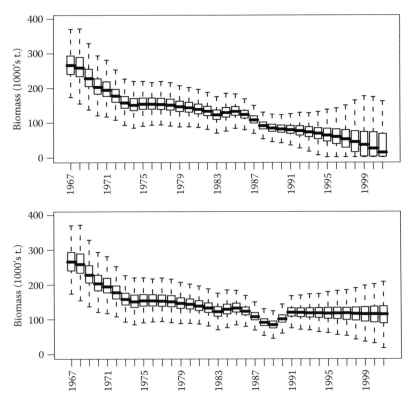

FIGURE 23.2
Boxplots of the posterior distribution of albacore biomass, fitted to data from 1967 to 1989, and projected to 2001. The top plot assumes catch equal to the Bayes estimate of maximum surplus production (19,400 tonnes) from 1990 onwards. The bottom plot assumes a moratorium on fishing in 1990 and 1991, and catch equal to maximum surplus production from 1992 onwards.

However, fish are not trees, and they do not remain still when the reaper calls. Bigger fish tend to be faster and more durable swimmers, and they may be able to swim away from the trawl before it engulfs them. In addition, some bottom fish may swim sufficiently high that they go over the top of a ground trawl. Smaller fish may be able to go under a ground trawl because the footrope of this gear typically rides over the sea floor on rollers and does not make direct contact with the sea bed. Also, some rougher bottom habitats are not amenable to trawling (due to the chance of ripping the trawl mesh, or losing the entire trawl if it gets stuck on rocks) and so trawls may not be viable in habitats of potentially higher abundance. These factors would result in the swept-area estimate of biomass tending to underestimate true biomass. On the other hand, the warps and bridles of trawl gears (Figure 23.3) can herd fish into the path of a trawl, so that it catches fish that are outside of its swept path, in which case the swept-area estimate of abundance could overestimate true biomass.

Swept-area estimates of rockfish biomass are calculated from research trawls off the West Coast of the USA. However, due to the factors noted above, these are regarded as a relative index of biomass. They are typically modeled according to Equation 23.5, with y_t being the swept-area estimate of biomass in year t, and parameter Q now called the *bulk* catchability. To combine information about bulk catchability across similar rockfish species Millar and Methot (2002) used the ADMB software (Section 23.3.2) to implement an age-structured meta-analysis of six West Coast rockfish species. Information required for the

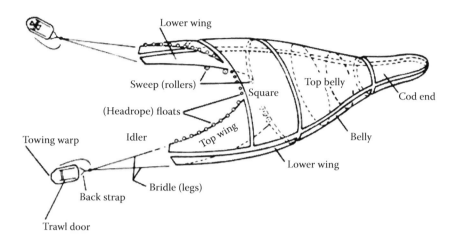

FIGURE 23.3
Sketch of an otter trawl. Reproduced from Figure 7.4B in (Hayes et al., 1996, p. 199) by permission of the American Fisheries Society.

model included catch-at-age data and knowledge about weight-at-age. This information was taken from the most recent stock assessment for each species. These assessment documents also provided a relative index of recruitment in each year. For each stock, the biomass B_t was obtained by summing the product of numbers-at-age and weight-at-age, over all relevant ages.

The hierarchical model of Millar and Methot (2002) assumed exchangeability of Q across the different species of rock fish. Denoting $q = \log(Q)$, it was assumed that

$$q \sim N(\mu_q, \sigma_q^2).$$

Two different hyperpriors were used:

$$\mu_q \sim N(0, 1), \quad \sigma_q^2 \sim \Gamma^{-1}(1, 2),$$

and

$$\mu_q \sim N(-1, 1), \quad \sigma_q^2 \sim \Gamma^{-1}(1, 1),$$

where $\Gamma^{-1}(\alpha, \beta)$ denotes an inverse gamma distribution with the associated gamma distribution having mean $\alpha\beta$ and variance $\alpha\beta^2$. The first induces a vague prior for Q with median slightly in excess of unity and an extremely long and slowly decaying right tail. The second prior is mildly informative and reflects a higher prior belief that catchability is below unity (Figure 23.4).

The posterior distribution of Q was reasonably insensitive to the choice of prior. In particular, the prior probabilities of Q being in excess of unity were 0.51 and 0.27 under the vague and mildly informative priors respectively, and the corresponding posterior probabilities were 0.06 and 0.05.

23.4.3 Hierarchical Modeling of Stock–Recruitment Relationship

It is generally considered that the number of recruits (young fish) joining the stock in any given year is highly dependent on conditions prevailing during their early life history,

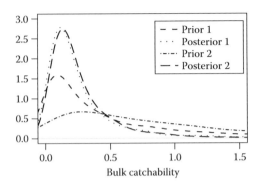

FIGURE 23.4
Prior and posterior densities for bulk catchability, Q.

especially the availability of food and extent of predation during the larval stage. These conditions can be highly variable, and this high variability often obscures the relationship between the size of the spawning parent stock and the number of resulting recruits. Nonetheless, if this relationship is not taken into consideration then the stock could suffer "recruitment over-fishing," that is, reach the point where the reduction in spawners results in substantial loss of recruitment. Proper risk assessment of fisheries policies must include this possibility.

In addition to large annual variability in recruitment (possibly with considerable temporal autocorrelation), estimates of spawning stock size and recruitment generally have high estimation uncertainty. Not surprisingly, the stock and recruitment estimates from a single stock often show little clear pattern of any relationship. To address this issue, a publicly available worldwide stock–recruit database, now containing approximately 700 data sets, was set up by the late Professor Ram Myers in the early 1990s to facilitate meta-analysis of this relationship (www.mathstat.dal.ca/~myers/welcome.html).

In a stock–recruit meta-analysis of US West Coast rockfish, Dorn (2002) considered the two most commonly employed stock–recruit curves, the Beverton–Holt and the Ricker curves. Here, attention will be confined to the Beverton–Holt curve, which can be expressed in the form

$$R = \frac{aS}{b + S}$$

where S is the spawning stock size (usually expressed as biomass) and R is the resulting number of recruits. This curve has recruitment asymptote at a, and b is the spawning stock biomass at which recruitment is $0.5a$.

It would not be sensible in a stock–recruit meta-analysis to assume that either of parameters a or b could be assumed exchangeable over different rockfish stocks because these parameters depend on the overall size of the stock, which in turn depends on habitat range and suitability. Instead, the Beverton–Holt stock–recruit curve can be uniquely determined by specifying the point (S_0, R_0), where S_0 and R_0 are the spawner size and recruitment of the stock in the absence of fishing, and the so-called "steepness" parameter, h, where hR_0 is the recruitment when the stock is at 20% of S_0 (Figure 23.5). The steepness parameter necessarily lies in the interval $(0.2, 1)$, with a value close to unity corresponding to a stock with recruitment that is robust to fishing, and a value close to 0.2 corresponding to a near proportional decrease in R with S. It is this parameter that is assumed exchangeable across different stocks of the same or similar species. In practice, if there is sufficient knowledge

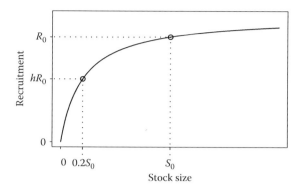

FIGURE 23.5
Beverton–Holt stock–recruit curve for a stock with steepness h.

about the rates of growth, (natural) mortality and fecundity of the stock, then the ratio between S_0 and R_0 can be determined, say $S_0 = \phi R_0$. Then, the Beverton–Holt curve can be expressed as a function of (R_0, h),

$$R = \frac{0.8 R_0 h S}{0.2 \phi R_0 (1 - h) + (h - 0.2) S}. \tag{23.6}$$

The meta-analysis of Dorn (2002) was implemented in ADMB and used stock–recruit data from 11 rockfish species. It used vague priors on individual values of R_0 for each stock, and a hierarchical prior on the scaled logit (mapping the interval $(0.2, 1)$ to $\mathbb{R}$) of h. That is,

$$\beta = \log\left(\frac{h - 0.2}{1 - h}\right) \sim N(\mu, \tau^2),$$

with relatively uninformative priors on hyperparameters μ and τ. For simplicity, temporal structure was not included in the model, spawner biomasses were assumed known without error, and ϕ was also assumed known for each stock. Conditional on (R_0, h), recruitment was modeled as lognormal, with mean given by Equation 23.6.

In the case of black rockfish, stock–recruit data were extremely limited, with only seven data points (Figure 23.6). For this species, $\phi = 1.21$ was calculated from biological information. If fitted to the black rockfish data only, the maximum likelihood estimate of h is close to unity, corresponding to a stock that appears immune to recruitment overfishing. In contrast, within the meta-analysis, the posterior modal value of h for black rockfish is shrunk to 0.68 (Figure 23.6).

Dorn (2002) incorporated the results of the stock–recruit meta-analysis into a reevaluation of the harvest policies for US West Coast rockfish. He concluded that, notwithstanding limitations due to simplifying assumptions used in the model, the fishing mortality of US West Coast rockfish generally exceeded the limits established by the Pacific Fishery Management Council to meet the requirements of the Magnuson–Stevens Fishery Conservation and Management Act (www.nmfs.noaa.gov/sfa/magact/).

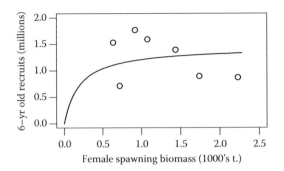

FIGURE 23.6
Beverton–Holt stock–recruit curve for black rockfish from a meta-analysis including 10 other similar stocks, using the posterior modal values of parameters h and R_0 (obtained from ADMB) for this stock.

23.5 Concluding Remarks

The above three examples give just a flavor for the application of MCMC to fisheries modeling. In particular, state–space models are natural tools for the temporal modeling of the unknown biomass of a stock, and MCMC has provided a viable computational framework for their general implementation. More generally, in an age-structured model the unknown state would be a vector of the numbers at different ages. The notion of deterministic population trajectories has no place in the sustainable management of fisheries resources.

Relatively simple examples were used here for readability. In comparison, a full Bayesian stock assessment model for an important commercial species may use data of many different types and can contain several hundred parameters, many of which will be random effects used to quantify variability in the population dynamics of the stock and uncertainties in measurements. For such highly parameterized models, the use of somewhat informative priors is often necessary. The hierarchical analyses in the last two examples demonstrated typical application of Bayesian meta-analysis to obtain such knowledge from similar stocks.

There is now a feeling among some fisheries scientists that the last few decades have been the Golden Age of fishery modeling (Quinn, 2003). MCMC has been prominent in the last part of this age and has allowed modelers to use more realism, and to incorporate prior knowledge. Nonetheless, the models cannot parsimoniously include all sources of relevant uncertainty (Cotter et al., 2004), some of which may be the dominant factors affecting the fishery. Examples of these rogue uncertainties include the response of fishers to regulation change (which often differs from that intended), the extent of poaching (often by organized crime), predator/prey shifts (due to halting of seal culls, say) or the effects of environmental change in the medium (e.g. El Niño and La Niña oscillations) or long term (e.g. global warming). To this end, much effort is now being directed at embedding fisheries modeling as a step within the wider conceptual framework of fisheries management evaluation (Hilborn, 2003). This will be a natural arena for application of decision theory, and it will be interesting to see what role Bayesian methods and MCMC eventually play in this larger scheme of things.

Acknowledgment

Comments from Mark Maunder (Inter-American Tropical Tuna Commission and ADMB fan) were gratefully received.

References

Browman, H. I. and Stergiou, K. I. 2004. Theme session: Perspectives on ecosystem-based approaches to the management of marine resources. *Marine Ecology Progress Series*, 274:269–303.

Cotter, A. J. R., Burt, L., Paxton, C. G. M., Fernandez, C., Buckland, S. T., and Pan, J.-X. 2004. Are stock assessment methods too complicated? *Fish and Fisheries*, 5:235–254.

Dorn, M. W. 2002. Advice on west coast rockfish harvest rates from Bayesian meta-analysis of stock-recruit relationships. *North American Journal of Fisheries Management*, 22:280–300.

Gelman, A., Carlin, J. B., Stern, H. S., and Rubin, D. B. 2003. *Bayesian Data Analysis*, 2nd edn. Chapman & Hall/CRC, Boca Raton, FL.

Gilks, W. R., Best, N. G., and Chan, K. K. C. 1995. Adaptive rejection Metropolis sampling within Gibbs sampling. *Applied Statistics*, 44:455–472.

Griewank, A. 2003. A mathematical view of automatic differentiation. *Acta Numerica*, 12:321–398.

Harley, S. J., Myers, R. A., and Dunn, A. 2001. Is catch-per-unit-effort proportional to abundance? *Canadian Journal of Fisheries and Aquatic Sciences*, 58:1760–1772.

Hayes, D. B., Ferreri, C. P., and Taylor W. W. 1996. Active fish capture methods. In B. R. Murphy and D. W. Willis (eds), *Fisheries Techniques*, 2nd edn, pp. 193–220. American Fisheries Society, Bethesda, MD.

Hilborn, R. 2003. The state of the art in stock assessment: Where we are and where we are going. *Scientia Mariana*, 67(Suppl. 1):15–21.

Hilborn, R., Pikitch, E. K., and McAllister, M. K. 1994. A Bayesian estimation and decision analysis for an age-structured model using biomass survey data. *Fisheries Research*, 19:17–30.

Hilborn, R. and Walters, C. J. 1992. *Quantitative Fisheries Stock Assessment: Choice, Dynamics and Uncertainty*. Chapman & Hall, New York.

Hilden, M. 1988. Errors of perception in stock and recruitment studies due to wrong choices of natural mortality rate in virtual population analysis. *Journal du Conseil International pour l'Exploration de la Mer*, 44:123–134.

ICES 2006. Report of the Workshop on Advanced Fish Stock Assessment (WKAFAT) 23–28 February, ICES CM 2006/RMC:01. ICES Headquarters, Copenhagen.

Kurlansky, M. 1997. *Cod: A Biography of the Fish that Changed the World*. Walker and Co., New York.

Liermann, M. and Hilborn, R. 1997. Depensation in fish stocks: A hierarchic Bayesian meta-analysis. *Canadian Journal of Fisheries and Aquatic Sciences*, 54:1976–1984.

Mangel, M., Marinovic, B., Pomeroy, C., and Croll, D. 2002. Requiem for Ricker: Unpacking MSY. *Bulletin of Marine Science*, 70: 763–781.

McAllister, M. K. and Ianelli, J. N. 1997. Bayesian stock assessment using catch-age data and the sampling-importance resampling algorithm. *Canadian Journal of Fisheries and Aquatic Sciences*, 54:284–300.

McAllister, M. K., Pikitch, E. K., Punt, A. E., and Hilborn, R. 1994. A Bayesian approach to stock assessment and harvest decisions using the sampling/importance resampling algorithm. *Canadian Journal of Fisheries and Aquatic Sciences*, 51:2673–2687.

Meyer, R. and Millar, R. B. 1999a. Bayesian stock assessment using a state–space implementation of the delay difference model. *Canadian Journal of Fisheries and Aquatic Sciences*, 56:37–52.

Meyer, R. and Millar, R. B. 1999b. BUGS in Bayesian stock assessment. *Canadian Journal of Fisheries and Aquatic Sciences*, 56:1078–1087.

Millar, R. B. and Methot, R. D. 2002. Age-structured meta-analysis of U.S. West Coast rockfish (Scorpaenidae) populations and hierarchical modeling of trawl survey catchabilities. *Canadian Journal of Fisheries and Aquatic Sciences*, 59:383–393.

Millar, R. B. and Meyer, R. 2000. Non-linear state space modelling of fisheries biomass dynamics by using Metropolis-Hastings within-Gibbs sampling. *Applied Statistics*, 49:327–342.

Myers, R. A. and Cadigan, N. G. 1995. Was an increase in natural mortality responsible for the collapse of northern cod? *Canadian Journal of Fisheries and Aquatic Sciences*, 52:1274–1285.

Otter Research Ltd 2007. An introduction to AD Model Builder Version 8.0.2 for use in nonlinear modeling and statistics. Otter Research Ltd, Sidney, BC, Canada. http://otter-rsch.com/admodel.htm

Patterson, K. R. 1999. Estimating uncertainty in harvest control law catches using Bayesian Markov chain Monte Carlo virtual population analysis with adaptive rejection sampling and including structural uncertainty. *Canadian Journal of Fisheries and Aquatic Sciences*, 56:208–221.

Polacheck, T., Hilborn, R., and Punt, A. E. 1993. Fitting surplus production models: Comparing methods and measuring uncertainty. *Canadian Journal of Fisheries and Aquatic Sciences*, 50:2597–2607.

Punt, A. E., Butterworth, D. S., and Penney, A. J. 1995. Stock assessment and risk analysis for the South Atlantic population of albacore *Thunnus alalunga* using an age-structured production model. *South African Journal of Marine Science*, 16:287–310.

Punt, A. E. and Hilborn, R. 1997. Fisheries stock assessment and decision analysis: The Bayesian approach. *Reviews in Fish Biology and Fisheries*, 7:35–63.

Quinn II, T. J. 2003. Ruminations on the development and future of population dynamics models in fisheries. *Natural Resource Modeling*, 16:341–392.

Raftery, A. E., Givens, G. H., and Zeh, J. E. 1995. Inference for a deterministic population dynamics model for bowhead whales (with discussion). *Journal of the American Statistical Association*, 90:402–430.

Ricker, W. E. 1975. Computation and interpretation of biological statistics of fish populations. *Bulletin of the Fisheries Research Board of Canada*, No. 191. Ottawa, Ontario, Canada.

Schaefer, M. B. 1954. Some aspects of the dynamics of populations important to the management of the commercial marine fisheries. *Bulletin of the Inter-American Tropical Tuna Commission*, 1:25–56. Reproduced (1991) in *Bulletin of Mathematical Biology*, 53(1-2):253–279.

Schnute, J. T. 1994. A general framework for developing sequential fisheries models. *Canadian Journal of Fisheries and Aquatic Sciences*, 51:1676–1688.

Shelton, P. A. 2005. Did over-reliance on commercial catch rate data precipitate the collapse of northern cod? *ICES Journal of Marine Science*, 62(6):1139–1149.

Sullivan, P. J. 1992. A Kalman filter approach to catch-at-length analysis. *Biometrics*, 48:237–257.

Walters, C. and Ludwig, D. 1981. Effects of measurement errors on the assessment of stock-recruitment relationships. *Canadian Journal of Fisheries and Aquatic Sciences*, 38:704–710.

24

Model Comparison and Simulation for Hierarchical Models: Analyzing Rural–Urban Migration in Thailand

Filiz Garip and Bruce Western

24.1 Introduction

Sociologists often argue that social context matters. Features of the social context, not just the characteristics of individuals, help produce aggregate outcomes such as the distribution of economic rewards, or paths of development. Multilevel designs where individuals are nested within social contexts provide a strong design for observing both contextual effects and the aggregate outcomes those effects might produce.

We present an analysis of migration in rural Thailand, in which survey respondents are nested within villages, providing annual reports on migration for the 1980s and 1990s. Rural–urban migration has propelled economic development as rural migrants remit their earnings back to their villages and return with news of economic opportunities for friends and family members. Though our data describe thousands of individual migration decisions, our interest focuses on aggregate differences across villages. The rural northeast of Thailand varies tremendously in the degree to which villages are integrated into the urban economies further south. The evolution of inequality in migration across villages is thus important for our understanding patterns of poverty and development in the rural areas of countries experiencing rapid growth.

Hierarchical models provide a valuable tool for studying multilevel sociological data such as the Thai migration surveys (Mason et al., 1983; Western, 1999). In sociology and demography, panel surveys of individuals and households, survey data from many countries, and pooled time series data from US states and cities have all been analyzed with hierarchical models (DiPrete and Forristal, 1994). Sometimes sociological applications have studied the heterogeneity of parameters across units, though more commonly hierarchical models offer a way to account for clustering in inferences about fixed parameters. In these cases, random effects are a nuisance, integrated out for correct inference.

Hierarchical models are common in sociology, but applied research often neglects two important topics. First, sociological analysis of hierarchical models rarely provides a detailed examination of model fit. In our analysis of the Thai migration data we study the fit of several alternative models by comparing the deviance information criterion (DIC) and posterior predictive statistics. Model fit is an important applied topic because sociological theory is often indifferent to alternative specifications of random effects. The structure of random effects may also have important implications for substantive conclusions. In particular, substantively important aggregate outcomes that are not directly modeled—like inequality in a response across units or response variable quantiles—may be sensitive to the specification of random effects. A second limitation of applied sociological research with hierarchical models is that these aggregate implications of model estimates typically go unexamined. Our analysis of rural–urban migration in Thailand examines several hierarchical models. In our analysis, Markov chain Monte Carlo (MCMC) computation for hierarchical models provides a convenient framework for studying aggregate patterns of variation by simulating migration given different hypothetical distributions of covariates.

24.2 Thai Migration Data

The Thai migration data are based on the Nang Rong Survey* of men and women aged 13–41 from 22 villages in the Nang Rong district of northeastern Thailand (Curran et al., 2005). We combine data from two waves (1994 and 2000) of the life history survey. The 1994 wave begins with men and women aged 13–35 in 1994, and asks about respondents' migration experiences since the age of 13. This design is replicated in 2000: men and women aged 18–41 are asked about their migration behavior starting at the age of 13. Some respondents were living away from the village at the time of the survey, and they were followed up and interviewed.[†] We merge these data with household censuses conducted in 1984, 1994, and 2000 to obtain household and village characteristics. The resulting data contain information on migration of 6768 respondents nested within 22 villages over a 16-year time period from 1984 to 2000 ($N = 93,914$).

Our interest focuses on how the level of migration in a village might subsequently promote more migration among individuals. Figure 24.1 shows the distribution of village migration rates, $\bar{y}_{jt} = \sum_i y_{ijt}/n_{jt}$, from 1984 to 2000. The survey data are retrospective, and the age distributions vary over time. The figure displays the migration rates for men and women aged 18–25, the age group that we observe every year. Migration rates generally increase until 1996. In 1984, around a quarter of young residents in Nang Rong left their district for at least two months. By 1996, the migration rate for the region had increased to about 50%. In 1996, the Asian financial crisis precipitated recession in Thailand. Migration rates declined over the next four years. In some villages, migration declines were particularly steep, with migration rates falling to around 10%. Trends for a high-migration and low-migration village are also shown in the plot. These trends share some common features, such as the increase in migration in the first decade and the decline from 1996.

Part of our substantive interest focuses on how the accumulation of migration experiences within villages is associated with an individual's likelihood of migration. Migration for an individual may become more likely if they live in a village in which many others have migrated. This phenomenon, called the cumulative causation of migration, occurs because prior migration generates resources or influence that make individuals more likely

---

* The Nang Rong Survey is a collaborative effort between investigators at the Carolina Population Center, University of North Carolina at Chapel Hill, and investigators at the Institute for Population and Social Research (IPSR), Mahidol University, Salaya, Thailand. It is partially funded by Grant R01-HD25482 from the National Institute of Child Health and Human Development to the Carolina Population Center. Information about the survey and the data analyzed in this chapter are available at http://www.cpc.unc.edu/projects/nangrong.

† Related project manuscripts report that success in finding migrants was relatively high (Rindfuss et al., 2007) On average, about 44% of the migrants were successfully interviewed at some point in the six months following the village surveys.

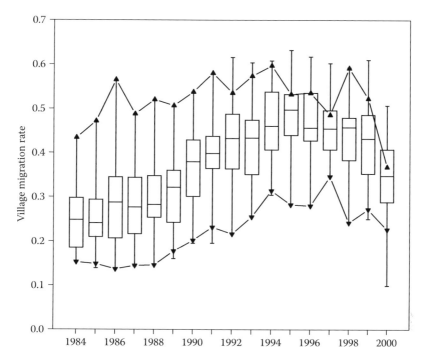

FIGURE 24.1

Boxplots of annual village migration rates for men and women aged 18–25, Nang Rong, Thailand, 1984–2000. Migration rates for villages with the largest and smallest migration rates in 1984 are shown by the trend lines.

to migrate (Massey, 1990). Extensive empirical evidence documents how past migration becomes a primary engine for future migration flows, eventually diminishing the importance of alternative explanations (Garip, 2008; Massey and Espinosa, 1997; Massey and Zenteno, 1999).

We study the effect of social context by a constructing a "village trips" variable that records the number of trips taken in a village in the years preceding the current year. A scatterplot of village trips and annual village migration rates for the 1984–2000 period is shown in Figure 24.2. In any given year, villages with the highest migration rates have a history of high levels of migration. This pattern is not surprising, but it remains an open empirical question whether a village's history of migration is associated with an individual's likelihood of migration, after accounting for their own history of migration, their family's migration history, and other covariates.

To study the effect of village trips for these multilevel data we write several hierarchical logistic regression models. For respondent i ($i = 1, \ldots, n_{tj}$) in village j ($j = 1, \ldots, 22$) in year t ($t = 1984, \ldots, 2000$), y_{ijt} denotes the binary migration outcome, taking the value 1 if the respondent travels away from the village for more than two months in the year, and 0 otherwise. Individual- and village-level covariates are collected in vectors, $\mathbf{x}_{ijt}$ and $\mathbf{z}_{jt}$. In each of the following logistic regressions, y_{ijt}, conditional on fixed and random effects collected in the vector $\boldsymbol{\theta}$, is assumed to be Bernoulli, $P(y|\boldsymbol{\theta}) = p^y(1-p)^{1-y}$, with expectation $E(y) = p$ and likelihood $L(\boldsymbol{\theta}; \mathbf{y}) = \prod P(y_{ijt}|\boldsymbol{\theta})$.

If we consider only the panel aspect of the data design, we can fit a respondent-level random effect, α_i, to allow for the correlation of observations for the same respondent,

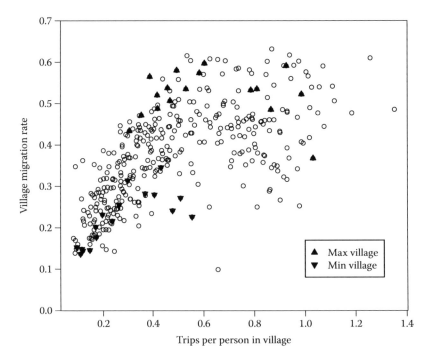

FIGURE 24.2
Scatterplot of village trips and village migration rates for men and women aged 18–25, Nang Rong, Thailand, 1984–2000. Villages with the smallest and largest migration rates in 1984 are indicated separately.

yielding the logistic regression:

$$\text{logit}(p_{ijt}) = \alpha_i + \delta_t + \boldsymbol{\beta}_1' \mathbf{x}_{ijt} + \boldsymbol{\beta}_2' \mathbf{z}_{jt}, \tag{24.1}$$

for $\mathbf{x}_{ijt}$ and $\mathbf{z}_{jt}$ as described in the previous paragraph, with corresponding fixed effects $\boldsymbol{\beta}_1$ and $\boldsymbol{\beta}_2$. This specification also includes a time effect, δ_t, that captures the common trend in migration across villages. The two levels of clustering, by respondent and village, could be modeled with separate effects, where a village effect, γ_j, captures a migration propensity that is common to all residents of the same village:

$$\text{logit}(p_{ijt}) = \alpha_i + \gamma_j + \delta_t + \boldsymbol{\beta}_1' \mathbf{x}_{ijt} + \boldsymbol{\beta}_2' \mathbf{z}_{jt}. \tag{24.2}$$

Finally, heterogeneity in village effects over time can be captured with a village-by-year effect, γ_{jt}:

$$\text{logit}(p_{ijt}) = \alpha_i + \gamma_{jt} + \delta_t + \boldsymbol{\beta}_1' \mathbf{x}_{ijt} + \boldsymbol{\beta}_2' \mathbf{z}_{jt}. \tag{24.3}$$

Given the observed variability in migration trends, this last model seems most realistic. It is shown as a directed acyclic graph in Figure 24.3. The parameters, μ and σ^2, are the means and variances of the hyperdistributions from which the random effects are drawn. Boxes and ovals denote covariates and parameters, respectively. Solid arrows indicate probabilistic dependencies, whereas dotted arrows are deterministic relationships. The clustered structure of the data (individuals within villages for each year) is denoted by stacked sheets.

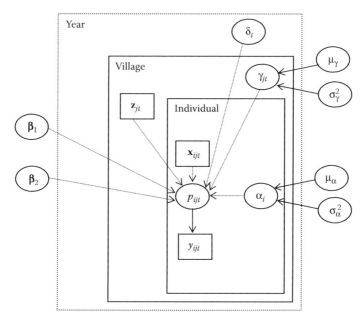

FIGURE 24.3
Three-level logit model with individual, village, and year effects on individual migration outcome.

In this figure, the year sheet is dotted, indicating that year-specific effects will induce correlations among observations from same time point, though individuals and villages are not nested within years.

The full Bayesian specification requires hyperdistributions for the random effects, and proper priors for their hyperparameters. In our analysis, the random effects for our three models are each given a normal distribution. The means are given diffuse normal prior distributions. The standard deviations are given uniform distributions. The priors, displayed in Table 24.1, are intended to be uninformative so the sample data dominates estimation of the hyperparameters (Gelman, 2006). We experimented with several alternative priors and obtained essentially the same results as those reported here.

TABLE 24.1

Hyperdistributions and Prior Distributions for Hierarchical
Logistic Regression Models of Thai Migration

Model	Random Effects	Prior Distributions
(24.1)	$\alpha_i \sim N(\mu_\alpha, \sigma_\alpha^2)$	$\mu_\alpha \sim N(0, 10^6)$
		$\sigma_\alpha \sim U(0, 1000)$
(24.2)	$\alpha_i \sim N(\mu_\alpha, \sigma_\alpha^2)$	$\mu_\alpha \sim N(0, 10^6)$
		$\sigma_\alpha \sim U(0, 1000)$
	$\gamma_j \sim N(\mu_\gamma, \sigma_\gamma^2)$	$\mu_\gamma \sim N(0, 10^6)$
		$\sigma_\gamma \sim U(0, 1000)$
(24.3)	$\alpha_i \sim N(\mu_\alpha, \sigma_\alpha^2)$	$\mu_\alpha \sim N(0, 10^6)$
		$\sigma_\alpha \sim U(0, 1000)$
	$\gamma_{jt} \sim N(\mu_\gamma, \sigma_\gamma^2)$	$\mu_\gamma \sim N(0, 10^6)$
		$\sigma_\gamma \sim U(0, 1000)$

24.3 Regression Results

We can easily explore the model fit and run simulation experiments with draws from the posterior obtained by MCMC simulation. The results below are based on 10,000 iterations from parallel chains, after a burn-in of 2500 iterations. Convergence diagnostics, including that of Gelman and Rubin (1992), for parallel chains indicate convergence for all parameters (results available upon request).

Posterior means and standard deviations for the regression coefficients are reported in Table 24.2. All variables are standardized to have zero mean and unit variance. The results show the positive association of the village history of migration with an individual's migration decision in a given year. A standard deviation difference in the trips per village nearly doubles the odds of migration for an individual ($e^{0.644} \approx 1.9$). A household's and individual's history of migration are also strongly associated with migration. All these effects are consistent across model specifications. Unsurprisingly, individual trips are estimated to have the strongest effect on individual migration. Less expected, however, is the relatively strong effect of the village level of migration. Covariate effects are also similar across models. Men, the unmarried, and the more educated are all somewhat more likely to migrate.

Most of the point estimates for the coefficients are insensitive to alternative specifications of the random effects, though some models may still fit the data better than others. The DIC statistic, proposed by Spiegelhalter et al. (2002), is readily calculated from MCMC output. The DIC is based on the usual deviance statistic, $D(\mathbf{y}, \boldsymbol{\theta}) = -2 \log L(\boldsymbol{\theta}; \mathbf{y})$, evaluated at the simulated values of the parameters. Like the deviance, better-fitting models have lower DIC statistics. DIC statistics are virtually the same for the individual and village random effects models. The DIC statistic for the village–year model, which includes random effects for each village in each year, is 137 points lower.

TABLE 24.2

Logistic Regression Coefficients (Standard Errors) for Hierarchical Models of Migration, Nang Rong, Thailand, 1984–2000

	Individual		Village		Village–Year	
Village trips	0.644	(0.006)	0.663	(0.063)	0.681	(0.073)
Household trips	0.115	(0.021)	0.118	(0.021)	0.114	(0.022)
Individual trips	1.457	(0.022)	1.454	(0.023)	1.462	(0.022)
Age	−0.248	(0.040)	−0.228	(0.039)	−0.243	(0.041)
Male	0.266	(0.069)	0.250	(0.067)	0.265	(0.070)
Married	−1.183	(0.038)	−1.185	(0.037)	−1.188	(0.038)
Education	0.756	(0.031)	0.774	(0.032)	0.763	(0.031)
Land	−0.056	(0.019)	−0.060	(0.019)	−0.052	(0.019)
σ_i	2.600	(0.036)	2.578	(0.036)	2.610	(0.037)
σ_v	—	—	0.431	(0.092)	—	—
σ_{vy}	—	—	—	—	0.190	(0.019)
DIC	0		−14		−137	
p_D	4955		4946		5094	

Note: $N = 93{,}914$ for 6768 individual respondents in 22 villages. DIC is adjusted by a constant ($-61{,}811$) to equal zero for the individual model.

A component of the DIC statistic, the p_D, is given by the difference between the posterior mean deviance and the deviance evaluated at the posterior mean and has been proposed as a measure of the effective number of parameters of a Bayesian model. The village–year model is parametrically the most complex and this is reflected in the relatively high p_D statistic. The village–year model includes an additional 352 random effects over the village model, an effective addition of 139 new parameters according to the p_D.

24.4 Posterior Predictive Checks

The DIC statistic is an omnibus measure of fit, and the p_D can yield odd results in some applications. An alternative approach, tailored to the substantive objectives of the research, examines model predictions for quantities of key substantive interest (Gelman et al., 1996). The posterior predictive distribution is the distribution of future data, $\tilde{y}$, integrating over the posterior parameter distribution for a given model:

$$p(\tilde{\mathbf{y}}|\mathbf{y}) = \int p(\tilde{\mathbf{y}}|\boldsymbol{\theta})p(\boldsymbol{\theta}|\mathbf{y})d\boldsymbol{\theta}.$$

To study the posterior predictive distribution the researcher must define a test statistic which can be calculated from the observed data. Because we are interested in the inequality in migration across villages and over time, we define the test statistic in year t as

$$R_t = \frac{\max(\bar{y}_{jt})}{\min(\bar{y}_{jt})},$$

the ratio of the largest to the smallest annual village migration rate. A well-fitting model should yield posterior predictions that track the observed trend in village inequality in migration.

Figure 24.4 compares the observed trend in R_t to the 95% posterior predictive confidence interval for R_t under the individual, village, and village–year models. For the individual-level model with respondent-level random effects, Figure 24.4a shows that the predictive distribution generally captures the U-shaped trend in inequality in village migration rates. In most years, the observed level of inequality falls within the predictive interval, indicating that the data are not extreme under the model. Several of the most extreme observations, however, fall well outside the predictive interval.

The village model adds time-invariant random effects for each village to the individual model that includes only respondent random effects. Figure 24.4b shows the posterior predictive interval for the village model. Adding village-level random effects does little to improve the model's fit to longitudinal patterns of inequality in village migration rates. As for the individual model, several extreme values at the ends of the time series are poorly predicted under the village model.

Finally, the village–year model adds a random effect for each village in each year. The posterior predictive distribution in this case covers the observed trend in inequality in all years but one. The flexibility of the village–year model is reflected in the relatively wide predictive distribution displayed in Figure 24.4c. Accounting for yearly differences in village effects adds significantly to predictive uncertainty about possible migration rates.

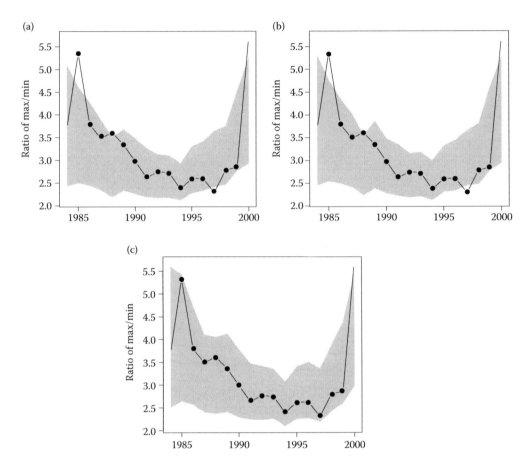

FIGURE 24.4
Inequality in village migration, R_t, and the 95% confidence region for the posterior predictive distribution of the (a) individual, (b) village, and (c) village–year models.

As a consequence, however, the observed trend in inequality is relatively likely under the village–year model.

24.5 Exploring Model Implications with Simulation

The posterior predictive check allows us to study the fit of the model, but we have not yet examined the implications of model estimates for understanding aggregate patterns. We explore the implications of the estimated model for inequality in village migration rates using simulations. Coefficient estimates show the strong effect of village trips on individuals' migration probabilities. Those living in villages with a high number of prior trips are more likely to migrate. In those villages, more trips accumulate over time, further increasing the likelihood of migration. This phenomenon, the cumulative causation of migration, suggests a dynamic mechanism of stratification in migration patterns across villages (Massey, 1990).

Due to cumulative causation, small initial differences in village trips may lead to large inequalities in village migration rates over time (Garip, 2008). Our model does not account for the initial distribution of village trips. The observed distribution of village trips in the data is one among many possible configurations. To observe the full extent of the implications of our model for inequality in village migration, we use a simulation exercise.

Keeping the aggregate trips constant, we alter the initial distribution of trips across villages in the data. We simulate the migration patterns from 1985 to 2000 using the following procedure. For each year, we compute individuals' predicted migration probabilities from our estimated model. We randomly assign migrants based on that probability. We then update the cumulative individual, household and village trips, and compute individuals' expected migration probabilities for the next year. We repeat this procedure many times ($N = 1000$), and compute average village migration rates over repetitions. In simulation runs, we take random draws from the MCMC-generated posterior distribution of the parameters to simulate inter-village inequality in migration. By drawing from the whole posterior distribution, simulation results reflect posterior uncertainty about parameters.

If we collect all the covariates and indicators for the random effects and the fixed time effect in the matrix $\mathbf{X}$, and the regression coefficients and random and fixed effects are in the vector $\boldsymbol{\beta}$, so the logistic regression in Equation 24.3 is written $\text{logit}(\mathbf{p}) = \mathbf{X}\boldsymbol{\beta}$, then the pseudo-algorithm is as follows:

1. Distribute the initial number of village trips, V_{t_0}, across villages $j = 1, \ldots, J$, according to scenario S such that $\sum_{j=1}^{J} v_{jt_0} = V_{t_0}$.

2. Sample parameters, $\hat{\boldsymbol{\beta}}$, from the MCMC-generated posterior distribution.

3. From the fitted model, $\text{logit}(\hat{\mathbf{p}}) = \mathbf{X}\boldsymbol{\beta}$, obtain predicted probabilities $\hat{p}_{ijt}$ for all i, j at time period t.

4. Simulate data $\mathbf{y}^*$ from the fitted model, that is, $y_{ijt+1}^* \sim \text{Binomial}(1, \hat{p}_{ijt})$ for all i, j.

5. Update cumulative independent variables (individual, household, and village trips), $x_{ijt+1} = x_{ijt} + f(y_{ijt+1}^*)$, where $f(\cdot)$ is a function transforming migration in $t + 1$ into trips for all i, j, yielding an updated covariate matrix, $\mathbf{X}^*$.

6. Compute predicted probabilities from the fitted model $\text{logit}(\mathbf{p}^*) = \mathbf{X}^* \hat{\boldsymbol{\beta}}$ using the updated independent variables.

7. Increment time period t to $t + 1$.

8. Repeat steps 3–7 T times, that is, generate a path of fitted values for T time periods.

9. Repeat steps 2–8 M times independently.

10. Compute typical values (e.g. means) of the predicted probabilities over the M replications.

This algorithm is repeated for each scenario S of the initial distribution of village trips.

Figure 24.5 shows the average migration rate observed in simulations under two scenarios. With minimum initial inequality, we distribute the aggregate number of trips equally across villages in 1984. With maximum initial inequality, we assign the total number of trips to one randomly selected village, giving all other villages zero initial trips. The minimum initial inequality case leads to slightly lower average migration until 1990, and the two scenarios are indistinguishable thereafter.

Figure 24.6 displays the observed ratio of the largest to the smallest annual village migration rates, R_t, and compares these to series under the two simulation scenarios. In the

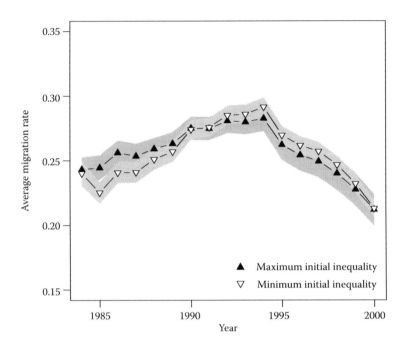

FIGURE 24.5
Annual migration rates and the 95% confidence region in simulations with maximum and minimum initial inequality in the distribution of village trips.

minimum initial inequality case, since all villages start at the same point, inequality in village migration rates does not grow over time. In this case, the cumulative mechanism identified in the model does not lead to increasing inequality in village migration. By contrast, with maximum initial inequality, initial inequality increases at a high rate after 1995. The observed inequality in the data, as expected, falls between the minimum and maximum inequality cases. The two extreme case scenarios provide upper and lower bounds for the potential inequality outcomes.

This simulation exercise thus links our estimates from the individual-level model to aggregate patterns of inequality between villages. Depending on the initial distribution of village trips, in a period of 16 years, the cumulative mechanism identified in our model could sustain or double inequality in village migration rates.

24.6 Conclusion

Hierarchical models are commonly used in sociology chiefly to study the effects of social context on individual outcomes. In our application, we examined the effects of households and villages on rural–urban migration in northern Thailand. With survey data on individuals at many points in time, individuals also formed contexts for migration decisions in any particular year. In data with this structure, we could specify as many as four hierarchies of random effects: at the individual, household, village, and village–year levels.

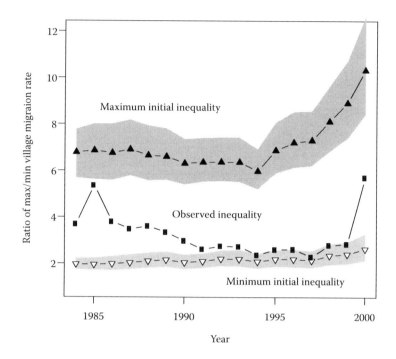

FIGURE 24.6
Inequality in village migration rate, R_t, and the 95% confidence region in simulations with maximum and minimum initial inequality in the distribution of village trips.

The nesting of observations within layers of social context creates data-analytic and substantive challenges. From the viewpoint of data analysis, a variety of equally plausible models can be specified to capture the multilevel structure of the data. From a substantive viewpoint, individual outcomes may aggregate to reshape the contexts in which the actions of individuals are determined. Though hierarchical models are common in sociology, the data-analytic problem of model comparison and the substantive problem of the aggregative effects of individual outcomes are often ignored.

Our analysis takes advantage of MCMC methods to fit hierarchical models, compare alternative models, and study the aggregate implications of the models. The problem of model fit was studied with both DIC statistics and posterior predictive checks. Both approaches yielded similar answers. Migration models including individual and village random effects fitted similarly well, but both were inferior to a model that allowed village effects to vary over time. The DIC statistic indicated the superior fit of the village–year model, and posterior predictive checks showed that this model better captured the observed trend in inequality in migration across villages.

We conducted a simulation exercise to help interpret the model parameters. The simulation experiment showed how the initial inequality in patterns of migration across villages influenced inequality in migration 16 years later. Inequality in migration nearly doubled where the initial distribution of migration was highly unequal. Had the initial distribution been equal across villages, this distribution would have remained largely unchanged.

In sum, MCMC computation for hierarchical provides an enormously flexible tool for analyzing contextual data. Far beyond the problems of estimation and inferences, posterior simulation with MCMC provides an important basis for data analysis and model interpretation. Though MCMC methods have so far seen relatively little application, they hold enormous promise for the analysis of hierarchical models in sociology.

References

Curran, S. R., Garip, F., Chung, C., and Tangchonlatip, K. 2005. Gendered migrant social capital: Evidence from Thailand. *Social Forces*, 84:225–255.

DiPrete, T. A. and Forristal, J. D. 1994. Multilevel models: Methods and substance. *Annual Review of Sociology*, 20:331–357.

Garip, F. 2008. Social capital and migration: How do similar resources lead to divergent outcomes? *Demography*, 45(3):591–617.

Gelman, A. 2006. Prior distributions for variance parameters in hierarchical models. *Bayesian Analysis*, 1(3):515–534.

Gelman, A. and Rubin, D. 1992. Inference from iterative simulation using multiple sequences. *Statistical Science*, 4:457–511.

Gelman, A., Meng, X.-L., and Stern, H. 1996. Posterior predictive assessment of model fitness via realized discrepancies. *Statistica Sinica*, 6:733–807.

Mason, W. M., Wong, G. Y., and Entwisle, B. 1983. Contextual analysis through the multilevel linear model. *Sociological Methodology*, 14:72–103.

Massey, D. S. 1990. Social structure, household strategies, and the cumulative causation of migration. *Population Index*, 56:3–26.

Massey, D. S. and Espinosa, K. 1997. What's driving Mexico-U.S. migration? A theoretical, empirical, and policy analysis. *American Journal of Sociology*, 102(4):939–999.

Massey, D. S. and Zenteno, R. 1999. The dynamics of mass migration. *Proceedings of the National Academy of Sciences of the USA*, 96(9):5328–5335.

Rindfuss, R. R., Kaneda, T., Chattopadhyay, A., and Sethaput, C. 2007. Panel studies and migration. *Social Science Research*, 36:374–403.

Spiegelhalter, D. J., Best, N., Carlin, B. P., and van der Linde, A. 2002. Bayesian measures of model complexity and fit. *Journal of the Royal Statistical Society, Series B*, 64(4):583–616.

Western, B. 1999. Bayesian analysis for sociologists: An introduction. *Sociological Methods and Research*, 28:7–34.

Index

A

Absolute catchability, fishing gear, 555
Acceptance rate, 31, 101, 150. *See also* Optimal
    scaling
    asymptotic, 97, 102
    HMC, 114
    jump updates, 308
    LMC, 150
    for MALA, 99
    optimal, 96, 141
    scaling and, 95, 96, 97
    within-component updates, 308
Acceptance/rejection method, 227, 236
    algorithm, 237
    running time, 237–238
ACF. *See* Autocorrelation function (ACF)
Adaptive MCMC methods. *See* Markov chain
        Monte Carlo (MCMC) methods,
        adaptive
Adaptive Metropolis algorithm (AM algorithm),
        104. *See also* Markov chain Monte Carlo
        (MCMC) methods, adaptive
    first coordinate trace plot, 105
    inhomogeneity factor trace plot, 106
    proposal distribution, 104, 105
    RAMA, 107
ADMB. *See* Automatic Differentiation Model
        Builder (ADMB)
Advance underlying state to time 0, 240
Aerometric Information Retrieval Service
        (AIRS), 494. *See* Air Quality System
        (AQS)
Age Structured Assessment Program (ASAP),
        550, 551
Air Quality System (AQS), 494
AIRS. *See* Aerometric Information Retrieval
        Service (AIRS)
Alleles, 341. *See also* Locus
Alzheimer's disease, 365. *See also* Bayesian
        model; Functional magnetic resonance
        imaging (fMRI)
    disease symptoms, 365, 366
    fMRI paradigm, 366
    image acquisition area, 366
AM algorithm. *See* Adaptive Metropolis
        algorithm (AM algorithm)
Antimonotone distributions, 203, 204
Antithetic perfect sampling, 220–221. *See also*
        Perfect sampling; Swindles

Antithetic variates, 220. *See also*
        Swindles—antithetic perfect sampling
Approximate Bayesian computation. *See*
        Likelihood-free (LF) computation
AQS. *See* Air Quality System (AQS)
AR(1) process, 9–10
Arbitrary time-dependence model, 422
ARE. *See* Asymptotic relative efficiency (ARE)
Areal (regionally aggregated) data. *See* Lattice
        data
ASAP. *See* Age Structured Assessment Program
        (ASAP)
Asymptotic relative efficiency (ARE), 268
Asymptotic variance, 6
    consistent overestimate, 16
    delayed-rejection schemes, 77
    estimation methods, 6, 8
    estimators, 16
    formula, 8
Attractive distributions. *See* Monotone
        distributions
Augmented sampler, 414
    approximation to true posterior, 321
    formulation, 414
    joint distribution, 321
    likelihood-free posterior approximation, 320
    model adequacy assessment, 321
    model errors, 321
    posterior distributions, 332
    pseudo-prior, 320
    sequence, 415
    univariate error distributions, 321
Autocorrelation function (ACF), 9, 12, 436
Autocovariance function, 9
Automatic Differentiation Model Builder
        (ADMB), 550
    automatic differentiation, 551–552
    features, 551, 553
    Metropolis–Hastings implementation,
        552–553
Autoregressive AR(1) process, 9
    autocorrelation plot, 13
    batch mean plot, 15
    digression, 10
    running averages plot, 12
    technical report, 11
    time series plot, 11
Auxiliary variable, 222
    in augmented formulation, 414
    finite mixtures, 214